M000236697

Radio Frequency and Microwave Electronics Illustrated

Matthew M. Radmanesh
Department of Electrical
& Computer Engineering
California State University,
Northridge

Prentice Hall PTR
Upper Saddle River, NJ 07458
www.phptr.com

Library of Congress Cataloging-in-Publication Data

Radmanesh, Matthew M., 1955–
Radio frequency and microwave electronics illustrated / Matthew M. Radmanesh
p. cm.
Includes index.
ISBN 0-13-027958-7
1. Microwave circuits. 2. Radio circuits. 3. Electronic circuit design. I. Title.

TK7876.R357 2001
621.381'3--dc21

00-046494

Editorial/Production Supervision: *Benchmark Productions, Inc.*
Acquisitions Editor: *Bernard Goodwin*
Cover Design Director: *Jerry Votta*
Cover Design: *Nina Scuderi*
Manufacturing Manager: *Alexis R. Heydt*
Editorial Assistant: *Michelle Vincenti*
Marketing Manager: *Dan DePasquale*
Project Coordinator: *Anne Trowbridge*

Prentice-Hall, Inc.
Upper Saddle River, NJ 07458

Prentice Hall books are widely used by corporations and government agencies for training, marketing, and resale.

The publisher offers discounts on this book when ordered in bulk quantities.
For more information, contact: Corporate Sales Department, Phone: 800-382-3419;
Fax: 201-236-7141; E-mail: corpsales@prenhall.com; or write: Prentice Hall PTR,
Corp. Sales Dept., One Lake Street, Upper Saddle River, NJ 07458.

Cover Illustration: The design on the cover shows the conceptual journey in engineering: zero to infinity, cause to effect, postulate to application, simplicity to complexity; from zero of an idea at the center to infinity of creation at the perimeter.

Printed in the United States of America

10 9 8 7 6 5 4 3 2 1

ISBN 0-13-027958-7

Prentice-Hall International (UK) Limited, *London*
Prentice-Hall of Australia Pty. Limited, *Sydney*
Prentice-Hall Canada Inc., *Toronto*
Prentice-Hall Hispanoamericana, S.A., *Mexico*
Prentice-Hall of India Private Limited, *New Delhi*
Prentice-Hall of Japan, Inc., *Tokyo*
Pearson Education Asia Pte. Ltd.
Editora Prentice-Hall do Brasil, Ltda., *Rio de Janeiro*

Dedicated to many generations of scientists, engineers and practical-minded philosophers, who have helped to enlighten the lost, to expand awareness and to create a higher potential survival for mankind in this universe.

Contents

Foreword

The field of RF and Microwaves has undergone a significant paradigm shift in the last decade. From being a technology that had its greatest utilization in defense and military-based applications, it is currently in the forefront as a fundamental technology in Wireless Communications and other industrial, medical, and commercial applications. As applications of RF and microwaves continue to evolve and as this technology becomes a common factor in the scientific and engineering communities it is imperative that university students and practicing scientists and engineers are thoroughly familiar with the measurement principles, electronics, and design fundamentals underlying this technology.

Several books on RF and microwave technology have been published since the landmark research and development effort that went into the development of radar and related techniques during World War II. Perhaps the most noteworthy and seminal publication in this area was the MIT Radiation Laboratory series which still stands out as an example of a major contribution to the scientific and engineering communities by a consortium of some of the world's leading scientists and engineers and as a memorial to the unnamed hundreds and thousands of engineers and scientists who actually carried out the research and development work, the results of which were presented in this series. Since the publication of this series, numerous books on RF and microwave technology have been published. These books have accommodated the advancements in the field and have provided rigorous mathematical analysis underlying the fundamental scientific principles. While these books have served their purposes as university texts and reference guides for practitioners, there was a significant gap in the available literature that as yet remained to be addressed. There has existed a need for a textbook that, through the use of graphical illustrations, rather than rigorous mathematical analysis, serves as a foundational treatise for graduate students and practicing engineers and scientists. Dr. Radmanesh's book *Radio Frequency and Microwave Electronics Illustrated* is an attempt directed at bridging this gap.

The book is organized into 5 distinct parts comprised of 21 chapters. At the end of each chapter are a list of symbols, sets of problems to test the reader's comprehension of the subject matter presented in the chapter, and a list of references for further reading or investigating a particular topic in greater detail.

Part I is comprised of four chapters. Chapters 1 and 2 introduce the reader to the fundamental postulates and axioms of science and engineering and the basic concepts and laws in electrical and electronics engineering. The presentation of these series of postulates and axioms is extremely helpful, since it lays the foundation for any of the engineering sciences and is particularly noteworthy, in that it has never been presented in any scientific text on RF and microwaves. Chapters 3 and 4 discuss the basic circuit mathematics and low frequency concepts. In Chapter 3, the groundwork is laid for basic mathematical concepts including phasors, basic circuit elements, Ohm's law, and its generalized version, Kirchoff's voltage and current laws, basic network theorems, and decibel scale. Chapter 4 introduces the reader to the behavior of the transistor circuits at DC and low frequencies. Even though microwave transistors are built differently and have a higher frequency range of operation, analysis of transistor circuits at DC and low frequencies provides a sound foundation for analysis and design of transistor circuits at RF and microwave frequencies.

Part II (Chapters 5–8) introduces the reader to the concept of wave propagation via basic concepts of Radio Frequency (RF) and microwave (MW) and their applications (Chapter 5), RF electronic concepts (Chapter 6), and fundamental concepts in wave propagation (Chapter 7). Chapter 8 is devoted strictly to RF/Microwave device and circuit characterization. Building on the concepts and fundamentals provided in previous chapters, Chapter 8 discusses the concepts of a two-port network. The reader is introduced to the characterization of two-part networks and its representation in terms of a set of parameters that can be cast into a matrix form.

Part III (Chapters 9–11) leads the reader to impedance matching concepts and passive circuit design. In Chapter 9, the "Smith Chart" (one of the most valuable and pervasive graphical tools in all of microwave engineering) is presented to the reader. The chart, as a reflection coefficient-to-impedance/admittance converter or vice versa, is discussed together with its ability to simplify the analysis of complex design problems involving transmission lines or lumped elements. Chapter 10 discusses the applications of the Smith Chart in three distinct categories: a) circuits containing primarily distributed elements, particularly transmission lines (TLs), b) circuits containing lumped elements, and c) circuits containing distributed and lumped elements in combination. Having developed a deep understanding of the Smith Chart and its applications, the reader is now introduced to the design of impedance matching networks in Chapter 11.

Part IV (Chapters 12–14) deals with the design considerations of microwave amplifiers. The chapters are systematically broken down to focus the reader on each of the important design parameters: stability concepts for a two-port network (Chapter 12), gain concepts in amplifiers (Chapter 13), and noise concepts in amplifier design (Chapter 14).

The readers conceptual journey is transformed to hard-core design examples of linear and nonlinear circuits and their applications in Part V (Chapters 15–21). The design examples include design of small-signal, narrow-band, low-noise, and multistage amplifiers (Chapter 15); design of large-signal, high-power amplifiers (Chapter 16); microwave transistor oscillator design, using the negative-resistance device concept in Chapter 17; microwave rectifier and detector design (Chapter 18); microwave mixer design (Chapter 19); and microwave control circuits, specifically switches, phase shifters, and attenuators (Chapter 20). The book concludes with Chapter 21, which is devoted entirely to RF and Microwave Integrated Circuits (MICs). The two classes of microwave integrated circuits, i.e., hybrid microwave integrated circuits and monolithic microwave integrated circuits (MMIC), are discussed in detail addressing the materials, processing techniques, reliability, and advantages of each type.

A comprehensive glossary of technical terms at the end of the book is a great aid in understanding RF and Microwaves and makes this book invaluable for anyone aspiring to master this field of study. The appendix section provides a list of all symbols used in the book, information on many physical constants, mathematical identities, and generally known laws and makes them sufficiently accessible for easy reference. The Computer-Aided design (CAD) examples provide actual design data and methodology, which the reader should find extremely helpful in many practical situations.

In summary, this book has some unique strengths which make it different from prior literature and an attractive reference for the reader.

1. The presentation of a series of scientific postulates and axioms at the start of the book lays the foundation for any of the engineering sciences and is unique to this book compared with similar RF and Microwave texts.
2. The presentation of classical laws and principles of electricity and magnetism, all inter-related, conceptually and graphically.
3. There is a shift of emphasis from rigorous mathematical solutions of Maxwell's equations, and instead has been aptly placed on simple yet fundamental concepts that underlie these equations. This shift of emphasis will promote a deeper understanding of the electronics, particularly at RF/Microwave frequencies.
4. Low-frequency electronics unlike most RF/Microwave texts has been amply treated, which makes an easy transition to RF/Microwave principles and prevents a gap of knowledge in the reader's mind.
5. New technical terms are precisely defined as they are first introduced, thereby keeping the subject matter in focus and preventing misunderstanding, and finally the abundant use of graphical illustrations and diagrams brings a great deal of clarity and conceptual understanding, enabling difficult concepts to be understood with ease.

I believe Dr. Radmanesh has addressed the literature gap that I discussed during the introduction of this foreword and has presented some unique aspects, which I believe the graduate student, the practicing engineer/scientist, and the university professor should find extremely beneficial and stimulating. The book should serve as a valuable readable reference for the RF and Microwave industry.

Dr. Asad M. Madni
C. Eng., Fellow IEEE, Fellow IEE, FIAE, FNYAS
President and Chief Operating Officer
BEI Technologies, Inc.

Preface

Education in the science of RF and microwave engineering consists of guiding the reader along a gradient of known data, with the highest attention to the basic concepts that form the foundation of this field of study. The basic concepts presented in this book are far more fundamental than the mother sciences of engineering (i.e., physics and mathematics) and cover the essential truth about our physical universe in which we live. These basic truths convey a much deeper understanding about the nature of the physical universe than has ever been discussed in any RF and microwave, or for that matter any scientific textbook.

These basic truths set up a background of discovered knowledge by mankind, against which a smaller sphere of information (i.e., RF and microwave engineering) can be examined. Many of the principles that appear in microwave books are easily describable and thus understood much better once the basic underlying concepts are grasped.

While studying sciences and engineering at the university, the author always looked for simplicity, a higher truth, and a deeper level of understanding in all of the rigorous mathematics and many of the physical laws that were presented. Upon further investigation, the underlying principles that form the backbone of all extant physical sciences have finally emerged and are presented as the fundamentals of physical sciences in Chapter 1 of this work.

A summary of philosophical formation of this work is presented in the form of a pyramid in Chapter 1. From this pyramid, we can see that workable knowledge is like a pyramid, where from a handful of common denominators efficiently expressed by a series of basic postulates, axioms, and natural laws, which form the foundation of a science, an almost innumerable number of devices, circuits, and systems can be thought up and developed. The plethora of the mass of devices, circuits, and systems generated is known as the application mass, which practically approaches infinity in sheer number. This is an important point to grasp, because the foundation portion never changes (a static) while the base area of the pyramid is an ever-changing and evolving arena (a kinetic) where this evolution is in terms of new implementation techniques and technologies.

Following this brief introduction, the fundamental laws and basic principles of electrical engineering, which most advanced textbooks take for granted, are discussed. The reason for their presentation at this early stage, is that in dealing with the subject of RF and microwave engineering, it has been found that a lack of deeper understanding of these fundamentals leads to a shallow perspective and a lack of appreciation of electrical engineering basics, which will eventually lead to serious miscomprehension and misapplication of the subject.

This book is written with emphasis on fundamentals and for this reason all new technical terms are thoroughly defined in the body of the text as they are introduced. This novel approach is based upon the results obtained in recent investigations and research in the field of education, which has shown that the lack of (or the slightest uncertainty on) the definition of terms poses as one of the most formidable obstacles in the reader's mind in achieving full comprehension of the material. A series of uncomprehended or misunderstood technical terms will block one's road to total comprehension and mastery of the subject. This undesirable condition will eventually lead to a dislike and total abandonment of the subject.

The initial motivation was to bring the basics to the forefront and orient the reader in such a way that he or she can think with these fundamentals correctly. This eventually led to writing the first manuscript several years ago and then the final preparation of this book at present.

In preparing this book, the emphasis was shifted from rigorous and sophisticated mathematical solutions of Maxwell's equations and instead has been aptly placed on RF and microwave circuit analysis and design principles using simple concepts while emphasizing the basics all the way.

This book is intended to be used in a 2-semester course in microwave electronics engineering for senior-level or graduate students and should serve as an excellent reference guide for the practicing RF and microwave engineer in the field as well.

The current work starts from very general postulates, considerations and laws and, chapter by chapter, narrows the focus to very specific concepts and applications, culminating in the design of various RF and microwave circuits. The book, divided into five parts and 21 chapters, develops and presents these chapters with the progressive development of concepts following the same pattern as presented in the pyramid of knowledge in Chapter 1, which is:

Part I The Highest Fundamentals
: Chapters 1–4 form the foundation of electronics.

Part II Wave Propagation in Networks
: Chapters 5–8 present the basics of RF and Microwave science, wave propagation, and network characterization concepts.

Part III Passive Circuit Design
: Chapters 9–11 deal with the Smith Chart and its numerous applications to matching circuits.

Part IV Basic Considerations in Active Networks
: Chapters 12–14 discuss the basic considerations of circuit design.

Part V Active Networks: Linear and Nonlinear Design
: Chapters 15–21 provide detailed analysis and design methodologies of linear and nonlinear active circuits.

A list of symbols used in each chapter and a series of problems are included at the end of each chapter to help the reader gain a fuller understanding of the presented materials. The book ends with a glossary of technical terms and several important appendixes. These appendixes cover physical constants and other important data needed in the analysis or design process, with one appendix fully devoted to several design examples of practical active circuits using computer-aided design techniques based on the "Libra/touchstone"® software Ver. 6.1 from HPEEsof.

ACKNOWLEDGMENTS

The author wishes to thank many generations of students who studied the first manuscript of this work and several of its revisions to bring the current work to its fruition. The author is grateful to Jeff Quin of Litton Guidance Center whose assistance in preparation of several segments of this book has been invaluable. Thanks are also due to Lyman Hayes of Hayes Associates, Ed Skochinsky of Edwards Air Force Base technical division, Oskar Ulloa of Peter Green design for graphic illustration, Chris Savage for many valuable discussions of graphical design and layout as well as numerous helpful suggestions throughout the entire work, and Kimo Watanaby for using superb technical skills in programming all the numerical examples in Visual Basic in the Microsoft Excel® Window environment. Many thanks are also due to Bernard Goodwin, Prentice-Hall Chief editor/publisher for the valuable hints, advice, and brilliant guidance along the way, Anne Trowbridge (PTR) for excellent coordination, and to Dmitri Nerubenko, Greg deZarn O'Hare (Benchmark Productions) on an expedient and well-organized job on the production of the final copy of the book.

The author also wishes to thank many of his colleagues, particularly Dr. Asad Madni (CEO/COO of BEI Technologies, Inc.), a valuable colleague and a great friend, Dr. E. S. Gillespie (California State University, Northridge, CA), Harvey Endler (JPL), Phillip Arnold (HP), Dr. George Haddad and Dr. C. M. Chu (University of Michigan, Ann Arbor, MI), Dr. M. Torfeh, Dr. H. Hizroglu and Dr. B. Guru (Kettering University, Flint, MI), Dr. Charles Alexander and Dr. M. E. Mokari (Ohio University, Athens, OH), who provided support and collegiality through the years.

Finally, my deep gratitude belongs to my wife, Jane Marie, my son, Jeremy William, for making life pleasant during this work, and to my parents, Mary and Dr. G.H. Radmanesh, for their love, encouragement, and unconditional support throughout my life.

Matthew M. Radmanesh, Ph.D.
California State University, Northridge,
Dept. of Electrical and Computer Engineering
18111 Nordhoff St., Northridge, CA 91330
November 2000

PART I

THE HIGHEST FUNDAMENTALS

CHAPTER 1

Fundamental Concepts of Science and Engineering

1.1 INTRODUCTION

In this chapter, we will deal with materials that underlie all of science and engineering. The materials that will be presented in this chapter will deal with the fundamental considerations of science and engineering. These fundamental concepts have been ignored, neglected, and even omitted from most scientific texts, much to the detriment of the students. This chapter presents a first-hand glimpse of these vital scientific basics and attempts to enhance the depth of understanding of physical sciences in general and electronics in particular.

1.2 KNOWLEDGE AND SCIENCE: DEFINITIONS

Because we are embarking on a road to knowledge, it is imperative that we define it at the outset of this work:

DEFINITION-KNOWLEDGE: *A body of facts, principles, data, and conclusions (aligned or unaligned) on a subject, accumulated through years of research and investigation, that provides answers and solutions in that subject.*

This is an important definition because it lays the groundwork for what comes in later chapters (see Figure 1.1). As students of knowledge, we must understand that every time we study a subject, we travel on the road to knowledge. Therefore the source of knowledge and his perspective on the subject play an important role both in terms of the degree of accuracy of the presented materials and of one's learning and guidance along this road to knowledge. This point should not be taken lightly because lack of a good and reliable source of knowledge, and sometimes the presence of a false source, is the downfall of many bright minds who, with proper guidance, would have otherwise been very productive and brilliant in that field of study.

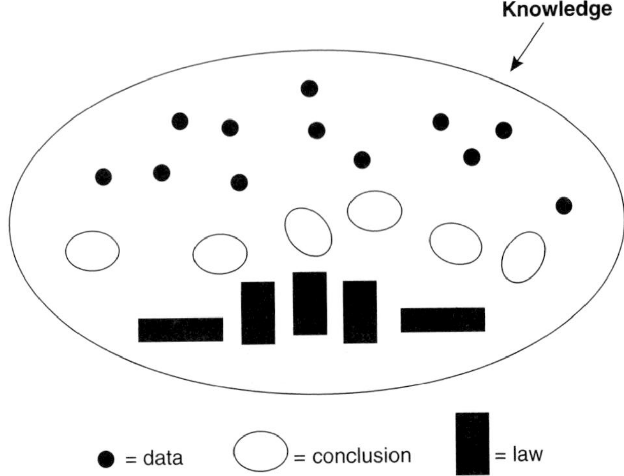

FIGURE 1.1 Definition of knowledge.

In our modern age we are surrounded with many fields of study, each with its own sphere of knowledge. The most common approach presented at universities today is through a science, as defined here:

DEFINITION-SCIENCE: *A branch of study concerned with establishing, systematizing, and aligning laws, facts, principles, and methods that are derived from hypotheses, observation, study, and experiments.*

From these two definitions, we can see that, in contrast to knowledge, a science is an aligned body of data, facts, and principles based on natural laws that have similarity in application (see Figure 1.2).

1.3 STRUCTURE OF A SCIENCE

We can observe that any science can be roughly divided into two divisions:

- **Considerations.** This division is an extremely important part of any science and primarily deals with all the theories, facts, research findings, postulates, hypotheses, design methodologies, technology, manufacturing techniques, discovered principles of operation, natural laws, and so on.
- **Application mass.** This division deals with all the related masses that are connected and/or obtained as a result of the application of the science. This includes all physical devices, machines, experimental setups, and other physical materials that are directly or indirectly derived as a result of the application of the science.

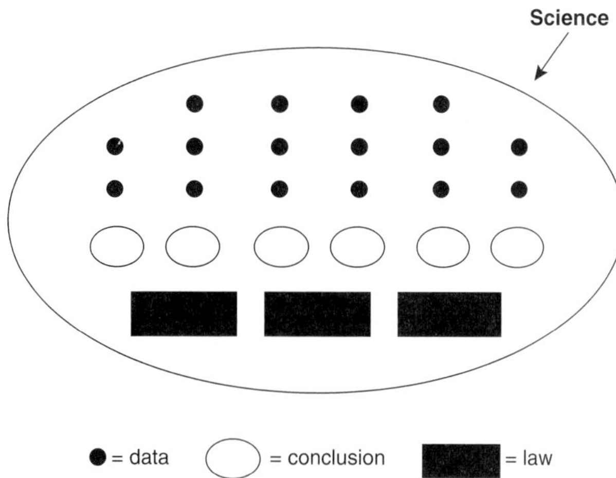

FIGURE 1.2 Definition of a science.

For example, for the field of electrical engineering, this division includes all electrical devices, circuits, motors, machinery, transmission lines, antennas, computers, assemblies, subsystems, systems, and networks, that are designed or directly obtained utilizing the electrical engineering theory and principles.

It is vital to note that of the two divisions, the former has much higher importance and seniority over the latter, even though the latter is much more voluminous and more abundant in practice. Therefore, this work's primary focus is on the considerations related to the field of electronics, and the discussions on application mass connected with the subject are all relegated to manufacturer's handbooks, datasheets, and other sources.

1.4 CONSIDERATIONS BUILT INTO A SCIENCE

In the field of considerations, we can observe that not every piece of data has the same level of importance as others. In fact, a few rank above all others and form the foundation of the science. In order of importance, the major considerations in any science can be summarized as follows:

1. Fundamental scientific postulates
2. Natural laws
3. Set of Nomenclature
4. Fundamental theorems
5. Analysis and theory of operation
6. Charts, diagrams, tables, etc.
7. Design methodology and procedures
8. Technology
9. Manufacturing techniques
10. Computer analysis and application software

These concepts are diagrammed in Figure 1.3. Due to the importance of these terms and their repeated use we now define a few of them briefly.

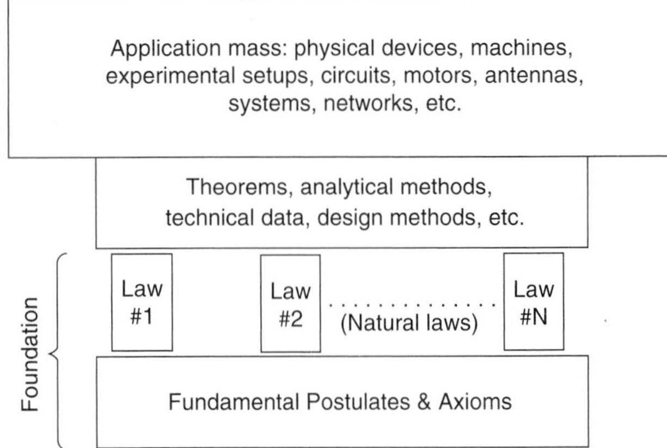

FIGURE 1.3 The foundation of the science of electrical engineering.

DEFINITION-SCIENTIFIC POSTULATE: *A scientific assumption that requires no proof, and is assumed to be true unconditionally and for all times, used as a basis for reasoning.*

DEFINITION-AXIOM: *A self-evident truth accepted without proof.*

DEFINITION-NATURAL LAWS: *A body of workable laws considered as derived solely from reason and study of nature.*

DEFINITION-THEOREM: *A proposition that is not self-evident but can be proven from accepted premises and, so is established as a principle.*

DEFINITION-TECHNOLOGY: *The application of knowledge for practical ends.*

1.4.1 Foundation of a Science

From the viewpoint of considerations, any extant science has a foundation on which it is built. This is very similar to a building structure whose footing and foundation are of essential importance because they support the weight of all the upper floors and their contents. The foundation of a science consists of two parts:

1. The fundamental postulates
2. Natural laws

On this solid foundation rest all theoretical research, extrapolations and design methodologies, application mass, and all future explorations, inventions and discoveries. The natural laws are discovered by observation and study but nevertheless have a lot in common with the fundamental postulates of a science.

It can be observed that the fundamental postulates of a science form the bedrock on which all natural laws rest. This means that the postulates and the discovered natural laws together form the foundation of a science. It is an important concept that is omitted in the majority of scientific texts. All the remaining considerations, such as scientific conclusions, technical data, design methods and rules, and the entire application mass of the subject rest on top of the foundation. Figure 1.4 shows the pyramid of knowledge in a workable science.

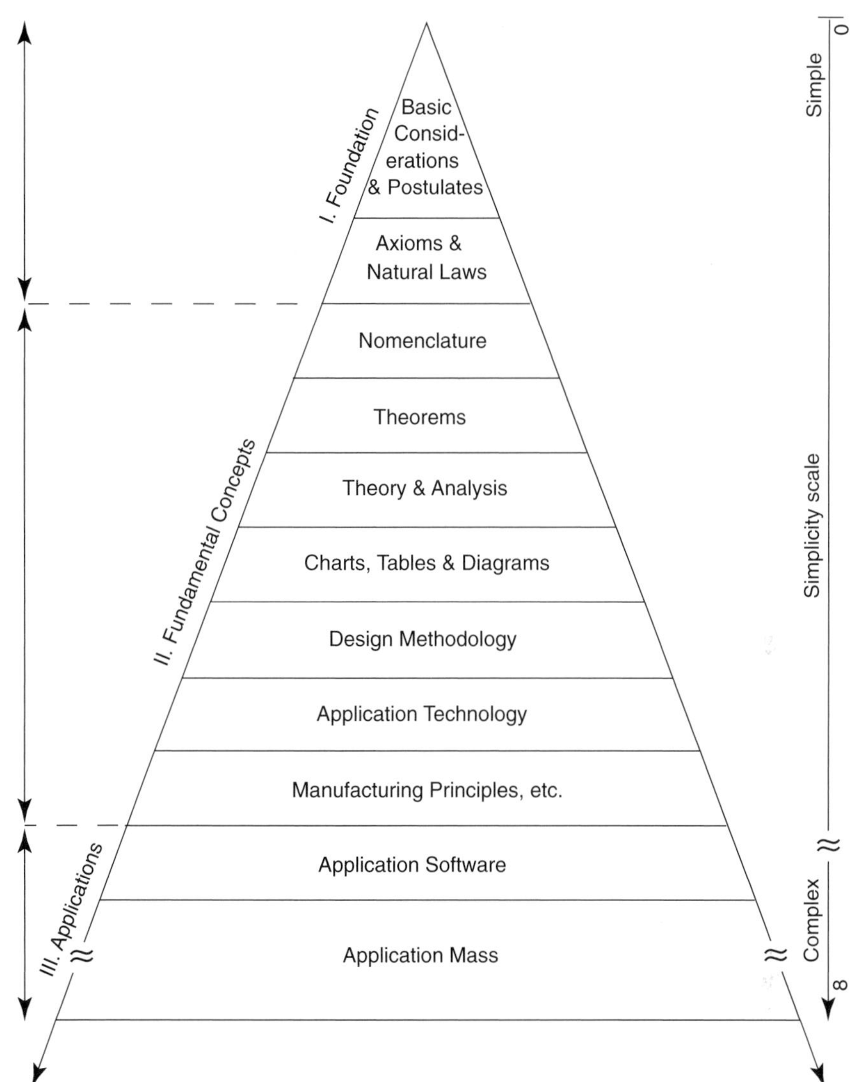

FIGURE 1.4 Pyramid of knowledge in a workable science.

Thus, the closer the laws are to simplicity and the basic postulates, the higher the workability of the science.

Furthermore, by observation we can generalize the pyramid of knowledge to include all of the extant sciences. This generalization would take the shape of a circle (see Figure 1.5). From Figure 1.5, we can see that the entire field of knowledge of mankind about the physical universe takes on the form of a pie, where each workable science is a slice of the pie. At the center of the circle lie the fundamental postulates of the physical universe, which are held in common by all of the sciences.

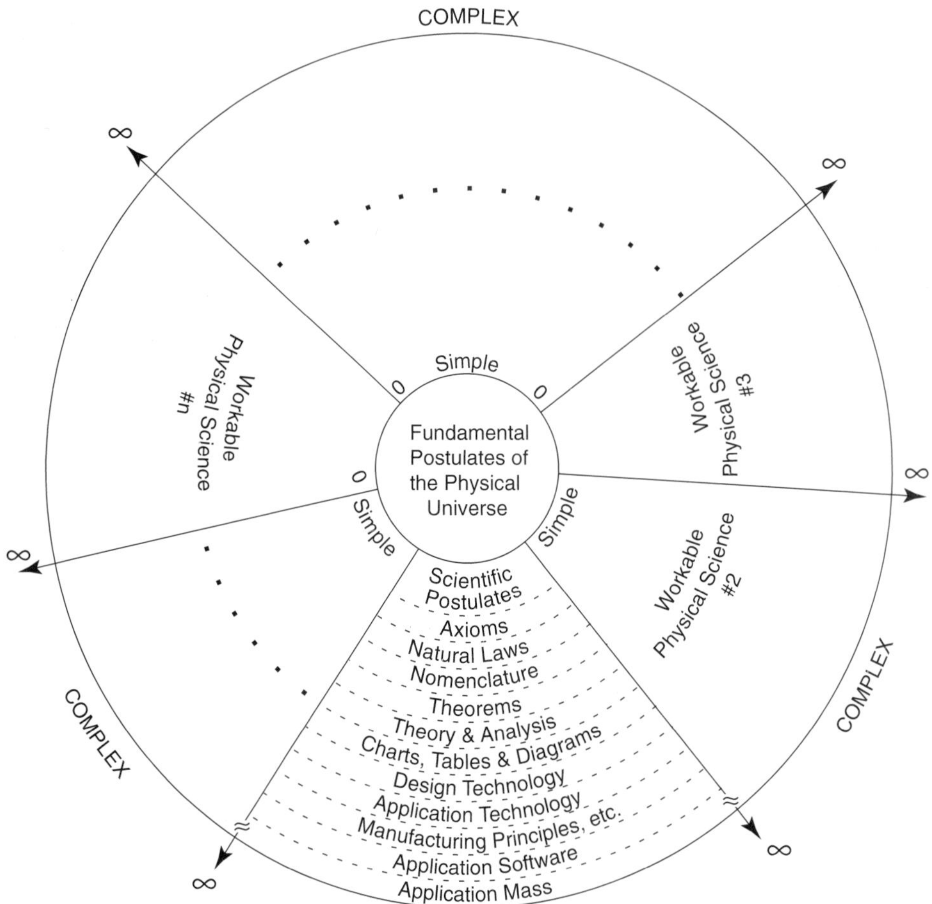

FIGURE 1.5 The structure of all sciences and their shared commonality in origin.

This common area, at the center of the circle, forms an interrelatedness among all sciences. The bulk of this chapter is devoted to discussing and examining the commonality and the interrelatedness of concepts in a science.

EXAMPLE 1.1

What is the foundation of the science of physics?

Solution:

We will first define what is meant by physics and then extract its fundamental postulates by deductive logic:

DEFINITION-PHYSICS: *The science dealing with energy and matter and the relationships between them, such as their interactions, motion, forces, properties, and changes in energy states excluding biological and chemical changes. Physics is divided* *into two fields: classical physics, which itself is further subdivided into electricity, magnetism, optics, acoustics, mechanics, and heat, and quantum physics, which is subdivided into atomic physics, nuclear physics, and solid-state physics.*

Physics is not focused on any specialized form of matter, as is chemistry (which focuses on the makeup of the different kinds of matter and the changes from one kind to another) or biology (which focuses on matter that is imbued with life force). Matter, in physics, is anything that has mass and occupies space, and for a very long time, it was considered to be distinctly separate and different from energy. With the increased knowledge of the nature of matter through several important discoveries, it has been shown that matter is fundamentally a condensed form of energy. In fact, through Einstein's formula, $E = mc^2$, we can see that the mass of matter (m) and its energy content (E) are equivalent through a proportionality constant (c^2). Therefore, higher physics and chemistry have become interrelated.

With the above preamble, now we can express the fundamental postulates of physics as follows:

Postulate #1: *Existence of energy, on a classical and/or quantum level.*

Postulate #2: *Existence of matter particles.*

Postulate #3: *Principle of conservation of energy to be held valid under all circumstances and for all times.*

NOTE 1: *The previous three postulates presuppose the existence of a continuous and linear space, in which all the energy and matter particles can be placed. Thus, a primary postulate should be added to the list:*

Postulate #4: *Existence of a continuous and linear space.*

NOTE 2: *The previous four postulates in action (i.e., particles in motion relative to one another in a linear space) instantly give birth to a new phenomenon: time. Thus, a secondary and fifth postulate can also be added to the list:*

Postulate #5: *Existence of a continuous and linear index of motion (or change) commonly known as time.*

To compare and measure all other motions relative to one another, a linear time reference is needed. Such a linear time reference is best furnished by having a number of particles that are moving relative to one another at a definite and fixed rate of motion. For example, rotation of the planets and their associated moons in the solar system (e.g., earth and its moon) around the sun at a regular and constant rate of motion establishes the necessary time standard against which all other motions can be measured. This will be further discussed in Chapter 2, *Fundamental Concepts in Electrical and Electronics Engineering*.

EXAMPLE 1.2

What forms the foundation of the science of electrical engineering?

Solution:
By observation, the foundation of electrical engineering consists of four scientific postulates and four classical natural laws.

These are given by:

1. Postulates of electrical engineering
   - Existence of a region of space that is continuous and behaves linearly.
   - Existence of charged particle(s) in the region of space. Each charged particle has a positive or a negative charge.
   - Existence of moving particles relative to one another at a fixed rate to establish a linear time yardstick or reference.
   - Principle of conservation of energy to be held valid unconditionally for all times.

Now given these four fundamental postulates, the field of electrical engineering is established. Through observation, study, and experimentation with this established field, the following natural laws can be and, in fact, have been derived:

2. Natural laws of electrical engineering (classical physics)
   - Faraday's law
   - Ampere's law
   - Gauss's law (electric)
   - Gauss's law (magnetic)

These four classical laws are commonly known as Maxwell's equations and are discussed in more depth in Chapter 2. Figure 1.6 shows the pyramid of knowledge in classical electrical engineering.

1.5 COMMONALITY AND INTERRELATEDNESS OF CONSIDERATIONS

Having examined and defined the classes of considerations that go into building a field of knowledge or a science, now we need to study the interrelations among these in greater depth.

1.5.1 Scientific Terminology

It could be observed that knowledge or information on a subject, when put into a language form, becomes quantized. In other words, knowledge is not continuous but rather made up of information quanta. The basic building blocks (or quanta) of information of any science are its technical terms or terminology. Each technical term (or quantum of information) carries with it an exact package of information that is embodied in its definition. The set of specific technical terms that is an inseparable part of any scientific subject is also commonly referred to as its nomenclature.

This concept indicates that in any topic, or generally in any body of knowledge or any scientific study, specific and exact definitions of terms are necessary in order to comprehend and communicate to others the laws, observations, assumptions, problems, solutions, and other relevant facts and conclusions (see Figure 1.7).

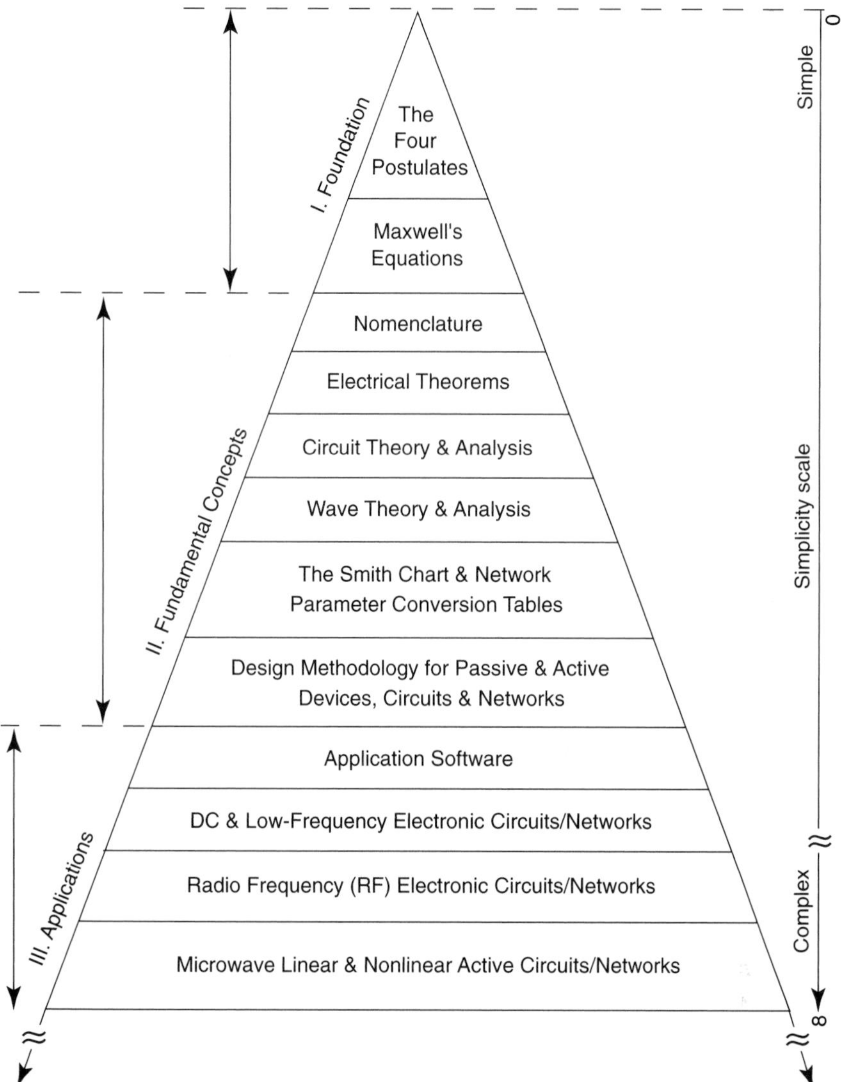

FIGURE 1.6 Pyramid of knowledge in electrical engineering (classical viewpoint).

Furthermore, it is a well-known fact that having an inadequate comprehension of the terminology is one of the leading causes of confusion about and misunderstanding of the subject (see Figures 1.8 and 1.9). Therefore, the terminology of a science forms an important part of the subject, and mastery of any subject requires mastery of its terminology, along with its accurate definitions. This means that a student, when faced with a stream of information regarding a scientific concept, needs to grasp each quantum of information fully (i.e., each technical term), with the exact definition that was originally intended by its author, before he or she can grasp the full scientific concept.

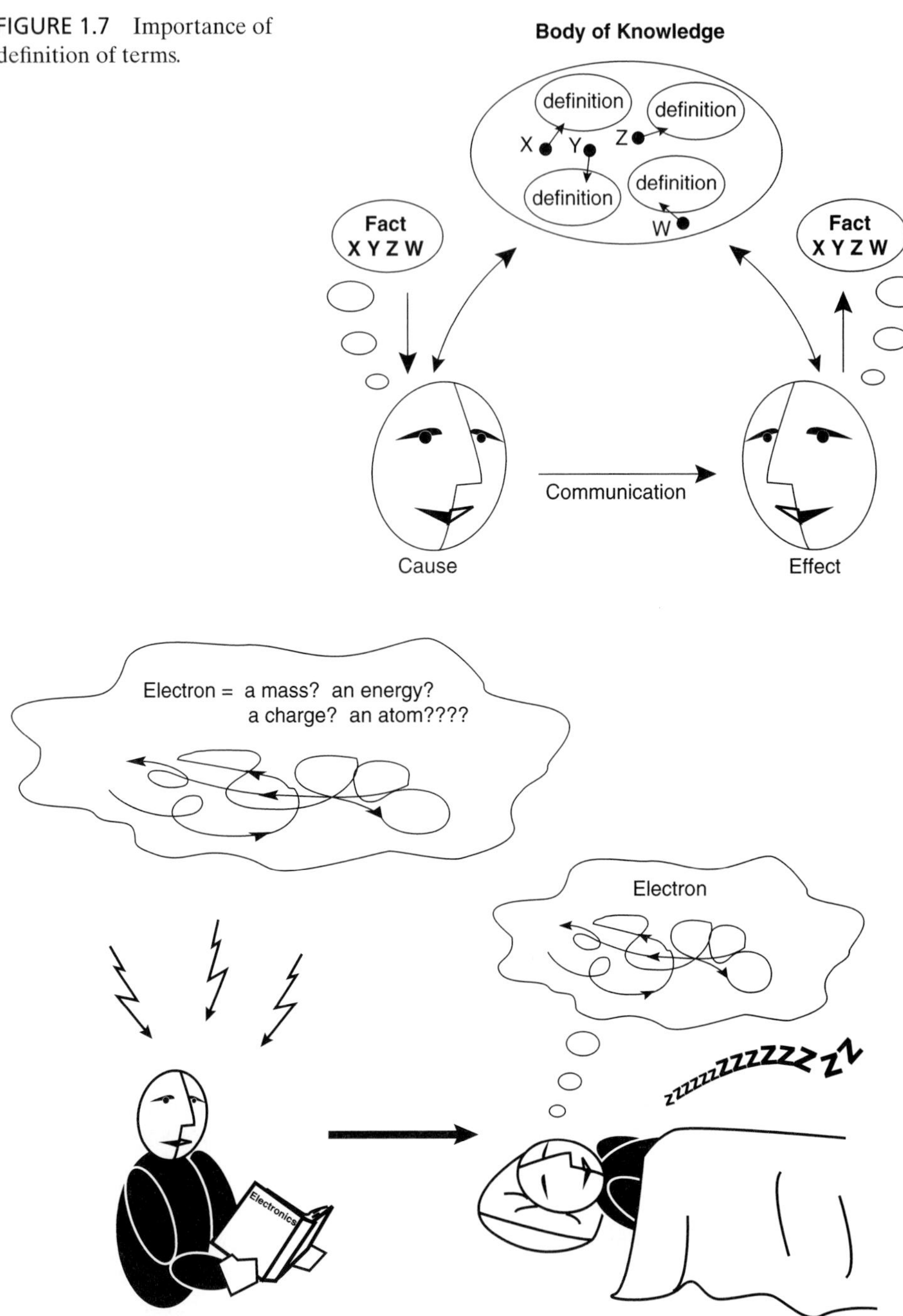

FIGURE 1.7 Importance of definition of terms.

FIGURE 1.8 Example of a confusion.

Knowing well that terminology plays a superior role in one's comprehension of a subject, we have taken special care throughout this work to define new terms as they

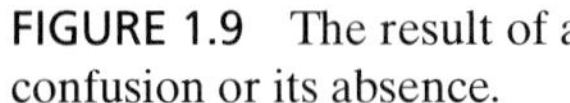

FIGURE 1.9 The result of a confusion or its absence.

are introduced. Additionally, and for easy reference, a relatively comprehensive glossary of important technical terms is provided at the end of the book.

A technical term may be defined in several ways. For example, the term "resistor" can be defined in any one of the following ways, which are all correct:

1. A device that is composed of a highly resistive material (such as carbon) designed to have a definite amount of resistance (descriptive definition).
2. A device that absorbs electric energy very much like a damping device used in door openers to absorb mechanical energy or in cars for shock absorption (associative definition).
3. A device used for dissipating energy, unlike a capacitor that is used for storing electric energy (comparative definition).
4. A device used in circuits to limit and create opposition to current flow or to provide voltage drop (action definition).

This approach can be applied to all terms in the field of electrical engineering with tremendous benefit.

1.5.2 Relativity of Knowledge

From the study of physics, we can see that a temperature of zero on the absolute scale cannot be reached but only asymptotically approached. A temperature of zero degrees Kelvin (on the absolute temperature scale) designates a state of "no energy" or "no motion" (i.e., a "complete static"), which is a mathematical absolute. Such a condition indicates a state of zero motion or energy (a nothingness) that is impossible to reach in the physical universe—a universe composed of constant interaction and motion of matter and energy relative to each other. It is also an obvious fact that a temperature of infinity, which is a state of infinite energy, cannot be reached. These simple observations bring us to the conclusion that the energy in the physical universe is bounded between two absolute energy levels: zero and infinity.

Due to the equivalence of matter and energy, and based on the same reasoning set forth for energy, a zero or infinity of matter are also impossible to attain. Furthermore, because any portion of the space in the physical universe is filled with either matter or energy[1], a zero or infinity of space cannot be achieved; such a condition leads to a zero or infinity of energy, which is impossible to attain as far as the physical universe is concerned.

From the discussion on the science of physics earlier, we observe that the main components of the physical universe are matter and energy (both finite), existing in a linear space (finite), and constantly interacting with each other, leading to time. Thus, we can conclude that the physical universe consists of three real components (matter, energy, and space) and a fourth apparent component (time). In other words, we are dealing with a universe where each of its components is finite. Thus, we can conclude that the physical universe itself as a whole is also a finite universe, and its total aggregate in terms of amount of energy, matter, and size is bounded between zero and infinity, that is:

$$0 < \text{the whole physical universe} < \infty$$

Having observed and concluded that the physical universe is a finite and bounded universe in energy, matter, and space, we can extrapolate or generalize this concept to any other entities in it (e.g., power, current, force, size, etc.), as well as any other related physical concepts concerning it (such as physical laws, facts, etc.). This means that all considerations and knowledge derived from or concerning the physical universe are perforce finite and do not achieve zero or infinity in importance or validity, which means that they have only relative value or importance.

This observation means that any and all data can be plotted on a graduated scale of importance or seniority, extending from zero to infinity, but not touching zero or infinity because these two are absolutes. Using this graduated scale of plotted data, we can observe that any datum (law, fact, theory, principle, etc.) about the physical universe has only relative validity or importance. Furthermore, there is no datum that has infinite importance or is true unconditionally at all times unless it is designated so by means of an agreement, and thus it must be expressed as a postulate.

---

[1] It should be noted that an absolute vacuum of space (i.e., space void of matter or energy) is impossible to attain because, even in outer space where there is no air, all types of particles such as space dust, photons, and cosmic rays still are contained in every region of space under consideration.

The idea of using a graduated or gradient scale is nothing new in mathematics and, in fact, has been used extensively in all of algebra and calculus. For example, "real positive numbers" are mapped onto points lying on a coordinate axis on a "one-to-one correspondence" basis. The coordinate axis extends from 0 to $+\infty$ with all real positive numbers graduated and marked for each point on the line, as shown in Figure 1.10.

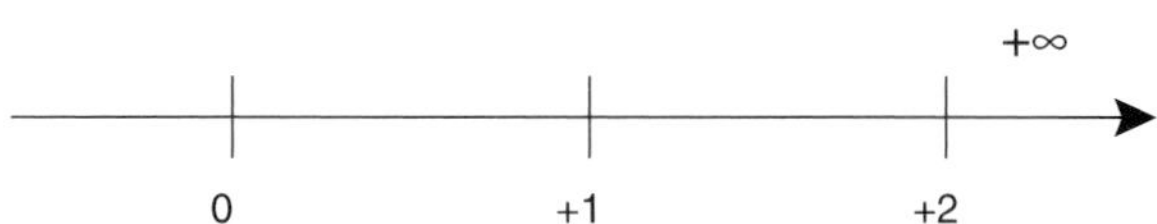

FIGURE 1.10 Gradient scale of real positive numbers in mathematics.

This gradient concept in mathematics, which applies only to real numbers, can now be generalized to any and all data concerning the physical universe quantities (e.g., force, mass, length, etc.) to plot out their order of importance, validity, and so on, as shown in Figure 1.11.

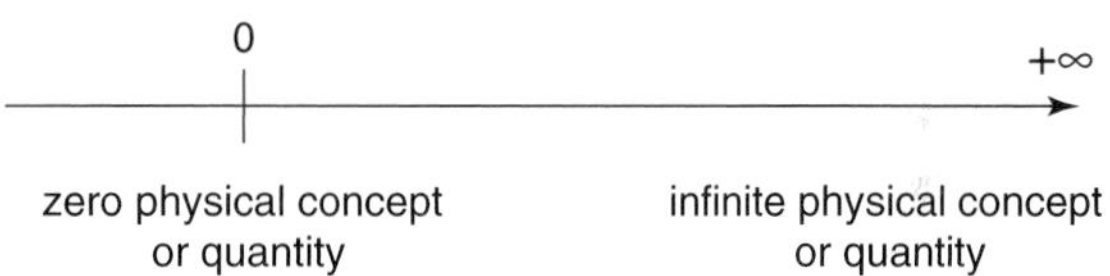

FIGURE 1.11 Gradient scale of physical concepts or physical quantities in the physical universe.

Because this graduated scale is based on the data derived from the finiteness of the physical universe, it would not permit any of the absolutes at either end (0 or ∞), to be achieved, only approached. Now, we need to know exactly what absolute means.

DEFINITION-ABSOLUTE: *(a) That which is without reference to anything else and, thus, not comparative or dependent on external conditions for its existence or for its specific nature, size, etc. (as opposed to relative), (b) that which is free from any limitations or restrictions and is unconditional.*

Thus, in learning a field of study, such as RF/microwave electronics, a presented fact or datum has only relative meaning and lies somewhere on a gradient scale and therefore, should never be considered to be an absolute!

From this graduated scale of plotted values for any quantity, we can see that "Zero (0)" and "Infinity (∞)," being at the extreme ends, exist only as mathematical absolutes—purely theoretical. In practice and in actuality, nobody has ever been able to achieve absolutes in force, mass, size, temperature, power, or energy, and no one ever will. For example, infinite positive charge or infinite voltage is impossible to achieve in the physical universe, no matter how many capacitors or batteries are connected in series to create the positive charge or voltage.

NOTE: *The dichotomy of real positive numbers is "real negative numbers," which extend from 0 to $-\infty$. If we juxtapose the two semi-infinite scales for positive and negative real numbers, we get a totally infinite scale extending from $-\infty$ to $+\infty$, encompassing all real numbers. By applying this concept, we can plot the negative* *charge on the semi-infinite scale to obtain the full spectrum of charged objects, as shown in Figure 1.12.*

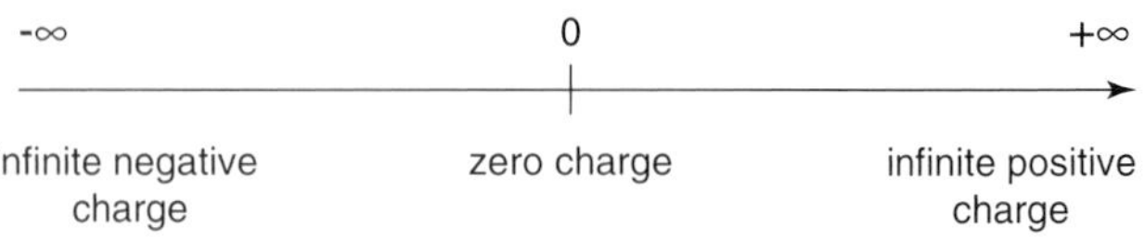

FIGURE 1.12 Gradient scale of charge.

EXAMPLE 1.3

Can a voltage value of 0.00 volts be reached?

Answer:

A voltage value of 0.00 volts (or any other value such as 10.00 V, 13.89 V, etc.), as an absolute, can not be reached and maintained by a source at all times without any further qualifications. Furthermore, such a voltage, even if it exists, can not be maintained indefinitely under all environmental conditions. Then there is the existence of either a measurement or calibration error that makes all measured values valid within a certain error range, making it impossible to achieve such a value. Thus, contrary to theory, such an absolute voltage value does not exist in practice.

This is a direct consequence of working within a finite physical universe. Even if randomly one achieves 0.00 volts, it can never be held at that value unconditionally and over an unlimited period of time without any further qualifications. In fact, this applies to any and all absolutes, as shown in Figure 1.13.

From this simple example, we can conclude the following:

CONCLUSION: **All of the physical laws (each having only a relative value), as cast into an exact mathematical form, are absolutes and can never be achieved in practice (unconditionally and for all times), but they can only be approached. Thus, they serve only as guidelines in practical applications—to find absolute solutions to idealized problems!**

EXAMPLE 1.4

Are Kirchhoff's voltage and current laws always valid?

Answer:

Kirchhoff's voltage and current laws are relatively true and apply at lower frequencies only. At higher frequencies (e.g., RF/microwaves) electromagnetic theory and Maxwell's equations must be used. Still at higher frequencies (e.g., optics) optical laws should be used (see Figure 1.14).

NOTE: *Example 1.4 clearly indicates that all of the laws in electrical engineering do not have a monotone order of importance. Each law has a different weight that should be evaluated properly relative to other laws before it is applied to a practical situation.*

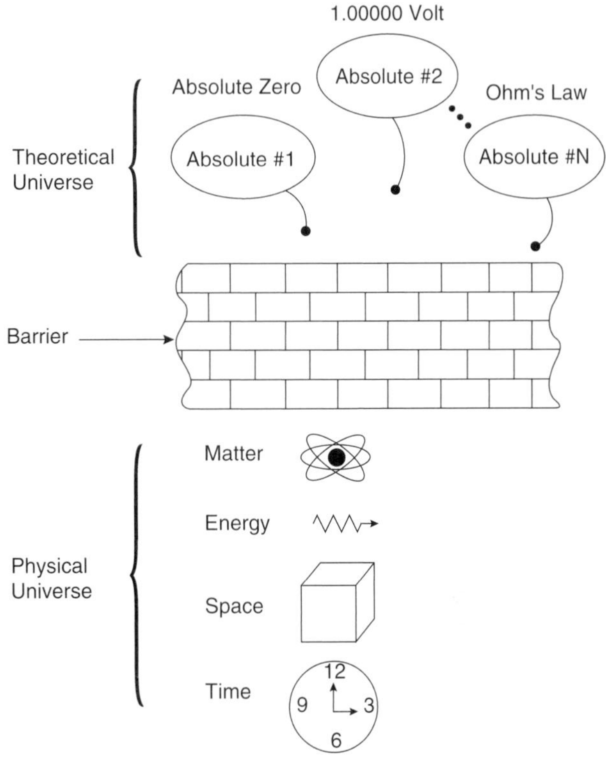

FIGURE 1.13 Relationship of absolutes to the physical universe.

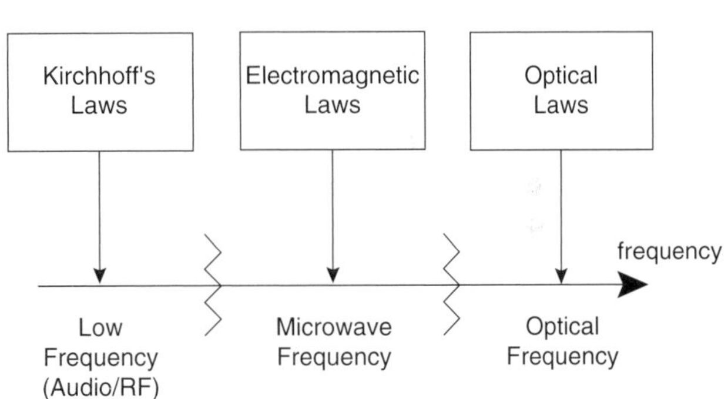

FIGURE 1.14 Relative truths of electrical laws on the frequency scale.

EXAMPLE 1.5

Let's consider the following two statements:

Datum #1: Ohm's law applies to all resistors.

Datum #2: Ohm's law is valid only at low frequencies.

Which one is truer?

Answer:

If we place these two data on a scale of validity, we can see that datum #2 is more valid than #1, as shown in Figure 1.15.

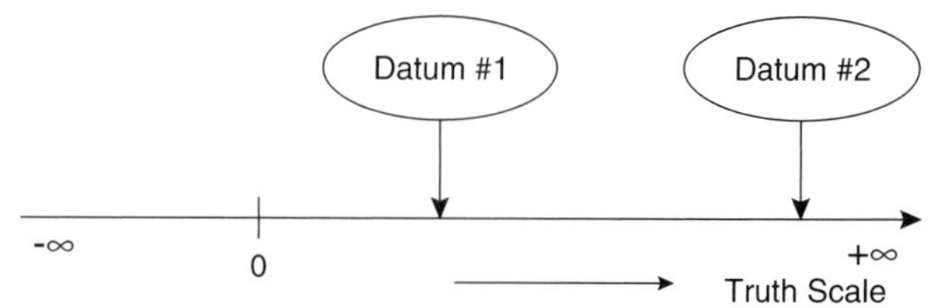

FIGURE 1.15 Relative truth of data.

1.5.3 Data Evaluation

It should be noted that when dealing with physical laws or engineering concepts (or, in general, any scientific datum), we need to find their worth and value for ourselves by examining them very closely and making up our minds about their usefulness before they become a valuable piece of information. This is the evaluation process as we know it, and it should precede any call to judgment.

The method commonly used to establish the value of any scientific data is determining the amount of alignment it provides to other known and existing facts and data in one's mind. Obviously, it goes without saying that the assigned value given to any scientific datum is wholly dependent on, determined, and modified by the viewpoint of the observer of the data. This factor alone, since the beginning of civilization, has created more debate and controversy about new and untried concepts than we can shake a stick at—putting it mildly!

EXAMPLE 1.6

Maxwell's equations embrace all of classical electromagnetic theory at any frequency, and they are considered to be a very valuable set of laws because they create alignment. By alignment we mean that all other low- or high-frequency laws can be derived from Maxwell's equations. How does the point of view of the observer modify Maxwell's equations?

Answer:

The assigned value for Maxwell's equations depends on the observer. For example, the same set of laws are valued differently by a microwave (or antenna) engineer, a power supply engineer, and a computer engineer, as noted here:

Observer	**Assigned value**
Microwave or antenna engineer	very high
Power supply engineer	some
Computer engineer	none

The cases of the microwave engineer and the computer engineer are depicted in Figure 1.16.

NOTE: *It is interesting to note that zero assignment of value to electromagnetic laws by computer engineers and computer engineering curricula has led to many poor designs, electromagnetically speaking! As a result, a whole new field has been*

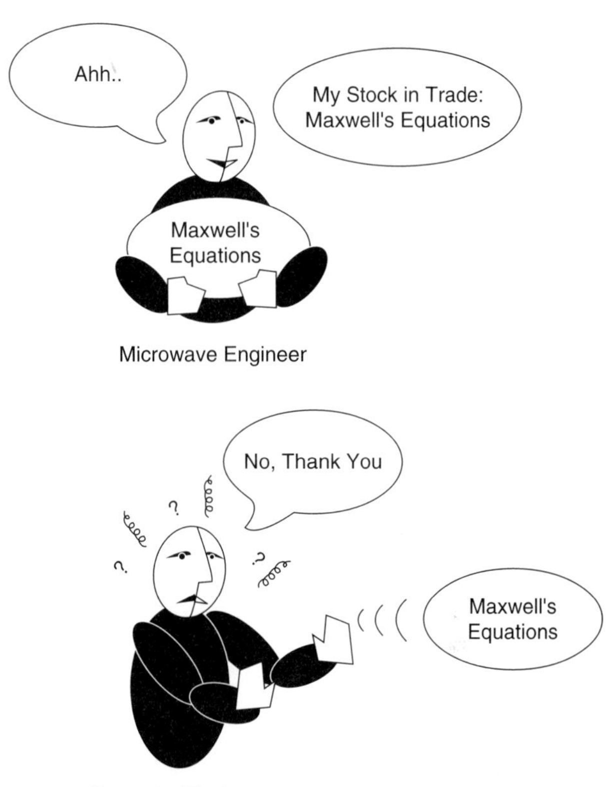

FIGURE 1.16 Assigned value is modified by the observer.

created to combat this problem and its serious ramifications in practice. This new field is aptly called "Electromagnetic Interference (EMI) or electromagnetic compatibility (EMC)," necessitated by assignment of no importance or value to principles of electromagnetic interference and radiation susceptibility by computer designers.

1.5.4 Physical Universe Dichotomy

By observation, we can conclude that physical universe material and the associated concepts come in dichotomies or opposite pairs. This observation leads to the "duality principle," a very important concept in electrical engineering that will be discussed in Chapter 3, *Mathematical Foundation for Understanding Circuits*. For example, (positive and negative), (kinetic and static), (electron and hole), ($+\infty$ and $-\infty$) are pairs of concepts in dichotomy.

By observing and understanding one of the items of a pair, the properties of the other item in the pair can be predicted and grasped. Thus, we can gain a deeper understanding about each item or the whole dichotomy by this process. We can apply this principle to any opposite pairs of concepts, particularly to a "known and unknown pair" or in general to any "two comparable data" having the same order of magnitude. This principle helps us understand the unknown by using the known. Utilizing this principle, we can gain a deeper understanding about the unknown datum than is usually possible (see Figure 1.17).

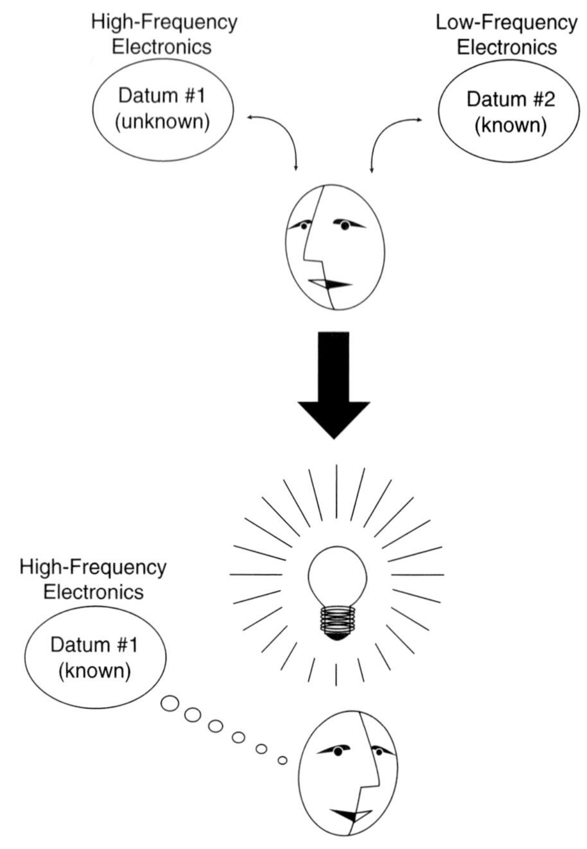

FIGURE 1.17 Understanding of an unknown occurs in the presence of a datum of known and comparable magnitude.

EXAMPLE 1.7

Is the charge of a single electron (a negative charge) by itself meaningful?

Answer:

A negative charge by itself is meaningless unless we compare it with its opposite—that is, a positive charge resident, for example, in a proton. Only then can we understand the concept of "charge."

EXAMPLE 1.8

Can we fully comprehend RF/microwave electronics without any other information?

Answer:

We understand RF/microwave electronics only when its low-frequency counterparts such as electrostatics and magnetostatics are fully understood. In the low-frequency counterpart, the circuit lengths are very small compared to the wavelength (unlike the high-frequency signals); thus, the delay-time in signal travel is negligible and all the low-frequency laws (Ohm's law, Kirchhoff's laws) apply. As will be seen later, as the frequency is increased beyond certain levels, these popular low-frequency laws utterly fail and need to be replaced by more exact laws applicable only at these high frequencies. It remains a fact that familiarity with the low-frequency laws will greatly enhance understanding and evaluation of high-frequency electronics concepts (see Figure 1.18).

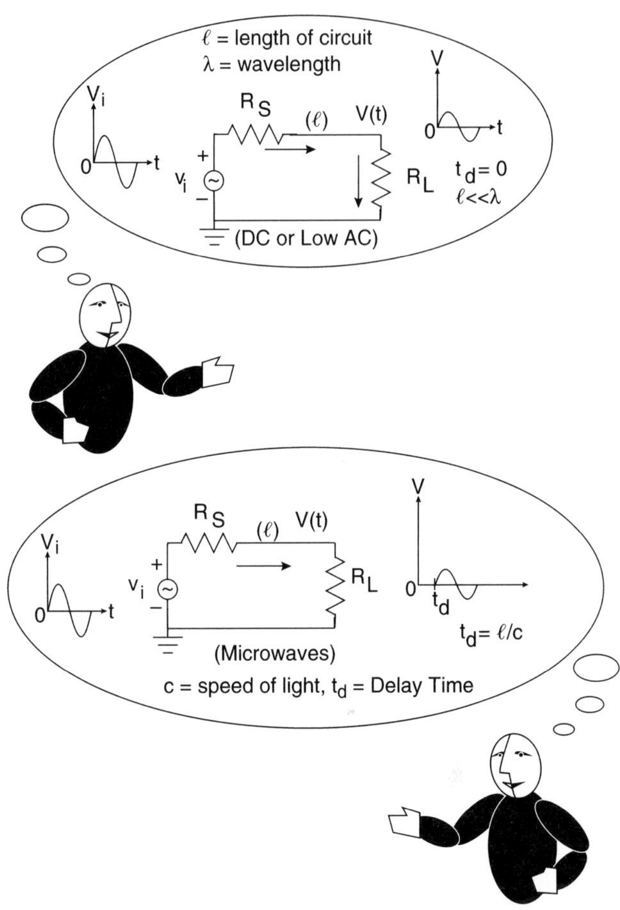

FIGURE 1.18 Low-frequency versus high-frequency circuits.

1.5.5 Solutions to Problems

The field of engineering is concerned primarily with solving problems in a systematic manner. Thus, we need to know in general how any problem can be resolved. The technique is very simple and applies to any problem.

STEP 1. Dissect the problem into sections of similar and related data.

STEP 2. Compare each area to already known natural laws.

STEP 3. Using the natural laws, obtain a solution for each section.

STEP 4. Resolve the remainder that cannot be known immediately by using the known part (and its solution) to arrive at the final answer (see Figure 1.19).

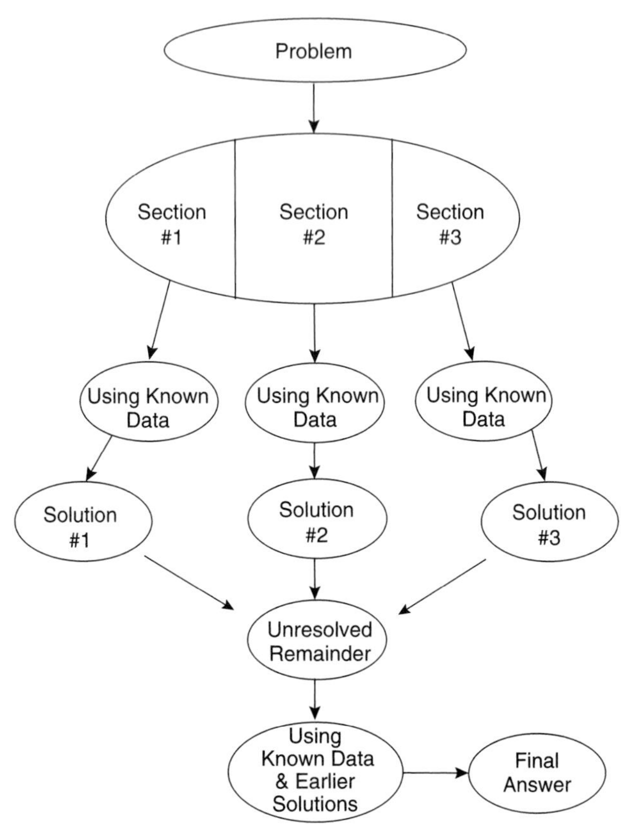

FIGURE 1.19 Solutions to problems.

EXAMPLE 1.9

Let's take the problem of designing an amplifier. How would a design engineer approach this problem?

Answer:

This design problem can roughly be sectioned into the following two areas: DC circuit design and AC circuit design.

The result is shown in Figure 1.20.

It can be observed that if unnecessary or arbitrary factors and assumptions, which do not derive from natural laws and axioms but only from misunderstanding or authoritarian opinion, are introduced into a problem (or a solution) the problem (or solution) will worsen. The following example illustrates this point.

FIGURE 1.20 Solution to amplifier design.

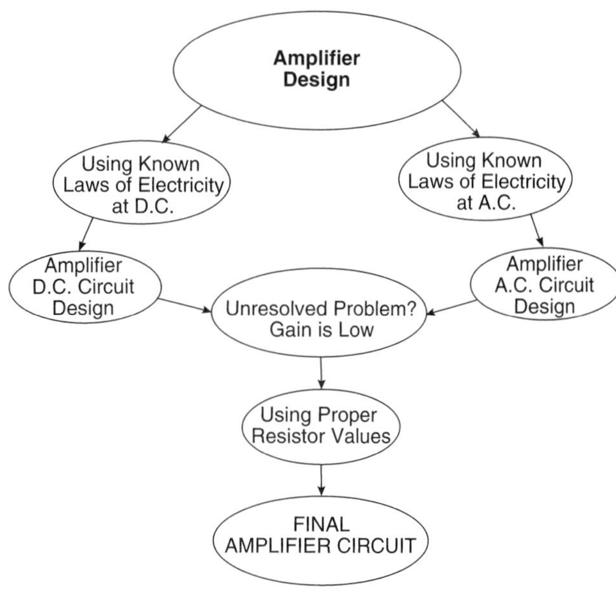

EXAMPLE 1.10

How could we introduce an unnecessary factor into a simple problem such as poor reception in a TV set?

Answer:

If we introduce an unnecessary factor (e.g., a hammer), which is not based on any natural law, we expect to get a worsening of the reception and have to miss our favorite football game altogether (see Figure 1.21).

FIGURE 1.21 Introduction of an unnecessary factor to a TV set.

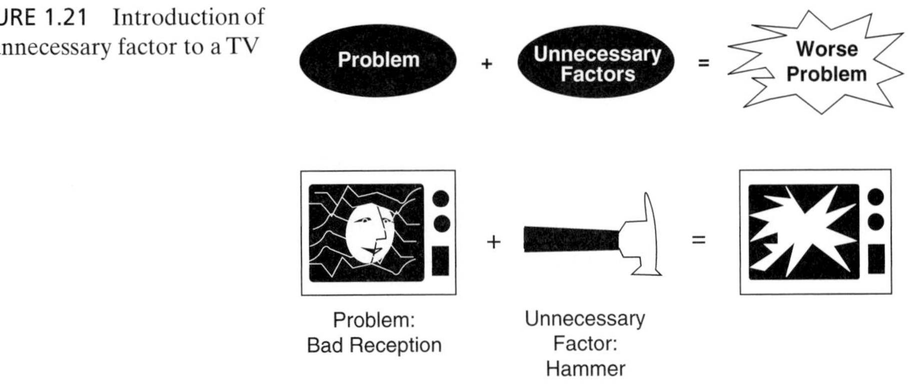

It should be noted that not all unnecessary factors are humorous or as glaring as the one presented in Example 1.10. In fact, the majority are very subtle and involve deviations from known laws, but all can be generally classified as arbitrary factors, defined as any false datum, or interjected rule or decision that does not fit or is unnecessary to a situation. The arbitrary factors in any subject form a general class of data that is derived from mere opinion or personal preference and usually violates natural laws.

It can be further observed that the introduction of an arbitrary factor (mostly generated by misunderstandings or given by authoritative sources) to a problem (or a solution) will invite the further introduction of arbitrary factors into the problem (or the solution) for its proper handling or resolution (see Figure 1.22). This is where the old maxim "lies beget more lies" comes to life and becomes the ruling principle. This concept, interestingly enough, can be seen to be true on a much wider scale of existence and thus, is applicable to all sciences and engineering. Example 1.11 illustrates this concept.

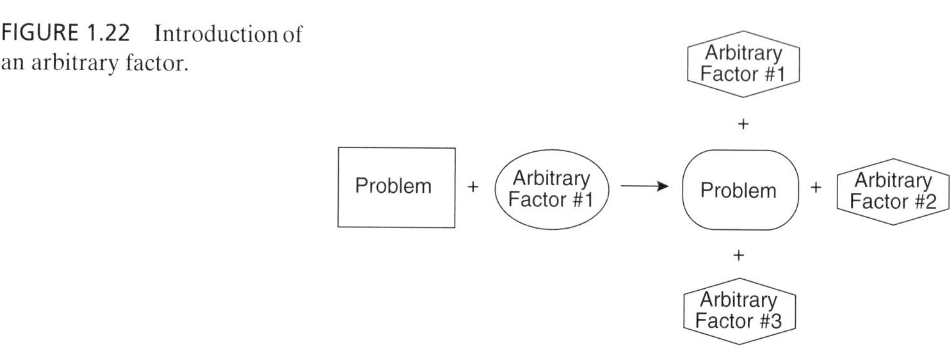

FIGURE 1.22 Introduction of an arbitrary factor.

EXAMPLE 1.11

Let's say that a microwave engineer in the process of designing a microwave amplifier decides to add an arbitrary factor such as "always use bipolar transistors." How would this affect the circuit design steps?

Answer:

Now having added this arbitrary factor to the problem of the design, we can see that the engineer has to add a few other arbitrary factors to keep the circuit functional, such as these:

"Always use high resistor values to increase input impedance"
"Always cool down the circuit to reduce noise"
"Always adjust resistors to get high gain"

These arbitrary factors are added to make the design process more effective and to produce a more functional circuit—all because of one added arbitrary factor, as shown in Figure 1.23.

We can extrapolate this concept of "adding arbitrary factors" further and conclude that those fields of study (e.g., arts, politics, etc.) that mostly depend on false data or authoritative opinions for their source of data contain the lowest number of natural laws (see Figure 1.24).

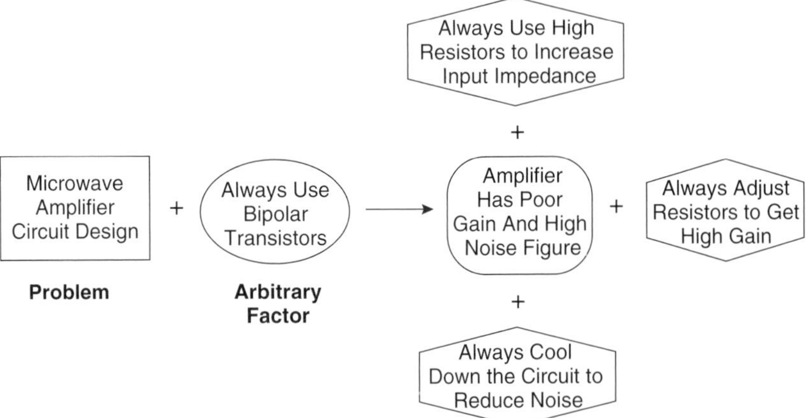

FIGURE 1.23 Example of adding an arbitrary factor.

FIGURE 1.24 Authoritarian opinions.

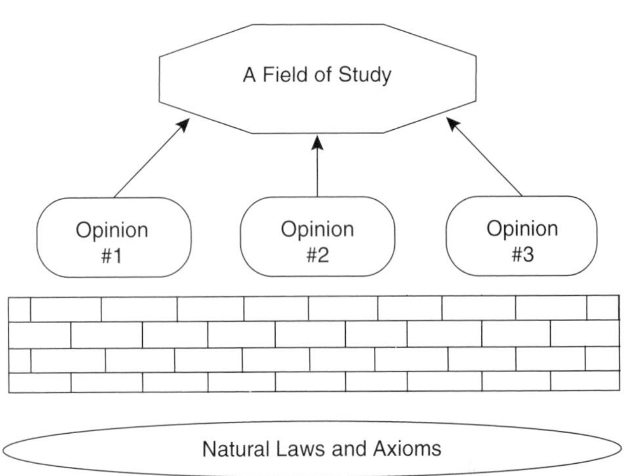

EXAMPLE 1.12

How does this concept apply to the field of "arts"?

Answer:

The field of arts has been bolstered with opinions of newspapers, directors, actors, and reviewers to a point where it is completely void of any of the natural laws of the field of arts (see Figure 1.25).

Uniqueness axiom. The previous discussion and examples bring a much wider sphere of knowledge into focus. This is the realm of "solutions to problems" as a whole class of data. Before we can analyze this class of data, we need to know its most central datum, "the unique solution." It is a well-known axiom in engineering that:

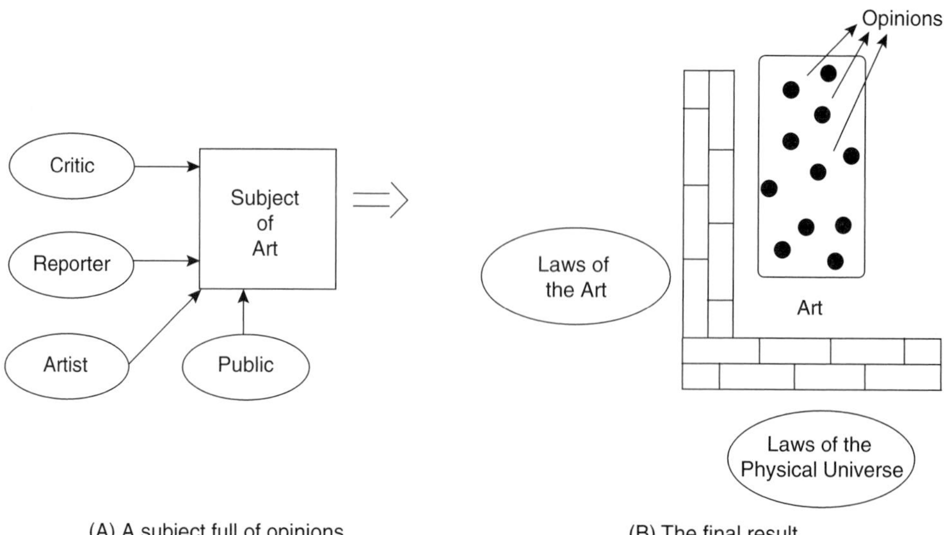

FIGURE 1.25 An example of an authoritarian field.

Given an exact set of necessary and sufficient initial or boundary conditions, any problem in the physical universe has a unique and exact solution.

When we say an axiom we mean a self-evident truth that is universally accepted as true without any proof or exceptions. The important point to remember here with regard to the uniqueness axiom is that it only applies when the problem under consideration exists in the "physical universe" with all necessary assumptions clearly specified.

This is unlike the social sciences and humanities where there may be many solutions to one problem; all of the proposed solutions seem equally valid and provide an equivalent and satisfactory answer to the social problem. This is where modern sciences and engineering fields have diverged greatly from social sciences and humanities and, with the help of uniqueness axiom, have assumed a commanding dominance and presence in our today's society or workaday world.

Unique vs. Arbitrary Solutions. Any unique and exact solution is surrounded by an almost infinite number of "arbitrary solutions," where all are deviations from the central datum. These deviations are all based on an introduction of an "arbitrary factor" at some point along the way in the solution process. Thus, we can conclude that:

Use of arbitrary factors produces arbitrary solutions which are all deviations from the exact and unique solution of any given physical universe problem.

By observation, it could be said that at this time the physical sciences with the help of an exact mathematics and axiomatic methods of analysis, have achieved the highly exalted status of "*unique solution type of sciences*" and systematically can and do obtain unique solutions to any given engineering or physical universe problems.

NOTE: *It is important to note that the uniqueness axiom only applies to "analysis" type problems and excludes all "design" type problems. For example, to fix a broken radio set, we need to troubleshoot the set first to find the exact nature of the problem. Once the problem has been identified to be, let's say, a bad component then only a new component with an exact part or model number would solve that problem (analysis problem—unique solution!). On the other hand, a skilled and knowledgeable circuit designer could easily design several models of a radio set, where all perform the same function and satisfy the same set of specifications but are structurally and electrically different (design problem—myriads of solutions!).*

1.5.6 Workability of a Scientific Postulate

The degree of workability of a scientific postulate is established by the degree to which it explains existing scientific phenomena already known, the degree to which it predicts new and undiscovered phenomena that when looked for will be found to exist, and the degree to which it does not require that imaginary phenomena (i.e., adding an arbitrary factor) be called into existence for its explanation (see Figure 1.26).

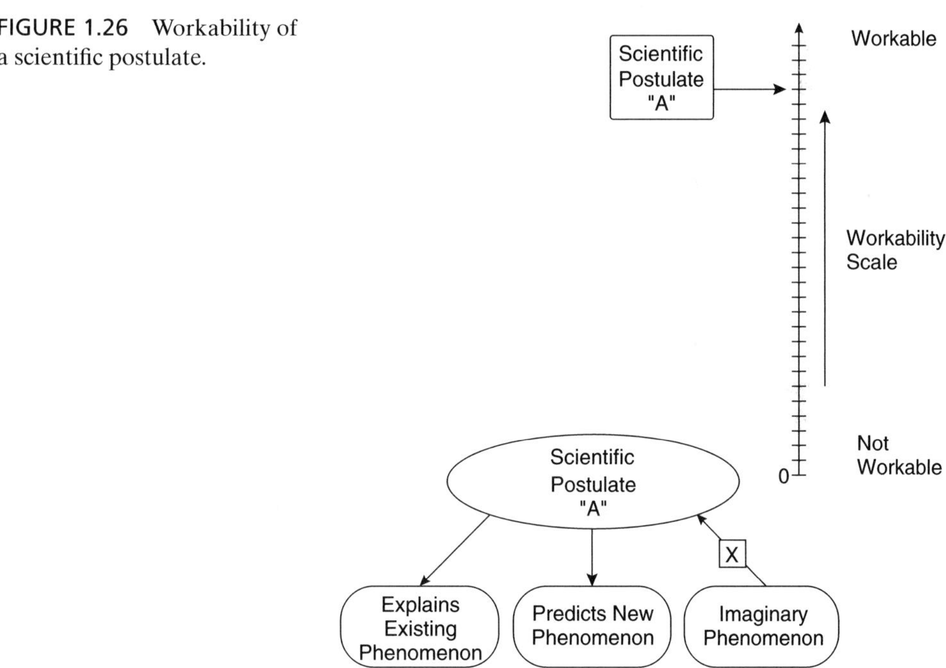

FIGURE 1.26 Workability of a scientific postulate.

EXAMPLE 1.13

How would you describe the practicality and workability of Ohm's law as an electrical circuit postulate?

Answer:
Ohm's law is a major postulate (or assumption) for low-frequency electronics; it states that all resistor elements always behave linearly even though it is not true due to the resistor's composition, geometry, higher-frequency effects, etc. So, the postulate of "resistor linearity" is valuable and workable only so long as it is applied at low frequencies and within a certain allowable tolerance band as designated on the resistor itself (see Figure 1.27).

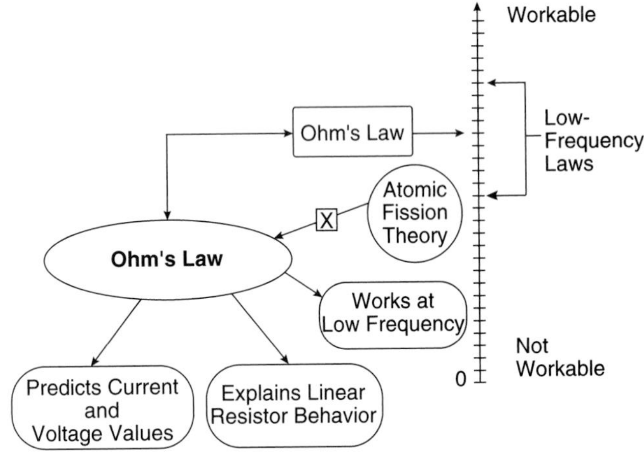

FIGURE 1.27 Ohm's law as a workable postulate.

EXAMPLE 1.14

How would you analyze the field of "arithmetic" in terms of its basic postulates?

Answer:
The field of "arithmetic" has four main postulates: addition, subtraction, multiplication, and division. These postulates and their associated rules are assumed to be true at all times and cannot be proven. As long as we are dealing with real numbers or the decimal number system, these postulates are workable and very valuable. But the moment we deviate, for example, into the binary number system, these postulates become unworkable or useless and have to be replaced by the postulates of "Boolean algebra."

EXAMPLE 1.15

How workable are Maxwell's equations?

Answer:
James Clerk Maxwell had to modify his original set of equations by adding a "displacement current density" term to Ampere's law. This modification created a self-consistent set of equations that theoretically was correct. Years later, Heinrich Hertz conducted many radio experiments whereby he discovered electromagnetic waves and proved the

correctness of Maxwell's equations experimentally. In fact, the addition of the displacement current density term, which was purely a theoretical conjecture at the time, was critical in predicting the wave propagation phenomena well in advance of its discovery by Hertz. This self-consistent classical set of equations is highly workable for all frequencies; it forms the backbone of all modern advances in the field of radar, telecommunications, navigation, telemetry, global positioning systems, and more.

1.6 THE ROLE OF MATHEMATICS

An interesting commentary on the subject of mathematics needs to be addressed at this early stage of our work in this chapter. Mathematics at this stage of development is so interwoven with the sciences that it has almost become an inseparable partner, or so it seems. On further investigation, we can make the following observation:

Mathematics are short-hand methods of stating, analyzing, or resolving real or abstract problems and expressing their solutions by symbolizing data, decisions, conclusions, and assumptions (see Figure 1.28).

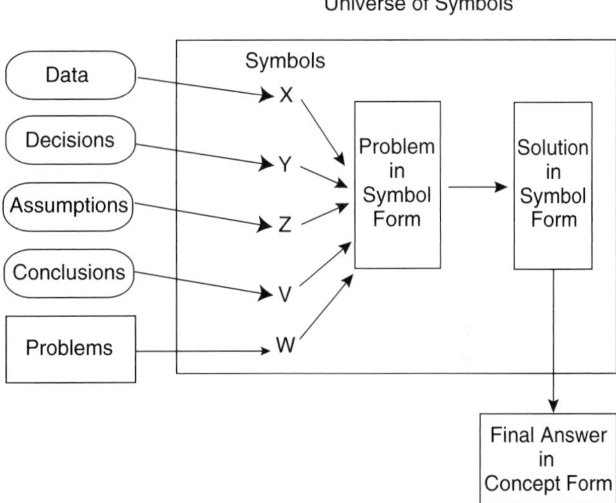

FIGURE 1.28 Use of mathematics.

This means that a science exists as an organized body of data expressed as a series of laws, concepts, and so on, with mathematics only as a servo-mechanism to expedite its organization, conveyance, and communication through symbolic representation. In other words, mathematics is junior in importance to the science it serves.

POINT OF CAUTION: *Utilizing exact mathematics in expressing natural laws or solving problems related to the physical universe introduces an absoluteness into the final solution that contradicts rule of non-absoluteness of the physical universe data. This contradiction can be resolved by realizing that we should expect only an* *approximate correlation between the actual measured quantities and the final solutions obtained through the use of exact mathematics.*

EXAMPLE 1.16

How would you use mathematics to express the concept of slope at a point on a curve in shorthand?

Answer:

The slope of a curve at a point, $f(x)$ at point x_0, is defined to be the rate of change of the function at that point, or the change in the value of the function for an infinitesimal change in the variable value (x). To express this concept mathematically, we note that the slope of a curve is the same as the slope of the tangent line at $x = x_0$; thus, we can write:

$$\frac{df}{dx} = \lim_{\Delta x \to 0}\left(\frac{f(x_0 + \Delta x) - f(x_0)}{\Delta x}\right) = \lim_{\Delta x \to 0}\left(\frac{\Delta f}{\Delta x}\right) \tag{1.1}$$

This concept is shown in Figure 1.29.

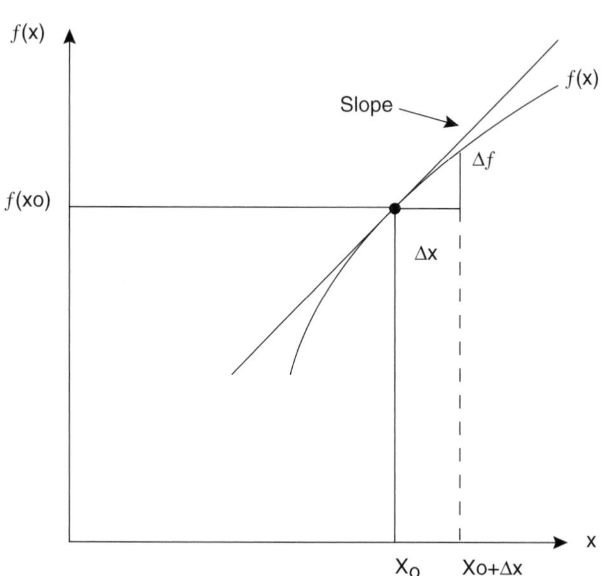

FIGURE 1.29 Slope of a curve.

EXAMPLE 1.17

How would an engineer express his income mathematically?

Answer:

On a first order of approximation, we may consider the engineer's income (I) to be proportional to and depend on:

His work ethics and honesty (H)
Correct actions (A)
Hard work (W)
Knowledge of the subject (K)

Thus, we can write the following relation as a shorthand:

$$I = H.A.W.K \tag{1.2}$$

This simple equation will serve as a shorthand for all the preceding information. It shows a linear relationship between income and the parameters on a first-order of approximation, and it can be used as an abbreviated version of the datum, but never to replace the datum itself (see Figure 1.30).

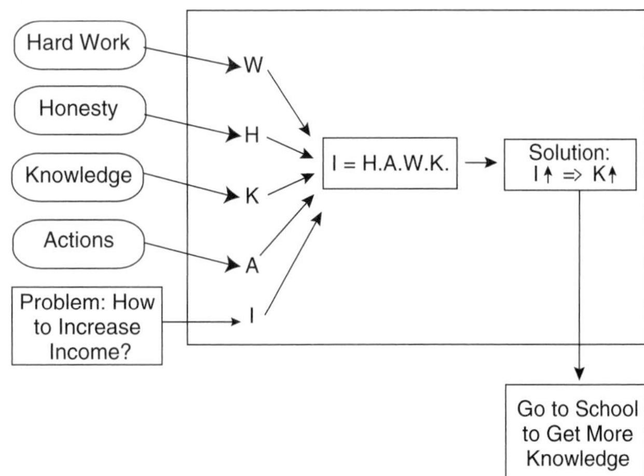

FIGURE 1.30 Use of mathematics in daily life.

EXAMPLE 1.18

How would you express Ohm's law mathematically?

Answer:

Ohm's law is an empirical law that states that the voltage across a resistor equals the amount of resistance of the element multiplied by the current through it. This may sound like a mouthful and hard to remember or carry around. Therefore by assigning the following symbols, we can write a shorthand notation and symbolize Ohm's law as follows.

Let:

$V \equiv$ voltage across a resistor
$I \equiv$ current through the resistor element
$R \equiv$ resistance value

Thus, Ohm's law can be written as a shorthand as:

$$V = RI \tag{1.3}$$

This is shown in Figure 1.31.

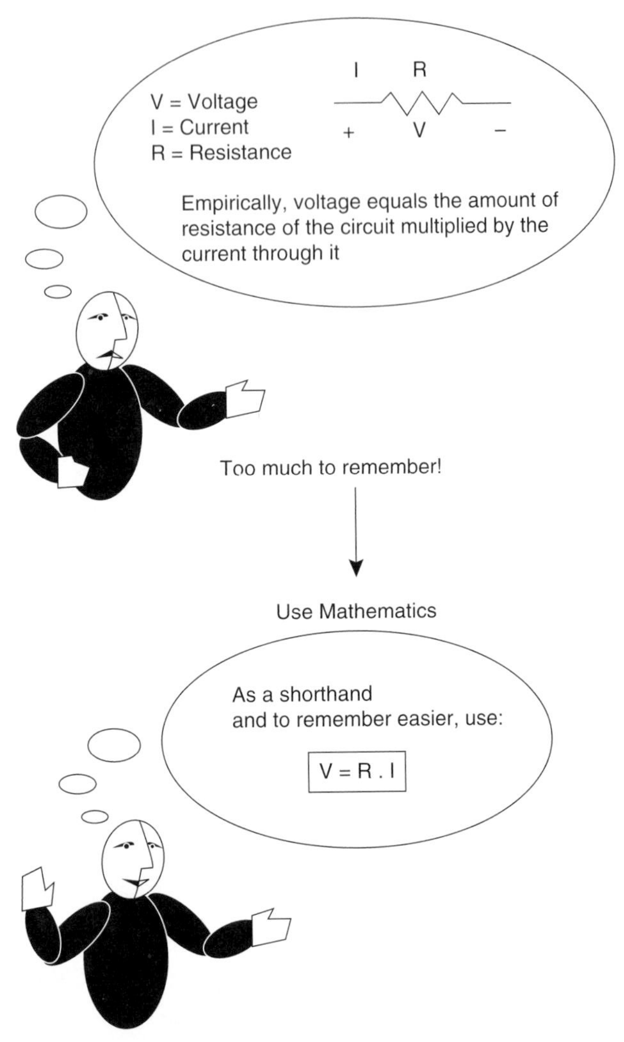

FIGURE 1.31 Use of mathematics in Ohm's law.

1.7 PHYSICAL SCIENCES: CLARIFICATION AND DEFINITION

It should be noted that the principle of conservation of energy applies only to an existing form of energy and can predict its future form or magnitude; however, it makes no determination or prediction about the origin of energy, its initial source, or how it comes about in the first place. It discusses only what happens to it once it exists in the physical universe. In this regard, this could be a limiting factor and a major shortcoming of this principle that is built into all of our extant physical sciences.

For this reason, the principle of conservation of energy applies only to inanimate matter and fails completely when applied to animate forms of life. In other words, it does not apply to animate matter (such as human beings, animals and plants) because these are endowed with life force (which exists in the thought universe), a parameter completely outside the scope of the physical sciences at this time. Thus, in a stretch of imagination we should never attempt to apply the principles presented in this work or in physics particularly (which are purely physical universe concepts) to the life force of life organisms that exist in a broader aspect of existence.

A general living organism can be roughly divided into two components: the life force component and the body or the physical component. This is a very rough division. Even though these two components are closely intertwined, they are distinctly different, with their own set of laws. The science of physics, as defined earlier, does not encroach on the life force component or the biological aspects; it studies the physical component purely as a mass composed of atoms and molecules.

Through rigorous clinical experiments, it has been shown and proven that the principles of conservation of mass or energy are null and void when it comes to the life force or the thought component of life. Thus, the laws in the science of physics and other important physical sciences fail to predict the behavior of a biological organism.

Therefore, at the outset of this work we need to clearly define the legitimate province of physical sciences, particularly physics, engineering, chemistry, and so on.

DEFINITION-A PHYSICAL SCIENCE: *The study and analysis of one or several inanimate components of the physical universe (consisting of matter and energy operating in linear space with the resultant creation of time), using exact mathematics with the intent of using the principles and results thus found for the betterment (or otherwise) of the conditions of mankind or all life organisms that are symbionts to existence.*

The above definition of a physical science is well in agreement with what English essayist/philosopher Francis Bacon (1561–1626 A.D.), who was a champion in the rebirth of scientific thinking and methodology, once wrote about the goal of science, "The sure and lawful goal of science is none other than this: that human life be endowed with new discoveries and power." Up to about the year 1500 A.D., knowledge as revealed by religious and philosophical scholars, was often accepted as final authority on matters of nature and the universe. Therefore, this statement written approximately four centuries ago, was in the teeth of authoritarianism, and in essence, sets the basis and pace for all of our modern physical sciences at this time.

In this regard, the field of "electronics" focuses on the behavior of electrical energy in space under boundary conditions set forth by matter and energy. This study is nothing but a subset of a much bigger field called physics, which is a subset of a vast arena called the physical universe. The latter itself is a subset of an extremely large sphere of existence called life and livingness, as shown in Figure 1.32. This means that in electronics, we are not studying an isolated subject but a very narrow field of study that is a subset of several other subsets.

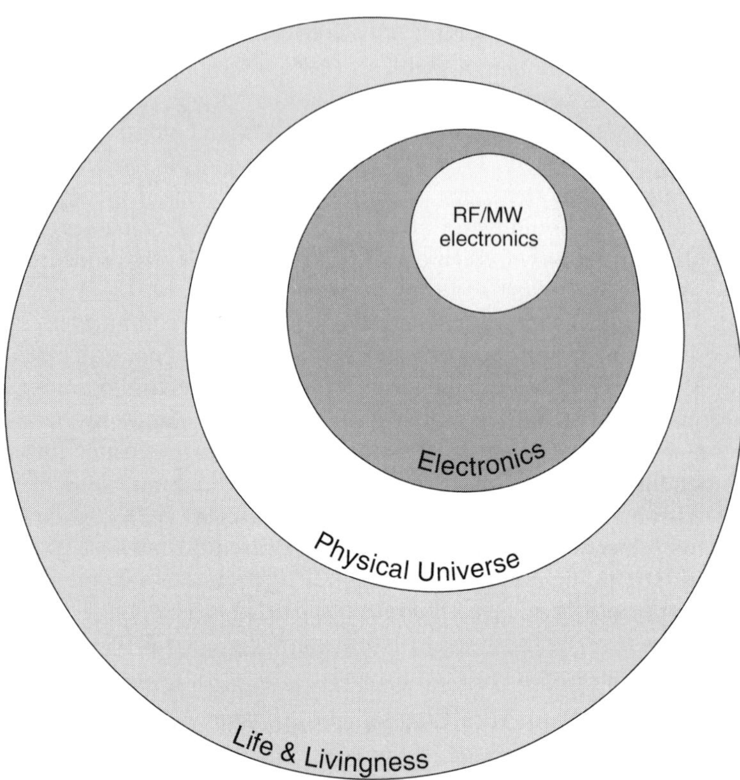

FIGURE 1.32 Field of "electronics" in correct perspective with regard to other fields.

NOTE 1: *As might have been expected, there are commonality and shared concepts between principles of electronics and laws of the physical universe. For example, the commonly known principle of inertia (i.e., a moving object tends to remain in motion, or a stationary object tends to stay motionless unless acted on by a force) may be seen to have application in the current flow in an inductor wherein the current tends to flow continuously and resists sudden shifts in magnitude or changes in its direction of flow (as explained by Faraday's law). This will be studied in more depth in the next chapter.*

NOTE 2: *It is interesting to note that a vague parallelism exists between the physical universe and the life force or the thought universe (such as the concept of space, energy, etc.), but it ceases to carry forward into a complete one-for-one correspondence between the two, as many scientists may have observed. Some attempts have been made by workers in the field (such as sociologists, psychologists, etc.) to apply the principles discovered in the physical sciences to the life force of an organism. The results have been sporadic and mixed. The reason is that the study of the life force or the thought component is a specialized field of study with its own set of laws, which* *are well beyond the scope of this work or any known physical science at this time. Therefore, the two components should never be confused.*

1.8 SUMMARY AND CONCLUSIONS

As part of the author's goal, the presentation of a number of these fundamental concepts is merely an effort to bring about the following:

- To set the reader on the right "thinking track" and eventually lead to a better appreciation of sciences and engineering, in general.
- To establish a scientific framework by which a science is presented and the concepts are interrelated in a logical and hierarchical fashion. This scientific framework allows the students of science to organize their thoughts logically!
- To enable the reader to select out the more salient aspects of a subject correctly and, therefore, be able to think with the data rather than memorizing it. This is what normally we refer to as the process of "selection of importances," which is crucial and essential to the full grasp of a science. This process prevents making the common mistake of assigning a monotone degree of importance to all scientific data.
- To promote a better and deeper understanding of the scientific methodology and improve one's study habits and comprehension rate.

These missing fundamentals from the field of science/engineering as well as education have caused great turmoil in the minds of engineering and science students all over the world. The brief presentation of these basics in this chapter is expected to bring about a renaissance in the field of engineering as a whole, particularly in the field of high-frequency electronics, a rapidly advancing field.

The author firmly believes, that in order to become a competent and proficient RF/microwave electronics engineer, one needs to master only a few basics including (but not limited to) the following:

- Fundamental postulates of the physical universe
- Fundamental axioms and the scientific methodology
- Essential electrical laws of electricity and magnetism
- Exact definitions for all basic and related electronics nomenclature
- The "low-frequency electronics" concepts
- The "high-frequency electronics" concepts

Armed with these tools, one is well equipped to handle the problems of the workaday world of electronics, particularly analysis and design, and will have a higher probability of success in becoming a professional RF/microwave engineer.

LIST OF SYMBOLS/ABBREVIATIONS

A symbol/abbreviation will not be repeated once it has been identified and defined with its definition remaining unchanged.

AC	Alternating current
DC	Direct current
c	Speed of light
I	Current
ℓ	Length of circuit
R	Resistance
RF	Radio frequency
R_L	Load resistance
R_S	Source resistance
t	Time
t_d	Delay time
TV	Television
V	Voltage
V(t)	Voltage with respect to time
HAWK	Honesty, actions, work, knowledge

PROBLEMS

1.1 What are the two major subdivisions of any science? Describe each. Which one of the two is more important? Why?

1.2 What are the main ingredients (in terms of knowledge, skill, tools, etc.) that would help to develop a competent engineer?

1.3 Why does a student need to know exact definitions of terms and symbols? Explain with an example.

1.4 What is meant by an absolute? Explain in terms of physical quantities and include an example of why an absolute is impossible to achieve but can only be approached.

1.5 Using an example, describe why any datum or "piece of information" has only relative importance, value, or seniority?

1.6 What is meant by the dichotomy in the physical universe material? Give an example.

1.7 Describe what happens when we add factors unnecessarily and arbitrarily to a problem. Give an example.

1.8 How do you determine the value of a postulate (or assumption)? Give an example.

1.9 Describe the steps used in solving a problem in general. Give an example.

1.10 Give an example of a field of study that is filled with opinions and theories. How many scientific laws can you find in that field?

1.11 Describe what mathematics is (in your own words) and what it does to scientific concepts. Give an example.

REFERENCES

[1.1] Boas, M. *The Scientific Renaissance.* New York: Harper, 1965.

[1.2] Buckley, H. *A Short History of Physics.* London: Methuen, 1927.

[1.3] Cajori, F. *A History of Physics.* New York: Dover, 1962.
[1.4] Dampier, W. C. *A History of Science.* Cambridge: Cambridge University Press, 1966.
[1.5] Gillmore, C. S. *Coulomb and the Evolution of Physics and Engineering in the 18th Century.* Princeton, NJ: Princeton University Press, 1971.
[1.6] Hart, I. *Makers of Science.* New York: Oxford University Press, 1923.
[1.7] Hawking, S. W. *A Brief History of Time.* New York: Bantam Books, 1988.
[1.8] Lipson, H. *Great Experiments in Physics.* Edinburgh: Oliver & Boyd, 1965.
[1.9] Mackenzie, A. E. E. *The Major Advancements in Science.* Cambridge: Cambridge University Press, 1960.
[1.10] Marion, J. V. *A Universe of Physics.* New York: John Wiley & Sons, 1970.
[1.11] Pledge, H. T. *Science Since 1500.* London: Science Museum, 1966.
[1.12] Radmanesh, M. M. *Obstacles to Comprehension of Engineering Sciences*, 2nd Annual IEE Engineering Conference Digest, Sacramento, CA, April, 1992.
[1.13] Radmanesh, M. M. *Creativity in Engineering Education for Higher Student Retention,* ASEE Pacific Southwest (PSW) Engineering Conference Digest, Flagstaff, AZ, October 1993.
[1.14] Runes, D. D. *A Treasury of World Science.* New York: Philosophical Library, 1961.
[1.15] Schwartz, G. and P. W. Bishop. *Moments of Discovery.* New York: Basic Books, 1962.
[1.16] Taton, R. *History of Science*, Vols. I–IV. New York: Basic Books, 1964.
[1.17] Wolff, A. *A History of Science, Technology and Philosophy in the 18th Century.* London: Ruskin House, 1952.

CHAPTER 2

Fundamental Concepts in Electrical and Electronics Engineering

2.1 INTRODUCTION

Very basic concepts underlie all education in electrical or electronics engineering. Knowing these basic concepts gives us a much deeper grasp of the more sophisticated concepts and prevents confusion and discouragement in studying this subject.

The science of electrical engineering sits squarely on the premises of the science of physics and its fundamental postulates. The science of physics deals with the fundamental building blocks of the whole physical universe, which has been discovered to be composed of "matter" and "energy" in constant interaction with themselves and with one another, all taking place in a medium called linear space. This constant interaction leads to another factor: time. These components will be specifically defined so that the reader may gain familiarity with and understand their nature and properties. For example, take a transistor operating in a circuit. This transistor has a mass (semiconductor materials, metal, etc., used in its fabrication) and handles electrical energy (current, voltage, etc.). Furthermore, it is presupposed that it occupies a certain space and has existed for a certain length of time (since its inception) during which there has been interaction of various energy forms with its structure.

To proceed further in our study of electronics we have to have specific definitions of the building blocks of the science of physics: energy and matter.

2.2 ENERGY

Energy is encountered in all aspects of existence, particularly in all physical scientific arenas. It is one of the foremost entities and deserves an exact definition.

DEFINITION-ENERGY: *The capacity or ability of a body to perform work. Energy of a body is either potential motion (called potential energy) or actual motion (called kinetic energy).*

From this definition, we can observe that the subject of energy deals with either **potential** or **actual motion**. Thus, energy could be subdivided into two categories: potential energy and kinetic energy, as discussed briefly in the text that follows. For example, a flying object has both potential energy (due to its height above the ground) and kinetic energy (due to its motion).

2.2.1 Potential Energy

DEFINITION-POTENTIAL ENERGY: *Any form of stored energy that has the capability of performing work when released.*

This motion is brought about because of proximity of particles relative to one another. Therefore, this type of energy could also be called "proximity energy."

The release of energy first translates into a quantity of force followed by work being performed. For example, two charges in close proximity to each other (as in a battery) have a stored potential energy (see Figure 2.1); this means that energy stored in the two terminals of a battery is potential energy until released in the form of force as an intermediary agent to perform the work at hand, such as starting an engine. Electrical forms of potential energy include the following:

- Static electric charge (i.e., nonmoving charge)
- Static electric force (or static electric field)
- Voltage (or electrical potential difference)

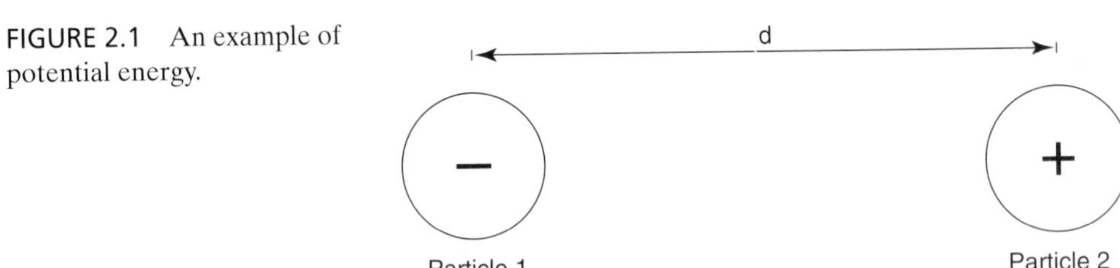

FIGURE 2.1 An example of potential energy.

2.2.2 Kinetic Energy

DEFINITION-KINETIC ENERGY: *The energy of a particle due to its motion.*

As an example, a moving electron in a conductor has a kinetic energy (see Figure 2.2), which could be converted into another form of energy such as in an incandescent filament to give off light or it could give up its kinetic energy to a highly resistive medium and create heat (such as in an electric oven). Electrical forms of kinetic energy include the following:

- Time-varying electric charge
- Dynamic electric field (i.e., time-varying field)
- Electric current (or moving charges in a region)
- Magnetic force (or magnetic field)
- Electromagnetic waves

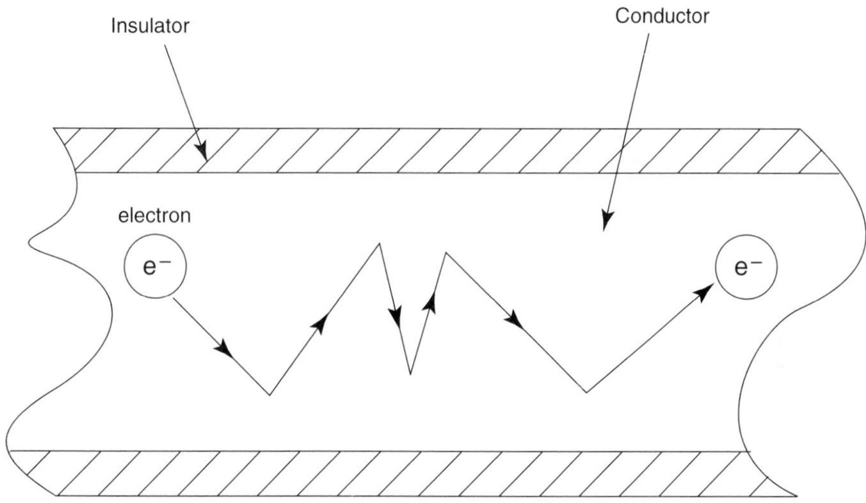

FIGURE 2.2 An example of kinetic energy.

2.2.3 Classical versus Quantum Physics

Before we proceed any further we need to define two important terms at this juncture.

DEFINITION-CLASSICAL PHYSICS (ALSO CALLED NONQUANTIZED OR CONTINUUM PHYSICS): *The branch of physics based on concepts established before quantum physics in conformity with Newton's mechanics and Maxwell's electromagnetic theory. In this branch, the distribution of energy, matter, and other physical quantities are studied under circumstances where their discrete or quantum nature is unimportant, and they may be regarded as continuous functions of position.*

DEFINITION-QUANTUM PHYSICS (ALSO CALLED QUANTUM MECHANICS OR QUANTUM THEORY): *The study of atomic structure that states an atom or molecule does not radiate or absorb energy continuously. Rather, it does so in a series of steps, each step being the emission or absorption of an amount of energy packet (E) called a quantum. The energy in each quantum is directly proportionate to the frequency (υ) of radiation or absorption, i.e., $E = h\upsilon$, where h is Planck's constant. It is the modern theory of matter, electromagnetic radiation, and their* *interaction with each other; and it differs from classical physics, which it generalizes and supersedes, mainly in the realm of atomic and subatomic phenomena.*

Based on the previous two definitions, and from the standpoint of motion, we can subdivide kinetic energy of a particle into two approximate divisions.

2.2.4 Types of Motion

There are two types of motion that need to be considered:

Quantum-physics type motion deals with motion of particles on atomic or subatomic scale that is governed by quantum physics. This type of motion could also be called "small motion." For example, consider an atom. It is composed of many electrons and a nucleus. Each electron has a spinning motion about itself while it also moves in an orbit around the nucleus. This particular subdivision of kinetic energy leads to study of atomic and subatomic particles, a subject called quantum mechanics, which was described earlier and is not covered in this book. It is mentioned only for completeness of the presented ideas.

Classical-physics type motion (also called large motion) deals with motion of particles of matter much greater in size than an atom and at speeds much slower than the speed of light and energy fields that appear to be continuous functions of positions, all governed by the laws of classical physics. This type of motion could also be called "large motion."

The classical-physics type motion breaks down into three classes of motion:

- A flow
- A divergence
- A standing wave

Classical-physics type motion and its associated kinetic energy is the area of primary focus of the current text. Each subdivision of classical-physics type motion is described next.

A **flow** is a transfer of energy particles, objects, or waves from point A to point B, such as electrical flow or heat flow. See Figure 2.3.

A **divergence** is the generalization of the concept of a flow and is defined to be the net outflow (or inflow) of particles from a common point. This is very similar in concept to but not synonymous with what is commonly referred to as "divergence of a vector quantity" in physics, which is a vector operation measuring the net outflux (or the influx) of a vector quantity from a point in space.

EXAMPLE 2.1

What are several examples of energy divergence?

Solution:

The following are several examples of energy divergence, which can be observed:

Explosion from a point (Figure 2.4a)

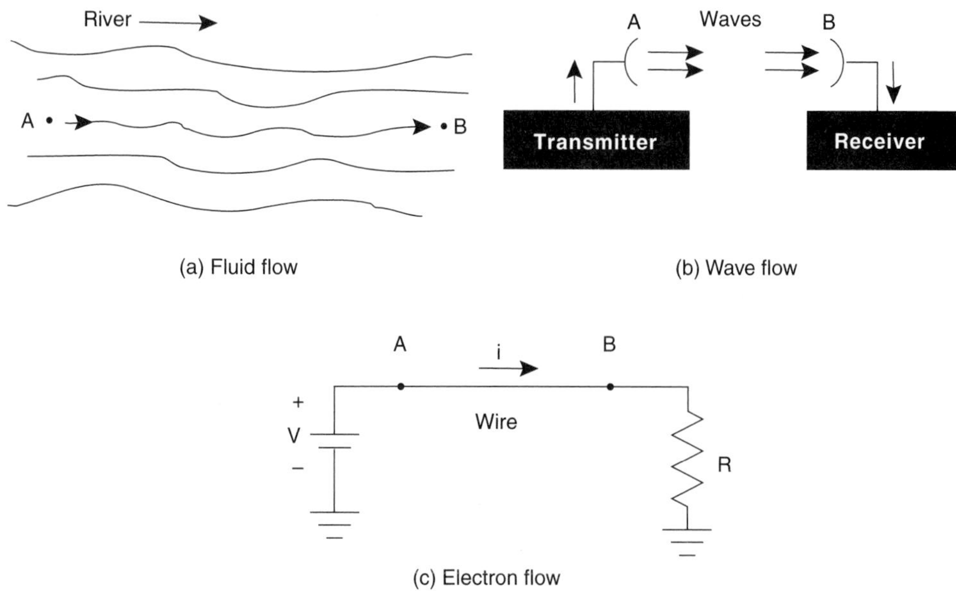

FIGURE 2.3 Examples of flow.

Implosion into a point (Figure 2.4b)
Radiation from an antenna (Figure 2.4c)
Electric Field (E-field) of a time-varying charge (Figure 2.4d)

A **standing** wave is caused by two energy flows, impinging against one another, with comparable characteristics to cause a suspension of energy particles in space, enduring with a duration longer than the duration of the flows themselves (see Figure 2.5).

It is important to note that characteristics of the two flows must be comparable in all aspects such as type, frequency, and amplitude. This means that, for example, light waves and microwaves do not form a standing wave (different frequencies), sound waves and microwaves do not lead to a standing wave either (different types), and a 1 Watt/meter^2 beam of microwaves at 1 GHz would not form a standing wave with another microwave beam of the same frequency but with only 1 milliWatt/m^2 of power density (different amplitudes).

It should be noted that unlike flows, which have direction, standing waves have location only. The frequency associated with a standing wave corresponds to the frequency of one of the flows before the impingement.

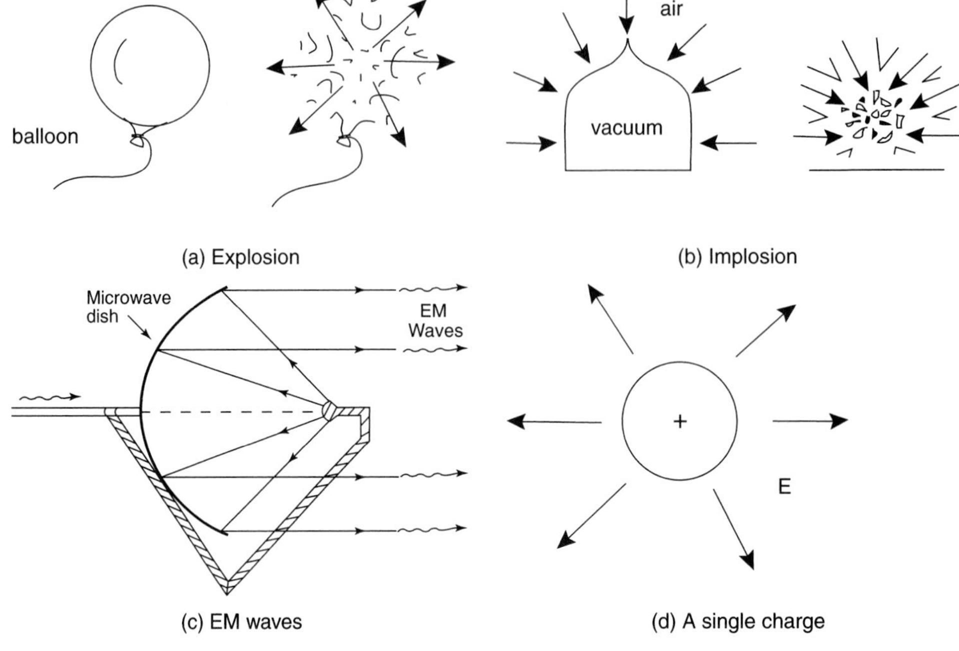

FIGURE 2.4 Examples of a divergence (or a dispersal).

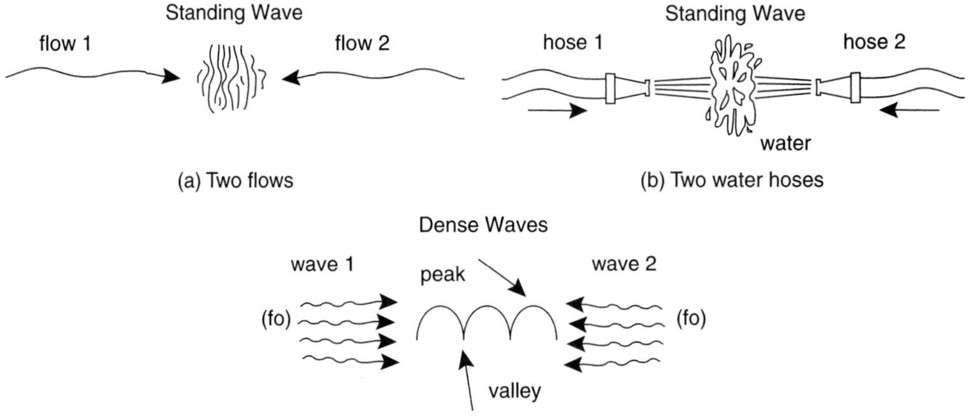

FIGURE 2.5 Example of standing waves.

2.3 MATTER

As described in Chapter 1, *Fundamental Concepts of Science and Engineering*, matter and energy were considered separate entities until recent discoveries and advances showed their equivalence. Thus, based on this understanding we need to define matter at this point.

DEFINITION-MATTER: *Matter particles are the result of bringing energy particles into close proximity where they occupy a very small volume.*

For example, we need to bring electrons into close proximity with a nucleus to create a matter particle called an atom. This concept is shown in Figure 2.6a. Conversely, matter becomes energy if dispersed or decompressed. For example, a nuclear blast converts matter into energy (see Figure 2.6b).

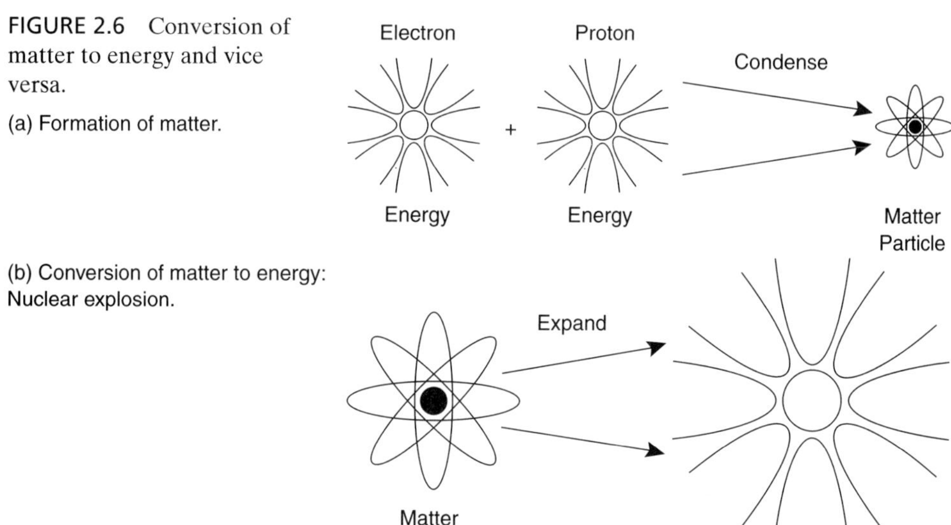

FIGURE 2.6 Conversion of matter to energy and vice versa.
(a) Formation of matter.
(b) Conversion of matter to energy: Nuclear explosion.

NOTE: *Mass is a fundamental property of a body that determines the acceleration of the body when a given force is applied. Mass, a measure of the quantity of matter, is only one aspect that describes the matter quantitatively and is usually considered to be a constant although it varies with the velocity of the body. Mass and energy are interchangeable by Einstein's relationship* $E = mc^2$*, which states that energy (E) is equal to the mass multiplied by a proportionality factor of* c^2.

POINT OF INTEREST: *Conversion of mass to energy is not the same as "combustion" as occurs in most modern engines. Combustion of fuel with oxygen releases the energy stored in chemical bonds, forming new chemicals with total mass at the end of the process remaining almost the same.*

Actual conversion of the molecules of the fuel into energy has no byproducts with the final mass being zero. For example, converting one kilogram of mass produces approximately 9×10^{16} joules of energy, which could light up Los Angeles for more than 10 years! This fact shows our current inefficient methods of creating energy, which gives us an energy crisis every few years and a polluted planet.

OBSERVATION: *We can observe that any piece of matter, on a classical-physics level of observation, could be considered to be a standing wave in its last stage of suspension in space (see Figure 2.7). Standing waves have a quite long duration and can exist long after the flow itself has ceased to exist. This is analogous to memory recordings of an incident in one's memory banks where one can recall and find out what happened long after the force of the impact has disappeared!*

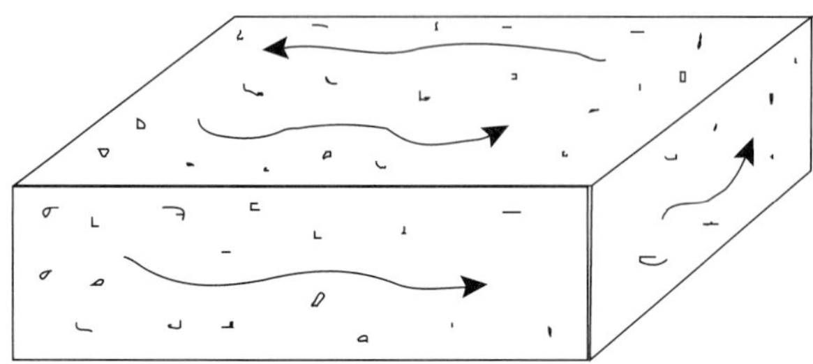

FIGURE 2.7 Matter as a standing wave with net motion equal to zero.

2.4 ADDITIONAL CONSIDERATIONS IMPLICIT IN PHYSICS

As it turns out, two additional considerations are immediately associated with the science of physics. These two additional considerations are intimately intertwined with the subject, and they need special attention as follows:

- Space is a consideration that is implicit but actually not embraced in the science of physics because it is one step behind the existence of matter and energy and needs to exist before matter and energy can exist.
- Time is a consideration that is a result of matter and energy and proceeds from their interactions.

Now we define and delineate each of these two considerations in more detail.

2.4.1 Space

The study of the existence of matter and energy as done in physics presupposes the existence of linear space in which we are placing these entities. Linear space is the first consideration behind the science of physics that is implicitly alluded to in its definition but not directly addressed. Thus, it is essential that we regress one step back from energy and matter and examine space in order to build the science of physics from the ground up—metaphorically speaking!

DEFINITION-SPACE (ALSO CALLED CREATED SPACE): *The continuous expanse extending in all directions that precedes existence of matter and energy and within which all things exist.*

Even though by definition there are an infinite number of dimensions extending out from a point of view in all directions, however in sciences, this situation has been greatly simplified by the use of a reference point and a set of three graduated axes for each of the principal dimensions (length, width, and depth) to bring about a "coordinate system of measurement." There are several types of coordinate systems:

- Rectangular (or Cartesian) coordinate systems
- Cylindrical coordinate systems
- Spherical coordinate systems

For example, a Cartesian coordinate system is shown in Figure 2.8.

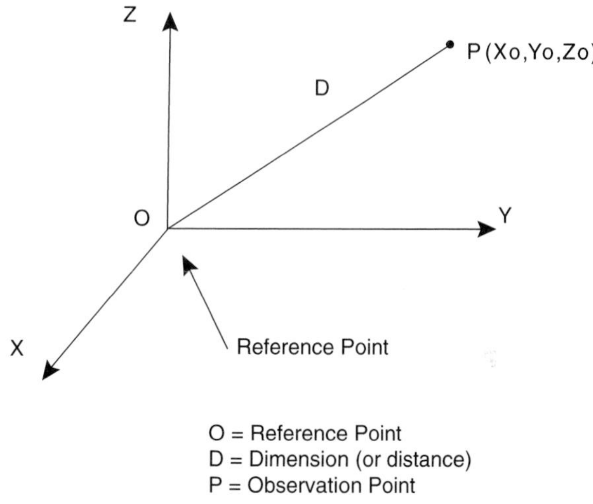

FIGURE 2.8 A Cartesian coordinate system.

The use of a coordinate system as a standard and indispensable tool in all of scientific analysis is based on the fact that any observation point can be uniquely located in space once all of the three coordinates of that point are specified.

The concept of space brings about the idea of a coordinate system, which exists in all aspects of physical sciences and mathematics. Mathematics attempts to provide ideal solutions to abstract or idealized problems posed by a science. These problems involve primarily space and then other variables, to which mathematics provides answers, through abstraction and symbolization of physical entities. The use of a coordinate system interwoven throughout the physical and engineering sciences is so inevitable and essential to the understanding and analysis of problems and design of new structures that without it most "sciences" become unworkable and highly speculative.

Thus, as can be seen from the previous observations, space assumes a senior position in construction of any universe (particularly, the physical universe) because it must exist first before other entities, such as matter and energy, can be located within it.

2.4.2 Time

As discussed, space is behind or precedes the existence of matter and energy; however, time is in front of or proceeds from them. The constant and continuous interaction of matter and energy with each other brings about the concept of time. The basis of time originates in altering the position of a particle in a region of space relative to another particle. Thus, in order to define time, the following requirements need to be met:

1. A region of linear space needs to be created.
2. Matter particles (or objects) are to be placed in this region of space.
3. These matter particles must have sufficient energy to cause their motion relative to one another.

With this preamble, time comes into view as defined next.

DEFINITION-TIME: *That characteristic of the physical universe at a given location (on a macroscopic and/or microscopic level) that orders the sequence of events. It proceeds from the interaction of matter and energy and is merely an index, which is used to keep track of change of particle location. The fundamental unit of time measurement is supplied by the earth's rotation on its axis while orbiting the sun.*

So, time is the action of energy in space, and its keynote concept is **"change in location relative to."** Thus, to have time, the particle's location or its relationship in space should change relative to its starting point, ending point, and other particles (see Figure 2.9). Time signifies change. For example, constant change of location and orientation of earth (orbital motion and rotation) relative to the sun and other planets gives us time that can be measured electronically on a digital clock or mechanically by a watch or an hourglass.

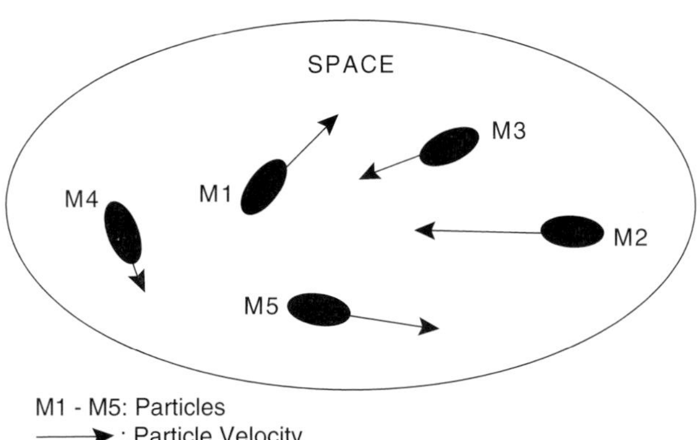

FIGURE 2.9 Definition of time.

NOTE 1: *On a lighter note, we could consider space as "the old ancestor or predecessor," while matter and energy are "the parents" with time as their direct "offspring." Thus, the phrase "Father Time" is a misnomer!*

NOTE 2: *It is also interesting to note that even though time apparently seems to be a measurable quantity, it is an abstract concept representing "change." Existence of time depends purely on the concept of "change of location in space" of particles relative to one another at a certain and known rate.*

2.5 THE FIELD OF ELECTRONICS

Electronics is a subset of the science of physics and needs to be defined at this juncture.

DEFINITION-ELECTRONICS: *The field of science and engineering concerned with the behavior of electrons (or lack thereof) in devices and the utilization of such devices.*

With this definition in mind we can see that electronics is, in essence, a field of science that deals with the study, control, and conduction (through different media—semiconductors, conductors, insulators, gases, a vacuum, etc.) of electricity, which is electrical energy in the form of electrons.

Knowing this definition and the previous preamble on the basics of the physical universe, we need to know the essential facts of the subject of electronics to understand and deal with it correctly.

We need to be thoroughly familiar with several fundamental concepts of the subject of electronics before we go very far in this field of study. These concepts can be briefly summarized as follows:

Fundamental concept #1—Electric charge. The basic building block of electrical energy is electric charge (in motion or not).

Fundamental concept #2—Byproducts of charge. Dealing with electric charge, we need to be thoroughly familiar with all of its different phenomena, byproducts, and the ramifications that it creates. These different aspects of existence of charge could be summed up as follows:

Byproducts of charge (electron, proton, etc.)

Byproduct #1: Generation of forces and fields

- ***a.*** *Electric force*
  - *Electric field*
- ***b.*** *Electric current*
  - *Magnetic force*
  - *Magnetic field*

Byproduct #2: Performing work

- ***a.*** *Work*
- ***b.*** *Voltage (or electrical potential difference), which is Work per unit charge*
- ***c.*** *Power, which is Work per unit time*

Byproduct #3: Transfer of electrical energy

- Electromagnetic waves

Fundamental concept #3—Electrical energy. The field of electronics deals with electrical energy that is either in the form of charge, which is either flowing (current) or is being stored (voltage) in various devices, transmission lines, elements, etc., or in the form of one of the byproducts of charge, such as electric field, magnetic field, or electromagnetic waves.

All of the electrical and/or magnetic laws on a classical level of observation deal with the flow or storage effect of charge or its byproducts under idealized conditions. This is a sweepingly general statement that applies to all known natural laws in the field of electrical engineering. Thus, our primary target in this text is the study of electrical energy flow (or lack thereof) through a region of space created by a device, a transmission line, an element, and so on.

Fundamental concept #4—Mathematical models. The use of exact mathematics to describe various forms of electric energy and its related laws and equivalent theorems greatly simplifies the analysis of idealized problems and assists the engineering design process.

The utilization of exact mathematical models (under ideal conditions) to describe processes and flows in this finite, nonideal, and imprecise physical universe can only be justified by noting that the mathematical answers obtained should never be interpreted as the exact numerical values that we are going to encounter in measurements in the real world, but only the best approximation that we can hope for.

These concepts exist regardless of the type or form or the degree of sophistication of mathematics involved. The building blocks of any subject are the "fundamental concepts," with mathematics serving only as an aid in expressing various concepts, problems, and solutions in shorthand notation. Therefore, this book is written in such a way that concepts are the primary focus, with the mathematics only as a servomechanism (or an aid or crutch) in arriving at or expressing the final answer in shorthand notation.

2.6 BASIC ELECTRICAL QUANTITIES, DEFINITIONS OF

Before we get involved in discussing the laws governing the field of electrical engineering, we need to establish the exact definitions and meanings of several technical terms that are essential to understanding this field of study. The definitions of these basic terms, along with their corresponding mathematical descriptions, are treated in the next several sections.

2.6.1 Electric Charge

The rock-bottom foundation of the whole subject of electronics is based on electric charge, which is a form of energy. Charges in motion create electric current (a flow) between two points, while nonflowing charges lead to static electricity and charge storage concepts.

2.6.2 Definition of Electric Charge

Before we proceed further, we need to define what we mean by electric charge.

DEFINITION-ELECTRIC CHARGE (OR CHARGE): *A basic property of elementary particles of matter (e.g., electron, protons, etc.) that is capable of creating a force field in its vicinity. This built-in force field is a result of stored electric energy. The charge of an object is the algebraic sum of the charges of its constituents (such as electrons, protons, etc.) and may be zero, a positive number, or a negative number. Electric charge is a form of electrical energy.*

To clear up a general misconception about electrons, it is worthwhile to note that an electron is a charge carrier; it inherently has a negative charge but itself is not charge but a particle with a mass. As such, it has its own mechanical type of energy (such as kinetic and/or potential). This is in addition to its charge, which is, in fact, electric energy due to the built-in electric force field.

The concept of charge leads to all other phenomena and fields of study as shown in Figure 2.10. From this diagram several concepts need to be expanded because they are essential to the understanding of the whole field of electronics.

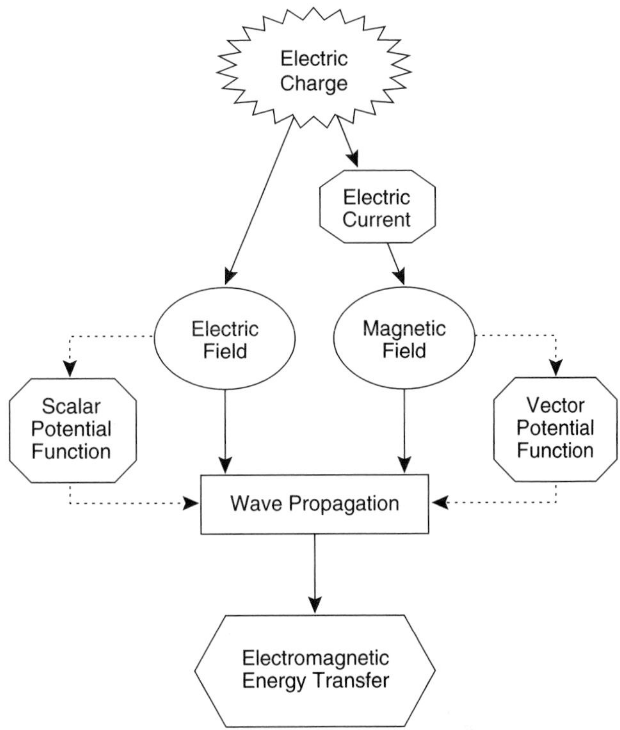

FIGURE 2.10 Diagram showing the electric charge as the cause of all electric and magnetic phenomena.

2.6.3 Electric Current

We defined the concept of "charge" earlier and described its properties and that it can be stationary (nonflowing) or in motion (flowing). Now, let us consider the concept of charges in motion, which leads to the idea of "electric current" or "current."

DEFINITION-ELECTRIC CURRENT (OR CURRENT): *The net transfer of electric charges (Q) across a surface per unit time. Electric current is a form of electrical energy.*

This concept, as shown in Figure 2.11, can be written mathematically as:

$$I = \frac{dQ}{dt} \tag{2.1}$$

Or, the total charge transferred is:

$$Q = \int_0^t I dt \tag{2.2}$$

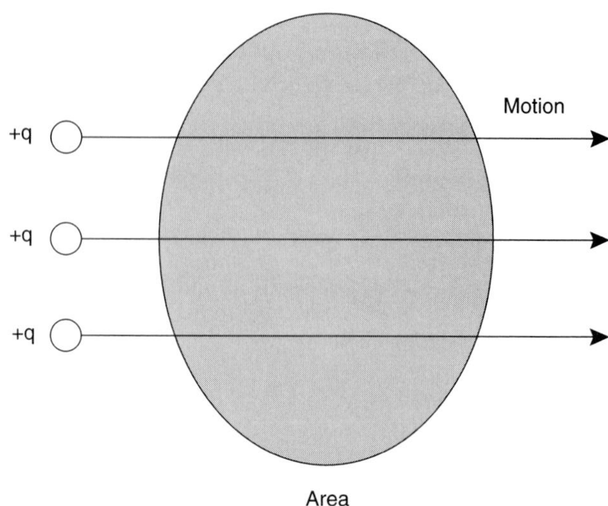

FIGURE 2.11 Definition of current.

2.6.4 Electrical Energy

Having defined "current" as a flow or a transfer of energy between two points, we are now ready to revisit another important concept: electrical energy. As defined earlier, energy *is the ability or capacity of a particle or a body to perform work or a task.*

Electric energy is a special class of the broad field of energy where electricity either directly or indirectly performs the intended task. Electrical energy could be subdivided into two categories. Each category of electrical energy has its own particular characteristics and peculiarities as briefly discussed here.

Electrical potential energy is stored electrical energy that has the capability of performing work when released. Examples include static charge, static electric field, and voltage.

Electrical kinetic energy is the energy of an electron (or many electrons) in motion, which is capable of performing a desired task directly or indirectly. Examples include electric current (direct motion of electrons), dynamic electric field (indirect), magnetic field (indirect), and electromagnetic waves (indirect).

2.6.5 Force and Field

The concept of energy brings about the idea of force and work, which we need to define.

DEFINITION-FORCE: *The agency that accomplishes work.*

Force can also be thought of as a form of energy that causes a particle or an object to move (if at rest) or change acceleration (if in motion).

Quantitatively, Force ($\overline{F}$) on a body is a vector (having a magnitude and a direction) equal to the time rate of change of momentum; that is,

$$\overline{F} = \frac{d(m\overline{V})}{dt}$$

where m is the mass and $(\overline{V})$ is the velocity of the object.

Having defined the concept of force, we need to define another basic concept related to force and widely used in all of engineering and particularly in electronics. That basic concept, of course, is "field."

DEFINITION-FIELD: *(a) A volume of space in which force is operative, (b) An entity that acts as an intermediary agent in interactions between particles, is distributed over a region of space, and whose properties are a function of space and time in general.*

Using definition (b), we can see that "field" is a vector quantity because it exists in a volume of space capable of exerting a force on any and all particles in that volume of space. Examples include an electric field, magnetic field, gravitational field, sound field, and so on.

There is also a shorthand definition for "field" given as follows.

A SHORTHAND DEFINITION-FIELD: *Cause (or action) at a distance.*

Of particular interest to the field of electronics, are two field quantities defined and described as follows:

DEFINITION-ELECTRIC FIELD INTENSITY ($\overline{E}$): *The electric force on a stationary positive unit charge at a point in an electric field (also called electric field strength, electric field vector, and electric vector).*

From this definition, we can see that electric field intensity ($\overline{E}$) is actually electric force per unit charge, distributed in a region of space.

DEFINITION-MAGNETIC FIELD INTENSITY ($\overline{H}$): *The force that a magnetic field would exert on a unit magnetic pole placed at a point of interest, which expresses the free space strength of the magnetic field at that point (also called magnetic field strength, magnetic intensity, magnetic field, magnetic force, and magnetizing force).*

From this definition, we can see that magnetic field intensity ($\overline{H}$) is magnetic force per unit pole, distributed in a region of space.

2.6.6 Work

DEFINITION-WORK: *The advancement of the point of application of a force on a particle.*

Therefore if a particle of charge (q) exists in an electrical field (E) and the particle moves a total distance (ℓ), we can write the amount of work performed mathematically as the summation of the dot product of electric force at each point ($q\overline{E}$) with the displacement vector ($\overline{d\ell}$) over a distance ℓ as:

$$W = q\int_0^{\ell} \overline{E} \cdot \overline{d\ell} \tag{2.3}$$

where $\overline{E}$ is the electric field vector defined as the electrical force exerted on a unit of charge and $\overline{d\ell}$ is an infinitesimal displacement vector. Differentiating Equation 2.3 with respect to q gives the differential form:

$$\frac{dW}{dq} = \int_0^{\ell} \overline{E} \cdot \overline{d\ell} \tag{2.4}$$

Equation 2.4 in essence gives the equation for the work performed per unit charge between two points, which basically is equivalent to the concept of voltage.

If the particle moves in the same direction as the applied force, then the equation for work simplifies into $W = qE\ell$, which simply states that work accomplished is the applied force (qE) multiplied by the distance (ℓ).

NOTE: *By close examination of the definition of "work," we can observe that it is the result of energy in action and represents "spent energy."*

2.6.7 Electrical Potential Differential (or Voltage)

The flow of an electric current between two points is caused by the electric potential difference (or voltage), which is defined next.

DEFINITION-ELECTRICAL POTENTIAL DIFFERENCE (OR VOLTAGE): *The electrical pressure or force between any two points caused by accumulation of charges at one point relative to another, which has the capability of creating a current between the two points.*

Obviously, voltage between two points can exist whether there is a current flow or not; however, in order to have a flow of current the medium between the two points should have a nonzero electrical conductivity (see Figure 2.12).

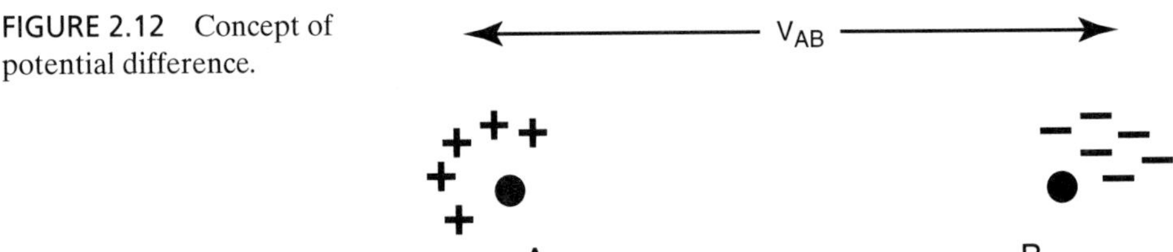

FIGURE 2.12 Concept of potential difference.

Having defined equations for energy, force, and work, we are now ready to define the potential difference mathematically. **Potential difference between two points (A and B) is defined to be the amount of work done against an electric field ($\overline{E}$) in order to move a unit charge ($q = 1C$) from A to B** (see Figure 2.13).

The force required to perform the task for a charge (q) in general, is:

$$\overline{F} = -(q\overline{E}) \tag{2.5}$$

Equation 2.5 gives the equation for the counteracting force that is required to move the charge from A to B. The work performed is given by:

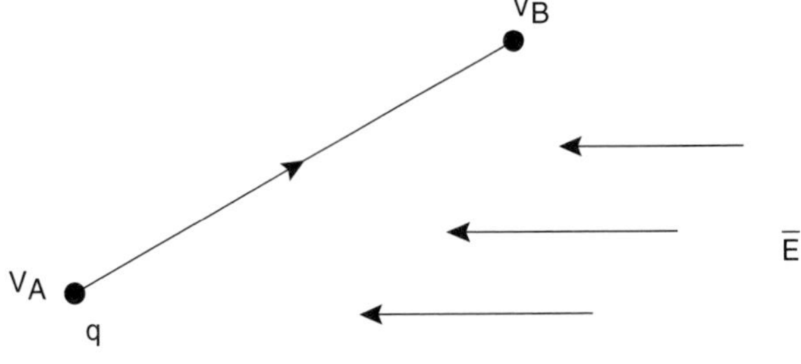

FIGURE 2.13 Definition of potential difference (or voltage).

$$W_{BA} = \int_A^B (-q\bar{E}) \cdot \overline{d\ell} \tag{2.6}$$

The voltage (or potential difference) is defined in terms of the work performed and is mathematically expressed as:

$$V_{BA} = V_B - V_A = \frac{W_{BA}}{q} = -\int_A^B \bar{E} \cdot \overline{d\ell} \tag{2.7}$$

Or, in differential form:

$$V = \frac{dW}{dq} \tag{2.8}$$

NOTE: *Theoretically speaking, the reference point (or ground) at which the potential function (V) is assumed to be zero is at an infinite distance away from the point of measurement (i.e., at infinity); however, in practice and in actual circuit analysis the ground can be assumed to be any designated point in the circuit, purely by prior agreement.*

2.6.8 Power

Even though the concept of energy as discussed here seems to be complete, often engineers are interested in the rate of transfer of energy between two points, which brings about the concept of power.

DEFINITION–POWER: *The rate at which work is performed, i.e., the rate at which energy is being either generated or absorbed:*

$$P = \frac{dW}{dt} \tag{2.9}$$

Or, in integral form:

$$W = \int_0^t P dt \tag{2.10}$$

Using the chain rule, Equation 2.8 can also be written as:

$$P = \frac{dW}{dt} = \frac{dW}{dq} \times \frac{dq}{dt} = VI \tag{2.11}$$

Thus, the total work performed or total energy transferred (or spent) is:

$$W = \int_0^t VIdt \tag{2.12}$$

2.6.9 The Governing Laws of Electrical Engineering

There are certain laws that electrical energy follows. To discover these laws, great minds have been at work for centuries. Therefore these laws represent the cumulative knowledge of mankind about electricity, electric charge flow, and magnetism, and they should not be regarded lightly.

One point of interest should be brought forth at this time: These laws have been cast into a mathematical format to ease their communication, simplification, and manipulation, and by themselves they represent a theoretical and an ideal yardstick—an absolute!

Because absolutes are unobtainable, we should expect only an approximate correlation between the actual measurements and the ideal answers that these laws provide.

Some of the most important laws, which are needed greatly in comprehending electrical engineering, are presented in chart form as shown in Figure 2.14 (also see Appendix F for mathematical descriptions of each of the laws or equations). From Figure 2.14, we can see that the primary laws of electronics in order of importance are as follows:

1. *Principle of conservation of energy*
2. *Primary laws*

   Maxwell's equations
   - **a.** *Ampere's law*
   - **b.** *Faraday's law*
   - **c.** *Electric Gauss's law*
   - **d.** *Magnetic Gauss's law*
3. Secondary laws (or equations)
   - **a.** *Scalar Poisson's and Laplace's equations*
   - **b.** *Vector Poisson's equation*
   - **c.** *General wave equations*
   - **d.** *Transmission line and waveguide equations*
   - **e.** *Helmholtz's equations*
   - **f.** *Electric and magnetic Coulomb's laws*
   - **g.** *Electric and magnetic Ohm's laws*
   - **h.** *Electric and magnetic Kirchhoff's laws*

Even though these laws will be discussed throughout this book on a "need to know" basis, we assume that the reader has a certain amount of familiarity to facilitate comprehension.

Before we discuss these laws and equations in the next several sections, it is imperative that we understand the basic underlying assumption held in common with any area of study. That commonality, of course, is the principle of conservation of

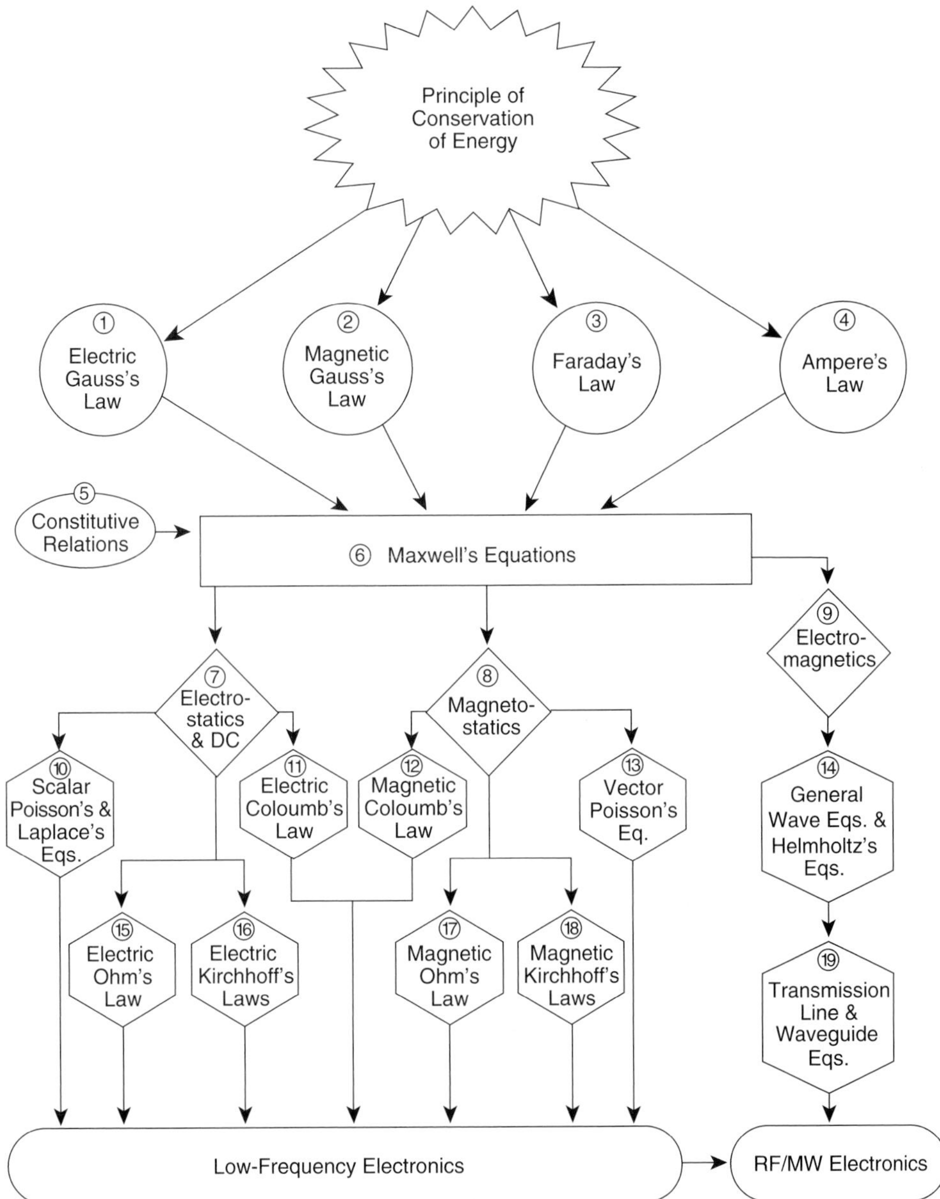

FIGURE 2.14 All fundamental classical laws of electricity and magnetism as they apply to the field of electronics.

energy, which deserves to be precisely defined at this early stage with all of its main features and frailties fully delineated.

2.7 PRINCIPLE OF CONSERVATION OF ENERGY

This principle is one of the most fundamental axioms in all of science, which needs special attention and must be considered and defined precisely at this point:

DEFINITION-PRINCIPLE OF CONSERVATION OF ENERGY (EXCLUDING ALL METAPHYSICAL SOURCES AND CAUSES OF ENERGY): *This fundamental law simply states that any form of energy in the physical universe can neither be created nor destroyed but only converted into another form of energy.*

In simple terms, we can say that in the physical universe "*energy comes from somewhere and goes somewhere.*" This means that in order to have energy, in a particular form and for a certain task, we need to use another form of energy (or matter) and convert it into the desired form first and then use it to perform the required task. For example, to light up a room by means of a flashlight, we need optical energy. To obtain this form of energy (i.e., optical), we use electrical energy. To obtain electrical energy, we need to use a battery to convert chemical energy into electrical energy, which then can be used in a flashlight to produce light.

Thus, any form of energy in the physical universe can be obtained only from the conversion of another source into the desired form of energy. This could also be called the *Principle of "immortality of energy."*

There is another closely related principle, the principle of "*conservation of matter,*" which follows the principle of conservation of energy through Einstein's relation $E = mc^2$. This relation states that mass is equivalent to energy through a proportionality constant c^2. The law of conservation of mass can be seen to be a corollary to the law of conservation of energy because matter is a condensed form of energy.

OBSERVATION: *It is interesting to observe that any and all physical laws are built on the principle of conservation of energy and all express the fact that one form of energy is equivalent to another form.*

For example, let us examine Ohm's law ($V = RI$). This law simply states that voltage (V) across a resistor (as one form of energy, i.e., potential energy) is equivalent to current (I) as another form energy (kinetic energy), through a proportionality factor called resistance (R). Another example is Faraday's law. Through simple observation of this law, we can conclude that it is merely stating a basic energy conversion fact: electrical energy in the form of electric field (E) equals another in the form of magnetic energy (H), through the use of spatial and time operators (Please see *Glossary of Technical Terms* for mathematical form of Faraday's law.).

Of course, this approach may be an oversimplification of the physical laws, but it helps us to understand their meaning without being overwhelmed by either their theoretical description or their mathematical complexity.

POINT OF CAUTION: *The principle of conservation of energy is a basic postulate that has been uniformly adopted as an underlying fundamental for all physical sciences practiced today. It should be noted, however, that the principle of conservation of energy applies only to an existing form of energy and can predict its future form or magnitude. It makes no determination or prediction about the origin of energy or its initial source. It discusses only what happens to it once it exists* *in the physical universe. In this regard, this fact could be a limiting factor and a major shortcoming, built into all of our extant physical sciences.*

2.8 MAXWELL'S EQUATIONS

Having defined all fundamental terms and principles, it is appropriate to introduce the most important classical equations in all of electrical engineering. Every known natural law in the field of electronics and electrical engineering can be derived from these equations.

Between 1864 and 1873, James Clerk Maxwell put forth a series of four advanced classical equations that describe the behavior of electromagnetic fields and waves in all practical situations. They relate the vector quantities for electric and magnetic fields as well as electric charges existing at any point, and they set forth stringent requirements that the fields must satisfy. These celebrated equations are described as follows:

NATURAL LAW #1: AMPERE'S LAW. *Current flowing in a wire (conduction current) or due to time rate of change of the electric displacement vector (displacement current) generates a magnetic flux that encircles the current in a clockwise direction when the current is moving away from the observer. The direction of the generated magnetic field follows the right hand rule. (Law of magnetic field generation)*

NATURAL LAW #2: FARADAY'S LAW. *When a magnetic field cuts a conductor, or when a conductor cuts a magnetic field, an electrical current will flow through the conductor if a closed path is provided over which the current can circulate. The induced electromotive force (emf) equals the negative of the rate of change of the magnetic flux linking the circuit. (Law of electromagnetic induction)*

NATURAL LAW #3: GAUSS'S LAW (ELECTRIC). *The summation of the normal component of the electrical displacement vector over any closed surface is equal to the electric charge within the surface, which in essence means that the source of the electric flux lines is the electric charge. (Law of electric charges)*

NATURAL LAW #4: GAUSS'S LAW (MAGNETIC). *The summation of the normal component of the magnetic flux density vector over any closed surface is equal to zero, which in essence means that the magnetic flux lines have no source or origin. (Law of magnetic charges)*

From these equations Maxwell predicted the existence of electromagnetic waves whose later discovery made radio possible. He showed that where a varying electric field exists, it is accompanied by a varying magnetic field induced at right angles, and vice versa, and the two form an electromagnetic field that could propagate as a transverse wave in free space.

He calculated that in a vacuum, the speed of the wave was given by $1/\sqrt{(\varepsilon_o \mu_o)}$, where ε_o and μ_o are the permittivity and permeability of vacuum. The calculated value for this speed was in remarkable agreement with the measured speed of light, and Maxwell concluded that light is propagated as electromagnetic waves. This remarkable discovery related the field of optics (a separate and isolated field of study at that time) to the field of electricity and magnetism.

Using Maxwell's equations, we can observe that the field of electrical engineering can be roughly subdivided into three distinct fields of study:

- Electrostatics & DC
- Magnetostatics
- Electromagnetics

These three areas are shown in Figure 2.14 and are briefly described next.

2.8.1 Electrostatics & DC

The field of electrostatics & DC is a subset of Maxwell's equations. To be specific, we define this field as follows.

DEFINITION-DC (DIRECT CURRENT): *A constant current which always flows in one direction.*

DEFINITION-ELECTROSTATICS: *The branch of electricity concerned with electrical charges at rest and their corresponding byproducts and effects such as electric fields, potential function, and so on. Such a stationary distribution of charge produces a static electric field. An example of such a case would be the fields associated with a fully charged capacitor at steady state.*

POINT OF CAUTION: *It should be noted that such an assumption of stationary distribution of charge is possible only at a classical level of observation (i.e., macroscopic or large-scale level relative to atomic dimensions). It can never be used at the quantum level of observation, where every charged particle is in constant motion due to the presence of several factors, such as thermal energy, coulombic forces, potential energy distribution in the material, and so on.*

In this field of study, we find many of the commonly used laws that are referred to in many popular texts. These laws can be briefly summarized as follows:

- Scalar Poisson's equation (potential function in a region filled by a charge distribution)
- Laplace's equation (potential function in free space)
- Electric Coulomb's law (forces between charged particles)
- Electric Ohm's law (linear resistor law)
- Electric Kirchhoff's laws (voltage and current laws)

Each one of these laws can be derived from Maxwell's equations as a special case. For further details and exact definitions of each of these laws, consult the glossary at the end of this text, Appendix F, *Classical Laws of Electricity, Magnetism and Electromagnetics*, or any basic text on field theory.

2.8.2 Magnetostatics

The second subset of Maxwell's equations is the field of magnetostatics.

DEFINITION-MAGNETOSTATICS: *The study of magnets and magnetic fields that are neither moving nor changing directions. Such a field could be produced by a stationary magnetic pole or by a DC current flowing in a stationary conductor.*

The main laws governing this field of study can be briefly summarized as follows:

- Vector Poisson's equation (magnetic potential function caused by a current distribution)
- Magnetic Coulomb's law (forces between magnetic poles)
- Magnetic Ohm's law (linear reluctance law)
- Magnetic Kirchhoff's laws (magnetic flux and mmf laws)

These laws can be derived from Maxwell's equations with relative ease and are relegated to more basic texts for a more detailed coverage. The reader is encouraged to find a brief mathematical description of these laws in Appendix F.

2.8.3 Electromagnetics

As the frequency of the signals in a circuit, component, or system is increased beyond DC, the fields of electricity and magnetism gradually become interwoven and inseparable. As a result, a new field of study was born to encompass this much wider sphere of activity. James Clerk Maxwell put the beginning of this vital field on solid ground in his celebrated Maxwell's equations (1864–1873). We now define this field of study as follows.

DEFINITION-ELECTROMAGNETICS: *The study of charges in motion resulting in electric and magnetic fields that are interdependent and interrelated, where an electric phenomenon cannot exist by itself without magnetic effects and vice versa. Examples include light emission from the sun and stars, thunderstorms and lightning, radio and television waves, power lines, and radar.*

This is quite contrary to the previous two fields of study where electric and magnetic phenomena can exist completely separate. It can be shown with relative ease that "electromagnetics" is a field of study that includes both fields of electrostatics and magnetostatics. Due to the exact nature of its formulation and precise nomenclature employed to describe its various concepts, it has created a world of enormous possibilities and a myriad of present and potential applications. It could easily be said that the field of electromagnetics put man on the moon and opened up the solar system to mankind's present and future space explorations, and it will enable us someday to conquer the whole galaxy!

The field of electromagnetics has several subdivisions, such as electrodynamics, physical optics, and so forth. One of the vital subdivisions of this exciting field is "electromagnetic wave propagation," which has tremendous applications in the field of RF and microwaves. The governing equations concerning electromagnetic wave propagation can be roughly divided into three subdivisions, with each more general than the next:

- General wave equations (dealing with general time-varying fields and the resulting waves)
- Helmholtz equations (dealing with time-harmonic fields and waves)

- Waves on a transmission line and waveguide (dealing with time-harmonic waves on a transmission line or waveguide)

The governing equations for any of these three subdivisions are briefly described mathematically at the back of the text in Appendix F. For further details and an in-depth treatment of this subject, the reader may consult other basic texts in this field.

2.9 SYSTEM OF UNITS

In mathematics, because we are dealing with abstract concepts represented by symbols, there is no need for units. If we are going to assign each of the mathematical symbols a specific physical quantity that is actual and measurable, then we need to know their units. This is an essential part of any scientific methodology where precise units are needed to obtain meaningful and finite answers to the real life problems that are encountered in the process of mankind's survival on this planet.

All scientific discoveries and the corresponding laws are most often cast into mathematical format to serve present and future generations and for no other purpose. Abstract mathematical formulas are understood clearly when meaningful and practical units of measurements are used.

One of the most useful units of measurement is the SI (international system of units). SI is built on four fundamental components of the physical universe: Matter, Energy, Space, and Time. The fundamental units of measurement are defined as follows:

- For mass, we use kilograms (kg)
- For energy, we use amperes (A)–(unit for current)
- For space, we use meters (m)
- For time, we use seconds (s)

These four units in SI are actually borrowed from a former system of measurement, which was commonly referred to as "MKSA system." Other units used in electrical and electronics engineering are derived from these four units and are expressed in terms of meters (m), kilograms (kg), seconds (s), and amperes (A). For example, the unit for charge is the coulomb (C), which can be expressed as (A–s).

NOTE: *In general, to be exact there are seven "base units" in the International system of units (SI): meter, kilogram (unit of macro-amount of substance), second, ampere, kelvin (unit of temperature), mole (unit of micro-amount of substance) and candela (unit of luminous intensity). In addition to the seven base units, a second class of SI units exist that are called "derived units." The derived units are formed by combining base units according to the algebraic relations linking the corresponding quantities, and each is usually named after a prominent scientist. Examples of derived units include Ohm, Farad, Volt, and Joule. There exists a third class of SI units, called "supplementary units," such as radian and steradian. These three classes of SI units form a coherent and useful system of measurement for all known quantities in the physical universe.*

Based on the previous discussion, it is essential and imperative that we present brief definitions of some of the most commonly used units in physics and electrical engineering for future reference:

Ampere (A). The unit of electric current defined as the flow of one coulomb of charge per second. Alternately, it can also be defined as the constant current that would produce a force of 2×10^{-7} Newton per meter of length in two straight parallel conductors of infinite length, of negligible cross section, placed one meter apart in a vacuum.

Celsius (°C). $1/100^{th}$ of the temperature difference between the freezing point of water (0°C) and boiling point of water (100°C) on the Celsius temperature scale.

Coulomb (C). The unit of electric charge defined as the charge transported across a surface in one second by an electric current of one ampere. An electron has a charge of $1.602\text{X}10^{-19}$ coulomb.

Fahrenheit (°F). $1/180^{th}$ of the temperature difference between the freezing point of water (32°F) and boiling point of water (212°F) on the Fahrenheit temperature scale.

Farad (F). The unit of capacitance in the MKS system of units equal to the capacitance of a capacitor that has a charge of one coulomb when a potential difference of one volt is applied.

Gauss. The unit of magnetic induction (also called magnetic flux density) in the CGS system of units equal to one flux line per square centimeter, which is the magnetic flux density of one Maxwell per square centimeter or 10^{-4} Tesla.

Gilbert (Gi). The unit of magnetomotive force (mmf) in the CGS system of units, equal to the magnetomotive force of a closed loop of one turn in which there is a current of $1/4\pi$ abampere (1 abampere = 10 A). Thus, one Gilbert = $10/4\pi$ ampere-turn.

Henry (H). The unit of self and mutual inductance in the MKS system of units equal to the inductance of a closed loop that gives rise to a magnetic flux of one Weber for each ampere of current that flows through.

Hertz (Hz). The unit of frequency equal to the number of cycles of a periodic function that occur in one second.

Joule (J). The unit of energy or work in the MKS system of units, equal to the work performed as the point of application of a force of one Newton moves through one meter of distance in the direction of the force.

Kelvin (K). The unit of absolute temperature, equal to a value of 1/273.16 of the absolute temperature of the triple point of water (which is a particular temperature, 273.16 K, and pressure point at which three different phases of water, vapor, liquid and ice, can coexist in equilibrium).

Maxwell (Mx). The unit for magnetic flux in the CGS system of units, equal to 10^{-8} Weber.

Newton (N). The unit of force in the MKS system of units equal to the force that imparts an acceleration of one meter per second to a mass of one kilogram.

Oersted (Oe). The unit of magnetic field in the CGS system of units equal to the field strength at the center of a plane circular coil of one turn and 1-cm radius when there is a current of $10/2\pi$ ampere in the coil.

Ohm (Ω). The unit of resistance in the MKS system of units equal to the resistance between two points on a conductor through which a current of one ampere flows as a result of a potential difference of one volt applied between the two points.

Siemens (S) (also called mho, inverse of Ohm). The unit for conductance, susceptance, and admittance in the MKS system of units; it is equal to the reciprocal of the resistance of an element that has a resistance of one ohm.

Tesla (T). The unit of magnetic field in the MKS system of units equal to one Weber per square meter.

Volt (V). The unit of potential difference (or electromotive force) in the MKS system of units equal to the potential difference between two points for which one coulomb of charge will do one Joule of work in going from one point to the other.

Watt (W). The unit of power in MKS system of units defined as the work of one Joule done in one second.

Weber (Wb). The unit of magnetic flux in the MKS system of units equal to the magnetic flux that, linking a circuit of one turn, produces an electromotive force of one volt when the flux is reduced to zero at a uniform rate in one second.

NOTE: *"CGS" is Centimeter-Gram-Second system of units.*

POINT OF INTEREST: THE MEN BEHIND THE UNITS *The units used in physics or electrical engineering are usually named after major contributors or inventors who have advanced this field of study materially and substantially. In this section, we briefly introduce the scientists behind the electric, magnetic, and other important units in chronological order:*

TABLE 2.1 The Men Behind the Units

Quantity	Unit Name	Scientist	Contribution
Magnetomotive Force (CGS)	Gilbert (Gi)	William Gilbert (1544–1603), English royal physician	Earth as a giant magnet & magnetism
Force	Newton (N)	Sir Isaac Newton (1642–1727), English mathematician and philosopher	Laws of gravity & motion
Temperature	Fahrenheit (°F)	Gabriel Daniel Fahrenheit (1686–1736), German physicist	Mercury thermometer
Temperature	Celsius (°C)	Anders Celsius (1701–1744), Swedish astronomer	Centigrade thermometer
Electric Charge	Coulomb (C)	Charles Augustin Coulomb (1736–1806), French engineer and physicist	Law of charges
Power	Watt (W)	James Watt (1736–1819), Scottish engineer/inventor	Birth of steam power
Electric Potential	Volt (V)	Alessandro Volta (1745–1827), Italian physicist	Electric batteries

TABLE 2.1 The Men Behind the Units *(Continued)*

Quantity	Unit Name	Scientist	Contribution
Electric current	Ampere (A)	Andre Marie Ampere (1775–1836), French physicist and mathematician	Circulation of current
Magnetic Field Intensity (CGS)	Oersted (Oe)	Hans Christian Oersted (1777–1851), Danish physicist	Connection between electricity and magnetism
Magnetic Field Strength (CGS)	Gauss	Karl Friedrich Gauss (1777–1855), German mathematician and astronomer	Terrestrial magnetics
Resistance	Ohm (Ω)	George Simon Ohm (1787–1854), German physicist and astronomer	Resistor law
Capacitance	Farad (F)	Michael Faraday (1791–1867), English scientist	Induction and laws of electrolysis
Inductance	Henry (H)	Joseph Henry (1797–1878), U.S. physicist	A glimpse of electromagnetic waves
Magnetic Flux	Weber (Wb)	Wilhelm Eduard Weber (1804–1891), German physicist	Terrestrial magnetics
Magnetic Flux (CGS)	Maxwell (Mx)	James Clerk Maxwell (1804–1891), Scottish physicist	Celebrated equations
Conductance	Siemens (S)	Werner (1816–1892, German) and William (1823–1883, British) Siemens (brothers), engineers	World telegraphy (Werner), the practical dynamo (William)
Work (or Energy)	Joule (J)	James Prescott Joule (1818–1889), English physicist	Thermodynamics and electricity
Temperature	Kelvin (K)	William Thomson Kelvin (1824–1907) British physicist and mathematician	Heat and electricity
Magnetic Flux Density	Tesla (T)	Nikola Tesla (1856–1943), U.S. inventor (born in Croatia)	High voltage/many great inventions
Frequency	Hertz (Hz)	Heinrich Rudolph Hertz (1857–1894), German physicist	Birth of radio

NOTE: *CGS is Centimeter-Gram-Second system of units.*

LIST OF SYMBOLS/ABBREVIATIONS

A symbol will not be repeated again once it has been identified and defined in an earlier chapter, as long as its definition remains unchanged.

C	Coulomb, unit of measure for charge
c	Speed of light
E	Energy
E-Field	Electric field

EM	Electromagnetic
F	Force
I	Current
K.E.	Kinetic energy
ℓ	Length
m	Mass
mmf	Magnetomotive force
MKSA	System of measurement represented by meter, kilogram, second, and ampere
P	Power
P.E.	Potential energy
Q	Total charge
q	Unit charge
SI	International system of units
$\overline{V}$	Velocity
W	Work

PROBLEMS

2.1 What are the basic components of the physical universe? Define each and describe the order of importance of each relative to the other.

2.2 What are the two subdivision of energy?

2.3 What are the main characteristics of energy as a large motion? Describe each.

2.4 Describe how mass and energy are derivable from and convertible into each other?

2.5 What are the prerequisites for existence of time? What is the keynote of time?

2.6 Why is time, in the strictest sense of the word, a consideration?

2.7 What is the single fundamental on which the whole field of electronics is built? Define it.

2.8 What are the byproduct phenomena of "charge"?

2.9 Describe what is the principle of conservation of energy?

2.10 What is meant by electric potential difference, conceptually and mathematically?

2.11 Describe what part of a life organism, in general, the laws of electronics directly apply to? Why?

2.12 Why are units important and what is one of the most useful systems of units?

REFERENCES

[2.1] Bordeau, S. B. *Volts to Hertz ... The Rise of Electricity.* Minneapolis: Burgess Publishing Co., 1982.

[2.2] Buckley, H. *A Short History of Physics.* London: Metheun, 1927.

[2.3] Dorf, R. C. and J. A. Savoboda. *Introduction to Electric Circuits.* New York: John Wiley & Sons, 1996.

[2.4] Flanagan, W. N. *Handbook of Transformer Design and Application.* Blue Ridge Summit: McGraw-Hill, 1993.

[2.5] Gibilisco, S. *The Illustrated Dictionary of Electronics.* Blue Ridge Summit: Tab Books, 1994.

[2.6] Gillmore, C. S. *Coulomb and the Evolution of Physics and Engineering in the 18th Century.* Princeton, NJ: Princeton University Press, 1971.

[2.7] Hawking, S. W. *A Brief History of Time.* New York: Bantam Books, 1988.

[2.8] Lapedes, D. N. *McGraw-Hill Dictionary of Physics and Mathematics.* New York: McGraw-Hill, 1978.

[2.9] Lincoln, E. S. *A Chronological History of Electrical Developments.* New York: National Electrical Manufacturer's Association, 1964.

[2.10] Lipson, H. *Great Experiments in Physics.* Edinburgh: Oliver & Boyd, 1965.

[2.11] Marion, J. V. *A Universe of Physics.* New York: John Wiley & Sons, 1970.

[2.12] Meyer, H. W. *A History of Electricity and Magnetism.* Cambridge, MA: MIT Press, 1972.

[2.13] Pledge, H. T. *Science Since 1500.* London: Science Museum, 1966.

[2.14] Runes, D. D. *A Treasury of World Science.* New York: Philosophical Library, 1961.

[2.15] Silsbee, F. B. *Systems of Electrical Units.* U.S. Government Printing Office: National Bureau of Standards, Monograph 56, 1962.

[2.16] Taton, R. *History of Science,* Vols. I-IV. New York: Basic Books, 1964.

[2.17] Tricker, R. A. R. *The Contributions of Faraday and Maxwell to Electrical Science.* New York: Pergamon Press, 1966.

[2.18] Wightman, W. P. D. *The Growth of Scientific Ideas.* New Haven, CT: Yale University Press, 1969.

CHAPTER 3

Mathematical Foundation for Understanding Circuits

3.1 INTRODUCTION

In this chapter, we will lay the groundwork for basic mathematical concepts including phasors, basic circuit elements, Ohm's law, Kirchhoff's voltage and current laws, basic network theorems, and the decibel scale. The main purpose of presenting these concepts is to help the reader better understand the ensuing materials and enhance the analysis and design of electronic circuits, which will be the main focus of this work.

3.1.1 The Phasor Concept

The following conditions should be present before the phasor concept can be used effectively:

- The circuit or system under consideration is linear.
- All independent sources are sinusoidal.
- Only the steady-state response is desired.

If these three strict conditions are met, then phasors can be employed with considerable ease in the analysis or design process of any circuit.

With the help of Euler's identity (which relates the polar representation of a complex number to its rectangular representation), we can write:

$$e^{\pm j\theta} = \cos\theta \pm j\sin\theta, \tag{3.1}$$

where $j = \sqrt{-1}$ is a unity imaginary number.

We will see shortly that the existence of a phasor is based on the concept of a mathematical transformation from one domain to another, which is a change in the mathematical description of a physical variable in order to facilitate computation (see Figure 3.1).

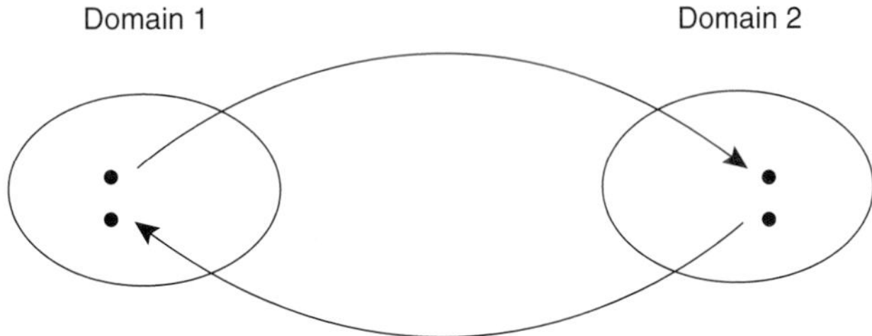

FIGURE 3.1 Transformation from domain 1 to domain 2.

3.2 PHASOR TRANSFORM

Consider a sinusoidal waveform of a voltage, current, or an electromagnetic (EM) wave given by:

$$x(t) = A_m \cos(\omega t + \varphi) \tag{3.2}$$

Using Euler's identity, $x(t)$ can be written as:

$$x(t) = Re[A_m\, e^{j\varphi}\, e^{j\omega t}] \tag{3.3}$$

The coefficient, $A_m e^{j\varphi}$, of the exponential term ($e^{j\omega t}$) is a complex number that carries the amplitude and phase angle of the given sinusoidal function. This complex number is defined to be the phasor representation [or the phasor transform (A)] of the given sinusoidal waveform:

$$A = A_m \cdot e^{j\varphi} \tag{3.4}$$

Thus, a phasor transforms the sinusoidal waveform from the time domain to the complex number domain (or to the frequency domain even though we have suppressed the exponential frequency factor $e^{j\omega t}$), as shown in Figure 3.2.

DEFINITION-PHASOR: *Is a result of a mathematical transformation of a sinusoidal waveform (voltage, current, or EM wave) from the time domain into the complex number domain (or frequency domain) whereby only the magnitude and phase angle information of the sinusoid is retained.*

FIGURE 3.2 Concept of a phasor.

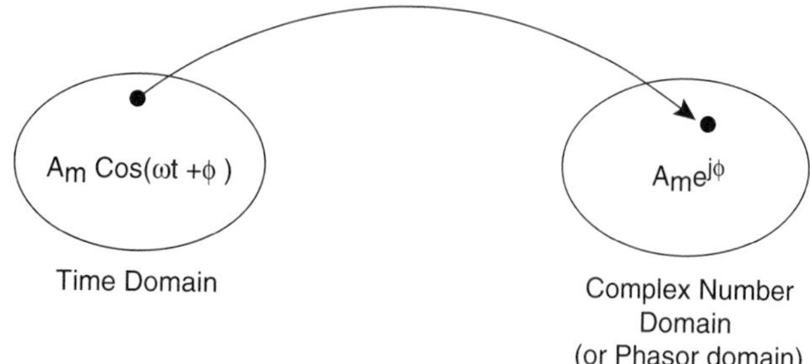

NOTE: *The value of ω cannot be deduced from a phasor quantity because it carries only the amplitude and phase information, though the frequency information is suppressed.*

3.3 INVERSE PHASOR TRANSFORM

The reverse operation, going from the phasor transform $A = A_m e^{j\varphi}$ to the time domain, is called the inverse phasor transform. It is obtained by multiplying the phasor by $e^{j\omega t}$ and taking its real part:

$$\text{Phasor } A \Rightarrow \text{ Time: } Re[Ae^{j\omega t}] = Re\,[A_m e^{j\varphi}\, e^{j\omega t}] = A_m \cos(\omega t + \varphi) = x(t) \qquad (3.5)$$

This is shown in Figure 3.3.

FIGURE 3.3 Inverse phasor transform.

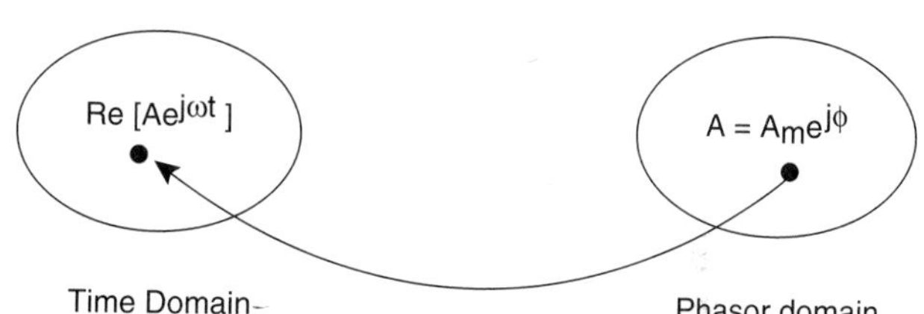

3.4 REASONS FOR USING PHASORS

The reasons for using phasors can be summarized in terms of the following facts:

FACT #1. The algebraic sum of any number of sinusoids of the same angular frequency (ω) and any number of their derivatives of any order is also a sinusoid of the same angular frequency (ω).

This suggests that we can treat sinusoids by algebraic methods using phasors to ease computation. The problem of finding the amplitude and phase angle of the steady state sinusoidal response is reduced to the problem of algebra of complex numbers.

FACT #2. Knowledge of a phasor (representing a sinusoid) determines the amplitude and the phase but not the frequency. When performing phasor calculations, it is important to keep in mind the frequency of operation (ω).

FACT #3. Summing sinusoids (voltages, currents, etc.) of the same frequency, very common in circuit theory and EM wave analysis, is easily done using phasors. Phasor transforms are very useful in this operation, considering the following:

$$\text{Let } X_i(t) = a_i \cos(\omega t + \varphi_i), \quad i = 1,\dots\dots\dots,n$$

Then the sum of n sinusoids, $X(t)$, can be written in the time domain as:

$$X(t) = X_1(t) + X_2(t) + \dots\dots\dots\dots\dots\dots\dots + X_n(t) \tag{3.6}$$

The phasor transform of the sum sinusoid, A, can be written as:

$$A = A_1 + A_2 + \dots\dots\dots\dots\dots\dots\dots\dots\dots + A_n \tag{3.7}$$

Where $A_i = a_i e^{j\varphi}$ $(i = 1,\dots\dots,n)$ is the corresponding phasor for each individual sinusoid $X_i(t)$. From Equations 3.6 and 3.7 we can write:

$$X(t) \Rightarrow A$$

Therefore, as shown above, the phasor representation of the sinusoidal sum is the sum of the phasors of the individual terms. Knowing the sum phasor (A) we can find the total sinusoid [$X(t)$] in the time domain and vice versa. Example 3.1 further illustrates this point:

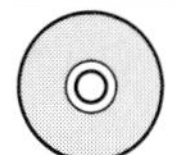

EXAMPLE 3.1

Determine the total sinusoidal function, $X(t)$, when it is given by:

$$X(t) = 20\cos(10t - 30°) + 40\cos(10t + 60°)$$

Solution:

$$A = 20e^{-j30°} + 40e^{j60°} = 20(\cos 30° - j\sin 30°) + 40(\cos 60° + j\sin 60°)$$

$$A = 37.3 + j24.6 = 44.7e^{j33°}$$

Once the sum phasor (A) is known, the corresponding sum sinusoid in the time domain [$X(t)$] can be easily found by the inverse transform:

$$X(t) = Re[Ae^{j10t}] = 44.7\cos(10t + 33°)$$

FACT #4. Differentiating a time domain sinusoid ($dX(t)/dt$) amounts to the multiplication of the corresponding phasor (A) by $j\omega$ as shown here:

$$dX(t) / dt = d\,[Re(A_m e^{j\omega t + \varphi})] / dt = Re\,[d(A_X e^{j\omega t}) / dt] = Re\,[A_X\, j\omega\, e^{j\omega\tau}] \quad (3.8)$$

Where $A_X = A_m e^{j\varphi}$ is the phasor for $X(t)$.
Therefore:

$$dX(t) / dt \Rightarrow j\omega A_X \quad (3.9)$$

FACT #5. Integrating a sinusoid in a time domain corresponds to dividing the phasor by $j\omega$ (with the use of Equation 3.9) as follows:

$$Y(t) = \int X(t)dt + Y_0 \Rightarrow X(t) = dY(t)/dt \Rightarrow A_X = j\omega A_Y \quad (3.10)$$

Therefore:

$$A_Y = (1 / j\omega)A_X \quad (3.11)$$

Where A_X and A_Y are the phasors for $X(t)$ and $Y(t)$, respectively.

3.5 LOW-FREQUENCY ELECTRICAL ENERGY CONCEPTS

At low frequencies, the laws governing electricity and magnetism are, in fact, an approximation and a simplified version of the higher frequency laws, allowing circuit analysis to be performed with considerable ease.

At these frequencies the signals have excessively long wavelengths compared to the circuit lengths, and all circuit components can be considered to be lumped elements with negligible time delay in signal propagation. This means that signals appear simultaneously at any and all points in the circuit.

Let's first define an important term:

DEFINITION-LUMPED ELEMENT: *A self-contained element that offers one particular electrical property throughout the frequency range of interest.*

Examples of lumped elements include resistors, capacitors, and inductors. In each of these elements, the associated property (resistance, capacitance, or inductance), resides at only one locale.

Let's consider the three basic lumped elements in the next section.

3.6 BASIC CIRCUIT ELEMENTS

At low electrical frequencies, three basic lumped passive components need to be considered. These are a resistor, a capacitor, and an inductor. They are described as follows.

3.6.1 Resistor

A resistor is a lumped bilateral and linear element that impedes the flow of current, $i(t)$, through it when a potential difference, $v_{12}(t)$, is imposed between its two terminals. This impeding property (represented by the symbol R) is defined to be the ratio of $v_{12}(t)$ over $i(t)$:

$$R \equiv v_{12}(t) \,/\, i(t) \tag{3.12}$$

Where $v_{12}(t) = \Delta v = v_1(t) - v_2(t)$ is the potential difference (or voltage) between the two resistor terminals as shown in Figure 3.4.

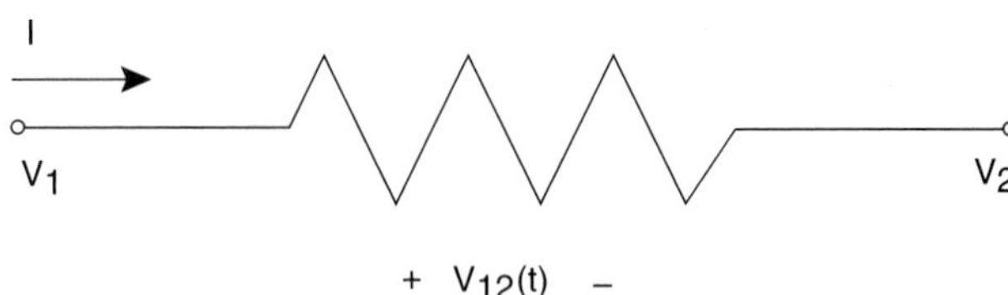

FIGURE 3.4 Circuit symbol for a resistor.

By bilateral, it is meant that the element property is independent of the direction of flow of current; indeed the same ratio would have been obtained had the voltage polarity been reversed, causing a current flow in the opposite direction.

3.6.2 Capacitor

A capacitor is a device consisting essentially of two conducting surfaces separated by an insulating material (or a dielectric), such as air, paper, or mica. By introducing capacitance, a circuit is capable of storing electrical energy, blocking totally the flow of direct current and partially the flow of alternating current.

Furthermore, the capacitance of a capacitor is defined to be the ratio of the magnitude of the charge (Q) on one of the conducting surfaces (there being an equal and opposite charge on the other conductor) to the voltage difference, $v_{12}(t)$, between the two conductors.

$$C = Q \,/\, v_{12}(t) \tag{3.13}$$

Where $v_{12}(t) = \Delta v = v_1(t) - v_2(t)$ is the potential difference (or voltage) as shown in Figures 3.5a and 3.5b.

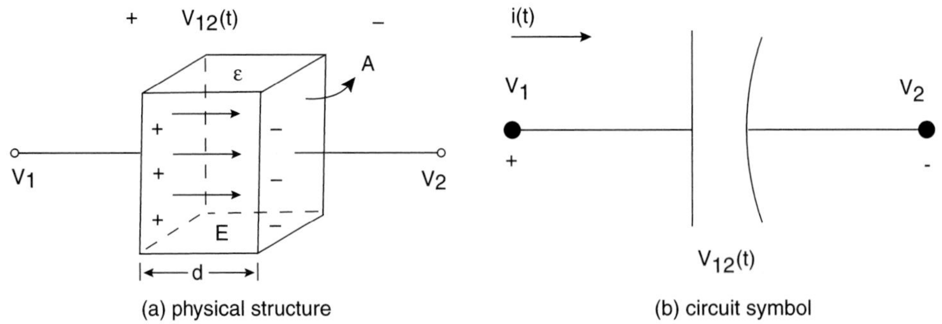

FIGURE 3.5 A parallel plate capacitor.

Quantitatively, capacitance is a measure of the ability of a device to store electrical energy in the form of separated charges, which leads to the concept of electric field storage. For example, in a parallel plate capacitor (see Figure 3.5), the capacitance (C) is found to be:

$$C = \varepsilon A / d \tag{3.14}$$

Where A is the common surface area shared between the two conducting surfaces, d is the distance between the two surfaces, and ε is the dielectric constant.

If we represent $v_{12}(t)$ in Equation 3.13 by a single variable, $v(t)$, we get:

$$Q = C\,v(t) \tag{3.15}$$

Differentiating Equation 3.15 yields:

$$i(t) = dQ/dt = Cdv(t)/dt \tag{3.16}$$

Equation 3.16 represents the current-voltage relationship for a lossless model of a capacitor. This is a linear relationship, which means that the current, $i(t)$, through a capacitor is proportional to the rate of change (or slope) of the voltage with time, as shown in Figure 3.6.

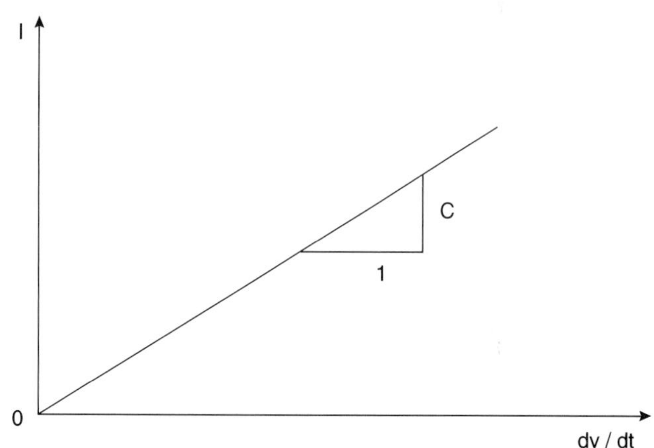

FIGURE 3.6 Current voltage relationship of a capacitor.

To sum up, a capacitor is a linear and bilateral element for time-harmonic signals. It should be noted, however, that because there are capacitors (such as electrolytic capacitors) that depend on a certain DC voltage polarity for proper operation, the bilateral property may not always be valid for signals having a net DC value.

For a time-harmonic signal (i.e., a signal that is varying sinusoidally) at frequency (ω), we can use the phasor concept to find the phasor domain relation of a capacitor from Equation 3.16. But first we need to define briefly the concept of impedance:

Impedance of a Capacitor.

DEFINITION-ELECTRICAL IMPEDANCE (ALSO KNOWN AS IMPEDANCE, Z): *The total opposition that a circuit presents to an AC signal, a complex number equal to the ratio of the voltage phasor(V) to the current phasor (I).*

$$Z = V / I = R + jX \tag{3.17}$$

The real part of impedance (R) is called "Resistance" while its imaginary part (X) is called "Reactance." The concept of impedance will be discussed later in full detail.

In the case of a lossless capacitor, the impedance is purely imaginary (i.e., totally reactive) and can be found by transforming Equation 3.16 into phasor domain as follows:

$$I = j\omega CV \tag{3.18}$$

Thus:

$$Z_C = \frac{V}{I} = jX_C = \frac{1}{j\omega C} \Rightarrow X_C = \frac{-1}{\omega C} \tag{3.19}$$

Thus, the reactance of a capacitor (X_C) is negative and diminishes as either the frequency (ω) or the capacitance value (C) is increased.

3.6.3 Inductor

An inductor is a device consisting of a coil of wire wound according to various designs, with or without a magnetic core, to store magnetic energy to a higher degree than a straight piece of wire would, as shown in Figure 3.7.

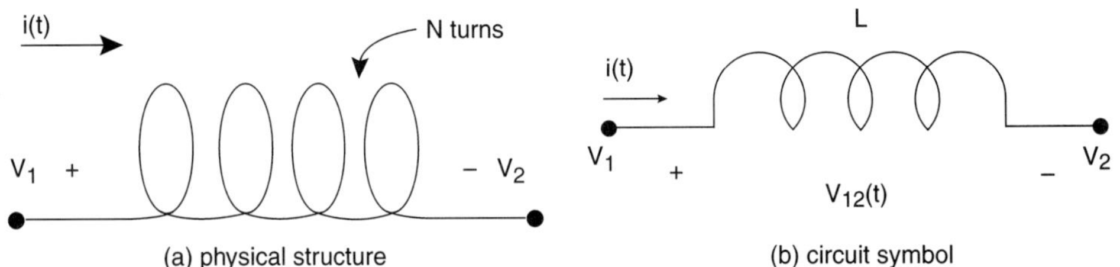

FIGURE 3.7 An inductor.

An inductor introduces the concept of inductance, which is defined as follows:

DEFINITION-INDUCTANCE (L): *The inertial property (caused by an induced reverse voltage) of an element that opposes the flow of current when a voltage is applied; it opposes a change in current that has been established. Presence of inductance is felt in a circuit only when the current is changing. It is a measure of the ability of a device in general to store energy in the form of magnetic field.*

If a current [$i(t)$] in an N-turn coil of wire creates a magnetic flux [$\Phi(t)$] in each turn, then the total flux is [$N\Phi(t)$], and the inductance (L) can be written as:

$$L = \frac{N\Phi(t)}{i(t)} \tag{3.20}$$

According to Faraday's law, the induced voltage, $v_{12}(t)$, across the inductor's terminals is equal to the rate of change of the magnetic flux in each turn multiplied by the number of turns (N), that is:

$$v_{12}(t) = N\frac{d\Phi(t)}{dt} = \frac{d[N\Phi(t)]}{dt} \tag{3.21}$$

Using Equation 3.20 and replacing $v_{12}(t)$ with $v(t)$, we can write Equation 3.21 as:

$$v(t) = L\frac{di(t)}{dt} \tag{3.22}$$

Equation 3.22 represents the current-voltage relationship for an inductor in the time domain illustrated in Figure 3.8.

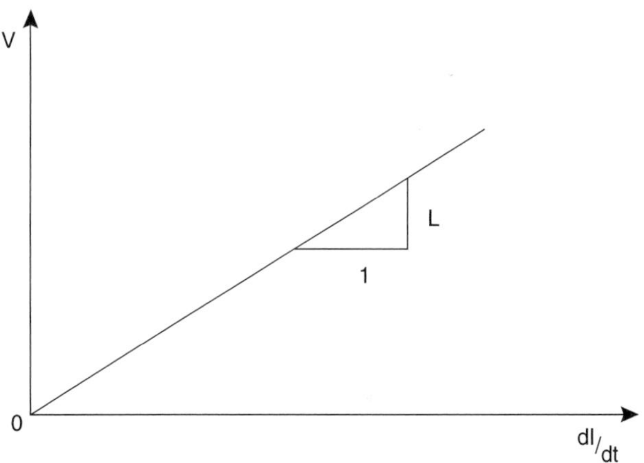

FIGURE 3.8 I-V characteristic of an inductor.

Figure 3.8 shows that an inductor is a linear element. Furthermore, because it is made of a piece of conductor, it is a bilateral element if the core material is air or a linear material. To increase the inductance greatly, however, the core is most often filled with a highly permeable material that has a high magnetic conductivity (such as ferromagnetic materials: iron, nickel, etc.). Because these materials are highly nonlinear, the inductor would have a preferred direction of voltage polarity and would no longer be considered a bilateral element.

Impedance of an Inductor. For a time-harmonic signal with frequency (ω), we can use the phasor transformation of Equation 3.22 into frequency domain to find the inductor's reactance as follows:

$$i(t) \Rightarrow I \text{ (phasor)}$$

$$v(t) \Rightarrow V \text{ (phasor)}$$

$$V = L \cdot j\omega I$$

Thus:

$$Z_L = V/I = jX_L = j\omega L \Rightarrow X_L = \omega L \quad (3.23)$$

From Equation 3.23 we can see that the reactance of an inductor (X_L) is positive and increases as the frequency or the inductance is increased.

3.7 SERIES AND PARALLEL CONFIGURATIONS

In actual practice, lumped elements are connected either in series or in parallel. In each case, because we are dealing with lumped elements, we can replace the entire circuit with an equivalent lumped element. Using this concept of equivalence, the following can be shown to be valid for N lumped elements. Let:

Resistor element values: R_i, $i = 1,............,N$

Capacitor element values: C_i, $i = 1,............,N$

Inductor element values: L_i, $i = 1,............,N$

Then the "equivalent lumped element" value for the series or parallel configuration would be as follows.

3.7.1 Series Configuration

This configuration is shown in Figure 3.9 with N passive elements connected in series. The total or the **"equivalent"** lumped values for the three kind of elements are summarized here as:

$$R_{eq} = \sum_{i=1}^{N} R_i \quad (3.24)$$

$$C_{eq} = \sum_{i=1}^{N} \frac{1}{C_i} \quad (3.25)$$

$$L_{eq} = \sum_{i=1}^{N} L_i \quad (3.26)$$

3.7.2 Parallel Configuration

This configuration is shown in Figure 3.10, with N passive elements connected in parallel. The total or the "equivalent" lumped values for the three kinds of elements are summarized here as:

$$\frac{1}{R_{eq}} = \sum_{i=1}^{N} \frac{1}{R_i} \quad (3.27)$$

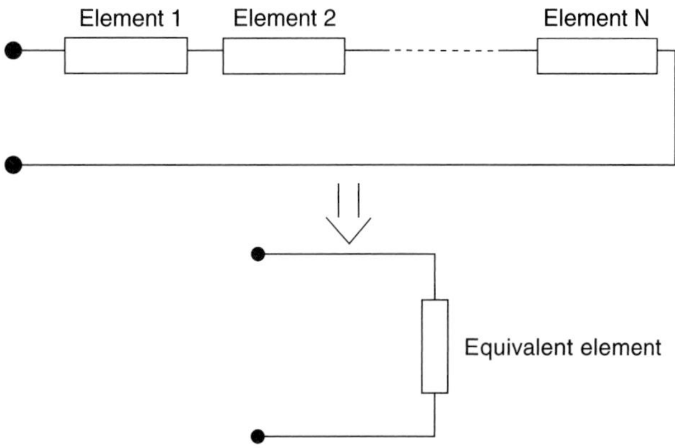

FIGURE 3.9 Series configuration.

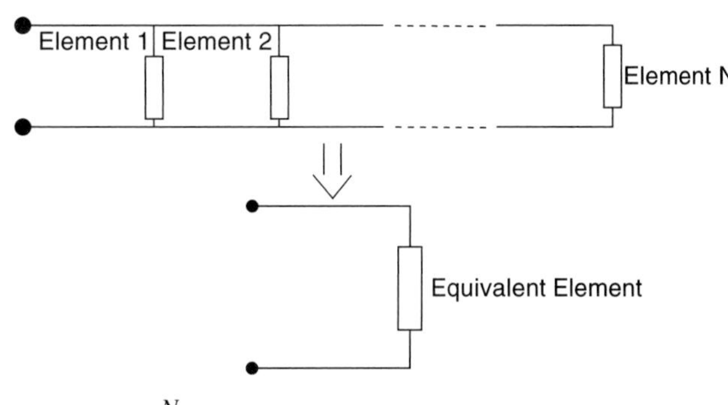

FIGURE 3.10 Parallel configuration.

$$C_{eq} = \sum_{i=1}^{N} C_i \tag{3.28}$$

$$\frac{1}{L_{eq}} = \sum_{i=1}^{N} \frac{1}{L_i} \tag{3.29}$$

3.8 CONCEPT OF IMPEDANCE REVISITED

In circuit theory, when dealing only with time-harmonic functions, we often transform the current and voltage (I–V) relationship of a linear element into a phasor for simplicity and ease of further analysis.

We earlier defined the electrical impedance (Z) of an element as the ratio of the voltage phasor (V) to current phasor (I).

$$Z \equiv V/I \text{ (measured in Ohms)}$$

Impedance, a complex number, is a measure of the total opposition that an element presents to an alternating current.

Because Z is a complex number, it may be written as follows:

$$Z = |Z|\, e^{j\theta} \text{ (exponential form)}$$

$$= |Z|\angle\theta \quad \text{(polar form)}$$

$$= R + jX \quad \text{(rectangular form)}$$

$|Z|$, θ, R, and X are the magnitude, phase angle, resistance, and reactance of the impedance, respectively. Figure 3.11 shows a graphical representation of impedance.

$$|Z| = \sqrt{R^2 + X^2} \tag{3.30a}$$

$$\theta = \tan^{-1}(X/R) \tag{3.30b}$$

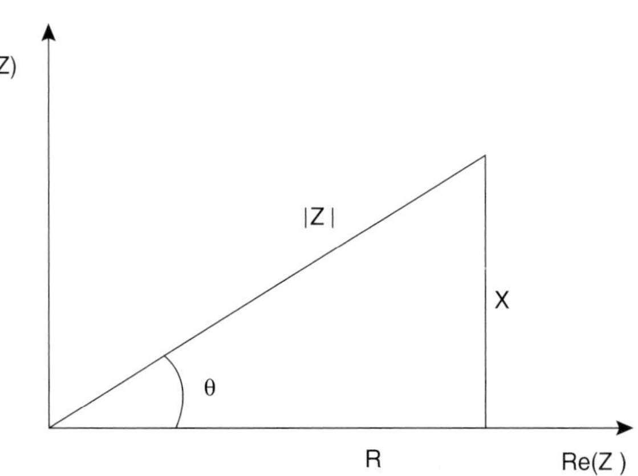

FIGURE 3.11 Graphical representation of the concept of impedance.

Thus, impedance is a generalized concept of resistance (or reactance) of an element, where an element may generally contain both a resistance and a reactance. For example, a lossy capacitor or inductor has an impedance that is not purely reactive; it has a resistive component in addition to its known reactance.

The inverse of impedance (Z) is admittance (Y), which is a measure of how readily an element will allow an alternating current to flow through it. Admittance (Y) can be written as:

$$Y = \frac{1}{Z} = \frac{1}{|Z|}e^{-j\theta} = \frac{1}{|Z|}\angle -\theta \tag{3.31a}$$

or

$$Y = \frac{1}{R + jX} = \frac{R - jX}{R^2 + X^2} = G + jB \tag{3.31b}$$

where

$$G = R/D,$$

$$B = -X/D$$

and

$$D = R^2 + X^2$$

G and B are called conductance and susceptance, respectively. Y is measured in mhos or Siemens (S).

3.9 LOW-FREQUENCY ELECTRICAL LAWS

The electrical laws that apply at "low" frequencies are a simplification of the general laws discussed in Chapter 2, *Fundamental Concepts in Electrical and Electronics Engineering*. These laws apply particularly at low frequencies where the size of the element is much larger than the wavelength and each component property (such as resistance, capacitance, and inductance) is separate from the other and is localized at one specific locale in the circuit. This means that a resistor will always exhibit lossy behavior, a capacitor will always store electric field, and an inductor will always hold a magnetic field throughout the frequency range of interest at their corresponding physical location.

This specific condition leads to treating all elements as lumped elements, and thus the following laws solely apply to all lumped element networks. Reciting these laws herein is only for contrast with the higher-frequency electromagnetic counterpart wherein the same element could potentially exhibit capacitive, magnetic, or resistive properties at different frequencies.

3.9.1 Kirchhoff's Current Law (KCL)

KCL is based on the law of conservation of charge and states the following:

For any lumped-element network, for any of its nodes (or junction points) and at any time, the net sum of currents in all circuit branches is equal to zero:

$$\sum_{n=1}^{N} i_n(t) = 0 \tag{3.32}$$

Where N is the total number of branches and $i_n(t)$ is the current in the nth branch, as shown in Figure 3.12.

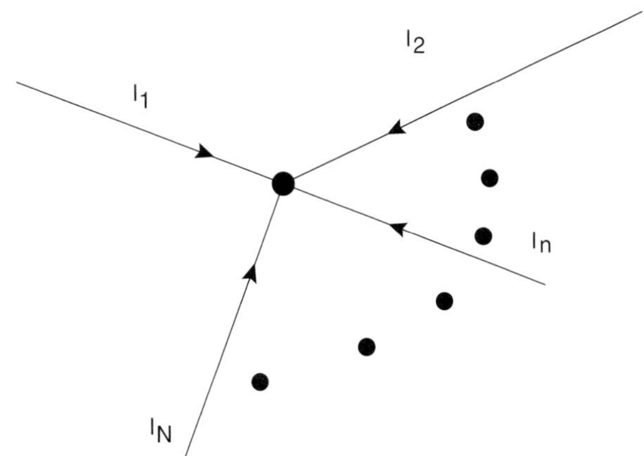

FIGURE 3.12 A node with N branches and N currents.

Several assumptions are built into this law:

ASSUMPTION #1. KCL assumes that the frequencies of interest are low enough such that there is no radiation at any of the nodes.

ASSUMPTION #2. KCL imposes a linear constraint on the branch currents.

ASSUMPTION #3. KCL as expressed in Equation 3.32 is valid both in time and phasor domain.

3.9.2 Kirchhoff's Voltage Law (KVL)

KVL is based on the law of conservation of energy and states the following:

In any lumped-element network, for any of its loops (or meshes) and at any time, the net sum of branch voltages around the loop equals zero.

$$\sum_{m=1}^{M} v_m(t) = 0 \tag{3.33}$$

Where M is the total number of branches in the loop and $v_m(t)$ is the branch voltage in the mth branch, as shown in Figure 3.13.

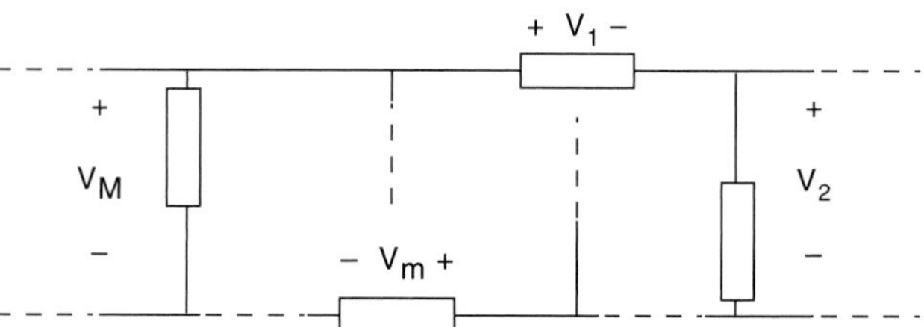

FIGURE 3.13 A loop with M branches.

KVL is based on the following assumptions:

ASSUMPTION #1. KVL assumes that all electric fields involved are "conservative fields," i.e., the work done on a particle in moving it from one point to another depends only on the particle's initial and final positions and not on its path of travel.

ASSUMPTION #2. KVL imposes a linear constraint between branch voltages of a loop.

ASSUMPTION #3. KVL as expressed in Equation 3.33 is valid for both time and phasor (frequency) domain.

Both of Kirchhoff's current and voltage laws apply to any and all *lumped networks*; it does not matter whether the circuit elements are linear, nonlinear, passive, active, etc. In other words, KCL and KVL are independent of the nature of the elements.

3.9.3 Ohm's Law

The third basic law is Ohm's law, which applies only to resistive elements (see Figure 3.14) and states the following:

The voltage (v) across any lumped constant-value resistor is equal to the current (i) through the element multiplied by the resistance (R).

$$v(t) = Ri(t) \tag{3.34}$$

NOTE: *Ohm's law applies only to linear resistors and thus is useless for nonlinear elements such as diodes.*

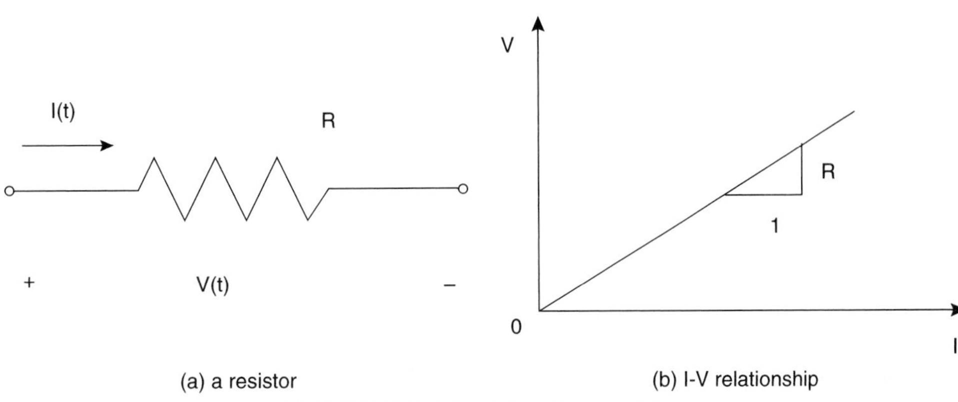

FIGURE 3.14 Ohm's law for a resistor.

3.9.4 Generalized Ohm's Law

When dealing with linear circuits under the influence of *time-harmonic signals*, Ohm's law can be restated under the *steady-state condition* in the phasor domain as follows:

> The phasor voltage drop (V) across any lumped linear element is equal to the element's impedance (Z) multiplied by the phasor current (I) through the element.

$$V = Z \cdot I \tag{3.35}$$

Where, $Z = R + j(\omega L\text{-}1/\omega C)$ in general.

Using these laws, we can easily derive and thus verify the validity of Equations 3.24–3.29.

EXERCISE 3.1

Using KVL, KCL, Ohm's law or Generalized Ohm's law, prove Equations 3.24–3.29.

Table 3.1 summarizes all three basic laws in time and phasor domain.

TABLE 3.1 Summary of Time and Frequency Domain Relationship

Property Name	Time Domain	Frequency Domain
Voltage	$v(t)$	$V(\omega)$
Current	$i(t)$	$I(\omega)$

TABLE 3.1 Summary of Time and Frequency Domain Relationship *(Continued)*

Property Name	Time Domain	Frequency Domain
Capacitor	$i(t) = C\,dv(t)/dt$	$I = j\omega CV$
Inductor	$v(t) = L\,di(t)/dt$	$V = j\omega LI$
Ohm's Law	$v(t) = Ri(t)$	$V = RI$
Generalized Ohm's Law	$v(t) = Ri + 1/C\int_0^t i\,dt + L\,di/dt$	$V(\omega) = Z(\omega)\,I(\omega)$, $Z = R + j(\omega L - 1/\omega C)$
KCL	$\sum_{n=1}^{N} i_n(t) = 0$	$\sum_{n=1}^{N} I_n(\omega) = 0$
KVL	$\sum_{m=1}^{M} v_m(t) = 0$	$\sum_{m=1}^{M} V_m(\omega) = 0$

3.10 FUNDAMENTAL CIRCUIT THEOREMS

Several fundamental circuit theorems need to be considered to simplify circuit analysis and design. These are Thevenin's, Norton's, Duality, Superposition, and Miller's theorems.

By definition, theorems are mathematical statements of identity, which can be proven rigorously. Therefore, the use of these theorems facilitates the solution to circuit problems by presenting an equivalent statement of the truth that is simpler and more workable.

Because all of these theorems apply only to linear networks, it behooves us to define this term at the outset of this section.

DEFINITION-LINEAR NETWORK: *A network in which the parameters of resistance, inductance, and capacitance of the lumped elements are constant with respect to current or voltage, and in which the voltage or current sources are independent of or directly proportional to other voltages and currents or their derivatives in the network.*

Furthermore, we need to note that all of these theorems allow calculation of the performance of a network from only its terminal properties without ever being concerned about what happens inside the network.

With this preamble we are ready to embark on the description of the three main theorems.

3.10.1 Thevenin's Theorem (also known as Helmholtz's Theorem)

Consider Figure 3.15 where the current (I_L) in the load (Z_L) is desired to be determined. This theorem states that:

At any given frequency, the current (I_L) that will flow through a load impedance (Z_L) when connected to any two terminals of a linear network is equal to the open-circuit voltage (V_{oc} or V_T), with the load (Z_L) removed, and divided by the sum of

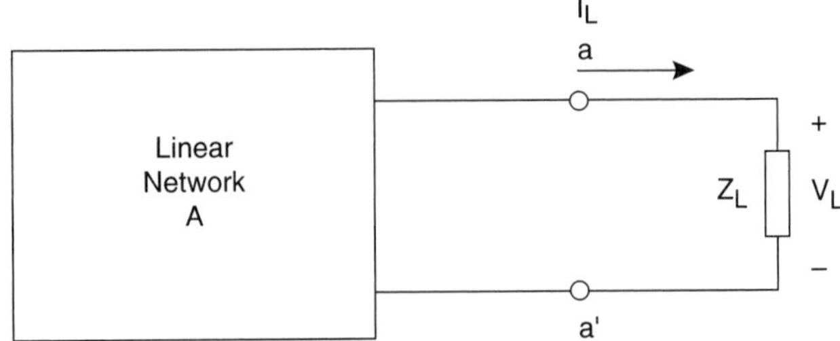

FIGURE 3.15 Linear network connected to a load.

the load impedance and the impedance (Z_T) obtained by looking back from the open terminals into the network with all independent sources reduced to zero (i.e., replacing each independent source by its internal impedance) as shown in Figure 3.16.

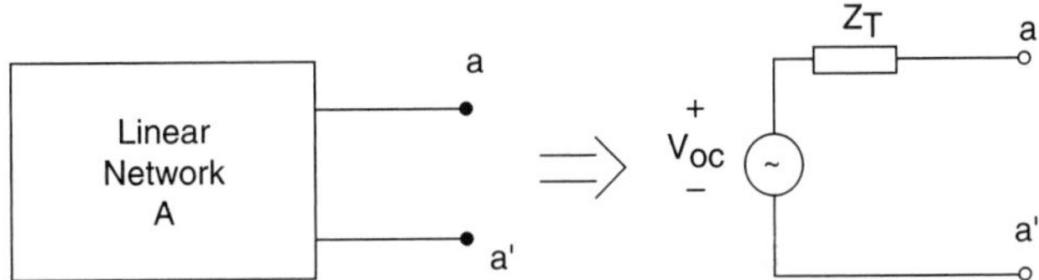

FIGURE 3.16 A linear network's equivalent circuit.

This valuable theorem in solving network problems allows calculation of the performance of a network (or device) only from its terminal properties. Thus according to this theorem, the linear network $[A]$ simplifies into the following circuit, as shown in Figure 3.17.

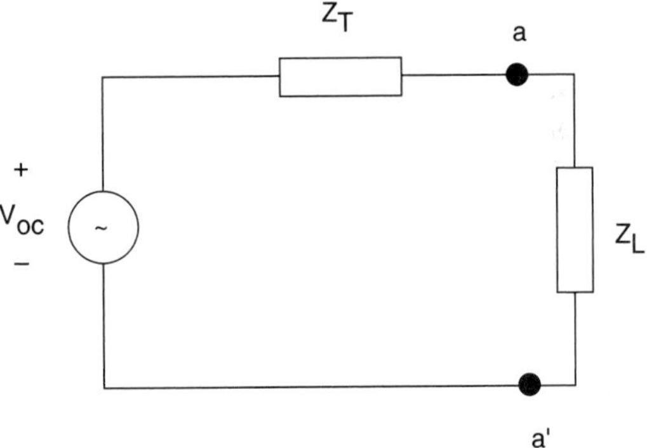

FIGURE 3.17 Thevenin's equivalent circuit.

Connecting the load back to the Thevenin's equivalent, the current through the load can easily be found to be:

$$I_L = \frac{V_{OC}}{Z_T + Z_L} \tag{3.36}$$

3.10.2 Norton's Theorem

Norton's theorem is the dual of Thevenin's theorem and states that:

> *The voltage across a load element (Z_L) that is connected to the two terminals of a linear network is equal to the short-circuit current (I_{sc}) between these terminals (in the absence of the load element) divided by the sum of the load admittance (Y_L) and the admittance (Y_T) of the network when looking back into these terminals, while setting all independent sources to zero (as shown in Figures 3.18 and 3.19).*

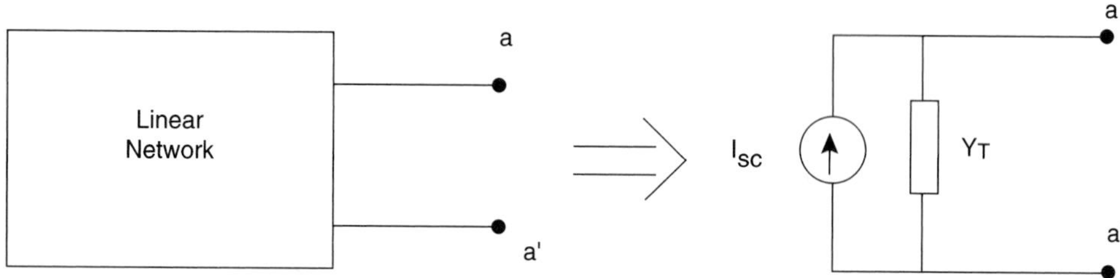

FIGURE 3.18 Norton's equivalent circuit.

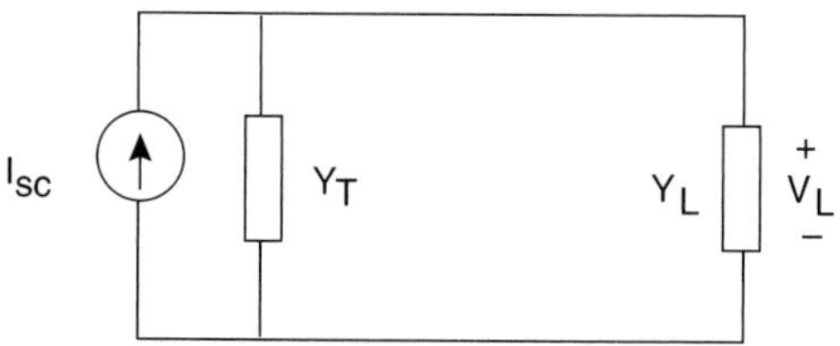

FIGURE 3.19 Norton's equivalent connected to a load.

By observation:

$$Y_T = \frac{1}{Z_T}$$

> **NOTE:** *Norton's theorem results from a more general principle, the duality principle, which is based on the duality theorem discussed next.*

3.10.3 Duality Theorem

> *This theorem states that when a theorem or statement is true, it will remain true if each quantity and operation is replaced by its dual quantity and operation. In circuit theory, the dual quantities are "voltage and current" and "impedance and admittance" and the dual operations are "series and parallel" and "meshes and nodes."*

Using the duality principle and the fact that Thevenin's equivalent and its corresponding Norton's equivalent represent the same network, the terminal voltages for Norton's and Thevenin's theorem must be the same. Thus, we obtain:

$$V_T = Z_T \cdot I_{sc}$$

In actuality, we can observe that the duality principle is based on a much more fundamental truth, which is for every datum or concept, there must be another one of comparable magnitude (see Chapter 1, *Fundamental Concepts of Science and Engineering*). In other words, a datum cannot exist all by itself (i.e., an absolute). For example, for a positive charge there is negative charge; for KVL there is KCL; for Thevenin's theorem there is Norton's theorem, and so on. A summary of dual quantities in electrical engineering is presented in Tables 3.2 and 3.3.

TABLE 3.2 Dual Quantities in the Subject of Electric Circuit Theory

Quantity or Operation	Dual Quantity or Operation
Electron	Hole
Current	Voltage
KCL	KVL
Node	Mesh
Inductor	Capacitor
Parallel elements	Series elements
Current source	Voltage source
Norton's theorem	Thevenin's theorem

In a broader arena of electromagnetics, we can see that numerous quantities, laws, and theorems in the field of electricity have duals in magnetism. These are briefly listed in Table 3.3.

TABLE 3.3 Dual Laws and Theorems in the Field of Electricity and Magnetism

Quantity, Law, or Theorem	Dual Quantity, Law, or Theorem
Electric charge	Magnetic pole
Electric field	Magnetic field
Electrostatics	Magnetostatics
Permittivity	Permeability
Electric current	Magnetic flux
Electromotive force (emf)	Magnetomotive force (mmf)
Resistance	Reluctance
Scalar potential function	Vector potential function
Electric Ohm's law	Magnetic Ohm's law
Electric Kirchhoff's law	Magnetic Kirchhoff's law
Electric Coulomb's law	Magnetic Coulomb's law
Scalar Poisson's equation	Vector Poisson's equation
Electric Gauss's law	Magnetic Gauss's law
Faraday's law	Ampere's law

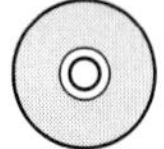

EXAMPLE 3.2

Given the following circuit (as shown in Figure 3.20):

a. *Find the voltage across the load (Z_L) using Norton's theorem*
b. *Find the Thevinin's equivalent*

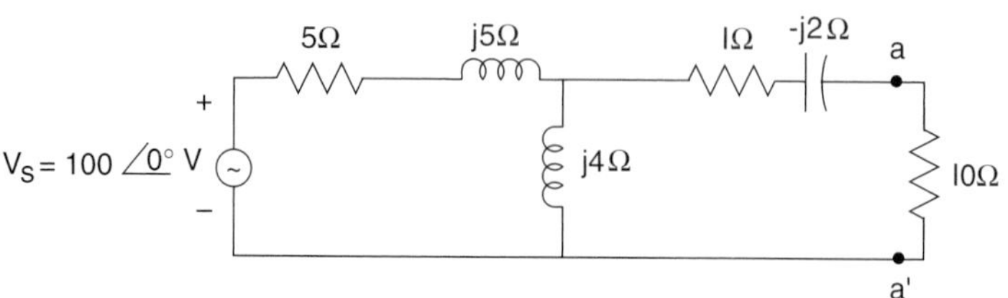

FIGURE 3.20 Circuit for Example 3.2.

Solution:

To find the load voltage and the Norton's equivalent, we do the following steps:

STEP 1. Disconnect the load and find $Y_T = 1 / Z_T$ by shorting the voltage source. Let:

$$Z_1 = 5 + j5, \quad Z_2 = j4, \quad Z_3 = 1 - j2.$$

Thus, Z_T is given by:

$$Z_T = Z_3 + (Z_1 \parallel Z_2) = (93 + j34)/53 = 1.75 + j0.64\ \Omega$$

STEP 2. We now find I_{sc} by shorting the terminal a–a' as shown in Figure 3.21:

$$I = V_s/(Z_1 + Z_2 \parallel Z_3)$$

$$I_{sc} = [(Z_2/(Z_2 + Z_3)]\ I = 20.5 + j3.24\ A$$

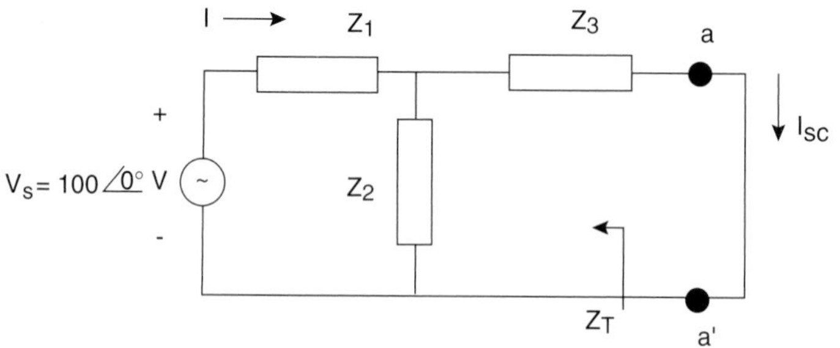

FIGURE 3.21 Circuit for step 2 of solution.

Step 3. The Norton equivalent is as shown in Figure 3.22. The current through the load is found by using the current division rule:

$$I_L = I_{SC}\frac{Z_T}{Z_T + Z_L}$$

$$I_L = (20.5 + j3.24) \times (1.75 + j0.64)/(11.75 + j0.64) = 3.28\angle 26° \text{ A}$$

$$V_L = 10\, I_L = 32.8\angle 26° \text{ V}$$

Using the duality principle, we obtain:

$$V_{OC} = Z_T\, I_{sc} = (1.75 + j0.64)(20.5 + j3.24) = 1.86\angle 20° \times 20.75\angle 9° = 38.6\angle 29°$$

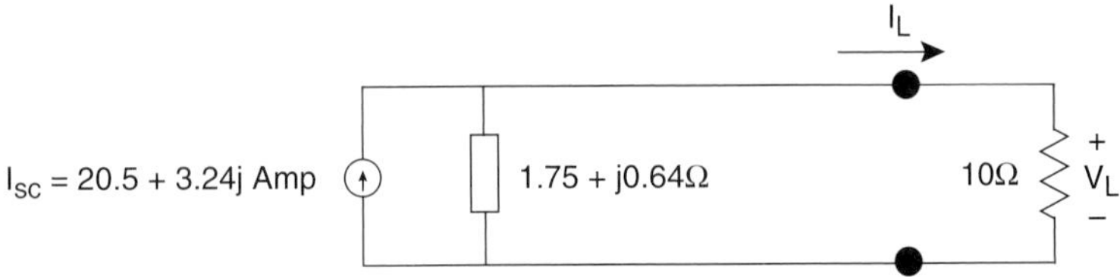

FIGURE 3.22 Circuit for step 3 of solution.

Thevenin's equivalent is shown in Figure 3.23.

FIGURE 3.23 Thevenin's equivalent circuit.

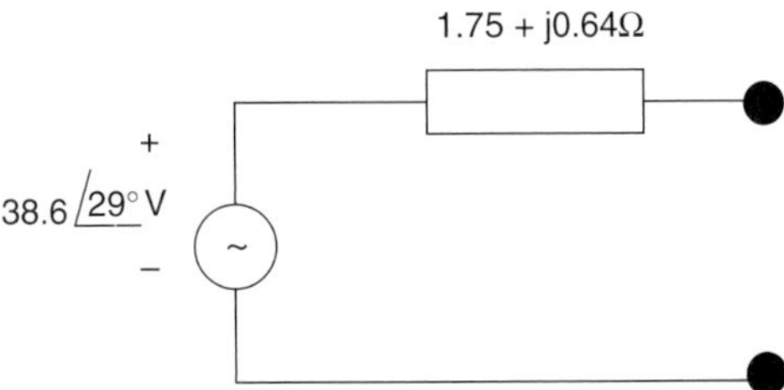

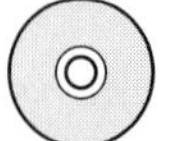

EXAMPLE 3.3

Given the circuit shown in Figure 3.24, find its corresponding dual.

Solution:

We know that the dual quantities are:

Voltage ↔ current

Series ↔ parallel

Thus, the voltage source is transformed into a current source (I_p) connected in parallel with a resistor R_p, as shown in Figure 3.25.

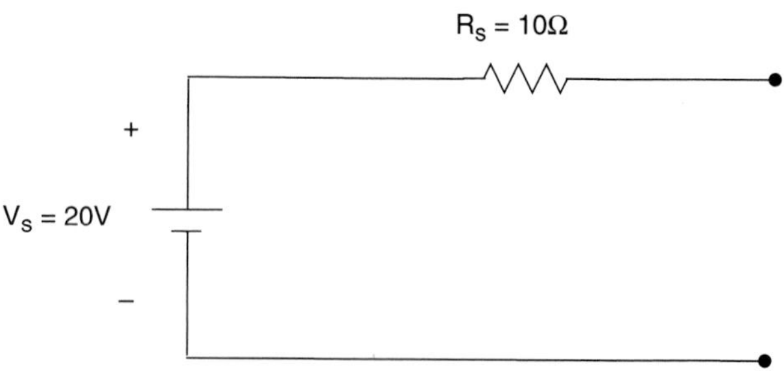

FIGURE 3.24 Circuit for Example 3.3.

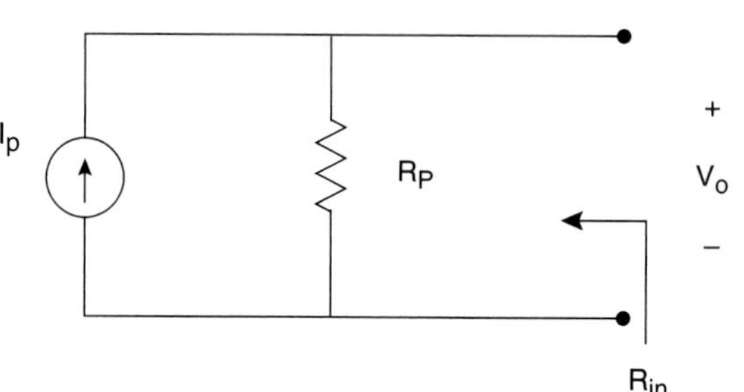

FIGURE 3.25 The corresponding dual.

In order to have equivalent performance, the output voltage (V_o) must be equal to $V_S = 20$V and the input impedance equal to $R_S = 10\ \Omega$. Thus, we have:

$$R_P = R_S = 10\ \Omega$$

$$V_S = I_P\, R_P = 20 \Rightarrow I_P = 2\ \text{A}$$

The final circuit is shown in Figure 3.26.

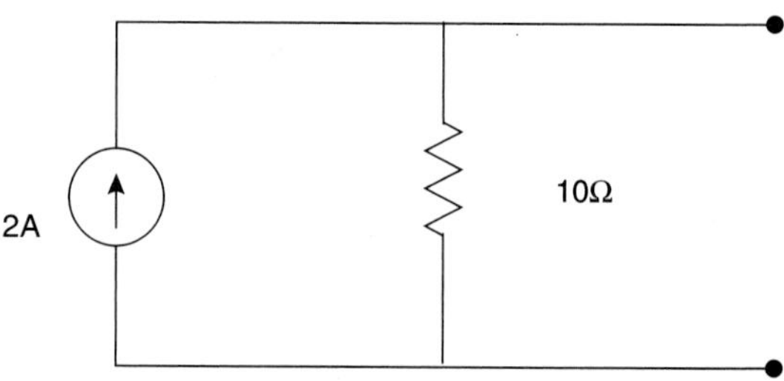

FIGURE 3.26 The final dual circuit.

3.10.4 The Superposition Theorem

Superposition applies to a situation where multiple causes (or independent sources) create one combined response at the output. With the help of this theorem, we can take apart this complicated response into its smaller components, which are easy to calculate. This theorem states that:

In a linear network, the voltage or current in any element resulting from several sources acting together is the sum of the voltages or currents resulting from each source acting alone, while all other independent sources are set to zero.

$$f(v_1 + v_2 + \ldots\ldots\ldots\ldots + v_n) = f(v_1) + f(v_2) + \ldots\ldots\ldots\ldots + f(v_n) \tag{3.37}$$

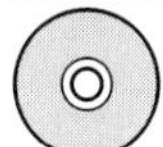

EXAMPLE 3.4

Find the current (I) in the 6Ω resistor, as shown in Figure 3.27.

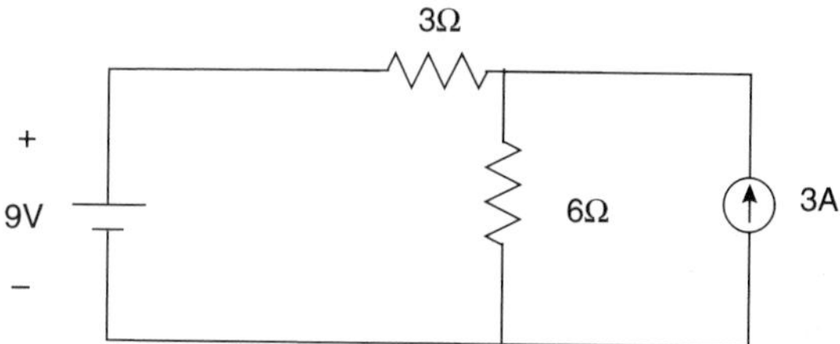

FIGURE 3.27 Circuit for Example 3.4.

Solution:

STEP 1. First set the current source to zero, as shown in Figure 3.28.

$$I_1 = 9/9 = 1\text{ A}$$

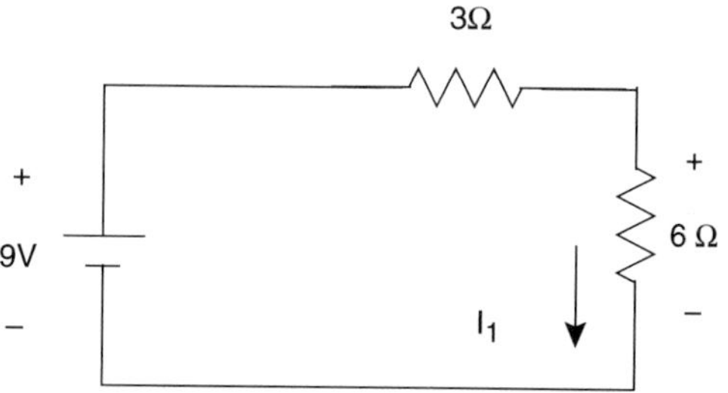

FIGURE 3.28 Circuit for step 1 of solution.

STEP 2. Set the voltage source to zero, as shown in Figure 3.29.

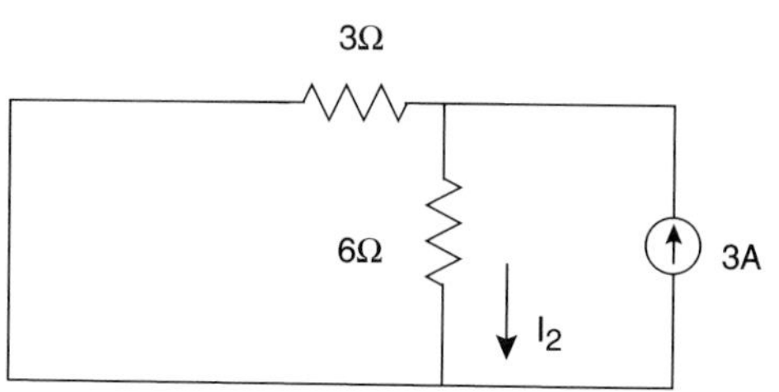

FIGURE 3.29 Circuit for step 2 of solution.

$$I_2 = 3 \times 3/9 = 1 \text{ A}$$

STEP 3. Thus, the total current (I) is given by:

$$I = I_1 + I_2 = 2 \text{ A}$$

NOTE: *All three theorems apply equally to time domain as well as the phasor domain problems and solutions, as long as the network remains linear.*

3.11 MILLER'S THEOREM

As will be seen in the next chapter, at higher frequencies there are transistor circuits where a passive element (usually a capacitor) bridges the output to the input by appearing in the feedback path. Such a feedback complicates the circuit analysis at high frequencies, as shown in Figure 3.30.

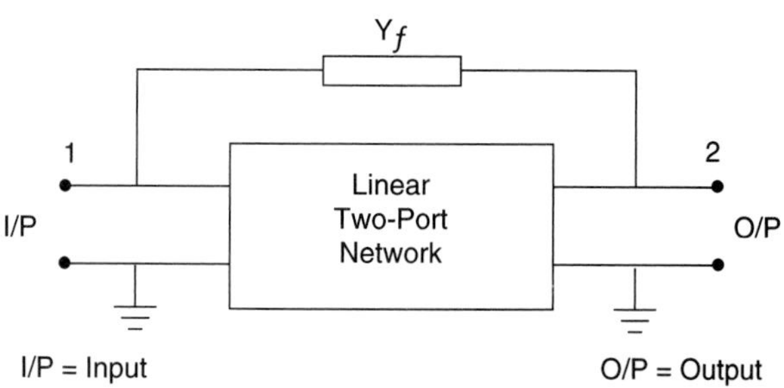

FIGURE 3.30 A linear two-port network with feedback.

There is a circuit theorem that replaces such a bridging element (Y_f) with two grounded elements. This replacement greatly simplifies circuit analysis, but more importantly it shows the effect of the bridging parasitic elements on the frequency response of a circuit at high frequencies.

Consider a linear two-port network (as shown in Figure 3.30) where the feedback (or bridging) element with admittance (Y_f) is connected from the output to the input. The overall voltage gain $(v_0/v_i = K)$ is expected to be known already through independent means.

With this assumption in mind, Miller's theorem states that:

The feedback element can be equivalently replaced by two admittances, Y_i and Y_o, as follows:

a. *Y_i between input node (1) and ground having a value of $Y_i = Y_f(1 - K)$*

b. *Y_o between output node (2) and ground with a value of $Y_o = Y_f(1 - \frac{1}{K})$*

The equivalent circuit is shown in Figure 3.31.

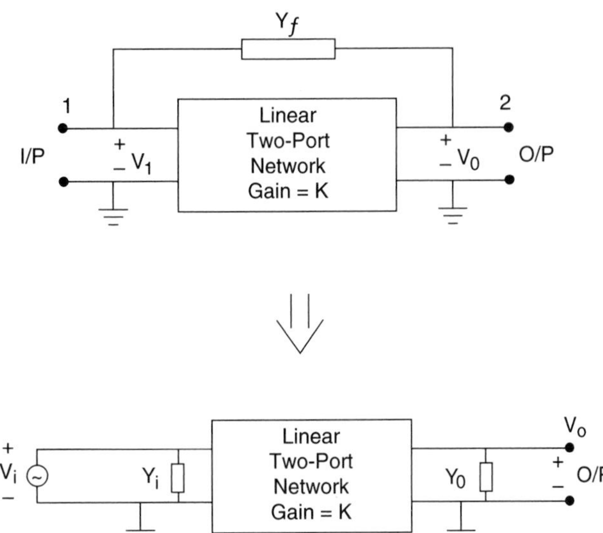

FIGURE 3.31 Miller's theorem.

This theorem, in essence, shows the deterioration of the effective input impedance caused by the presence of a feedback from the output port to the input port of a linear network.

POINTS OF CAUTION:

1. *Miller's theorem assumes that the voltage gain (K) can be determined by independent means beforehand.*
2. *Miller's equivalent circuit (see Figure 3.31) is valid as long as none of the conditions that existed in the original network (when K was determined) are changed. Therefore, Miller's theorem can be used only in the input impedance calculations of the network when the voltage gain is known. This means that this theorem cannot be used to calculate the output impedance because doing so requires that the input source be eliminated and be replaced with a source at the output terminals. Such an action would change the value of K in general, thus making Miller's equivalent circuit invalid.*

EXAMPLE 3.5

A broadband amplifier has an input impedance of $R_{in} = 100\ \text{k}\Omega$ at low frequencies. Calculate the total input impedance at a higher frequency of $f = 1$ MHz. The feedback capacitor's value is $C = 1\text{pF}$, and the overall gain is measured to be:

$$K = -100 \text{ at } f = 1 \text{ MHz (see Figure 3.32).}$$

Solution:

Miller's equivalent circuit is shown in Figure 3.33 where:

$$C_i = [1 - (-100)]C = 101 \text{ pF}$$

$$C_o = (1 - (-1/100)]C = 1.01 \text{ pF}$$

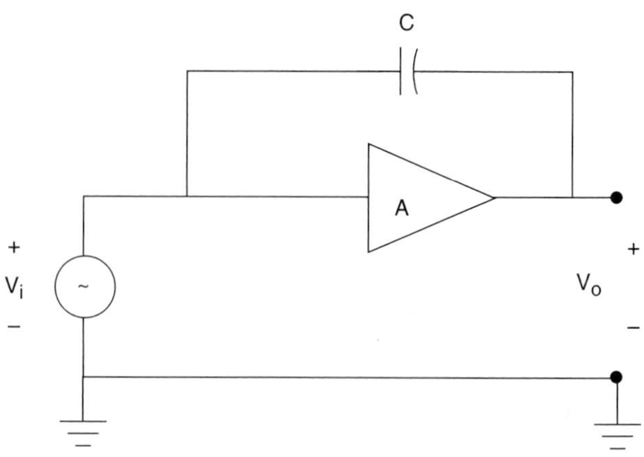

FIGURE 3.32 The amplifier circuit.

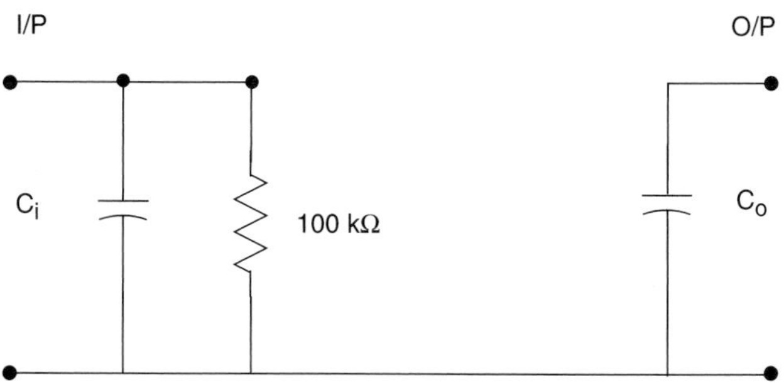

FIGURE 3.33 Miller's equivalent circuit for Example 3.5.

This indicates that the feedback capacitor appearing at the input is multiplied more than 100 times (known as Miller's effect) with a reactance of :

$$jX_C = 1/(j2\pi \times 10^6 \times 101 \times 10^{-12}) = -j1.57 \text{ k}\Omega$$

Therefore, the overall input impedance is substantially reduced from its original value because:

$$Z_{in} = 100 \,||\, -j1.57 = 0.025 - j1.55 \text{ k}\Omega$$

This substantial reduction in the input impedance is equivalent to "shorting out the input," which leads to tremendous deterioration of the high-frequency performance of the amplifier.

3.11.1 Relationship of Circuit Theorems to Circuit Laws

A summary of all four theorems discussed so far is depicted in Figure 3.34, which shows that by applying KCL, KVL, and the generalized Ohm's law we can analyze a circuit and predict its behavior under different signal conditions.

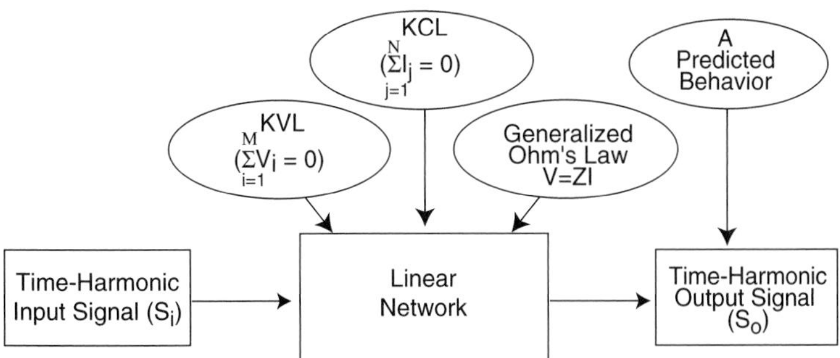

FIGURE 3.34 Low-frequency circuit analysis.

For a linear network, we can first simplify the circuit and obtain an equivalent circuit by applying one or more of the above four theorems (Thevenin's, Norton's, Superposition, or Miller's theorems) and then proceed to use KCL, KVL, and generalized Ohm's law to obtain the desired output signal, as shown in Figure 3.35.

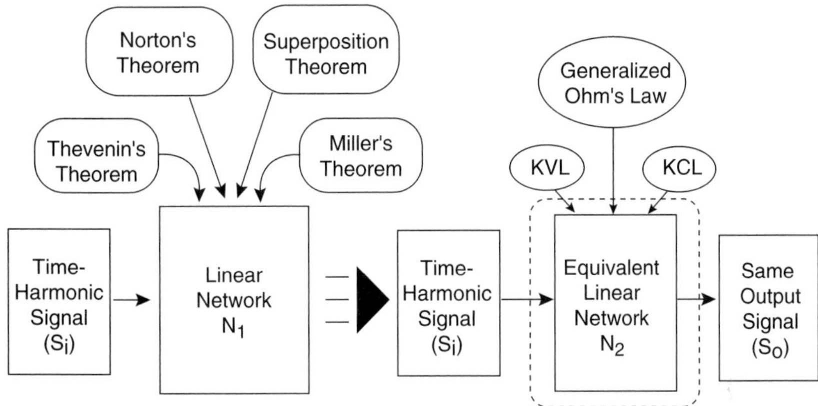

FIGURE 3.35 Relationship of theorems to actual circuit analysis.

3.12 POWER CALCULATIONS IN SINUSOIDAL STEADY STATE

In Chapter 2, power was generally defined as:

$$P(t) = v(t)i(t) \tag{3.38}$$

We now use this equation to define several specific power terms for signals in the sinusoidal steady state for a general circuit, as shown in Figure 3.36.

Let

$$v(t) = V_m \cos(\omega t + \varphi_v)$$

$$i(t) = I_m \cos(\omega t + \varphi_i)$$

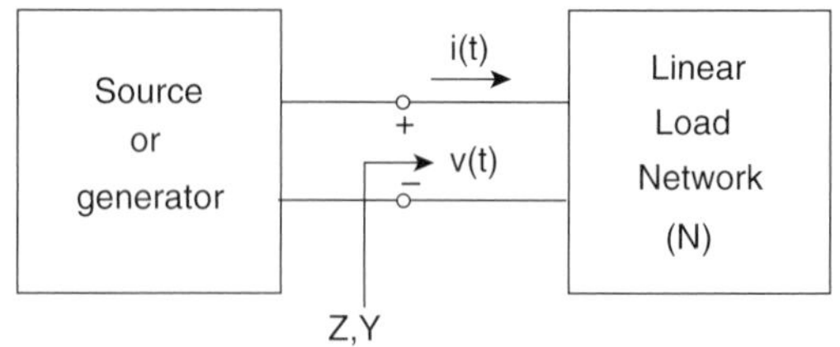

FIGURE 3.36 Block diagram of source and load network.

Where $\omega = 2\pi f = 2\pi/T$ is the frequency of operation, with f and T being the frequency and period of the sinusoid, respectively.

We now define concepts of instantaneous power, average power, and complex power for sinusoidal steady state.

3.12.1 Instantaneous Power, *p*(*t*)

Instantaneous power, $p(t)$, is defined as:

$$p(t) = v(t)i(t)$$

$$= I_m V_m \cos(\omega t + \varphi_v)\cos(\omega t + \varphi_i) \tag{3.39}$$

$$= \frac{1}{2}[I_m V_m \cos(\varphi_v - \varphi_i)] + \frac{1}{2}[I_m V_m \cos(2\omega t + \varphi_v + \varphi_i)]$$

3.12.2 Average Power, P_{av}

Average power is defined as:

$$P_{av} = \frac{1}{T}\int_0^T p(t)dt = \frac{1}{T}\int_0^T v(t)i(t)dt \tag{3.40}$$

We know that the integral of any sinusoid over any number of its periods is zero.

$$\int_0^T \cos(n\omega t + \phi) = 0, \qquad n = 1,2,3,\ldots$$

Thus, Equation 3.40 gives:

$$P_{av} = \frac{1}{2}[I_m V_m \cos(\varphi_v - \varphi_i)] \tag{3.41}$$

Figure 3.37 shows the relationship of the P_{av} and $p(t)$ to $v(t)$ and $i(t)$.

NOTE 1: *For a linear resistor, $\varphi_v = \varphi_i = 0$, thus Equation 3.41 can be written as:*

$$P_{av} = \frac{1}{2}[I_m V_m] = \frac{1}{2}[RI_m^2] = \frac{1}{2}[V_m^2/R] \tag{3.42a}$$

NOTE 2: *For a DC signal ($\omega = 0$) because $\varphi_v = \varphi_i = 0$ and $V_m = V_{DC}$ and $I_m = I_{DC}$, then the average power delivered to a resistive load (R) is simply given by:*

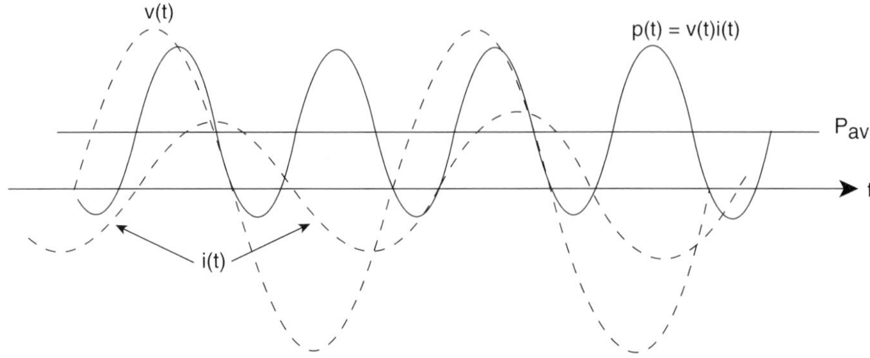

FIGURE 3.37 Relationship of $p(t)$ and P_{av} with respect to $i(t)$ and $v(t)$.

$$P_{av} = V_{DC} I_{DC} = RI_{DC}^2 = V_{DC}^2/R \tag{3.42b}$$

3.12.3 Complex Power, *P*

Complex power can now be defined as:

$$P = \frac{1}{2}[VI^*] \tag{3.43}$$

Where V and I are phasors for $v(t)$ and $i(t)$, respectively, and (*) represents the complex conjugate operation. V and I are defined as:

$$V = V_m e^{j\varphi_v}$$

$$I = I_m e^{j\varphi_i}$$

Thus, Equation 3.43 can be written as:

$$P = \frac{1}{2}[I_m V_m e^{j(\varphi_v - \varphi_i)}] = \frac{1}{2}\{I_m V_m[\cos(\varphi_v - \varphi_i) + j\sin(\varphi_v - \varphi_i)]\} \tag{3.44}$$

OBSERVATION #1:

$$P_{av} = Re(P) = \frac{1}{2}Re(VI^*) \tag{3.45}$$

OBSERVATION #2:

Letting $Z(j\omega)$ or $Y(j\omega)$ be the input impedance (or input admittance) of the load network (see Figure 3.36), we have:

$$Z(j\omega) = \frac{1}{Y(j\omega)} = \frac{V}{I} = \frac{V_m}{I_m}e^{j(\varphi_v - \varphi_i)} = Z_m e^{j\varphi_z}$$

Where

$$Z_m = V_m/I_m,$$

$$\varphi_z = \varphi_v - \varphi_i$$

$$P_{av} = \frac{1}{2}(I_m V_m \cos\varphi_z) = \frac{1}{2}[I_m^2 ReZ(j\omega)] = [V_m^2 ReY(j\omega)] \tag{3.46}$$

Which indicates that the average power delivered to a load can be changed by merely changing the phase of the load without changing its magnitude.

3.12.4 Superposition of Average Powers

Now let's consider a case where the electrical source is composed of the sum of n sinusoids at different frequencies ($\omega_1, \omega_2, \ldots\ldots, \omega_n$). With the network in the steady state, we can see that because the load network is linear, the total output is the sum of all individual outputs due to each one of the inputs (see Superposition theorem). The instantaneous power, being a nonlinear function, does not follow the Superposition theorem. It can easily be shown, however, that superposition holds valid for average power (P_{av}) as follows:

$$(P_{av})_{total} = P_{av,\omega 1} + P_{av,\omega 2} + \ldots\ldots + P_{av,\omega n} \tag{3.47}$$

3.12.5 Effective or Root-Mean-Square (rms) Value of a Periodic Signal

The concept of an "effective value" comes from a desire to have a periodic signal (voltage or current) deliver to a resistor load (R) the same average power (P_{av}) as an equivalent DC signal (voltage or current) would. Thus, the effective value of a signal (V_{eff} or I_{eff}) is a measure of its effectiveness in delivering real power to a load resistor (R).

To find the effective value of a signal, we need to find an equivalent DC value (V_{eff} or I_{eff}) that will deliver the same average power (P_{av}) to a resistor load (R) as would be delivered by the periodic signal itself. Thus, from Equation 3.40 and Equation 3.42b, we can write the following:

For a DC signal

$$(P_{av})_{DC} = Ri_{eff}^2, \tag{3.48}$$

And, for a periodic signal

$$P_{av} = \frac{1}{T}\int_0^T v(t)i(t)dt \tag{3.49}$$

substituting for $v(t) = Ri(t)$ and equating Equation 3.48 with 3.49, we have:

$$P_{av} = \frac{1}{T}\int_0^T Ri(t)^2 dt = Ri_{eff}^2 \tag{3.50}$$

Or

$$I_{eff} = \left[\frac{1}{T}\int_0^T i(t)^2 dt\right]^{1/2} \tag{3.51}$$

From Equation 3.51 we see that I_{eff} is the square root of the mean of the squared (or root-mean-square) value of the current. Thus, the effective current (I_{eff}) is commonly referred to as the root-mean-square current (I_{rms}):

$$I_{eff} = I_{rms}$$

Similarly, V_{eff} is found as follows:

$$P_{av} = \frac{1}{T}\int_0^T \frac{v(t)^2}{R}dt = \frac{V_{eff}^2}{R} \tag{3.52}$$

$$V_{eff} = \left[\frac{1}{T}\int_0^T v(t)^2 dt\right]^{1/2} = V_{rms} \tag{3.53}$$

Summary: *Because* $V_{rms} = RI_{rms}$, *we can write:*

$$P_{av} = RI_{rms}^2 = V_{rms}^2/R$$

and

$$P_{av} = I_{eff}V_{eff} = I_{rms}V_{rms} \tag{3.54}$$

Special Case—Sinusoidal Signals. When the periodic signal is sinusoidal [e.g., $i(t) = I_m\cos\omega t$], I_{rms} (or I_{eff}) can be calculated as follows:

$$i(t) = I_m\cos\omega t$$

$$I_{rms} = \left[\frac{1}{T}\int_0^T I_m^2\cos^2\omega t dt\right]^{1/2} = \left[\frac{I_m^2}{T}\int_0^T \frac{1+\cos 2\omega t}{2}dt\right]^{1/2} = \left[\frac{I_m^2}{T}\frac{T}{2}\right]^{1/2}$$

Or

$$I_{rms} = I_{eff} = \frac{I_m}{\sqrt{2}} \tag{3.55a}$$

For $i(t) = I_m\sin\omega t$ we get the identical result as follows:

$$I_{rms} = \left[\frac{1}{T}\int_0^T I_m^2\sin^2\omega t dt\right]^{1/2} = \left[\frac{I_m^2}{T}\int_0^T \frac{1-\cos 2\omega t}{2}dt\right]^{1/2} = \left[\frac{I_m^2}{T}\frac{T}{2}\right]^{1/2}$$

Or

$$I_{rms} = I_{eff} = \frac{I_m}{\sqrt{2}} \tag{3.55b}$$

NOTE:

$$\cos\omega^2 t = \frac{1+\cos 2\omega t}{2}$$

$$\sin\omega^2 t = \frac{1 - \cos 2\omega t}{2}$$

and

$$\int_0^T \cos 2\omega t = 0.$$

Similarly, for sinusoidal voltage:

$$v(t) = V_m \cos\omega t,$$

Or,

$$v(t) = V_m \sin\omega t$$

The effective (or rms) value is given by

$$V_{rms} = V_{eff} = \frac{V_m}{\sqrt{2}} \tag{3.56}$$

NOTE 1: *Similar to average powers, superposition also holds valid for the rms values of several sinusoids as follows:*

$$(P_{rms})_{total} = P_{rms,\omega 1} + P_{rms,\omega 2} + \dots\dots + P_{rms,\omega n} \tag{3.57}$$

NOTE 2: *For a general load (Z), from Equation 3.41, the average power delivered can now be written in terms of I_{rms} and V_{rms} as:*

$$Z = R + jX = |Z|e^{j\phi_z}$$

$$P_{av} = \frac{I_m V_m}{2}\cos(\phi_v - \phi_i)$$

Or,

$$P_{av} = I_{rms} V_{rms} \cos(\varphi_z) \tag{3.58}$$

The concept of effective (or rms) value is used in many AC voltmeters and ammeters where the readings are given in terms of rms values. This means that to obtain the amplitude or peak value, the rms value reading must be multiplied by ($\sqrt{2} \approx 1.414$). For example, the domestic line voltage in the United States is V_{rms} = 110 V (220 V in Europe), which corresponds to a peak value of $V_m \approx 155.5$ V (311 V in Europe).

3.13 THE DECIBEL UNIT (dB)

The decibel is a standard unit for describing transmission gain (or loss) and relative power levels, which we need to define at the outset:

DEFINITION-DECIBEL: *The ratio of two powers or intensities or the ratio of a power to a reference power. It is one-tenth of an international unit known as the "Bel." The Bel, named after Alexander Graham Bell, was used to measure attenuation in telephone cables. "Bel" is defined to be the logarithm to the base 10 (also known as common logarithm) of the power ratio:*

$$\text{Bel} = \log_{10}(P_2/P_1) \tag{3.59}$$

Therefore, if P_1 (usually the input power) and P_2 (usually the output power) are two amounts of power, the latter is said to be N **decibels** greater than the first (for a positive N value), where:

$$N(\text{dB}) = 10 \log_{10}(P_2/P_1) \tag{3.60}$$

If $N < 0$ then P_2 is said to be N decibels smaller than P_1.

To convert from dB to power ratios, we use Equation 3.60 to obtain:

$$\frac{P_2}{P_1} = 10^{N(\text{dB})/10}$$

3.13.1 Decibels Above or Below 1 Watt (dBW)

If the chosen reference level is P_1 = one Watt (1 W) then the term dBW is used, which is defined by:

$$N(\text{dBW}) = 10 \log(P_2/1\text{ W}) \tag{3.61 a}$$

3.13.2 Decibels Above or Below 1 Milliwatt (dBm)

If the reference level is chosen to be P_1 = one milliWatt (10^{-3} Watt or 1 mW) then the term dBm is commonly used, which is defined by:

$$N(\text{dBm}) = 10 \log(P_2/1\text{ mW}) \tag{3.61 b}$$

3.13.3 Decibels Above or Below 1 Microwatt (dBμW)

If the chosen reference level is P_1 = one microWatt (10^{-6} Watt or 1 μW) then the term dBμW is used, which is defined by:

$$N(\text{dB}\mu\text{W}) = 10 \log(P_2/1\ \mu\text{W}) \tag{3.61 c}$$

NOTE: *The decibel (dB) is "the logarithm of a power ratio" and not a unit of power; however, dBμW, dBm and dBW are units of power in the logarithmic system of numbers.*

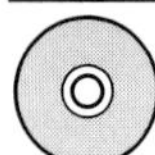

EXAMPLE 3.6

Convert the following into dBm or dBW:

$P = 1$ mW, $P(\text{dBm}) = ?$
$P = 10$ mW, $P(\text{dBm}) = ?$
$P = 0.1$ mW, $P(\text{dBm}) = ?$
$P = 1$ W, $P(\text{dBW}) = ?$
$P = 0.1$W, $P(\text{dBW}) = ?$
$P = 10$ W, $P(\text{dBW}) = ?$

Solution:

$P = 1$ mW $\Rightarrow$ $P = 0$ dBm
$P = 10$ mW $\Rightarrow$ $P = 10$ dBm
$P = 0.1$ mW $\Rightarrow$ P = –10 dBm
$P = 1$ W, $P(\text{dBW}) = 0$ dBW
$P = 0.1$W, $P(\text{dBW}) = -10$ dBW
$P = 10$ W, $P(\text{dBW}) = 10$ dBW

Therefore, the dBm and dBW units measure "Above or below" one milliwatt and one Watt, respectively.

3.13.4 Voltage or Current Gain

At times, relative voltages or currents in terms of decibels are desired. In this case, if the resistance level is the same at the points where both power levels are measured, the relative currents or voltages at the output (P_{out}) relative to the input (P_{in}), are expressed as:

$$P_{in} = V_{in}^2 / R = I_{in}^2 R$$

$$P_{out} = V_{out}^2 / R = I_{out}^2 R$$

Thus, in terms of voltage we have:

$$G(\text{dB}) = 10 \log_{10}(P_{out}/P_{in}) = 10 \log_{10}(V_{out}/V_{in})^2 = 20 \log_{10}(V_{out}/V_{in}) \quad (3.41)$$

Or, in terms of current

$$G(\text{dB}) = 10 \log_{10}(I_{out}/I_{in})^2 = 20 \log_{10}(I_{out}/I_{in}) \quad (3.62\text{a})$$

NOTE 1: *Loss in ratio is defined as the inverse of the gain in ratio.*

$$\text{Loss (ratio)} = P_{in}/P_{out} = 1/\text{Gain (ratio)}$$

Or, in dB, loss is the negative value of gain.

$$L(\text{dB}) = 10 \log_{10}(P_{in}/P_{out}) = -G(\text{dB}) \quad (3.62\text{b})$$

NOTE 2: *For a cascaded network consisting of several gain stages (G_1, G_2,...,G_k in dB) and several loss stages (L_1, L_2,..., L_m in dB), the total gain (G_{total}) in dB is the sum of all the gain stages minus all of the loss stages.*

$$G_{total} = (G_1 + G_2, ... + ... + G_k) - (L_1 + L_2 + ... + L_m), \text{ dB} \tag{3.63}$$

3.13.5 Neper (Np)

Neper is defined to be a unit of attenuation used for expressing the ratio of two currents, voltages, or fields by taking the natural logarithm (logarithm to base e) of this ratio. If voltage V_1 is attenuated down to V_2 so that:

$$V_2/V_1 = e^{-N}$$

Then N is attenuation in Nepers (always a positive number) and is defined by:

$$N\ (\text{Np}) = \log_e(V_2/V_1)^{-1} = -\ln(V_2/V_1) \tag{3.64}$$

NOTE: *The unit Neper is named after John Napier, a Scottish scientist and inventor of natural logarithms.*

In circuits matched in impedance, the following conversion between Neper and dB can be derived:

$$1\ \text{Np} = -\ln(V_2/V_1) \Rightarrow V_2/V_1 = 1/e$$

This gives Np in terms of dB as:

$$1\ \text{Np} = 20 \log_{10}[1/(V_2/V_1)] = 20 \log_{10}(e) = 8.686 \text{ dB} \tag{3.65}$$

Thus 1 Np is a larger unit than dB by a factor of 8.86. Conversely:

$$1 \text{ dB} = 0.115 \text{ Np} \tag{3.66}$$

As noted, values expressed in dB can take on positive or negative values while Np values can take on only positive values. Example 3.7 illustrates this point.

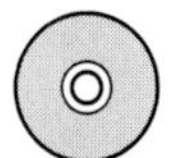

EXAMPLE 3.7

If a voltage at the input of a circuit attenuates from 1 to 0.5 volt when it reaches the output, what is the voltage ratio in dB and Nepers?

Solution:

$V_o/V_i = 0.5$
$N(\text{dB}) = 20 \log_{10}(0.5) = -6 \text{ dB}$
Or,
$N(\text{Np}) = 6 \text{ dB} \times 0.115 = 0.69 \text{ Np}$

LIST OF SYMBOLS/ABBREVIATIONS

A symbol or abbreviation will not be repeated once it has been identified and defined in an earlier chapter, as long as its definition remains unchanged.

A	Used to represent amplitude of a function
B	Susceptance
Bel	International unit for measuring attenuation
C	Capacitance
C_{eq}	Equivalent lumped value capacitance
dB	Decibel
dBm	Decibels referenced to 1 milliwatt
G	Conductance
I_L	Current through the load
I_m	The imaginary portion of a complex number
I_{SC}	Short-circuit current
i(t)	Current with respect to time
j	An imaginary number where $j = \sqrt{-1}$
K	Overall voltage gain
KCL	Kirchhoff's Current
KVL	Kirchhoff's Voltage Law
L	Inductance
L_{eq}	Equivalent lumped value inductance
N	Used to represent the number of turns of wire in an inductor
Np	Used to represent the unit Neper that defines attenuation
pF	pico-Farad
Re	The real portion of a complex number
R_{eq}	Equivalent lumped value resistance
R_i	Used to represent a lumped element resistance
$v_{12}(t)$	Notation used to denote a difference in voltage between two designated points, 1 and 2, in a circuit
V_{OC}	Open circuit voltage
X	Reactance
Y	Admittance
Y_T	Thevenin admittance
Z	Impedance
Z_L	Load impedance
Z_T	Thevenin impedance
$\angle$	Angle
$\|Z\|$	Uprights used to denote absolute value of the variable inside
ω	Frequency given in radians/second
Φ	Magnetic flux
φ	Phase of a function
θ	Phase angle of a complex number

PROBLEMS

3.1 Define a phasor and describe its applications.

3.2 What are the reasons for using phasors?

3.3 Determine the phasors that represent the following real-valued time functions:

a. $A(t) = 10\cos(2t + 30°) + 5\sin 2t$

b. $B(t) = \sin(3t - 90°) + 10\sin(3t + 45°)$

c. $C(t) = \cos(t) + \cos(t + 30°) - \cos(t + 60°)$

3.4 Find the steady-state voltage, $v(t)$, represented by the phasor:

a. $V = 10\angle -140°$ volts

b. $V = 80 + j75$ volts

3.5 Express the following summations into one single sinusoid, $(A\cos\omega t)$:

a. $A(t) = 2\cos(6t + 120°) + 4\sin(6t - 60°)$

b. $B(t) = 5\cos 8t + 10\sin(8t + 45°)$

c. $C(t) = 2\cos(2t + 60°) - 4\sin 2t + d/dt(2\sin 2t)$

3.6 A current in an element is $2\cos 100t$. Find the steady-state voltage $v(t)$ across the element if the element is:

a. A resistor: $R = 20\ \Omega$

b. An inductor: $L = 20$ mH

c. A capacitor: $C = 20$ mF

3.7 Find the input impedance and admittance at terminals a – b for the circuit shown in Figure P3.7 where $\omega = 100$ rad/s.

FIGURE P3.7

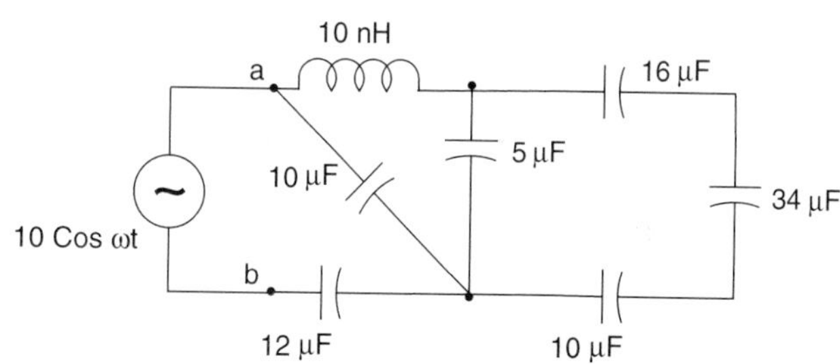

3.8 Find the steady-state voltage (v) when $I_s = 20\cos\omega t$, $\omega = 1000$ rad/s for the circuit shown in Figure P3.8.

3.9 Find the steady-state current, $i(t)$ when $V_S = 10\cos 3t$ for the circuit shown in Figure P3.9.

3.10 In the circuit shown, determine the voltage across the inductor, $V_L(t)$, when $V_{S1} = 20\cos\omega t$, $V_{S2} = 30\cos(\omega t - 90°)$, and $\omega = 1000$ rad/s (see Figure P3.10).

3.11 Find the current through the load ($Z_L = 8 - j4\Omega$) in which $I_1 = V_{OC}/10$, where V_{OC} is the voltage across the output terminals (a – b)for a disconnected load (see Figure P3.11). Determine I_L using Thevenin's theorem and Norton's theorem.

FIGURE P3.8

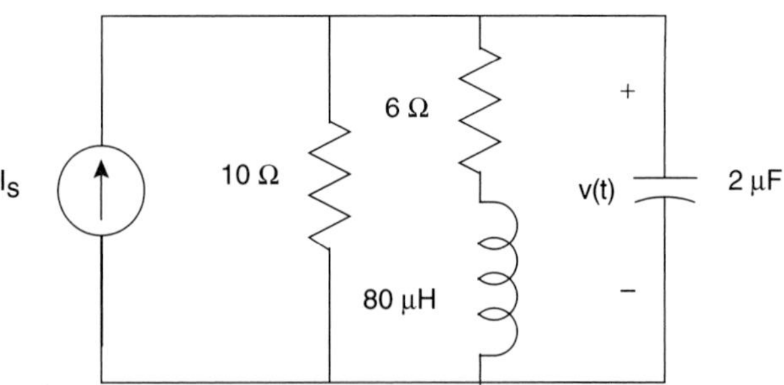

FIGURE P3.9

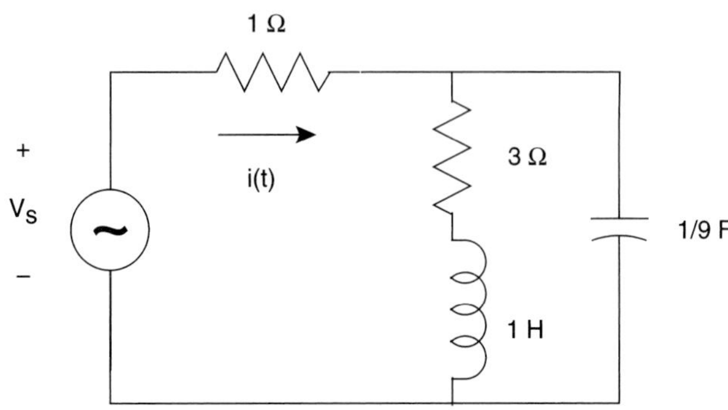

FIGURE P3.10

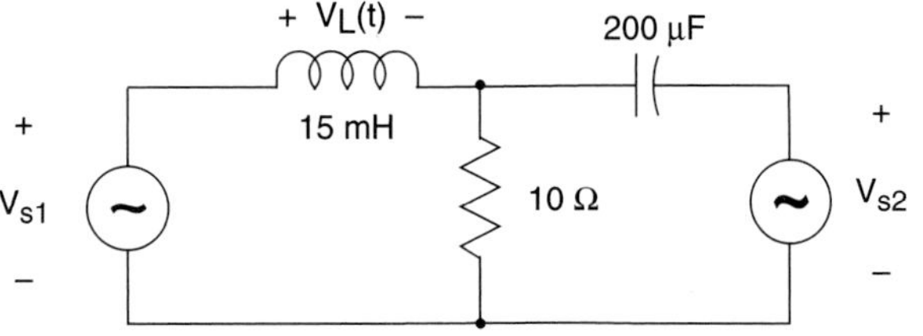

FIGURE P3.11

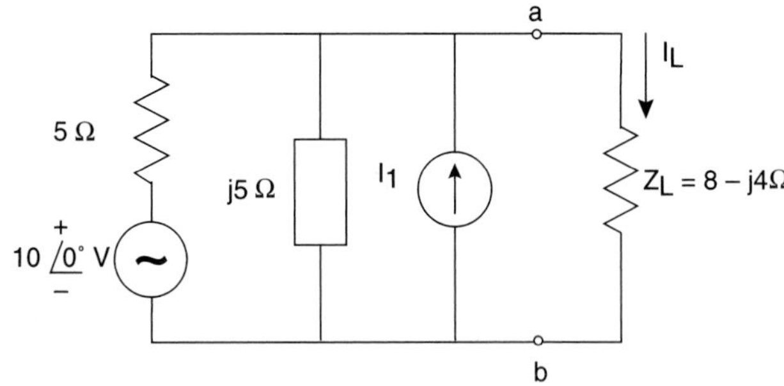

3.12 Using the Superposition theorem, determine *i(t)* in the circuit shown in Figure P3.12 when $V_S(t) = 10\cos 10t$.

FIGURE P3.12

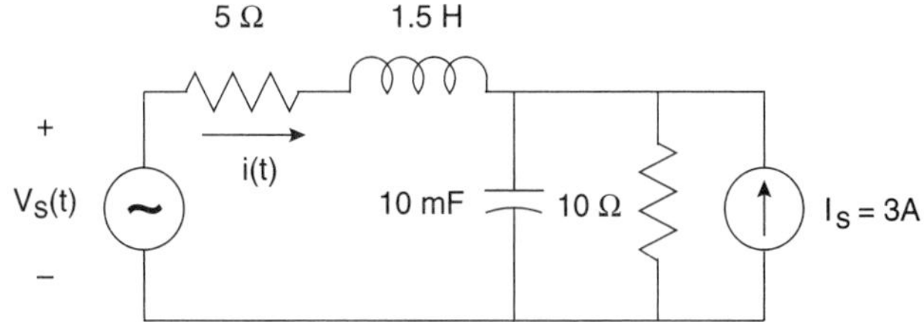

3.13 Calculate the power ratios for the following decibel values:

a. 5

b. 25

c. 50

d. 75

3.14 The input signal to a three-stage system is -35dBm. The first stage has a gain of 28 dB, the second stage has a loss of 3 dB, and the third stage has a gain of 7 dB. What is the power output at each stage in milliwatts?

3.15 Find the rms value of the voltage (v) for:

a. $v = 2 - 4\cos 2t$ V

b. $v = 3\sin \pi t + 2\cos \pi t$ V

c. $v = 2\cos 2t + 4\cos(2t + \pi/4) + 12\sin 2t$ V

3.16 Find the instantaneous power, the average power and the complex power for the circuit shown in Figure P3.16.

FIGURE P3.16

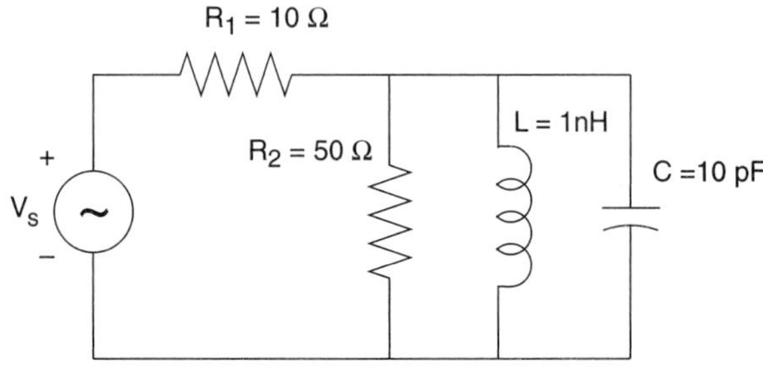

REFERENCES

[3.1] Desor, C. A. and E. S. Kuh. *Basic Circuit Theory.* Tokyo: McGraw-Hill, 1969.

[3.2] Dorf, R. C. and J. A. Savoboda. *Introduction to Electric Circuits.* New York: John Wiley & Sons, 1996.

[3.3] Dorf, R. C. *Electrical Engineering Handbook.* Boca Raton, FL: CRC Press, 1993

[3.4] Sedra, A. S. and K. C. Smith. *Microelectronic Circuits.* New York: Oxford University Press, 1998.

CHAPTER 4

DC and Low-Frequency Circuits Concepts

4.1 INTRODUCTION

Modern electronic circuits commonly use transistors. To understand higher-frequency concepts, it is imperative that we first study the behavior of transistor circuits at DC and low frequencies. Even though microwave transistors are built differently and have a higher-frequency range of operation, analysis of transistor circuits at DC and low frequencies will provide a sound foundation for analysis and design of transistor circuits at much higher frequencies. In fact, we must be able to perform this lower-frequency analysis with relative ease before we can master higher-frequency circuit analysis and design.

The analysis presented in this chapter requires familiarity with different types of transistors and their operation. We assume that the reader has a certain amount of knowledge about the basics of transistors and their function and behavior. As a refresher, some of the basics about diodes and transistors are discussed first before we proceed with actual circuit analysis.

4.2 DIODES

A diode is one of the most basic semiconductor devices, and it serves as a building block for the construction of a transistor. It is the simplest and most fundamental nonlinear element, and a certain amount of familiarity with this device is assumed.

DEFINITION-SEMICONDUCTOR DIODE (OR CRYSTAL DIODE): *A two-electrode device consisting of n-type and p-type semiconductor materials that makes use of the rectifying properties of a p-n junction, namely it passes current in the forward direction (from anode to cathode) and blocks current in the reverse direction.*

The I–V characteristic of a diode is plotted in Figure 4.1.

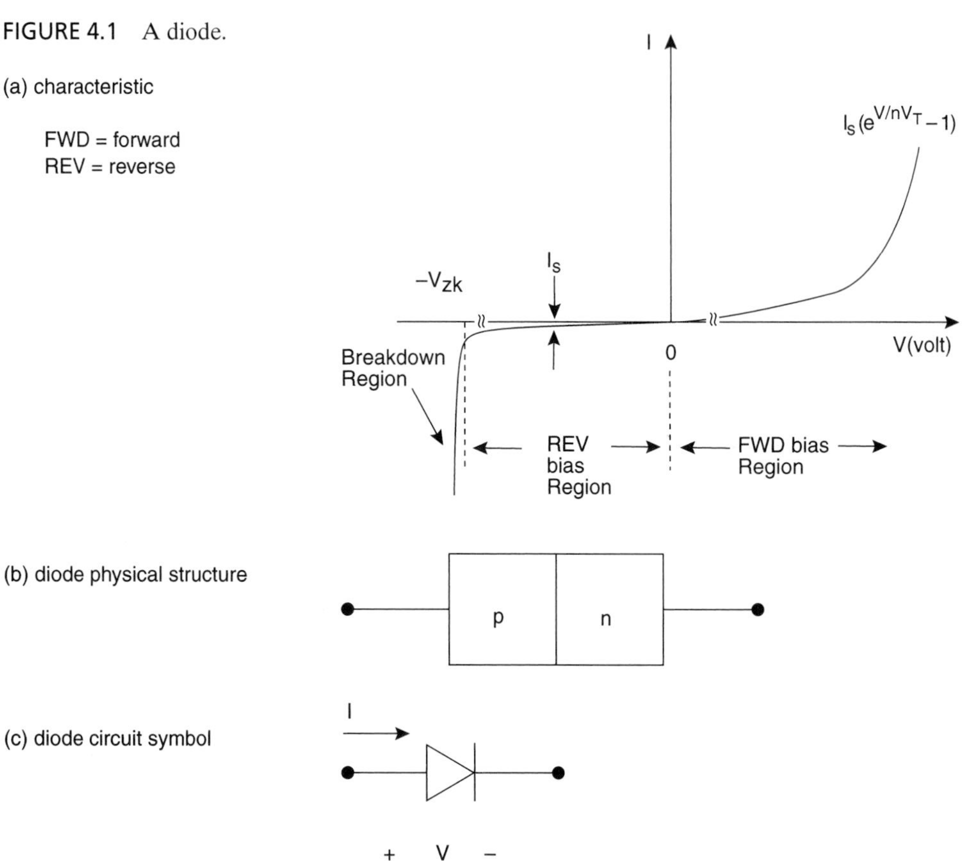

FIGURE 4.1 A diode.

4.2.1 Diode's Three Modes of Operation

The *I–V* characteristic of a diode (see Figure 4.1) consists of three distinct regions:

1. The forward-bias region: $V > 0$
2. The reverse-bias region: $V < 0$
3. The breakdown region: $V < -V_{ZK}$

These three regions are described briefly as follows:

1. The **forward-bias region** is characterized by:

$$i = I_S(e^{v/nV_T} - 1) \qquad v > 0 \tag{4.1}$$

When $i >> I_S$, that is, when there is appreciable current in the forward direction, then Equation 4.1 can be approximated by:

$$i \approx I_S e^{v/nV_T} \tag{4.2a}$$

Or, Equation 4.2a can alternately be expressed as:

$$v = nV_T \ln\frac{i}{I_S} \tag{4.2b}$$

Where:

a. I_S is the reverse saturation current, which is a constant for a diode at a certain temperature. The current (I_S) is also called "the scale current" because it, being directly proportional to the cross-sectional area of the diode, would double the value of current (i) for a given voltage (v) if the junction area were doubled. The value of I_S is a very strong function of temperature (T) and will approximately double in value for every 5°C rise in temperature. Typical values of I_S for a low-power diode is in the order of 10^{-9} A to 10^{-15} A.

b. V_T is a constant called the thermal voltage and is given by:

$$V_T = kT/q \tag{4.3a}$$

k = Boltzmann's constant = 1.38×10^{-23} J/K
T = The absolute temperature of the diode in Kelvin (K), (K = 273.15 + °C)
q = The magnitude of the charge of an electron (1.6×10^{-19} C)

Using these values, Equation 4.3a can be written as:

$$V_T = \frac{T}{11600} \tag{4.3b}$$

At room temperature:

$$T = 293° \text{ K (or 20 °C)} \Rightarrow V_T \approx 25 \text{ mV.}$$

This value of V_T will be used throughout this book for rapid circuit calculations.

c. n is a constant with a value between 1 and 2 (i. e., $1 \le n \le 2$) depending on the material and physical structure of the diode. For example, $n = 1$ applies to diodes that are fabricated using standard integrated circuit technology whereas $n = 2$ applies to discrete two-terminal diodes. In this book, for simplicity of analysis, we will use $n = 1$ throughout unless otherwise specified.

NOTE: *Because both V_T and I_S are strong functions of temperature, the forward I-V characteristic of a diode as given by Equation 4.1 varies with temperature. An increase in temperature (T) causes an increase in (I_S), causing a decrease in the value of voltage (v), which is required if the original value of current (i) is to be maintained. For temperatures in the vicinity of T = 293° K, it is found that:*

$$\frac{dv}{dT} = -2.2 \text{ mV/°C} \tag{4.4}$$

This decrease in voltage with temperature is shown in Figure 4.2. For example, a 30°C increase in room temperature ($T = 323°$K) will decrease a 0.7 V drop across a diode, to approximately 0.64 V.

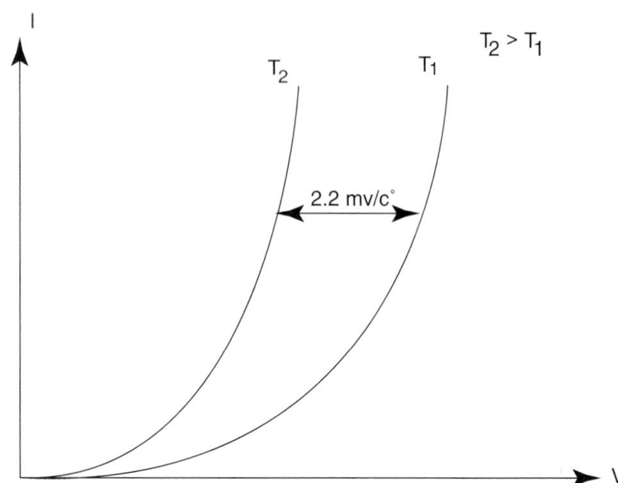

FIGURE 4.2 Temperature dependence of diode *I–V* curve.

2. The **reverse-bias region** is the region where $v < 0$. Under this condition, the current through the diode is $i \approx -I_S$. The reverse current increases somewhat with an increase in temperature or reverse voltage; this effect can be neglected in this simple model.
3. The **breakdown region** is entered when the magnitude of the reverse voltage exceeds a threshold value (V_{ZK}), i.e., $V < -V_{ZK}$.

   In this region, the reverse current increases rapidly with a very small increase in voltage. The diode breakdown is not destructive as long as the reverse current does not exceed the maximum safe value as specified in the device data sheets.

   There are two possible breakdown mechanisms that occur separately or in combination, depending on the breakdown voltage as follows:

   - Zener breakdown when $V_{ZK} < 5\ V$
   - Avalanche breakdown when $V_{ZK} > 7\ V$
   - A combination of Zener and Avalanche breakdown when $5 < V_{ZK} < 7\ V$

4.2.2 DC Analysis of Forward-Biased Diode Circuits

In this section, we will focus on diodes operating in the forward-bias region. Let's consider a diode circuit as shown in Figure 4.3. Using the diode's characteristic equation and applying KVL, we have:

$$I_D \approx I_s e^{V_D/nV_T} \tag{4.5}$$

$$V_{DD} = RI_D + V_D \tag{4.6}$$

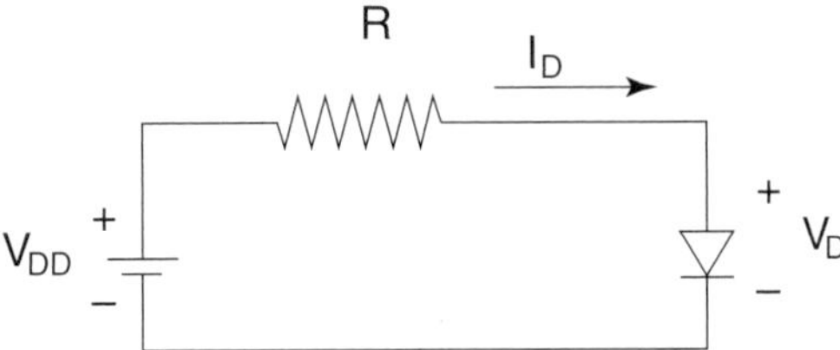

FIGURE 4.3 Simple diode circuit.

From Equations 4.5 and 4.6, we can see that we have two equations with two unknowns (I_D,V_D) that need to be solved for. There are two methods of analysis:

- Graphical method
- Mathematical method using the iterative analytical technique

These methods are discussed next:

Graphical Method. This method is an invaluable conceptual tool in diode circuit analysis but is seldom used in practice. It is important to understand this method because it gives us a visual understanding of DC analysis.

In this method we plot both equations to scale on a graph in the $I_D - V_D$ plane and obtain the solution as the coordinates of the point of intersection, as shown in Figure 4.4.

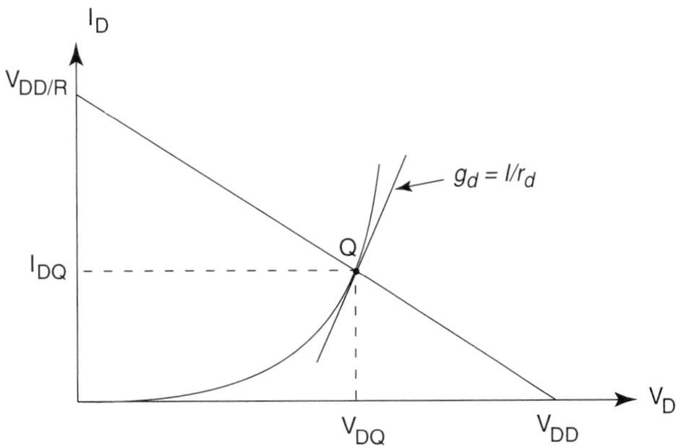

FIGURE 4.4 Graphical representation of small-signal analysis.

Equation 4.6 expresses a linear relationship between I_D and V_D and is often referred to as the "load line." The intersection point (often referred to as the quiescent point or Q-point) has coordinates I_{DQ} and V_{DQ}, which represent the solution to the circuit of Figure 4.3.

Mathematical Method. In this method, the solution to the diode circuit is found through calculation. Depending on the desired degree of accuracy of the final solution for voltage and current, we can have two possible diode models: first-order DC model and second-order DC model.

Diode First-Order DC Model A glance at the *I–V* characteristic of a silicon diode in the forward-bias region indicates that the current is negligibly small for $V < 0.5$ V (called the cut-in voltage). Furthermore, for a fully conducting diode, the voltage drop lies around 0.6 to 0.8 *V*, which gives rise to a simple first-order diode DC model.

A conducting diode has an approximate 0.7 V voltage drop across it (see Figure 4.5).

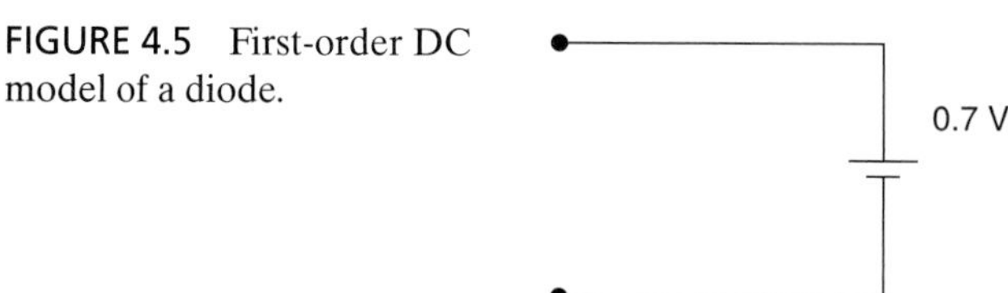

FIGURE 4.5 First-order DC model of a diode.

Diodes with different cross-sectional areas (different current ratings) exhibit the 0.7 V drop at different current values. For example, a low-power diode when fully conducting ($V_D \approx 0.7$ V) has $I = 1$ mA whereas a high-power diode, having a larger cross sectional area, has $I = 1$ A when fully conducting with $V_D \approx 0.7$ V.

NOTE: *The above applies to a real diode. For an ideal diode, the voltage drop across the diode is zero.*

Diode Second-Order DC Model Because Equation 4.5 is a nonlinear equation, we need to use an iterative procedure to obtain the solution (I_{DQ}, V_{DQ}). Before we start this iterative process, let's derive an important relation between two distinct points on the diode characteristic curve as follows:

Point #1 (I_{D1},V_{D1}): $$I_{D1} \approx I_S e^{V_{D1}/nV_T} \quad (4.7\text{a})$$

Point #1 (I_{D2},V_{D2}): $$I_{D2} \approx I_S e^{V_{D2}/nV_T} \quad (4.7\text{b})$$

Dividing Equation 4.7b by 4.7a and simplifying, we obtain:

$$\frac{I_{D2}}{I_{D1}} = \frac{e^{V_{D2}/nV_T}}{e^{V_{D1}/nV_T}} \Rightarrow \Delta V_D = V_{D2} - V_{D1} = nV_T \ln\left(\frac{I_{D2}}{I_{D1}}\right) \quad (4.8\text{a})$$

Or, in terms of base 10 logarithms, Equation 4.8a becomes:

$$\Delta V_D = V_{D2} - V_{D1} = 2.3nV_T \log_{10}\left(\frac{I_{D2}}{I_{D1}}\right) \quad (4.8\text{b})$$

If we repeat this process for *m* points, we obtain:

$$\Delta V_{Dm-1} = nV_T \ln\left(\frac{I_{Dm}}{I_{Dm-1}}\right) \tag{4.8 c}$$

Based on Equation 4.8, we can develop a second-order DC model, as shown in Figure 4.6. In Figure 4.6, V_{Dm} is the diode voltage for the mth point (which is the final converging point) in the iteration process with its corresponding voltage shift of ΔV_{Dm-1}. The following example will illustrate the two DC models outlined previously.

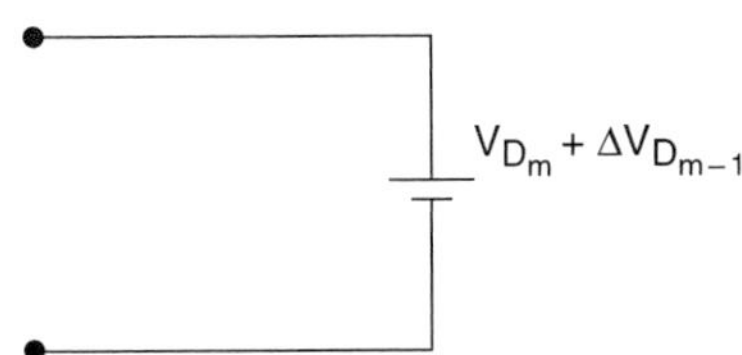

FIGURE 4.6 Second-order DC model of a diode.

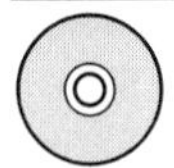

EXAMPLE 4.1

Consider the circuit shown in Figure 4.3 with V_{DD} = 10 V and R = 1 kΩ. The diode has a current of 1 mA for a voltage drop of 0.7 V (assume n = 1).

Determine the Q-point of the circuit by using:

1. *First-order model*
2. *Second-order model*

Solution:

1. Using the first-order model, we have:

$$V_{DD} = RI_D + V_D$$

$$I_D = \frac{V_{DD} - V_D}{R} = \frac{10 - 0.7}{1k} = 9.3\,\text{mA}$$

2. Using the second-order model, we obtain a more accurate solution. To begin the iteration procedure, we use the iterative analytical technique by assuming point #1 is at:

$$(I_{D1}, V_{D1}) = (1\text{ mA}, 0.7\text{V})$$

Iteration #1: Using KVL, we obtain I_{D2} and then use Equation 4.8c to calculate V_{D2} as follows:

$$I_{D2} = (V_{DD} - V_{D1})/R = (10 - 0.7)/1\text{k} = 9.3\text{ mA}$$

$$\Delta V_{D1} = nV_T \ln(I_{D2}/I_{D1}) = 1 \times 25 \times \ln(9.3/1) = 55.8\text{ mV}$$

Thus, V_{D2} = 700 + 55.8 = 755.8 mV

Iteration #2: Using KVL, we obtain I_{D3} and then use Equation 4.8c to calculate V_{D3} as follows:

$$I_{D3} = (V_{DD} - V_{D2})/R = (10 - 0.7558)/1\text{k} = 9.244\text{ mA}$$

$$\Delta V_{D2} = nV_T \ln(I_{D3}/I_{D2}) = 1 \times 25 \times \ln(9.244/9.3) = -0.15 \text{ mV}$$

Thus, $V_{D3} = 755.8 - 0.15 = 755.65$ mV

Because ΔV_{D2} in the second iteration is close to zero, we can see that the iteration has converged, and we have our final solution as:

$$(I_{DQ}, V_{DQ}) = (9.244 \text{ mA}, 755.65 \text{ mV})$$

4.2.3 Diode AC Small-Signal Analysis

Symbol convention: For the sake of clarity of notation and distinction between DC, AC, and total (DC + AC) values of any variable, the following convention is adopted for use throughout the rest of this book:

DC signal variable: Use capital letter with capital letter subscript, e.g., V_D, I_D, etc.
AC signal variable: Use small letter with small letter subscript, e. g., v_d, i_d, etc.
Total signal variable: Use small letter with capital letter subscript, e.g.,
$v_D(\text{total}) = V_D(\text{DC}) + v_d(\text{AC})$, $i_D = I_D + i_d$, etc.

Before we get deeply involved in mathematical analysis, we will define a few important terms first:

DEFINITION-SMALL SIGNAL: *A low-amplitude signal that covers such a small part of the operating characteristic curve of a device that operation is nearly always linear.*

DEFINITION-SMALL SIGNAL ANALYSIS: *Consideration of only small excursions from the no-signal bias, so that a device can be represented by a linear equivalent circuit.*

AC Models of a Diode. Let's consider the conceptual diode circuit as shown in Figure 4.7. A DC voltage (V_{DD}) as well as a time-varying voltage [$v_d(t)$] is applied across the diode.

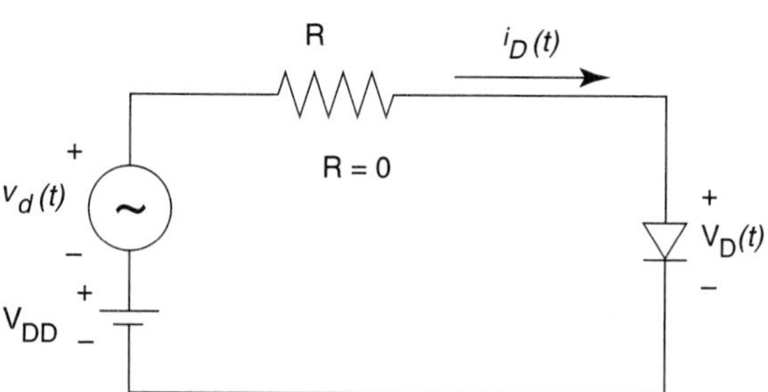

FIGURE 4.7 A conceptual diode circuit.

The object of this analysis is to understand the behavior of the diode when small signals are applied. To this end, we will find the current under two sets of voltage bias conditions:

a. DC Bias

$$v_d(t) = 0,$$

Thus:

$$v_D(t) = V_{DD}$$

$$I_{DQ} = I_S e^{V_{DD}/nV_T} \tag{4.9}$$

b. DC Bias and AC Small Signal

$$v_D(t) = V_{DD} + v_d(t) \tag{4.10}$$

$$\begin{aligned} I_D &= I_S e^{v_D/nV_T} \\ &= I_S e^{[V_{DD} + v_d(t)]/nV_T} = I_S e^{V_{DD}/nV_T} e^{v_d(t)/nV_T} = I_{DQ} e^{v_d(t)/nV_T} \end{aligned} \tag{4.11}$$

To consider $v_d(t)$ a small signal for linear operation, we require the following condition to be satisfied in Equation 4.11:

$$\left|\frac{v_d(t)}{nV_T}\right| \ll 1 \tag{4.12}$$

This is because we know that the exponential term $e^{v_d(t)/nV_T}$ can be expanded (using Taylor series expansion, see Appendix J) and greatly simplified by the condition imposed by Equation 4.12 as follows:

$$e^{v_d(t)/nV_T} \approx 1 + v_d(t)/nV_T \quad \text{for} \quad \left|\frac{v_d(t)}{nV_T}\right| \ll 1 \tag{4.13}$$

Using Equation 4.13, we can write Equation 4.11 as:

$$i_D \approx I_{DQ}[1 + v_d(t)/nV_T] = I_{DQ} + i_d \tag{4.14}$$

where

$$i_d = g_d v_d(t) \tag{4.15 a}$$

and

$$g_d = 1/r_d = (I_{DQ}/nV_T) \tag{4.15b}$$

Equation 4.15a relates the small-signal current (i_d) to small-signal voltage (v_d) with a proportionality constant (g_d), called the diode small-signal conductance. The inverse of g_d, called the diode small-signal resistance or incremental resistance (r_d), is inversely proportional to the DC bias current (I_{DQ}) as given by:

$$r_d = 1/g_d = nV_T/I_{DQ} \tag{4.16}$$

NOTE: *From Equation 4.12, we can write:*

$$V_T = 25 \text{ mV}$$

$$|v_d(t)| << 50 \text{ mV}, \quad n \le 2$$

We can see that in order to satisfy Equation 4.12, the magnitude of the small signal $|v_d(t)|$ *should not exceed more than approximately 5 to 10 mV in order to have the small-signal analysis hold valid.*

From a graphical point of view, g_d can be considered to be equal to the slope of the I–V curve at the operating point (Q-point) as shown in Figure 4.4.

EXERCISE 4.1

Show by direct derivation of the diode equation that g_d is given by:

$$g_d = \left.\frac{di(t)}{dv}\right|_{i = I_{DQ}} \tag{4.17}$$

First-Order AC Model (Low Frequency). From Equation 4.15a we can see that as a first-order AC model, a diode can be represented as a resistor, as shown in Figure 4.8. This model applies primarily at low frequencies where the diode's resistive effect (i.e., loss or dissipation effect) dominates all other charge transport effects.

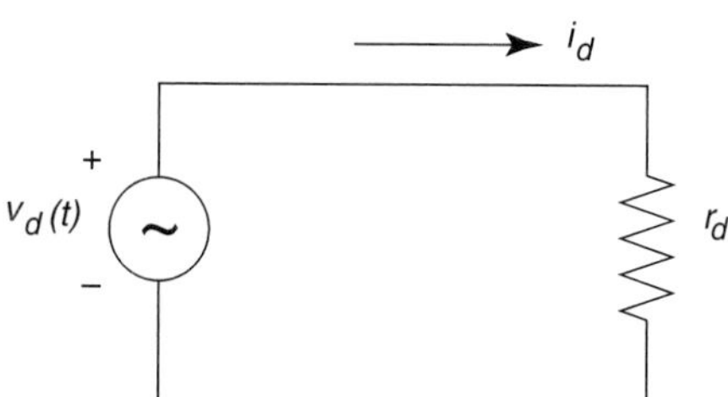

FIGURE 4.8 First-order AC model of a diode (low frequency).

Second-Order AC Model (High Frequency). The first-order model, composed of a single resistor, is commonly used at low frequencies where the diode's charge-storage (or capacitive) effects are negligible. At higher frequencies, however, there are two diode capacitances that become important and need be considered as follows:

- The depletion layer capacitance or junction capacitance (C_j)
- The diffusion capacitance (C_d)

Including these two capacitances (C_j, C_d) in parallel with the small-signal resistance (r_d), results in the second-order model, as shown in Figure 4.9.

4.2.4 Diode Model in the Reverse Breakdown Region

One of the important applications of a diode operating in the reverse breakdown region is in the design of voltage regulators. Such diodes (called breakdown diodes or Zener diodes) have a very steep I–V characteristic curve, as shown in Figure 4.10.

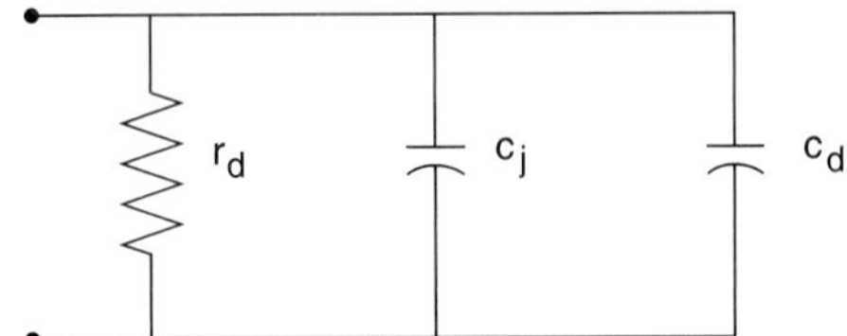

FIGURE 4.9 Second-order AC model of a diode (high frequency).

From Figure 4.10, we can see that for voltages and currents greater than the knee values $(-V_{zk}, -I_{zk})$, the I–V characteristic curve is almost a straight line with a slope $1/r_z$ for $|V| > V_{zk}$. r_z, the incremental resistance (also called the dynamic resistance) of the Zener diode, is the inverse of the slope of the I–V curve in the breakdown region and has a typical value of a few to a few tens of Ohms, usually specified in the data sheet.

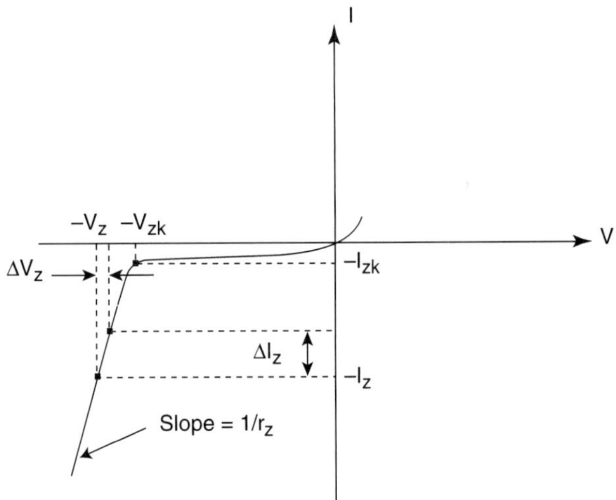

FIGURE 4.10 A Zener diode in the reverse breakdown region.

For $|V| > V_{zk}$, we can write:

$$\Delta V_z = r_z \Delta I_z \tag{4.18}$$

Zener Diode DC Model. Based on Equation 4.18, a Zener diode in the breakdown region can be modeled as a simple resistor (r_z) in series with a constant battery (V_{zk}), as shown in Figure 4.11.

Zener Diode AC Model. Based on Equation 4.18, a Zener diode in the breakdown region can be modeled as a simple resistor (r_z) as shown in Figure 4.12.

A Special Case—Ideal Zener Diode. An ideal Zener diode has $r_z = 0$, which corresponds to an I–V curve that is a straight vertical line at $V_z = V_{zk}$. In this case, we can write:

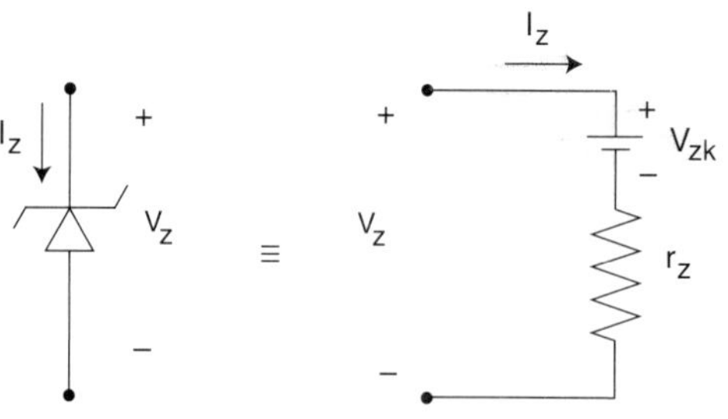

FIGURE 4.11 DC model of a Zener diode.

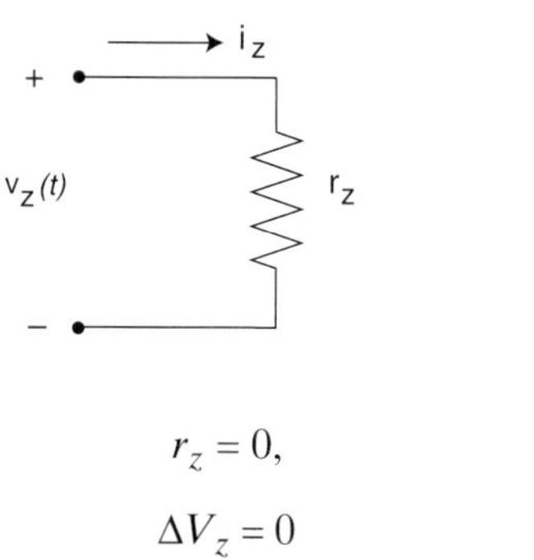

FIGURE 4.12 AC model of a Zener diode in breakdown.

$$r_z = 0,$$
$$\Delta V_z = 0$$

Here the diode's small-signal resistance is a simple short-circuit. This corresponds to a perfect voltage regulator that can hold $V_z = V_{zk}$ voltage regardless of the amount of current that passes through it. This ideal situation is impossible to achieve in practice but can be approached with some success for a small current range.

EXAMPLE 4.2

A string of two diodes is used in the circuit shown in Figure 4.13. Calculate the change in the output voltage:

a. *If the DC input voltage changes by* ± 1 V.

b. *If* $R_L = 2\ k\Omega$ *is connected to the output terminals.*

Assume each diode has $n = 2$.

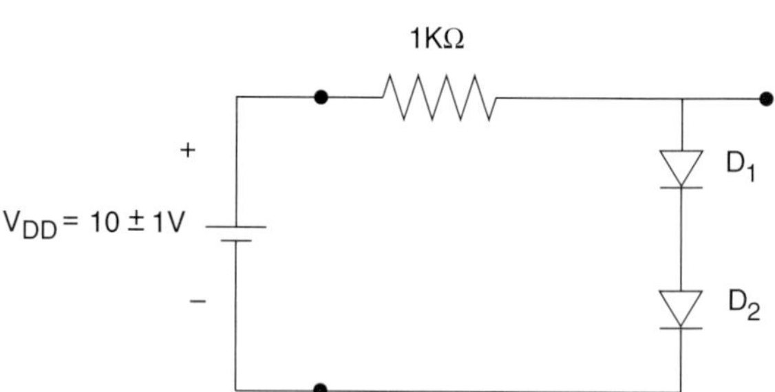

FIGURE 4.13 Circuit for Example 4.2.

Solution:

a. In this first part, we will do a DC as well as an AC analysis.

DC analysis

Because both diodes are forward biased, we can see that for the first order of approximation, the voltage drop across each diode is 0.7 V, causing a DC output voltage of 1.4 V, that is,

$$(V_o)_{DC} = 1.4\text{ V}$$

$$I_{DQ} = (10 - 1.4)/1\text{k} = 8.6\text{ mA}$$

AC analysis

±1 V change in the input voltage acts as an AC signal: $|v_{dd}(t)| = 1$ V.
The diode's incremental resistance is found as:

$$r_d = nV_T/I_{DQ} = 2 \times 25/8.6 = 5.8\ \Omega$$

Using the small-signal model, we get the AC equivalent circuit as shown in Figure 4.14:

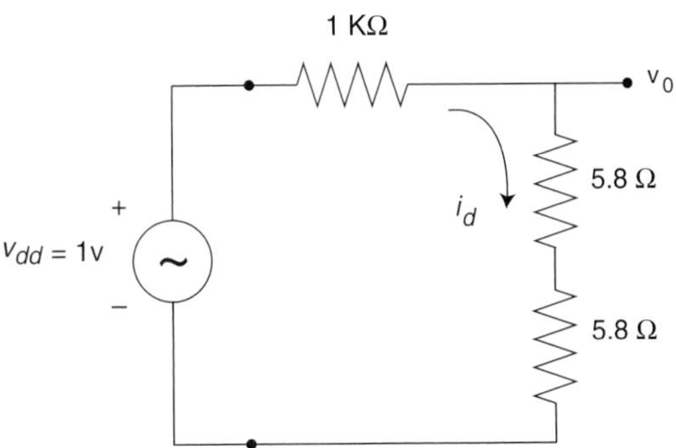

FIGURE 4.14 AC equivalent circuit.

$$i_d = 1/(1000 + 2 \text{ x } 5.8) = 0.99\text{ mA}$$

$$(v_o)_{AC} = 2r_d i_d = 2 \times 5.8 \times 0.99 = 11.5\text{ mV}$$

Thus, the output voltage will change by ±11.5 mV for a change of ±1 V.
Furthermore, because ±11.5 mV total voltage variation corresponds to ±5.75 mV (±11.5/2) for each diode, use of the small-signal model is justified, that is,

$$v_d/nV_T = 5.75/50 = 0.115 << 1$$

b. If $R_L = 2\text{ k}\Omega$ is connected to the output, the variation can be calculated as follows (see Figure 4.15):

$$I_L = (V_o)_{DC}/R_L = 1.4/2\text{k} = 0.7\text{ mA}$$

This current reduces the two diodes' biasing current by:

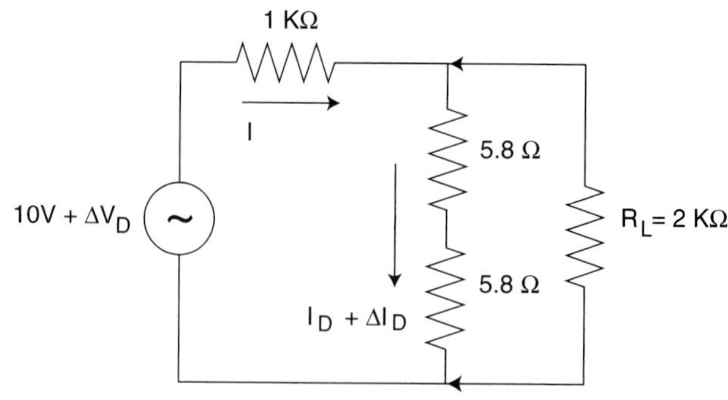

FIGURE 4.15 Connecting R_L to the output terminals.

$$\Delta I_D = -0.7 \text{ mA}$$

The corresponding incremental change in voltage can be calculated as:

$$\Delta V_D = 2r_d \Delta I_D = 2 \times 5.8 \times (-0.7) = -8.1 \text{ mV}$$

Because the voltage variation is -4.05 ($-8.1/2$) mV per diode, use of small-signal analysis is justified.

4.3 TRANSISTORS

At the outset of this section, we will first define an important term.

DEFINITION-TRANSISTOR: *A nonlinear three-terminal, active semiconductor device where the flow of electrical current between two of the terminals is controlled by the third terminal. The name is an acronym for transfer resistor.*

In simple terms we can see that operationally, a transistor transfers a large resistance at its input (with a very small current through it) to a small output resistance with a much larger output current flowing through the load. Transistors may be used in circuits as amplifiers, oscillators, detectors, switches and so on. Transistors may be subdivided into two general classes:

- Bipolar Junction Transistors (BJTs)
- Field Effect Transistors (FETs)

We will analyze each type of transistor separately in the next few sections.

4.4 BIPOLAR JUNCTION TRANSISTORS (BJTs)

These transistors consist of three layers of semiconductors and two junctions. The semiconductor layers may be alternate N-type and P-type or vice versa. Thus, BJTs are either NPN or PNP. The three terminals are called emitter, base, and collector, and the two junctions are the emitter-base junction (EBJ) and the collector-base junction (CBJ), as shown in Figures 4.16 and 4.17.

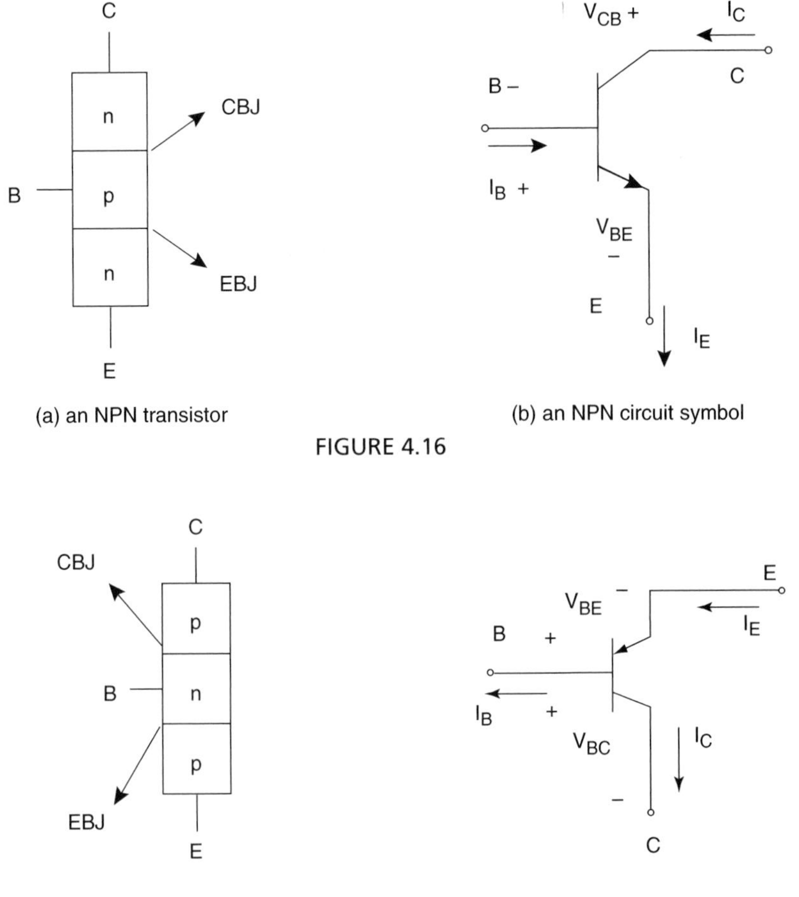

(a) an NPN transistor (b) an NPN circuit symbol

FIGURE 4.16

(a) a PNP transistor (b) a PNP transistor circuit symbol

FIGURE 4.17

4.4.1 Graphical Representation of BJT Characteristics

It is often useful to describe a BJT graphically in terms of its I–V characteristics. Consider the following conceptual circuit (shown in Figure 4.18) where the transistor is connected in the common-emitter configuration (emitter connected to ground). In this setup, the DC voltage sources are independent variables and create three dependent variables (I_C, I_B, V_{CE}). To obtain a single transistor characteristic curve that is plotted in the $I_C - V_{CE}$ plane with I_B as a parameter, the DC voltage source is set to a voltage (e.g., V_{BE1} with corresponding base current I_{B1}) and then the V_{CE} voltage source setting is theoretically varied from 0 to infinity while measuring the corresponding collector current (I_C). This procedure yields one characteristic curve; if repeated for several base DC-voltage settings, then the result is a family of characteristic curves with I_B as a parameter, as shown in Figure 4.19.

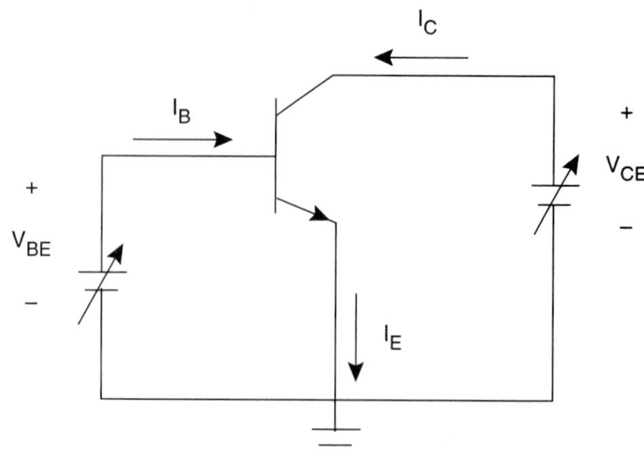

FIGURE 4.18 A conceptual circuit.

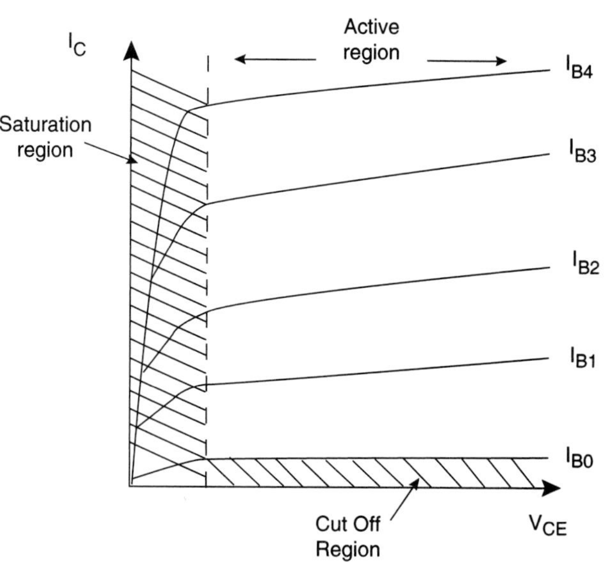

FIGURE 4.19 $I_C - V_{CE}$ characteristics of a BJT.

The $I_C - V_{CE}$ family of characteristic curves can be subdivided into several regions: active, saturation, and cut-off. To understand these three regions, we need to look into the various bias conditions under which the transistor is operating.

4.4.2 BJT DC Biasing

The DC biasing of a transistor plays a major role in the operation and proper functioning of an active circuit. It is best to define this term before proceeding any further:

> **DEFINITION-DC BIASING:** *Is the setting of the DC voltages at each of the two transistor junctions (EBJ or CBJ) such that the transistor will perform in a stable fashion in the intended mode (e.g., active mode for amplifiers, etc.).*

Setting the proper values for each of the two transistor junction voltages can be translated equivalently into terminal current values such as emitter or collector currents, which can alternatively be used to specify the DC bias values of the transistor. These currents, in conjunction with the bias voltage values of junctions, make the DC bias specification of a transistor complete.

4.4.3 BJT Modes of Operation

Depending on the bias conditions on each of the two junctions (EBJ or CBJ), there can be four modes of operation. Assuming the following notations,

$$\text{FWD} \equiv \text{forward bias}$$

$$\text{REV} \equiv \text{reverse bias}$$

these four modes of operation are shown in Figure 4.20.

FIGURE 4.20 Four modes of operation of a BJT.

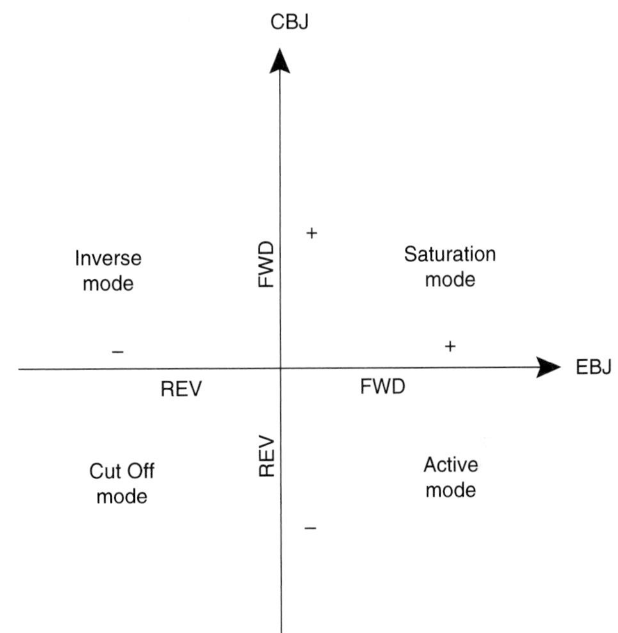

Each mode can be defined as follows:

Saturation mode is the mode in which both EBJ and CBJ are forward biased. In this mode an increase in base current (I_B) produces no further increase in collector current (I_C), i.e., collector current is almost independent of the base current.

Cut-off mode is the mode where EBJ and CBJ are both reverse biased (or have zero bias); thus, there is no current of any kind through the circuit, i.e.,

$$I_E = 0,$$

$$I_C = 0,$$

and

$$I_B = 0.$$

Active mode is the mode where EBJ is forward biased and CBJ is reverse biased. In this mode the collector current (I_C) is proportional to the base current:

$$I_C = \beta I_B, \tag{4.19}$$

Where β is the "common-emitter current gain."
Kirchhoff's current law (KCL) gives:

$$I_E = I_B + I_C = (1+1/\beta)\, I_C$$

$$I_C = \frac{\beta}{1+\beta} I_E = \alpha I_E \tag{4.20a}$$

where

$$\alpha = \frac{\beta}{1+\beta} \tag{4.20b}$$

α is called the "common-base current gain."
A first-order model for the operation of a transistor in the active mode can be represented by the hybrid-π equivalent circuit shown in Figure 4.21:

$$I_C = I_S e^{V_{BE}/V_T} \tag{4.21}$$

Where I_S is the reverse saturation current, V_T is the thermal voltage defined earlier to be $V_T = kT/q$, where k is Boltzmann's constant (1.38×10^{-23} *J/K*), T is the absolute temperature in Kelvin [T(K) = 273.15 + T(°C); T(°C) is temperature in degrees Celsius or centigrade] and q is the magnitude of electronic charge (= 1.602×10^{-19} C).
At room temperature ($T \approx 20$°C or 293 °K), the value of V_T is approximately 25 millivolts (mV). Even though V_T changes slightly with temperature, nevertheless as discussed earlier, for the sake of simplicity and to promote rapid circuit analysis, we shall use $V_T = 25$ mV throughout this book.
In this first-order model of the active mode (see Figure 4.21a), the forward voltage (V_{BE}) causes an exponentially related collector current (I_C) to flow, and as long as the CBJ remains reverse biased (i.e., $V_{CB} \geq 0$), the collector terminal behaves as a nonlinear voltage-controlled current source depending exponentially on V_{BE}. As defined earlier, β is the common-emitter current gain and ordinarily is much larger than unity for a good transistor, that is, β >> 1 (e.g., 100 or more). As a result, I_B is seen to be much smaller than I_C. Furthermore, because α is very close to unity for a good transistor (e.g., $\alpha \approx 0.99$), the collector current (I_C) is approximately equal to the emitter current (I_E): $I_C \approx I_E$.

FIGURE 4.21 Two possible large-signal equivalent circuit models of an NPN BJT in active mode.

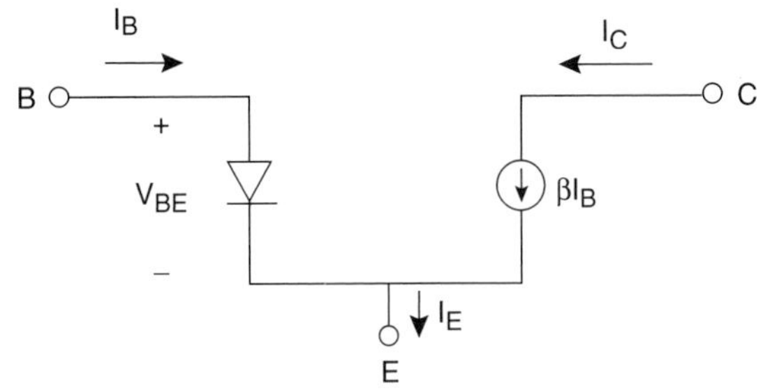

(b)

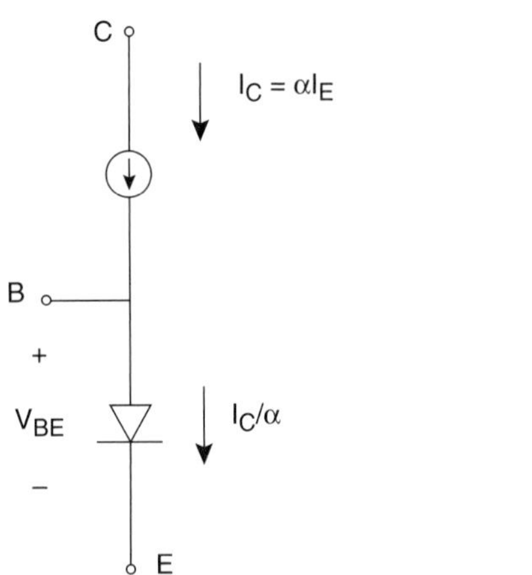

NOTE: *The use of NPN transistors in the previous active mode description is justified by the fact that at RF/microwave frequencies, these transistors (rather than PNP transistors) are used primarily due to their superior performance and higher speed. Therefore, all transistors used throughout the rest of this text will be assumed to be NPN.*

A mathematically equivalent way of drawing the above hybrid-π circuit for the active mode is the use of T-model as shown, in Figure 4.21b. This alternate equivalent circuit (the T-model) in some instances simplifies the analysis of a complex circuit greatly.

Because of its importance in amplifier design, we shall discuss this mode in further depth in the next few sections.

Inverse mode is a mode in which the EBJ is reverse biased and CBJ forward biased; that is, the emitter's and collector's roles are reversed. This mode may theoretically be used in the same manner as the active mode. In practice, though, the functions of the collector and emitter are switched very rarely. In fact, it is almost never done except for special design situations, such as in transistor-transistor logic (TTL) circuits. Inverse mode is seldom used in practical applications and certainly not in this text.

4.4.4 Active Mode Simplification

Mathematically, I_C is an exponential function of V_{BE}, which is sketched as shown in Figure 4.22:

$$V_T \approx 25 \text{ mV} \tag{4.22a}$$

$$I_C \approx I_S e^{40V_{BE}} \tag{4.22b}$$

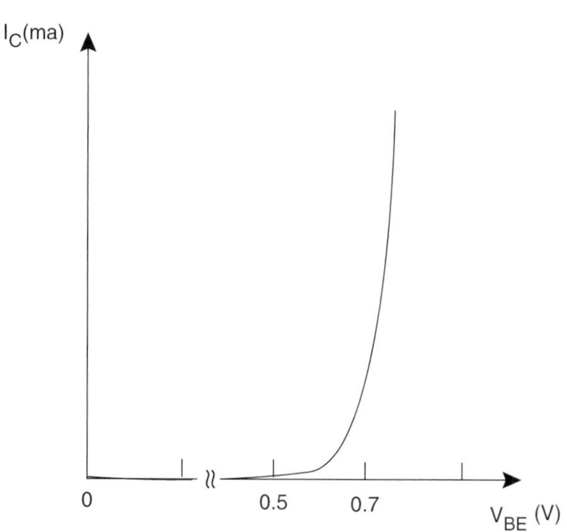

FIGURE 4.22 Exponential relation of I_C with V_{BE}.

Because the constant factor (40) in the exponential term is very high, the current (around 0.7V) rises very sharply for a small increase of base voltage. Furthermore, we notice that between the base and emitter lies a diode, which requires only a small increase of 17.3 mV in V_{BE} in order to double the collector current as proved below:

Proof:

$$I_C \approx I_S e^{40V_{BE}} \tag{4.23a}$$

We need to find an increase of ΔV_{BE} above the V_{BE} value that causes the current to double:

$$2I_C \approx I_S e^{40(V_{BE} + \Delta V_{BE})} \tag{4.23b}$$

Divide Equation 4.23b by Equation 4.23a and upon simplification, we obtain:

$$\Delta V_{BE} = 17.3 \text{ mV} \tag{4.24}$$

This means that I_C is too sensitive to changes in the V_{BE} voltage, and thus V_{BE} is not an accurate parameter to represent the control of BJTs. Instead we use I_B as the prime parameter of control.

Furthermore, because V_{BE} varies in the range of 0.6 to 0.8 volts over most of the normal current range for a silicon transistor, it is safe to assume that $V_{BE} \approx 0.7$ (assume silicon transistors) for all rapid first-order DC calculations; this is very similar to the diode analysis.

4.4.5 A Simple BJT Circuit Analysis at DC

The following simple transistor circuit, even though not used in practice, is presented purely to illustrate transistor operation and analysis.

Consider the circuit shown in Figure 4.23. We wish to determine the transistor's operating currents and voltages (I_{BQ}, I_{CQ}, I_{EQ}, V_{CEQ}, and V_{CBQ}) with a known β. Assuming the transistor to be in active mode and $V_{BE} \approx 0.7$ V, we have the following:

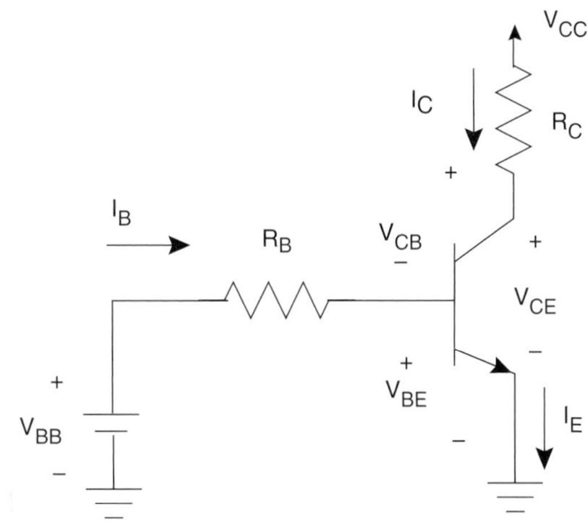

FIGURE 4.23 A theoretical DC circuit for a BJT.

Mathematical Analysis.

$$V_{BEQ} \approx 0.7\,\text{V}$$

$$I_{BQ} = (V_{BB} - 0.7)/R_B \tag{4.25a}$$

$$I_{CQ} = \beta\, I_{BQ} \tag{4.25b}$$

$$= \alpha\, I_{EQ} \tag{4.25c}$$

$$V_{CEQ} = V_{CC} - R_C\, I_{CQ} \tag{4.25d}$$

Kirchhoff's voltage law (KVL) gives:

$$-V_{CEQ} + V_{BEQ} + V_{CBQ} = 0 \tag{4.25e}$$

Therefore:

$$V_{CBQ} = V_{CEQ} - 0.7 \tag{4.26}$$

Equations 4.25 and 4.26 form the basic equations that must be considered in analyzing transistor circuits operating in active mode.

Graphical Analysis. From Equation 4.25, the operating point is given by:

$$V_{BB} = R_B I_B + V_{BE} \Rightarrow I_{BQ} = (v_{BB} - 0.7)/R_B \tag{4.27}$$

$$V_{CEQ} = V_{CC} - R_C I_{CQ} \Rightarrow I_{CQ} = V_{CC}/R_C - V_{CEQ}/R_C \quad \text{(load line)} \tag{4.28}$$

Equation 4.28 is the equation of a straight line that can be written in a general form as:

$$Y = mX + b, \tag{4.29}$$

where:

$$Y = I_{CQ},$$

$$X = V_{CEQ},$$

$$m = -1/R_C$$

and

$$b = V_{CC}/R_C$$

Equation 4.28 is known as the "equation of a load line." Drawing the load line equation on the transistor characteristic curves and finding its intersection with the branch that corresponds to the base currents (I_{BQ}) will provide the operating point (or the Quiescent point, Q-point) of the transistor, as shown in Figure 4.24. The coordinates of the Q-point give the DC values of V_{CEQ} and I_{CQ} that can be read from the graph.

NOTE: *For Class A amplifiers (classes of amplifiers will be explained in Chapter 15, and in the Glossary of Technical Terms at the end of the text) and maximum voltage swing, it is best to have the Q-point in the middle of the active region.*

EXAMPLE 4.3

In the circuit shown in Figure 4.25, design the bias network to establish a Q-point having an emitter current of 1 mA using a power supply of V_{CC} = 12 V. Assume the BJT has a $\beta = 100$ ($\alpha = 0.99$).

Solution:

As a design rule of thumb, V_B and V_{CEQ} are chosen to be one-third of the power supply voltage value; that is:

$$V_B = 12/3 = 4\text{ V}$$

$$V_{CEQ} = 12/3 = 4\text{ V}$$

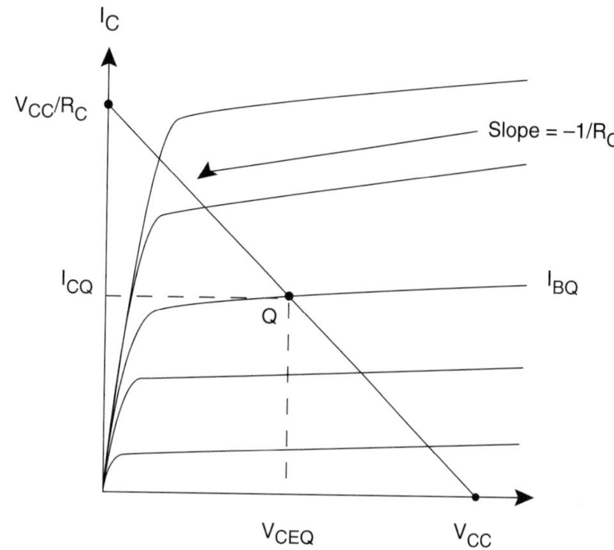

FIGURE 4.24 Graphical analysis and use of load line in a simple DC circuit for a BJT.

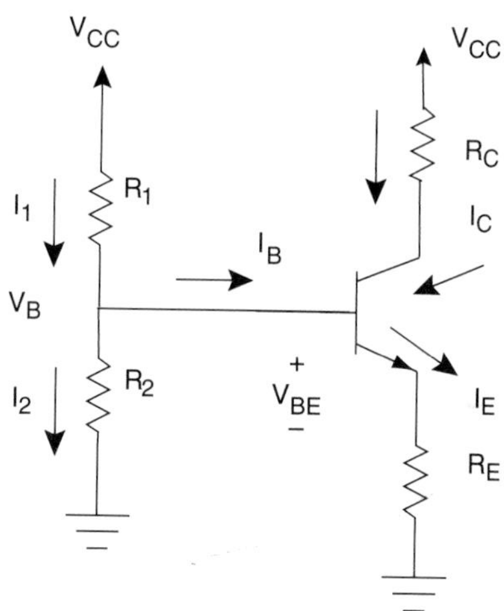

FIGURE 4.25 Circuit for Example 4.3.

Assuming active mode (to be verified later) and $V_{BE} \approx 0.7$ V, we obtain:

$$I_{CQ} = \alpha I_{EQ} = 0.99 \text{ mA}$$

$$I_{BQ} = I_{CQ}/\beta = 9.9\ \mu\text{A}$$

$$\text{KVL:}\quad V_B = V_{BE} + R_E\, I_{EQ} \;\Rightarrow R_E = (4 - 0.7)/10^{-3} = 3.3\ \text{K}\Omega$$

$$V_{CC} = R_C I_{CQ} + V_{CEQ} + R_E I_{EQ} \;\Rightarrow\; R_C = (V_{CC} - V_{CEQ} - R_E I_{EQ})/I_{CQ} \;\Rightarrow R_C = 4.7\ \text{K}\Omega$$

We notice that if $I_{BQ} << I_1$ then it can be safely assumed that $I_1 \approx I_2$. Thus, to satisfy the condition of $I_{BQ} << I_1$, we require that I_1 be at least 10 times I_{BQ}, that is:

$$I_1 = 10\, I_{BQ} = 99\ \mu\text{A}$$

Setting $I_{BQ} << I_1$ allows us to use the voltage division rule to determine R_1 and R_2:

$$V_B = V_{CC} \times R_2/(R_1 + R_2) \Rightarrow R_2/(R_2 + R_1) = 1/3 \Rightarrow R_1/R_2 = 2$$

$$V_B = I_1 R_2 \Rightarrow R_2 = 4/(99 \times 10^{-6}) = 40.4\ \text{K}\Omega$$

$$R_1 = 2\, R_2 = 80.8\ \text{K}\Omega$$

Active mode assumption verification:

$$V_{BEQ} = 0.7\ \text{V} > 0 \quad \text{(emitter-base junction is forward biased)}$$

$$V_{CBQ} = V_{CEQ} + V_{BEQ} = 4 + 0.7 = 4.7\ \text{V} > 0 \ \text{(collector-base junction is reverse biased)}$$

Because EBJ is forward biased and CBJ is reverse biased, the transistor is indeed in active mode, and we are justified in using it in all of our calculations.

4.4.6 The AC Analysis of a BJT

To use a transistor as an amplifier or oscillator, it must be biased in the active region. Proper biasing will establish the Q-point of the transistor and will provide a constant DC current in the emitter (or collector). This current should be a known value and insensitive to variations of β, temperature, etc. Predictability and constancy of the bias current are major requirements in the circuit design because the operation of the transistor as an amplifier for RF/microwave signals is heavily influenced by the values of the bias current and voltages.

Symbol convention: Similar to diode analysis and for the sake of clarity of notation and distinction between DC, AC, and total values of any variable, the following conventions are adopted for use throughout the rest of this book:

DC signal variable: Use capital letter with capital letter subscript, e.g., V_{BE}, I_B, etc.
AC signal variable: Use small letter with small letter subscript, e,g, v_{be}, i_b, etc.
Total signal variable: Use small letter with capital letter subscript, e.g.,
$v_{BE}(\text{total}) = V_{BE}(\text{DC}) + v_{be}(\text{AC}),\ i_B = I_B + i_b$, etc.

4.4.7 Small-Signal AC Analysis

Using the same definition of small-signal and its analysis technique presented and employed earlier in the diode section, the following first- and second-order AC models for a transistor can be developed:

(A) BJT First-Order Models. Let's consider a conceptual amplifier circuit with the BJT in the active mode (as shown in Figure 4.26) for which we will derive two simple small-signal models for the first order of approximation.

(1) *BJT Hybrid-π Model* When a small AC signal (v_{be}) is applied to the base of the transistor that is already biased to have a value of V_{BEQ}, the corresponding collector current based on our first-order model becomes:

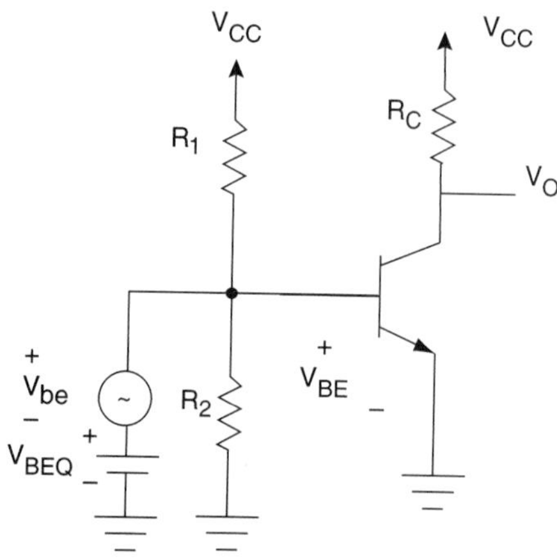

FIGURE 4.26 A conceptual BJT amplifier circuit.

DC Bias ($v_{be} = 0$)

$$V_{BE} = V_{BEQ}$$

$$I_{CQ} = I_S e^{V_{BEQ}/V_T} \tag{4.30}$$

AC Small Signal ($v_{be} \neq 0$)

$$v_{BE} = V_{BEQ} + v_{be}$$

$$i_C = I_S e^{(V_{BEQ} + v_{be})/V_T} = I_{CQ} e^{v_{be}/V_T} \tag{4.31}$$

To consider v_{be} a **small signal** for linear operation, we require the following condition to be satisfied:

$$v_{be} << V_T \tag{4.32}$$

Equation 4.32 states that v_{be} should be much smaller than 25 mV in order to be considered a small signal. This condition is obtained by using Taylor expansion for e^x (around $x = 0$) to obtain:

$$e^x \approx 1 + x \quad \text{for} \quad x << 1 \tag{4.33}$$

Therefore, based on the small-signal approximation ($x = v_{be}/V_T << 1$) and using Equation 4.33, we have the total collector current (I_C) expressed as:

$$e^{v_{be}/V_T} \approx 1 + v_{be}/V_T \quad \text{for} \quad (v_{be}/V_T) << 1 \tag{4.34}$$

Thus:

$$i_C = I_{CQ}\,(1 + v_{be}/V_T) = I_{CQ} + I_{CQ}\, v_{be}/V_T \tag{4.35 a}$$

or,

$$i_C = I_{CQ} + i_c \tag{4.35b}$$

where

$$i_c = I_{CQ}\, v_{be}/V_T \tag{4.35 c}$$

is the AC small-signal current in the collector terminal.

Equation 4.35c relates the AC small signal current in the collector to the corresponding AC base-emitter voltage that causes the current in the first place. The proportionality factor is aptly called the transconductance (g_m):

$$i_c = g_m\, v_{be} \tag{4.36}$$

where g_m is defined by:

$$g_m = I_{CQ}/V_T \tag{4.37}$$

We observe that the transconductance of a BJT is directly proportional to the collector bias current (I_{CQ}). To obtain a constant value for g_m, we require a constant and predictable collector bias current (I_{CQ}). For BJTs, g_m is typically around 40 mA/V.

From a graphical point of view, g_m can be considered to be equal to the slope of the $I_C - V_{BE}$ characteristic curve at I_{CQ}, because:

$$i_C = I_{CQ} e^{v_{be}/V_T} \tag{4.38}$$

Now differentiating Equation 4.38, we obtain:

$$di_C/dv_{be} = (I_{CQ}/V_T)\ e^{v_{be}/V_T} = g_m e^{v_{be}/V_T}$$

At the Q-point, $v_{be} = 0$ (i.e., no voltage variation). Thus, we have:

$$g_m = di_C/dv_{be} \text{ (at the Q-point).}$$

This concept is shown in Figure 4.27.

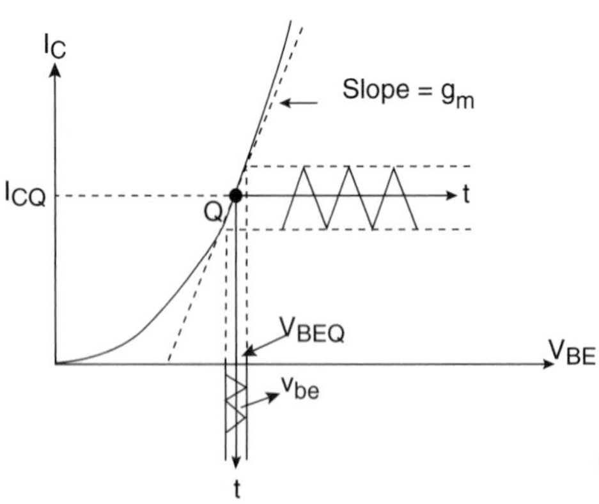

FIGURE 4.27 Graphical analysis showing amplification of a small signal by a BJT.

Furthermore, because the transistor is in active mode, the AC base current (i_b) is β times smaller than the collector current:

$$i_b = i_c/\beta = (g_m/\beta)v_{be}$$

Thus, looking into the base, the input resistance (denoted by r_π) is given by:

$$r_\pi = v_{be}/i_b = \beta/g_m = \beta/(I_{CQ}/V_T) = V_T/I_{BQ} \tag{4.39}$$

This small-signal analysis shows that for a first-order model and small AC signals ($|v_{be}| << V_T$), a BJT can be considered to be a simple resistor (r_π) at the input terminal (base). At the output terminal (collector), a BJT appears to be a voltage-controlled current source. This model is known as the hybrid-π model and is shown in Figure 4.28.

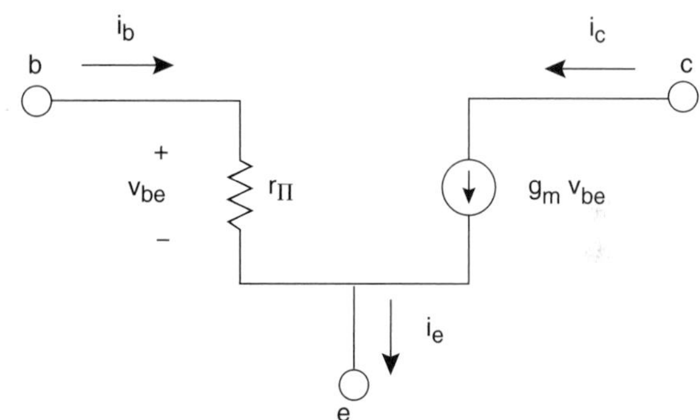

FIGURE 4.28 First-order AC model of a BJT (hybrid-π model).

As can be seen, the values of g_m and r_π in this small-signal analysis depend heavily on the current bias value, I_{CQ}, which is determined by the location of the Q-point in the $I_C - V_{CE}$ (or $I_C - V_{BE}$) characteristic curve.

In this first version of hybrid-π model, the BJT is considered to be a voltage-controlled current source with v_{be} as the control voltage. Based on this model, a few equivalent models can be derived rather easily as follows.

(2) *Alternate Hybrid-π Model of a BJT* This is a slightly different equivalent circuit model in which a BJT is modeled as a current-controlled current source. The collector current (i_c) can be written as:

$$i_c = g_m\, v_{be} \tag{4.40}$$

where,

$$v_{be} = r_\pi\, i_b$$

thus,

$$i_c = g_m\,(r_\pi\, i_b) = g_m\,(\,\beta i_b/g_m\,) = \beta i_b \tag{4.41}$$

The result is shown in Figure 4.29 with the control base current (i_b).

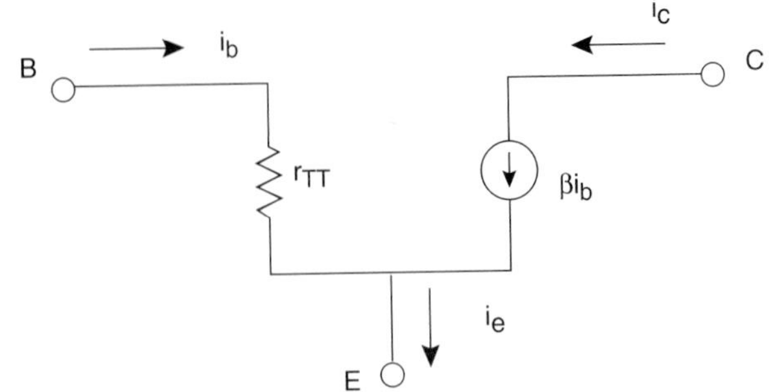

FIGURE 4.29 Alternate first-order AC model for a BJT (hybrid-π model).

(3) *The T-Model of a BJT* The T-model is an alternative to the hybrid-π model. The T-model in some situations (such as in the common-emitter configuration) becomes a more convenient model to work with. This model is shown in Figure 4.30. Resistance (r_e) in this model needs to be determined in terms of previously known parameters. From this model, we have:

$$v_{be} = r_e i_e \tag{4.42}$$

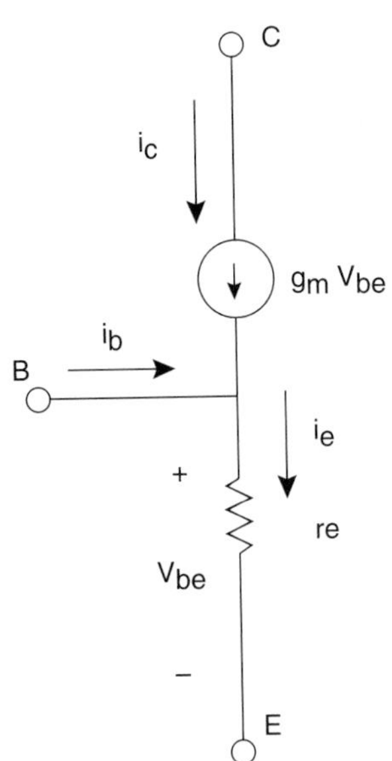

FIGURE 4.30 T-model of a BJT (first-order model).

To find r_e, from Equation 4.39, we have:

$$v_{be} = r_\pi i_b = r_\pi (i_c/\beta) = r_\pi(\alpha i_e/\beta) = [r_\pi/(1+\beta)]\, i_e \tag{4.43}$$

Thus, comparing Equations 4.42 and 4.43, we obtain:

$$r_e = r_\pi/(1+\beta) \tag{4.44}$$

From Equation 4.39, it can easily be shown that:

$$r_e = V_T / I_{EQ} \tag{4.45}$$

Figure 4.30 shows the T-model that represents BJT as a voltage-controlled current source with the control voltage being v_{be}.

(4) *Alternate T-Model of a BJT* In this model, a BJT can be represented as a current-controlled current source with the control signal being i_e. Note that the current source ($g_m\, v_{be}$) can be converted in terms of i_e by:

$$g_m v_{be} = g_m\,(r_e i_e) = g_m\,(\alpha / g_m) i_e = \alpha i_e \tag{4.46}$$

Using Equation 4.46, the alternate T-model is obtained and is shown in Figure 4.31.

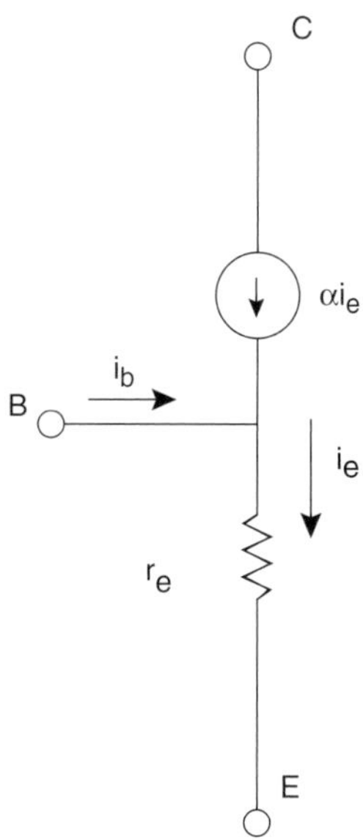

FIGURE 4.31 Alternate T-model of a BJT (first-order model).

(B) BJT Second-Order Hybrid-π Model. The hybrid-π as well as the T-model were developed for use as first-order approximations to a BJT's AC small-signal behavior. Adding more components will upgrade this model to a second-order approximation. Depending on the frequency range of operation we can have the following two cases:

(1) *BJT Low-Frequency AC Model* At low frequencies, there are two additional nonfrequency-dependent elements that need to be added to our first-order model: r_o and r_μ.

r_o In the first-order AC model of a BJT, the collector current is independent of the collector voltage; i.e., the output resistance is assumed to be infinity. In practice, it has a finite value (r_o). This effect can be modeled by a shunt resistance (r_o) between the collector and emitter terminals.

$\boldsymbol{r_\mu}$ There is a small dependence of i_b on v_{cb} that can be modeled by the addition of a very large resistor (r_μ) between the collector and base, to the first-order hybrid-π model. The final low-frequency second-order model with these additional elements is shown in Figure 4.32.

Please note that the resistance r_x is added to the model to represent the loss in the base region of the transistor.

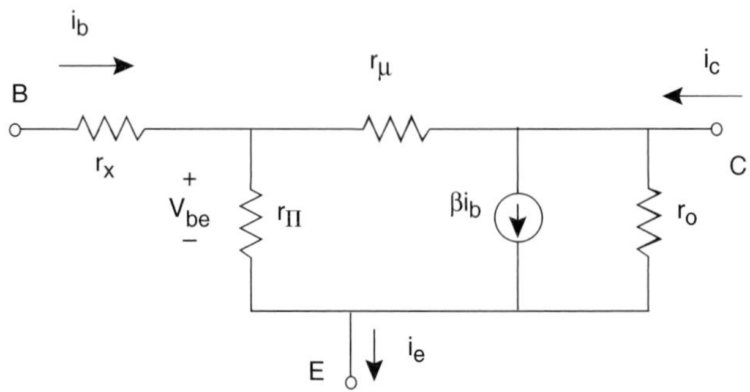

FIGURE 4.32 Low-frequency second-order AC model.

In a typical BJT at low frequencies, the following considerations apply:

- The resistance (r_μ) is very large and in most cases can be ignored, preserving the transistor's unilateral character and simplifying the analysis greatly.
- The resistance r_x is much smaller than r_π (r_x is on the order of tens of ohms while r_π is on the order of KΩ) and so can be ignored. The effect of r_x, while negligible at low frequencies, becomes significant at high frequencies where the input impedance becomes highly capacitive. These capacitive effects can short out the resistor r_π, leaving behind only the resistor r_x to account for the frequency response of the transistor at high frequencies.
- The output resistance (r_o), which accounts for the finite slope of the $I_C - V_{CE}$ characteristic curves, should be added to the first-order hybrid-π model. Output resistance can be calculated using the following approximate relation:

$$r_o \approx V_A / I_C \tag{4.47}$$

where V_A (called the "Early voltage") is the voltage obtained by extrapolating the characteristic curves until they meet at a point on the negative V_{CE} axis at $V_{CE} = -V_A$ and I_C is the collector current at the edge of the active region.

V_A is a positive number, as shown in Figure 4.33. Typically, V_A is in the range of 50 to 100 volts.

(2) *BJT High-Frequency AC Model* The high-frequency model includes all the elements of the low-frequency second-order model and adds two capacitors:

$\boldsymbol{C_\pi}$ C_π is the capacitance between the emitter-base terminals. Its value is usually on the order of a few to tens of pico-Farads.

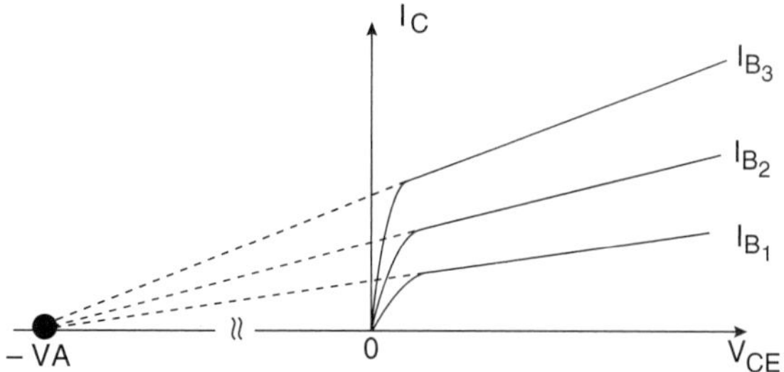

FIGURE 4.33 Extrapolation of the characteristic curves to find Early voltage.

C_μ C_μ is the capacitance between the collector and base terminals, and is usually in the range of a fraction to a few pico-Farads. This capacitance plays a major role in the high-frequency performance of a BJT, particularly for common-emitter configurations. This second-order model is shown in Figure 4.34.

NOTE 1: *The resistance r_μ can easily be ignored because it is in parallel with the capacitance (C_μ), which has a much smaller reactance value than r_μ's resistance value. At high frequencies, the capacitor C_μ becomes very important because it appears at the input but multiplied many times in value (the Miller effect), effectively shorting out the input signal with consequent deterioration of the frequency response of the transistor.*

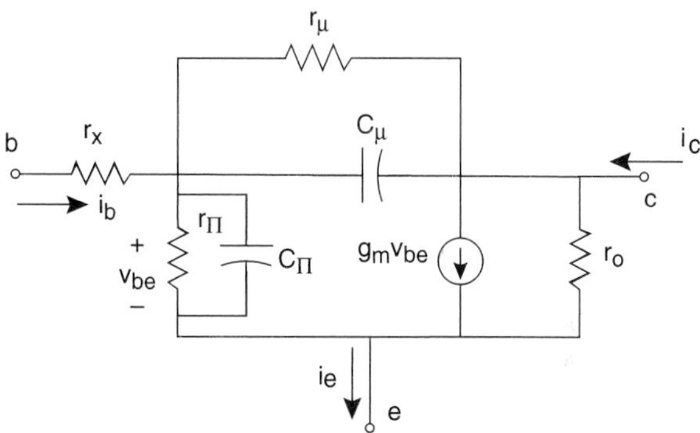

FIGURE 4.34 High-frequency second-order AC model of a BJT.

EXAMPLE 4.4

In the amplifier circuit shown in Figure 4.35, determine the voltage gain and the input impedance. Assume the transistor has a $\beta = 200$ ($\alpha = 0.995$) and the input signal is given by $v_i(t) = 0.01 \cos\omega_o t$ V, $\omega_o = 2\pi \times 10^4$ rad/sec.

The element values are as follows:

$$R_1 = 100\ \text{K}\Omega,\ R_2 = 100\ \text{K}\Omega,\quad R_C = 4.3\ \text{K}\Omega,\quad R_{E1} = R_{E2} = 3.4\ \text{K}\Omega,$$

$$R_S = 100\ \Omega,\ R_L = 4.3\ \text{K}\Omega,\quad V_{CC} = 15\ \text{V},\ C_1 = C_2 = C_3 = 10\ \mu\text{F}$$

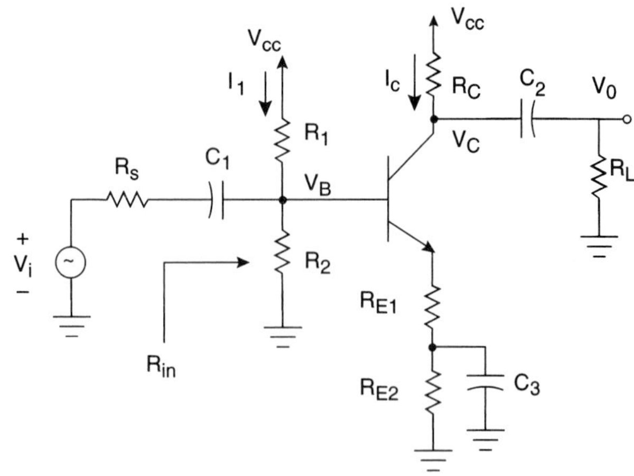

FIGURE 4.35 Circuit for Example 4.4.

Solution:

C_1 and C_2 are coupling capacitors intended to block the DC while allowing the AC signal through. C_3 is the bypass capacitor intended to increase the gain of the amplifier circuit by shorting out the resistor R_{E2} in order to increase the gain, as will be seen later.

DC analysis: In this case, all capacitors act like an open circuit, and the analysis is similar to Example 4.3.

Assume transistor is in active mode (this will be verified later). Thus, $V_{BEQ} \approx 0.7$ V.

Because β is very large I_B is very small and the voltage division rule can be applied (see Figure 4.36):

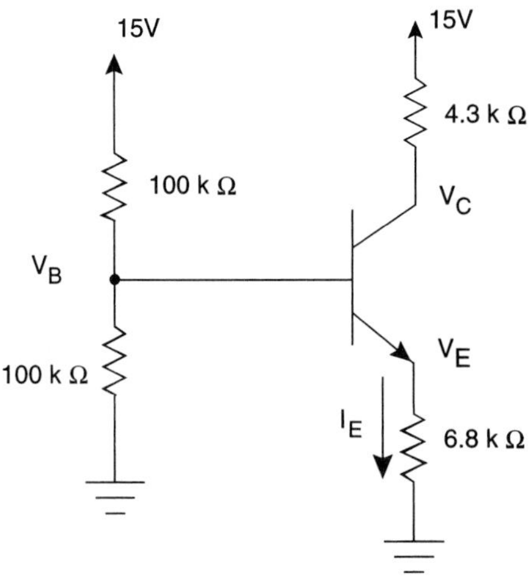

FIGURE 4.36 DC circuit for Example 4.4.

$$V_B = V_{CC} \times R_2/(R_1 + R_2) = 7.5\text{ V}$$

$$V_{EQ} = 7.5 - 0.7 = 6.8\text{ V}$$

Thus: $I_{EQ} = 6.8/6.8\text{k }\Omega = 1\text{ mA},$

$$I_{CQ} = \alpha\, I_{EQ} = 0.995 \text{ mA}$$

$$I_{BQ} = I_{CQ}/\beta = 5 \ \mu\text{A}$$

Knowing the current values enables us to find the voltage values:

$$V_{CQ} = 15 - 0.995 \times 4.3 = 10.7 \text{ V}$$

$$V_{CEQ} = 10.7 - 6.8 = 3.9 \text{ V}$$

Now check the active mode assumption:

$$V_{CBQ} = 10.7 - 7.5 = 3.2 \text{ V}$$

NOTE 1: *Because $V_{CBQ} > 0$, therefore the collector-base junction is reverse-biased and the transistor is indeed in active mode.*

NOTE 2: *The current (I_1) in the biasing resistor (R_1) is given by:*

$$I_1 = V_{CC}/(R_1 + R_2) = 15/200\text{k} = 75 \ \mu\text{A}$$

Because $I_1 >> I_B = 5\mu A$, the voltage division rule used at the base to obtain V_B is justified.

AC analysis: At $\omega_o = 2\pi \times 10^4$ rad/sec, each capacitor has a reactance of:

$$X_C = 1/\omega C = 1/(2\pi \times 10^4 \times 10 \times 10^{-6}) = 1.59 \ \Omega$$

With a reactance value of 1.59 Ω, each capacitor can be considered to be a short circuit compared to all resistor values in the circuit.

Using a T-model for a BJT (preferred for a common-emitter configuration), the equivalent circuit, as shown in Figure 4.37, can be used to find the gain as follows:

$$r_e = V_T/I_{EQ} = 25/1 = 25 \ \Omega$$

$$v_o = -(R_C \,\|\, R_L)\, \alpha i_e = -2139 \, i_e$$

$$v_i = v_b = (r_e + R_{E1}) i_e = 3425 \, i_e$$

Thus, $v_o/v_i = -\alpha(R_C \,\|\, R_L) / (r_e + R_{E1}) = -2139 / 3425 = -0.62$

$$R_{in} = 100\text{K} \,\|\, 100\text{K} \,\|\, 201 \times 3.425\text{K} = 46.6 \text{ K}\Omega$$

NOTE 3: *The general equation for the gain of a common-emitter amplifier can be seen from this example to be given by:*

$$v_o/v_i = -(R_C \,\|\, R_L) / (r_e + R_{E1})$$

To increase the gain further, the emitter resistance (R_{E1}) should be completely eliminated at AC frequencies by a bypass shunt capacitor, leading to:

$$v_o/v_i = -2130/25 = -85.2$$

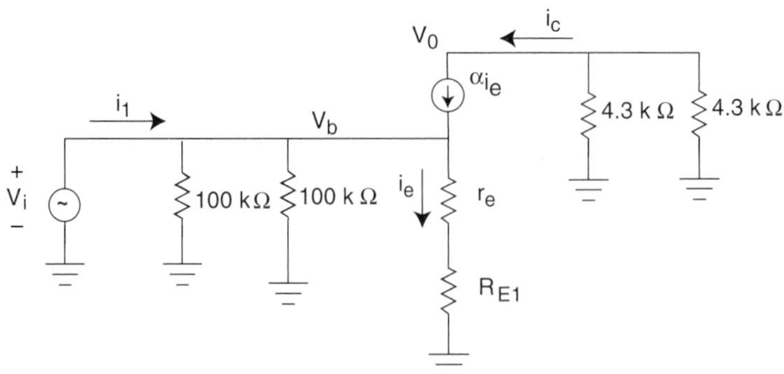

FIGURE 4.37 AC equivalent circuit for Example 4.4.

NOTE 4: *Reducing the emitter resistance (R_E) to zero would reduce the input impedance to approximately 5025 Ω (201 × 25 Ω), which is not desirable. Thus a compromise needs to be made to keep the gain high while maintaining adequate input impedance.*

4.5 FIELD EFFECT TRANSISTORS (FETs)

A field effect transistor by definition consists of three regions: source, drain, and gate wherein the resistance of the current path from source to drain is controlled by applying a voltage to the gate electrode. The gate voltage varies the depletion layer under the gate area and thus reduces or increases the conductance of the path. In this type of transistor, the current is conducted by only one type of carrier (electrons or holes) which gives the FET an alternate name: the **Unipolar Transistor.**

Another major difference between an FET and a BJT is the fact that an FET's gate input impedance is very large and that the current through the gate, for all practical purposes, can be considered to be zero.

There are several types of FETs. The Metal-Oxide Semiconductor FET (MOSFET) and the Junction FET (JFET) are used for lower frequencies and digital circuits. The Metal-Semiconductor FET (MESFET), using Gallium Arsenide or Indium phosphide material, is used at high RF, microwave and millimeter-wave frequencies.

Due to the diversity of the type of devices that can exist for FETs, we will discuss a "general version" of the device for DC biasing and modeling under small-signal conditions.

4.5.1 $I_D - V_{DS}$ Characteristic Curves of an FET

At higher frequencies, the device of choice is an N-Channel type FET because electrons as main charge carriers (compared to holes) have higher mobility and thus higher speed. Therefore, in the following analysis, we will focus on N-channel FETs only.

Consider a conceptual circuit with voltages V_{DS} and V_{GS} applied (see Figure 4.38a). This circuit can be used to measure $I_D - V_{DS}$ characteristics, which are a family of curves each measured at a constant V_{GS} (as shown in Figure 4.38b). This is very similar to a BJT's $I_C - V_{CE}$ characteristic curves, described earlier.

These devices have a threshold gate voltage (V_t) that must be overcome before the device is functional; that is, a channel is induced by applying a bias voltage to the gate such that $V_{GS} \geq V_t$. Otherwise, the device is in cut-off for $V_{GS} < V_t$.

FIGURE 4.38

(a) An FET conceptual circuit

(b) FET characteristic curves

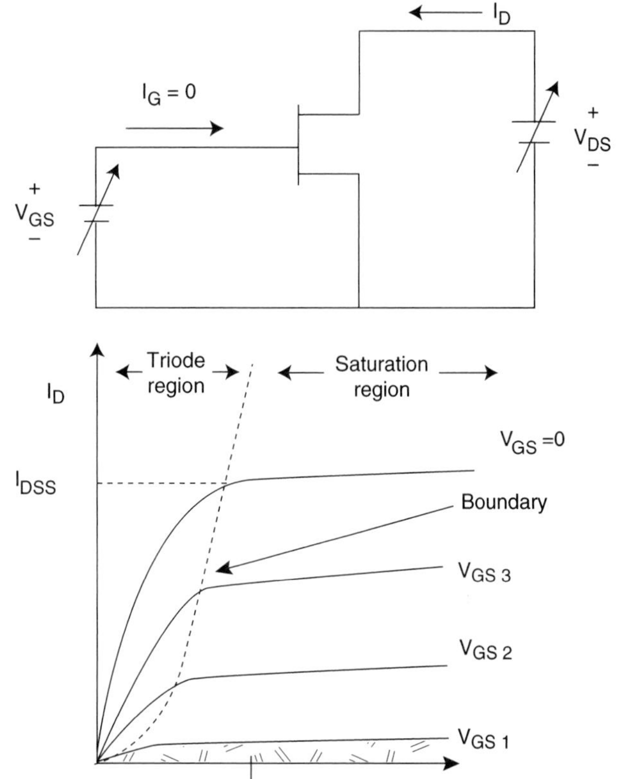

From Figure 4.38, three distinct regions can be identified: the cut-off, the triode, and the saturation regions. This is very similar to the BJT operation, where to work as a switch the cut-off and the triode regions are utilized, while the saturation region is used if the FET is to operate as an amplifier or oscillator. The FET circuit symbol is shown in Figure 4.39.

A brief description of each of the three regions is presented next (see Figure 4.40a).

1. **The Cut-Off Region.** To operate in this region, we need to have $V_{GS} < V_t$ where V_t is the threshold voltage.

   DEFINITION-THRESHOLD VOLTAGE (ALSO REFERRED TO AS PINCH-OFF VOLTAGE, V_P, FOR JFETs): *The minimum gate voltage required to turn the transistor "on."*

2. **The Triode Region.** To operate the FET in the triode region, we must have the following two conditions met:

   a. First, turn the FET on by overcoming the threshold voltage (V_t):

$$V_{GS} \geq V_t$$

$$V_{GD} > V_t$$

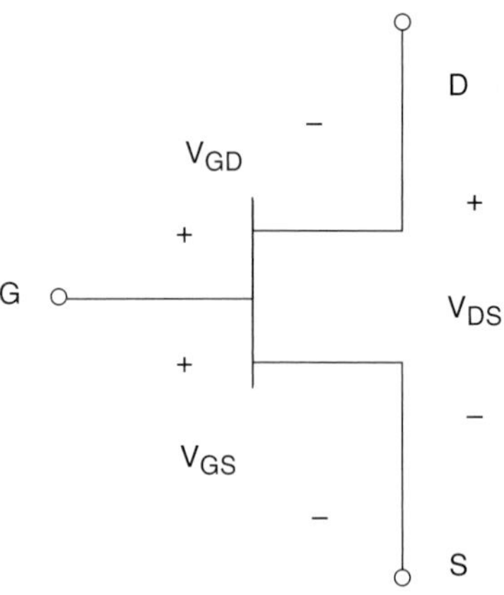

FIGURE 4.39 Circuit for an N-channel FET.

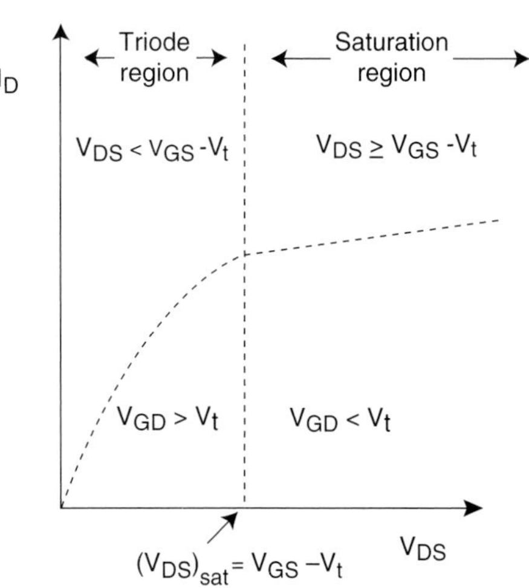

FIGURE 4.40

(a) FET characteristic curve

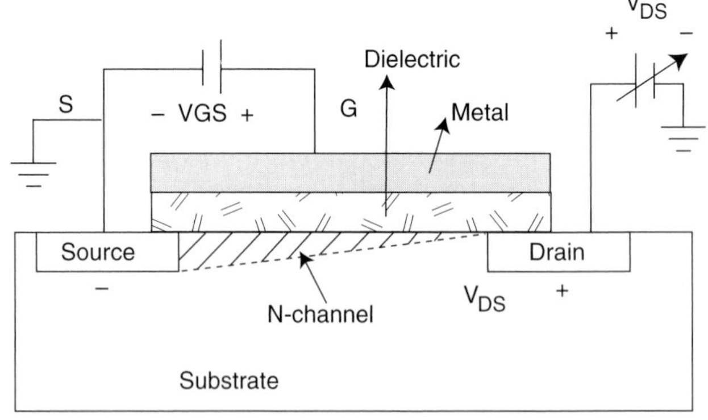

(b) A general FET N-channel device in pinch-off

b. Then, keep V_{DS} small enough so that the device operates below the pinch-off condition.

DEFINITION-PINCH-OFF CONDITION: *The condition in which the current flow between the source and the drain (I_D) no longer increases with further increase of V_{DS}.*

When all of the channel between these two electrodes is completely depleted of carriers by the gate voltage, (as shown in Figure 4.40b), the pinch-off condition prevails.

To prevent the device from going into the pinch-off condition, we require the gate-to-drain voltage (V_{GD}) to be bigger than the threshold voltage to keep the channel continuous, i.e., $V_{GD} > V_t$.

Under these two conditions, the $I_D - V_{DS}$ characteristics can be described approximately by the relationship:

$$I_D = K\,[2(V_{GS} - V_t)V_{DS} - V_{DS}^2] \tag{4.48}$$

where K is a device parameter dependent on the device geometry with units of A/V^2. (Note: In some existing literature for MESFETs, β is used instead of K, which has no relationship to the β of a BJT.)

Approximation. If V_{DS} is sufficiently small so that the second-order term can be neglected, we obtain a linear relationship:

$$I_D \approx 2K(V_{GS} - V_t)\,V_{DS} \tag{4.49}$$

This relationship clearly shows that for small V_{DS} values, an FET can be used as a linear resistor (R_{DS}) controllable by V_{GS}:

$$R_{DS} = V_{DS}/I_D = 1/[2\,K(V_{GS} - V_t)] \tag{4.50}$$

NOTE: *As mentioned earlier, the current through the gate is practically zero (i.e., $I_G = 0$) due to its very high input impedance (unlike BJTs), and thus the only controlling factor for the current through the device from drain to source is the gate voltage, as can be seen throughout all of these equations.*

3. **The Saturation Region.** To operate a FET in the saturation region (also called **pinch-off region** for JFETs) the following conditions have to be met:

   a. A channel needs to be established by applying a gate voltage (larger than V_t), which is the same condition earlier discussed for the triode region.

   $$V_{GS} \geq V_t$$

   b. V_{DS} must be large enough to pinch off the channel at the drain end, resulting in the gate-to-drain voltage falling below V_t.

   $$V_{GD} < V_t \tag{4.51}$$

   Thus, the **boundary** between the triode and saturation regions is given by:

$$V_{GD} = V_t \Rightarrow V_{GS} - V_{DS} = V_t \quad (4.52)$$

Or

$$V_{DS} = V_{GS} - V_t \quad (4.53)$$

The value of V_{DS} as given by Equation 4.53 provides the minimum V_{DS} value needed to place the device at the onset of the saturation region, i.e.,

$$(V_{DS})_{min} = V_{GS} - V_t \quad (4.54)$$

Because saturation region current, $(I_D)_{sat}$, remains approximately constant for values of $V_{DS} \geq (V_{DS})_{min}$, its value can be found from the triode region by simply substituting $(V_{DS})_{min}$ for V_{DS} in Equation 4.48, which yields:

$$(I_D)_{sat} = K\,(V_{GS} - V_t)^2 \quad (4.55)$$

Thus, in saturation, Equation 4.55 provides a drain current value that is independent of the drain voltage (V_{DS}) and is a function of gate-to-source voltage (V_{GS}) according to a square-law relationship.

NOTE 1: *These equations represent a first-order of approximation for the triode and saturation regions. If "channel length modulation" exists (as in MESFETs), there is a slight linear dependence of $(I_D)_{sat}$ on V_{DS} in the saturation region that can be accounted for by incorporating a factor of $(1 + \lambda V_{DS})$ in the $(I_D)_{sat}$ equation as follows:*

$$(I_D)_{sat} = K\,(V_{GS} - V_t)^2\,(1 + \lambda V_{DS}) \quad (4.56)$$

where λ is a positive number. Similar to the BJT, we observe that when the sloped $I_D - V_{DS}$ characteristic curves in saturation are extrapolated, the V_{DS} axis is intercepted at a point given by:

$$(V_{DS})_{intercept} = -1/\lambda = -V_A \quad (4.57)$$

where V_A is a positive voltage similar to the Early voltage in a BJT. Typically, we have:

$$0.005 \leq \lambda \leq 0.03\ V^{-1}, \text{ and}$$

$$\text{correspondingly, } 30 \leq V_A \leq 200\ \text{V}.$$

NOTE 2: *The value of the drain current obtained at the onset of saturation with $V_{GS} = 0$ is denoted by I_{DSS} and is given by:*

$$I_{DSS} = K V_t^2 \quad (4.58)$$

(For JFETs, $V_t = V_P$ and, thus, $I_{DSS} = K V_P^2$).

From this equation, if I_{DSS} and V_t are known, K can be found as follows:

$$K = I_{DSS}/V_t^2 \quad (4.59)$$

4.5.2 Comparison of FET Devices

The channel in an FET can either be created (or induced) by a voltage at the gate or physically exist due to an already implanted doped semiconductor. Therefore, based on the physical structure of the channel between the source and the drain, an FET can generally be classified into two categories. These are *enhancement type* (also called normally off) and *depletion type* (also called normally on) devices. Considering N-channel FETs exclusively, we can see that in enhancement type devices the channel does not ordinarily exist and has to be induced by applying a positive voltage at the gate above the threshold value. In N-channel enhancement type FETs, the threshold voltage is always positive (i.e., $V_t > 0$). On the other hand, depletion type devices already have a physically implanted channel and control of the channel is obtained by applying a positive or negative gate voltage so as to enhance or deplete the channel of charge carriers (i.e., electrons). In N-channel depletion type FETs, the threshold voltage is always negative (i.e., $V_t < 0$).

For comparison, typical $I_D - V_{GS}$ characteristics of enhancement type and depletion type FETs with their threshold values are shown in Figure 4.41. As with regard to $I_D - V_{DS}$ characteristics, both type of devices have similar curves except that V_t values have opposite signs.

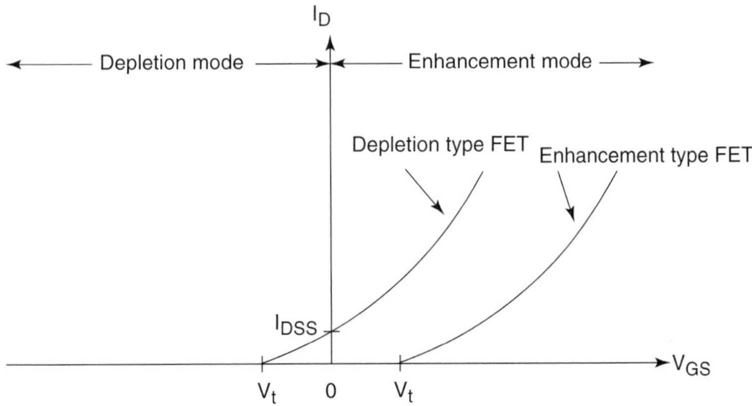

FIGURE 4.41 $I_D - V_{GS}$ characteristics for N-channel FETs.

In an N-channel depletion type device the channel can either be enhanced (by applying $V_{GS} > 0$ and thus attracting more electrons into it) or depleted (by applying $V_{GS} < 0$ and thus repelling the existing electrons from it). Therefore we can see that a depletion type device can be operated in two possible modes: "enhancement mode" and "depletion mode." This is in contrast to enhancement type devices in which their channel can not be depleted and perforce can operate in only one mode. This mode is aptly called the "enhancement mode."

N-channel MOSFETs and MESFETs can generally be built as enhancement type ($V_t > 0$) or depletion type ($V_t < 0$) devices. However, due to their physical structure, JFETs are strictly depletion type ($V_P < 0$) devices (As noted earlier in JFETs, the threshold voltage is called the pinch-off voltage and is denoted by V_P).

4.5.3 FET Circuits at DC

Having considered the $I_D - V_{DS}$ characteristics of a general N-channel FET device, we now present a few simple DC circuit examples to show the application of these principles.

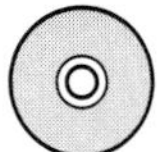

EXAMPLE 4.5 (Design)

Design the circuit shown in Figure 4.42 by determining R_S and R_D such that the FET operates at

$$I_D = 0.4 \text{ mA} \quad \text{and} \quad V_D = +1 \text{ V}$$

The FET has $V_t = 2V$ and $K = 0.4\ mA/V^2$. Neglect channel length modulation (i.e., $\lambda = 0$).

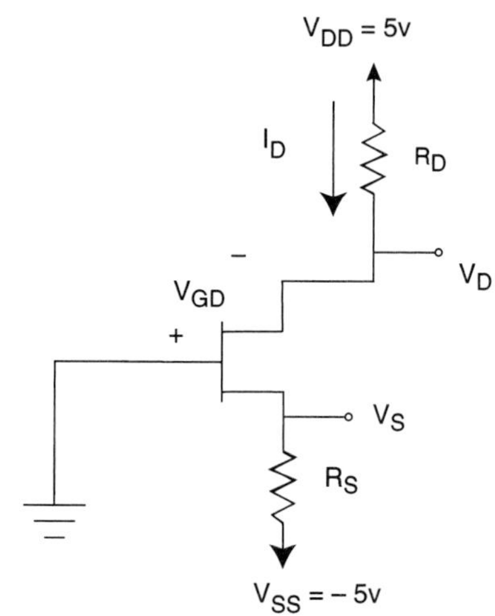

FIGURE 4.42 Circuit for Example 4.5.

Solution:

$$R_D = (V_{DD} - V_D)/I_D = 10 \text{ K}\Omega$$

To find R_S we note that $V_{GD} = V_G - V_D = -1 < V_t$ (because $V_t = 2$ V).

Thus, the transistor is operating in the saturation region and V_{GS} is given by:

$$I_D = K(V_{GS} - V_t)^2 \Rightarrow 0.4 = 0.4(V_{GS} - 2)^2$$

Two possible solutions for V_{GS} are given by:

$$V_{GS} = 1 \text{ V}$$

$$V_{GS} = 3 \text{ V}$$

The first solution puts the FET in cut-off (because $V_{GS} < V_t$) and this fact contradicts the FET being in saturation region.

The second solution allows the transistor to be turned on because $V_{GS} > V_t$ and thus is acceptable. Because the gate is grounded ($V_G = 0$) we have:

$$V_{GS} = V_G - V_S = 3 \Rightarrow V_S = -3\text{V}$$

Thus:

$$R_S = (V_S - V_{SS})/I_D = (-3 + 5)/0.4 = 5\ \text{K}\Omega.$$

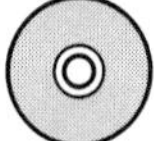

EXAMPLE 4.6 (Design)

Find the values of resistance R_D and voltage V_D when a current $I_D = 0.4$ mA flows through the drain of an FET circuit as shown in Figure 4.43. The FET has a threshold voltage $V_t = 2$ V and $K = 0.1$ mA/V^2 (assume $\lambda = 0$).

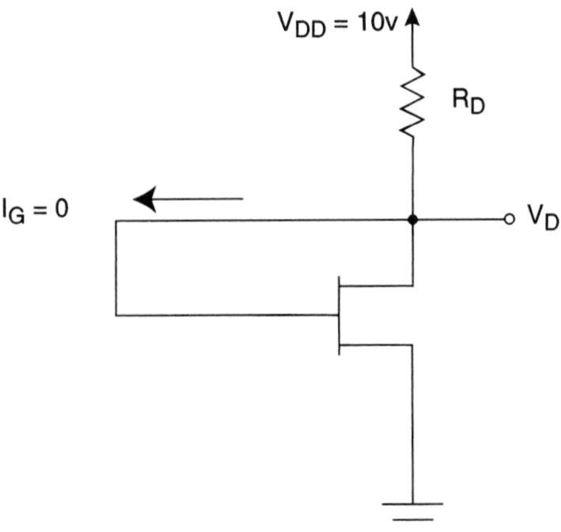

FIGURE 4.43 Circuit for Example 4.6.

Solution:

Because $V_{GD} = 0 < V_t$, the device is always in saturation. Therefore, we have:

$$I_D = K\,(V_{GS} - V_t)^2$$

$$0.4 = 0.1(V_{GS} - 2)^2$$

which gives two solutions:

$$V_{GS} = 0\,\text{V} < V_t \quad \Rightarrow \quad \text{FET is in cut-off (not acceptable).}$$

$$V_{GS} = 4\text{V} > V_t \quad \Rightarrow \quad \text{FET is in saturation, i.e., "on" (acceptable)}$$

because $V_{GD} = V_{GS} - V_{DS} = 0 \quad \Rightarrow \quad V_{DS} = V_D = 4\,\text{V}$

$$R_D = (V_{DD} - V_D)/I_D = (10 - 4)/0.4 = 15\ \text{K}\Omega.$$

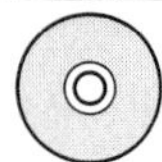

EXAMPLE 4.7 (Analysis)

Analyze the circuit shown in Figure 4.44 by determining the voltages at all nodes and the current through all branches. The FET has $V_t = 1$ V, $K = 0.5$ mA/V^2. Neglect channel length modulation (i.e., $\lambda = 0$).

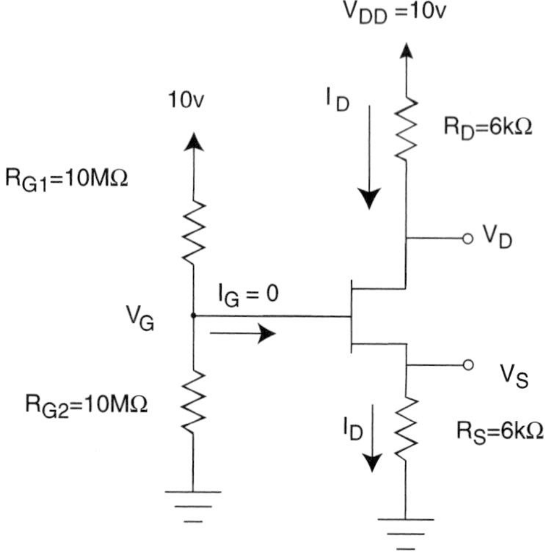

FIGURE 4.44 Circuit for Example 4.7.

Solution:

Using the voltage division rule we have:

$$V_G = R_{G2}[V_{DD}/(R_{G1} + R_{G2})] = 10\ (10/20) = 5\text{V}$$

Also,

$$V_S = 6\ I_D$$

Thus, $V_{GS} = V_G - V_S = 5 - 6\ I_D$

Because it is not clear in what region the transistor is operating, we assume saturation-region operation, solve the problem, and then check the validity of our assumption. If our original assumption turns out to be incorrect, then triode-region operation should be considered using the same procedure until the original assumption can be correctly verified.

Assume saturation:

$$I_D = K\ (V_{GS} - V_t)^2 \Rightarrow I_D = 0.5\ [(5 - 6\ I_D) - 1]^2 = 0.5\ (4 - 6I_D)^2$$

$$18\ I_D^2 - 25\ I_D + 8 = 0$$

Solving this quadratic equation yields two solutions:

$$I_D = 0.89 \text{ mA}$$

$$I_D = 0.5 \text{ mA}$$

Using the first solution $V_S = 6\,I_D = 5.34\text{ V} > V_G = 5\text{ V}$, which is impossible!
Using the second solution $V_S = 6\,I_D = 3.0\text{ V} < V_G = 5\text{ V}$, which is acceptable!
$V_{GS} = 5 - 3 = 2\text{V} > V_t = 1\text{V}$, which means the device is on. Using the second solution we obtain:

$$V_S = 6\,I_D = 3\text{ V}$$

$$V_D = 10 - 6\,I_D = 7\text{ V}$$

$$V_{DS} = 7 - 3 = 4\text{V}$$

NOTE: *Now we need to verify the saturation assumption:*

$$V_{GD} = V_G - V_D = 5 - 7 = -2 < V_t$$

$$(V_{DS})_{min} = V_{GS} - V_t = 5 - 1 = 4\text{ V}$$

because $V_{DS} = (V_{DS})_{min}$, therefore, the FET is indeed at the onset of saturation, and our initial assumption has been proved correct!

4.5.4 The DC Operating Point of an FET

Similar to a BJT, the Q-point of an FET should be chosen in the middle of the saturation region, allowing maximum voltage swing for class-A type amplifiers.

A typical and yet popular biasing arrangement very similar to the circuit of Example 4.3 is shown in Figure 4.45.

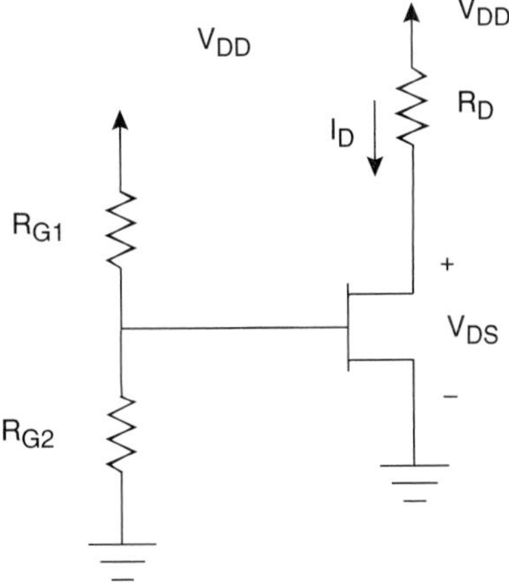

FIGURE 4.45 A typical circuit for FET biasing.

To obtain the Q-point, we use the $I_D - V_{DS}$ characteristic curves of the FET as shown in Figure 4.46 with the plotted load line, which is given by the following equation:

KVL:

$$V_{DD} = V_{DS} + R_D I_D \Rightarrow I_D = V_{DD}/R_D - V_{DS}/R_D \tag{4.60}$$

The intersection of this load line with a suitable curve will provide a useful Q-point in the middle of the saturation region, allowing for the required signal swing without the device ever entering the triode region (see Figure 4.46).

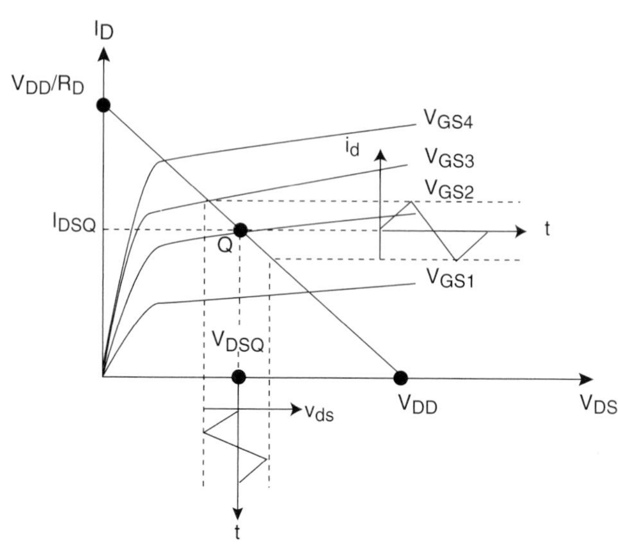

FIGURE 4.46 FET characteristic curves, the load line, and the Q-point.

4.5.5 FET AC Analysis

For an FET to operate as a linear amplifier or an oscillator (which is an amplifier with positive feedback), it must be operated in the middle of the saturation region and must remain in this region for proper operation at all times, as discussed in the previous sections.

Small-Signal AC-Analysis. As discussed earlier, if the amplitude of the input voltage is small, we can develop an approximate first-order model for any device or, in this case, the FET's behavior in the saturation region.

Consider the conceptual amplifier circuit (shown in Figure 4.47) with the FET in the saturation region. The total drain current based on a first-order model can be calculated as follows:

- Under DC bias (i.e., $v_{gs} = 0$):

$$V_{GS} = V_{GSQ} \tag{4.61 a}$$

$$I_{DQ} = K\,(V_{GSQ} - V_t)^2 \tag{4.61 b}$$

- Under small AC signal (i.e., $|v_{gs}| > 0$):

$$v_{GS} = V_{GSQ} + v_{gs} \tag{4.62 a}$$

$$i_D = K(V_{GSQ} + v_{gs} - V_t)^2 = K\,(V_{GSQ} - V_t)^2 + 2K\,(V_{GSQ} - V_t)\,v_{gs} + K v_{gs}^2 \tag{4.62 b}$$

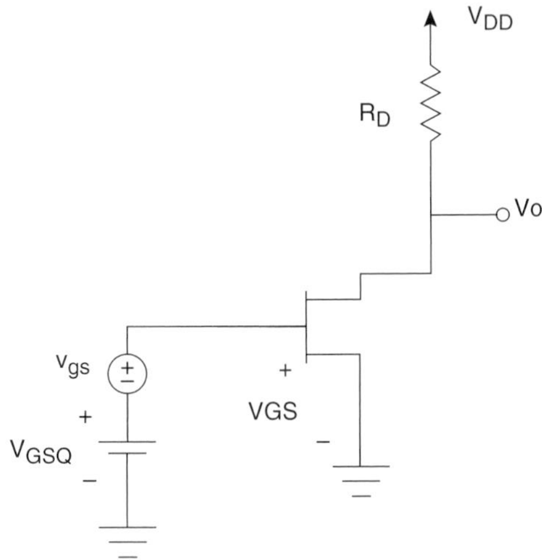

FIGURE 4.47 A conceptual FET amplifier circuit.

Equation 4.62b is nonlinear; however, with the help of small-signal assumption, we can linearize it. To be specific about the size of small AC signal (v_{gs}), we now set the following small-signal condition in Equation 4.62b:

$$Kv_{gs}^2 << 2K(V_{GSQ} - V_t)v_{gs} \Rightarrow v_{gs} << 2\,(V_{GSQ} - V_t) \tag{4.63}$$

Using the small-signal condition (as defined by Equation 4.63), Equation 4.62b simplifies into:

$$i_D = I_{DQ} + i_d \tag{4.64}$$

where i_d, the small AC signal, is given by:

$$i_d = 2\,K(V_{GSQ} - V_t)v_{gs} \tag{4.65}$$

Similar to BJT analysis, the small-signal transconductance (g_m) can be defined as:

$$g_m = i_d/v_{gs} = 2K(V_{GSQ} - Vt) \tag{4.66}$$

Alternately, Equation 4.66 can also be expressed as:

$$g_m = 2\,I_{DQ}/(V_{GSQ} - V_t) \tag{4.67}$$

As observed from this analysis, the value of g_m in this AC first-order model depends, in a major way, on the bias values (V_{GSQ}, I_{DQ}, and V_{DSQ}), which are determined by the location of the Q-point in the $I_D - V_{DS}$ plane.

4.5.6 First-Order AC Model of an FET

Based on Equations 4.65 through 4.67, the first-order hybrid-π model of an FET can be considered to be an open circuit at the gate, in conjunction with a simple current source at the output drain terminal, as shown in Figure 4.48.

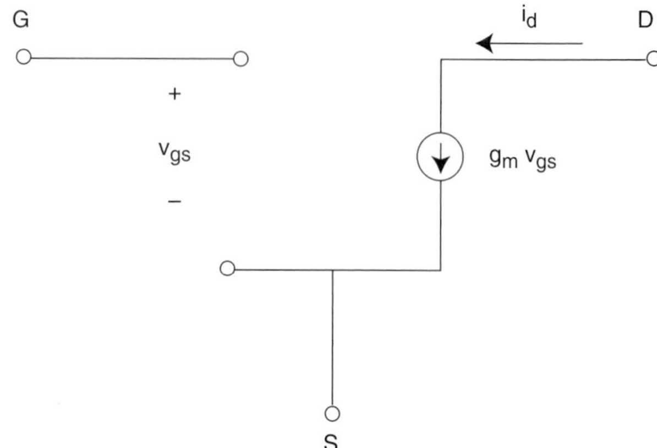

FIGURE 4.48 First-order hybrid-π model of an FET (low frequency).

4.5.7 Second-Order AC Models of an FET

Adding more elements to the first-order model will upgrade it to the second-order approximation as follows:

Low-Frequency AC Model. In the previous first-order model, the output resistance (r_o) is assumed to be infinite. In practice, though, the output resistance is finite and should be added at the output to create a more accurate second-order model, as shown in Figure 4.49.

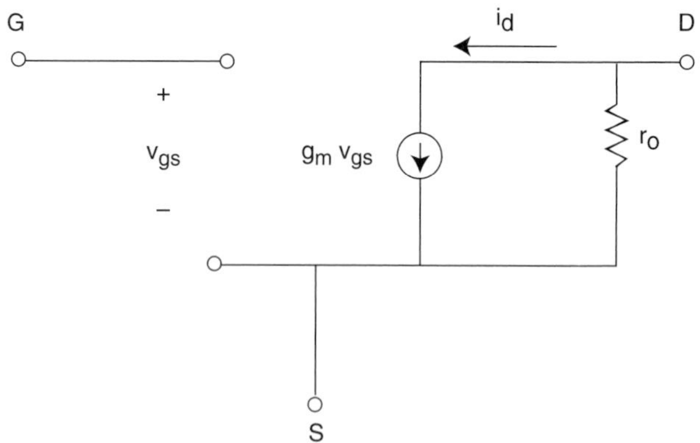

FIGURE 4.49 Second-order hybrid-π model of an FET in saturation (low frequency).

Please note that the output resistance (r_o) creates a linear dependence of i_d with an increase in v_{ds}, which is a more realistic model for an FET.

High-Frequency AC Model. The second-order model of an FET at high frequencies takes into account the capacitive effects existing at the junctions of the device. Such a model consists of the low-frequency second-order model with two additional capacitors (C_{gs} and C_{gd}), as shown in Figure 4.50.

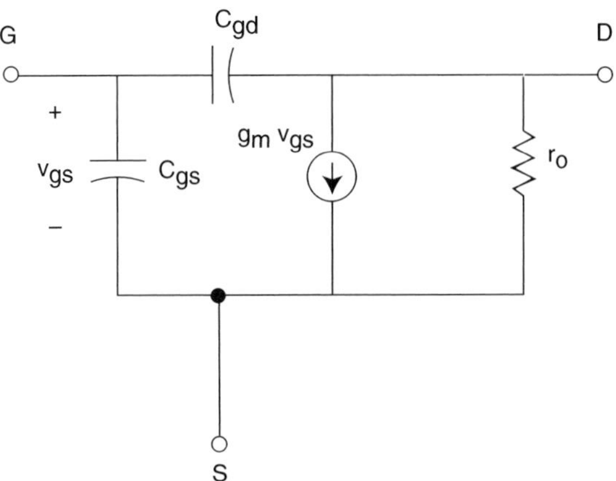

FIGURE 4.50 Second-order high-frequency hybrid-π model of an FET in saturation.

Similar to a BJT's output resistance, r_o is given by:

$$r_o = V_A / I_D$$

where $V_A = 1/\lambda$ is an FET's early voltage parameter and I_D is the drain current near the boundary region for a given V_{GSQ}.

From the hybrid-π model, we can observe that when an FET is connected in the common-source configuration, the capacitor C_{gd} appears in the feedback path, which effectively bridges the output terminal (or drain) to the input terminal (or gate), very similar to the capacitor (C_μ) in BJTs.

Using Miller's theorem, the feedback capacitor (C_{gd}) can be replaced with two grounded and magnified capacitors: one at the input and another at the output terminals.

This replacement not only simplifies the circuit analysis but also clearly indicates the magnified role of C_{gd} at the input terminals causing an eventual deterioration of the frequency response of the amplifier.

EXAMPLE 4.8

In the amplifier circuit shown in Figure 4.51, calculate the input impedance and the voltage gain of the amplifier. Assume the FET has $K = 0.125\ mA/V^2$, $V_t = 1.5\ V$ and $V_A = 50\ V$ and all coupling capacitors (C_1, C_2) to be large enough as to act as short circuits at the signal frequencies of interest.

Solution:

To determine small signal parameters, we first need to do a DC analysis:

DC analysis ($v_i = 0$)

Because $I_{GQ} = 0 \Rightarrow V_{GD} = 0 < V_t \Rightarrow$ FET is always in the saturation region, and we can write:

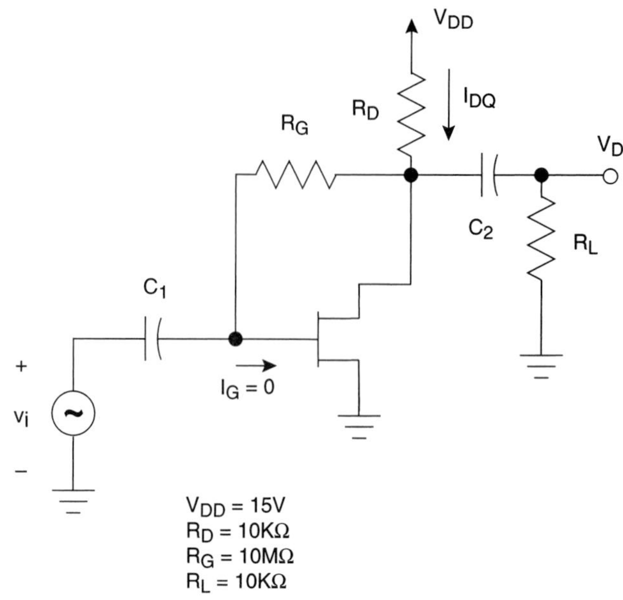

FIGURE 4.51 Circuit for Example 4.8.

$$V_{GS} = V_D$$

Also,

$$I_{DQ} = 0.125\,(V_{GS} - 1.5)^2 \tag{4.68a}$$

using KVL, we have:

$$V_{DD} = R_D I_{DQ} + V_D \Rightarrow 15 = 10 I_{DQ} + V_{GS} \tag{4.68b}$$

There are two possible solutions to Equations 4.68a and b as follows:

Solution 1 $I_{DQ} = 1.72$ mA, $V_{GS} = V_D = -2.2$ V
Solution 2 $I_{DQ} = 1.06$ mA, $V_{GS} = V_D = 4.4$ V

Because V_{GS} is always positive (i.e., $V_{GS} > 0$), *Solution 2* provides the operating point as:

$$I_{DQ} = 1.06 \text{ mA}$$

$$V_{GS} = V_D = 4.4 \text{ V}$$

AC analysis ($\mathbf{v}_i \neq 0$)

$$g_m = 2K(V_{GS} - V_t) = 2 \times 0.125 \times (4.4 - 1.5) = 0.725 \text{ mA/V}$$

$$r_o = V_A / I_D = 50/1.06 = 47 \text{ k}\Omega$$

The total voltage gain is calculated from the small equivalent circuit (see Figure 4.52): Because $R_G = 10 \text{ M}\Omega >> 1 \Rightarrow i_g \approx 0$

$$v_o = -g_m v_{gs}\,(R_D \,\|\, R_L \,\|\, r_o), \quad v_i = v_{gs}$$

Thus: $v_o/v_i = -g_m\,(10\text{k} \,\|\, 10\text{k} \,\|\, 47\text{k}) \Rightarrow v_o/v_i = -3.3$

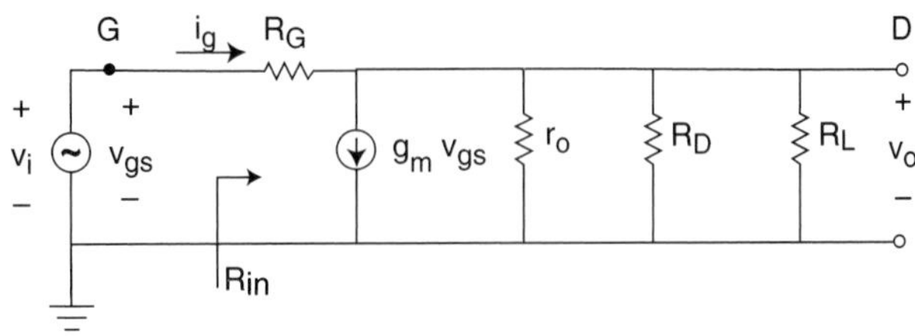

FIGURE 4.52 Small signal equivalent circuit for Example 4.8.

The input impedance is given by: $R_{in} \equiv v_i / i_g$

$$i_g = (v_i - v_o)/R_G = v_i(1 - v_o/v_i)/R_G = 4.3\ v_i/R_G$$

Thus: $R_{in} = R_G/4.3 \Rightarrow R_{in} = 2.33\ \text{M}\Omega$

NOTE: *Generalizing the results from Example 4.8, we can develop a basic equation for the overall gain of a common-source FET amplifier configuration, where the source is not shorted to ground (i.e., $R_S \neq 0$) as follows:*

$$v_o/v_i = -g_m(R_D \,\|\, R_L \,\|\, r_o)/(1 + g_m R_S)$$

4.6 HOW TO DO AC SMALL-SIGNAL ANALYSIS

Having examined a p-n junction diode and both types of transistors (BJTs and FETs) and performed analysis/design on several active circuits, it is time for us to summarize these methods and findings into a workable technique we can draw upon for all of our future analytical or design needs. The complex problem of small-signal analysis of a general active circuit, as shown in Figure 4.53, breaks down into two major subdivisions (see Section 1.5.5, *Solutions to Problems* in Chapter 1, *Fundamental Concepts of Science and Engineering*):

- DC analysis
- AC analysis

Each one of the two subdivisions has several exact steps that must be followed to obtain the desired AC quantities, such as gain, input impedance, and so on.

4.6.1 DC Analysis

Because all of the AC small-signal models and the ultimate performance of any active circuit depend directly on the value of the DC current flowing through the transistor (i.e., the device's Q-point), the first logical step in the analysis is to perform a DC analysis.

As noted earlier, the purpose of biasing is to establish a constant DC current in the device that is relatively stable and insensitive to external variations (e.g., temperature) or internal changes (e.g., β of a transistor). Given an active circuit, the steps required to perform the DC analysis are as follows:

STEP 1. Open-circuit all capacitors in the circuit.

STEP 2. Short-circuit all inductors in the circuit.

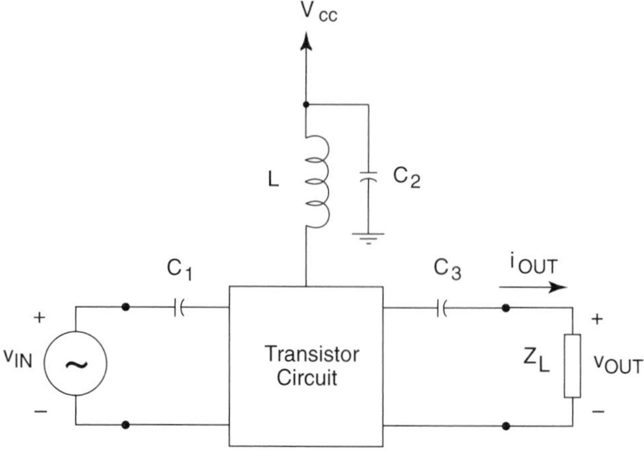

FIGURE 4.53 Block diagram of a transistor circuit with power supply connection as shown.

C_1, C_3 coupling capacitors
C_2 filtering capacitor

STEP 3. Set all AC sources to zero, i.e.,
- **a.** Short-circuit all AC voltage sources
- **b.** Open-circuit all AC current sources

STEP 4. Use KVL, KCL, and Ohm's law as well as the device's DC I–V relationship (e.g., $V_{BE} \approx 0.7$ V for a BJT) to obtain the device's Q-point.

STEP 5. Calculate all small-signal parameter values (e.g., g_m, r_e, etc.).

4.6.2 AC Small-Signal Analysis

The goal here is to find the desired AC quantities (e.g., gain, input impedance, etc.) based on the DC values obtained in the previous section. To this end, we need to establish and draw the "AC small-signal equivalent circuit," as delineated in the following steps:

STEP 1. Calculate the impedance values of all capacitors ($Z_C = -j/\omega C$) and, if these values are negligible compared to the lowest impedance value remaining in the circuit, we can then safely replace them by a short-circuit. Otherwise, they remain in the circuit if the frequency is not high enough to justify the short-circuit.

STEP 2. Calculate the impedance values of all inductors ($Z_L = j\omega L$) and if these values are much larger compared to the highest impedance value remaining in the circuit, we can then safely open-circuit them. Otherwise, they remain in the circuit if the frequency is not high enough to justify the open circuit.

STEP 3. Replace each active device by its small-signal model (e.g., hybrid-π model, T-model, etc.) as appropriate.

STEP 4. Set each independent DC voltage and current source to zero value, i.e.,
- **a.** DC voltage source → replace with a short circuit,
- **b.** DC current source → replace with an open circuit.

STEP 5. Use KVL, KCL, Ohm's law, and circuit theorems to obtain AC branch currents and AC node voltages.

STEP 6. Use proper mathematical relationships to obtain the desired AC quantities (e.g., V_{out}/V_{in}, $R_{in} = V_{in}/I_{in}$, $R_{out} = V_{out}/I_{out}$, etc.).

NOTE: *To find R_{out} (= V_{out}/I_{out}), we need to connect an AC source (V_{out}) at the output terminals, set the AC input source to zero, obtain a new AC equivalent circuit, calculate I_{out} and then find the ratio of V_{out}/I_{out}. Special attention should be paid here because setting the input AC source to zero and connecting another AC source at the output, changes the AC equivalent circuit.*

4.7 SUMMARY AND CONCLUSIONS

In this chapter, we have presented the general operation and structure of diodes as the fundamental building blocks for transistors. We discussed that there are two basic types of transistors: BJT and FET. The general equations presented for each type of transistor apply to all of the transistor circuits and their variations in that category. Not so much for BJTs, but this is applicable particularly when it comes to the various types of FET devices (such as JFET, MOSFET, and MESFET), which are structurally different and yet satisfy the same general equations presented in this chapter.

Knowing this chapter well lays the groundwork for all the future chapters and familiarizes the reader with the very basic device knowledge needed to handle higher frequency devices and circuits comfortably.

LIST OF SYMBOLS/ABBREVIATIONS

A symbol will not be repeated again once it has been identified and defined in an earlier chapter, as long as its definition remains unchanged.

B	Base
BJT	Bipolar junction transistor
C	Coulomb
CBJ	Collector Base junction of a transistor
C_π	Hybrid-π model capacitance between emitter-base terminals
C_μ	Capacitance between collector and base in Hybrid-π model
E	Emitter
EBJ	Emitter Base junction of a transistor
FET	Field Effect Transistor
g_m	Transconductance
I_B	Base current for a transistor
I_C	Collector current for a transistor
I_{CQ}	Collector current at the quiescent point
I_D	Drain current, FET
I_{DSQ}	Drain to source quiescent current, FET
I_{DSS}	Drain current at the onset of saturation when $V_{GS} = 0$, FET
I_E	Emitter current for a transistor
I_G	Gate current, FET
I_S	Saturation current
K	Boltzmann's constant (= 1.38×10^{-23} $J/°K$)
NPN	Refers to the type of doping used to construct a transistor; n-type collector, p-type base, and n-type emitter

PNP	Refers to the type of doping used to construct a transistor; p-type collector, n-type base, and p-type emitter
Q	Quiescent point
q	Magnitude of electric charge (=1.602×10^{-19} C)
R_B	Base resistor
R_C	Collector resistor
r_o	Shunt resistance of Hybrid-π model
r_π	Hybrid-π model input resistance
r_μ	Resistance between collector and base in Hybrid-π model (BJT)
T	Temperature (use degrees Kelvin for computation)
V_{BE}	Voltage from the base to emitter for a transistor
V_{CC}	Collector supply voltage for a transistor
V_{CE}	Collector to emitter voltage for a transistor
V_{CEQ}	Collector to emitter voltage at the quiescent point
V_D	Drain voltage, FET
V_{DD}	Drain supply voltage, FET
V_{DS}	Drain to source voltage, FET
V_{DSQ}	Drain to source quiescent voltage, FET
V_G	Gate voltage, FET
V_{GS}	Gate to source voltage, FET
V_{GSQ}	Gate to source quiescent voltage, FET
V_S	Source voltage, FET
V_T	Thermal voltage defined as $V_T = kT/q$
V_{ZK}	Voltage at the zener knee of the *I-V* characteristic of a diode
α	Common-base current gain
β	Common-Emitter current gain
λ	A positive number describing $(I_D)_{sat}$ linear relation with V_{DS}

NOTE: *Symbol convention uses all capital letters for DC analysis. Small-signal AC analysis uses similar designations for transistor voltages and currents but with small letters and small letter subscripts. Total signal variables use small letters with capital letter subscripts.*

PROBLEMS

4.1 Assuming the ideal diodes are fully conducting as shown in Figure P4.1, find the values of I and V.

4.2 Find the value of an ideal current source I required to obtain an output voltage $V_o = 2$V for the circuit shown in Figure P4.2. Assume diodes are fabricated using integrated circuit technology with $I_S = 10^{-14}$ A.

4.3 A regulator uses a Zener diode with an incremental resistance of 5Ω with a series resistor of 100 Ω (see Figure P4.3). If the supply voltage changes by 2 V, find the corresponding change in the regulated output voltage.

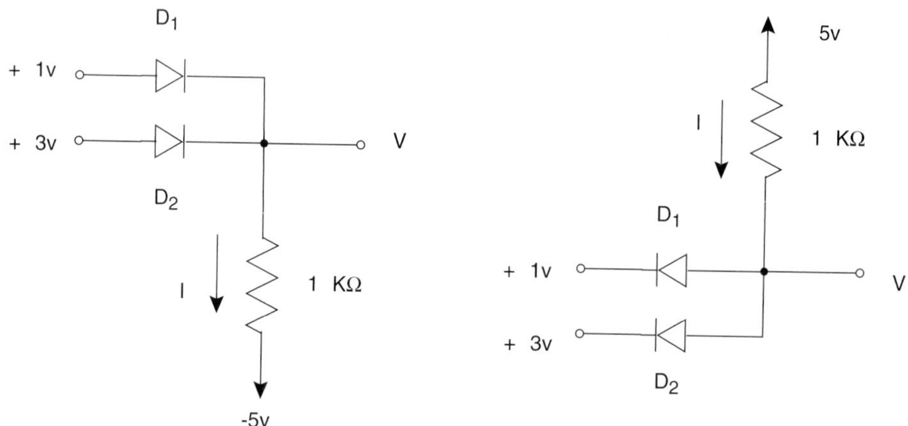

FIGURE P4.1

FIGURE P4.2

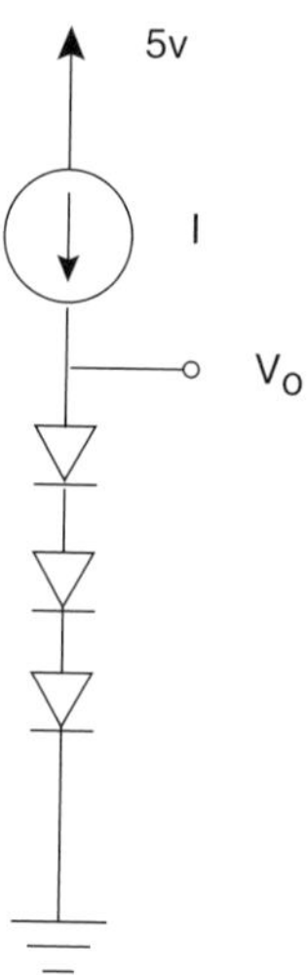

FIGURE P4.3

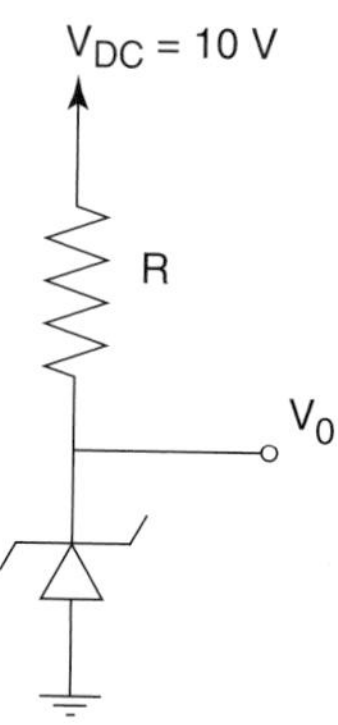

4.4 For the circuit shown in Figure P4.4, assume that the transistor has a $\beta = 100$. Find V and I.

FIGURE P4.4

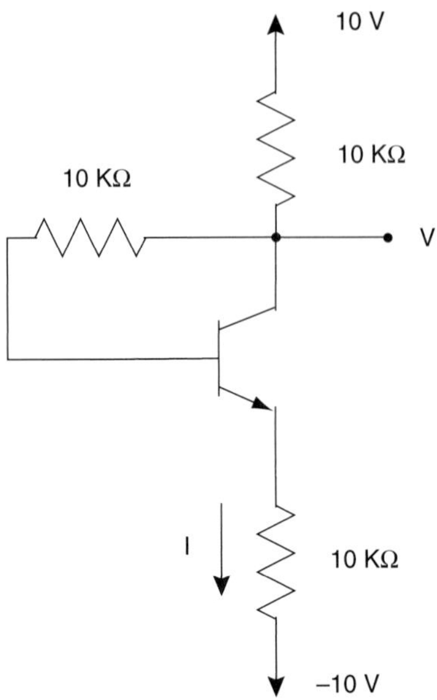

4.5 For the emitter-follower circuit shown in Figure P4.5, the BJT has a $\beta = 200$ and $V_S = 0.5\cos 2\pi \times 10^7 t$ Volts. Calculate:

a. I_{EQ}, V_B, and V_E (all DC values)
b. The input impedance R_{IN}
c. The voltage gain v_o/v_i

FIGURE P4.5

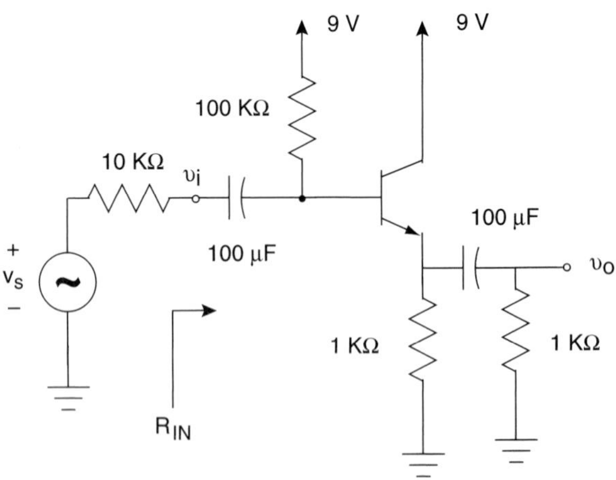

4.6 For the common-emitter amplifier shown in Figure P4.6, calculate the DC bias current (I_{EQ}), R_{IN}, R_{OUT}, and voltage gain (v_o/v_i). Assume: V_{CC} = 9V, R_1 = 27$k\Omega$, R_2 = 15 $k\Omega$, R_E = 1.2 $k\Omega$, R_C = 2.2 $k\Omega$, $C_1 = \infty$, β = 100 and V_A = 100 V.

FIGURE P4.6

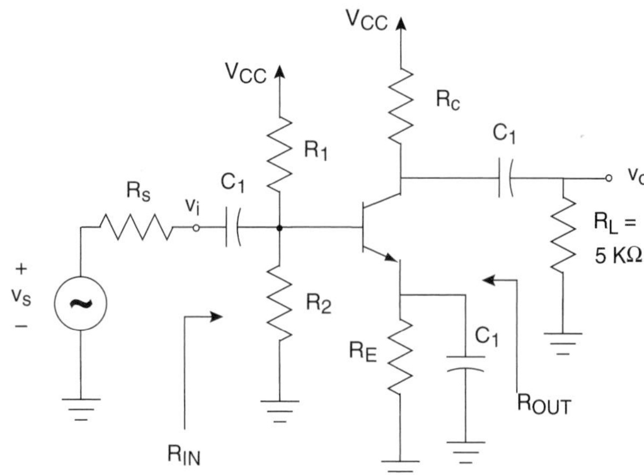

4.7 For an FET device, the threshold voltage is V_t = –2V. Find the minimum value of V_{DS} required to operate in the saturation region when V_{GS} = +1V. What is the corresponding value of I_D? Assume I_{DSS} = 1 mA.

4.8 Design the FET circuit shown in Figure P4.8 to obtain a current I_D = 0.4 mA. Determine R and V_D. Assume the FET has V_t = 2V, K = 0.5 mA/V^2. Neglect channel length modulation (i.e., λ = 0).

FIGURE P4.8

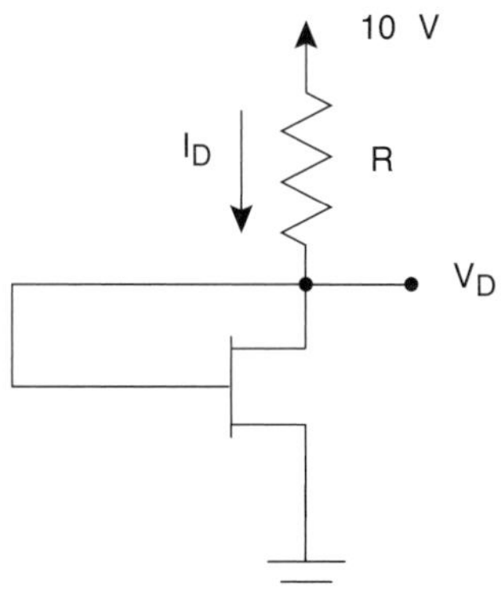

4.9 A common-source FET amplifier having an open-circuit voltage gain of –9 and an output resistance of 10 Ω, is connected to a load resistance of 10 kΩ. What is the resulting voltage gain?

4.10 The FET amplifier shown in Figure P4.10 has V_t = –4V and I_{DSS} = 12 mA. At I_D = 12 mA, the output resistance is r_o = 25 $k\Omega$.

a. Determine the DC bias quantities: V_{GQ}, I_{DQ}, V_{GSQ}, and V_{DQ}.
b. Determine R_{IN}, v_o/v_i and v_o/v_s ($\omega = 2\pi \times 10^8$ rad/s)

FIGURE P4.10

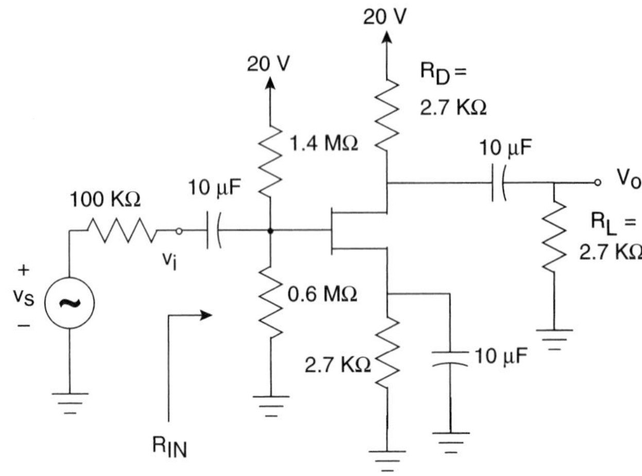

REFERENCES

[4.1] Burns, S. B. and P. R. Bond. *Principles of Electronic Circuits*, St. Paul: West, 1987.

[4.2] Gray, P. R. and C. L. Searle. *Electronic Principles*, 3rd Edition, New York: John Wiley & Sons, 1993.

[4.3] Hambley, A. R. *Electronics*, 2nd Edition, Upper Saddle River: Prentice Hall, 2000.

[4.4] Hamilton, D. J. and W. G. Howard. *Basic Integrated Engineering*, New York: McGraw-Hill, 1975.

[4.5] Hodges, D. A. and H. G. Jackson. *Analysis and Design of Digital Integrated Circuits*, Second Edition, New York: McGraw-Hill, 1988.

[4.6] Horowitz, P. and W. Hill. *The Art of Electronics*. Cambridge University Press, 1983.

[4.7] Johns, A. and S. E. Butner. *Analog Integrated Circuit Design*. New York: John Wiley & Sons, 1997.

[4.8] Lackey, J. E., J. L. Massey, and M. D. Hehn. *Solid State Electronics*. New York: Holt, Rinehart and Winston, 1986.

[4.9] Millman, J. A. and A. Grabel. *Microelectronics*, 2nd ed. New York: McGraw-Hill, 1987.

[4.10] Navon, H. *Semiconductor Microdevices and Materials*. New York: Holt, Rinehart and Winston, 1986.

[4.11] Sedra, A. S. and K. C. Smith. *Microelectronic Circuits*. Oxford University Press, 1998.

[4.12] Soclof, S. *Analog Integrated Circuits*, Englewood Cliffs, NJ: Prentice-Hall, 1985.

[4.13] Streetman, B. J. *Solid State Electronic Devices*, 4th Edition, Englewood Cliffs, NJ: Prentice-Hall, 1995.

PART II

WAVE PROPAGATION IN NETWORKS

CHAPTER 5

Introduction to Radio Frequency and Microwave Concepts and Applications

5.1 INTRODUCTION

This chapter lays the foundation for understanding higher-frequency wave phenomena and divides the task of active circuit design for RF/MW frequencies into specific concept blocks. These concept blocks create a gradual approach to understanding and designing RF/MW circuits and represent specific realms of knowledge that need to be mastered to become an accomplished designer.

Before we describe and analyze these types of waves we need to consider why RF/microwaves as a subject has become so important, that it is placed at the forefront of our modern technology. Furthermore, we need to expand our minds to the many possibilities that these signals can provide for peaceful practices by exploring various commercial applications useful to mankind.

5.1.1 A Short History of RF and Microwaves

Circa 1864–1873, James Clark Maxwell integrated the entirety of man's knowledge of electricity and magnetism by introducing a set of four coherent and self-consistent equations that describe the behavior of electric and magnetic fields on a classical level. This was the beginning of microwave engineering, as presented in a treatise by Maxwell at that time. He predicted, purely from a mathematical standpoint and on a theoretical basis, the existence of electromagnetic wave propagation and that light was also a form of electromagnetic energy—both completely new concepts at the time.

From 1885 to 1887, Oliver Heaviside simplified Maxwell's work in his published papers. From 1887 to 1891, a German physics professor, Heinrich Hertz, verified Maxwell's predictions experimentally and demonstrated the propagation of electromagnetic waves. He also investigated wave propagation phenomena along transmission lines and antennas and developed several useful structures. He could be called the first microwave engineer.

Marconi tried to commercialize radio at a much lower frequency for long-distance communications, but as he had a business interest in all of his work and developments, this was not a purely scientific endeavor.

Neither Hertz nor Heaviside investigated the possibility of electromagnetic wave propagation inside a hollow metal tube because it was felt that two conductors were necessary for the transfer of electromagnetic waves or energy. In 1897, Lord Rayleigh showed mathematically that electromagnetic wave propagation was possible in waveguides, both circular and rectangular. He showed that there are infinite sets of modes of the TE and TM type possible, each with its own cut-off frequency. These were all theoretical predictions with no experimental verifications.

From 1897 to 1936, the waveguide was essentially forgotten until it was rediscovered by two men, George Southworth (AT&T) and W. L. Barron (MIT), who showed experimentally that a waveguide could be used as a small bandwidth transmission medium, capable of carrying high power signals.

With the invention of the transistor in the 1950s and the advent of microwave integrated circuits in the 1960s, the concept of a microwave system on a chip became a reality. There have been many other developments, mostly in terms of application mass, that have made RF and microwaves an enormously useful and popular subject.

Maxwell's equations lay the foundation and laws of the science of electromagnetics, of which the field of RF and microwaves is a small subset. Due to the exact and all-encompassing nature of these laws in predicting electromagnetic phenomena, along with the great body of analytical and experimental investigations performed since then, we can consider the field of RF and microwave engineering a "mature discipline" at this time.

5.1.2 Applications of Maxwell's Equations

As indicated earlier in Chapter 2, *Fundamental Concepts in Electrical and Electronics Engineering*, standard circuit theory can neither be used at RF nor particularly at microwave frequencies. This is because the dimensions of the device or components are comparable to the wavelength, which means that the phase of an electrical signal (e.g., a current or voltage) changes significantly over the physical length of the device or component. Thus use of Maxwell's equations at these higher frequencies becomes imperative.

In contrast, the signal wavelengths at lower frequencies are so much larger than the device or component dimensions, that there is negligible variation in phase across the dimensions of the circuit. Thus Maxwell's equations simplify into basic circuit theory, as covered in Chapter 3, *Mathematical Foundation for Understanding Circuits*.

At the other extreme of the frequency range lies the optical field, where the wavelength is much smaller than the device or circuit dimensions. In this case, Maxwell's equations simplify into a subject commonly referred to as geometrical optics, which treats light as a ray traveling on a straight line. These optical techniques may be applied successfully to the analysis of very high microwave frequencies (e.g., high millimeter wave range), where they are referred to as "quasi-optical." Of course, it should be noted that further application of Maxwell's equations leads to an advanced field of optics called "physical optics or Fourier optics," which treats light as a wave and explains such phenomena as diffraction and interference, where geometrical optics fails completely.

The important conclusion to be drawn from this discussion is that Maxwell's equations present a unified theory of analysis for any system at any frequency, provided we use appropriate simplifications when the wavelengths involved are much larger, comparable to, or much smaller than the circuit dimensions.

5.1.3 Properties of RF and Microwaves

An important property of signals at RF, and particularly at higher microwave frequencies, is their great capacity to carry information. This is due to the large bandwidths available at these high frequencies. For example, a 10 percent bandwidth at 60 MHz carrier signal is 6 MHz, which is approximately one TV channel of information; on the other hand 10 percent of a microwave carrier signal at 60 GHz is 6 GHz, which is equivalent to 1000 TV channels.

Another property of microwaves is that they travel by line of sight, very much like the traveling of light rays, as described in the field of geometrical optics. Furthermore, unlike lower-frequency signals, microwave signals are not bent by the ionosphere. Thus use of line-of-sight communication towers or links on the ground and orbiting satellites around the globe are a necessity for local or global communications.

A very important civilian as well as military instrument is radar. The concept of radar is based on radar cross-section which is the effective reflection area of the target. A target's visibility greatly depends on the target's electrical size, which is a function of the incident signal's wavelength. Microwave frequency is the ideal signal band for radar applications. Of course, another important advantage of use of microwaves in radars is the availability of higher antenna gains as the frequency is increased for a given physical antenna size. This is because the antenna gain being proportional to the electrical size of the antenna, becomes larger as frequency is increased in the microwave band. The key factor in all this is that microwave signal wavelengths in radars are comparable to the physical size of the transmitting antenna as well as the target.

There is a fourth and yet very important property of microwaves: the molecular, atomic, and nuclear resonance of conductive materials and substances when exposed to microwave fields. This property creates a wide variety of applications. For example, because almost all biological units are composed predominantly of water and water is a good conductor, microwave technology has tremendous importance in the fields of detection, diagnostics, and treatment of biological problems or medical investigations (e.g., diathermy, scanning, etc.). There are other areas in which this basic property would create a variety of applications such as remote sensing, heating (e.g., industrial purification and cooking) and many others that are listed in a later section.

5.2 REASONS FOR USING RF/MICROWAVES

Over the past several decades, there has been a growing trend toward use of RF/microwaves in system applications. There are many reasons among which the following are prominent:

- Wider bandwidths due to higher frequency
- Smaller component size leading to smaller systems
- More available and less crowded frequency spectrum
- Better resolution for radars due to smaller wavelengths
- Lower interference due to lower signal crowding
- Higher speed of operation
- Higher antenna gain possible in a smaller space

On the other hand, there are some disadvantages to using RF/microwaves, such as: more expensive components, availability of lower power levels, existence of higher signal losses, and use of high-speed semiconductors (such as GaAs or InP) along with their corresponding less-mature technology (relative to the traditional silicon technology, which is now quite mature and less expensive).

In many RF/microwave applications the advantages of a system operating at these frequencies outweigh the disadvantages and propel engineers to a high-frequency design.

5.3 RF/MICROWAVE APPLICATIONS

The major applications of RF/microwave signals can be categorized as follows:

5.3.1 Communication

This application includes satellite, space, long-distance telephone, marine, cellular telephone, data, mobile phone, aircraft, vehicle, personal, and wireless local area network (WLAN), among others. Two important subcategories of applications need to be considered: TV and radio broadcast, and optical communications.

TV and Radio Broadcast. In this application, RF/microwaves are used as the carrier signal for audio and video signals. An example is the Direct Broadcast System (DBS), which is designed to link satellites directly to home users.

Optical Communications. In this application, a microwave modulator is used in the transmitting side of a low-loss optical fiber with a microwave demodulator at the other end. The microwave signal acts as a modulating signal with the optical signal as the carrier. Optical communication is useful in cases where a much larger number of frequency channels and less interference from outside electromagnetic radiation are desired. Current applications include telephone cables, computer network links, low-noise transmission lines, and so on.

5.3.2 Radar

This application includes air defense, aircraft/ship guidance, smart weapons, police, weather, collision avoidance, and imaging.

5.3.3 Navigation

This application is used for orientation and guidance of aircraft, ships, and land vehicles. Particular applications in this area are as follows:

- Microwave Landing System (MLS), used to guide aircraft to land safely at airports
- Global Positioning System (GPS), used to find one's exact coordinates on the globe

5.3.4 Remote Sensing

In this application, many satellites are used to monitor the globe constantly for weather conditions, meteorology, ozone, soil moisture, agriculture, crop protection from frost, forests, snow thickness, icebergs, and other factors such as monitoring and exploration of natural resources.

5.3.5 Domestic and Industrial Applications

This application includes microwave ovens, microwave clothes dryers, fluid heating systems, moisture sensors, tank gauges, automatic door openers, automatic toll collection, highway traffic monitoring and control, chip defect detection, flow meters, power transmission in space, food preservation, pest control, and so on.

5.3.6 Medical Applications

This application includes cautery, selective heating, heart stimulation, hemorrhage control, sterilization, and imaging.

5.3.7 Surveillance

This application includes security systems, intruder detection, and Electronic Warfare (EW) receivers to monitor signal traffic.

5.3.8 Astronomy and Space Exploration

In this application, gigantic dish antennas are used to monitor, collect, and record incoming microwave signals from outer space, providing vital information about other planets, stars, meteors, and other objects and phenomena in this or other galaxies.

5.3.9 Wireless Applications

Short-distance communication inside as well as between buildings in a local area network (LAN) arrangement can be accomplished using RF and microwaves. Connecting buildings via cables (e.g., coax or fiber optic) creates serious problems in congested metropolitan areas because the cable has to be run underground from the upper floors of one building to the upper floors of the other. This problem, however, can be greatly alleviated using RF and microwave transmitter/receiver systems that are mounted on rooftops or in office windows (see Figure 5.1). Inside buildings, RF and microwaves can be used effectively to create a wireless LAN in order to connect telephones, computers, and various LANs to each other. Using wireless LANs has a major advantage in office rearrangement where phones, computers, and partitions are easily moved with no change in wiring in the wall outlets. This creates enormous flexibility and cost savings for any business entity.

A summary of RF and microwave applications is shown in Table 5.1.

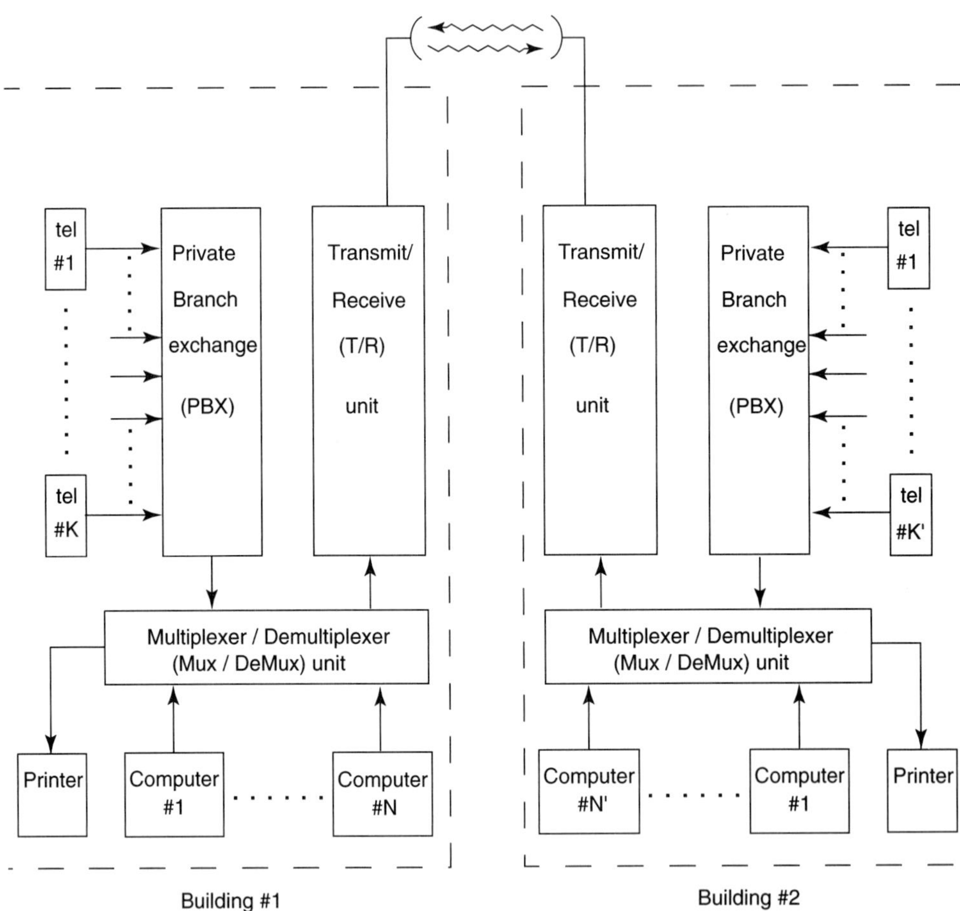

FIGURE 5.1 A typical local area network (LAN) for connectivity using microwaves.

TABLE 5.1 Summary of Applications of RF and Microwaves

Category of Application	Description
Astronomy and space exploration	Deep space probes
	Galactic explorations
Communication	Optical communications
	Telephone systems
	Computer networks
	Low-noise transmission media
	TV and radio broadcast
	Direct broadcast satellite
	High-definition TV
Domestic & industrial applications	Agriculture
	Moisture detection and soil treatment
	Pesticides
	Crop protection from freezing
	Automobiles
	Anti-theft radar or sensor
	Automotive telecommunication
	Blind spot radar
	Collision avoidance radar
	Near-obstacle detection
	Radar speed sensors
	Road-to-vehicle communication
	Vehicle-to-vehicle communication
	Highway
	Automatic toll collection
	Highway traffic control and monitoring
	Range and speed detection
	Structure inspection
	Vehicle detection
	Microwave Heating
	Home microwave ovens
	Microwave clothes dryer
	Industrial heating
	Microwave Imaging
	Hidden weapon detection
	Obstacle detection & Navigation
	Office
	Mail sorting
	Wireless phones and computers
	Power
	Beamed power propulsion
	Power transmission in space
	Preservation
	Food preservation
	Treated manuscript drying
	Production control
	Etching system production
	Industrial drying
	Moisture control

TABLE 5.1 Summary of Applications of RF and Microwaves *(Continued)*

Category of Application	Description
Medical applications	Cautery Heart stimulation Hemorrhaging control Hyperthermia Microwave imaging Sterilization Thermography

Radar	Air defense
	Navigation & position information
	  Airport traffic control
	  Global positioning system (GPS)
	  Microwave landing system (MLS)
	Police patrol (velocity measurement)
	Smart weapons
	Tracking
	Weather forecast

Remote sensing	Earth monitoring Meteorology Pollution control Natural resources and exploration

Surveillance	Security system
	  Intruder detection
	  Security system
	  Signal traffic monitoring

Wireless applications	Wireless local area networks (LANs)

5.4 RADIO FREQUENCY (RF) WAVES

Having briefly reviewed many of the current applications of RF/microwaves, we can see that this rapidly advancing field has great potential to be a fruitful source of many future applications.

As discussed earlier, electromagnetic (EM) waves are generated when electrical signals pass through a conductor. EM waves start to radiate from a conductor when the signal frequency is higher than the highest audio frequency, which is approximately 15 to 20 kHz. Because of this radiating property, signals of such or higher frequencies are often known as radio frequency (RF) signals.

5.4.1 RF Spectrum Bands

Because it is not practical either to design a circuit that covers the entire frequency range or to use all radio frequencies for all purposes, the RF spectrum is broken into various bands. Each band is used for a specific purpose and usually RF circuits are designed to be used in one particular band. Table 5.2 shows the most common assignment of RF commercial bands.

TABLE 5.2 Commercial Radio Frequency Band

Name of Band	Abbreviation	Frequency Range
Very low frequency	VLF	3–30 kHz
Low frequency	LF	30–300 kHz
Medium frequency	MF	300 kHz-3 MHz
High frequency	HF	3–30 MHz
Very high frequency	VHF	30–300 MHz
Ultra-high frequency	UHF	0.3–3 GHz
Super-high frequency	SHF	3–30 GHz
Extra-high frequency	EHF	30–300 GHz

Definition of Microwaves. When the frequency of operation starts to increase toward approximately 1GHz and above, a whole set of new phenomena occurs that is not present at lower frequencies. The radio waves at frequencies ranging from 1 GHz to 300 GHz are generally known as *microwaves*. Signals at these frequencies have wavelengths that range from 30 cm (at 1 GHz) to 1 millimeter (at 300 GHz). The special frequency range from 30 GHz to 300 GHz has a wavelength in the millimeter range; thus, it is generally referred to as millimeter-waves.

NOTE: *In some texts, the range 300 MHz to 300 GHz is considered the microwave frequency range. This is in contrast with the microwave frequency range defined previously, where the frequency range from 300 MHz to 1 GHz is referred as the RF range.*

Microwave Bands. The microwave frequency range consisting of the three main commercial frequency bands (UHF, SHF, and EHF) can further be subdivided into several specific frequency ranges, each with its own band designation. This band subdivision and designation facilitate the use of microwave signals for specific purposes and applications.

In electronics industries and academic institutions, the most commonly used microwave bands are set forth by the Institute of Electrical and Electronics Engineers (IEEE); they are shown in Table 5.3. In this table the "Ka to G" are the millimeter-wave (mmw) bands.

TABLE 5.3 IEEE and Commercial Microwave Band Designations

Band Designation	Frequency Range (GHz)
L Band	1.0-2.0
S band	2.0-4.0
C band	4.0-8.0
X band	8.0-12.0
Ku band	12.0-18.0

TABLE 5.3 IEEE and Commercial Microwave Band Designations *(Continued)*

Band Designation	Frequency Range (GHz)
K band	18.0-26.5
Ka band (mmw)	26.5-40.0
Q band (mmw)	33.0-50.0
U band (mmw)	40.0-60.0
V band (mmw)	50.0-75.0
E band (mmw)	60.0-90.0
W band (mmw)	75.0-110.0
F band (mmw)	90.0-140.0
D band (mmw)	110.0-170.0
G band (mmw)	140.0-220.0

5.5 RF AND MICROWAVE (MW) CIRCUIT DESIGN

Because of the behavior of waves at different frequencies, basic considerations in circuit design have evolved greatly over the last few decades and generally can be subdivided into two main categories:

- RF circuit design considerations
- Microwave (MW) circuit design considerations

Each category is briefly described next.

5.5.1 Low RF Circuit Design Considerations

Low RF circuits have to go through a three-step design process. In this design process the effect of wave propagation on the circuit operation is negligible and the following facts in connection with the design process can be stated:

1. The length of the circuit (ℓ) is generally much smaller than the wavelength, (i.e., $\ell << \lambda$)
2. Propagation delay time(t_d) is approximately zero (i.e., $t_d \approx 0$).
3. Maxwell's equations simplify into all of the low-frequency laws such as KVL, KCL, and Ohm's law. Therefore, at RF frequencies, the delay time of propagation (t_d) is approximately zero when $\ell << \lambda$ and all elements in the circuit can be considered to be lumped.

   The design process has the following three steps:

   STEP 1. The design process starts with selecting a suitable device and performing a DC design to obtain a proper Q-point.

   STEP 2. Next, the device will be characterized (through either measurement or calculations) to obtain its AC small signal parameters based on the specific DC operating point selected earlier.

   STEP 3. The third step consists of designing two matching circuits that transition this device to the outside world: the signal source at one end and the load at the other. Various design considerations and criteria such as stability, gain, and noise, are included at this stage and must be incorporated in the design of the final matching networks.

The design process for RF circuits is summarized and shown in Figure 5.2.

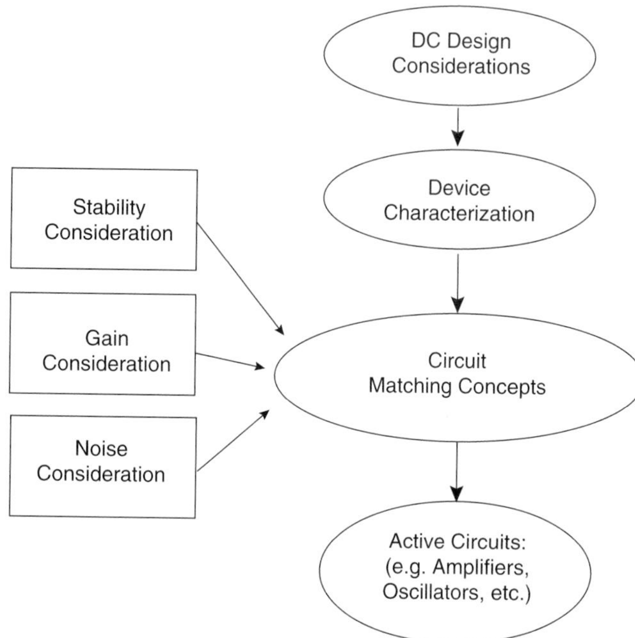

FIGURE 5.2 RF circuit design steps.

5.5.2 High RF and Microwave Circuits

To understand high RF and microwave circuits we should know that microwave circuits may have one or more lumped elements but should contain at least one distributed element. This last needs to be defined at this point:

> **DEFINITION-DISTRIBUTED ELEMENT:** *An element whose property is spread out over an electrically significant length or area of a circuit instead of being concentrated at one location or within a specific component.*

EXAMPLE 5.1

Describe what a distributed inductor is.

Answer:

A distributed inductor is an element whose inductance is spread out along the entire length of a conductor (such as self-inductance), as distinguished from an inductor whose inductance is concentrated within a coil.

EXAMPLE 5.2

Describe what a distributed capacitor is.

Answer:
A distributed capacitor is an element whose capacitance is spread out over a length of wire and not concentrated within a capacitor, such as the capacitance between the turns of a coil or between adjacent conductors of a circuit.

Working with distributed circuits, we need to know the following facts about them:

a. The wave propagation concepts as set forth by the Maxwell Equations fully apply.
b. The circuit has a significant electrical length, i.e., its physical length is comparable to the wavelength of the signals propagating in the circuit.

This fact brings the next point into view:

c. The time delay (t_d) due to signal propagation can no longer be neglected (i.e., $t_d \neq 0$).

To illustrate this point we will consider the following example.

EXAMPLE 5.3

How does a two-conductor transmission line (such as a coaxial line, etc.) behave at low and high frequencies?

Answer:
At low frequencies this transmission line is considered to be a short piece of wire with a negligibly small distributed resistance that can be considered to be lumped for the purpose of analysis (since $t_d \approx 0$).

At higher frequencies, however, the resistive, capacitive, and inductive properties can no longer be separated, and each infinitesimal length (Δx) of this transmission line exhibits these properties, as shown in Figure 5.3.

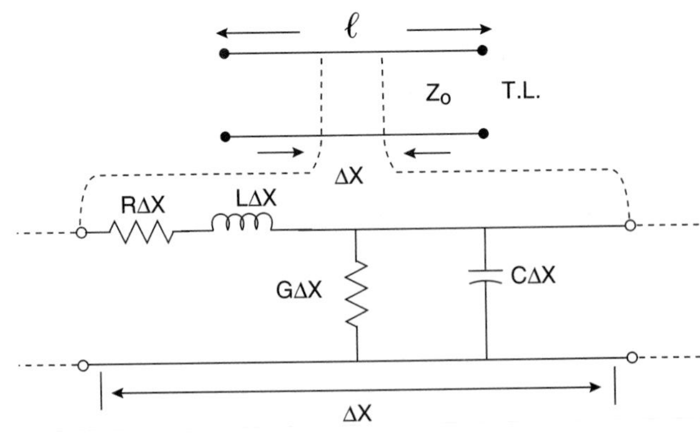

FIGURE 5.3 An infinitesimal portion of a transmission line (TL).

From this figure, we can see that the elements are: series elements (R, L) and shunt elements (G, C). These are defined as follows:

R = resistance per unit length in Ω/m
L = inductance per unit length in H/m
G = conductance per unit length in S/m
C = capacitance per unit length in F/m

This equivalent circuit is referred to as a distributed circuit model of a two-conductor transmission line and will be used in the next example to derive the governing differential equations for propagating voltage and current waves along a transmission line.

EXAMPLE 5.4

Using KVL and KCL derive the relationship between voltage and current in a transmission line at:

a. *Low frequencies*
b. *High frequencies (i.e., RF/microwave frequencies)*

Solution:

a. At low frequencies a transmission line (which can be lossy in general) can be represented as shown in Figure 5.4. In this Figure, "R" is the distributed loss resistance of the line, which can be modeled as a lumped element. The voltage and current relationship can be written as:

$$V_1 = V_2 + IR$$

NOTE: *If the line is lossless, then we have:*

$$V_1 = V_2$$

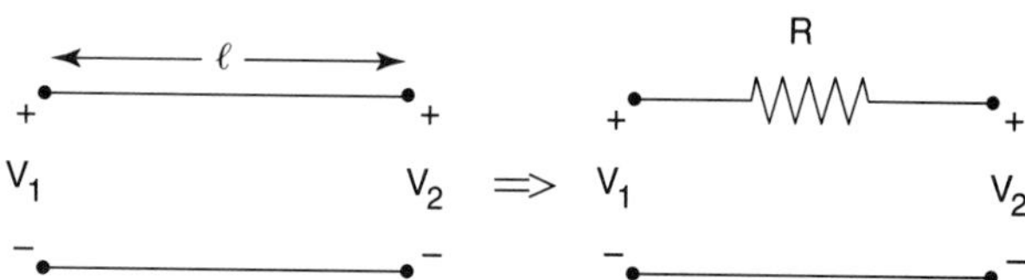

FIGURE 5.4 Equivalent circuit of a TL at low frequencies.

b. At high frequencies, based on Figure 5.3 a transmission line can be modeled as shown in Figure 5.5.

To develop the governing differential equations, we will examine one Δx section of a transmission line, as shown in Figure 5.6. Using KVL for the Δx section, we can write:

$$v(x,t) = i(x,t)\,R\Delta x + L\Delta x\,\partial i(x,t)/\partial t + v(x+\Delta x,t)$$

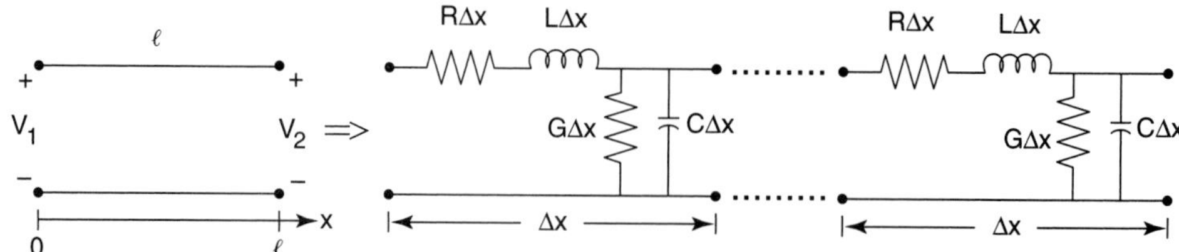

FIGURE 5.5 Equivalent circuit of a TL at high frequencies.

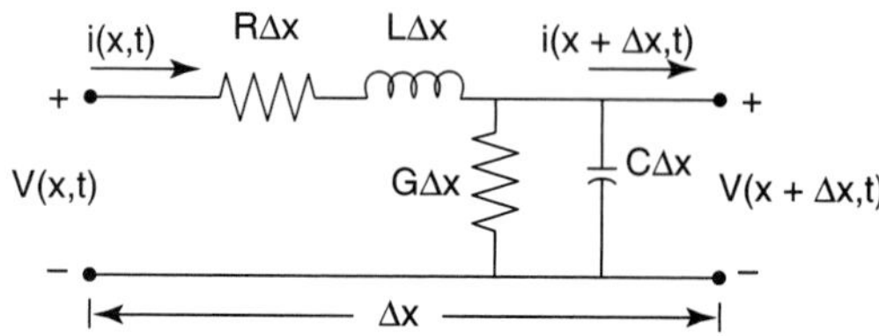

FIGURE 5.6 Voltage and current in an infinitesimal length of TL.

Upon rearranging terms and dividing both sides by Δx, we obtain:

$$-\frac{v(x+\Delta x, t) - v(x, t)}{\Delta x} = Ri(x, t) + L\frac{\partial i(x, t)}{\partial t}$$

Letting $\Delta x \to 0$, yields:

$$-\frac{\partial v(x, t)}{\partial x} = Ri(x, t) + L\frac{\partial i(x, t)}{\partial t} \tag{5.1}$$

Similarly, using KCL we can write:

$$i(x, t) = v(x+\Delta x, t)G\Delta x + C\Delta x\frac{\partial v(x+\Delta x, t)}{\partial t} + i(x+\Delta x, t)$$

Upon rearranging terms, dividing by Δx and letting $\Delta x \to 0$, we have:

$$-\frac{\partial i(x, t)}{\partial x} = Gv(x, t) + C\frac{\partial v(x, t)}{\partial t} \tag{5.2}$$

Equations 5.1 and 5.2 are two cross-coupled equations in terms of v and i. These two equations can be separated by first differentiating both equations with respect to x and then properly substituting for the terms, which leads to:

$$\begin{aligned}-\frac{\partial^2 v(x, t)}{\partial x^2} &= R\frac{\partial i(x, t)}{\partial x} + L\frac{\partial^2 i(x, t)}{\partial x \partial t}\\ &= -R\left(Gv(x, t) + C\frac{\partial v(x, t)}{\partial t}\right) - L\left(G\frac{\partial v(x, t)}{\partial t} + C\frac{\partial^2 v(x, t)}{\partial t^2}\right)\end{aligned}$$

or

$$\frac{\partial^2 v(x, t)}{\partial x^2} = LC\frac{\partial^2 v(x, t)}{\partial t^2} + (RC + LG)\frac{\partial v(x, t)}{\partial t} + RGv(x, t) \tag{5.3}$$

Similarly for i we can write:

$$\frac{\partial^2 i(x,t)}{\partial x^2} = LC\frac{\partial^2 i(x,t)}{\partial t^2} + (RC + LG)\frac{\partial i(x,t)}{\partial t} + RGi(x,t) \tag{5.4}$$

For sinusoidal signal variation for v and i, we can write the corresponding phasors as follows:

$$v(x,t) = Re[V(x)e^{j\omega t}]$$

$$i(x,t) = Re[I(x)e^{j\omega t}]$$

Using phasor differentiation results from Chapter 3, Equations 5.3 and 5.4 can be written as:

$$\frac{d^2 V(x)}{dx^2} - \gamma^2 V(x) = 0 \tag{5.5a}$$

$$\frac{d^2 I(x)}{dx^2} - \gamma^2 I(x) = 0 \tag{5.5b}$$

where

$$\gamma = \alpha + j\beta = \sqrt{(R + j\omega L)(G + j\omega C)} \tag{5.5c}$$

γ is the **propagation constant**, with real part (α) and imaginary part (β), called the **attenuation constant** (Np/m) and **phase constant** (rad/m), respectively.
The solution to the second-order differential equations, as given by Equations 5.5a and 5.5b, can be observed to be of exponential type format ($e^{\pm\gamma x}$). Thus, we can write the general solutions for $V(x)$ as follows:

$$V(x) = V^+e^{-\gamma x} + V^-e^{\gamma x} \tag{5.6a}$$

where the complex constants V^+ and V^- are determined from the boundary conditions imposed by the source voltage and the load value.
Similarly, $I(x)$ can be obtained from $V(x)$ (see Equation 5.1) as:

$$I(x) = \left(\frac{-1}{R + j\omega L}\right)\frac{dV(x)}{dx} = \frac{V^+e^{-\gamma x} - V^-e^{\gamma x}}{Z_o} \tag{5.6b}$$

where

$$Z_o = \sqrt{\frac{R + j\omega L}{G + j\omega C}} \tag{5.7}$$

is the characteristic impedance of the transmission line.

Special Case: A Lossless Transmission Line

For this case, we have $R = G = 0$. This yields the following simplifications:

$$\gamma = j\omega\sqrt{LC} = j\beta$$

$$Z_o = \sqrt{L/C}$$

where $\beta = \omega\sqrt{LC}$ is the phase constant.

In this case Equation 5.5 can be written as:

$$\frac{d^2V(x)}{dx^2} + \beta^2 V(x) = 0 \tag{5.8a}$$

$$\frac{d^2I(x)}{dx^2} + \beta^2 I(x) = 0 \tag{5.8b}$$

Similar to Equations 5.6a and 5.6b, the solutions to Equation 5.8 are given by:

$$V(x) = V^+ e^{-j\beta x} + V^- e^{j\beta x} \tag{5.9a}$$

$$I(x) = \frac{V^+ e^{-\gamma\beta x} - V^- e^{\gamma\beta x}}{Z_o} \tag{5.9b}$$

NOTE 1: *Transmission line Equations 5.5 and 5.8 could all have been derived using Maxwell's equations directly from the field quantities E and H, as delineated in Appendix F, General Laws of Electricity and Magnetism, under items 14 and 19.*

It will be seen in Chapter 7, *Fundamental Concepts in Wave Propagation*, that the term $e^{-\gamma x}$ [or $e^{-j\beta x}$] represents a propagating wave in the "$+x$" direction while $e^{\gamma x}$ [or $e^{j\beta x}$] represents a propagating wave in the "$-x$" direction on a transmission line. The combination of the two comparable waves propagating in opposite directions to each other forms a standing wave on the transmission line (as discussed earlier in Chapter 2). These will be all explored further in Chapter 7.

NOTE 2: *Based on a given source voltage ($x = 0, V = V_1$) and a known load voltage ($x = \ell, V = V_2$), the constants V^+ and V^- can easily be found from the following two equations:*

$$x = 0, \quad V_1 = V^+ + V^- \tag{5.10}$$

$$x = \ell, \quad V_2 = V^+ e^{-j\beta\ell} + V^- e^{j\beta\ell} \tag{5.11}$$

EXERCISE 5.1

a. Derive expressions for V^+ and V^- from Equations 5.10 and 5.11 in terms of V_1 and V_2.

b. Given the load value as $Z = Z_L$, find V^+ and V^- in terms of V_1 and Z_L [as in part (a)].

HINT: Use $V_2 = Z_L\,[V^+ e^{-j\beta\ell} - V^- e^{j\beta\ell}]/Z_o$

5.5.3 High RF and Microwave Circuit Design Process

Circuit design process at high RF and microwaves is very similar to the low RF circuit design except for the wave propagation concepts that should be taken into account. The design process has the following three steps:

STEP 1. The design process starts with the design of the DC circuit to establish a stable operating point.

STEP 2. The next step is to characterize the device at the operating point (Q-point), using electrical waves to measure the percentage of reflection and transmission that the device presents at each port.

STEP 3. The third step consists of designing the matching networks that transition the device to the outside world such that the required specifications such as stability, overall gain, etc., are satisfied.

Except for the fact that our familiarity with wave propagation concepts becomes crucial, the microwave circuit design process is similar to the RF circuit design steps delineated in Figure 5.2.

5.6 THE UNCHANGING FUNDAMENTAL VERSUS THE EVER-EVOLVING STRUCTURE

Before we get into specific analysis and design of RF and microwave circuits, it is worthwhile first to examine a general communication system in which each circuit or component has a specific function in a bigger scheme of affairs. In general, any communication system is based on a very simple and yet extremely fundamental truth, commonly referred to as the "universal communication principle."

The "universal communication principle" is a fundamental concept that is at the heart of a wide sphere of existence called "life and livingness" or, for that matter, any of its subsets particularly the field of RF/microwaves. This principle is intertwined throughout the entire field of RF/microwaves and thus plays an important role in our understanding of this subject. Therefore it behooves us well to define it at this juncture:

THE UNIVERSAL COMMUNICATION PRINCIPLE: *This principle states that communication is the process whereby information is transferred from one point in space and time (X_1, Y_1, Z_1, t_1, called the source point), to another point in space and time (X_2, Y_2, Z_2, t_2, called the receipt point), with the intention of creating an* *exact replica of the source information at the receipt point. Usually, the receipt point at location (X_2, Y_2, Z_2) is separated by a distance (d) from the source point location (X_1, Y_1, Z_1).*

The physical embodiment of the universal communication principle is a "communication system," which takes the information from the source point and delivers an exact replica of it to the receipt point (see Figure 5.7). Thus in general, it can be seen that any communication system can be broken down into three essential elements:

Source point: A point of emanation or generation of information.
Receipt point: A point of receipt of information.
Distance (or imposed space): The space existing between the "source point" and "receipt point" through which the information travels.

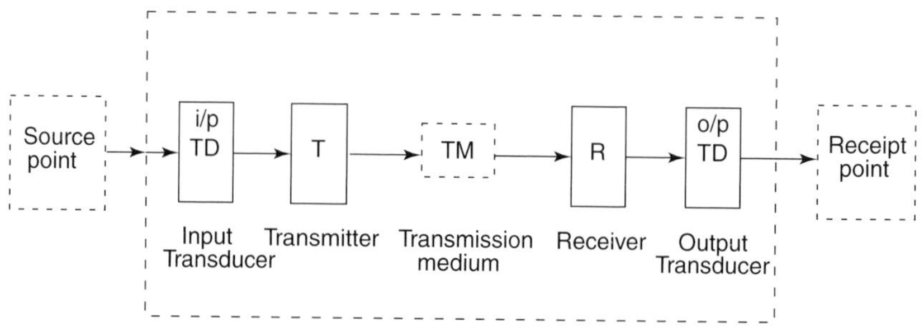

FIGURE 5.7 Depiction of a communication system.

Furthermore, it can be observed that in order to achieve effective communication between two systems, we need to have three more factors present: (a) There must be intention on the part of the source point and the receipt point to emit and to receive the information, respectively, (b) source and receipt points must have attention on each other (i.e., both being ready for transmission and reception), and (c) duplication (i.e., an exact replica) must occur at the receipt point of what emanated from the source point.

Use of the universal communication principle in practice creates a one-way communication system (such as radio and TV broadcast), and it forms one leg of a two-way communication system (such as CB radio or telephone), where this process is reversed to create the second leg of the communication action.

An important application of the universal communication principle is in a radar communication system where the source point (X_1, Y_1, Z_1) is at the same physical location as the receipt point (X_2, Y_2, Z_2), i.e., $X_1 = X_2$, $Y_1 = Y_2$, $Z_1 = Z_2$; however, the times of sending and reception are different ($t_1 \neq t_2$). Otherwise no communication would take place. This brings us to the obvious conclusion that we can not have a condition where the source and the receipt points are the same, simultaneously!

Based on this simple concept of communication, the most complex communication systems can be understood, analyzed, and designed. Figure 5.8 is a simple and yet very generalized block diagram of such a practical communication system in use today.

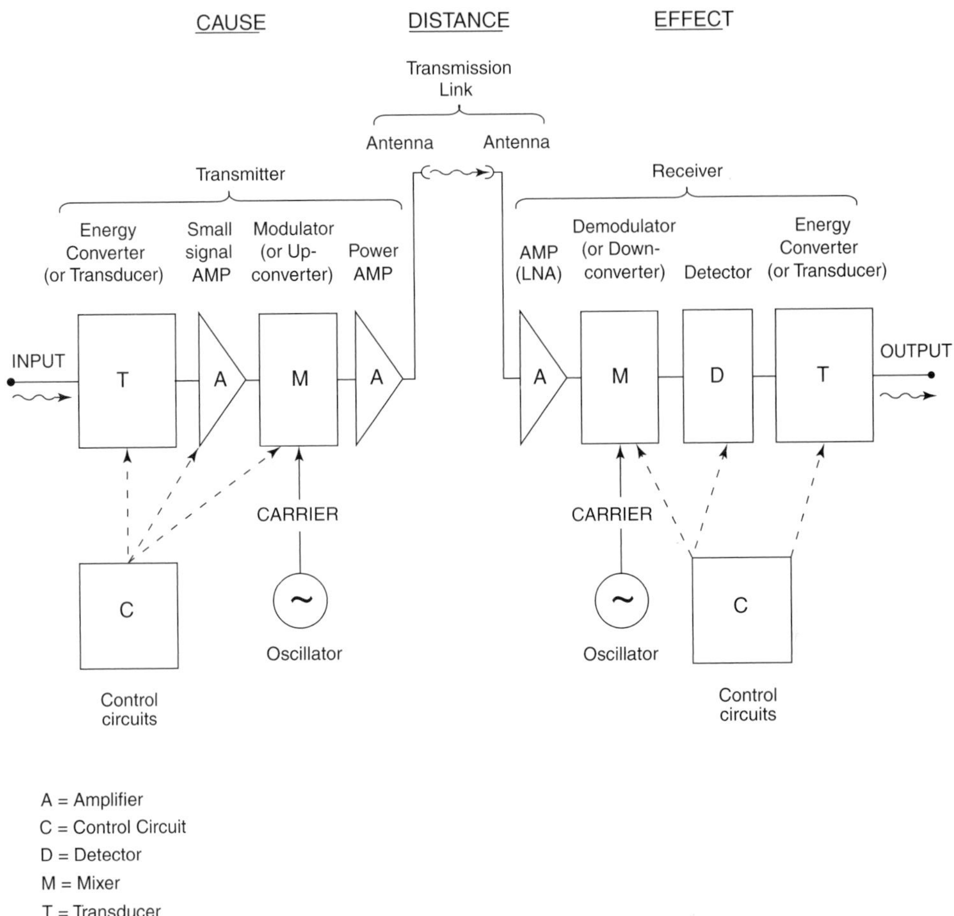

FIGURE 5.8 Block diagram of a general communication system.

It should be noted that the design and structure of this communication system can change and evolve into a more efficient system with time, whereas the universal communication principle will never change. Of course, this should be no surprise to workers in the field because the foundation (fundamental postulates, axioms, and natural laws) and basic concepts (theorems, analytical techniques, and theory of operation) of any science are far superior in importance to any designed circuitry, machinery, or network. This observation brings us to the following conclusion:

Fundamentals of any science are superior and dictate the designed forms, structures, or in general the entire application mass of that science, and not vice versa.

This is true in all aspects of design: While the underlying principle remains constant, the structure, which is the electronic circuit, constantly undergoes improvement with new designs and evolves in time toward a more efficient circuitry.

This can best be described as "engineering principle as a constant" versus "the application mass as a constantly evolving structure" that approaches closer and closer to a perfect embodiment of the underlying principle with each improvement.

Even though, rarely, new discoveries may bring new underlying fundamentals to the forefront, nevertheless the fundamentals, as a general rule, remain invariant.

For example, between 1864 and 1873 James Clerk Maxwell interrelated all of the known data about electricity and magnetism in classical laws of electromagnetics. Since that time, which is over a century, tremendous technological changes and advances have happened all over the globe, yet Maxwell's equations have not changed an iota. This set of celebrated equations has remained timeless!

Of course, it should be noted that quantum mechanics, dealing with subatomic particles, may be considered by some, to have generalized these equations and shown that energy is not continuous but quantized. Nevertheless, Maxwell's equations at the classical level of observation have not been surpassed and are still true today; they currently form the foundation of "electromagnetics" as a science—the backbone of electronics and electrical engineering.

Now to build a communication system in the physical universe that works and is practical, we must satisfy two conditions:

- First, it must be based on the fundamental concept of "the universal communication principle" and then Maxwell's equations—both in combination form a static that is unchanging!
- Second, it must follow and conform to the current state of technology in terms of manufacturing, materials, device fabrication, circuit size, and structure—a kinetic and constantly evolving! These two prerequisites, in essence, clearly demonstrate and confirm the interplay of *static versus kinetic*, which is interwoven throughout our entire world of science and technology.

The previous two steps of system design set up the blueprint for any general engineering system design. We must heed these points carefully before we go very far in the quest for workable knowledge.

5.7 GENERAL ACTIVE-CIRCUIT BLOCK DIAGRAMS

Considering Figure 5.8, we note several stages from left to right:

Energy conversion stage: This is a simple transducer causing the incoming energy (e.g., sound, etc.) to be converted to electrical energy. An example for this stage could be a microphone.

Amplification stage: This is a high-gain small-signal amplifier causing a higher signal to compensate for losses in the energy conversion stage.

Frequency conversion stage (also called modulation or up-conversion): This stage follows the amplification stage and causes a carrier wave to be modulated by the amplified signal. This is the stage that prepares the signal for transmission for long distance by increasing its frequency because higher-frequency signals travel further and require smaller antennas. A local oscillator is needed to produce the carrier wave before the modulation process can take place.

Power amplification stage: This is the stage where the signal power level is boosted greatly so that a higher range of reception is allowed.

Transmission link: This is the transmission media in which the modulated signal is transported from "cause or source point" to the "effect or receipt point."

Low-noise amplification stage: This is the first stage (or front-end) of the receiver wherein the modulated signal is amplified and prepared by a low-noise amplifier (LNA) in such a way that the effect of noise that could possibly be added to the signal by later stages, is minimized.

Frequency conversion stage (or demodulation or down-conversion): This stage demodulates the signal and brings the carrier frequency down to workable levels. Just as in the modulation stage, a local oscillator of a certain frequency is needed to make the demodulation process effective.

NOTE: *If the local oscillator is tunable, then the same receiver can be used to receive signals from other sources at other frequencies (a heterodyne receiver!).*

Detector stage: This stage removes the carrier wave and reconstructs the original signal.

Energy conversion stage: This stage converts the electrical signal back to its original form (e.g., sound). An example for this stage could be a speaker.

Control stage: This is where all the decisions with regard to circuit connection/ disconnection, routing, switching, and so on, take place. A control stage is present at both the source and the receipt points of the communication system.

In the remainder of this book, we will explore and focus on techniques for analysis and design of circuits used in all stages of an RF/microwave communication system. To gain a full conceptual understanding of different types of circuit designs we need to have an overall idea of "how different components fit together." To bring this point into a realm of practicality each specific type of microwave circuit has been cast into an exact block diagram that clearly depicts the relationship of the device with other circuit components and sections. The circuits considered for the purpose of the block diagram are defined as follows:

Amplifier: An electronic circuit capable of increasing the magnitude or power level of an electrical signal without distorting the wave-shape of the quantity. The block diagram for this circuit is shown in Figure 5.9.

Oscillator: An electronic circuit that converts energy from a DC source to a periodically varying electrical signal. The block diagram for this circuit is shown in Figure 5.10.

Mixer: An electronic circuit that generates an output frequency equal to the sum and difference of two input frequencies; (also called a frequency converter). The block diagram for this circuit is shown in Figure 5.11.

Detector: An electronic circuit concerned with demodulation; it extracts a signal that has modulated a carrier wave. The block diagram for this circuit is shown in Figure 5.12.

From these block diagrams we can see that the device forms the "heart" or "engine" of the circuit around which all other circuit components should be properly designed to control the input/output flow of signals and eventually obtain optimum performance. Furthermore, these four block diagrams show the irresistible fact that the knowledge gained in earlier chapters is essential in the design of these complicated circuits.

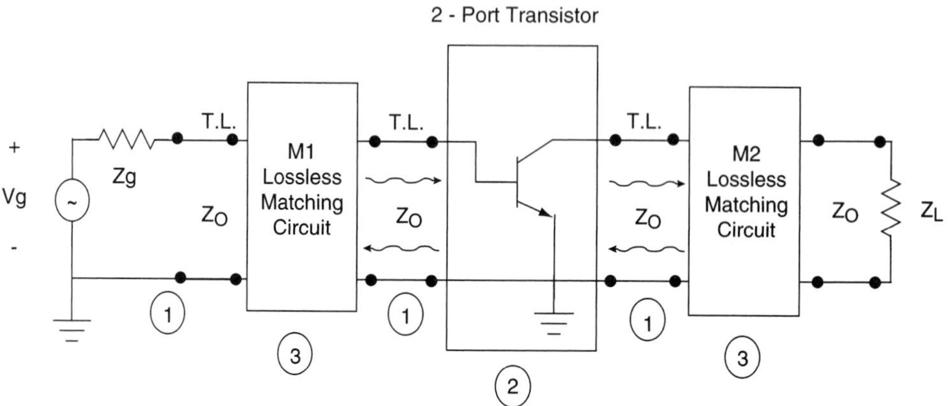

FIGURE 5.9 An amplifier circuit block diagram.

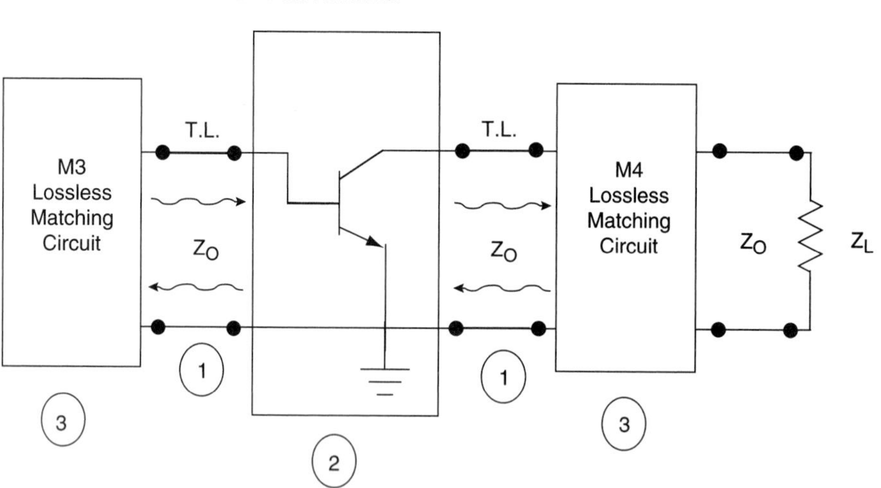

FIGURE 5.10 An oscillator circuit block diagram.

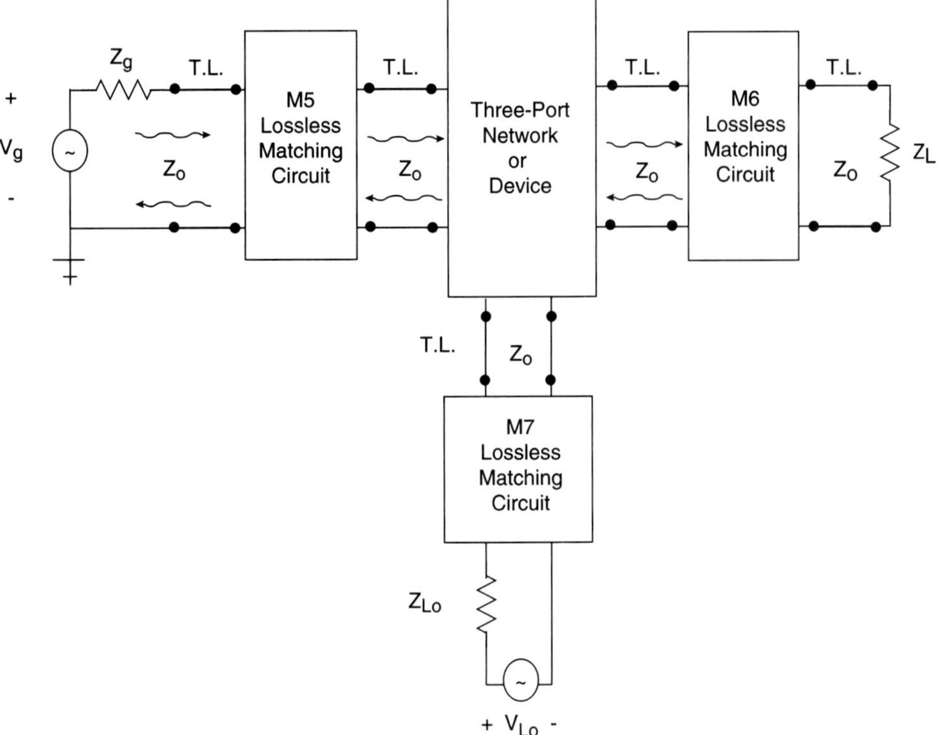

FIGURE 5.11 A mixer circuit block diagram.

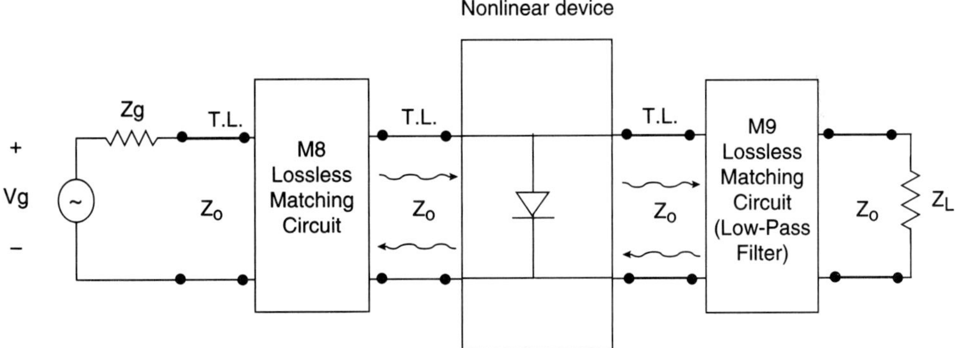

FIGURE 5.12 A detector circuit block diagram.

5.8 SUMMARY

To be proficient at higher frequency circuits (analysis or design), we need to master, on a gradient scale, all of the underlying principles and develop a depth of knowledge before we can be called a skilled microwave practitioner.

Figure 5.13 depicts the gradient scale of concepts that need to be fully understood to achieve a mastery of circuit design skills at higher frequencies. As shown in this figure, we start with the fundamental axioms of sciences, fundamental concepts in electronics, and we progress toward high-frequency electronic circuit design by learning the DC and low-frequency concepts first, then wave propagation concepts, device-circuit characterization, matching concepts, and we eventually arrive at the final destination of RF/MW active circuit design concepts, originally set forth as the goal of this book. Knowing this progressive series of concepts will enable us to design amplifiers, oscillators, mixers, detectors, control circuits, and integrated circuits with relative ease and proficiency at RF/MW frequencies.

LIST OF SYMBOLS/ABBREVIATIONS

A symbol will not be repeated again once it has been identified and defined in an earlier chapter, as long as its definition remains unchanged.

- ℓ Length of a component or circuit
- t_d Time delay
- λ Wavelength

PROBLEMS

5.1 What is the difference between a lumped element and a distributed element?

5.2 How many steps are required to design an RF circuit? A microwave circuit? Describe the steps.

5.3 What are the similarities and difference(s) between an RF and a microwave circuit design procedure?

5.4 Describe: (a) What is meant by "fundamentals versus application mass"? (b) What is meant by timelessness of a fundamental truth? Give an example. (c) What part of a system constantly evolves? (d) What are the prerequisites for any general system design?

5.5 What is at the heart of an amplifier, an oscillator, a mixer, or a detector block diagrams?

5.6 What are the main concepts we need to master to design an RF or a microwave circuit?

5.7 Why is it necessary to understand low-frequency electronics fully before trying to master RF/microwave electronics?

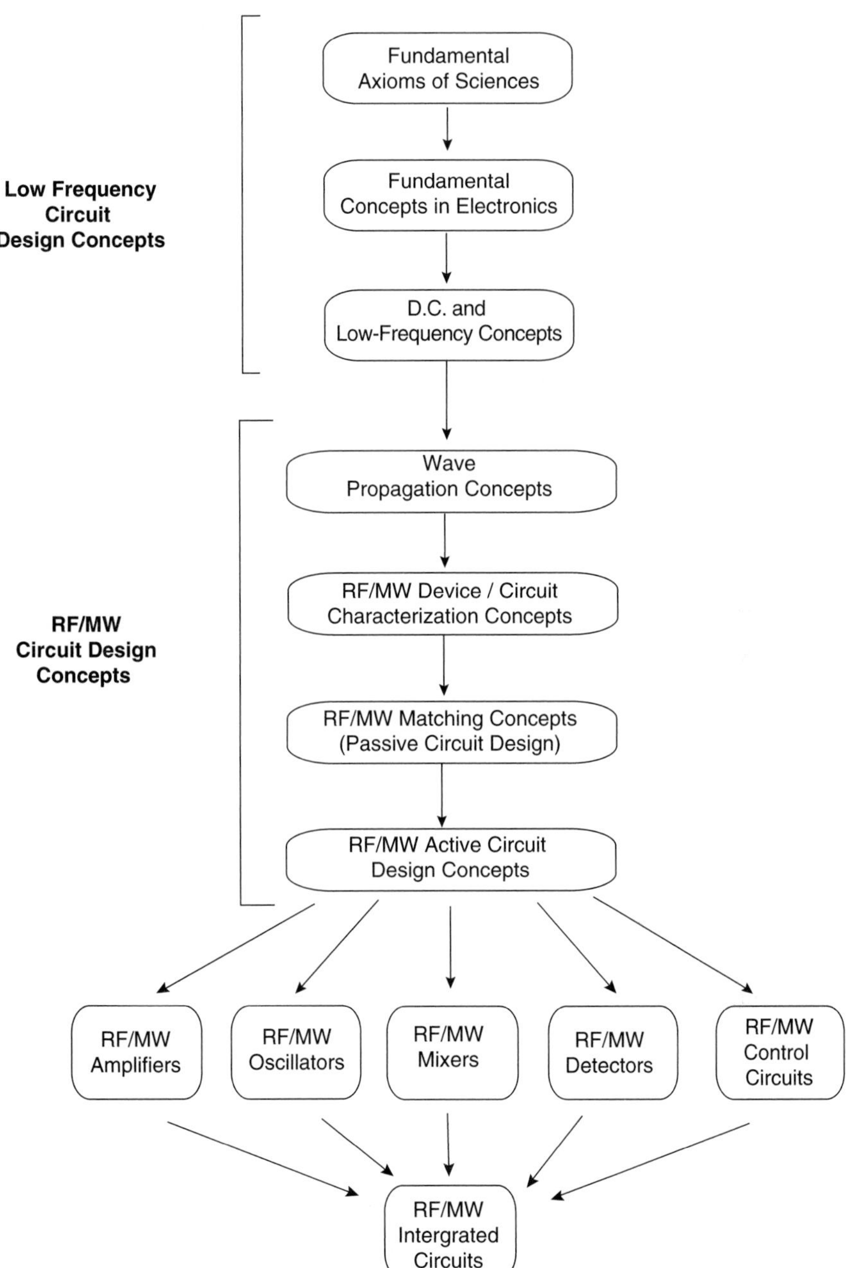

FIGURE 5.13 The gradient scale of concepts in RF/MW circuit design.

REFERENCES

[5.1] Carlson, A. B. *Communication Systems: An Introduction to Signals and Noise in Electrical Communication*. New York: McGraw-Hill, 1968.

[5.2] Cheung, W. S. and F. H. Levien. *Microwave Made Simple.* Dedham: Artech House, 1985.

[5.3] Gardiol, F. E. *Introduction to Microwaves.* Dedham: Artech House, 1984.

[5.4] Ishii, T. K. *Microwave Engineering.* 2nd ed., Orlando: Harcourt Brace Jovanovich, publishers, 1989.

[5.5] Laverahetta, T. *Practical Microwaves.* Indianapolis: Howard Sams, 1984.

[5.6] Lance, A. L. *Introduction to Microwave Theory and Measurements.* New York: McGraw-Hill, 1964.

[5.7] Radmanesh, M. M. *Applications and Advantages of Fiber Optics as Compared with other Communication Systems*, Hughes Aircraft Co., Microwave Products Div., pp. 1-11, April 1988.

[5.8] Radmanesh, M. M. *Radiated and Conducted Susceptibility Induced Current in Bundles: Theory and Experiment*, Boeing Co., HERF Div., pp. 1–115, Sept. 1990.

[5.9] Saad, T. *Microwave Engineer's Handbook*, Vols I, II. Dedham: Artech House, 1988.

[5.10] Scott, A. W. *Understanding Microwaves.* New York: John Wiley & Sons, 1993.

[5.11] Vendelin, G. D., A. M. Pavio, and Ulrich L. Rhode. *Microwave Circuit Design.* New York: John Wiley & Sons, 1990.

C H A P T E R 6

RF Electronics Concepts

6.1 INTRODUCTION

It is important to set the stage properly for the introduction of microwave circuits. To that end we will introduce RF circuit analysis and design to serve as a platform of fundamental information in order to catapult us into the world of microwave circuit design. Therefore this chapter primarily deals with the world of RF circuit design, with the intention of preparing the reader for a much broader field of study, namely, microwave circuit analysis and design, presented in the future chapters.

6.2 RF/MICROWAVES VERSUS DC OR LOW AC SIGNALS

There are several major differences between signals at higher radio frequency (RF) or microwaves (MW) and their counterparts at low AC frequency or DC. These differences, which greatly influence electronic circuits and their operation, become increasingly important as the frequency is raised. The following four effects provide a brief summary of the effects of RF/MW signals in a circuit that are not present at DC or low AC signals:

Effect #1. Presence of stray capacitance. This is the capacitance that exists:

- Between conductors of the circuit
- Between conductors or components and ground
- Between components

This effect is shown in Figure 6.1.

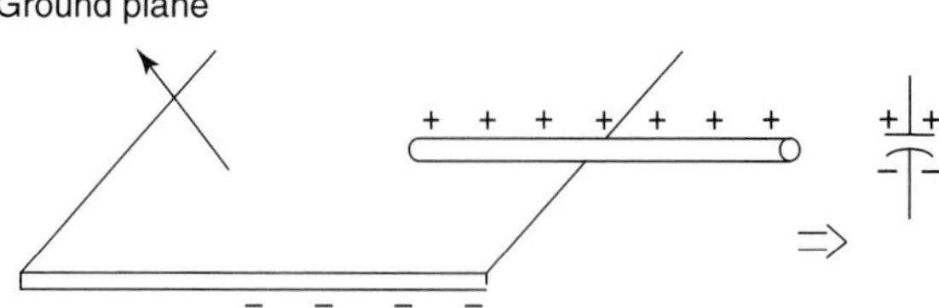

FIGURE 6.1 The stray capacitance effect in a circuit.

Effect #2. Presence of stray inductance. This is the inductance that exists due to:

- The inductance of the conductors that connect components
- The parasitic inductance of the components themselves

These stray parameters are not usually important at DC and low AC frequencies, but as frequency increases, they become a much larger portion of the total. This concept is shown in Figure 6.2.

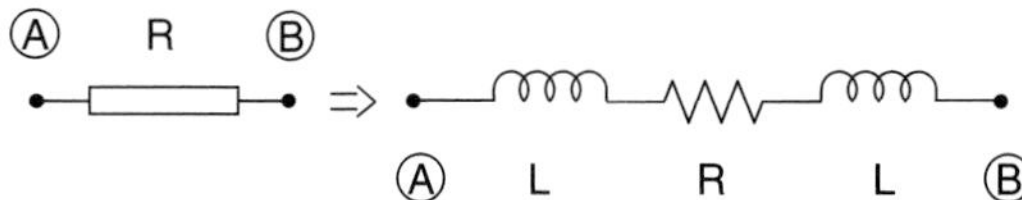

FIGURE 6.2 The stray inductance effect in a circuit.

Effect #3. Skin effect. This is due to the fact that AC signals penetrate a metal partially and flow in a narrow band near the outside surface of each conductor. This effect is in contrast to DC signals that flow through the whole cross-section of the conductor, as shown in Figures 6.3a and 6.3b.

For AC signals, the current density falls off exponentially from the surface of the conductor toward the center. At a critical depth (δ), called the skin depth or depth of penetration, signal amplitude is $1/e$ or 36.8 percent of its surface amplitude; see Figure 6.3c. The skin depth is given by:

$$\delta = \sqrt{\frac{1}{\pi f \mu \sigma}}$$

where μ is the permeability (H/m) and σ is the conductivity of the conductor.

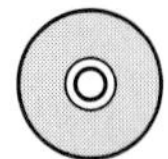

EXAMPLE 6.1

Considering copper as the conductive medium, what is the skin depth at 60 Hz and 1 MHz?

Solution:

For copper we have:

$\mu = 4\pi \times 10^{-7}$ H/m

$\sigma = 5.8 \times 10^{7}$ S/m

At $f = 60$ Hz $\Rightarrow \delta = (1/\pi \times 60 \times 4\pi \times 10^{-7} \times 5.8 \times 10^{7})^{1/2} = 0.85$ cm

While on the other hand for $f = 1$ MHz, we calculate δ to be :

$\delta = 0.007$ cm

which is a substantial reduction in penetration depth.

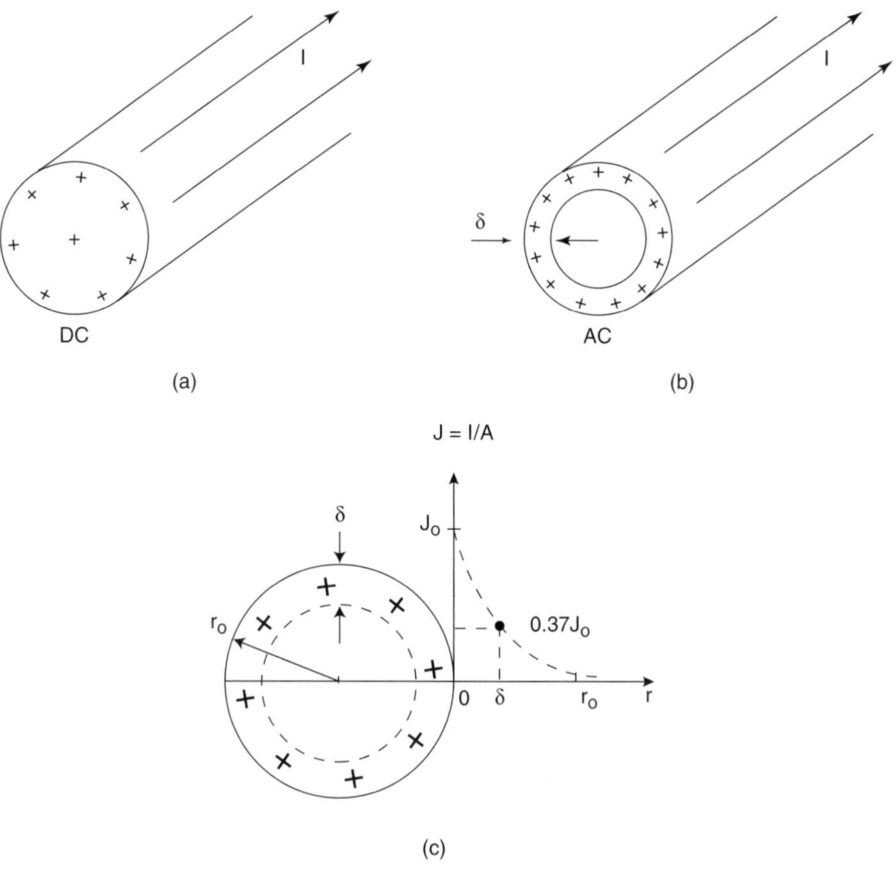

FIGURE 6.3 Skin effect and current flow for (a) DC signal, (b) AC signal, (c) skin effect.

As seen from Example 6.1, we can observe that as frequency increases, skin effect produces a smaller zone of conduction and a correspondingly higher value of AC resistance compared with DC resistance.

Effect #4. Radiation. This is caused by the leakage or escape of signals into the air. This, in essence, means that the signals bypass the conducting medium, and not all of the source energy reaches the load. Radiation can occur outside or within a circuit, as shown in Figure 6.4.

The radiation factor causes coupling effects to occur as follows:

- Coupling between elements of the circuit
- Coupling between the circuit and its environment
- Coupling from the environment to the circuit

"Electromagnetic interference" (EMI), also called "radio frequency interference" (RFI) or "RF-noise," is due to radiation of signals at RF/MW frequencies and is considered to be negligible in most low-frequency AC circuits and absent in DC circuits.

FIGURE 6.4 Radiation of a circuit (a) outside or (b) inside.

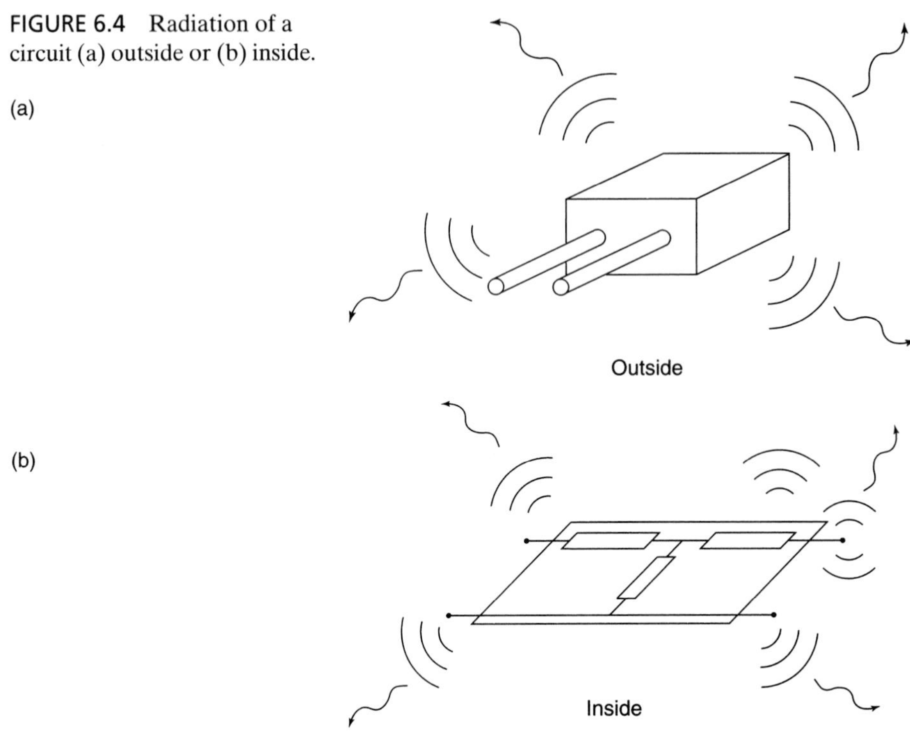

6.3 EM SPECTRUM

When an RF/MW signal radiates, it becomes an EM wave that is propagating through a medium such as air. The range of frequencies of electromagnetic waves known as the EM spectrum is shown in Figure 6.5.

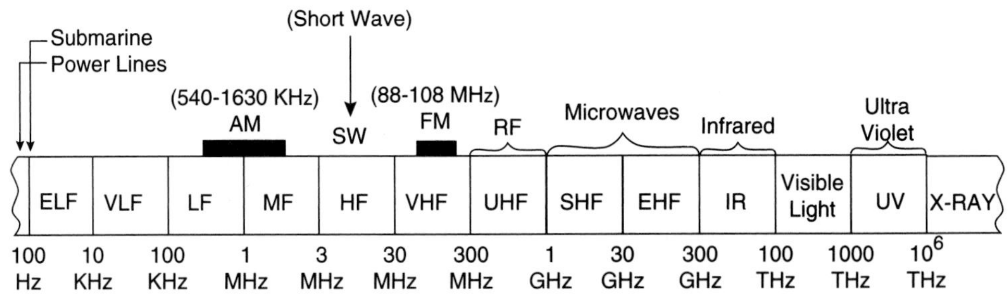

FIGURE 6.5 The EM spectrum.

Looking at this spectrum we may wonder, "How do microwaves differ from other EM waves?" The answer lies in the fact that microwaves is treated as a separate topic because at these frequencies the wavelength (λ) approximates the physical size of electronic components, as discussed in Chapter 5, *Introduction to RF and MC and Applications*. Therefore, components behave differently at microwave frequencies than they do at lower frequencies.

EXAMPLE 6.2

How does an ordinary resistor element behave at microwave frequencies?

Solution:

An ordinary carbon resistor at microwave frequencies (e.g., at $f = 10$ GHz) has a stray capacitor and a stray inductor as well as a higher resistance due to the skin effect (because the cross-section is reduced) and radiation (because part of the power is lost in the air). These factors are added into the equivalent circuit (as shown in Figure 6.6). Addition of the extra parasitic elements in the equivalent circuit is necessary because the combined length of the leads and the the component size itself is comparable to the wavelength.

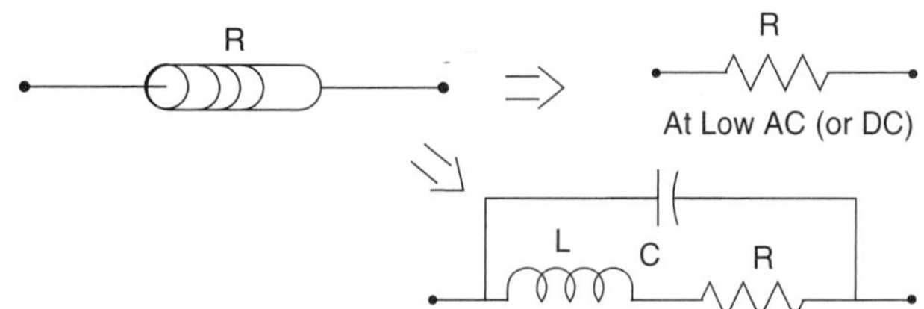

FIGURE 6.6 The equivalent circuit of an ordinary resistor at low AC and at microwave frequency.

6.4 WAVELENGTH AND FREQUENCY

When an electromagnetic wave with a certain oscillation frequency (f) propagates through the air or any other medium, it does so at a certain fixed speed or velocity [also known as the phase velocity (V_p)] with a corresponding fixed wavelength (λ), as shown in Figure 6.7. These three factors, f, V_p ,and λ , are not independent from each other and, in fact, are interrelated such that the product of frequency (f) and wavelength (λ) is equal to the velocity (V_p), i.e.,

$$\lambda f = V_p \tag{6.1}$$

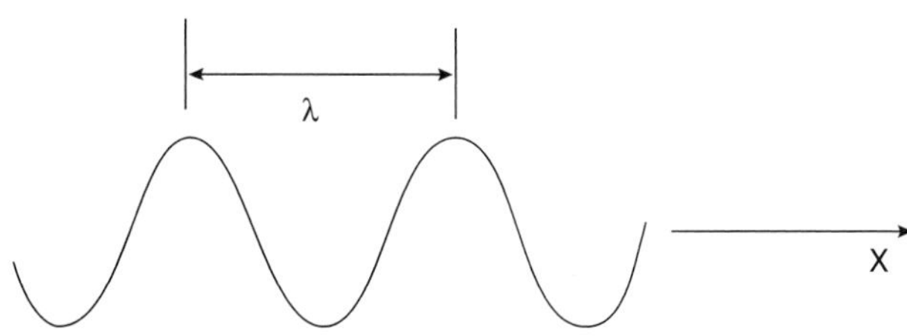

FIGURE 6.7 Wavelength of a wave.

It has been shown that the velocity of propagation for any and all EM waves through the air is approximately equal to the speed of light (c):

$$c = 3 \times 10^8 \text{ m/s}$$

If the medium is not air the speed is lower than "c" and can be shown to be:

$$V_p = c/\sqrt{\varepsilon_r} \tag{6.2}$$

where ε_r is the relative dielectric constant of the medium of propagation.

6.5 INTRODUCTION TO COMPONENT BASICS

In this section, the properties of resistors, capacitors, and inductors at high radio frequencies will be studied. But first, we take a brief look at the simplest component of all, a piece of wire. We consider this element first and examine its problems at radio frequencies.

6.5.1 Wire

A wire is the simplest element to study having a zero resistance, which makes it appear as a short circuit at DC and low AC frequencies. Yet at RF/MW frequencies it becomes a very complex element and deserves special attention, which will be given shortly. Wire in a circuit can take on many forms, such as the following:

- Wire wound resistors
- Wire wound inductors
- Leaded capacitors (see Figure 6.8)
- Element-to-element interconnect applications

(a) (b)

FIGURE 6.8 A leaded capacitor: (a) axial, (b) radial.

The behavior of a wire in the RF spectrum depends to a large extent on the wire's diameter and length. A system for different wire sizes is the American Wire Gauge (AWG) system. In this system, the diameter of a wire will roughly double for every six gauges.

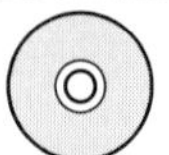

EXAMPLE 6.3

Given that the diameter of AWG 50 is 1.0 mil (0.001 inch), what is the diameter of AWG 14?

Solution:

Starting from AWG 50 we descend downward by 6 gauges until we reach AWG 14 as follows:

AWG 50 $\Rightarrow$ d = 1 mil,
AWG 44 $\Rightarrow$ d = 2 mils,
AWG 38 $\Rightarrow$ d = 4 mils,
AWG 32 $\Rightarrow$ d = 8 mils,
AWG 26 $\Rightarrow$ d = 16 mils,
AWG 20 $\Rightarrow$ d = 32 mils,
AWG 14 $\Rightarrow$ d = 64 mils.

Problems associated with a piece of wire

Problems associated with a wire can be traced to two major areas: skin effect and straight-wire inductance.

These two problems are discussed next.

Skin effect in a wire

As frequency increases, the electrical signals propagate less and less in the inside of the conductor. The current density increases near the outside perimeter of the wire and causes a higher impedance to be seen by the signal, as shown in Figure 6.9. This is because resistance of the wire is given by:

$$R = \frac{\rho \ell}{A},$$

and if the effective cross-sectional area (A) decreases, this leads to an increase in resistance (R).

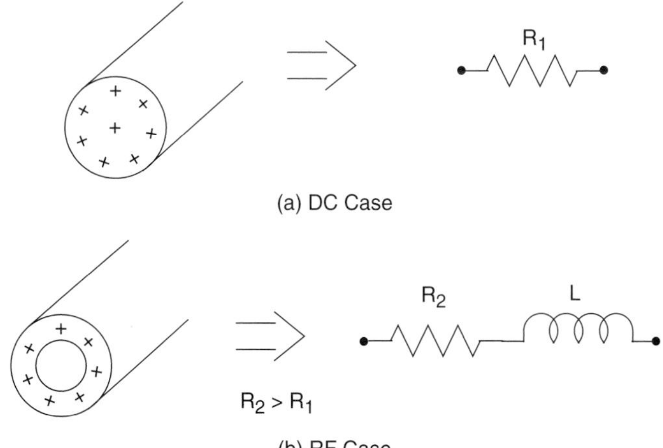

FIGURE 6.9 The skin effect in a wire: (a) DC, (b) RF.

Straight-wire inductance

In the medium surrounding any current carrying conductor, there exists a magnetic field. If the current (I) is AC, this magnetic field is alternately expanding and contracting (and even reversing direction if there is no DC bias present). This produces an induced voltage (as specified by Faraday's law) in the wire that opposes any change in the current flow. This opposition to change is called "self-inductance," as shown in Figure 6.10.

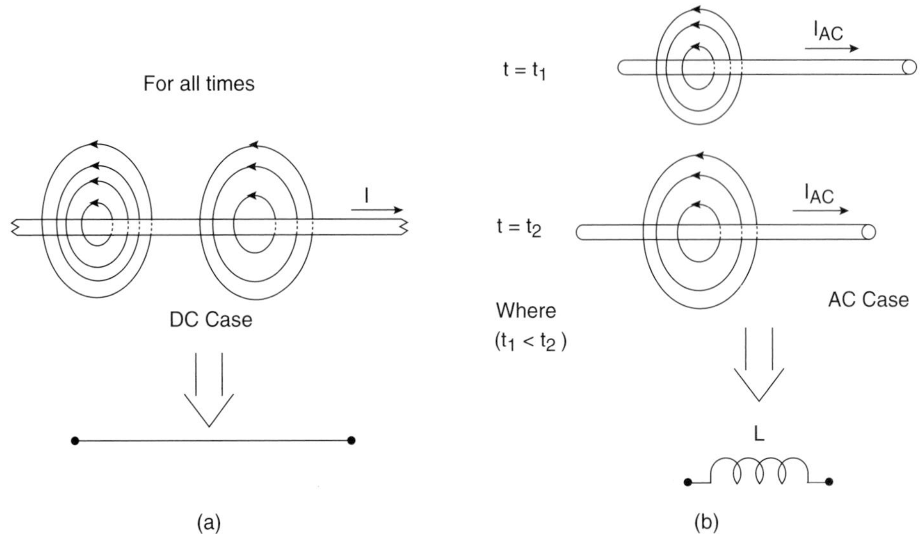

FIGURE 6.10 Interactive properties of a wire: (a) DC case, self-inductance not present; (b) AC case, self-inductance present.

The concept of inductance is important because at RF/MW, any and all conductors including hookup wires, capacitor leads, bonding wires, and all interconnects tend to become inductors and exhibit the property of inductance as shown in Figure 6.11.

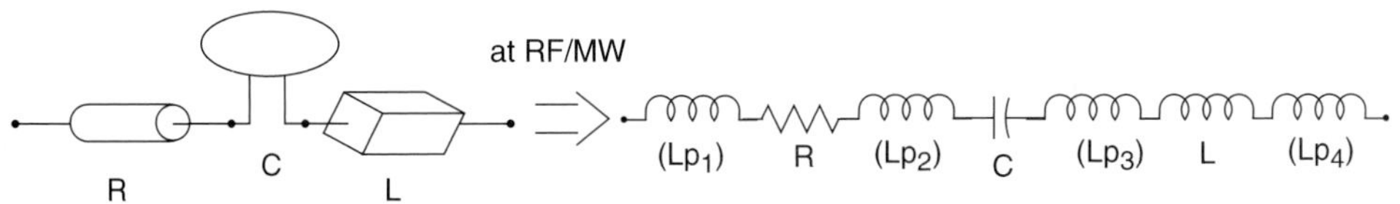

FIGURE 6.11 A simple RLC circuit at RF/MW frequency.

6.5.2 Resistors

DEFINITION-RESISTOR: *An element specializing in the resistance property of a material. The resistance of a material is a property whose value determines the rate at which electrical energy is converted into thermal energy when an electric current passes through it.*

Resistors are used in almost all circuits for different purposes, such as:

- In transistor bias networks, to establish an operating point
- In attenuators (also called pads), to control the flow of power
- In signal combiners, to produce a higher output power
- In transmission lines, to create matched conditions, etc.

Once we depart from the world of DC, resistors start to behave differently:

At DC: $V = RI$ (Ohm's law),
At low AC: $V \approx RI$,
At high RF/MW: $V \neq RI$.

At RF/MW frequencies, a resistor (R) appears as a combination of several elements, as shown in Figure 6.12.

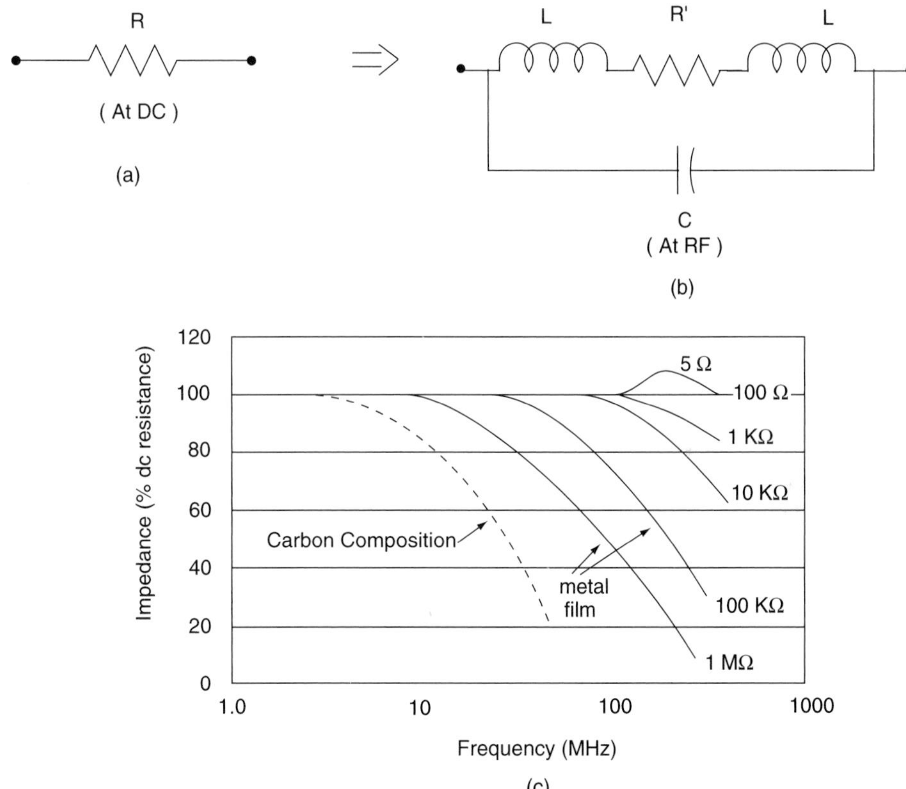

FIGURE 6.12 A simple resistor at (a) DC, (b) high RF/MW frequency, (c) frequency dependence of resistors. (adapted from Handbook of Components for Electronics, McGraw-Hill)

Figure 6.12a shows a simple resistor at DC. As frequency increases, the lead wire inductance (L), increased resistor value ($R' > R$) due to skin effect, and parasitic capacitances become prominent, as shown in Figure 6.12b. The net effect of all these parasitic elements, on the average, is a decrease in the resistor value as shown in Figure 6.12c for carbon-composition and metal resistors.

NOTE: *The 5 Ω resistor graph in Figure 6.12c shows a slight resonance due to the parallel combination of lead inductance and capacitance, which causes a small increase in the resistor value with a subsequent decline as frequency is increased further.*

There are several types of resistors, which can be briefly summarized as follows:

- Carbon composition type resistors, which have a high capacitance due to carbon granules' parasitic capacitance
- Wirewound resistors, which have high lead inductance
- Metal film resistors, which are usually made up of highly resistive films, such as NiCr, etc.
- Thin-film chip resistors that are produced on an Alumina or Beryllia substrate, which reduce the parasitic reactances greatly

These four types of resistors are shown in Figure 6.13a, b, c, and d.

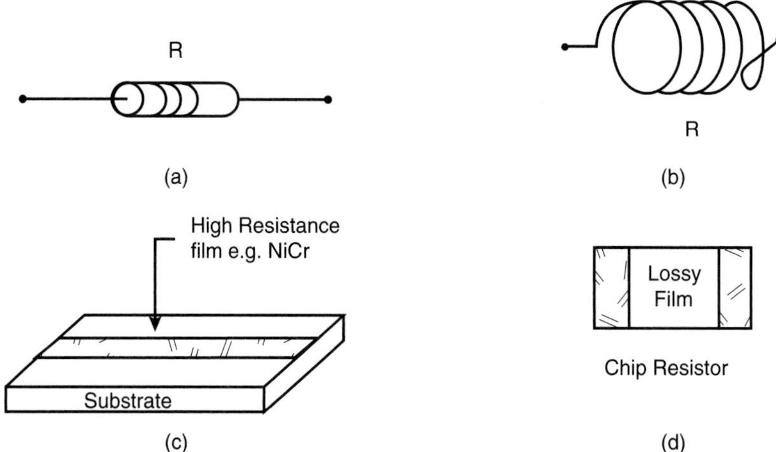

FIGURE 6.13 Various types of resistors: (a) carbon resistor, (b) wire wound resistor, (c) metal film resistor, and (d) thin film chip resistor.

6.5.3 Capacitors

DEFINITION-CAPACITOR: *A device that consists of two conducting surfaces separated by an insulating material or dielectric. The dielectric is usually ceramic, air, paper, mica, or plastic. The capacitance is the property that permits the storage of charge when a potential difference exists between the conductors. It is measured in Farads (F), (see Figure 6.14).*

The performance of a capacitor is primarily dependent on the characteristics of its dielectric. The dielectric determines the voltage and temperature range in which

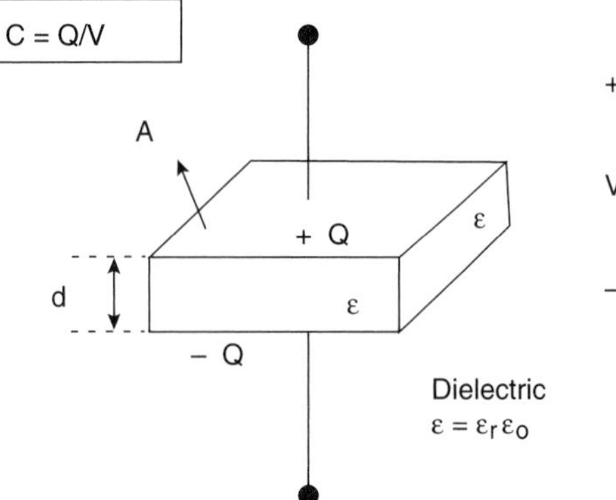

FIGURE 6.14 A parallel-plate capacitor.

the capacitor is operational. Any losses or imperfections in the dielectric have an enormous effect on the circuit operation. A few examples of different types of dielectric are shown in Figure 6.15.

FIGURE 6.15 Effect of dielectric on the capacitance value for (a) air, (b) paper.

A practical capacitor has several parasitic elements that become important at higher frequencies. The equivalent circuit of a real capacitor is shown in Figure 6.16.

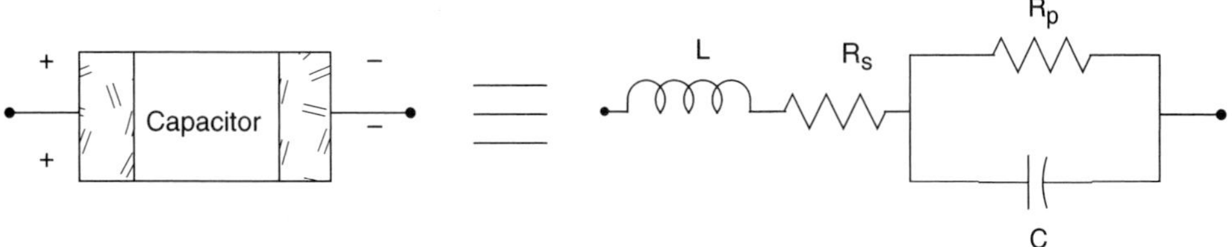

FIGURE 6.16 Parasitic elements in a capacitor.

In Figure 6.16 the elements are defined as follows:

C is the actual capacitance, L is the lead inductance, R_S is the series resistance, and R_P is the insulation resistance. Both resistances create heat and loss. The existence of parasitic elements, as shown in Figure 6.16, brings the concept of real-world capacitors to the forefront, which need further explanation:

Perfect Capacitors. In a perfect capacitor, current will lead the applied voltage in phase by 90 degrees. In phasor notation this can be written as:

$$I = j\omega CV = \omega CVe^{j90°} \tag{6.3}$$

Practical Capacitors. In a real-world capacitor, the phase angle (ϕ) will be less than 90 degrees (i.e., $\phi < 90°$) as shown in Figure 6.17. The reason $\phi < 90°$ is the existence of R_S and R_P (parasitic resistances shown in Figure 6.16), which combine into one equivalent resistor (R_{EQ}), as shown in Figure 6.18.

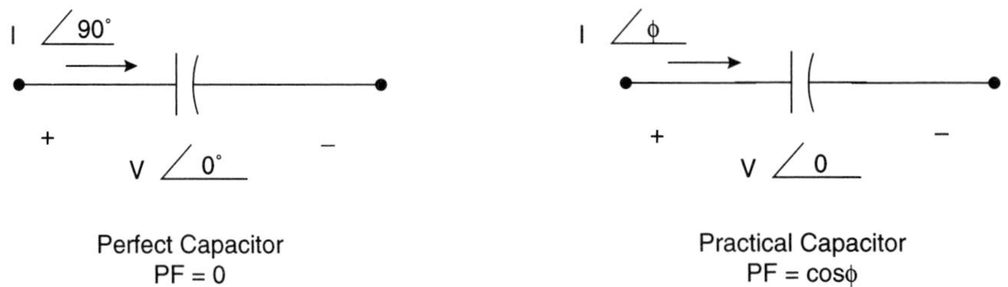

FIGURE 6.17 A capacitor current-voltage relationship for (a) a perfect capacitor, (b) a practical capacitor.

FIGURE 6.18 Phase angle ϕ (<90°).

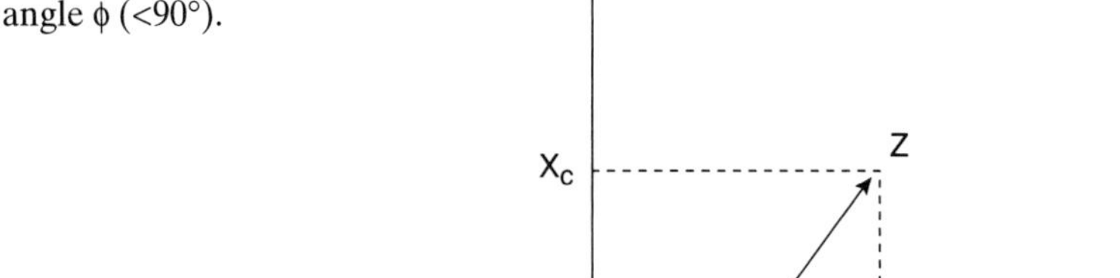

In a practical capacitor cos(ϕ), called the power factor (PF), can be written as (see Figure 6.18):

$$PF = \cos(\phi) = \frac{R_{EQ}}{\sqrt{X_C^2 + R_{EQ}^2}} \tag{6.4}$$

Usually $R_{EQ} << X_C$ where $X_C = 1/\omega C$. Therefore we can write:

$$PF = \cos(\phi) \approx R_{EQ}/X_C \tag{6.5}$$

An important factor in practical capacitors or in general any imperfect element is the Quality Factor(Q) :

DEFINITION-QUALITY FACTOR(Q): *A measure of the ability of an element (or circuit with periodic behavior) to store energy, equal to 2π times the average energy stored divided by the energy dissipated per cycle.*

Q is a figure of merit for a reactive element and can be shown to be the ratio of the element's reactance to its effective series resistance. For a capacitor, Q is given by:

$$Q = X_C / R_{EQ} = \frac{1}{\omega C R_{EQ}} \approx 1/PF \tag{6.6}$$

From Equation 6.6, we can observe that for a practical capacitor, as the effective series resistance (R_{EQ}) decreases, Q will increase until $R_{EQ} = 0$, which corresponds to a perfect capacitor having $Q = \infty$, i.e.,

$$R_{EQ} = 0 \Rightarrow PF = 0,\ Q = \infty \text{ (a perfect capacitor)} \tag{6.7}$$

The effect of these imperfections in a capacitor is shown in Figure 6.19.

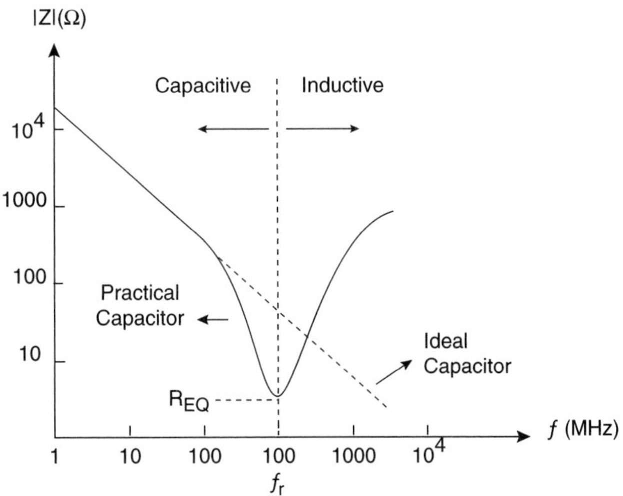

FIGURE 6.19 The behavior of a capacitor versus frequency.

From Figure 6.19, two distinct regions in the frequency response plot of a capacitor can be identified. These two regions straddle the resonance frequency (f_r) as follows:

a. $f < f_r$. In this region as frequency increases, the lead inductance's reactance goes up gradually, cancelling the capacitor's reactance and thus causing resonance (f_r).

b. $f > f_r$. In this region the capacitor acts like an inductor and is no longer performing its intended function.

From Figure 6.19 we can conclude that we need to examine the capacitor at RF/MW frequencies before its use in final design and production. The concept of capacitance behavior is shown in Figure 6.20, where the distinction between the low AC and RF/MW is clearly shown.

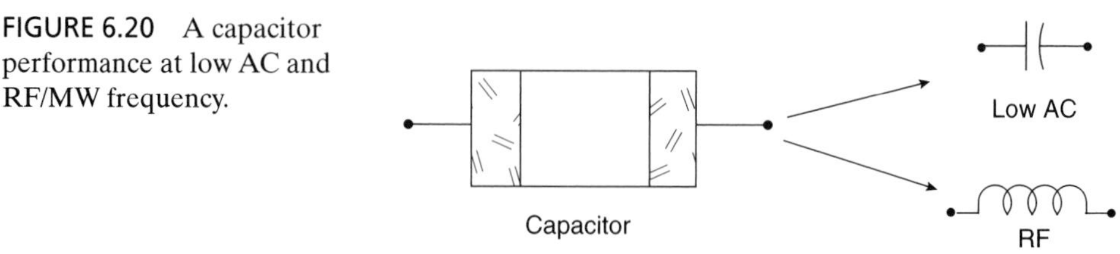

FIGURE 6.20 A capacitor performance at low AC and RF/MW frequency.

6.5.4 Inductors

DEFINITION-INDUCTOR: *A wire that is wound (or coiled) in such a manner as to increase the magnetic flux linkage between the turns of the coil. The increased flux linkage increases the wire's self-inductance, as shown in Figure 6.21.*

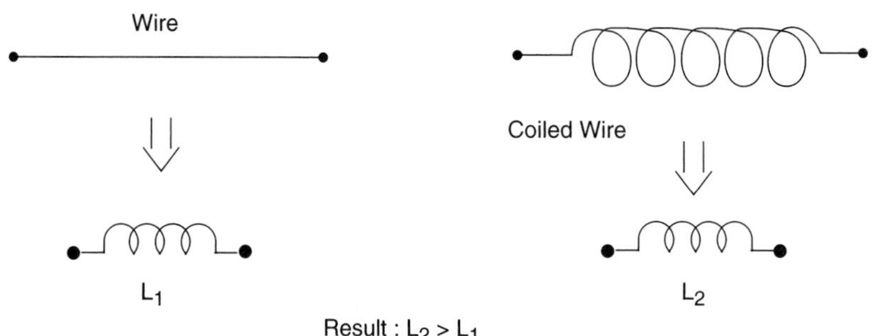

FIGURE 6.21 A simple piece of wire versus a coiled wire.

Inductors have a variety of applications in RF circuits such as in resonance circuits, filters, phase shifters, delay networks, and RF chokes, as shown in Figure 6.22.

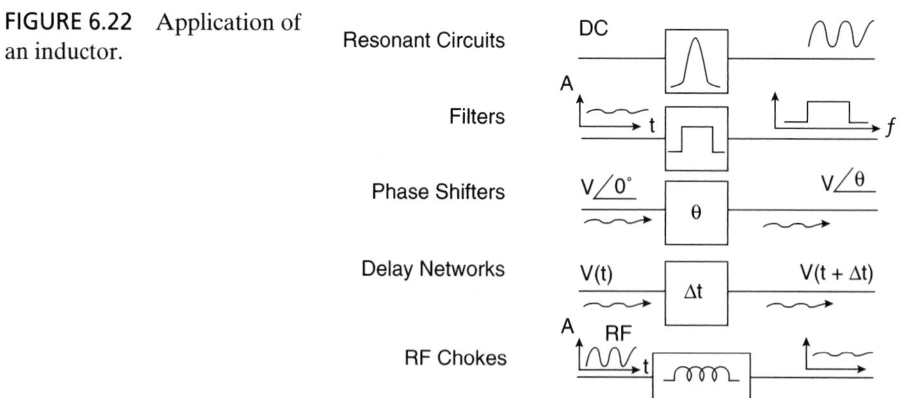

FIGURE 6.22 Application of an inductor.

There is no such thing as a perfect component. Among all components, inductors are most prone to very drastic changes over frequency. This is due to the fact that the distributed capacitance (C_d) and series resistance (R) in an inductor at RF/MW play a major role in the performance of an inductor, as shown in Figure 6.23.

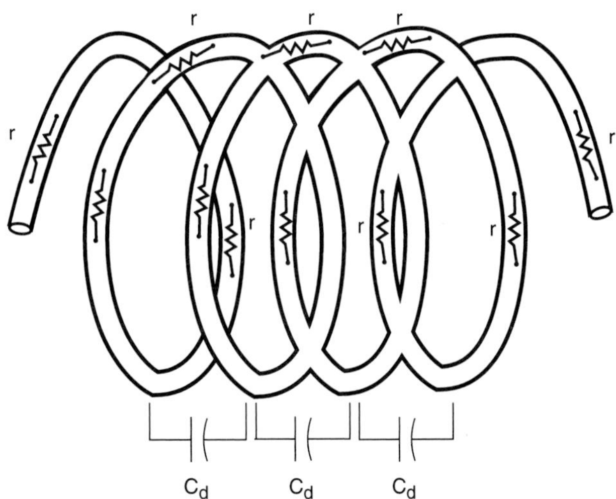

FIGURE 6.23 The distributed parasitic elements of an inductor.

From Figure 6.23, we can see that C_d exists due to a voltage drop in the coil caused by internal resistance. The voltage drop causes a voltage difference between two turns of the coil separated from each other (with air as the dielectric). The aggregate of all small C_d's and R's provides the equivalent circuit, as shown in Figure 6.24.

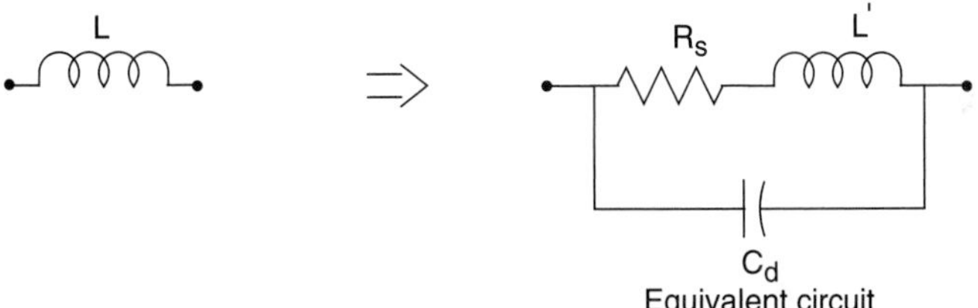

FIGURE 6.24 The equivalent circuit of an inductor at RF/MW frequencies.

The effect of C_d on an inductor's frequency response is shown in Figure 6.25. From this figure (just like a capacitor) there are two regions that straddle the resonant circuit. These two regions can be identified as follows:

a. $f < f_r$. In this region, the inductor's reactance ($X_L = \omega L$) increases as frequency is increased.

b. $f > f_r$. In this region the inductor behaves like a capacitor, and as the frequency is increased the reactance decreases.

At $f = f_r$ resonance takes place in an inductor (inductor's reactance is cancelled by parasitic distributed capacitor), and theoretically the inductor's reactance is infinity; however, in practice, the total impedance of the element is finite due to a nonzero series resistance.

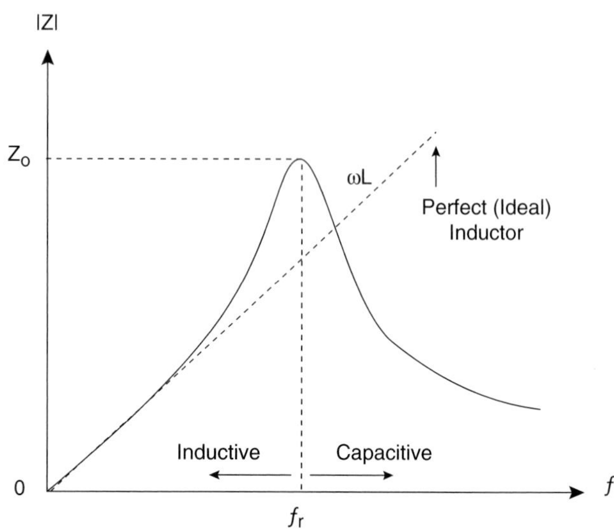

FIGURE 6.25 Effect of parasitic C_d on an inductor's reactance.

The quality factor (Q) of an inductor is defined to be:

$$Q = X_L/R_S = \omega L/R_S \tag{6.8}$$

For a perfect inductor the series resistance is zero; thus we have:

$$R_S = 0 \Rightarrow Q = \infty \quad \text{(A perfect inductor)} \tag{6.9}$$

At low frequencies, Q is very large because R_s is very small; however, as frequency increases the skin effect and winding distributed capacitor (C_d) begin to degrade the Q of an inductor, as shown in Figure 6.26.

From Figure 6.26, it can be seen that as frequency increases, Q will increase up to Q_o, which is at $f = f_o$. For frequencies $f_o < f < f_r$, R_s and C_d combine to decrease the Q of the inductor toward zero. At resonance ($f = f_r$), where the total reactance of the element is zero, the inductor is no longer useful.

To extend the frequency range of an inductor (by increasing its Q), we can use one of the following solutions:

- Use a larger diameter for the wire, which effectively reduces the resistance value.
- Spread the winding apart, which reduces the distributed capacitance (C_d) between the windings.
- Increase the inductance (L) by increasing the permeability of the flux linkage path by using a magnetic-core material.

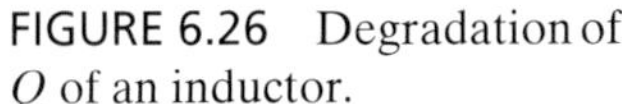
FIGURE 6.26 Degradation of Q of an inductor.

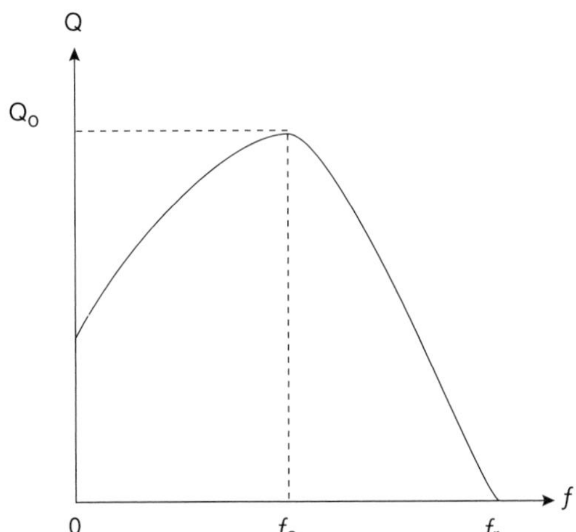

These solutions are shown in Figure 6.27. The third solution is the most effective and practical of the three.

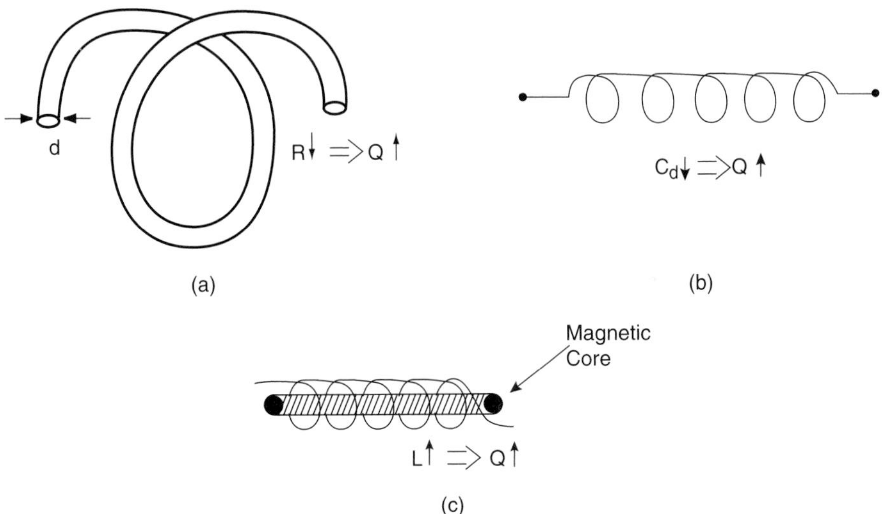

FIGURE 6.27 Three possible ways of increasing Q in an inductor.

6.6 RESONANT CIRCUITS

A resonant circuit (also called a filter) is certainly not new and, in fact, has been and is used in practically every transmitter, receiver, or piece of test equipment in existence. Its function is to pass selectively a certain frequency (or frequency range) from the source to the load, while attenuating all other frequencies outside of this passband, as shown in Figure 6.28.

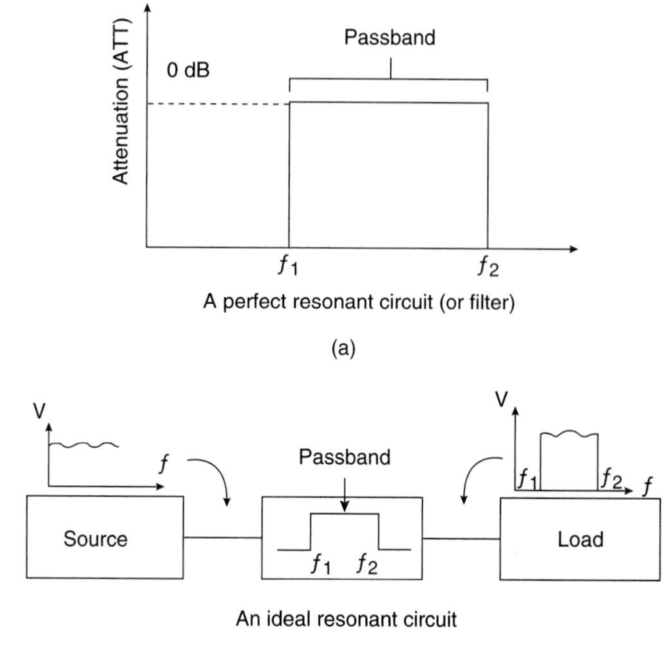

FIGURE 6.28 An ideal resonant circuit: (a) frequency response, (b) application.

Because there is no perfect component, a perfect resonant circuit does not exist and cannot be built (see Chapter 1, Section 1.5.2, *Relativity of Knowledge*). Knowing the mechanics of resonant circuits, we can tailor an imperfect resonant circuit (or filter) to suit our needs. A typical practical filter's frequency response is shown in Figure 6.29.

From this figure several important features need to be defined:

DEFINITION-ANGULAR BANDWIDTH (BW): *The difference between upper (ω_2) and lower (ω_1) angular frequencies at which the amplitude response is 3 dB below the passband response value (also called the half-power BW). Therefore we can write:*

$$BW = (\omega_2 - \omega_1).$$

DEFINITION-CIRCUIT Q: *The ratio of center angular frequency (ω_o) to the bandwidth(BW), i.e.,*

$$Q = \omega_o/BW = f_o/(f_2 - f_1) = f_o/\Delta f \tag{6.10}$$

where

$$BW = (\omega_2 - \omega_1) = 2\pi\Delta f,$$

$$\Delta f = f_2 - f_1,$$

$$\omega_o = 2\pi f_o,$$

$$\omega_2 = 2\pi f_2,$$

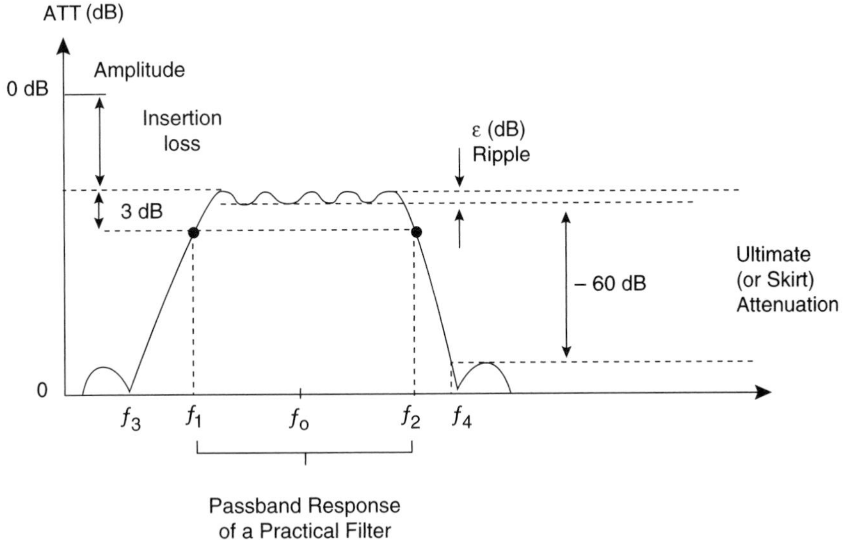

FIGURE 6.29 A practical filter frequency response.

and $$\omega_1 = 2\pi f_1.$$

It is important to note that "circuit Q" should not be confused with "component Q," which is a measure of component loss, while "circuit Q" is a measure of the selectivity of a resonance circuit. This means that as the angular bandwidth (BW) decreases, Q, as well as the selectivity of the resonant circuit, increases. Furthermore, it should be noted that the component Q does have an effect on the circuit Q but the reverse is not true.

DEFINITION-SHAPE FACTOR (SF) OF A RESONANT CIRCUIT: *Is the ratio of the 60-dB bandwidth to the 3-dB bandwidth, i.e.,*

$$SF = \frac{f_4 - f_3}{f_2 - f_1} \tag{6.11}$$

Shape factor (SF) is simply a measure of the steepness of the skirts. The smaller the SF number, the steeper the response skirts. A perfect filter has SF = 1; however, in practice SF is always greater than 1 (i.e., SF ≥ 1). When SF is less than 1(i.e., SF < 1), we have a physical impossibility, as shown in Figure 6.30.

DEFINITION-ULTIMATE ATTENUATION: *The final, maximum attenuation that the resonance circuit presents outside the specified passband. A perfect resonant circuit has an ultimate attenuation of infinity. If there are response peaks outside the passband, then this will detract from the ultimate attenuation specification of the circuit, as shown in Figure 6.31.*

DEFINITION-INSERTION LOSS: *The attenuation resulting from inserting a circuit between source and load (usually expressed in dB). This concept is depicted in Figure 6.32.*

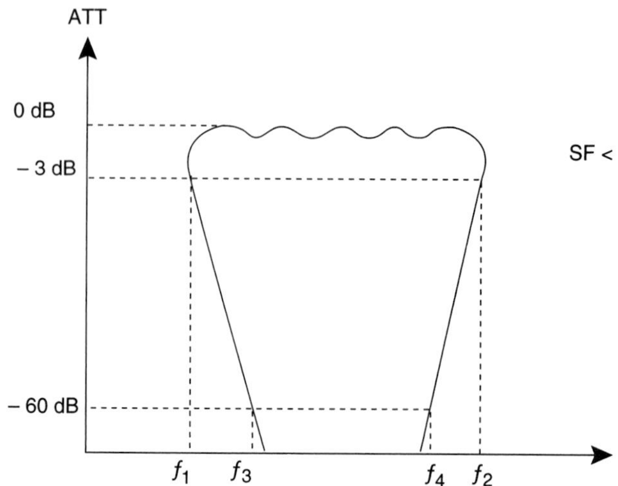

FIGURE 6.30 Physical impossibility when SF < 1.

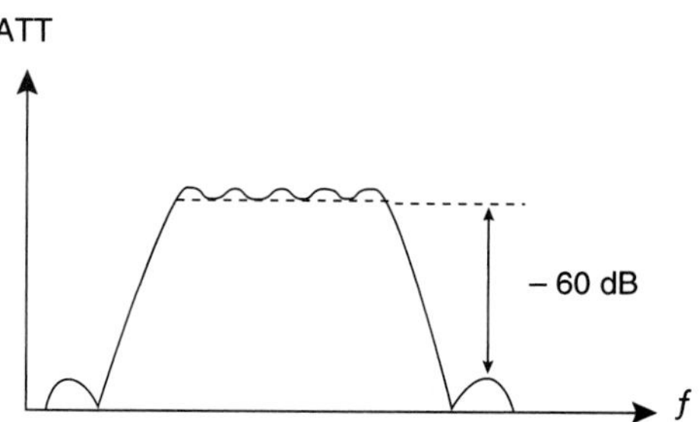

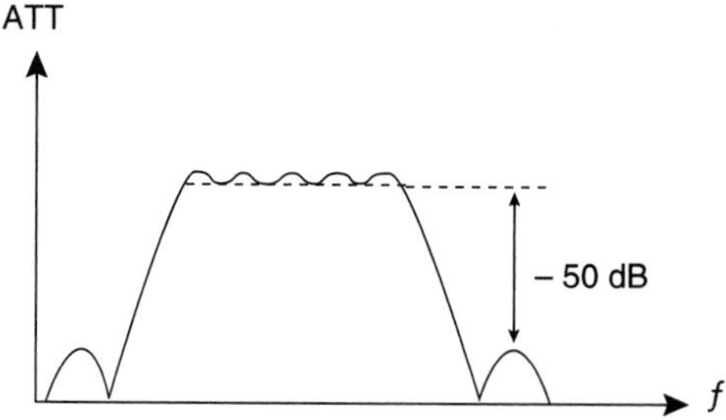

FIGURE 6.31 Response peaks outside the passband reduce ultimate attenuation.

DEFINITION-RIPPLE: *A measure of the flatness of the frequency response of the resonance circuit, defined as the attenuation difference (in dB) between the maximum value and the minimum in the passband, i.e.,*

$$\varepsilon = |\text{max. attenuation} - \text{min. attenuation}| \quad \text{(in dB)}$$

This concept is shown in Figure 6.33.

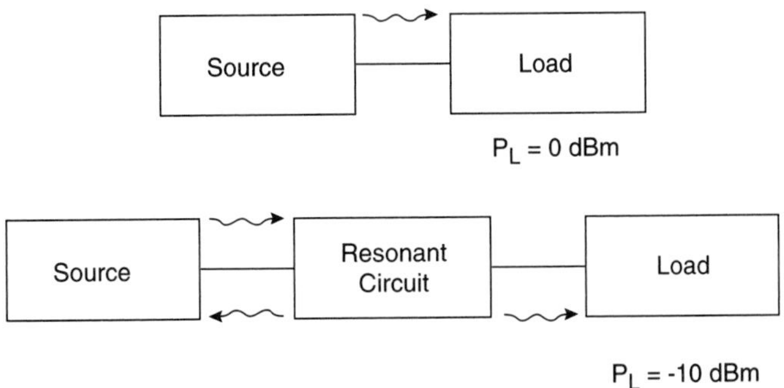

FIGURE 6.32 Concept of insertion loss.

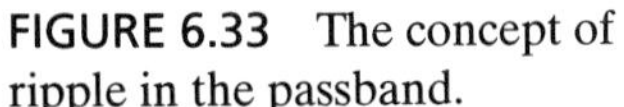

FIGURE 6.33 The concept of ripple in the passband.

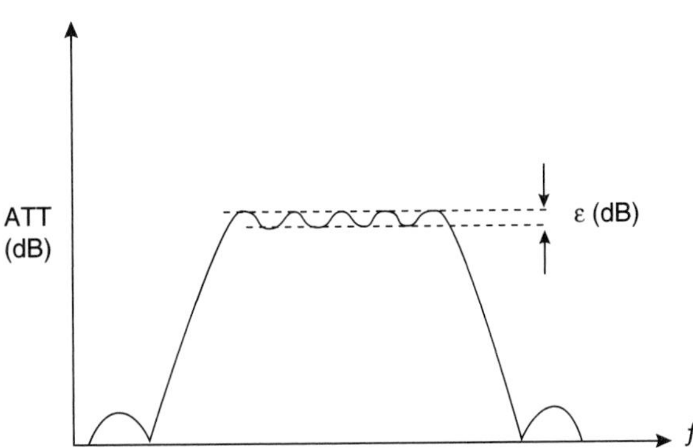

6.7 ANALYSIS OF A SIMPLE CIRCUIT IN PHASOR DOMAIN

Consider the circuit shown in Figure 6.34, which consists of a series resistance R_S and shunt element with impedance Z_P. We wish to calculate the total voltage gain (V_o/V_i) of the circuit. Using KVL in phasor domain, we have:

$$V_o = Z_P I_i, \tag{6.12}$$

and,

$$V_i = (R_S + Z_P)I_i \tag{6.13}$$

FIGURE 6.34 A simple series-shunt circuit.

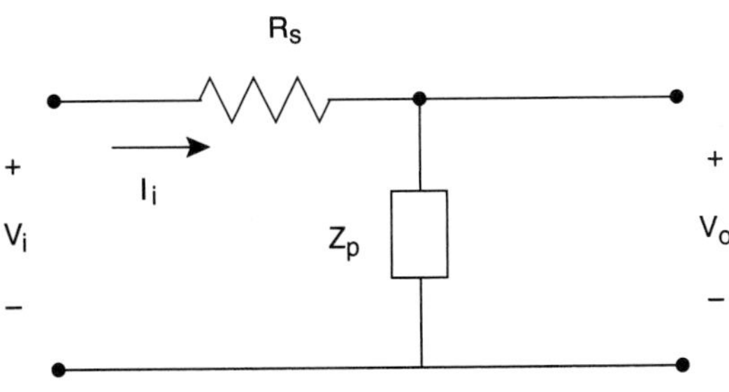

Dividing Equation 6.12 by 6.13 we obtain the total voltage gain as:

$$V_o/V_i = H(\omega) = |H|\, e^{j\phi} = |H|\ \angle\phi = \frac{Z_P}{R_S + Z_P} \tag{6.14}$$

The gain magnitude in dB would be given by:

$$20\log_{10}|H| = 20\log_{10}|V_o/V_i| = 20\log_{10}\left(\frac{Z_P}{R_S + Z_P}\right)\ \text{(dB)} \tag{6.15}$$

NOTE 1: *From Equation 6.15 we can observe that output voltage magnitude will always be less than or at best equal to the input voltage magnitude. This is true for all passive circuits. Therefore, we can write for all passive circuits:*

$$|V_o| \le |V_i| \tag{6.16}$$

NOTE 2: *If the shunt element (in Figure 6.34) contains a capacitor or an inductor then the impedance of the shunt element (Z_P) will be frequency dependent and so would the output voltage (V_o) or the total voltage gain (V_o/V_i) as the following examples illustrate.*

EXAMPLE 6.4

If the shunt element in Figure 6.34 is a perfect capacitor, calculate and plot the voltage gain magnitude and phase.

Solution:

From the circuit shown in Figure 6.35 and Equation 6.14, we can write the following:

$$Z_P = 1/j\omega C$$

$$H(\omega) = |H|\angle\phi = \frac{(1/j\omega C)}{(R + 1/j\omega C)} = \frac{1}{1 + j\omega RC} = \frac{1}{1 + (\omega RC)^2}\angle -\tan^{-1}(\omega RC)$$

or,

$$|H(\omega)|\ \text{(dB)} = 20\log_{10}|H| = -20\log_{10}[1 + (\omega RC)^2]\ \ \text{(in dB)}$$

and,

$$\phi = -\tan^{-1}(\omega RC)$$

The magnitude and the phase are plotted in Figures 6.36a, b. From the magnitude plot we can see that this circuit performs like a low-pass filter.

NOTE: *Attenuation is 6 dB for every octave increase of frequency (i.e., doubling the frequency). This is due to only a single reactive element. In general, for each significant reactive element added in the circuit, the slope will increase by an additional 6 dB.*

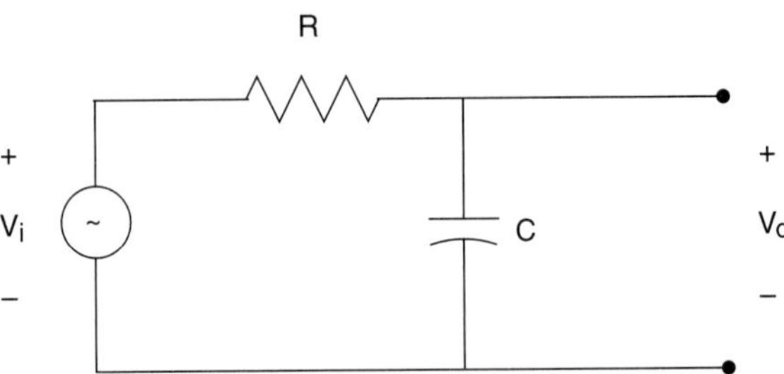

FIGURE 6.35 Circuit for Example 6.4.

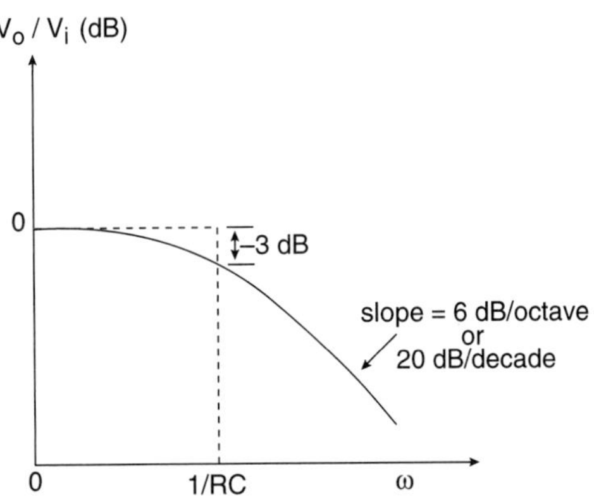

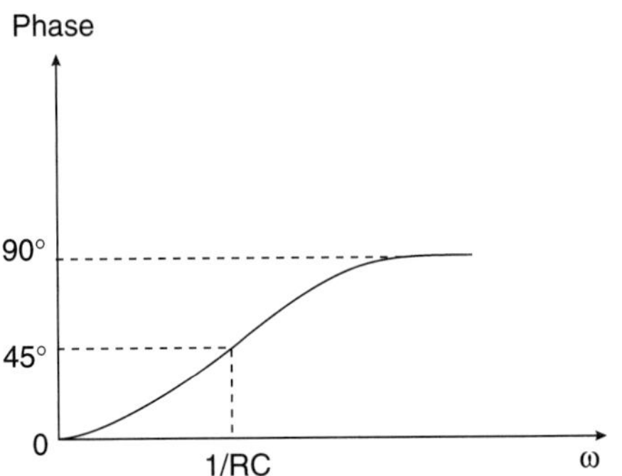

FIGURE 6.36 Frequency response: (a) magnitude plot, (b) phase plot.

(a)

(b)

EXAMPLE 6.5

If the shunt element in Figure 6.34 is a perfect inductor, calculate and plot the voltage gain magnitude and phase.

Solution:

From the circuit shown in Figure 6.37 and Equation 6.14, we can write the following:

$$Z_P = j\omega L$$

$$H(\omega) = \frac{j\omega L}{R_S + j\omega L} = \frac{1}{1 - jR_S/\omega L} = \left[\frac{1}{1 + (R_S/\omega L)^2}\right]^{\frac{1}{2}} \angle \tan^{-1}(R_S/\omega L)$$

or

$$|H|\ (\text{dB}) = 20 \log_{10}|H| = -10 \log_{10} [1 + (R_S/\omega L)^2] \quad (\text{dB})$$

and

$$\phi = \tan^{-1}(R_S/\omega L)$$

FIGURE 6.37 Circuit for Example 6.5.

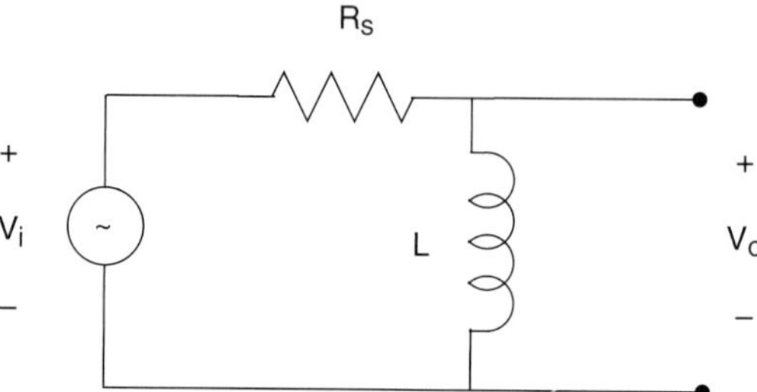

The magnitude and the phase are plotted in Figures 6.38a, b. From the magnitude diagram we can see that this circuit performs like a high-pass filter.

EXAMPLE 6.6

If the shunt element in Figure 6.39 is a combination of a perfect capacitor in parallel with a perfect inductor, calculate and plot the voltage gain magnitude and phase.

Solution:

From the circuit shown in Figure 6.39, we can write Z_P (from Equation 6.14) as the parallel combination of the capacitor and inductor as follows:

$$Z_P = j\omega L \parallel -j/\omega C = \frac{j\omega L}{(1 - \omega^2 LC)}$$

$$H(\omega) = \frac{Z_P}{R_S + Z_P} = \frac{j\omega L}{R_s - \omega^2 R_S LC + j\omega L} \quad (\text{dB})$$

$$|H| = 20\log_{10}|H| = 20\log_{10}\left(\frac{\omega L}{\sqrt{(R_S - \omega^2 LC)^2 + (\omega L)^2}}\right)^{\frac{1}{2}} \quad (\text{dB})$$

The magnitude is plotted in Figure 6.40.

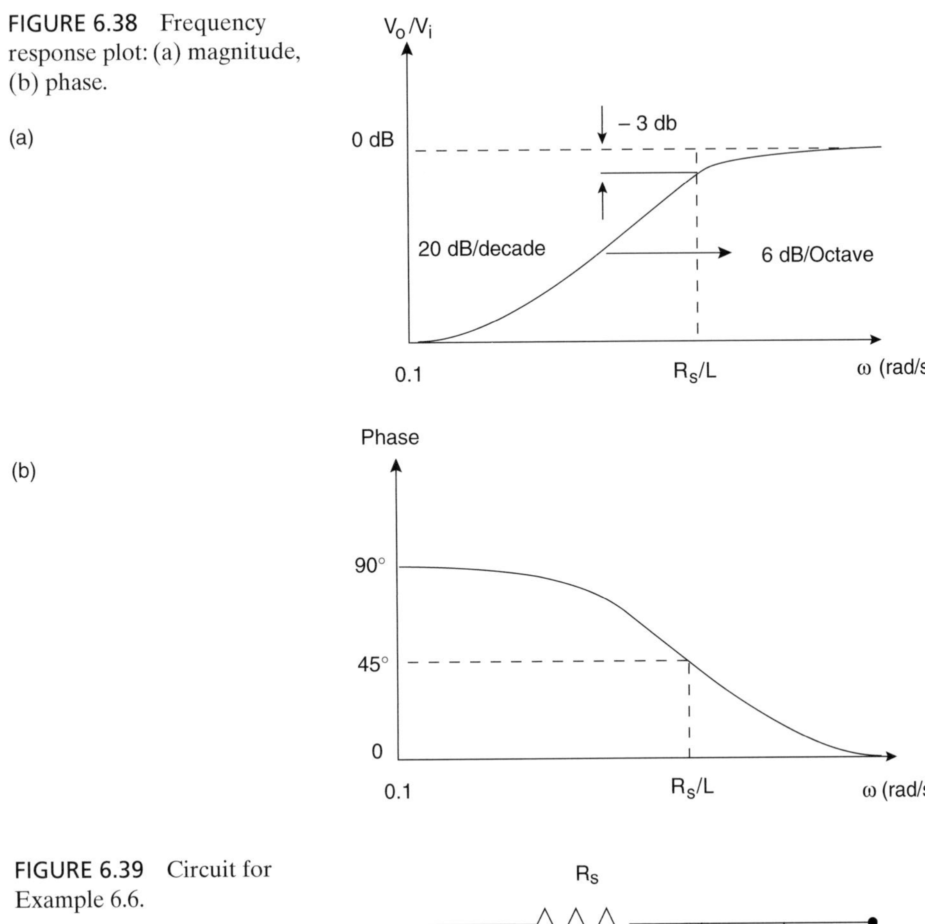

FIGURE 6.38 Frequency response plot: (a) magnitude, (b) phase.

FIGURE 6.39 Circuit for Example 6.6.

NOTE 1: *Near the resonance frequency of the tuned circuit, the slope of the resonance curve increases to 12 dB/octave because there are now two significant reactances present and each one is changing at the rate of 6 dB/octave (therefore 12 dB/octave slope).*

NOTE 2: *Away from resonance, only one reactance becomes significant, therefore there would be only a 6 dB/octave of slope in effect.*

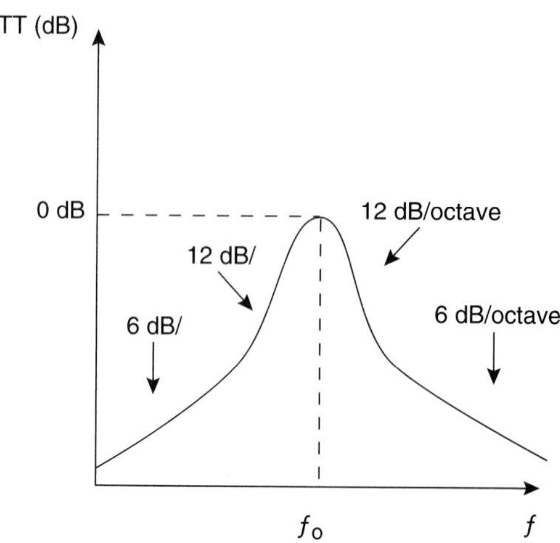

FIGURE 6.40 Magnitude plot of the frequency response of Example 6.6.

6.7.1 Loaded Q

The Q of a resonant circuit was defined earlier as:

$$Q = \frac{\omega_o}{BW}.$$

The "circuit Q" is often called the "loaded Q" because it describes the passband characteristics of the resonant circuit under actual "in-circuit" or "loaded" condition. In general, the "loaded Q" depends on three main factors as follows:

- The source resistance (R_S)
- The load resistance (R_L)
- The component Q (of each of the reactive elements)

Figure 6.41 shows a block diagram of a resonant circuit and its frequency performance.

From Figure 6.41 we can observe that:

1. The resonant circuit sees an equivalent resistance of R_S in parallel with R_L as its true load. This is shown in Figure 6.42. The loaded Q can be calculated by noting that:

$$R_P = R_S \parallel R_L, \tag{6.17}$$

X_P = the inductive or capacitive reactance of either of the reactive components (because they are equal at resonance!)
Therefore:

$$Q = R_P/X_P \tag{6.18}$$

2. If R_S or R_L increases, then the equivalent resistance increases, which will reduce the energy losses and will narrow the curve. This will increase the selectivity and as a result the "loaded Q."
3. For a fixed R_P, if X_P is decreased by choosing a smaller "L" or a larger "C," Q will increase. This point is illustrated in the next example.

FIGURE 6.41 A resonant circuit: (a) block diagram, (b) frequency response.

(a)

(b)

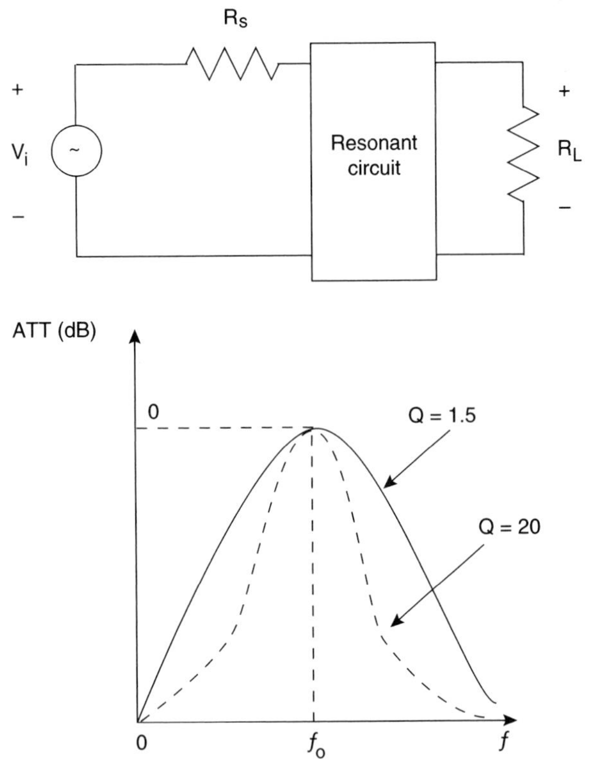

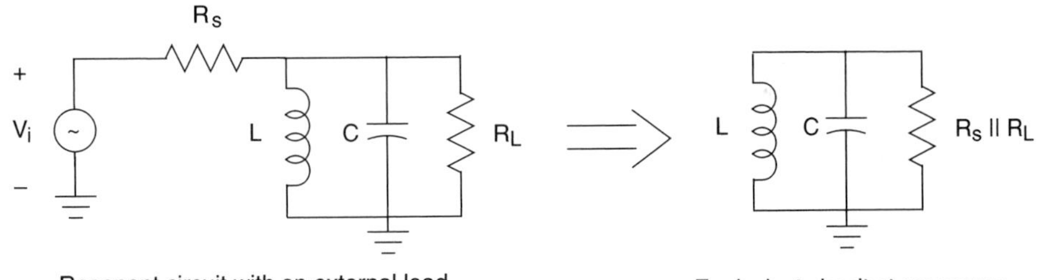

Resonant circuit with an external load

Equivalent circuit at resonance

FIGURE 6.42 A resonant circuit at resonance.

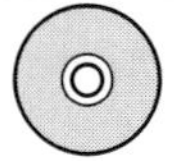

EXAMPLE 6.7

Design a resonant circuit with a loaded $Q = 1.1$ at $f = 142.4$ MHz that operates between a source resistance of 100 Ω and load resistance of 100 Ω. Discuss how to increase Q. Use perfect components.

Solution:

$$R_P = 100 \parallel 100 = 50\ \Omega$$

$$X_P = R_P/Q = 50/1.1 = 45.45\ \Omega = \omega L = 1/\omega C$$

$$\text{Thus, we obtain: } L = 50\text{ nH}, \quad C = 25\text{ pF}$$

Given fixed R_S and R_P, we can increase Q by 20 times through scaling up the capacitor value by 20 while scaling down the inductor value by 20, i.e.,

$$Q = 22 \Rightarrow C_{new} = 500\text{ pF}, \quad L_{new} = 2.5\text{ nH}$$

These two cases are shown in Figure 6.43.

Figure 6.43 shows two equivalent circuits for two different Qs obtained by scaling the inductor's and the capacitor's values appropriately.

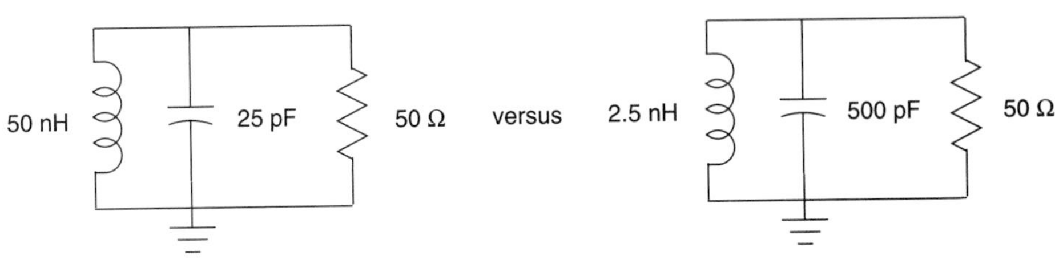

FIGURE 6.43 Changing the Q value by adjusting the inductor and capacitance value.

Therefore a circuit designer has two design approaches in designing resonant circuits:

- Select an optimal value of R_S and R_L to get the specified Q
- Given R_S and R_L, select component values of "L" and "C" to optimize Q

NOTE 1: *If poor quality reactive components (i.e., low Q) are used in highly selective resonant circuits, the net result is that we effectively place a low-value shunt resistor directly across the circuit, which will drastically reduce its loaded Q and increase the bandwidth.*

NOTE 2: *At resonance, an ideal LC parallel circuit has a very high (ideally infinite) total impedance, as shown in Figure 6.44.*

NOTE 3: *Usually we need to involve only the Q of the inductors in the loaded-Q calculations; the Qs of most capacitors are quite high over their useful frequency range, which means that they have a very small resistive passive element.*

6.7.2 Impedance Transformation

The most common type of impedance transformation is from the series elements to shunt elements, as shown in Figure 6.45.

We first define the component "Q" (represented by Q_C) as:

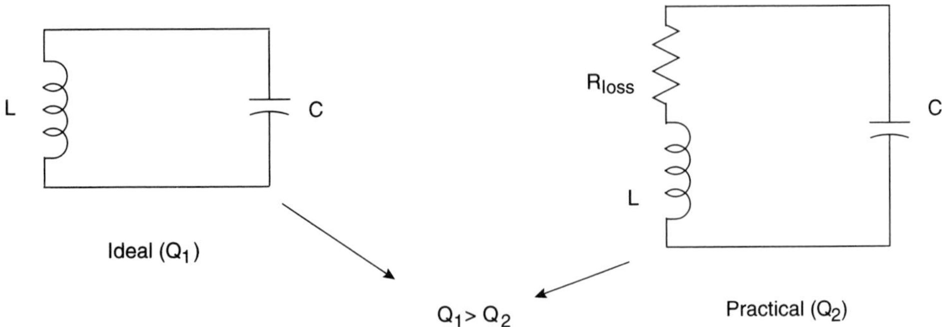

FIGURE 6.44 An ideal versus a practical resonant circuit.

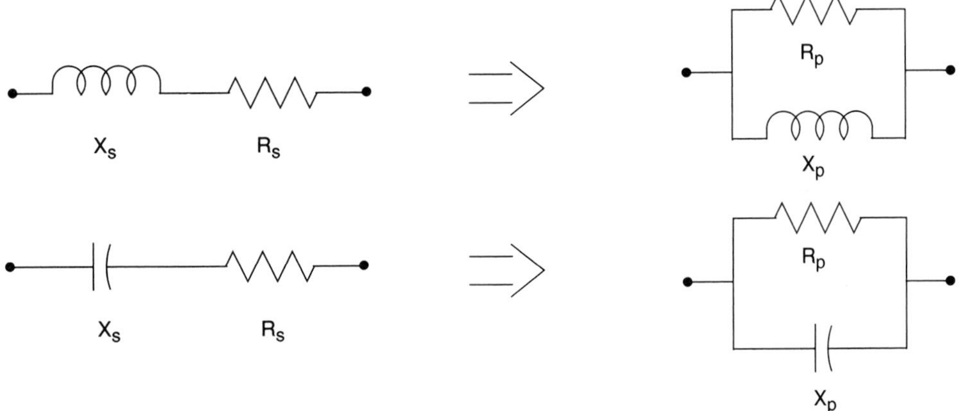

FIGURE 6.45 Impedance transformation.

$$Q_C = Q_S = Q_P, \quad \text{(component } Q\text{)} \tag{6.19}$$

Equation 6.19 states that the Q remains the same in the process of series-to-shunt transformation, which is a correct assumption because we are still dealing with the same element even though we are changing its equivalent circuit.

Through simple mathematical manipulation, we can write:

$$R_P = (Q_C^{\,2} + 1)\, R_S \tag{6.20}$$

and,

$$X_P = R_P / Q_C \tag{6.21}$$

Using Equation 6.21 for a shunt configuration, Q_C is defined as:

$$Q_C = R_P / X_P \text{ (shunt)} \tag{6.22}$$

which is in contrast with the series configuration earlier defined as:

$$Q_C = X_S / R_S \text{ (series)} \tag{6.23}$$

EXERCISE 6.1

It will be a worthwhile exercise to derive the above impedance transformation equations, i.e., Equation 6.20 and Equation 6.21.

HINT: *Set Z_{in}, of each circuit equal to each other.*

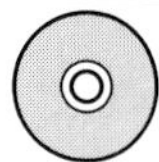

EXAMPLE 6.8

An imperfect inductor has an inductance of 50 nH with a series loss resistance of 10 Ω. Find the following:

a. *Q_C at 100 MHz*
b. *The equivalent parallel configuration at f = 100 MHz*

Solution:

$$Q_C = X_S/R_S = 2\pi fL/R_S = 2\pi \text{ x } 100 \times 10^6 \times 50 \times 10^{-9}/10 = 3.14$$

$$R_P = (Q_C^2 + 1)R_S = (3.14^2 + 1) \times 10 = 108.7\ \Omega$$

$$X_P = R_P/Q_C = 108.7/3.14 = 34.62\ \Omega$$

$$X_P = \omega L_P \Rightarrow L_P = X_P/\omega = 55.1 \text{ nH}$$

The equivalent circuit is shown in Figure 6.46.

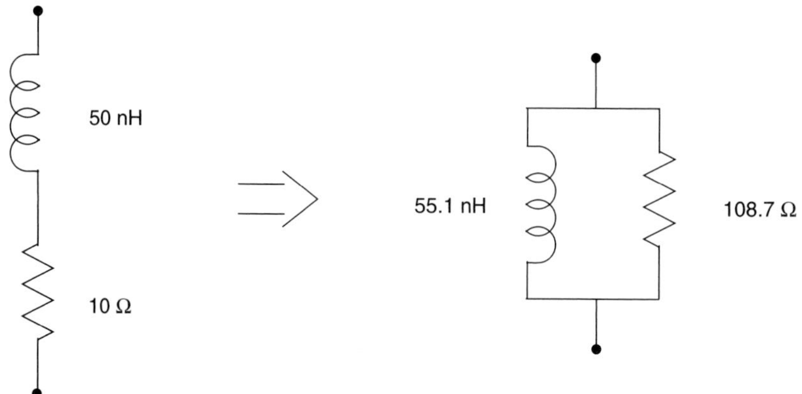

FIGURE 6.46 Circuit transformation for Example 6.8.

6.7.3 Insertion loss (IL)

If the inductor and the capacitor were perfect components with no internal loss, then the insertion loss for LC resonant circuits would be zero dB. In actuality, this is not the case, and insertion loss is a very critical parameter in specifying a resonant circuit, as shown in Figure 6.47. The following example will illustrate the concept of insertion loss further.

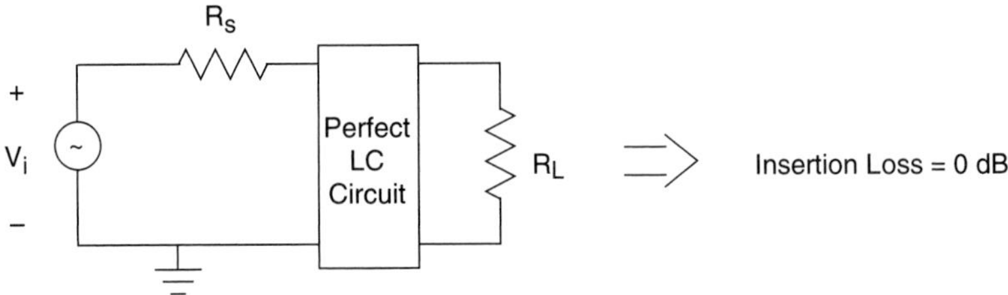

FIGURE 6.47 Insertion loss of a perfect LC circuit.

EXAMPLE 6.9

Calculate the insertion loss of the LC resonant circuit shown in Figure 6.48 at f = 1430 MHz.

$$R_S = R_L = 1\ \text{k}\Omega$$

$$\textit{Inductor: } L = 0.05\ \mu\text{H},\ Q_{C1} = 10$$

$$\textit{Capacitor: } C = 0.25\ \text{pF},\ Q_{C2} = \infty$$

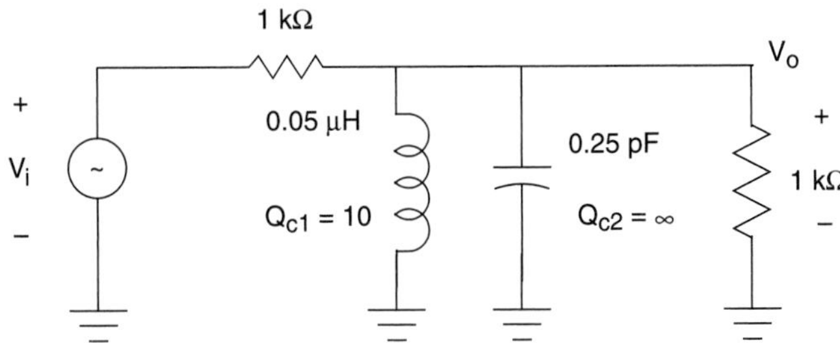

FIGURE 6.48 Circuit for Example 6.9.

Solution:

a. Removing the *LC* circuit gives:

$$V_o = 1000/(1000 + 1000) \times V_i = 0.5\ V_i$$

b. Next we convert the inductor's series configuration into parallel:

$$Q_{C1} = X_{SL}/R_{SL} \Rightarrow R_{SL} = 2\pi \times 1430 \times 10^6 \times 0.05 \times 10^{-6}/10 = 45\ \Omega$$

$$R_{PL} = (Q_{C1}^2 + 1)\ R_{SL} \Rightarrow R_{PL} = 4.5\ \text{k}\Omega$$

Therefore at resonance we have a circuit as shown in Figure 6.49:

$$(R_L)_{EQ} = 1 \text{ k} \parallel 4.5 \text{ k} = 820\ \Omega$$

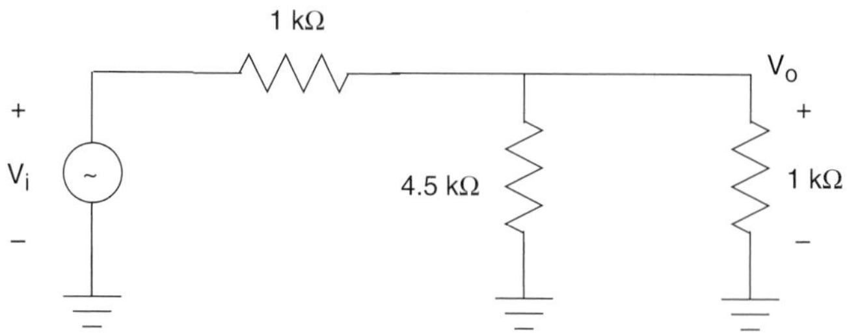

FIGURE 6.49 The equivalent circuit of Example 6.9.

Thus, the load voltage with resonant circuit in place is:

$$V'_o = 820/(1000 + 820)V_i = 0.45\ V_i$$

$$\text{Insertion Loss } (IL) = V'_o/V_o = 0.45/0.5 = 0.9$$

$$IL(\text{dB}) = 20\log_{10}(0.9) = 0.92 \text{ dB}$$

An insertion loss of 0.92 dB may not appear much but can add up very quickly if we cascade several resonant circuits together.

EXAMPLE 6.10

Design a simple parallel resonant circuit to work between a source resistance (R_S) of 1 kΩ and load resistance (R_L) of 1 kΩ to provide a 3dB bandwidth of 10 MHz at a center frequency (f_o) of 100 MHz. Also calculate the insertion loss. Assume all capacitors are perfect and the inductor has a Q of 85.

Solution:

Considering the circuit shown in Figure 6.50, we have:

$$Q = f_o/\Delta f = 100/10 = 10$$

For inductor:

$$Q_C = R_P/X_P = 85 \Rightarrow R_P = 85\ X_P \qquad (1)$$

The loaded Q for the circuit is:

$$Q = 10 = R_{tot}/X_P = (R_P \parallel 1 \text{ k} \parallel 1 \text{ k})/X_P \qquad (2)$$

Solving (1) and (2) for R_P and X_P we obtain:

$$X_P = 44.1\ \Omega = \omega_o L = 1/\omega_o C$$

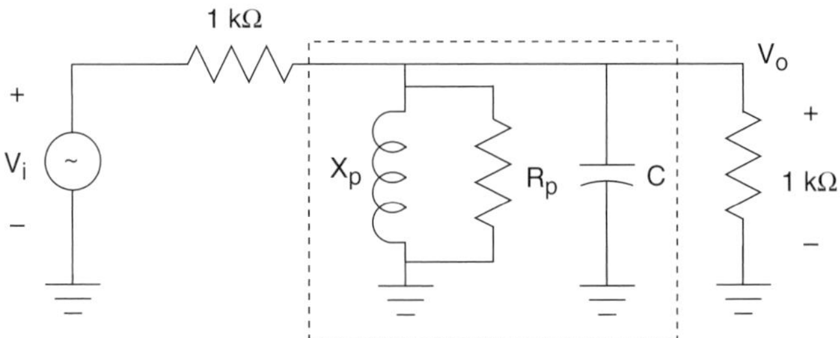

FIGURE 6.50 Circuit for Example 6.10.

$$R_P = 3.75 \text{ k}\Omega$$

$$L = X_P/\omega_o = 70 \text{ nH}$$

$$C = 1/\omega_o X_P = 36 \text{ pF}$$

To find the insertion loss we note two cases as follows:

a. Without the resonant circuit

$$V_o = 1\text{k}/(1\text{k}+1\text{k}) = 0.5\ V_i$$

b. With the resonant circuit in place

$$(R_L)_{EQ} = R_P \| R_L = 3.75\text{k} \| 1\text{k} = 789.5\ \Omega$$

$$V_o' = V_i \times 789.5/(1000 + 789.5) = 0.44\ V_i$$

$$\text{I.L.(dB)} = 20\log_{10}(0.44\ V_i/0.5\ V_i) = 1.1 \text{ dB}$$

6.8 IMPEDANCE TRANSFORMERS

By observation, it becomes apparent that low values of R_S and R_L tend to load down a given resonant circuit, leading to a decrease of its Q and broadening of its bandwidth. Thus it is very difficult to design a high-Q simple LC resonant circuit that would function well between two low values of R_L and R_S, as shown in Figure 6.51.

To solve this design problem, two types of impedance transformer may be used as shown in Figures 6.52 and 6.53. The impedance transformer placed between the load and the source in the circuit converts R_S or R_L to a much larger resistance. Thus the Q of the circuit is pushed to a higher value because the resonating circuit as a whole is presented with a higher resistance value. Figures 6.52 and 6.53 show two types of transformers:

- Tapped-C transformer
- Tapped-L transformer

These are described next.

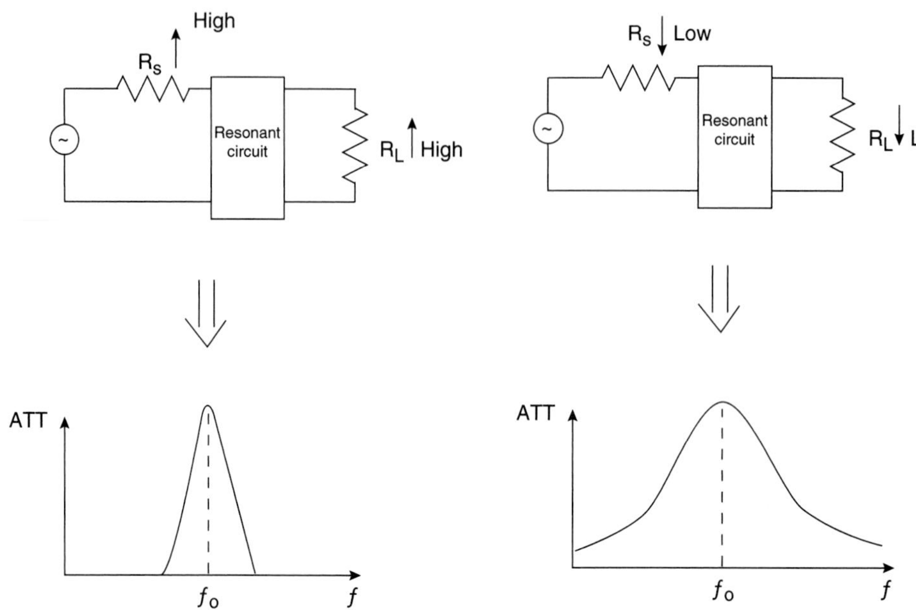

FIGURE 6.51 The comparison of frequency response: (a) high R_S and R_L, (b) low R_S and R_L.

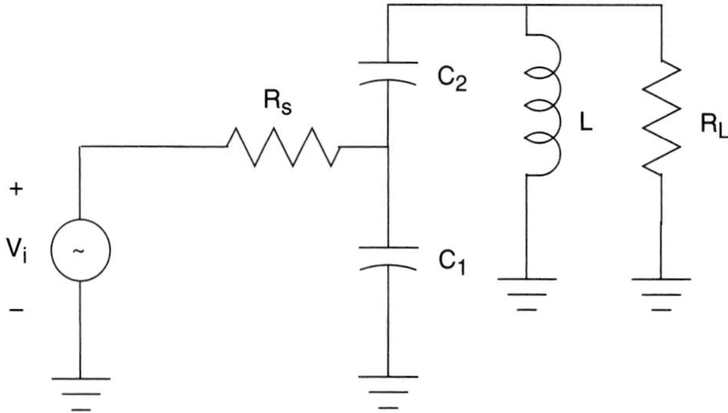

FIGURE 6.52 Tapped-C transformer.

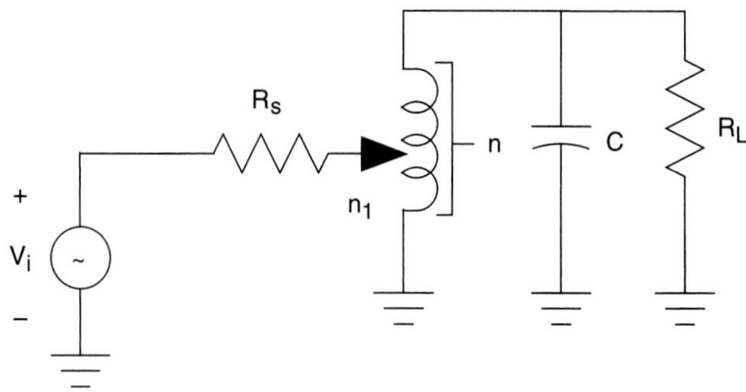

FIGURE 6.53 Tapped-L transformer.

6.8.1 Tapped-C Transformer

$$R'_S = R_S(1 + C_1/C_2)^2, \tag{6.24}$$

$$C_T = \frac{C_1C_2}{C_1 + C_2} \tag{6.25}$$

where "C_T" is the equivalent capacitance that resonates with "L."

6.8.2 Tapped-L Transformer

$$R'_S = R_S(n/n_1)^2 \tag{6.26}$$

$$L_T = L \tag{6.27}$$

The transformer circuits, tapped-C and tapped-L, present a much larger R_S and R_L that is actually present. For example, for a circuit with $R_S = 50\ \Omega$, a transformer could turn the 50 Ω into a 500 Ω and the circuit will be able to see a higher R_S and thus its "Q" would be much higher.

EXAMPLE 6.11

Design a resonant circuit that operates between $R_S = 50\ \Omega$ and $R_L = 2000\ \Omega$ with a $Q = 20$ at the center frequency $f_o = 100$ MHz. The inductor has a $Q_C = 100$ at 100 MHz. You may use a tapped-C transformer to achieve the desired Q (see Figure 6.54).

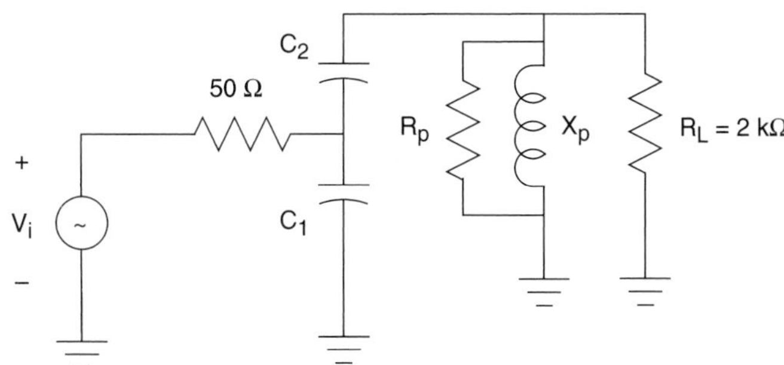

FIGURE 6.54 Circuit for Example 6.11.

Solution:

We will use a tapped-C transformer to step up $R_S = 50\ \Omega$ to 2000 Ω in order to match the load resistance for maximum power transfer.

$$R'_S = R_S(1 + C_1/C_2)^2 \Rightarrow 2000 = 50(1 + C_1/C_2)^2$$

Inductor: $$Q_C = R_P/X_P = 100 \Rightarrow R_P = 100\,X_P \tag{1}$$

$$Q = R_{tot}/X_P = (R'_S \parallel R_L \parallel R_P)/X_P = 20 \tag{2}$$

Using (1) and (2), we solve for R_P and X_P:

$$R_P = 4\ \text{k}\Omega$$

$$X_P = 40\ \Omega \Rightarrow L = X_P/\omega_o = 63.6\ \text{nH}$$

$$C_T = 1/X_P\omega_o = 39.8\ \text{pF}$$

$$C_T = 39.8 = \frac{C_1C_2}{C_1 + C_2} \Rightarrow C_1/C_2 = 5.3$$

If we select $C_1 = 250$ pF, then:

$$C_2 = 47\ \text{pF}$$

The final design is shown in Figure 6.55.

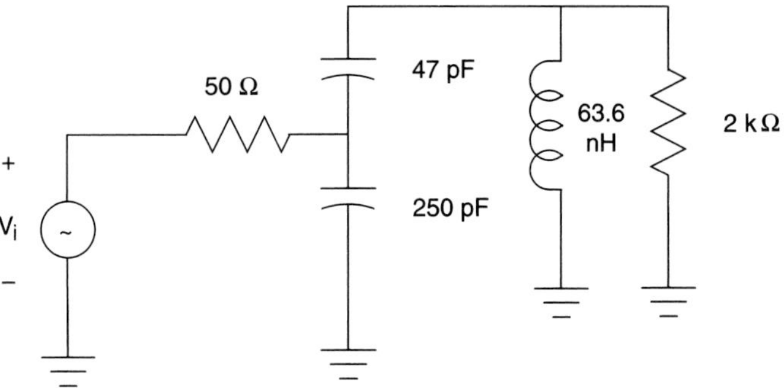

FIGURE 6.55 Final circuit design of Example 6.11.

6.9 RF IMPEDANCE MATCHING

Impedance matching is often necessary in the design of RF circuitry to provide maximum possible transfer of power between a source and its load, as shown in Figure 6.56.

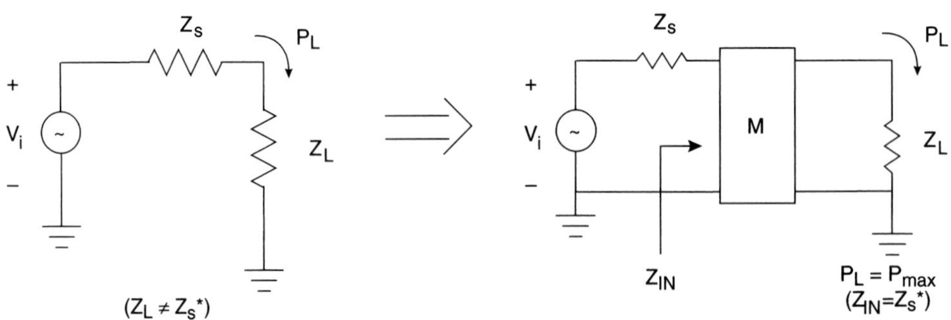

FIGURE 6.56 The concept of using an RF impedance matching network M.

One of the applications of an RF impedance matching network is in the front end of any sensitive receiver where the signal is extremely weak and none of it can be wasted due to mismatch. Therefore use of an appropriate matching network becomes crucial to the overall performance of the circuit.

The maximum power transfer theorem states that (see Chapter 11, *Design of Matching Networks*, for further details):

$$\text{for DC: } Z_S = R_S = Z_L = R_L \qquad \text{(i.e., there is no reactance)} \tag{6.28}$$

and,

$$\text{for AC: } Z_L = Z_S^* \tag{6.29a}$$

$$R_S = R_L \tag{6.29b}$$

$$X_S = -X_L \tag{6.29c}$$

NOTE: *$X_S = -X_L$ is valid at only one frequency (the frequency of resonance). Therefore, a perfect match can occur only at the resonant frequency; this poses a problem in broadband matching of circuits.*

At all other frequencies removed from the matching center frequency, the impedance match becomes progressively worse and eventually nonexistent, as shown in Figure 6.57.

FIGURE 6.57 Power value transfer to the load as a function of load impedance.

(a)

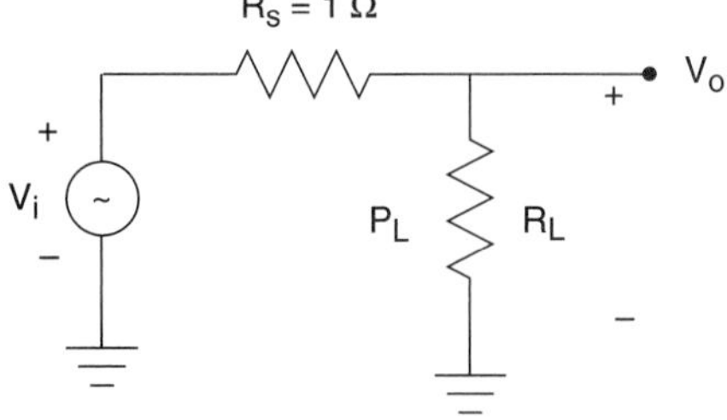

(b)

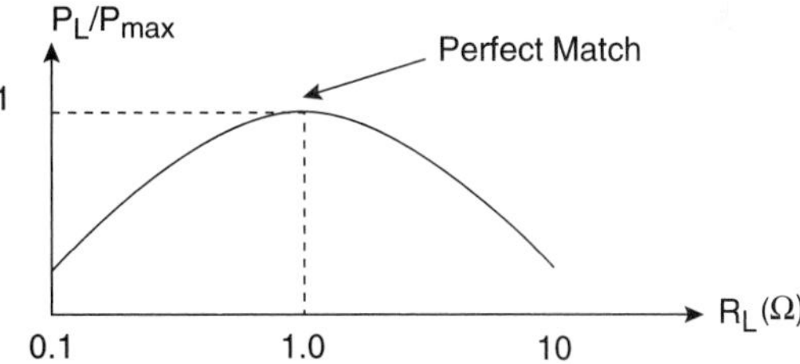

There are an infinite number of possible networks that could be used to perform the impedance matching function. For example, a circuit as simple as a two-element LC network or as elaborate as a seven-element matching network would work equally well. But first we will analyze a simple matching network, as illustrated in the following example.

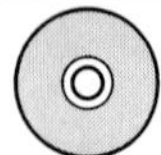

EXAMPLE 6.12

Analyze the LC matching network (shown in Figure 6.58), which transforms a source resistance $R_S = 100\ \Omega$ to a load of $R_L = 1000\ \Omega$.

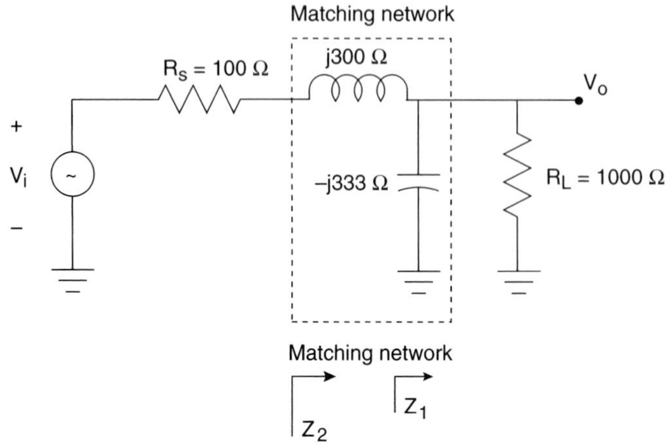

FIGURE 6.58 Circuit for Example 6.12.

Solution:

From Figure 6.59 we have:

$$Z_1 = 1000 \parallel -j333 = 100 - j300\ \Omega$$

$$Z_2 = j300 + (100 - j300) = 100\ \Omega$$

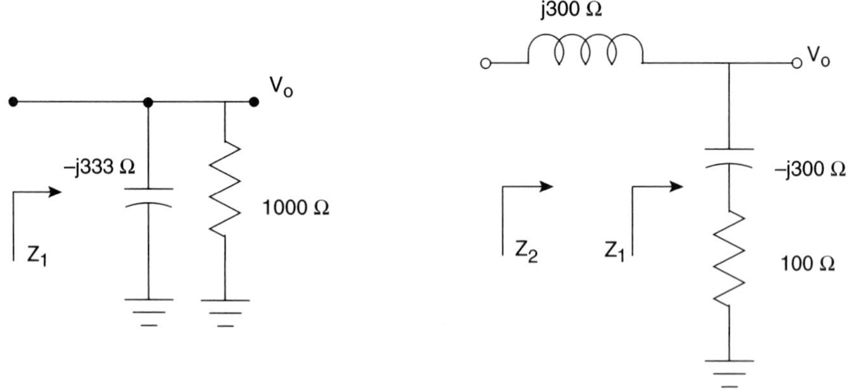

FIGURE 6.59 Circuit analysis for Example 6.12.

Therefore the source sees $R_{IN} = 100\ \Omega$ as the total load, which creates a matched condition with the source resistance ($R_S = 100\ \Omega$), and therefore maximum power transfer (for $V_i = 1$ V) occurs as follows:

$$V_o = 0.5V_i$$

$$(P_L)_{max} = 1/2\ (V_o^2\ /R_{IN})$$

$$= 1/2(0.5^2\ /100) = 1.25\ \text{mW} = 0.97\ \text{dBm}$$

These calculations bring us to an important question:

> How much power would have been transferred if matching were not placed between R_S and R_L?

From the diagram shown in Figure 6.60, we can write:

$$V_i = 1\ \text{V}$$

$$V_o = (1000/1100)V_i = 0.91\ V_i$$

$$P_L = (V_o^2\ /R_L)/2 = (0.91^2\ /1000)/2 = 0.41\ \text{mW} = -3.83\ \text{dBm}$$

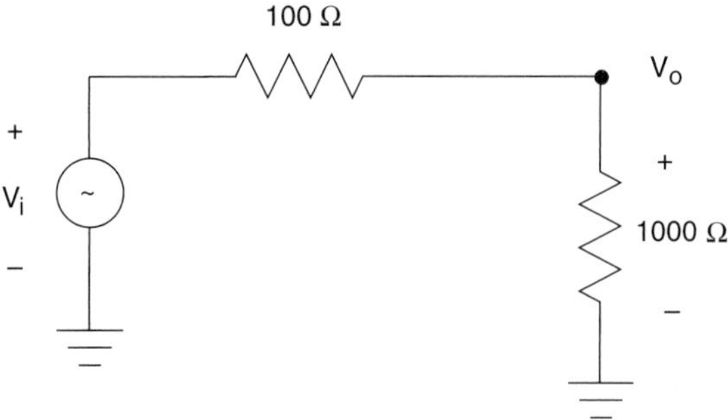

FIGURE 6.60 Simplified equivalent circuit.

Therefore compared to the matched case the power loss would have been:

$$\text{Power loss} = 0.41/1.25 = 0.328 = -4.89\ \text{dB}$$

We can see that only one-third of the available power is transferred and the two-thirds remaining is wasted (i.e., reflected back to the source) due to mismatch.

> **NOTE:** *The function of the shunt component is to transform a larger impedance down to a smaller value with a real part equal to the real part of the source impedance and the reactive component capable of resonating with (or cancelling out) the reactive part of the source impedance (see Figure 6.61).*

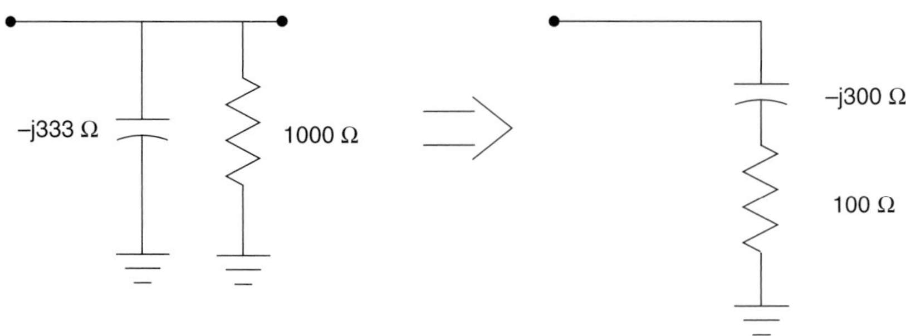

FIGURE 6.61 The result of adding a shunt element.

6.9.1 The L-network

The simplest and most widely used matching circuit for lumped elements is the *L*-network, as shown in Figures 6.62a, b, c, d. The generalized configuration is shown in Figure 6.63 where we can write:

$$R_P = (Q^2 + 1)R_S \tag{6.30}$$

$$Q = Q_S = Q_P \tag{6.31}$$

$$Q_S = Q_P = \sqrt{(R_P/R_S) - 1} \tag{6.32}$$

$$Q_S = X_S/R_S \tag{6.33}$$

$$Q_P = R_P/X_P \tag{6.34}$$

where Q_S, R_S, and X_S are for the series leg and Q_P, R_P and X_P are for the parallel leg.

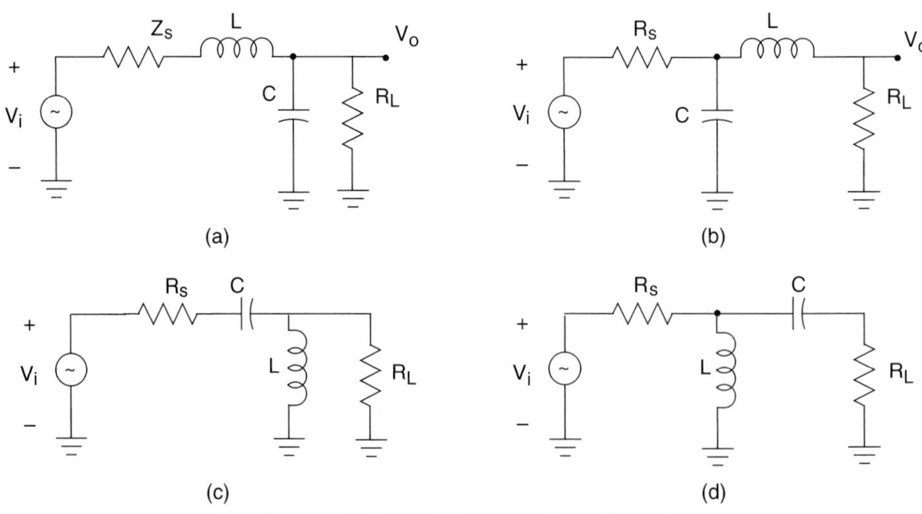

FIGURE 6.62 Several L-network configurations.

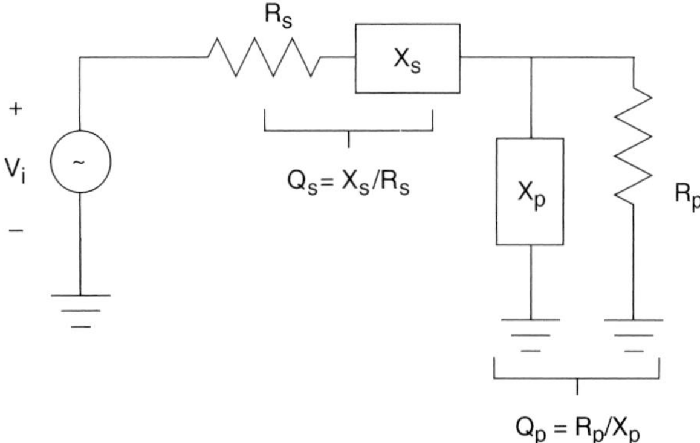

FIGURE 6.63 Generalized L-network configuration.

NOTE: *X_S and X_P may be either capacitive or inductive reactances but must be of opposite types. The following example illustrates the concept of L-networks more clearly.*

EXAMPLE 6.13

Design a circuit to match a 100 Ω source resistance to a 1000 Ω load resistance at f_o = 100 MHz. Assume that a DC voltage must also be transferred from the source to the load and all elements are perfect.

Solution:

The need for DC at the output dictates the need for an inductor in the series leg, as shown in Figure 6.64.

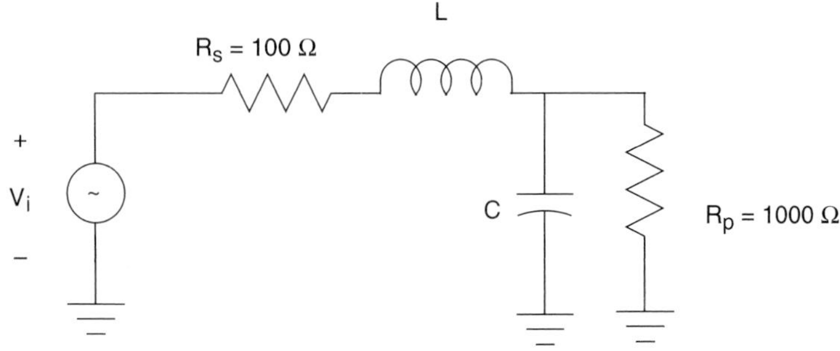

FIGURE 6.64 Design for Example 6.13.

$$R_S = 100\ \Omega, R_P = 1000\ \Omega$$

$$Q_S = Q_P = (1000/100 - 1)^{1/2} = 3$$

$$Q_S = X_S/R_S \Rightarrow X_S = 3R_S = 300\ \Omega$$

$$Q_P = R_P/X_P \Rightarrow X_P = R_P/3 = 1000/3 = 333$$

$$X_S = \omega L \Rightarrow L = X_S/\omega = 477 \text{ nH}$$

$$X_P = 1/\omega C \Rightarrow C = 1/\omega X_P = 4.8 \text{ pF}$$

NOTE: *This circuit was analyzed earlier in Example 6.12.*

The previous examples dealt with matching two real impedances. In actual practice, however, we deal with transistors, transmission lines, antennas, and so on. These all present complex input and output impedances, as shown in Figure 6.65.

FIGURE 6.65

(a) a transistor circuit

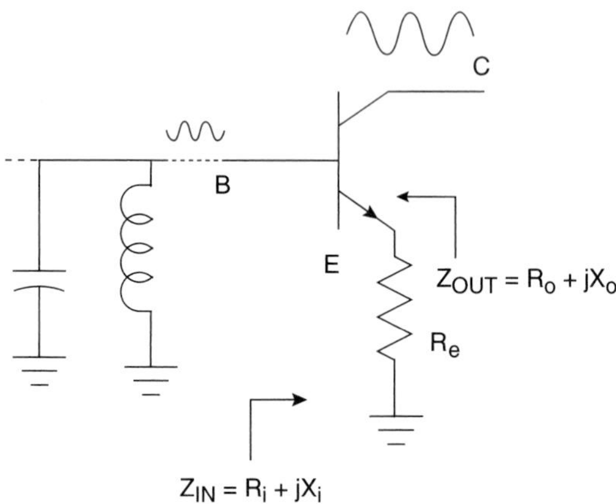

(b) the AC equivalent circuit with parasitic capacitance

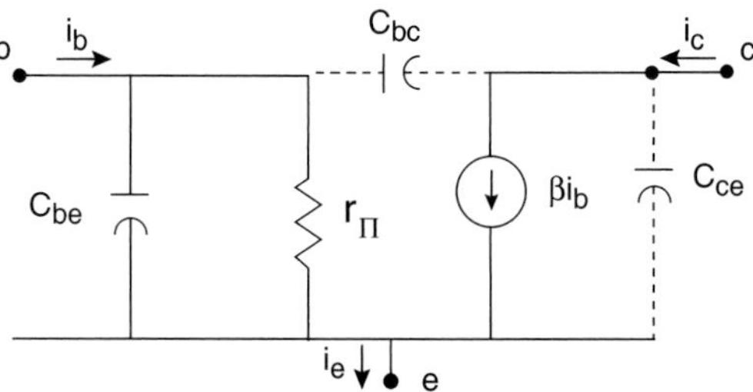

There are two basic approaches to handling complex impedances:

- The absorption method
- The resonance method

These two methods are explained next.

6.9.2 The Absorption Method

This is a method in which any stray reactances can be absorbed into the impedance-matching network by prudent placement of each matching element such that the following occurs:

- Element capacitor (C) is calculated and placed in parallel with stray capacitor (C_p)
- Element inductor (L) is calculated and placed in series with stray inductor (L_p)

- Next, the stray component values are then subtracted from the calculated element values to arrive at the final matching network. The new element values, C' and L', are given by:

$$C = C_p + C' \Rightarrow C' = C - C_p \tag{6.35}$$

$$L = L_p + L' \Rightarrow L' = L - L_p \tag{6.36}$$

The following example further illustrates this method.

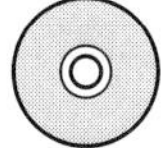

EXAMPLE 6.14

Use the absorption method to match the source (100 + j126 Ω) to a load (1000 || –j795.8 Ω) at 100 MHz, as shown in Figure 6.66.

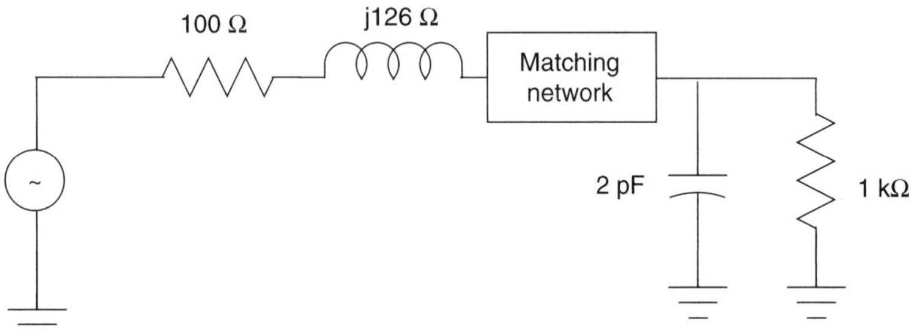

FIGURE 6.66 Circuit for Example 6.14.

Solution:

$$Z_C = -j795.8 = -j/\omega C \Rightarrow C = 1/(795.8 \times 2\pi \times 10^8) = 2 \text{ pF}$$

STEP 1. Totally ignore the reactances and simply match 100 Ω to 1000 Ω such that the inductors are in series and capacitors are in parallel. This step has already been done in Example 6.13, and we use the results for the matching network directly (see Figure 6.67).

$$L = 477 \text{ nH},$$

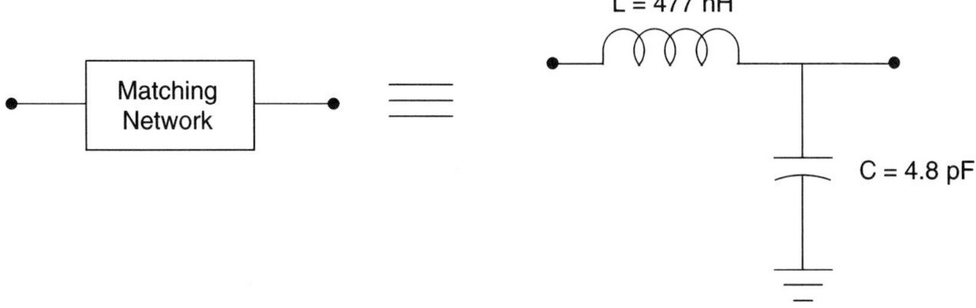

FIGURE 6.67 Matching network realization.

$$C = 4.8 \text{ pF}$$

STEP 2. The new elements are given by:

$$L' = L - L_p = 477 - 200 = 277 \text{ nH}$$

$$C' = C - C_p = 4.8 - 2 = 2.8 \text{ pF}$$

Based on these values, the final design is shown in Figure 6.68.

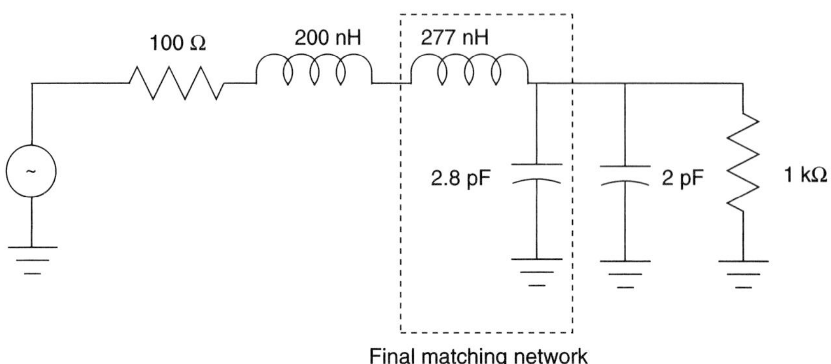

FIGURE 6.68 Final design for Example 6.14.

NOTE: *This method is workable* ***only*** *if the calculated element values (L,C) are higher than the stray values (L_p,C_p).*

6.9.3 The Resonance Method

The resonance method is a method in which any stray reactances are resonated with equal and opposite reactances at the frequency of interest. Once this is done, the design proceeds the same way as two pure resistances: one at the source and the other at load. Example 6.15 illustrates this method.

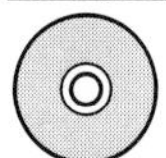

EXAMPLE 6.15

Design a matching network that will match a source resistance of 50 Ω to a capacitive load (at f_o = 75 MHz), as shown in Figure 6.69. The matching circuit should block the DC to the output. Use the resonance method.

Solution:

STEP 1. Resonate 40 pF with a shunt inductor (L_1) with the following value (see Figure 6.70):

$$\omega L_1 = 1/\omega C \Rightarrow L_1 = 1/C\omega^2 \Rightarrow L_1 = 112.6 \text{ nH}$$

STEP 2. Now we match 50 Ω to 600 Ω by the same technique as before (see Figure 6.71):

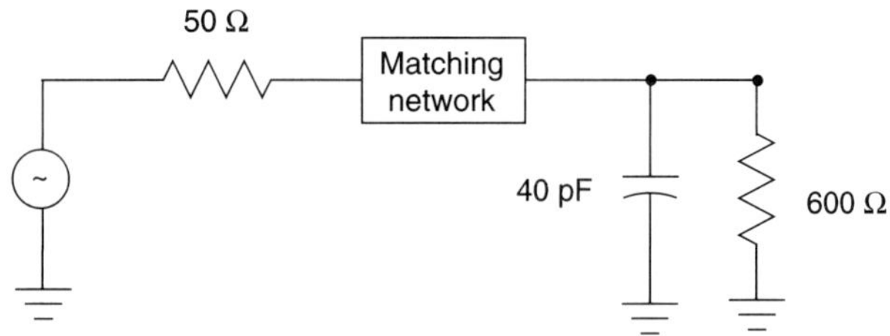

FIGURE 6.69 Circuit for Example 6.15.

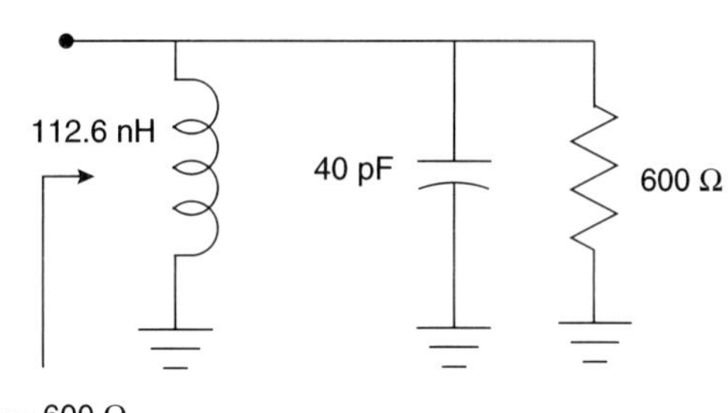

FIGURE 6.70 Resonating L = 112.6 nH with the capacitor.

$$Q_S = Q_P = \sqrt{(600/50) - 1} = 3.32$$
$$X_S = Q_S R_S = 50 \times 3.32 = 166\ \Omega$$
$$X_P = R_P/Q_P = 600/3.32 = 181\ \Omega$$

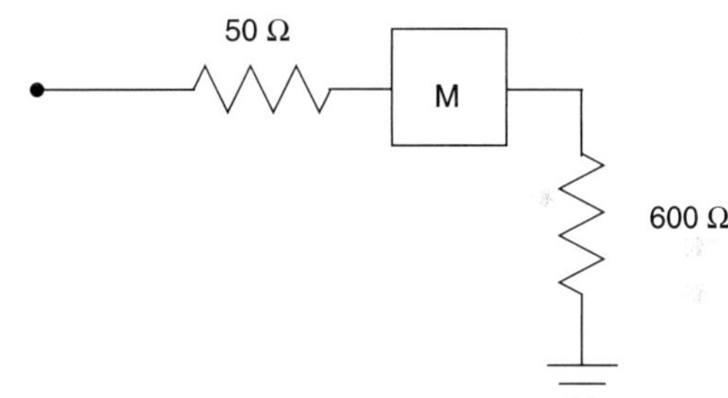

FIGURE 6.71 Matching 50 Ω to 600 Ω by using a matching network (M).

Using a series cap to block the DC, we have:

$$C = 1/\omega X_S = 12.8\ \text{pF}$$
$$L_2 = X_P/\omega = 384\ \text{nH}$$

Combining L_1 and L_2 in parallel (as shown in Figure 6.72) we obtain:

$$L_{tot} = \frac{L_1 L_2}{L_1 + L_2} = 87\ \text{nH}$$

The final circuit is shown in Figure 6.73.

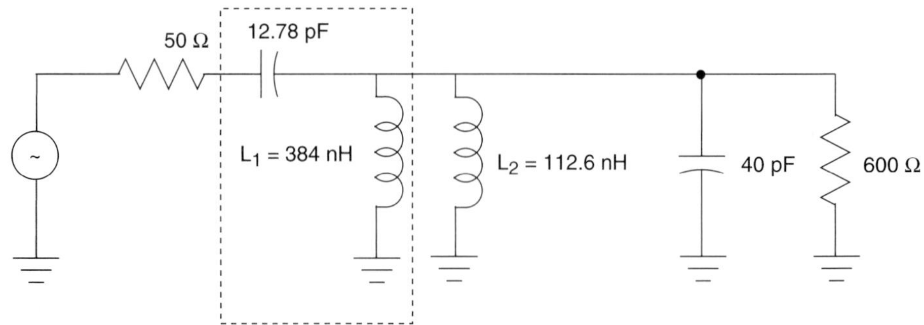

FIGURE 6.72 Using L_1 and L_2 to match a capacitive load to 50 Ω.

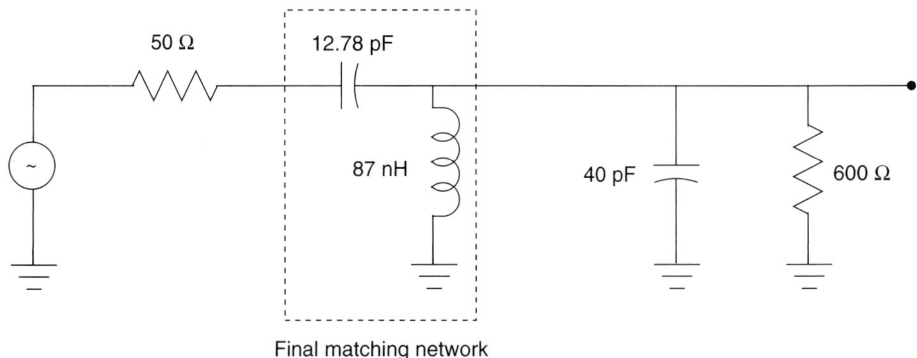

FIGURE 6.73 The final circuit design of Example 6.15.

6.10 THREE-ELEMENT MATCHING

We can observe with the L-networks that once R_s and R_p are given, the Q of the network is defined and the designer no longer has a choice over its value because:

$$Q = \sqrt{(R_P/R_s) - 1} \tag{6.37}$$

If a narrow bandwidth is desired, this will cause a design problem. An example of a circuit with a low Q is shown in Figure 6.74.

The minimum Q available is the circuit Q established with an L-matching network. However, three-element networks overcome this disadvantage and can be used for narrow-band high-Q applications. There are two types of three-element networks, as shown in Figure 6.75.

Using more than three elements in the matching network design would bring about a greater amount of design flexibility, which can lead to tedious mathematical equations. For this reason mathematical calculations of the reactive elements X_1, X_2, and X_3, which are too complicated, are omitted here. Instead we will show a simpler way to calculate these values later using a graphical tool called a Smith chart. As will be seen shortly, the use of a Smith chart greatly simplifies a very complex design process.

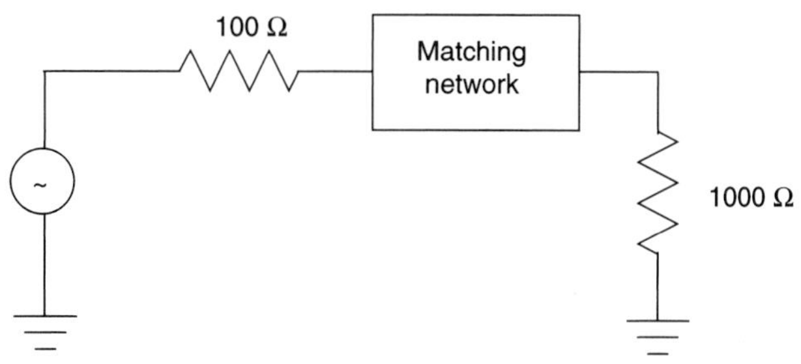

FIGURE 6.74 An example of a circuit with a low Q value.

FIGURE 6.75 Use of three-element networks.

(a) Pi-network

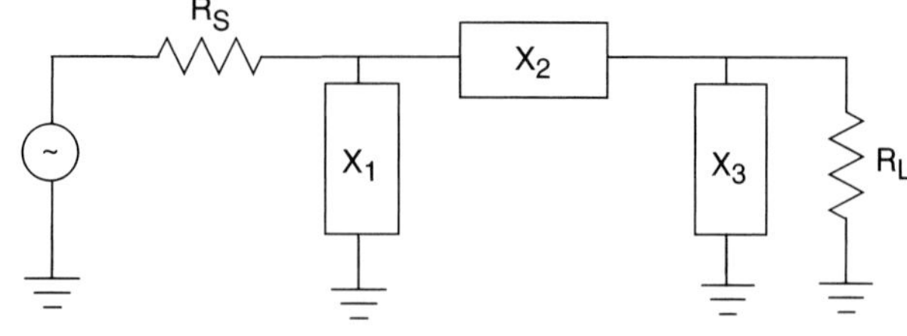

(b) T-network

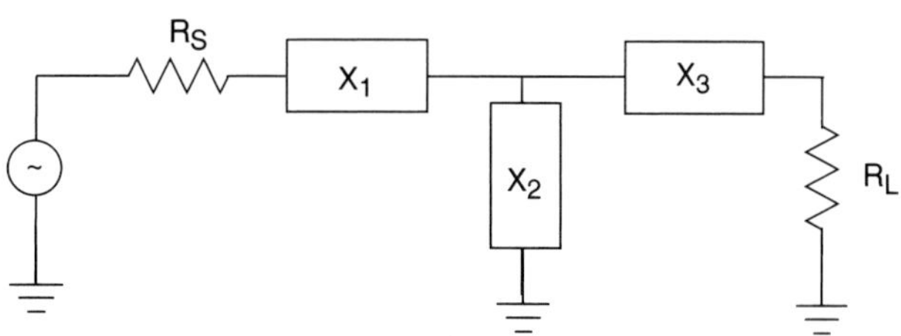

LIST OF SYMBOLS/ABBREVIATIONS

A symbol will not be repeated again, once it has been identified and defined in an earlier chapter, as long as its definition remains unchanged.

AWG	American Wire Gauge
BW	Bandwidth
C_d	Distributed capacitance
C_T	Total capacitance of a center tapped capacitor
f_o	Frequency where Q is maximum; center of pass band
f_r	Resonance frequency
$H(\omega)$	Voltage gain
IL	Insertion loss
PF	Power factor
μ	Permeability
Q	Quality factor
Q_o	Maximum quality factor where $f = f_o$

R_P	Insulation resistance; also parallel resistance
R_S	Series resistance
SF	Shape factor
V_P	Phase velocity
σ	Conductivity
δ	Skin depth or depth of penetration
ϕ	Phase angle
ε_o	Dielectric constant of free space
ε	Dielectric constant of a material
ε_r	Relative dielectric constant of a material
ρ	Resistivity

PROBLEMS

6.1 Design a resonant circuit with a loaded Q of 50 that operates between a source of 100 Ω and a load of 2000 Ω at a frequency of 100 MHz.

6.2 Transform a series configuration of a 250 nH inductor into an equivalent parallel configuration at 50 MHz.

6.3 Using the tapped-C method, design a resonant circuit with a loaded Q of 40 at a center frequency of 100 MHz that operates between a source resistance of 100 Ω an a load resistance of 3000 Ω. The capacitors are all lossless and the inductor has a Q of 100 at 100 MHz.

6.4 Design a simple parallel LC resonant circuit to provide a bandwidth of 10 MHz at a center frequency of 100 MHz. The resonant circuit is operating between a source and a load impedance of 2000 Ω each. The capacitor is lossless, and the Q of the inductor is 85. Calculate the insertion loss of the resonant circuit in operation.

FIGURE P6.4

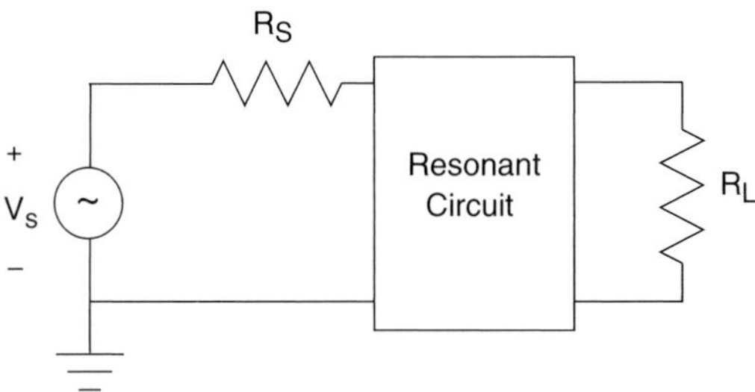

6.5 Using an L-network, design a circuit to match a 50 Ω source resistance to an 850 Ω load at 50 MHz. Assume that the DC must also be transferred from the source to the load.

6.6 Using the absorption method, design a matching network to match the source and the load at 50 MHz, as shown in Figure P6.6.

FIGURE P6.6

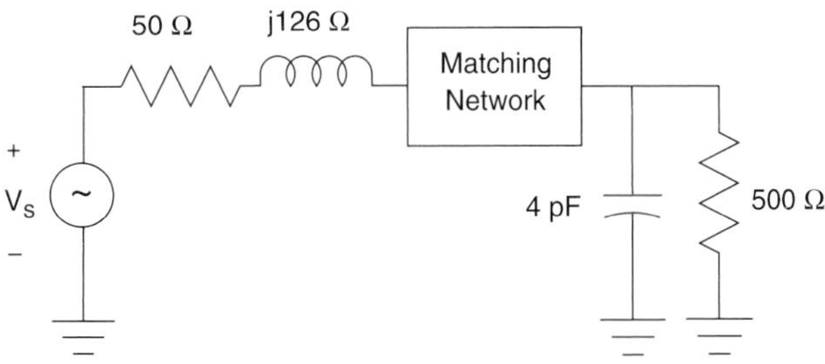

6.7 Using the resonance method, design an impedance matching network that will block the flow of DC from the load, as shown in Figure P6.7. Assume $f = 100$ MHz.

FIGURE P6.7

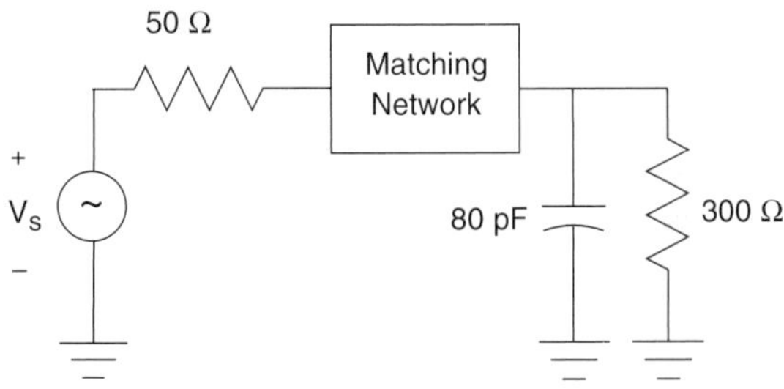

REFERENCES

[6.1] Bowick, C. *RF Circuit Design*. Carmel: SAMS-Prentice Hall, 1982.

[6.2] Carr, J. J. *Secrets of RF Circuit Design*. New York: McGraw-Hill, 1991.

[6.3] Gottlieb, I. W. *Practical RF Power Design Techniques*. New York: McGraw-Hill, 1993.

[6.4] Harsany, S. C. *Principles of Microwave Technology*. Upper Saddle River: Prentice Hall, 1997.

[6.5] Krauss, H. L., C. W. Bostian, and F. H. Raab. *Solid State Radio Engineering*. New York: John Wiley & Sons, 1980.

[6.6] Lenk, J. D. *Lenk's RF Handbook*. New York: McGraw-Hill, 1992.

[6.7] Matthaei, G., L. Young, and E. M. *Jones Microwave Filters, Impedance-Matching Networks, and Coupling Structures*. Dedham: Artech House, 1980.

[6.8] Scott, A. W. *Understanding Microwaves*. New York: John Wiley & Sons, 1993.

[6.9] Vizmuller, P. *RF Design Guide*. Norwood: Artech House, 1995.

CHAPTER 7

Fundamental Concepts in Wave Propagation

7.1 INTRODUCTION

The subject of RF/Microwaves primarily deals with electrical energy at high frequencies. Therefore, to know microwaves, we need to know the three qualities of energy in general. By quality we mean the character or nature of energy.

7.2 QUALITIES OF ENERGY

The following qualities apply to any and all types of energy (whether electrical, mechanical, thermal, or chemical), at high or low frequencies. Because we are dealing with electronics, we will narrow the following discussion to electrical energy and waves only.

7.2.1 Quality #1: Existing Characteristics

These characteristics can be divided into three classes:

1. **A flow** is the transfer of energy from one point to another. The energy in a flow can have any type of waveform. So a flow is a transfer. This is shown in Figure 7.1.
2. **A divergence (also referred to as a "dispersal")** is a generalized case of a "flow" where a number of flows travel from or to a common center as shown in Figure 7.2a and b.

   NOTE: *"A divergence" is similar in concept but different (in definition) from "divergence of a vector quantity," which is an exact mathematical operation measuring the net outflux (or influx) of a vector quantity.*

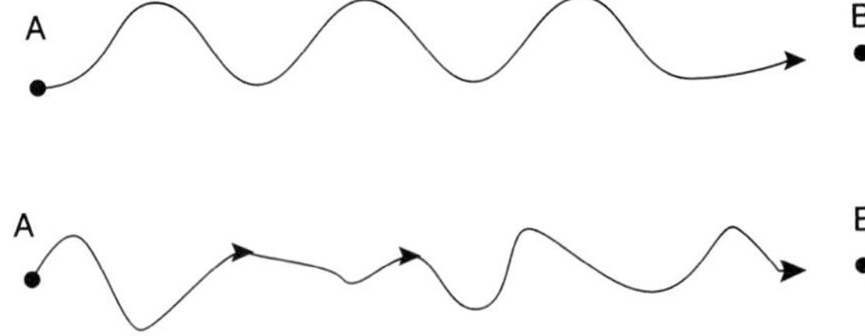

FIGURE 7.1 A flow.

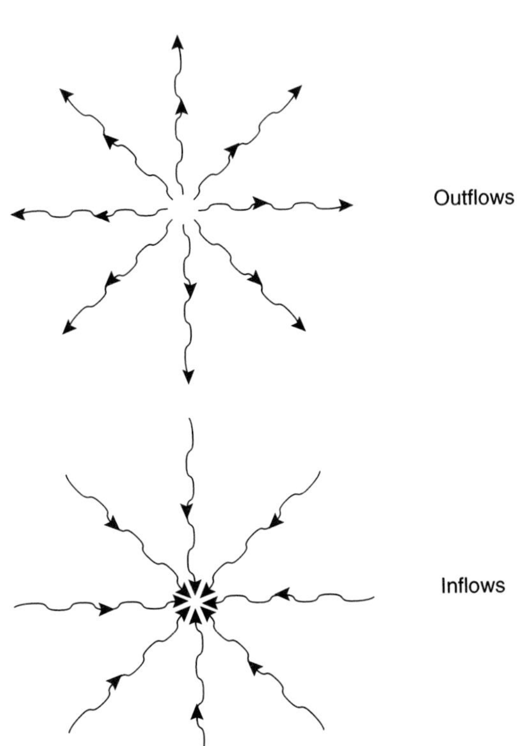

FIGURE 7.2 A divergence.

(a) Net outlow

(b) Net inflow

3. **A standing wave (also called a ridge of energy)** is energy suspended in space and comes about when two flows or divergences of approximately equal magnitude and exactly equal frequency impinge against one another with sufficient amplitude to cause an enduring state of energy, which may last after the flow itself has ceased. For example, a resonator or a cavity oscillator falls into the category of devices that generate this type of wave characteristic. A few examples are shown in Figure 7.3a and b.

7.2.2 Quality #2: Wavelength

Wavelength is a characteristic of an orderly flow of motion and describes its regular and repeated pattern by the distance between its peaks. Many motions are too random and too chaotic to have an orderly flow and thus have no wavelength.

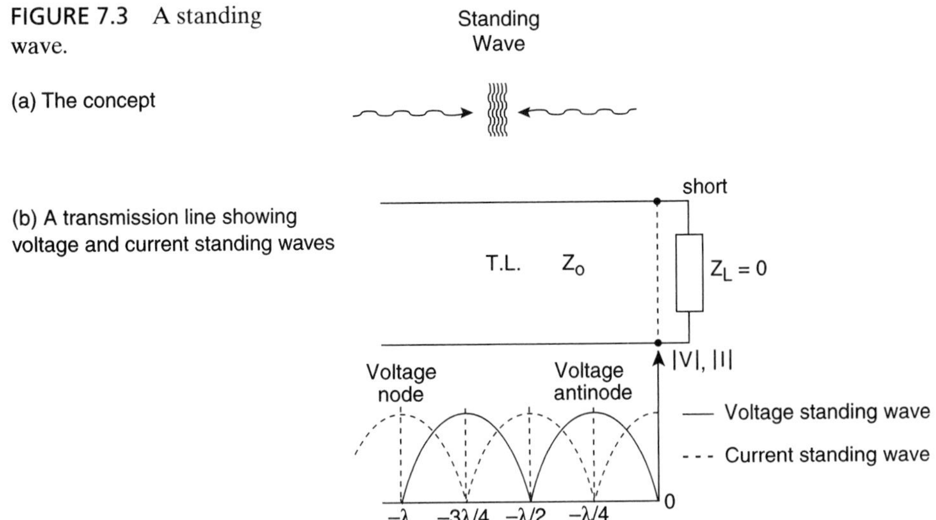

FIGURE 7.3 A standing wave.

(a) The concept

(b) A transmission line showing voltage and current standing waves

DEFINITION-WAVELENGTH: *is defined to be the physical distance between two points having the same phase in two consecutive cycles of a periodic wave along a line in the direction of propagation, as shown in Figure 7.4.*

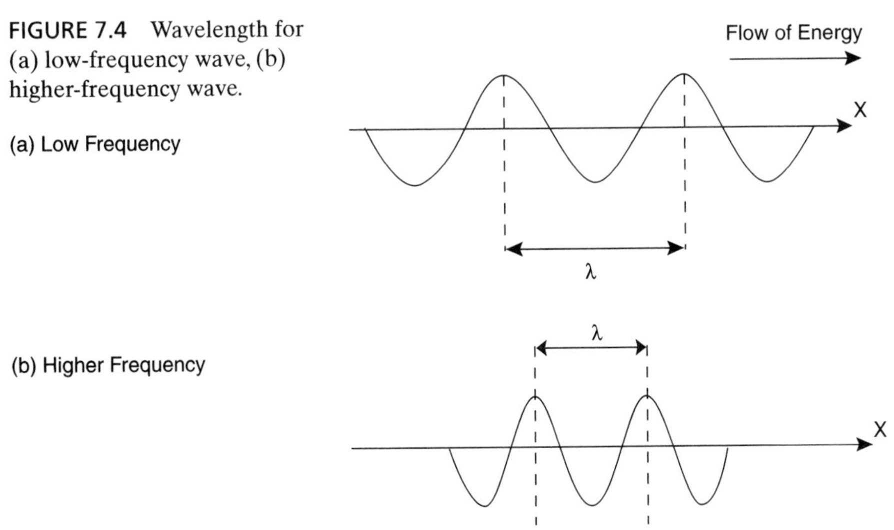

FIGURE 7.4 Wavelength for (a) low-frequency wave, (b) higher-frequency wave.

(a) Low Frequency

(b) Higher Frequency

As frequency increases, the wavelength (λ) decreases as can be observed. Thus, higher-frequency waves have shorter wavelengths, as already discussed in Chapter 5, *Introduction to Radio Frequency and Microwave Concepts and Applications* (see Figure 7.4).

Wavelength has no bearing on the wave characteristics (Quality #1) but applies to the repetition property of the wave flow. A standing wave has a potential flow when released; therefore it may be considered to have a wavelength even though it is not a flow or a wave in the truest sense of the word.

If a random wave is periodic, it can be considered to have a wavelength using the Fourier theorem. It can be proven mathematically that through the use of Fourier analysis, any wave can be decomposed into its Fourier harmonics, provided that the wave is continuously flowing and periodic, as shown in Figure 7.5.

FIGURE 7.5 Examples of wave patterns.

7.2.3 Quality #3: A Flow's Direction (or Absence Thereof)

This quality describes the direction or the absence of direction of flow. A few examples are shown in Figure 7.6. This quality is an important one because energy can have a flow with no net transfer of energy, that is, absence of the direction of flow. For example, a wave traveling from a transmitter to a receiver, or electrons moving in a wire under the influence of an electric field, is said to have a "direction of flow." Examples of absence of direction include a free electron moving in the lattice of a solid at equilibrium (i.e., when no external field is applied), which is a flow with an absence of direction, or an electron in an atom moving in an orbit around the nucleus. Both are flows without a net transfer of energy.

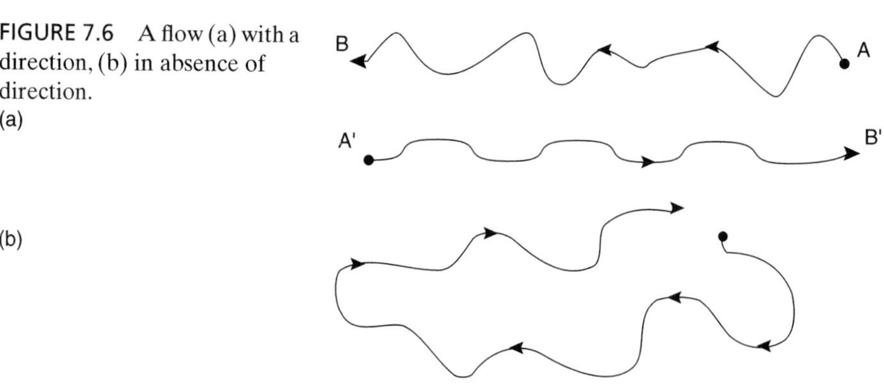

FIGURE 7.6 A flow (a) with a direction, (b) in absence of direction.

7.3 DEFINITION OF A WAVE

So far we have loosely used "a wave" to mean a special case of a flow of energy. Now we need to define it exactly:

> **DEFINITION-WAVE:** *A disturbance that propagates from one point in a medium to other points without giving the medium, as a whole, any permanent displacement.*

This general definition of a wave includes any and all disturbances that could be of electrical or nonelectrical origins. We further restrict our definition to a special class of waves that are of electrical origin. These waves are called electromagnetic (EM) waves. Now, we need to define an important term:

DEFINITION-ELECTROMAGNETIC (EM) WAVE: *A radiant energy flow produced by oscillation of an electric charge. In free space and away from the source (which is composed of vibrating electric charges), EM rays of waves consist of vibrating electric and magnetic fields that move at the speed of light (in a vacuum) and are at right angles to each other and to the direction of motion.*

The propagation of a simple electromagnetic wave in free space is shown in Figure 7.7. EM waves propagate with no actual transport of matter and grow weaker in amplitude as they travel farther in space. EM waves include radio waves, microwaves, infrared waves, visible/ultraviolet light waves, and X-, gamma-, and cosmic rays. (See the discussion of the electromagnetic spectrum in Chapter 6, *RF Electronics Concepts*.) These are all different types of electrical energy, and all follow the same principles that we have discussed so far in this chapter (see Figures 7.8 and 7.9).

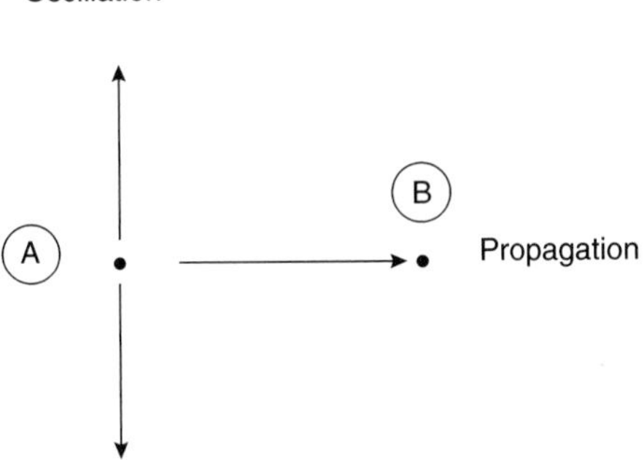

FIGURE 7.7 The propagation of a simple wave in space from A to B.

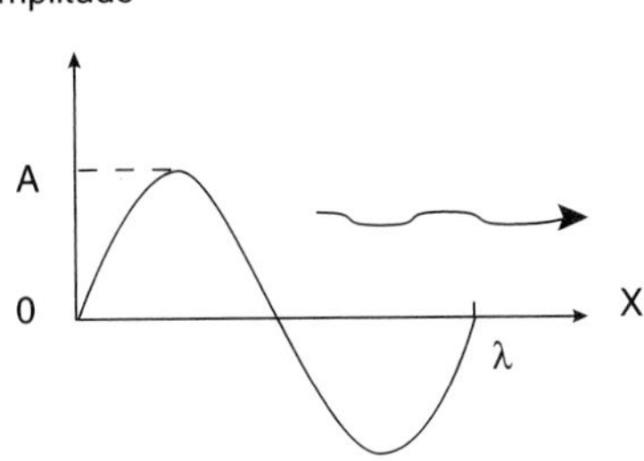

FIGURE 7.8 Wave amplitude in space.

On a larger view of things, we can observe that RF and microwaves are a special case of EM waves, which are themselves a subset of a larger field of study, i.e., waves. Of course, this last itself is a subset of a much larger sphere of existence known as "energy," as shown in Figure 7.10.

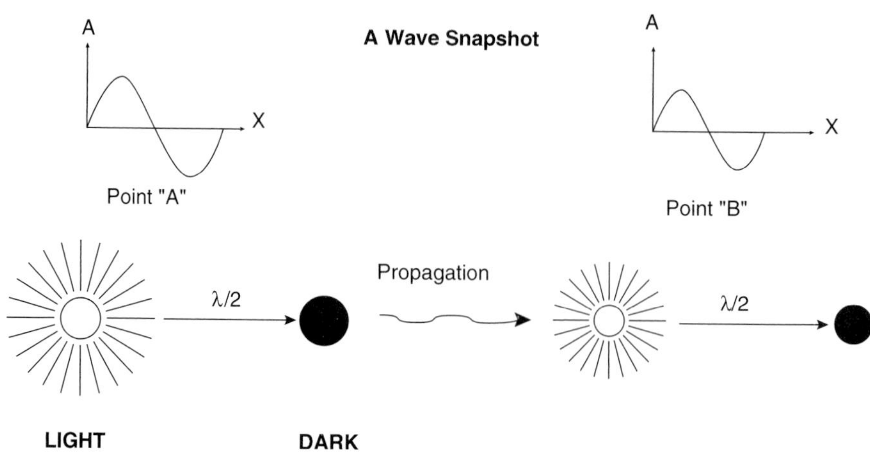

FIGURE 7.9 A reduction of wave amplitude as it propagates.

FIGURE 7.10 Relationship of RF and microwaves to other forms of energy.

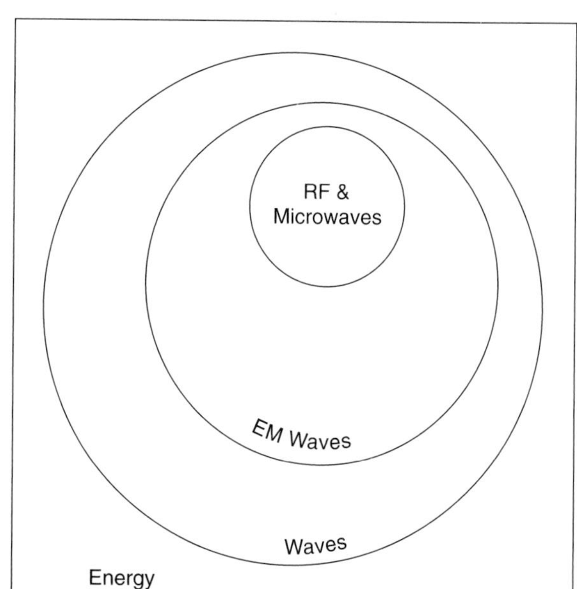

7.4 MATHEMATICAL FORM OF PROPAGATING WAVES

We know that $f(x - x_o)$ is the same function as $f(x)$ except shifted to the right a distance x_o along the $+x$ axis. If instead we consider $f(x - vt)$, then the function $f(x)$ is shifted to the right a distance $x_o = vt$, where v and t can be considered to be the velocity of motion and the elapsed time, respectively. The distance (x_o) increases as time elapses; therefore, the function is displaced continuously farther out along the $+x$ axis as time elapses.

7.4.1 An Important Special Case: Sinusoidal Waves

Assume $f(x)$ is a sinusoidal function:

$$f(x) = A\cos\beta x \tag{7.1}$$

Where A is the amplitude and β is the phase constant. Then a sinusoidal wave propagating in the $+x$ direction would be represented in the time and phasor domains by:

a. Time domain form:

$$f(x,t) = A\cos\beta(x - vt) = A\cos(\beta x - \omega t) \tag{7.2}$$

b. Phasor form:

$$F = Ae^{-j\beta x} \tag{7.3}$$

where ω $(= \beta v)$ is the angular frequency.

To find the wavelength (λ), we know that it is defined to be the physical distance between two peaks (or valleys). We note that at $t = 0$, the wave's peak is at $x = 0$. The next peak is at $x = \lambda$ and the sinusoidal wave has a phase of 2π, thus:

$$\beta\lambda = 2\pi \Rightarrow \lambda = 2\pi/\beta \tag{7.4}$$

For the wave propagating in the $-x$ direction, the following can be written:

a. Time domain:

$$f(x,t) = A\cos(\beta x + \omega t) \tag{7.5}$$

b. Phasor domain:

$$F = Ae^{j\beta x} \tag{7.6}$$

The phase velocity (V_P), which is defined to be the velocity at which the plane of the constant phase propagates, can be obtained from:

$$\beta x - \omega t = k, \tag{7.7}$$

where k is an arbitrary constant.

Differentiating Equation 7.7 with respect to time gives the phase velocity:

$$\beta dx/dt - \omega = 0 \Rightarrow V_p = dx/dt = \omega/\beta$$

In an unrestricted or "free" space, a plane wave travels at velocity V_P, which is given by:

$$V_p = 1/\sqrt{\mu_o\varepsilon_o} \tag{7.8}$$

where μ_0 and ε_0 are the permeability and permittivity of free space.

Equations 7.2 and 7.5 show a simple wave that keeps its size and shape while propagating at a constant velocity V_P. This type of propagation is said to be undistorted and unattenuated because it is propagating in free space (or a vacuum), which is a nondispersive medium. We now need to define a term:

DEFINITION-DISPERSIVE MEDIUM: *A medium in which the phase velocity (V_P) of an EM wave is a function of frequency.*

This means that a complex wave, consisting of several frequencies, travels through a dispersive medium at different velocities; that is, each frequency component travels at $V_p = \omega/\beta$ with different time delays. This would cause the wave to be distorted at the exit point. For example, a square-pulse waveform entering and traveling through a dispersive medium will lose its shape and will appear rounded at both of its edges when exiting the medium.

EM waves can have a "rise and fall" as well as an "advance and retreat" type of oscillation of the field quantity, as they propagate (see Figure 7.11).

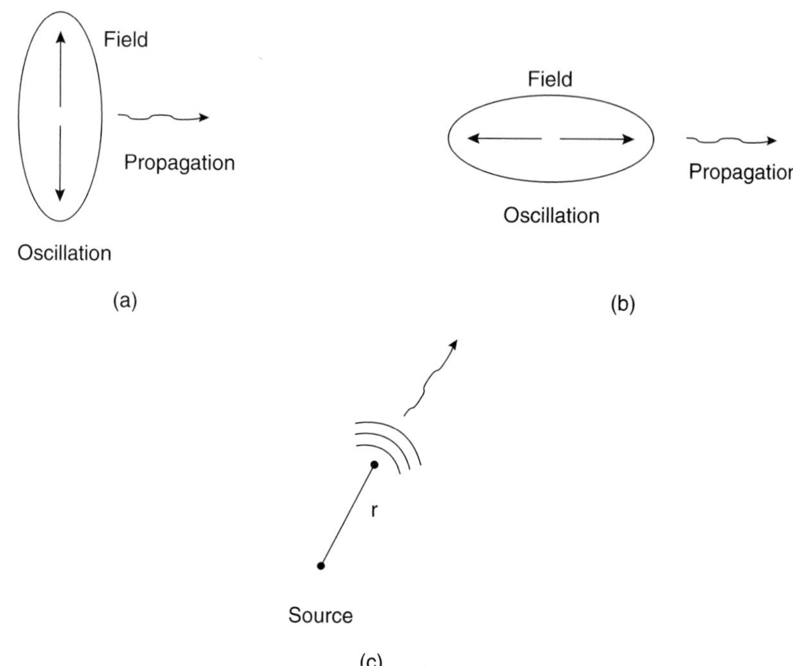

FIGURE 7.11 Propagation of an EM wave for different types of oscillation of fields: (a) transverse field, (b) longitudinal field, (c) wave a distance r from source.

7.4.2 Types of Waves

Waves are like fluids and propagate according to the medium in which they find themselves. If the medium is unrestricted, then it would be called "free-space wave propagation." When the source is a point and the medium of propagation is free space, waves have spherical wavefronts, as shown in Figure 7.12.

7.4.3 A Special Case: Plane Waves

When waves under consideration are at an infinite distance away from the source of disturbance, then the wavefront of each wave is a plane surface, and these waves are called plane waves (see Figure 7.13). These plane waves are in the TEM mode of propagation, as defined next.

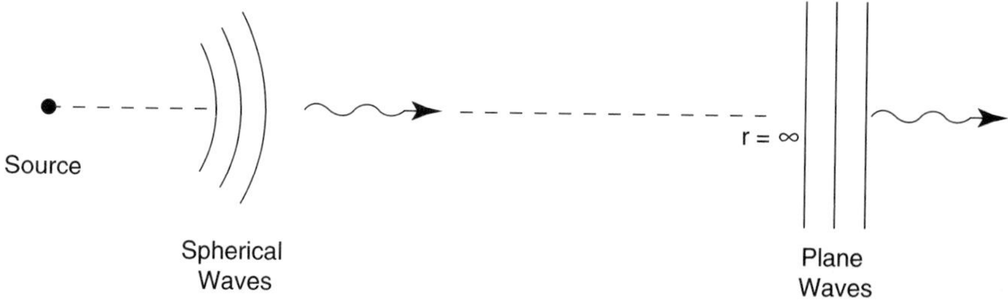

FIGURE 7.12 A spherical wave at an infinite distance from a source.

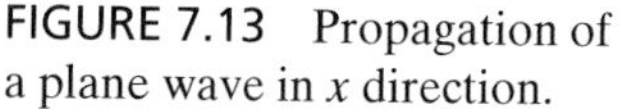

FIGURE 7.13 Propagation of a plane wave in x direction.

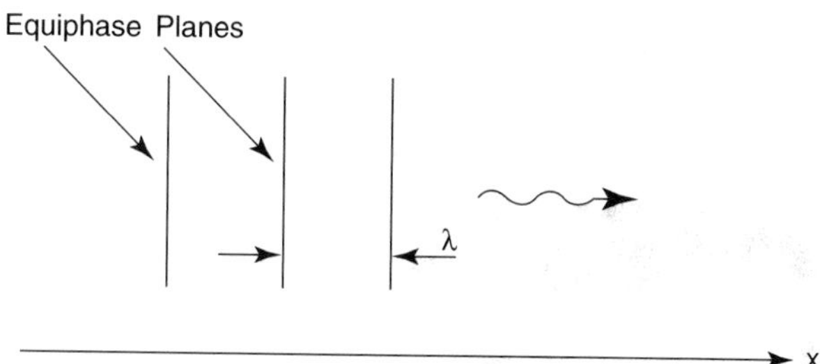

DEFINITION-TEM (TRANSVERSE ELECTRO-MAGNETIC) MODE: *Waves having their electric and magnetic fields perpendicular to each other and to the direction of propagation. These waves have no field components in the direction of propagation.*

A typical TEM wave in free space is shown in Figure 7.14.

FIGURE 7.14 A typical TEM wave in free space.

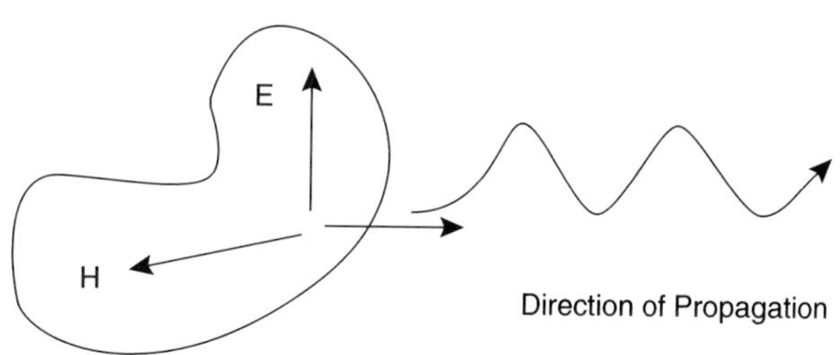

The mathematical expression, $a(x, t)$, for the plane wave propagation is defined in the text that follows.

1. General time domain form is given by:

$$a(x,t) = A_o f(\omega t - \beta x) \tag{7.9}$$

For a time harmonic wave, we can write:

$$a(x,t) = A_o \cos(\omega t - \beta x) = Re(A_o e^{-j\beta x} e^{j\omega t}) \tag{7.10}$$

2. In phasor domain, we have:

$$A(x) = A_o e^{-j\beta x}, \tag{7.11}$$

which is a plane wave propagating in the $+x$ direction, as shown in Figures 7.12 and 7.13.

7.5 PROPERTIES OF WAVES

By property we mean a distinctive attribute or feature of waves. Several properties of waves are worthy of consideration at the outset of this section:

7.5.1 Property #1: Flow Property

This property is in common with Quality #1 for energy. A wave is a flow. It goes from point A to point B, and in doing so a transfer of energy takes place, of course, with a reduced amplitude at the destination, as shown in Figure 7.15.

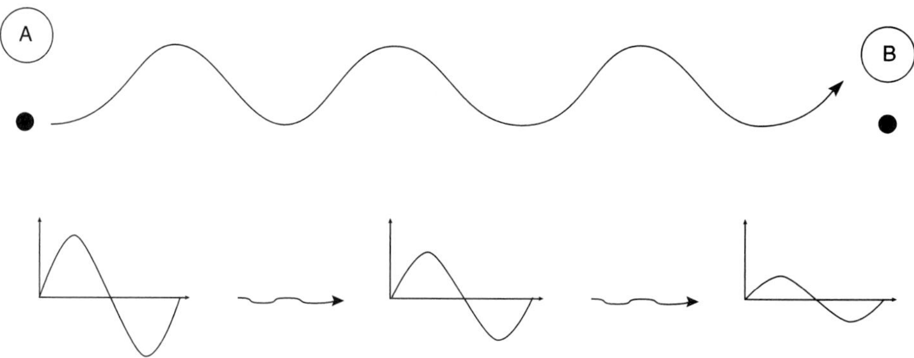

FIGURE 7.15 Wave as a flow of energy.

7.5.2 Property #2: Wavelength Property

This property was discussed as Quality #2 of energy. A wave with a regular and periodic (or repeating) waveform has a wavelength that is the physical distance between two peaks (or valleys) in two consecutive cycles, as defined earlier in a more precise way. This concept is shown in Figure 7.16.

In order to derive wavelength (λ), we know that the speed of propagation (v) is uniform, thus, the distance (λ) traveled in one wave period ($T = 1/f$) is:

$$\lambda = vT = v/f \tag{7.12}$$

At high RF and microwaves, the wavelength ranges from one meter to one millimeter corresponding to a frequency of 300 MHz to 300 GHz.

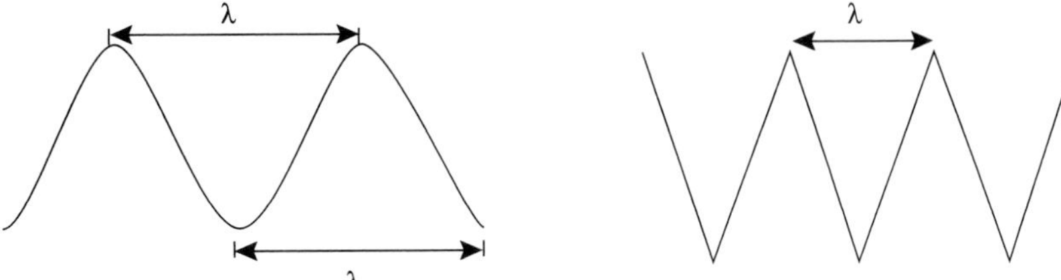

FIGURE 7.16 The concept of wavelength.

7.5.3 Property #3: Reflection and Transmission Property

When a wave encounters an obstacle or a different medium, some of it reflects back (called a reflected wave) and the rest of it transmits through (called a transmitted wave). This is true for any and all types of waves.

EXAMPLE 7.1 Perfect Reflection

How does a perfect mirror behave for an incident wave?

Solution:

For a perfect mirror we have perfect reflection, i.e., 100 percent of the incident wave reflects back and zero transmission takes place, as shown in Figure 7.17.

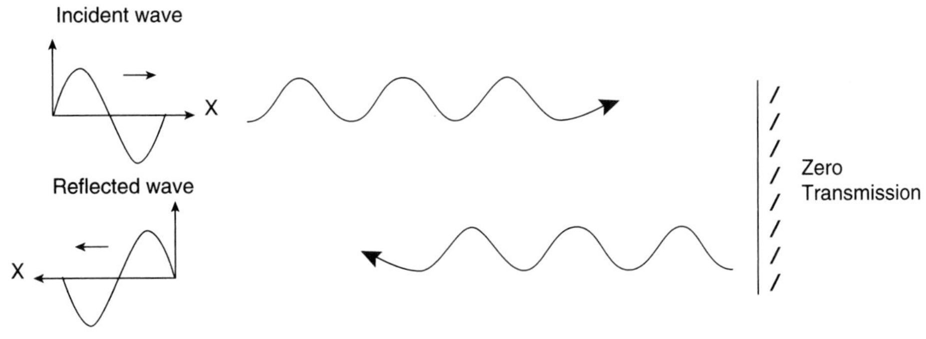

FIGURE 7.17 A perfect mirror.

EXAMPLE 7.2 Perfect Transmission

What would constitute a perfect transmission condition?

Solution:

For a perfect transmission, the two media have to be identical in their electrical properties (such as permittivity, permeability, etc.), as shown in Figure 7.18. This means that for this condition to occur, the second medium has to continue to behave electrically the same as the first.

FIGURE 7.18 A perfect transmission.

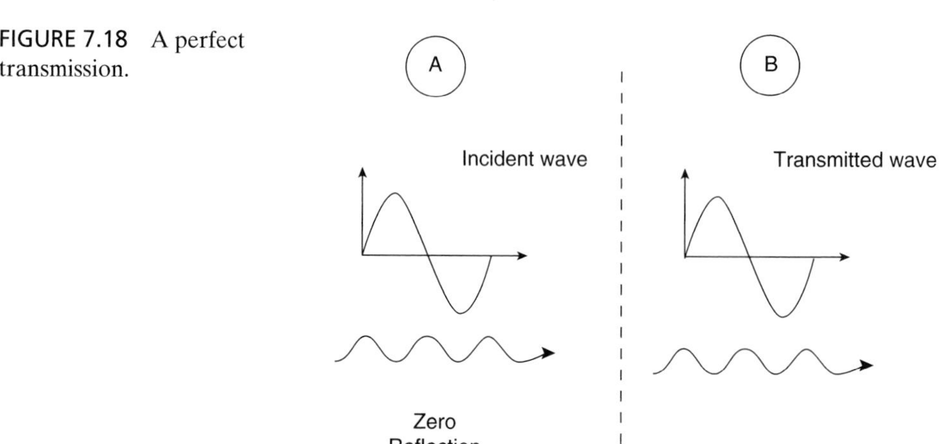

7.5.4 Property #4: Standing-Wave Property

When two waves of exactly the same magnitude and frequency travel opposite to each other, the result is not a wave but an "oscillation with no propagation" called a "standing wave," which has a fixed location, as shown in Figure 7.19.

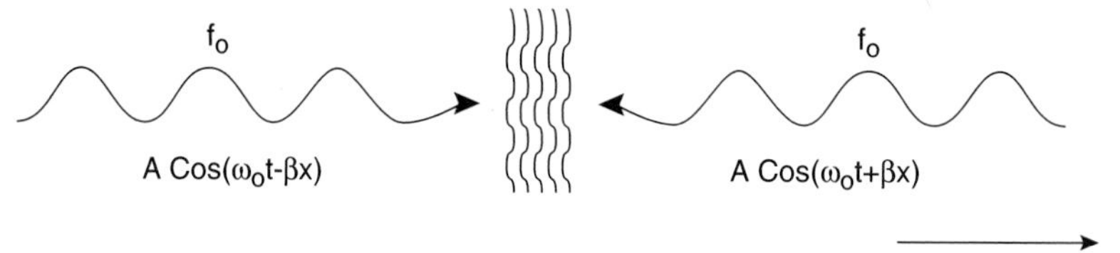

FIGURE 7.19 A standing wave.

The standing wave can be written mathematically in the following domains as:

1. Phasor domain:

$$Ae^{-j\beta x} + Ae^{+j\beta x} = 2A\cos\beta x \tag{7.13a}$$

2. Time domain:

$$2A\cos(\beta x)\cos(\omega t) \tag{7.13b}$$

Because Equation 7.13b is not of the form $f(\beta x - \omega t)$, it is not a wave but a pure oscillation at a fixed location!

NOTE: *A definite prerequisite for a standing wave is two opposite waves of exactly the same frequency. If their amplitudes are comparable but not equal, the result would be a standing wave plus a traveling wave and not a pure standing wave as described earlier.*

7.6 TRANSMISSION MEDIA

When waves are constricted to a limited transmission space (also called a line, guide, channel, etc.), then the waves take on different forms and patterns according to the shape of the guide, just like fluid flow in a pipe.

7.6.1 Types of Transmission Media

A few examples of the wave patterns in different transmission media are coaxial line, two-wire transmission line, a waveguide, a microstrip line, a parallel plate waveguide, and a stripline, as shown in Figure 7.20.

Generally, any and all of these five transmission media could be called "transmission lines," but the terminology has been made more specific to convey more exact concepts, thus:

1. (a), (b), (e) and (f) are generally labeled as transmission lines (TLs)
2. (c) is labeled a waveguide
3. (d) is labeled a microstrip line

(a), (b), (e) and (f) all will support propagation of transverse electromagnetic (TEM) waves and will be used specifically in this book. An example of a TEM wave was shown earlier in Figure 7.14, where the direction of propagation is perpendicular to the oscillating electric and magnetic fields.

NOTE 1: *Structure (d), a microstrip line, supports a quasi-TEM wave, which is a wave with a small axial field. This type of transmission line has gained tremendous popularity in microwave integrated circuits due to its planar structure and ease of fabrication using printed circuit technology. Microstrip lines will be discussed in detail in a later section.*

NOTE 2: *Structures (a) and (c), a coaxial line and a waveguide, are closed structures and are preferred because they have much lower radiation losses than the other three open structures.*

NOTE 3: *Structure (f), a stripline transmission line, can be thought of as a "flattened-out" coaxial line, where both have a center conductor that is enclosed by an outer ground conductor with a uniform dielectric material filling the space between the two.*

NOTE 4: *Other types of transmission media include slotline, dielectric waveguide, coplanar waveguide, and ridge waveguide. These transmission media have non-TEM modes of propagation and as such are beyond the scope of this book to be discussed.*

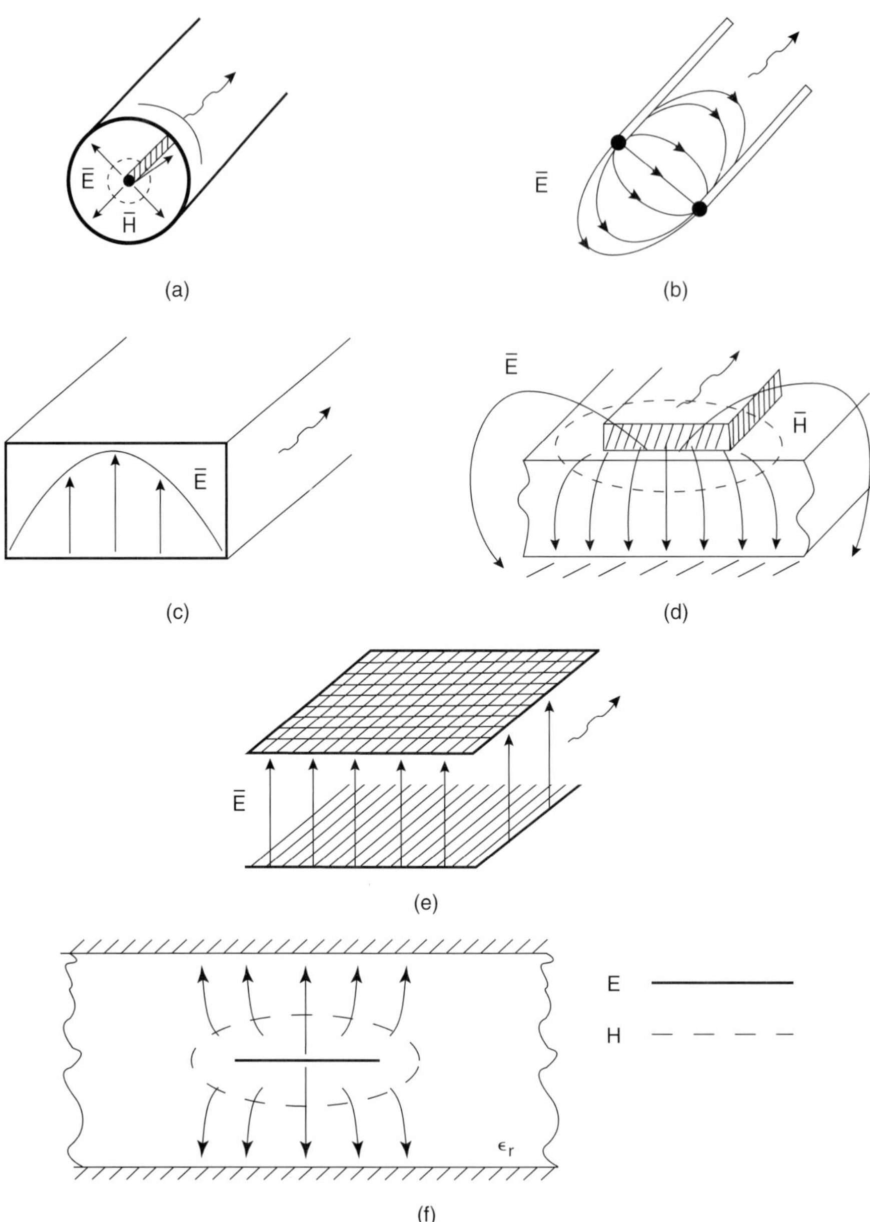

FIGURE 7.20 The electric and magnetic field patterns of a wave existing in different types of transmission media: (a) coaxial line, (b) two-wire transmission line, (c) a waveguide, (d) a microstrip line, (e) parallel plate waveguide, (f) stripline.

A summary of transmission media and their different characteristics is shown in Table 7.1. The comparison made in this table can be roughly divided into two important general areas:

- Electrical considerations: mode of propagation, dispersion, bandwidth, power loss, and power capacity (items 1 through 5)
- Mechanical considerations: physical size, ease of fabrication, and ease of integration with other elements and components (items 6 through 8)

TABLE 7.1 Comparison of Various Transmission Media

Feature	Coaxial	Stripline	Microstrip	Waveguide
1.) Propagating Mode	Main: TEM Other: TM, TE	Main: TEM Other: TM, TE	Main: Quasi-TEM Other: TM, TE	Main: TE_{10} Other: TM, TE
2.) Dispersion	None	None	Low	Medium
3.) Bandwidth	High	High	High	Low
4.) Power loss	Medium	High	High	Low
5.) Power capacity	Medium	Low	Low	High
6.) Size	Large	Medium	Small	Large
7.) Ease of fabrication	Medium	Easy	Easy	Medium
8.) Ease of integration	Hard	Fair	Easy	Hard

7.6.2 A Short History of Transmission Media

Waveguides were used for most microwave systems during the 1930s and 1940s, but they have a limited bandwidth and are bulky and expensive, even though they have the advantage of being able to handle the high powers needed for radar applications. During this same period, coaxial lines were also developed as a broadband and medium power transmission line, but they are difficult to integrate into or fabricate in integrated circuit technology, which is better suited to planar transmission lines.

Planar transmission lines received attention in 1950s. They are low cost, compact, and capable of being integrated with planar microwave integrated devices and circuits. Therefore, they play an important role in planar microwave technology for transmission of signals between devices, circuits, and networks. Examples of planar transmission lines include microstrip line (developed in 1952), stripline (developed circa 1955), and slotline (developed in 1969).

Other planar transmission lines (e.g., coplanar waveguides, finlines, etc.) have also been developed through time. Overall and among all planar transmission lines, none has proven as popular as microstrip line technology, which has gained tremendous interest in planar circuit applications. For this reason. microstrip lines are discussed and analyzed in depth in a later section in this chapter.

7.6.3 Waves on a Transmission Line (TEM Mode)

When we mention a "transmission line," it is commonly understood to be any system of conductors suitable for conducting TEM-mode electromagnetic waves efficiently between two or more terminals. Common examples of TEM-mode transmission lines include telephone lines, power lines, coaxial lines, and parallel plate lines.

At lower frequencies, the length of the line is much smaller than the signal wavelength, and so the transmission line can be treated as a "lumped element" with almost zero loss and no time delay for signal propagation between two points.

At high RF/microwave frequencies, the length of the line is comparable to the signal wavelength, and the time delay of propagation (and the corresponding signal phase shift) can no longer be ignored. Under these conditions, the "distributed circuit model" is used to analyze a transmission line (see Chapter 5). Such a model provides the governing differential equations for voltage and current waves propagating along a transmission line without a need to resort to Maxwell's equations to solve for the electromagnetic field quantities.

7.6.4 Sinusoidal Wave Propagation on a Transmission Line

Consider a transmission line as shown in Figure 7.21. Assuming a sinusoidal signal excitation, the propagating voltage and current waves on a transmission line are also sinusoidal and can be expressed as:

$$v(x,t) = Re[V(x)e^{j\omega t}] \tag{7.14a}$$

$$i(x,t) = Re[I(x)e^{j\omega t}] \tag{7.14b}$$

where complex quantities $V(x)$ and $I(x)$ are phasor quantities.

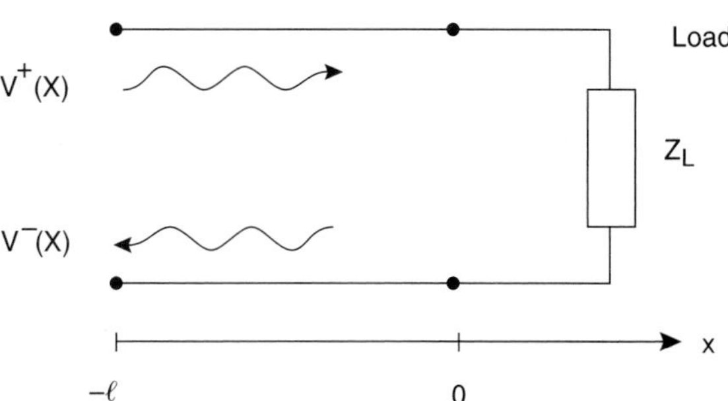

FIGURE 7.21 Incident and reflected waves on a transmission line.

Using the distributed circuit model of a transmission line and its corresponding equivalent circuit, the following differential equations for $I(x)$ and $V(x)$ can be derived (see Chapter 5, Example 5.4) for an infinitesimal length of a transmission line:

$$\frac{d^2V(x)}{dx^2} - \gamma^2 V(x) = 0 \tag{7.15a}$$

$$\frac{d^2I(x)}{dx^2} - \gamma^2 I(x) = 0 \tag{7.15b}$$

where γ is the complex propagation constant given by:

$$\gamma = \alpha + j\beta = [(R + j\omega L)(G + j\omega C)]^{1/2} \tag{7.16}$$

where:

α = attenuation constant (in Nepers/m)
β = phase constant (in radian/m)
R = resistance per unit length (in Ω/m)
L = inductance per unit length (in H/m)
G = conductance per unit length (in S/m)
C = capacitance per unit length (in F/m)

By observation, we notice that the general solution to the problem is of exponential form. Therefore, for the two possible solutions we can write:

$$V_1(x) = V_o^+ e^{-\gamma x} \Rightarrow I_1(x) = \frac{V_o^+}{Z_o} e^{-\gamma x} \tag{7.17a}$$

$$V_2(x) = V_o^- e^{\gamma x} \Rightarrow I_2(x) = -\frac{V_o^-}{Z_o} e^{\gamma x} \tag{7.17b}$$

where Z_o is the characteristic impedance of the transmission line; V_o^+ and V_o^- are complex constants in general.

As we are dealing with a linear system, the general solution for voltage and current is obtained using the superposition theorem as follows:

$$V(x) = V_1(x) + V_2(x) = V_o^+ e^{-\gamma x} + V_o^- e^{\gamma x} \tag{7.17c}$$

$$I(x) = I_1(x) + I_2(x) = \frac{V_o^+}{Z_o} e^{-\gamma x} - \frac{V_o^-}{Z_o} e^{\gamma x} \tag{7.17d}$$

From Equation 7.17c, d, we observe that voltage and current are a pair of waves coexisting and inseparable for a distributed circuit. Each solution for voltage or current consists of two waves that will be labeled as follows:

An incident wave: $$e^{-\gamma x} = e^{-\alpha x} e^{-j\beta x} \tag{7.18a}$$

A reflected wave: $$e^{\gamma x} = e^{\alpha x} e^{j\beta x} \tag{7.18b}$$

Where βx is referred to as the electrical length.

As already mentioned, each wave travels at the phase velocity (V_P) given by:

$$V_P = \omega/\beta = c \text{ (vacuum)}, \tag{7.19}$$

where c is the speed of light in vacuum given by:

$$c = 1/(\mu_o \varepsilon_o)^{1/2} = 2.9988 \times 10^8 \approx 3 \times 10^8 \text{ m/s.}$$

The time-average incident power propagating along a transmission line is given by (assuming Z_o is a real number):

$$P^+(x) = \frac{1}{2}[Re V^+(x) I^+(x)^*] = \frac{V_o^{+2}}{2Z_o} e^{-2\alpha x} \tag{7.20}$$

The same can be written for the reflected power propagating back to the source.

The law of conservation energy requires that the rate of decrease of propagating power $P(x)$ along the line should equal the average power loss per unit length (P_{loss}). Thus we can write:

$$-\frac{\partial P(x)}{\partial x} = P_{loss} = 2\alpha P(x)$$

$$\alpha = \frac{P_{loss}}{2P(x)} \approx \frac{-\Delta P/\Delta x}{2P} = \frac{-[P(x+\Delta x) - P(x)]/\Delta x}{2P(x)} \tag{7.21}$$

Equation 7.21 shows an interesting and yet very practical way to measure the actual attenuation constant (α). This method is particularly helpful if we are trying to establish the integrity of a faulty line because a simple comparison of the measured (α) with the nominal (α) would reveal the needed information. The following example elucidates this point further.

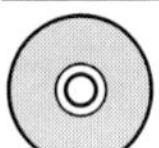

EXAMPLE 7.3

The microwave power at one point (P_1) on a transmission line is measured to be 10 mW. At a distance of d = 50 cm away, another power measurement (P_2) indicates a power of 7 mW. Determine the attenuation constant of the transmission line.

Solution:

$$\alpha = \frac{P_{loss}}{2P(x)} \approx \frac{-\Delta P/\Delta x}{2P} = \frac{-[P(x+\Delta x) - P(x)]/\Delta x}{2P(x)}$$

Where

$$\Delta P/\Delta x = (P_2 - P_1)/d$$

Thus, we have:

$$\alpha = \frac{-(P_2 - P_1)/d}{2P_1} = \frac{(10-7)/0.5}{2 \times 10} = 0.3 \text{ Np/m}$$

7.6.5 The Concept of the Reflection Coefficient

Any time an incident wave encounters a second medium different from the first, it is partly reflected (creating a reflected wave) while the remainder is transmitted through (creating a transmitted wave). The reflected wave encountering the incident wave forms a standing wave, as described earlier. We can see that there are four possible waves in a transmission line:

- An incident wave,
- A reflected wave,
- A transmitted wave, and
- A standing wave.

Let us now define an important term:

DEFINITION-REFLECTION COEFFICIENT: *The ratio of the reflected wave phasor to the incident wave phasor.*

In the special case of a *uniform* transmission line when the incident wave encounters a second medium such as a termination (load) or a discontinuity, then under these conditions, the ratio of the reflected wave phasor to the incident wave phasor is the reflection coefficient as defined above. To illustrate this concept, consider a transmission line circuit with a load (Z_L) located at $x = 0$ (the reference point), as shown in Figure 7.21. For this circuit, we can write:

$$V^+(x) = V_o^+ e^{-\gamma x} \tag{7.22a}$$

$$V^-(x) = V_o^- e^{\gamma x} \tag{7.22b}$$

$$\Gamma(x) = \frac{V^-(x)}{V^+(x)} = \frac{V_o^- e^{\gamma x}}{V_o^+ e^{-\gamma x}} = \frac{V_o^-}{V_o^+} e^{2\gamma x} = \Gamma_L e^{2\gamma x} \tag{7.23}$$

where $\Gamma_L = \Gamma(0) = \dfrac{V_o^-}{V_o^+}$ is the reflection coefficient at the load end of the transmission line.

Thus, the total voltage and current phasors $[V(x), I(x)]$ along the transmission line can be written as:

$$V(x) = V^+(x) + V^-(x) = V_o^+ e^{-\gamma x} + V_o^- e^{\gamma x}$$

or

$$V(x) = V_o^+ (e^{-\gamma x} + \Gamma_L e^{\gamma x}). \tag{7.24}$$

Similarly,

$$I(x) = I^+(x) - I^-(x) = V^+(x) / Z_o - V^-(x) / Z_o$$

or

$$I(x) = \frac{V_o^+}{Z_o} (e^{-\gamma x} - \Gamma_L e^{\gamma x}). \tag{7.25}$$

The input impedance, $Z_{IN}(x)$, at any point along the transmission line is obtained by dividing Equation 7.24 over 7.25 and is given by:

$$Z_{IN}(x) = \frac{V(x)}{I(x)} = Z_o \frac{e^{-\gamma x} + \Gamma_L e^{\gamma x}}{e^{-\gamma x} - \Gamma_L e^{\gamma x}} \tag{7.26}$$

NOTE: *At any junction point on a transmission line where two regions meet (e.g., at a load, at a second T.L., etc.), the total tangential E-field (or equivalently, the net voltage) on each side of the junction is continuous. Therefore, the transmitted voltage ($V_t = T V^+$) in the second region can be written as:*

$$V_t = T V^+ = V^+ + V^- = (1 + \Gamma) V^+ \Rightarrow T = 1 + \Gamma$$

where T is the transmission coefficient.

A Special Case. At the load end (where $x = 0$), the following is observed:

$$Z_{IN}(0) = Z_L = Z_o \frac{1+\Gamma_L}{1-\Gamma_L} \Rightarrow \Gamma_L = \frac{Z_L - Z_o}{Z_L + Z_o} \tag{7.27a}$$

Or, in general, at any arbitrary point (x) along a transmission line with an input impedance (Z_{IN}), we can generalize Equation 7.27a and write the reflection coefficient (Γ_{IN}) at that point as:

$$\Gamma_{IN}(x) = \frac{Z_{IN}(x) - Z_o}{Z_{IN}(x) + Z_o} \tag{7.27b}$$

Using Equation 7.27a and letting $x = -\ell$, Equation 7.26 can be written as:

$$Z_{IN}(-\ell) = Z_o \frac{Z_L + Z_o \tanh\gamma\ell}{Z_o + Z_L \tanh\gamma\ell} \tag{7.28}$$

7.6.6 Lossless Transmission Lines

Because most of the transmission lines at RF/microwave frequencies have negligible losses, we will focus exclusively on lossless transmission lines.

In a lossless transmission line, there is no series resistance (R) or shunt leakage conductance (G). Thus, the energy propagating on the line does not get attenuated in strength (or power). Considering Figure 7.22, the following simplifications can be made:

$$\alpha = 0,$$

$$\gamma = j\beta,$$

$$Z_o = (L/C)^{1/2}, \tag{7.29a}$$

$$V_P = \omega/\beta = 1/(LC)^{1/2} \tag{7.29b}$$

$$\lambda = V_P/f = 2\pi/\beta \tag{7.29c}$$

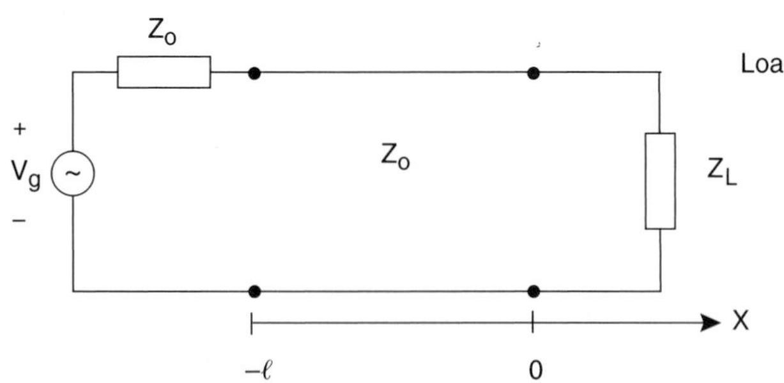

FIGURE 7.22 A lossless transmission line.

Using Equation 7.23, we can write:

$$\Gamma(x) = \frac{V^-(x)}{V^+(x)} = \frac{V_o^- e^{j\beta x}}{V_o^+ e^{-j\beta x}} = \frac{V_o^-}{V_o^+} e^{j2\beta x} = \Gamma_L e^{j2\beta x} \tag{7.30}$$

where V_o^+ and V_o^- are complex constants in general, and $\Gamma_L = V_o^-/V_o^+$.

Using Equations 7.24, 7.25 and 7.30, we can write the voltage and current on a lossless transmission line as:

$$V(x) = V^+(x) + V^-(x) = V_o^+ e^{-j\beta x} + V_o^- e^{j\beta x}$$

$$V(x) = V_o^+ e^{-j\beta x}\left(1 + \frac{V_o^-}{V_o^+} e^{j2\beta x}\right)$$

or

$$V(x) = V_o^+ e^{-j\beta x}[1 + \Gamma(x)] \tag{7.31 a}$$

Similarly, $I(x)$ can be written as:

$$I(x) = I^+(x) - I^-(x) = \frac{V_o^+}{Z_o} e^{-j\beta x} - \frac{V_o^-}{Z_o} e^{j\beta x}$$

$$I(x) = \frac{V_o^+}{Z_o} e^{-j\beta x}[1 - \Gamma(x)] \tag{7.31b}$$

At the source ($x = -\ell$), we have:

$$V(-\ell) = V_g - Z_o I(-\ell)$$

upon substitution for $V(-\ell)$ and $I(-\ell)$, we obtain:

$$V_o^+ = (V_g/2) e^{-j\beta\ell} \tag{7.32}$$

The input impedance at any point (x) on the transmission line can now be written as:

$$Z_{IN}(x) = Z_o \frac{1 + \Gamma(x)}{1 - \Gamma(x)} \tag{7.33 a}$$

At any point ($x = -\ell$) Equation 7.33a for Z_{IN} yields:

$$Z_{IN}(-\ell) = Z_o \frac{Z_L + jZ_o \tan\beta\ell}{Z_o + jZ_L \tan\beta\ell} \tag{7.33b}$$

The time-average incident and reflected power propagating along a transmission line is given by (assuming Z_o is a real number, $\alpha = 0$):

$$P^+(x) = \frac{1}{2} Re[V^+(x)\, I^+(x)^*] = \frac{|V_o^+|^2}{2Z_o} \tag{7.34 a}$$

$$P^-(x) = \frac{1}{2} Re[V^-(x)\, I^-(x)^*] = |\Gamma|^2 \frac{|V_o^+|^2}{2Z_o}, \tag{7.34b}$$

and

$$P_L = \frac{1}{2}|V_L|^2 \frac{Re(Z_L)}{|Z_L|^2} = P^+(x) - P^-(x) = (1 - |\Gamma|^2)\frac{|V_o^+|^2}{2Z_o} \tag{7.34 c}$$

Earlier, we defined the transmission coefficient (T) as: $V_L = T\ V^+$; thus, from Equations 7.34a–c we have:

$$|T|^2 = (1 - |\Gamma|^2)\frac{|Z_L|^2}{Z_o Re(Z_L)} \tag{7.34d}$$

7.6.7 Standing Wave Ratio

As described earlier, a standing wave results from two waves having the same frequency traveling in opposite directions on a transmission line. The meeting of these two waves produces a standing wave pattern of voltage and current on a transmission line.

The maximum value of voltage anywhere along the transmission line is given by:

$$V_{max} = |V(x)|_{max} = |V_o^+| + |V_o^-| = |V_o^+|(1 + |\Gamma_L|) \tag{7.35}$$

The minimum value of the voltage is given by:

$$V_{min} = |V(x)|_{min} = |V_o^+| - |V_o^-| = |V_o^+|(1 - |\Gamma_L|) \tag{7.36}$$

Similarly, for the current standing wave we have:

$$I_{max} = |I(x)|_{max} = |I_o^+| + |I_o^-| = \frac{|V_o^+|}{Z_o} + \frac{|V_o^-|}{Z_o} = \frac{|V_o^+|}{Z_o}(1 + |\Gamma_L|) \tag{7.37}$$

$$I_{min} = |I(x)|_{min} = |I_o^+| - |I_o^-| = \frac{|V_o^+|}{Z_o} - \frac{|V_o^-|}{Z_o} = \frac{|V_o^+|}{Z_o}(1 - |\Gamma_L|) \tag{7.38}$$

Equations 7.35 to 7.38 are used to define the standing wave ratio (SWR) (often referred to as voltage standing wave ratio, or $VSWR$), as follows:

$$VSWR = \frac{V_{max}}{V_{min}} = \frac{I_{max}}{I_{min}} = \frac{1 + |\Gamma_L|}{1 - |\Gamma_L|} \tag{7.39}$$

or

$$|\Gamma_L| = \frac{VSWR - 1}{VSWR + 1} \tag{7.40}$$

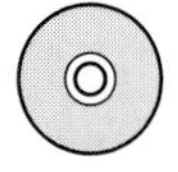

EXAMPLE 7.4

What is the VSWR for a matched transmission line ($Z_o = 50\ \Omega$)?

Solution:

$$Z_L = Z_o \Rightarrow \Gamma_L = 0$$

$$VSWR = (1 + 0)/(1 - 0) = 1$$

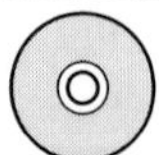

EXAMPLE 7.5

What is the VSWR for:

a. *An open load ($Z_L = \infty$),*

b. *A short load ($Z_L = 0$)? Assume $Z_o = 50\ \Omega$.*

Solution:

$$\text{a. } Z_L = \infty \Rightarrow \Gamma_L = \lim_{Z_L \to \infty} (Z_L - 50)/(Z_L + 50) = 1$$

$$VSWR = (1 + 1)/(1 - 1) = \infty$$

$$\text{b. } Z_L = 0 \Rightarrow \Gamma_L = (0 - 50)/(0 + 50) = -1 \Rightarrow |\Gamma_L| = 1$$

$$VSWR = (1 + 1)/(1 - 1) = \infty$$

CONCLUSION: From Examples 7.4 and 7.5, we can see that:

$$\mathbf{1 \leq VSWR \leq \infty} \tag{7.41}$$

EXAMPLE 7.6

What is the Z_{IN} of a TL at $x = -\ell$ for an open circuit load?

Solution:

For $Z_L = \infty$, Equation 7.33b can be written as:

$$Z_{OC} = Z_{IN}(-\ell) = \lim_{Z_L \to \infty} Z_o \frac{Z_L + jZ_o \tan\beta\ell}{Z_o + jZ_L \tan\beta\ell}$$

$$Z_{OC} = -jZ_o \cot\beta\ell \tag{7.42}$$

EXAMPLE 7.7

What is the Z_{IN} of a TL at $x = -\ell$ for a short circuit load?

Solution:

For $Z_L = 0$, Equation 7.33b can be written as:

$$Z_{SC} = Z_{IN}(-\ell) = Z_o \frac{0 + jZ_o \tan\beta\ell}{Z_o + 0}$$

$$Z_{SC} = jZ_o \tan\beta\ell \tag{7.43}$$

7.6.8 Quarter-Wave Transformers

The main functions of any transmission line at any frequency, are two-folded as follows:

1. Transmission of power, and/or
2. Transmission of information.

At RF/microwave frequencies, it becomes essential that all lines be matched to each other, to the source and finally to the load. This is due to the obvious fact that reflections due to mismatch or discontinuities (e.g., at a connection, at a junction, etc.) will result in echoes, will reduce the transmitted power and will distort the information carrying signal.

A simple method for matching a resistive load Z_L to a lossless feed line (having a real characteristic impedance Z_o) is the use of a quarter-wave transformer which is a piece of a transmission line having a $\lambda/4$ length and a characteristic impedance of (Z_T).

The characteristic impedance of the quarter-wave transformer (Z_T) terminated in a real load Z_L can be derived as follows:

Derivation of (Z_T)

$$\ell = \lambda/4 \Rightarrow \beta\ell = (2\pi/\lambda)(\lambda/4) = \pi/2 \Rightarrow \tan\beta\ell = \infty$$

Thus, the input impedance of the quarter-wave transformer from Equation 7.33b terminated in a real load Z_L, can be written as:

$$x = -\lambda/4$$

$$Z_{IN}(-\lambda/4) = \lim_{\ell \to \lambda/4} Z_T \frac{Z_L + jZ_T \tan\beta\ell}{Z_T + jZ_L \tan\beta\ell} = Z_T \frac{jZ_T \tan\beta\ell}{jZ_L \tan\beta\ell} \Rightarrow Z_{IN} = \frac{Z_T^2}{Z_L} \tag{7.44}$$

Or,

$$Z_T = \sqrt{Z_{IN} Z_L} \tag{7.45}$$

POINT OF CAUTION: *This simple method of matching is applicable only when both of the following conditions are met:*

a. *The feed transmission line is lossless (this leads to a characteristic impedance value which is a real number), and*

b. *The load is resistive.*

There are cases where the load is a complex number, and thus at fist glance a quarter-wave transformer does not seem to lend itself useful for matching purposes. However, in such a case the load should first be converted into a real number by adding a reactance having the same value as the load's reactance but with the opposite sign. The resultant load is resistive and can then be transformed to the feed line's characteristic impedance through the use of a quarter-wave transformer as described above (for further details on matching techniques, please see Chapter 11, *Design of Matching Networks*).

The following example further elucidates this simple method of matching.

EXAMPLE 7.8

What is the input impedance of a short circuited lossless TL if the length of the line is $\lambda/4$?

Solution:

From Equation 7.44, we have:

$$Z_{IN} = \frac{Z_T^2}{Z_L} = \frac{Z_T^2}{0} = \infty$$

NOTE: *Example 7.8 clearly shows why these types of shorted transformers are ideal for electrically isolating the RF circuitry from the DC bias source in an amplifier circuit as will be discussed later in Chapter 15, RF/Microwave Amplifiers I:*

Small-Signal Design. This is because the RF circuitry is connected at the input side of the transformer while DC bias source is at the short-circuited end of the transformer (of course, the short circuit is created by the use of a high-value capacitor to ground). In this fashion, the RF signals "see" an open circuit at the RF side and would not be able to travel to the DC bias source, while the DC bias "sees" a direct connection (i.e. a short circuit) to the RF circuitry.

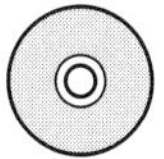

EXAMPLE 7.9

What is (Z_T) of a quarter-wave transformer to transform a load of 100 Ω to a 50 Ω feed line as shown in Figure 7.23?

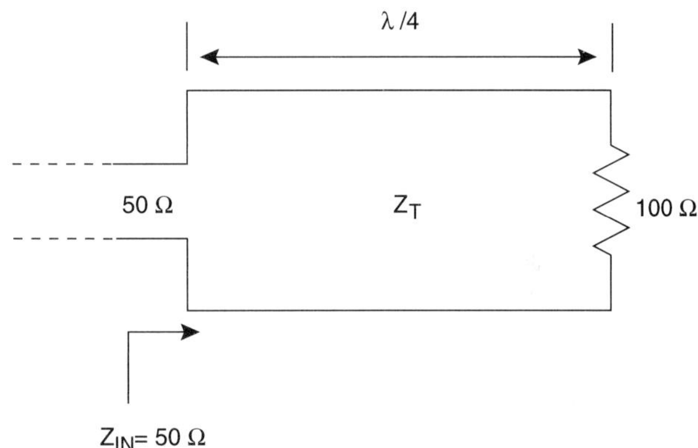

FIGURE 7.23 A quarter-wave transformer.

Solution:

To create a match, we require that Z_{IN} to be the same as the characteristic impedance of the feed line, i.e., $Z_{IN} = 50\ \Omega$. Using Equation 7.45, we can write:

$$Z_T = (Z_L Z_{IN})^{1/2}$$

Thus, we obtain:

$$Z_T = (100 \times 50)^{1/2} = 70.7\ \Omega$$

7.6.9 A Generalized Lossless Transmission Line Circuit

In the previous examples, the main focus has been on the effects of a load on the current and voltage waves traveling on a transmission line. The source of the waves, the generator located at the other end, plays an important role in the propagation of waves along a transmission line. Up to this point in our discussion, the generator's internal impedance has been a real number equal to the characteristic impedance of the transmission line. In effect, the generator was matched to the line and only the effects of mismatch of the load have been studied so far. Obviously, this is a special case. The most general case is having mismatches at both ends (i.e., at the generator and at the load ends), which will now be discussed in detail.

Analysis. Consider a finite lossless transmission line (T.L.) of length (ℓ) with a characteristic impedance (Z_o) driven by a generator (V_g) with an internal impedance (Z_g) at $x = -\ell$ and terminated in a load (Z_L) at $x = 0$, as shown in Figure 7.24.

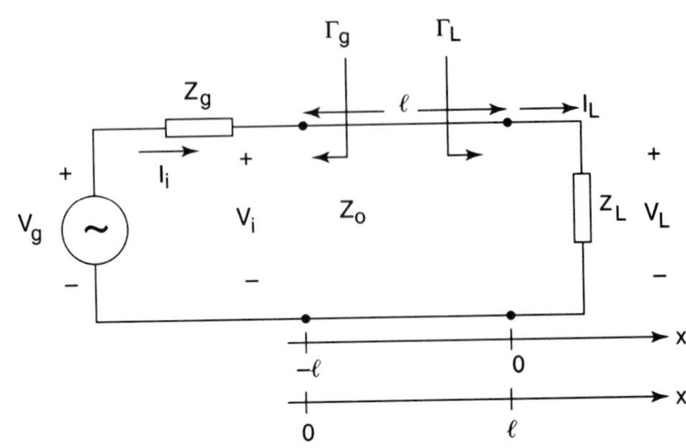

FIGURE 7.24 A general transmission line circuit. x'-axis is the shifted x-axis by $+\ell$ (i.e., $x' = x + \ell$).

The boundary condition (B.C.) at each end can be written as:

1. B.C. #1—Voltage and current at $x = -\ell$ is given by:

$$V_i = V_g - Z_g I_i$$

2. B.C. #2—Voltage and current at $x = 0$ is given by:

$$V_L = Z_L I_L$$

3. Voltage and current on the T.L. for $-\ell \le x \le 0$, from Equation 7.31 and Equation 7.31 is given by:

$$V(x) = V^+(x) + V^-(x) = V_o^+ e^{-j\beta x}[1 + \Gamma(x)] \tag{7.46a}$$

$$I(x) = I^+(x) - I^-(x) = \frac{V_o^+}{Z_o} e^{-j\beta x}[1 - \Gamma(x)] \tag{7.46b}$$

Where (from Equation 7.30), $\Gamma(x)$ is given by:

$$\Gamma(x) = \frac{V^-(x)}{V^+(x)} = \frac{V_o^- e^{j\beta x}}{V_o^+ e^{-j\beta x}} = \frac{V_o^-}{V_o^+} e^{2j\beta x} = \Gamma_L e^{2j\beta x} \tag{7.46c}$$

Applying the boundary condition given by B.C. #1, we can solve for $|V^+|$ as follows:

$$V_i = V(-\ell) = V_o^+ e^{j\beta\ell}[1 + \Gamma(-\ell)] = V_g - Z_g \frac{V_o^+}{Z_o} e^{+j\beta\ell}[1 - \Gamma(-\ell)] \tag{7.47}$$

where

$$\Gamma(-\ell) = \Gamma_L e^{-j2\beta\ell}$$

From Equation 7.47, we can solve for V_o^+ in terms of V_g to obtain:

$$V_o^+ = \frac{Z_o V_g}{Z_o + Z_g}\left(\frac{e^{-j\beta\ell}}{1 - \Gamma_L \Gamma_g e^{-j2\beta\ell}}\right) \tag{7.48a}$$

where

$$\Gamma_g = \frac{Z_g - Z_o}{Z_g + Z_o} \tag{7.48b}$$

Thus, $V(x)$ and $I(x)$ under this general condition may be obtained by substituting for V_o^+ from Equation 7.48a in Equations 7.46 as follows:

$$V(x) = \frac{Z_o V_g}{Z_o + Z_g} e^{-j\beta(x+\ell)}\left(\frac{1 + \Gamma_L e^{j2\beta x}}{1 - \Gamma_L \Gamma_g e^{-j2\beta\ell}}\right) \tag{7.49a}$$

$$I(x) = \frac{V_g}{Z_o + Z_g} e^{-j\beta(x+\ell)}\left(\frac{1 - \Gamma_L e^{j2\beta x}}{1 - \Gamma_L \Gamma_g e^{-j2\beta\ell}}\right) \tag{7.49b}$$

NOTE: *Equations 7.49a, b show phasor expressions for the voltage and current due to a sinusoidal voltage source (V_g) feeding a finite transmission line, which is terminated in a general load(Z_L).*

These equations represent the summation of infinite number of reflections from both ends of the transmission line, i.e.,

$$V(x) = V_1^+ + V_1^- + V_2^+ + V_2^- + \ldots = \sum_{i=1}^{\infty} (V_i^+ + V_i^-) \tag{7.50}$$

where

$$|V_1^+| = |V_o'^+|$$
$$|V_1^-| = |\Gamma_L||V_o'^+|$$
$$|V_2^+| = |\Gamma_g||\Gamma_L||V_o'^+|,$$
$$|V_2^-| = |\Gamma_g||\Gamma_L|^2|V_o'^+|, \text{ etc.}$$

and,

$$V_o'^+ = \left(\frac{Z_o}{Z_o + Z_g}\right)V_g$$

EXERCISE 7.1

Prove that the summation of infinite number of voltage reflections as shown by Equation 7.50 converges to Equation 7.49a.

HINT: *Note that:*

$$\sum_{n=0}^{\infty} x^n = 1 + x + x^2 + \ldots + x^n + \ldots = \frac{1}{1-x}, \ |x| < 1$$

Special Cases. From Equations 7.49a, b, we can derive several special useful cases as follows:

Case I: Matched at Both Ends

$$Z_g = Z_L = Z_o \text{ (see Figure 7.25a)}$$

$$\Gamma_g = \Gamma_L = 0$$

$$V(x) = \frac{V_g}{2} e^{-j\beta(x+\ell)} \tag{7.51a}$$

$$I(x) = \frac{V_g}{2Z_o} e^{-j\beta(x+\ell)} \tag{7.51b}$$

FIGURE 7.25 Three special cases of a transmission line circuit.

(a)

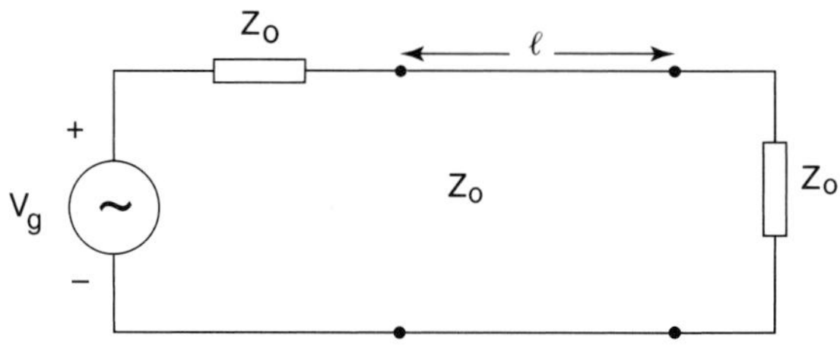

(b)

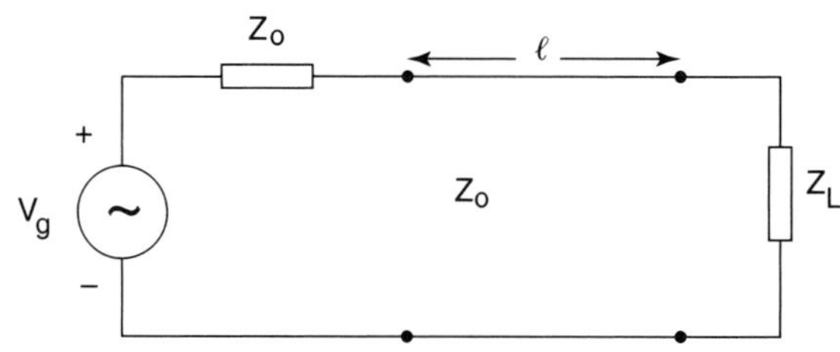

(c)

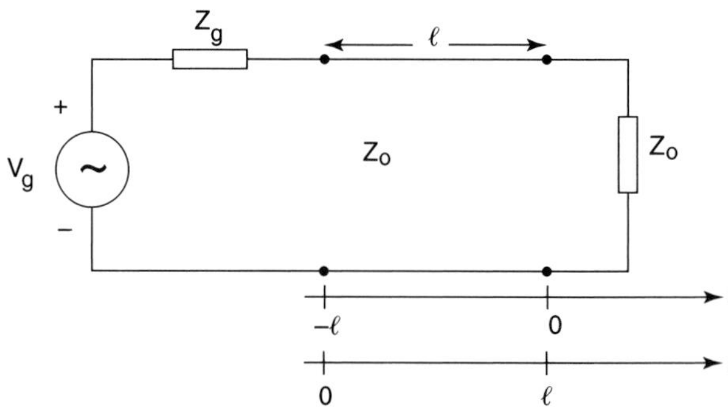

At the generator end ($x = -\ell$), we have:

$$x = -\ell$$

$$V(-\ell) = V_i = \frac{V_g}{2} \tag{7.51c}$$

$$I(-\ell) = I_i = \frac{V_g}{2Z_o} \quad (7.51\text{d})$$

and at the load end ($x = 0$), we can write:

$$V(0) = V_L = \frac{V_g}{2}e^{-j\beta\ell} \quad (7.51\text{e})$$

$$I(0) = I_L = \frac{V_g}{2Z_o}e^{-j\beta\ell} \quad (7.51\text{ f})$$

In this case there are no standing waves on the transmission line and the magnitude of the voltage and current are the same everywhere on the line, i.e.,

$$|V_o^+| = |V_i| = |V_L| = |V(x)| = \frac{V_g}{2} \quad (7.51\text{g})$$

$$|I_o^+| = |I_i| = |I_L| = |I(x)| = \frac{V_g}{2Z_o} \quad (7.51\text{h})$$

NOTE: *In some texts, the reference for length is located at the generator end rather than at the load end. This means that there is a shift in the x-axis (by $+\ell$), i.e.,*

$$x' = x + \ell \quad (7.51\text{ i})$$

Thus, Equations 7.51a, b can now be written in terms x' as:

$$V(x') = \frac{V_g}{2}e^{-j\beta x'} \quad (7.51\text{ j})$$

$$I(x') = \frac{V_g}{2Z_o}e^{-j\beta x'} \quad (7.51\text{k})$$

Case II: Matched at the Source End Only

$$Z_g = Z_o,\ Z_L \neq Z_o \text{ (see Figure 7.25b)}$$

$$\Gamma_g = 0, \Gamma_L \neq 0$$

$$V(x) = \frac{V_g}{2}e^{-j\beta(x+\ell)}(1 + \Gamma_L e^{j2\beta x}) \quad (7.52\text{a})$$

$$I(x) = \frac{V_g}{2Z_o}e^{-j\beta(x+\ell)}(1 - \Gamma_L e^{j2\beta x}) \quad (7.52\text{b})$$

In terms of the shifted axis ($x' = x + \ell$), we can write Equations 7.52a, b as:

$$V(x') = \frac{V_g}{2}e^{-j\beta x'}(1 + \Gamma_L e^{j2\beta(x'-\ell)}) \quad (7.52\text{c})$$

$$I(x') = \frac{V_g}{2Z_o}e^{-j\beta x'}(1 - \Gamma_L e^{j2\beta(x'-\ell)}) \quad (7.52\text{d})$$

At the generator end ($x = -\ell$), we have:

$$x = -\ell$$

$$V(-\ell) = V_i = \frac{V_g}{2}(1 + \Gamma_L e^{-j2\beta\ell}) \tag{7.52e}$$

$$I(-\ell) = I_i = \frac{V_g}{2Z_o}(1 - \Gamma_L e^{-j2\beta\ell}) \tag{7.52f}$$

And at the load end ($x = 0$), we have:

$$x = 0$$

$$V(0) = V_L = \frac{V_g}{2} e^{-j\beta\ell}(1 + \Gamma_L) \tag{7.52g}$$

$$I(0) = I_L = \frac{V_g}{2Z_o} e^{-j\beta\ell}(1 - \Gamma_L) \tag{7.52h}$$

Case III: Matched at the Load End Only

$$Z_g \neq Z_o,\ Z_L = Z_o \text{ (see Figure 7.25c)}$$

$$\Gamma_g \neq 0, \Gamma_L = 0$$

$$V(x) = \frac{Z_o V_g}{Z_o + Z_g} e^{-j\beta(x + \ell)} \tag{7.53a}$$

$$I(x) = \frac{V_g}{Z_o + Z_g} e^{-j\beta(x + \ell)} \tag{7.53b}$$

At the generator end ($x = -\ell$), we have:

$$x = -\ell$$

$$V(-\ell) = V_i = \frac{Z_o V_g}{Z_o + Z_g} \tag{7.53c}$$

$$I(-\ell) = I_i = \frac{V_g}{Z_o + Z_g} \tag{7.53d}$$

And, at the load end ($x = 0$) we have:

$$x = 0$$

$$V(0) = V_L = \frac{Z_o V_g}{Z_o + Z_g} e^{-j\beta\ell} \tag{7.53e}$$

$$I(0) = I_L = \frac{V_g}{Z_o + Z_g} e^{-j\beta\ell} \tag{7.53f}$$

This case reduces to a simple voltage division between the source internal impedance (Z_g) and the transmission line presenting a constant input impedance of (Z_o). The voltage and current on the transmission line have the same magnitude except for a phase shift with length, that is,

$$V_i = V_L e^{j\beta\ell} \tag{7.53g}$$

$$V(x) = V_L e^{-j\beta x} \tag{7.53h}$$

Similarly, we can write for current:

$$I_i = I_L e^{j\beta\ell} \tag{7.53 i}$$

$$I(x) = I_L e^{-j\beta x} \tag{7.53 j}$$

Equations 7.53a and 7.53b can be written in terms of the shifted axis ($x' = x + \ell$) as:

$$V(x') = V_L e^{-j\beta(x' - \ell)} = V_i e^{-j\beta x'} \tag{7.53k}$$

$$I(x') = I_L e^{-j\beta(x' - \ell)} = I_i e^{-j\beta x'} \tag{7.53 l}$$

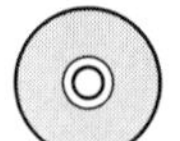

EXAMPLE 7.10

Consider a 50 Ω lossless transmission line of length $\ell = 1$ m, connected to a generator operating at $f = 1$ GHz and having $V_g = 10$ V with $Z_g = 50\ \Omega$ at one end and connected to a load Z_L=100 Ω at the other (see Figure 7.26). Determine:

a. *The voltage and current at any point on the transmission line*

b. *The voltage at the generator (V_i) and load (V_L) ends*

c. *The reflection coefficient and VSWR at any point on the line*

d. *The average power delivered to the load*

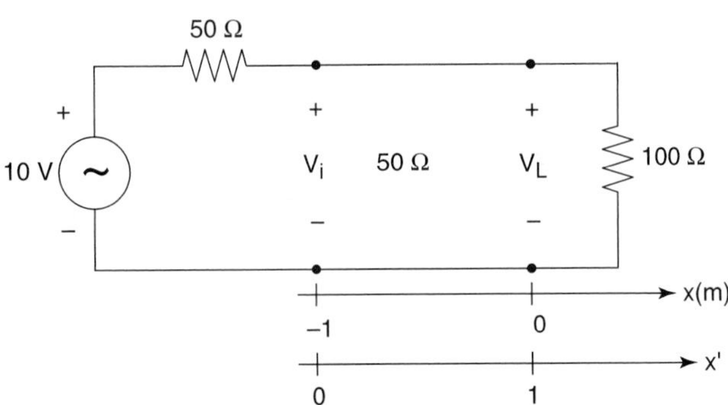

FIGURE 7.26 Circuit for Example 7.10.

Solution:

a.

$$Z_g = Z_o = 50\ \Omega \Rightarrow \Gamma_g = 0$$

Because $\Gamma_g = 0$, special case II applies here. Thus, we can write:

$$\beta = \omega/c = 2\pi \times 10^9/3 \times 10^8 = 20\pi/3 \text{ rad/m}$$

$$\Gamma_L = \frac{Z_L - Z_o}{Z_L + Z_o} = \frac{100 - 50}{100 + 50} = \frac{1}{3}$$

$$V(x) = \frac{V_g}{2}e^{-j\beta(x+\ell)}(1+\Gamma_L e^{j2\beta x}) = 5e^{-j20\pi(x+1)/3}\left(1+\frac{1}{3}e^{j40\pi x/3}\right)$$

$$I(x) = \frac{V_g}{2Z_o}e^{-j\beta(x+\ell)}(1-\Gamma_L e^{j2\beta x}) = \frac{10}{2\times 50}e^{-j20\pi(x+1)/3}\left(1-\frac{1}{3}e^{j40\pi x/3}\right)$$

$$= 0.1e^{-j20\pi(x+1)/3}\left(1-\frac{1}{3}e^{j40\pi x/3}\right)$$

b. At the generator end ($x = -1$ m), we have:

$$x = -1$$

$$V_i = V(-1) = 5\left(1+\frac{1}{3}e^{-j40\pi/3}\right) = -4.16 + j1.44$$

At the load end, we have:

$$x = 0$$

$$V_L = V(0) = 5e^{-j20\pi/3}\left(1+\frac{1}{3}\right) = \frac{20}{3}e^{-j20\pi/3}$$

c. The reflection coefficient and *VSWR* are as follows:

$$\Gamma(x) = \Gamma_L\, e^{j2\beta x} = \frac{1}{3}e^{j40\pi x/3}$$

$$VSWR = \frac{1+|\Gamma_L|}{1-|\Gamma_L|} = \frac{1+1/3}{1-1/3} = 2$$

d. The average power delivered to the load is ($\alpha = 0$):

$$P_L(x) = \frac{1}{2}Re[V_L(x)I_L^*(x)] = \frac{Re}{2}\left[V_L\frac{V_L^*}{Z_L^*}\right] = |V_L|^2\frac{Re[Z_L]}{2|Z_L|^2}$$

thus,

$$P_L(x) = \frac{|V_L|^2}{2Z_L} = \frac{|20e^{j20\pi/3}/3|^2}{2\times 100} = \frac{2}{9} = 0.22 \text{ W}$$

NOTE: *If the load were completely matched to the line, the power delivered to the load would be:*

$$Z_L = 50\ \Omega$$

$$|V_i| = |V_L| = V_g/2 = 5 \text{ V}$$

$$(P_{av})_{max} = \frac{|V_L|^2}{2Z_L} = \frac{5^2}{2\times 50} = 0.25 \text{ W}$$

Because there is no reflected power, $(P_{av})_{max}$ is also the incident power (P_i), which is higher than the (P_{av}) calculated earlier under unmatched conditions. The difference in the two powers is due to the reflected power back to the source:

$$P_r = |\Gamma_L|^2 P_i = (1/9)(0.25) = 0.03 \text{ W}$$

7.7 MICROSTRIP LINE

Among all planar transmission lines, microstrip line has gained much popularity and importance in microwave planar circuit technology, and so we will consider and analyze microstrip lines in this section. A microstrip line is a transmission line consisting of a strip of conductor of thickness (t), width (w), and a ground plane separated by a dielectric medium of thickness (h), as shown in Figure 7.27.

FIGURE 7.27 A microstrip transmission line: (a) geometry and (b) field configuration.

(a)

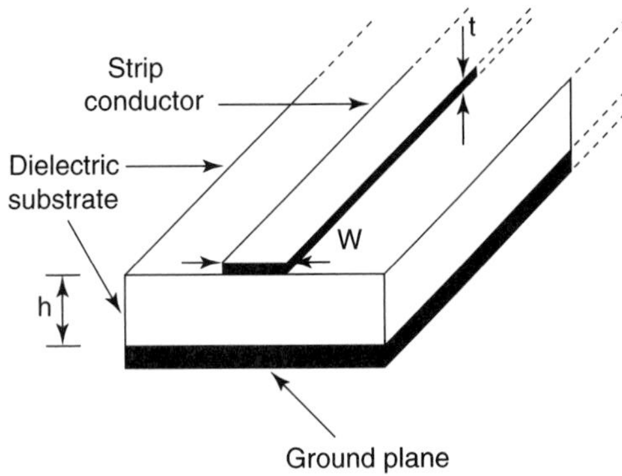

(b)

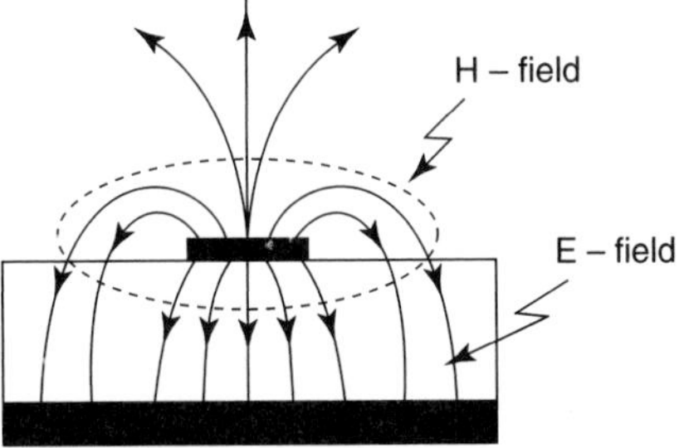

Because it is an open conduit for wave transmission, not all of the electric or magnetic fields will be confined in the structure. This fact, along with the existence of a small axial E-field, leads not to a purely TEM wave propagation, but to a quasi-TEM mode of propagation.

These types of transmission lines are very popular and are used extensively in microwave planar circuit design and microwave integrated circuit (MIC) technology. Use of printed circuit board technology and its simplicity of fabrication, along with ease of placement and interconnection of lumped elements and components, has made this type of transmission line very popular and much superior to other types of planar transmission lines.

7.7.1 Wave Propagation in Microstrip Lines

The dielectrics used in the fabrication of the microstrip line are characterized by a dielectric constant (ε_r) defined by:

$$\varepsilon_r = \varepsilon / \varepsilon_o, \tag{7.54}$$
$$\varepsilon_o = 8.854 \times 10^{-12} \text{ F/m}$$

where ε and ε_o are the dielectric's and vacuum's permittivity, respectively. The most popular dielectrics are: RT/Duroid 5880/6006/6010.5 ($\varepsilon_r = 2.23/6/10.5$), alumina 85%/96% ($\varepsilon_r = 8.0/8.9$), quartz ($\varepsilon_r = 3.7$), silicon ($\varepsilon_r = 11.7$), and Epsilam-10 ($\varepsilon_r = 10$).

The EM-wave propagation in a microstrip line is approximately nondispersive below a cut-off frequency (f_o), which is given by:

$$f_o(GHz) = 0.3\sqrt{\frac{Z_o}{h\sqrt{\varepsilon_r - 1}}} \tag{7.55}$$

where h is in centimeters.

The phase velocity for a quasi-TEM is given by:

$$V_P = c/\sqrt{\varepsilon_{ff}}$$

where c is the speed of light and ε_{ff} is the effective relative dielectric constant.

Because the field lines are not contained in the structure and some exist in the air (see Figure 7.27b), the effective dielectric constant satisfies the following relation:

$$1 < \varepsilon_{ff} < \varepsilon_r$$

In general, the effective dielectric constant (ε_{ff}) is a function of not only the substrate material (ε_r) but also of the dielectric thickness (h) and conductor width (W) as will be shown later.

The characteristic impedance (Z_o) is given by:

$$Z_o = \frac{1}{V_p C_o} \tag{7.56}$$

where C_o is the capacitance per unit length.

The wavelength (λ) of a propagating wave in the microstrip line is given by:

$$\lambda = V_P/f = \lambda_o/\sqrt{\varepsilon_{ff}} \tag{7.57a}$$

where $\lambda_o = c/f$ is the wavelength in free space.

NOTE: *The wavelength of a TEM wave (λ_{TEM}) propagating in the dielectric material is different from the wavelength (λ_o) of a propagating wave in free space as follows:*

$$\lambda_{TEM} = \lambda_o/\sqrt{\varepsilon_r} \tag{7.57b}$$

As can be seen from these equations, the characteristic impedance (Z_o) and the wavelength (λ) both vary and thus are functions of the geometry (W, h) of the microstrip line. This variation is shown in Figures 7.28 and 7.29.

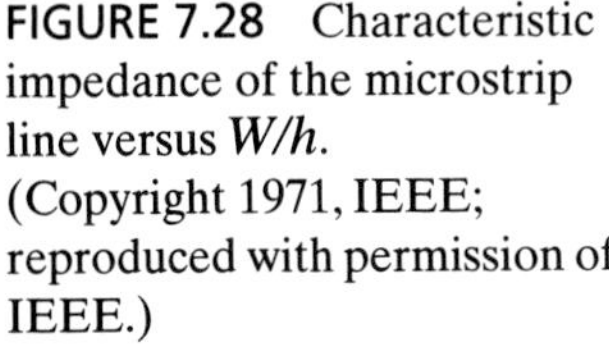

FIGURE 7.28 Characteristic impedance of the microstrip line versus W/h.
(Copyright 1971, IEEE; reproduced with permission of IEEE.)

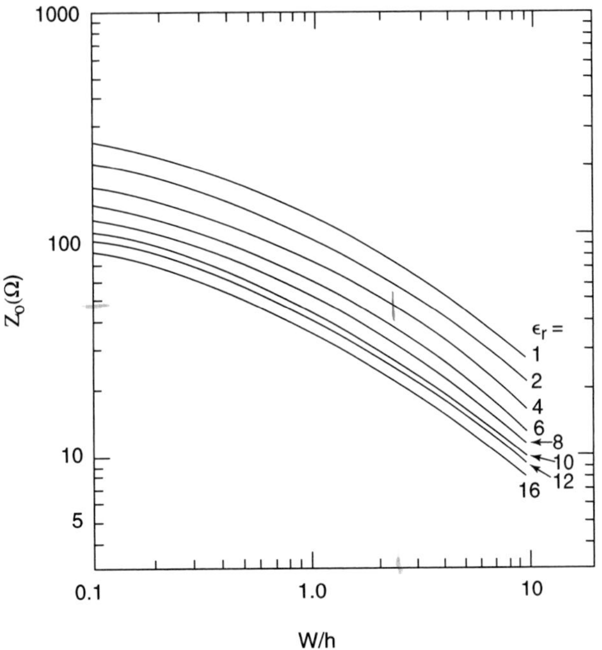

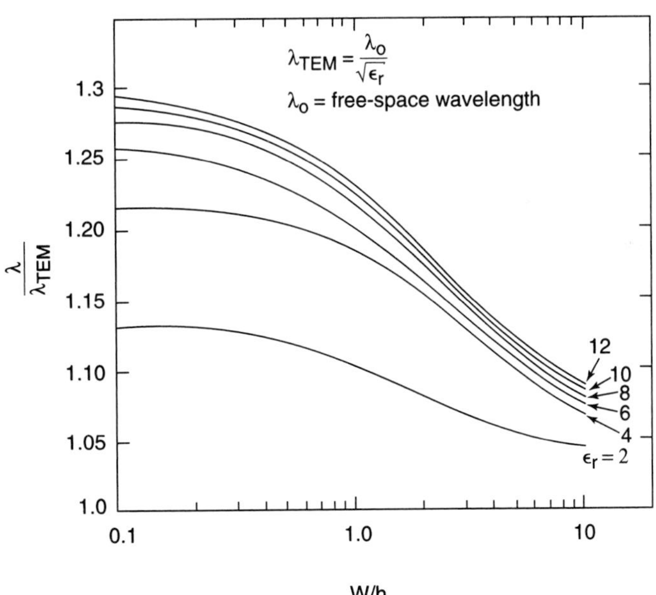

FIGURE 7.29 Normalized wavelength of the microstrip line versus W/h.
(Copyright 1971, IEEE; reproduced with permission of IEEE.)

7.7.2 Empirical Formulas for Effective Dielectric Constant (ε_{ff}), Characteristic Impedance (Z_o), Wavelength (λ), and Attenuation Factors (α_d, α_c)

The essential empirical formulas for a microstrip line can be categorized as follows (assuming zero or negligible thickness of the strip of metal on top of the dielectric, i.e., $t/h < 0.005$):

1. The effective dielectric constant (ε_{ff}) is given by [assuming that the dimensions of the microstrip line (W, h) are known]:
   For $W/h \le 1$:
$$\varepsilon_{ff} = \frac{\varepsilon_r + 1}{2} + \frac{\varepsilon_r - 1}{2}\left[\left(1 + 12\frac{h}{W}\right)^{-1/2} + 0.04\left(1 - \frac{W}{h}\right)^2\right], \tag{7.58}$$
   For $W/h \ge 1$:
$$\varepsilon_{ff} = \frac{\varepsilon_r + 1}{2} + \frac{\varepsilon_r - 1}{2}\left(1 + 12\frac{h}{W}\right)^{-1/2} \tag{7.59}$$
   The effective dielectric constant (ε_{ff}) can be thought of as the dielectric constant of a homogeneous medium that would fill the entire space, replacing air and dielectric regions.

2. The characteristic impedance is given by [assuming that the dimensions of the microstrip line (W, h) are given or known]:
   For $W/h \le 1$:
$$Z_o = \frac{60}{\sqrt{\varepsilon_{ff}}}\ln\left(\frac{8h}{W} + \frac{W}{4h}\right) \tag{7.60}$$
   For $W/h \ge 1$:
$$Z_o = \frac{120\pi}{\sqrt{\varepsilon_{ff}}[W/h + 1.393 + 0.667\ln(W/h + 1.444)]} \tag{7.61}$$

3. Assuming (ε_{ff}) and Z_o are given, then the microstrip dimensions (W, h) can be found as follows (a design problem):
   For $W/h \le 2$:
$$\frac{W}{h} = \frac{8e^A}{e^{2A} - 2} \tag{7.62}$$
   For $W/h \ge 2$:
$$\frac{W}{h} = \frac{2}{\pi}\left[B - 1 - \ln(2B - 1) + \frac{\varepsilon_r - 1}{2\varepsilon_r}\left\{\ln(B - 1) + 0.39 - \frac{0.61}{\varepsilon_r}\right\}\right] \tag{7.63}$$
   where
$$A = \frac{Z_o}{60}\sqrt{\frac{\varepsilon_r + 1}{2}} + \frac{\varepsilon_r - 1}{\varepsilon_r + 1}\left(0.23 + \frac{0.11}{\varepsilon_r}\right) \tag{7.64}$$

and

$$B = \frac{377\pi}{2Z_o\sqrt{\varepsilon_r}} \tag{7.65}$$

4. The wavelength in the microstrip line (λ) is given by:
For $W/h < 0.6$:

$$\lambda = \frac{\lambda_o}{\sqrt{\varepsilon_r}}\left[\frac{\varepsilon_r}{1 + 0.6(\varepsilon_r - 1)(W/h)^{0.0297}}\right]^{1/2} \tag{7.66}$$

For $W/h \geq 0.6$:

$$\lambda = \frac{\lambda_o}{\sqrt{\varepsilon_r}}\left[\frac{\varepsilon_r}{1 + 0.63(\varepsilon_r - 1)(W/h)^{0.1255}}\right]^{1/2} \tag{7.67}$$

5. Attenuation factors
Another characteristic of the microstrip line is its attenuation when signals travel on it. There are two types of losses in a microstrip line:

- Dielectric substrate loss due to dielectric conductivity
- Conductor ohmic loss due to skin effect

The loss factor (α) can be found by noting that the power carried along a transmission line in ($+x$ direction) in a quasi-TEM mode can be written as:

$$P^+(x) = \frac{1}{2}[V^+(x)I^+(x)^*] = \frac{[V^+(x)]^2}{Z_o} \tag{7.68}$$

where

$$V^+(x) = |V^+|e^{-\alpha x}\,e^{-j\beta x} \tag{7.69}$$

Thus, we have:

$$P^+(x) = \frac{|V^+|^2}{2Z_o}e^{-2\alpha x}e^{-j2\beta x} = |P^+|e^{-j2\beta x} \tag{7.70}$$

where

$$|P^+| = \frac{|V^+|^2}{2Z_o}e^{-2\alpha x}$$

and α is the total attenuation factor, which is composed of two components:

$$\alpha = \alpha_d + \alpha_c \tag{7.71}$$

where

α_d = Dielectric loss factor
α_c = Conductor loss factor

These two loss factors are discussed next:

(A) *Attenuation Due to Dielectric Loss* Attenuation due to dielectric loss, identified by "dielectric loss factor (α_d)" using the quasi-TEM mode of propagation, is given by the following.

For low-loss dielectric

$$\alpha_d = 27.3\frac{\tan\delta}{\lambda_o}\left(\frac{\varepsilon_r}{\varepsilon_r - 1}\right)\left(\frac{\varepsilon_{ff} - 1}{\sqrt{\varepsilon_{ff}}}\right)\text{(dB/cm)} \tag{7.72}$$

where ($\tan\delta$) is the loss tangent given by:

$$\tan\delta = \frac{\sigma}{\omega\varepsilon}$$

For high-loss dielectric

$$\alpha_d = 4.34\sigma\left(\frac{\mu_o}{\varepsilon_o}\right)^{1/2}\left(\frac{1}{\varepsilon_r - 1}\right)\left(\frac{\varepsilon_{ff} - 1}{\sqrt{\varepsilon_{ff}}}\right)\text{(dB/cm)} \tag{7.73}$$

where σ is the conductivity of the dielectric and $\mu_o = 4\pi \times 10^{-7}$ (H/m) is the permittivity of free space.

(B) *Attenuation Due to Conductor Loss* Attenuation due to the conductor identified by "conductor loss factor (α_c)" using the quasi-TEM mode of propagation (for $W/h \to \infty$), is given approximately by:

$$W/h \to \infty,$$

$$\alpha_c = \frac{R_s}{Z_o W}\text{(Np/m)} \tag{7.74}$$

where

$$R_s = \sqrt{\frac{\pi f \mu_o}{\sigma}} \tag{7.75}$$

is the surface resistivity of the conductor (due to skin effect). Usually conductor loss is more dominant than the dielectric loss in most microstrip lines, i.e.,

$$\alpha_c >> \alpha_d;$$

$$\Rightarrow \alpha = \alpha_c + \alpha_d \approx \alpha_c$$

There are some cases, though (such as in silicon substrates), where the dielectric loss factor (α_d) is of the same order or larger than the conductor loss factor (α_c).

EXAMPLE 7.11

A 50 Ω microstrip transmission line needs to be designed using a sheet of Epsilam-10® ($\varepsilon_r = 10$) with $h = 1.02$ mm. Determine W, λ, and ε_{ff} by:

a. *An exact method*

b. *An approximate method*

Solution:

a. Exact method

We will design a microstrip line with $W/h \le 2$. Thus, from Equation 7.62 we have:

$$\frac{W}{h} = \frac{8e^A}{e^{2A} - 2}$$

where

$$A = \frac{Z_o}{60}\sqrt{\frac{\varepsilon_r + 1}{2}} + \frac{\varepsilon_r - 1}{\varepsilon_r + 1}\left(0.23 + \frac{0.11}{\varepsilon_r}\right) = \frac{50}{60}\sqrt{\frac{10 + 1}{2}} + \frac{10 - 1}{10 + 1}\left(0.23 + \frac{0.11}{10}\right)$$

$$\Rightarrow A = 2.152$$

Thus, (W/h) is obtained to be:

$$\frac{W}{h} = 0.96$$

and

$$W = 1.02 \times 0.96 = 0.98 \text{ mm}$$

Because $W/h > 0.6$, we use Equation 7.66 to find λ and then use Equation 7.57a to find ε_{ff} as follows:

$$\lambda = \frac{\lambda_o}{\sqrt{\varepsilon_r}}\left[\frac{\varepsilon_r}{1 + 0.63(\varepsilon_r - 1)(W/h)^{0.1255}}\right]^{1/2} = \frac{\lambda_o}{\sqrt{10}}\left[\frac{\varepsilon_r}{1 + 0.63(10 - 1)(0.96)^{0.1255}}\right]^{1/2}$$

$$\Rightarrow \lambda = 0.39\lambda_o$$

and

$$\lambda = 0.39\lambda_o = \lambda_o/\sqrt{\varepsilon_{ff}} \Rightarrow \varepsilon_{ff} = (1/0.39)^2 = 6.6$$

b. Approximate method

Using Figure 7.28, we obtain W/h for $Z_o = 50\ \Omega$ and $\varepsilon_r = 10$ to be:

$$Z_o = 50 \Rightarrow W/h \approx 1 \Rightarrow W = h = 1.02 \text{ mm}$$

From Figure 7.29 for $W/h = 1$, we obtain:

$$\lambda/\lambda_{TEM} = 1.23$$

From Equation 7.57b, we have:

$$\lambda_{TEM} = \lambda_o/\sqrt{\varepsilon_r} = \lambda_o/\sqrt{10} = 0.316\lambda_o$$

Thus, λ is found to be:

$$\lambda = 1.23 \times 0.316\lambda_o = 0.39\lambda_o$$

and from Equation 7.57a, we have:

$$\lambda = \lambda_o/\sqrt{\varepsilon_{ff}} \Rightarrow \varepsilon_{ff} = (\lambda_o/\lambda)^2$$

$$\varepsilon_{ff} = (1/0.39)^2 = 6.6$$

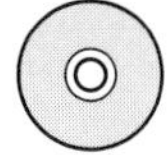

EXAMPLE 7.12

Design a 50 Ω transmission line that provides 90° phase shift at 2.5 GHz. Assume $h = 1.27$ mm and $\varepsilon_r = 2.2$.

Solution:

To find W, we assume that $W/h \geq 2$ and will verify this assumption later. From Equations 7.63 and 7.65, we find:

$$B = \frac{377\pi}{2Z_o\sqrt{\varepsilon_r}}$$

$$B = 7.985,$$

and

$$\frac{W}{h} = \frac{2}{\pi}\left[B - 1 - \ln(2B-1) + \frac{\varepsilon_r - 1}{2\varepsilon_r}\left\{\ln(B-1) + 0.39 - \frac{0.61}{\varepsilon_r}\right\}\right] \Rightarrow \frac{W}{h} = 3.08$$

$$W = 3.08 \times 1.27 = 3.91 \text{ mm}$$

The value of $W/h = 3.08$ is obviously greater than 2, which justifies our earlier assumption.

So far we have found the width of the line; now, we need to know the length of the line. Using the given phase shift of 90° yields:

$$\phi = \beta\ell = \omega\ell/V_p = 2\pi f\ell/(c/\sqrt{\varepsilon_{ff}}) = 2\pi f\ell\sqrt{\varepsilon_{ff}}/c = 90^\circ = \pi/2$$

$$\Rightarrow \ell = c/(4f\sqrt{\varepsilon_{ff}})$$

From the preceding equation we can see that in order to find ℓ, we need to find ε_{ff}. Using Equation 7.59, we obtain for $W/h \geq 1$:

$$\varepsilon_{ff} = \frac{\varepsilon_r + 1}{2} + \frac{\varepsilon_r - 1}{2}\left(1 + 12\frac{h}{W}\right)^{-1/2}$$

$$\varepsilon_{ff} = 1.87$$

Thus, the length of the transmission line is given by:

$$\ell = 3 \times 10^8/(4 \times 2.5 \times 10^9 \times \sqrt{1.87}) = 0.0219 \text{ m} = 2.19 \text{ cm}$$

LIST OF SYMBOLS/ABBREVIATIONS

A symbol will not be repeated again, once it has been identified and defined in an earlier chapter, as long as its definition remains unchanged.

C_o	Capacitance per unit length
EM	Electromagnetic
k	Arbitrary constant
TEM	Transverse Electromagnetic
TL	Transmission line
v	Speed of propagation

V_P	Phase velocity
V^+	Incident voltage
V^-	Reflected voltage
Z_o	Characteristic Impedance
Z_{OC}	Open-circuit impedance
Z_{SC}	Short-circuit impedance
Z_T	Characteristic impedance of a quarter-wave transformer
β	Phase constant
Γ	Reflection coefficient
Γ_L	Reflection coefficient at the load
$\Gamma(x)$	Reflection coefficient at location x
ε	Dielectric permittivity
ε_{ff}	Effective relative dielectric constant
ε_o	Permittivity of vacuum (8.85×10^{-12} F/m)
ε_r	Dielectric constant of a material
γ	Propagation constant
λ_o	Wavelength in free space
ω	Angular frequency

PROBLEMS

7.1 In the two-port network shown in Figure P7.1, assume that $(V_S)_{RMS} = 20 \angle 0°$ V, and $Z_L = 50 + j50\ \Omega$.

a. Find $V^+(0), V^+(\lambda/8), V^-(0), V^-(\lambda/8)$.

b. Calculate net voltages: $V(0), V(\lambda/8), I(0)$, and $I(\lambda/8)$.

c. Calculate the average input powers at $x = 0, \lambda/8$ and show: $P(0) = P(\lambda/8)$.

d. Find $Z_{IN}(0)$.

NOTE: *For parts a–d assume $Z_S = 50\ \Omega$.*

FIGURE P7.1

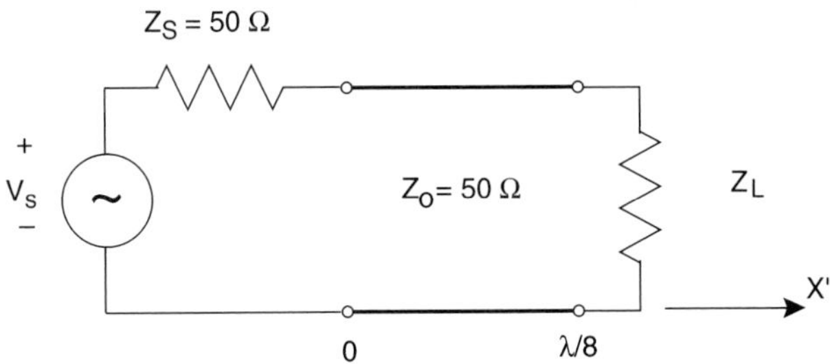

7.2 Find the input impedance and the reflected power at port (1) and the power delivered to the load at port (2) for the circuit shown in Figure P7.2. Assume $V_s = \cos 2\pi \times 10^9 t$.

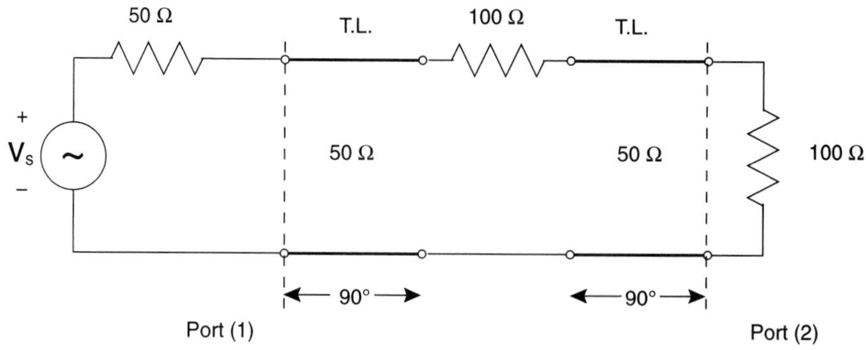

FIGURE P7.2

7.3 In the lossless transmission line circuit shown in Figure P7.3, calculate the incident power, the reflected power, and the power transmitted into the 75 Ω line. Show that: $P_{INC} = P_{REF} + P_{TRANS}$.

FIGURE P7.3

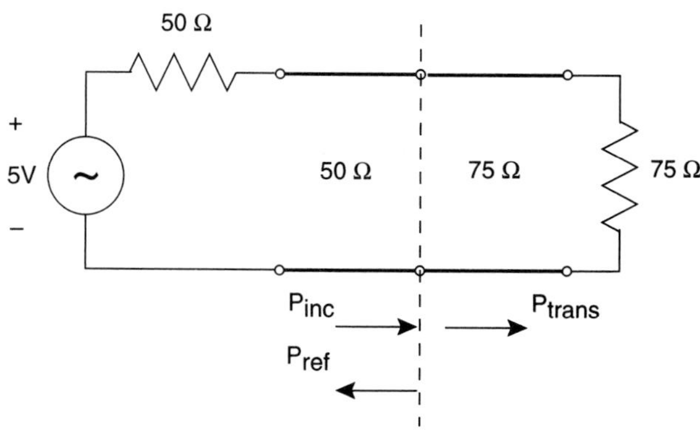

7.4 A lossless transmission line ($\ell = 0.6\lambda$, $Z_o = 50\ \Omega$) is terminated in a load impedance ($Z_L = 40 + j20\ \Omega$). Find the reflection coefficient at the load, the input impedance of the line, and the *VSWR* on the line.

7.5 In the circuit shown in Figure P7.5, calculate the reflection coefficient at the load and the *VSWR* on the line.

FIGURE P7.5

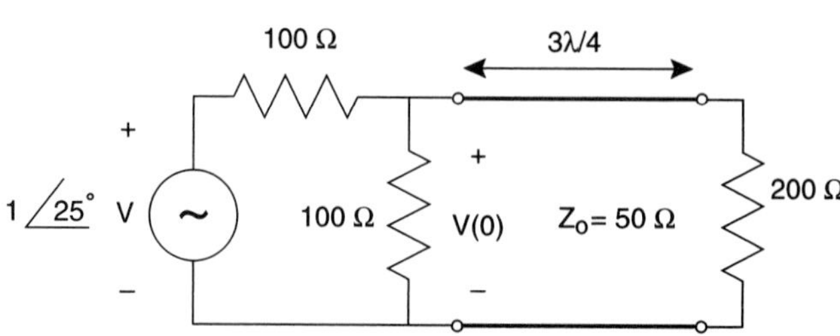

7.6 Consider the lossless transmission line circuit shown in Figure P7.6. Calculate:

a. The load impedance (Z_L)

b. The reflection coefficient at the input of the line

c. The *VSWR* on the line

FIGURE P7.6

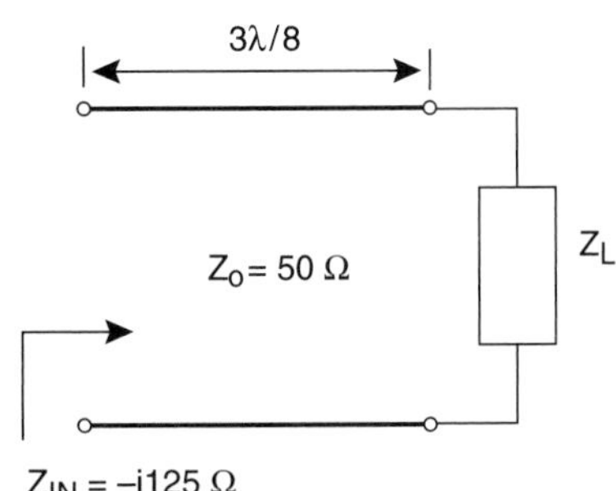

7.7 A lossless transmission line is terminated with a 200 Ω load. If the *VSWR* on the line is 2.0, find the possible values for the line's characteristic impedance.

7.8 For a lossless transmission line, terminated in a reactive load ($Z_L = jX$), find the reflection coefficient and the *VSWR*. What is $|\Gamma|$?

REFERENCES

[7.1] Cheng, D. K. *Fundamentals of Engineering Electromagnetics.* Reading: Addison Wesley, 1993.

[7.2] Cheung, W. S. and F. H. Levien. *Microwave Made Simple, Principles and Applications.* Dedham: Artech House, 1985.

[7.3] Collin, R. E. *Foundation For Microwave Engineering*, 2nd Ed., New York: McGraw-Hill, 1992.

[7.4] Edwards, T. C. *Foundations for Microstrip Circuit Design*. New York: John Wiley & Sons, 1981.

[7.5] Gardiol, F. *Microstrip Circuits.* New York: John Wiley & Sons, 1994.

[7.6] Gonzalez, G. *Microwave Transistor Amplifiers, Analysis and Design*, 2nd ed. Upper Saddle River: Prentice Hall, 1997.

[7.7] Kraus, J. D. *Electromagnetics*, 3rd Ed., New York: McGraw-Hill, 1984.

[7.8] Plonsey, R. and R. E. Collin. *Principles and Applications of Electromagnetic Fields*, 2nd Ed., New York: McGraw-Hill, 1982.

[7.9] Radmanesh, M. M. and B. W. Arnold, *Generalized Microstrip-Slotline Transitions, Theory and Simulation Vs. Experiment,* Microwave Journal, Vol. 36, No. 6, pp. 88–95, June 1993.

[7.10] Radmanesh, M. M. and B. W. Arnold, *Microstrip-Slotline Transitions: Simulation Versus Experiment*, EESof User's Group, IEEE MTT-S International Microwave Symposium, Albuquerque, New Mexico, June 1992.

[7.11] Shen, L. C. and J. A. Kong. *Applied Electromagnetism*, 2nd Ed., Boston: PWS Engineering, 1987.

[7.12] Stratton, J.A. *Electromagnetic Theory.* New York: McGraw-Hill, 1941.

[7.13] Wadell, B. C. *Transmission Line Design Handbook.* Norwood: Artech House, 1991.

[7.14] Zahn, M. *Electromagnetic Field Theory.* New York: John Wiley & Sons, 1979.

CHAPTER 8

Circuit Representations of Two-Port RF/Microwave Networks

8.1 INTRODUCTION

RF/microwave devices, circuits, and components can be classified as one-, two-, three-, or *N*-port networks. A majority of circuits under analysis are two-port networks. Therefore, we will focus primarily on two-port network characterization and will study its representation in terms of a set of parameters that can be cast into a matrix format.

DEFINITION-A TWO-PORT NETWORK: *A network that has only two access ports, one for input or excitation and one for output or response.*

The description of two-port networks from a circuit viewpoint can best be achieved both at low and high frequencies through the use of network parameters. These parameters are discussed in the upcoming sections.

8.2 LOW-FREQUENCY PARAMETERS

To characterize a linear network at low frequencies, several different sets of parameters are available, where one may be selected to fit the application to obtain optimum results.

Voltages and currents at each port provide us with four variables of interest: v_1, v_2, i_1, and i_2. There are six ways of picking two out of a set of four variables, but only four combinations (or sets) will yield nontrivial and unique parameters. These are called *Z*-, *Y*-, *h*-, and *ABCD*-parameters. A two-port network with terminal voltages and currents is shown in Figure 8.1. These four sets of parameters are defined as follows.

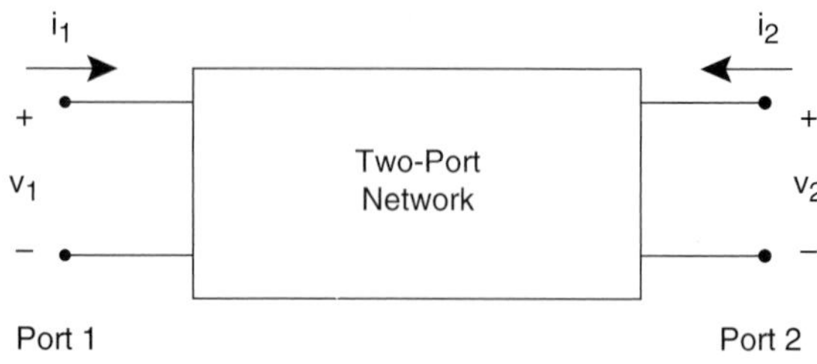

FIGURE 8.1 A block diagram of a two-port network.

8.2.1 Impedance or Z-Parameters

$$v_1 = z_{11}i_1 + z_{12}i_2 \tag{8.1a}$$

$$v_2 = z_{21}i_1 + z_{22}i_2 \tag{8.1b}$$

Or in matrix form:

$$[V] = [Z][I] \tag{8.2}$$

$$[V] = \begin{bmatrix} v_1 \\ v_2 \end{bmatrix} \tag{8.3a}$$

$$[I] = \begin{bmatrix} i_1 \\ i_2 \end{bmatrix} \tag{8.3b}$$

$$[Z] = \begin{bmatrix} z_{11} & z_{12} \\ z_{21} & z_{22} \end{bmatrix} \tag{8.4}$$

8.2.2 Admittance or Y-Parameters

Similarly, we can write the *Y*-parameters in matrix form as:

$$[I] = [Y][V] \tag{8.5}$$

where $[I]$ and $[V]$ are defined as before and $[Y]$ as follows:

$$[Y] = \begin{bmatrix} y_{11} & y_{12} \\ y_{21} & y_{22} \end{bmatrix} \tag{8.6}$$

8.2.3 Hybrid or h-Parameters

$$\begin{bmatrix} v_1 \\ i_2 \end{bmatrix} = \begin{bmatrix} h_{11} & h_{12} \\ h_{21} & h_{22} \end{bmatrix} \begin{bmatrix} i_1 \\ v_2 \end{bmatrix} \tag{8.7}$$

8.2.4 Transmission or ABCD-Parameters

$$\begin{bmatrix} v_1 \\ i_1 \end{bmatrix} = \begin{bmatrix} A & B \\ C & D \end{bmatrix} \begin{bmatrix} v_2 \\ -i_2 \end{bmatrix} \tag{8.8}$$

EXAMPLE 8.1

Find the [ABCD] matrix for a series impedance element (Z) as shown in Figure 8.2.

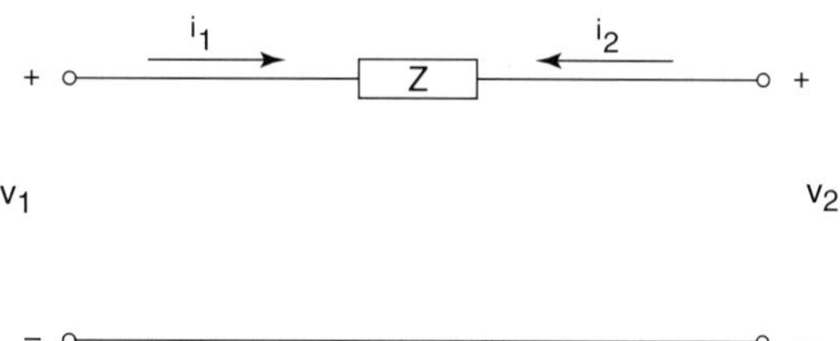

FIGURE 8.2 A series element (Example 8.1).

Solution:

Using KVL and KCL, the following can be written:

$$v_1 = v_2 - Zi_2 = Av_2 - Bi_2$$

$$i_1 = -i_2 = 0 - i_2 = Cv_2 - Di_2$$

Thus, the $[ABCD]$ matrix is given by:

$$\begin{bmatrix} A & B \\ C & D \end{bmatrix} = \begin{bmatrix} 1 & Z \\ 0 & 1 \end{bmatrix}$$

EXAMPLE 8.2

Find the [ABCD] matrix for a shunt element (Y) as shown in Figure 8.3.

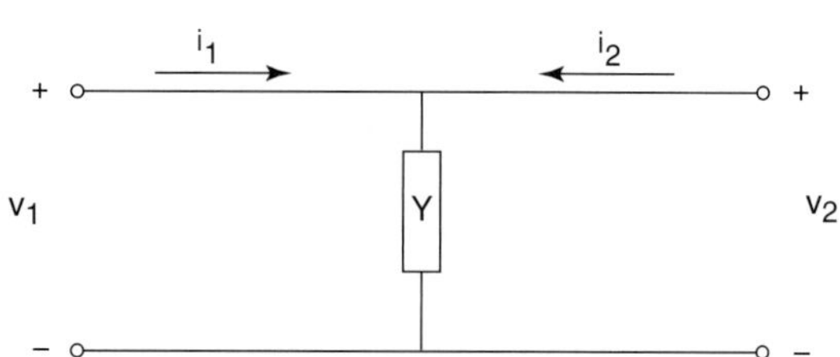

FIGURE 8.3 A shunt element (Example 8.2).

Solution:

Using KVL and KCL, the following can be written:

$$v_1 = v_2 = v_2 + 0 = Av_2 - Bi_2$$

$$i_1 = Yv_2 - i_2 = Cv_2 - Di_2$$

Thus, the $[ABCD]$ matrix is given by:

$$\begin{bmatrix} A & B \\ C & D \end{bmatrix} = \begin{bmatrix} 1 & 0 \\ Y & 1 \end{bmatrix}$$

EXAMPLE 8.3

Find the [ABCD] matrix for a circuit consisting of a series element (Z) and a shunt element (Y) as shown in Figure 8.4.

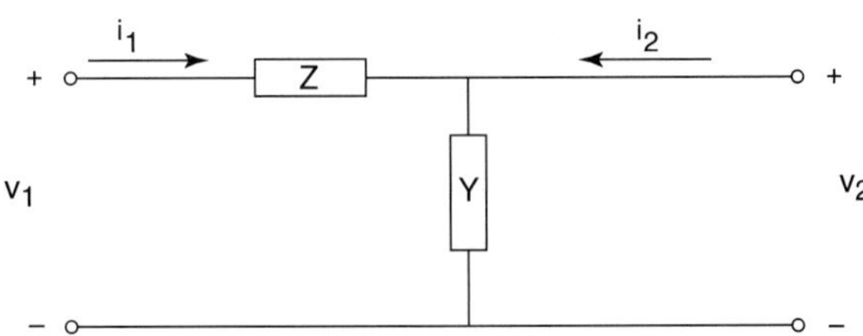

FIGURE 8.4 A series-shunt circuit (Example 8.3).

Solution:

The $[ABCD]$ matrix for the whole circuit, which is a cascade of a series and a shunt element, is a multiplication of the two matrices as follows:

$$\begin{bmatrix} A & B \\ C & D \end{bmatrix} = \begin{bmatrix} A_1 & B_1 \\ C_1 & D_1 \end{bmatrix} \begin{bmatrix} A_2 & B_2 \\ C_2 & D_2 \end{bmatrix} = \begin{bmatrix} 1 & Z \\ 0 & 1 \end{bmatrix} \begin{bmatrix} 1 & 0 \\ Y & 1 \end{bmatrix}$$

Thus, the $[ABCD]$ matrix is given by:

$$\begin{bmatrix} A & B \\ C & D \end{bmatrix} = \begin{bmatrix} 1 + ZY & Z \\ Y & 1 \end{bmatrix}$$

EXAMPLE 8.4

Find the [ABCD] matrix for a transformer as shown in Figure 8.5.

Solution:

Using the transformer voltage and current rule, which states that if the voltage is stepped down then in order to preserve the power flow the current must be proportionately stepped up, we have:

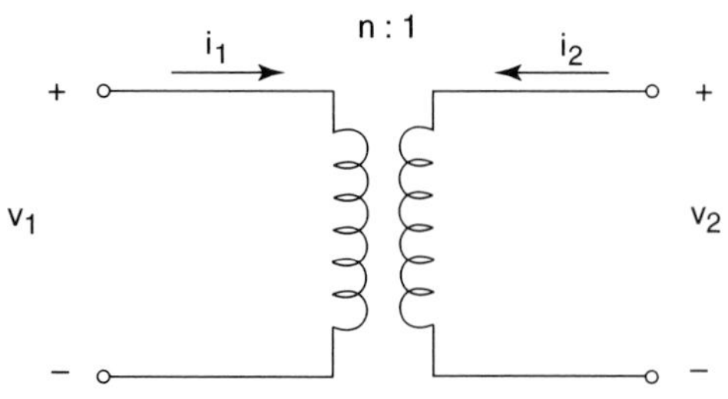

FIGURE 8.5 A transformer (Example 8.4).

$$v_1 = nv_2 = Av_2 - Bi_2$$

$$i_1 = -\frac{1}{n}i_2 = Cv_2 - Di_2$$

Thus the $[ABCD]$ matrix is given by:

$$\begin{bmatrix} A & B \\ C & D \end{bmatrix} = \begin{bmatrix} n & 0 \\ 0 & \frac{1}{n} \end{bmatrix}$$

EXAMPLE 8.5

Find the [ABCD] matrix for a lossless transmission line of length (ℓ) and characteristic impedance (Z_o) as shown in Figure 8.6.

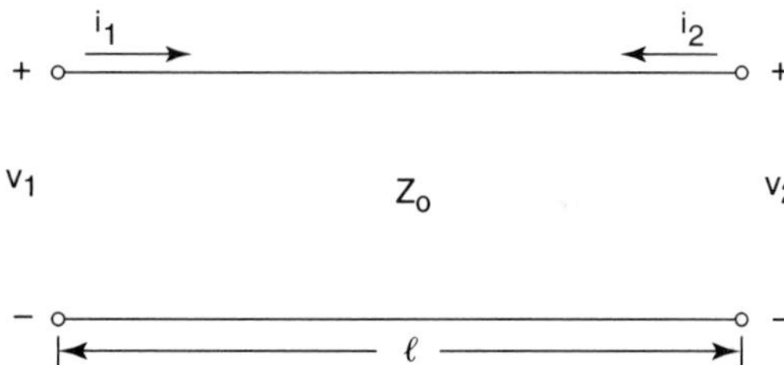

FIGURE 8.6 Lossless transmission line (Example 8.5).

Solution:

Using results obtained in Chapter 7, *Fundamental Concepts in Wave Propagation*, we know that the output voltage or current has the same magnitude as the input but lags behind in phase by $e^{-j\beta\ell}$. Thus, the following can be written:

$$v_2 = v_1\, e^{-j\beta l} \Rightarrow v_1 = v_2\, e^{+j\beta l} = v_2 \cos\beta\ell + jv_2 \sin\beta\ell$$

$$i_2 = -i_1\, e^{-j\beta l} \Rightarrow i_1 = -i_2\, e^{+j\beta l} = -i_2 \cos\beta\ell - ji_2 \sin\beta\ell$$

Because the load end is considered to be matched to the transmission line, we can write:

$$v_2 = -Z_o i_2$$

$$v_1 = (\cos\beta\ell)v_2 - (jZ_o \sin\beta\ell)i_2 = Av_2 - Bi_2$$

$$i_1 = (jY_o\sin\beta\ell)v_2 - (\cos\beta\ell)\, i_2 = Cv_2 - Di_2$$

Thus, the $[ABCD]$ matrix can be written as:

$$\begin{bmatrix} A & B \\ C & D \end{bmatrix} = \begin{bmatrix} \cos\beta\ell & jZ_o\sin\beta\ell \\ jY_o\sin\beta\ell & \cos\beta\ell \end{bmatrix}$$

8.3 HIGH-FREQUENCY PARAMETERS

We note that Z-, Y-, h-, and $ABCD$-parameters are based on the following considerations at each of the network ports:

- Net voltage (v) and net current (i)
- Short and open circuit terminations

Simple observations at high RF/microwave frequencies reveal the following:

- Shorts and open circuit terminations are difficult to implement over a broad range of frequencies and, thus, cannot be used to characterize networks.
- At high RF/microwave frequencies, the net voltage (or net current) is a combination of two or more voltage (or current) traveling waves.

Based on these observations, the Z-, Y-, h-, and $ABCD$-parameters cannot be accurately measured at these higher frequencies; therefore, we have to use the concept of propagating or traveling waves to define the network parameters.

The network representation of a two-port network at high RF/microwave frequencies is called "scattering parameters" (or "S-parameters" for short).

When cascading networks, a variation of S-parameters called chain scattering parameters (or T-parameters) are used to simplify the analysis.

These two types of high-frequency parameters are very popular and are primarily used at the high RF/microwave frequencies.

8.4 FORMULATION OF THE S-PARAMETERS

The high frequency S- and T-parameters are used to characterize high RF/microwave two-port networks (or N-port networks, in general). These parameters are based on the concept of traveling waves and provide a complete characterization of any two-port network under analysis or test at high RF/microwave frequencies.

In view of the linearity of the electromagnetic field equations and the linearity displayed by most microwave components and networks, the "scattered waves" (i.e., the reflected and transmitted wave amplitudes) are linearly related to the incident wave amplitude. The matrix describing this linear relationship is called the "scattering matrix," or $[S]$.

While the lower-frequency network parameters (such as *Z*- or *Y*-matrices, etc.) are defined in terms of net (or total) voltage and currents at the ports, these concepts are not practical at high RF/microwave frequencies where it is found that any set of parameters, to be meaningful, must be defined in terms of a combination of traveling waves.

To characterize a two-port network that has *identical characteristic impedances* (Z_o) at both the input and output ports, let us consider the incident and reflected voltage waves at each port, as shown in Figure 8.7.

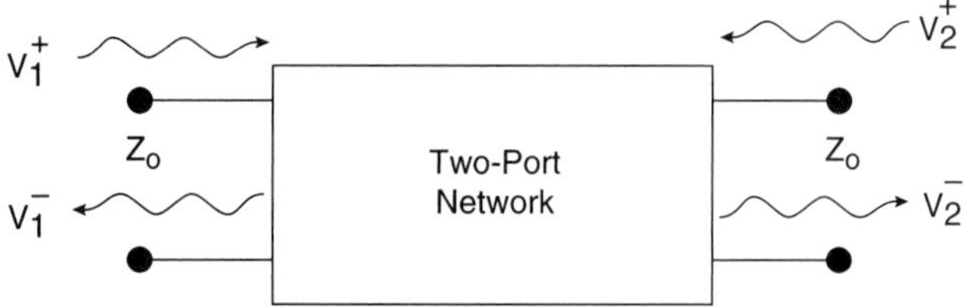

FIGURE 8.7 A two-port with incident and reflected waves at each port.

To define the *S*-parameters accurately, we will consider a voltage phasor $[V_i^+]$ incident on and a voltage phasor $[V_i^-]$ reflected from the terminals of a two-port network ($i = 1,2$) as shown in Figure 8.7.

The scattering matrix, $[S]$, is now defined to describe the linear relationship between the incident voltage wave phasor matrix $[V_i^+]$ and the reflected or transmitted wave phasor matrix $[V_i^-]$ at any of the two ports as follows:

$$V_1^- = S_{11}V_1^+ + S_{12}V_2^+$$

$$V_2^- = S_{21}V_1^+ + S_{22}V_2^+$$

Or, in matrix form we can write:

$$\begin{bmatrix} V_1^- \\ V_2^- \end{bmatrix} = \begin{bmatrix} S_{11} & S_{12} \\ S_{21} & S_{22} \end{bmatrix} \begin{bmatrix} V_1^+ \\ V_2^+ \end{bmatrix} \tag{8.9}$$

or,

$$[V^-] = [S][V^+]$$

where

$$[V^-] = \begin{bmatrix} V_1^- \\ V_2^- \end{bmatrix}$$

$$[V^+] = \begin{bmatrix} V_1^+ \\ V_2^+ \end{bmatrix}$$

and

$$[S] = \begin{bmatrix} S_{11} & S_{12} \\ S_{21} & S_{22} \end{bmatrix} \tag{8.10}$$

This linear relationship is expressed in terms of a ratio of two phasors that are complex numbers with the magnitude of the ratio always less than or equal to 1. Each specific element of the $[S]$ matrix is defined as:

$$S_{11} = \left.\frac{V_1^-}{V_1^+}\right|_{V_2^+=0} = \Gamma_{IN} \quad \text{Input reflection coefficient when output port is terminated in a matched load.} \tag{8.11}$$

$$S_{21} = \left.\frac{V_2^-}{V_1^+}\right|_{V_2^+=0} \quad \text{Forward transmission coefficient when output port is terminated in a matched load.} \tag{8.12}$$

$$S_{12} = \left.\frac{V_1^-}{V_2^+}\right|_{V_1^+=0} \quad \text{Reverse transmission coefficient when input port is terminated in a matched load.} \tag{8.13}$$

$$S_{22} = \left.\frac{V_2^-}{V_2^+}\right|_{V_1^+=0} = \Gamma_{OUT} \quad \text{Output reflection coefficient when input port is terminated in a matched load.} \tag{8.14}$$

S-parameters, as defined above, have many advantages at high RF/microwave frequencies that can be briefly stated as follows:

- *S*-parameters provide a complete characterization of a network, as seen at its two ports.
- *S*-parameters make the use of short or open (as prescribed at lower frequencies) completely unnecessary at higher frequencies. It is a known fact that the impedance of a short or an open varies with frequency which is one reason why they are not useful for device characterization at high RF/microwave frequencies. Furthermore, the presence of a short or open in a circuit can cause strong reflections (because $|\Gamma_L| = 1$), which usually lead to oscillations or damage to the transistor circuitry.
- *S*-parameters require the use of matched loads for termination and because the loads absorb all the incident energy, the possibility of serious reflections back to the device or source is eliminated.

S-parameters can be converted to low-frequency parameters as illustrated in the next example.

EXAMPLE 8.6

Given the [ABCD] matrix for a two-port network, derive its [S] matrix (see Figure 8.8).

Solution:

To obtain S_{11}, we terminate port 2 in a matched load and find the input reflection coefficient (Γ_{IN}) as follows:

$$S_{11} = \Gamma_{IN} = \frac{Z_{IN} - Z_o}{Z_{IN} + Z_o}$$

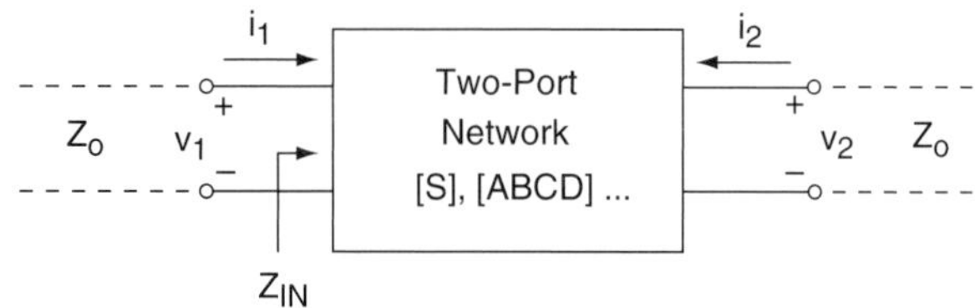

FIGURE 8.8 Network for Example 8.6.

Where

$$Z_{IN} = v_1/i_1$$

$$v_1 = Av_2 - Bi_2$$

$$i_1 = Cv_2 - Di_2$$

$$v_2 = -Z_o i_2$$

Substituting for v_1 and i_1 in terms of $[ABCD]$, we have:

$$Z_{IN} = \frac{v_1}{i_1} = \frac{Av_2 - Bi_2}{Cv_2 - Di_2} = \frac{AZ_o - B}{CZ_o - D}$$

Now, substituting for Z_{IN}, we can write S_{11} as:

$$S_{11} = \Gamma_{IN} = \frac{A + BY_o - CZ_o - D}{\Delta}$$

where Δ is given by:

$$\Delta = A + BY_o + CZ_o + D$$

Similarly, S_{12}, S_{21}, and S_{22} can be found as follows:

$$S_{12} = \frac{2(AD - BC)}{\Delta}$$

$$S_{21} = \frac{2}{\Delta}$$

$$S_{22} = \Gamma_{OUT} = \frac{-A + BY_o - CZ_o + D}{\Delta}$$

In general, using the same technique as demonstrated in Example 8.6, any set of network parameters can be converted into another set of parameters. Appendix H, *Conversion Among Two-Port Network Parameters,* shows the conversion relation between the z-, y-, h-, $ABCD$-, and the S-parameters.

The conversion among the three transistor configurations is an important relation that becomes useful in many practical design situations. Appendix I, *Conversion among the Y-Parameters of a Transistor (Three Configurations: CE, CB, and CC)*, shows the conversion relation between the y-parameters of a transistor in common-emitter, common-base, and common-collector configurations. If parameters other than Y-parameters (e.g., S-parameters) are needed, then Appendix H can be used effectively to convert Y- to S-parameters.

8.5 PROPERTIES OF S-PARAMETERS

The S-parameters of an N-port network, in general, have certain properties and interrelationships among the parameters themselves that are worth considering. In the following discussion, due to their popularity and frequent use, we limit our discussion solely to two-port networks. Depending on whether the network is reciprocal or lossless, the S-parameters will have different properties, as discussed next.

8.5.1 Reciprocal Networks

A reciprocal network is defined to be a network that satisfies the reciprocity theorem, which is defined as follows:

DEFINITION-RECIPROCITY THEOREM: *Is a theorem stating that the interchange of electromotive force at one point (e.g., in branch k, v_k) in a passive linear network, with the current produced at any other point (e.g., branch m, i_m) results in the same current (in branch k, i_k) when the same electromotive force is applied in the new location (branch m, v_m); that is,*

$$v_k/i_m = v_m/i_k \tag{8.15}$$

or

$$Z_{km} = Z_{mk}$$

As observed, this theorem applies only to passive networks having linear bilateral impedances. Networks that satisfy this condition include all passive networks that contain linear passive elements including resistors, capacitors, inductors, and transformers except independent or dependent sources, nonlinear elements, and/or active solid-state devices such as diodes, transistors, and so on.

It can be shown that for all reciprocal networks, the $[S]$ matrix is symmetrical:

$$S_{12} = S_{21} \tag{8.16a}$$

Generalizing the previous equation, it can be shown that for an N-port network:

$$S_{ij} = S_{ji} \quad \text{for } i \neq j \tag{8.16b}$$

where

$$i = 1,.....,N$$

$$j = 1,.....,N$$

A Special Case: A Symmetrical Reciprocal Network. A special case of a reciprocal network is a symmetrical network. These networks have identical size and arrangement for corresponding electrical elements in reference to a plane or line of symmetry.

Due to symmetry of the network topology and by observation, the input impedance obtained by looking into the input port is equal to the impedance looking into the output port. The equality of input and output impedances leads to the equality of input and output reflection coefficients in addition to equality of S_{12} and S_{21} as required by the reciprocity theorem stated earlier. Therefore, for symmetrical networks, we can always write:

$$S_{11} = S_{22} \tag{8.17a}$$

$$S_{12} = S_{21} \tag{8.17b}$$

Or, in general for any symmetrical passive N-port network, we can write:

$$S_{ii} = S_{jj} \tag{8.18a}$$

$$S_{ij} = S_{ji} \tag{8.18b}$$

Where $i \neq j$, and

$$i = 1,\ldots\ldots, N$$

$$j = 1,\ldots\ldots, N.$$

8.5.2 Lossless Networks

For a lossless passive network (i.e., one containing no resistive elements), the power entering the circuit will always be equal to the power leaving the network, i.e., the power is conserved. This condition will impose a number of restrictions on the S-parameters that give rise to the unity and zero properties as follow:

The Unity Property of [S] Matrix. This property states that for a passive lossless N-port network, the sum of the products of each term of any one row (or any one column) multiplied by its own complex conjugate is unity, i.e.,

$$\sum_{i=1}^{N} S_{ij}S_{ij}^{*} = 1\ , \qquad j = 1,2,\ldots.,N \tag{8.19}$$

where i and j are row and column numbers respectively.

For a two-port network, Equation 8.19 yields two equations:

$$S_{11}S_{11}^{*} + S_{21}S_{21}^{*} = 1 \tag{8.20a}$$

$$S_{12}S_{12}^{*} + S_{22}S_{22}^{*} = 1 \tag{8.20b}$$

Furthermore, if the lossless network is also reciprocal (i.e., $S_{12} = S_{21}$), these two equations are greatly simplified as follows:

$$S_{12} = S_{21} \tag{8.21a}$$

$$|S_{11}| = |S_{22}| \tag{8.21b}$$

$$|S_{11}|^2 + |S_{21}|^2 = 1 \tag{8.21c}$$

The Zero Property of [S] Matrix. This property states that for a passive lossless N-port network, the sum of the products of each term of any row (or any column) multiplied by the complex conjugate of the corresponding terms of any other row (or column) is zero:

$$\sum_{k=1}^{N} S_{ki}S_{kj}^{*} = 0 \quad \text{for } i \neq j, \ \& \ i,j = 1,2,....,N \tag{8.22}$$

where i and j are row and column numbers, respectively.

For a two-port network, this equation simplifies into two equations:

$$S_{11}S_{12}^{*} + S_{21}S_{22}^{*} = 0 \tag{8.23a}$$

$$S_{12}S_{11}^{*} + S_{22}S_{21}^{*} = 0 \tag{8.23b}$$

Furthermore, if the lossless network is also reciprocal (i.e., $S_{12} = S_{21}$), then the previous two equations simplify into:

$$S_{12} = S_{21}$$

$$S_{11}S_{21}^{*} + S_{21}S_{22}^{*} = 0 \tag{8.24a}$$

$$|S_{11}| = |S_{22}| \tag{8.24b}$$

NOTE: *A matrix satisfying the zero and unity property is called a unitary matrix.*

Analysis of Reciprocal Lossless Networks. From the zero and unity properties of the S-matrix, the S-parameters of a *reciprocal lossless network* are constrained by Equations 8.20, 8.21, and 8.24 as follows:

$$S_{21} = S_{12} \tag{8.25a}$$

$$|S_{11}| = |S_{22}| \tag{8.25b}$$

$$|S_{11}|^2 + |S_{21}|^2 = 1 \tag{8.25c}$$

$$S_{11}S_{21}^{*} + S_{21}S_{22}^{*} = 0 \tag{8.25d}$$

If we let:

$$S_{11} = |S_{11}|e^{j\theta_{11}},$$

$$S_{22} = |S_{11}|e^{j\theta_{22}}$$

and

$$S_{21} = |S_{21}|e^{j\theta_{21}}$$

Then, Equations 8.25c and 8.25d give:

$$|S_{21}| = (1 - |S_{11}|^2)^{1/2} \tag{8.26a}$$

$$|S_{11}|(1 - |S_{11}|^2)^{1/2}(e^{j(\theta_{11} - \theta_{21})} + e^{j(\theta_{21} - \theta_{22})}) = 0$$

which yields:

$$\left(e^{j(\theta_{11}-\theta_{21})}+e^{j(\theta_{21}-\theta_{22})}\right)=0 \Rightarrow e^{j(\theta_{11}-\theta_{21})}=e^{-j\pi}e^{j(\theta_{21}-\theta_{22})}$$

$$\Rightarrow \theta_{11}+\theta_{22}=2\,\theta_{21}-\pi\pm 2n\pi$$

or

$$\theta_{21}=\frac{\theta_{11}+\theta_{22}}{2}+\pi\left(\frac{1}{2}\mp n\right)\quad \text{For } n=0,1,2,\ldots \tag{8.26b}$$

Equations 8.26a and 8.26b provide the magnitude and phase of S_{21} (or S_{12}) in terms of magnitude and phase of S_{11} and S_{22}. Therefore, from a measurement knowledge of S_{11} and S_{22}, we can completely describe and specify a reciprocal lossless two-port network. This use of S-parameters in specifying a reciprocal lossless two-port network shows its usefulness and versatility. The following will illustrate the concept of S-parameters further.

EXAMPLE 8.7

What are the S-parameters of a series element (Z) as shown in Figure 8.9?

FIGURE 8.9 Circuit for Example 8.7.

a) A series element

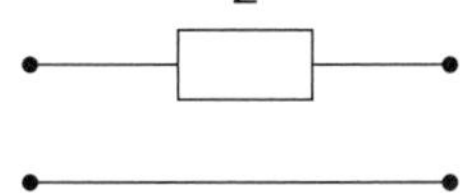

b) Circuit for S_{11}-parameter calculation

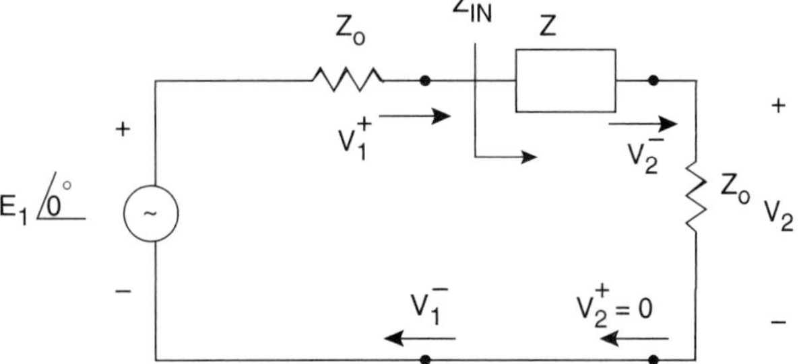

Solution:

Because this is a reciprocal and symmetrical network, we have:

$$S_{11}=S_{22},$$
$$S_{12}=S_{21}$$

So, we only need to find S_{11} and S_{21}.

NOTE: *This is not a lossless network because $Z = R + jX$ has a loss component!*

a.
$$S_{11}=\left.\frac{V_1^-}{V_1^+}\right|_{V_2^+=0} \tag{8.27}$$

According to Equation 8.27, S_{11} is the input reflection coefficient when the output is matched; that is,

$$S_{11} = \Gamma_{IN} = (Z_{IN} - Z_o)/(Z_{IN} + Z_o),$$

Where $Z_{IN} = Z + Z_o$, thus we have:

$$S_{11} = Z/(Z + 2Z_o) \tag{8.28}$$

b.

$$S_{21} = \left.\frac{V_2^-}{V_1^+}\right|_{V_2^+ = 0} \tag{8.29}$$

From Equation 8.29, we can see that S_{21} is the voltage gain (or loss) when the output is matched. Thus by applying a source voltage (E_1) at port 1, the voltage gain is found as follows:

$$I = E_1/(Z_o + Z_{IN})$$

$$V_2 = V_2^- + V_2^+$$

Because the load is matched, $V_2^+ = 0$. Thus, we have:

$$\Rightarrow V_2^- = Z_o I \tag{8.30a}$$

$$V_1 = V_1^+ + V_1^- = V_1^+(1 + S_{11}) = Z_{IN} I \Rightarrow V_1^+ = Z_{IN} I/(1 + S_{11}) \tag{8.30b}$$

Dividing Equation 8.30a by 8.30b, we have:

$$S_{21} = V_2^-/V_1^+ = Z_o(1 + S_{11})/Z_{IN} \Rightarrow S_{21} = 2Z_o/(Z + 2Z_o) \tag{8.31}$$

Therefore, the whole S-matrix can be written as:

$$S = \begin{bmatrix} \frac{Z}{Z + 2Z_o} & \frac{2Z_o}{Z + 2Z_o} \\ \frac{2Z_o}{Z + 2Z_o} & \frac{Z}{Z + 2Z_o} \end{bmatrix} \tag{8.32}$$

OBSERVATION: *For a "series Z" network, from Equations 8.28 and 8.31, we can see that:*

$$S_{21} = 1 - S_{11}$$

EXAMPLE 8.8

What are the S-parameters of a shunt element (Y) as shown in Figure 8.10?

Solution:

Similar to Example 8.1, this is a reciprocal and symmetrical network, thus:

$$S_{11} = S_{22},$$

and

$$S_{12} = S_{21}$$

so we only need to find S_{11} and S_{21}.

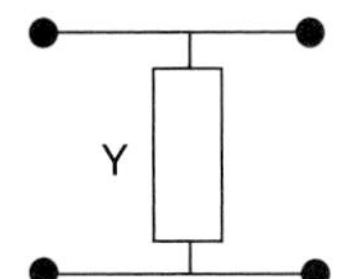

FIGURE 8.10 Circuit for Example 8.8.
a) A shunt element

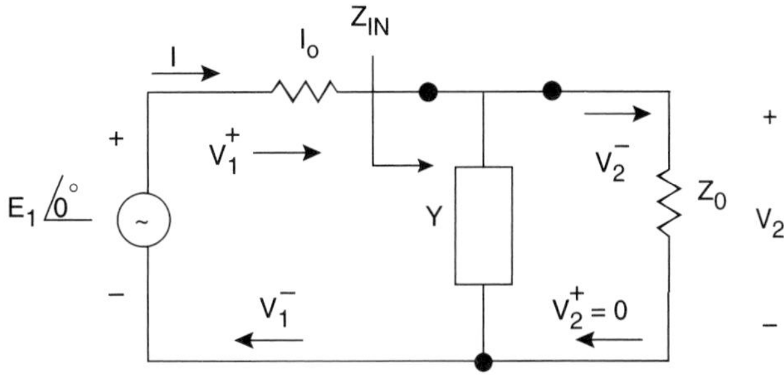

b) Circuit for S_{11}-parameter calculation

a.

$$S_{11} = \left.\frac{V_1^-}{V_1^+}\right|_{V_2^+ = 0}$$

$$S_{11} = (Z_{IN} - Z_o)/(Z_{IN} + Z_o)$$

$$Z_{IN} = Z_o + (1/Y \| Z_o)$$

Substituting for Z_{IN} in S_{11}, we obtain:

$$S_{11} = -Z_oY/(2 + Z_oY) \tag{8.33}$$

b.

$$S_{21} = \left.\frac{V_2^-}{V_1^+}\right|_{V_2^+ = 0}$$

By applying a source voltage E_1 to port 1, we obtain:

$$I = E_1/(Z_o + Z_{IN})$$

Because port 2 is terminated in a matched load (i.e., $V_2^+ = 0$), we can write:

$$V_2 = V_2^- = (1/Y \| Z_o)I \tag{8.34a}$$

$$V_1 = V_1^+ + V_1^- = V_1^+(1 + S_{11}) = Z_{IN}I \Rightarrow V_1^+ = Z_{IN}I/(1 + S_{11}) \tag{8.34b}$$

Dividing Equation 8.34a by 8.34b, we obtain:

$$S_{21} = V_2^-/V_1^+ = (1/Y \| Z_o)(1 + S_{11})/Z_{IN} \Rightarrow S_{21} = 2/(2 + Z_oY) \tag{8.35}$$

Therefore, the whole S-matrix can be written as:

$$S = \begin{bmatrix} \frac{-Z_oY}{2 + Z_oY} & \frac{2}{2 + Z_oY} \\ \frac{2}{2 + Z_oY} & \frac{-Z_oY}{2 + Z_oY} \end{bmatrix} \tag{8.36}$$

OBSERVATION: *For a "shunt Y" network, from Equations 8.33 and 8.35, we can see that:*

$$S_{21} = 1 + S_{11}$$

8.6 SHIFTING REFERENCE PLANES

The S-parameters relate amplitude and phase of traveling waves that are incident on, transmitted through, or reflected from a network terminal. Therefore, the location of the reference plane must be known precisely to calculate or measure the exact phase of the S-parameters.

Consider a two-port network in which the reference plane at port 1 has moved a distance ℓ_1 to port 1'. Similarly, the reference plane at port 2 has moved a distance ℓ_2 to port 2' as shown in Figure 8.11. The phasors for voltage waves at each new port (i.e., 1' and 2') can now be written as:

$$V'^{+}_{i} = V^{+}_{i} e^{j\theta_i}, \quad i = 1,2 \tag{8.37a}$$

$$V'^{-}_{i} = V^{-}_{i} e^{-j\theta_i}, \quad i=1,2 \tag{8.37b}$$

where $\theta_i = \beta\ell_i$ ($i = 1,2$) is the electrical length corresponding to the reference plane shift at each port.

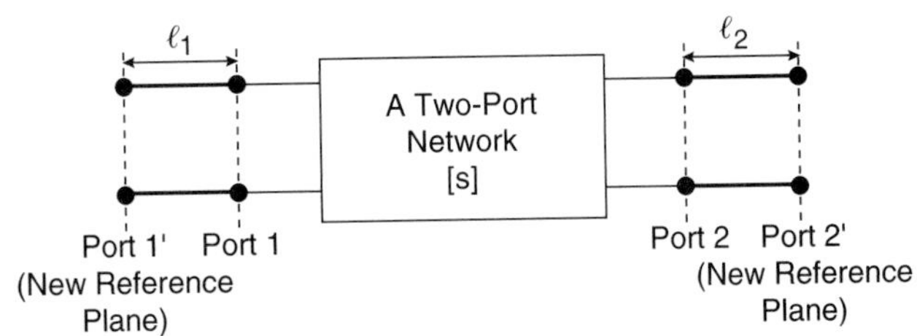

FIGURE 8.11 A two-port network with new reference planes.

Inverting Equation 8.37, we can write:

$$V^{+}_{i} = V'^{+}_{i} e^{-j\theta_i}, \quad i = 1,2 \tag{8.38a}$$

$$V^{-}_{i} = V'^{-}_{i} e^{j\theta_i}, \quad i = 1,2 \tag{8.38b}$$

Upon substitution of Equations 8.38 in:

$$[V^-] = [S][V^+]$$

and further mathematical manipulation, we obtain $[S']$, which is the shifted S-parameters as:

$$[V'^-] = [S'][V'^+], \qquad [S'] = \begin{bmatrix} S_{11}e^{-j2\theta_1} & S_{12}e^{-j(\theta_1+\theta_2)} \\ S_{21}e^{-j(\theta_1+\theta_2)} & S_{22}e^{-j2\theta_2} \end{bmatrix} \tag{8.39}$$

or conversely

$$[S] = \begin{bmatrix} S'_{11}e^{j2\theta_1} & S'_{12}e^{j(\theta_1+\theta_2)} \\ S'_{21}e^{j(\theta_1+\theta_2)} & S'_{22}e^{j2\theta_2} \end{bmatrix} \tag{8.40}$$

To summarize this analysis, we note that:

$$S'_{ii} = S_{ii}e^{-j2\theta_i}, \quad i = 1,2 \tag{8.41}$$

$$S'_{ij} = S_{ij}e^{-j(\theta_i+\theta_j)}, \quad i \neq j, i = 1,2 \tag{8.42}$$

Equation 8.41 shows that the phase of S_{ii} is shifted by twice the electrical length because the incident wave travels twice over this length upon reflection. On the other hand, at port i ($i = 1,2$), Equation 8.42 shows that S_{ij} ($i \neq j$) is shifted by the sum of the electrical lengths because the incident wave must pass through both lengths in order to travel from one shifted port to the other.

8.7 TRANSMISSION MATRIX

The following discussion in general applies to a cascade of N-port networks. For the sake of simplicity, however, we limit our analysis to two-port networks only. When cascading a number of two-port networks in series, a more useful network representation is needed to facilitate the calculation of the overall network parameters.

This new representation should relate the output quantities in terms of input quantities. Using such a representation will enable us to obtain a description of the complete cascade by simply multiplying together the matrices describing each network.

At low frequencies, the transmission matrix (also known as the *ABCD* matrix) is defined in terms of the net input voltage and current as the independent variables and output net voltage and current as the dependent variables.

At high RF and microwave frequencies, however, the transmission matrix [T] is expressed in terms of the input incident and reflected waves as the independent variables and the output incident and reflected waves as the dependent variables.

Using the latter definition at RF/microwave frequencies, the transmission matrix formulation becomes very useful when dealing with multistage circuits (such as filters, amplifiers, etc.) or infinitely long periodic structures such as those used in circuits for traveling wave tubes, etc.

The transmission matrix (also called chain scattering parameters or scattering transfer parameters) for a two-port network, as shown in Figure 8.12, is defined as:

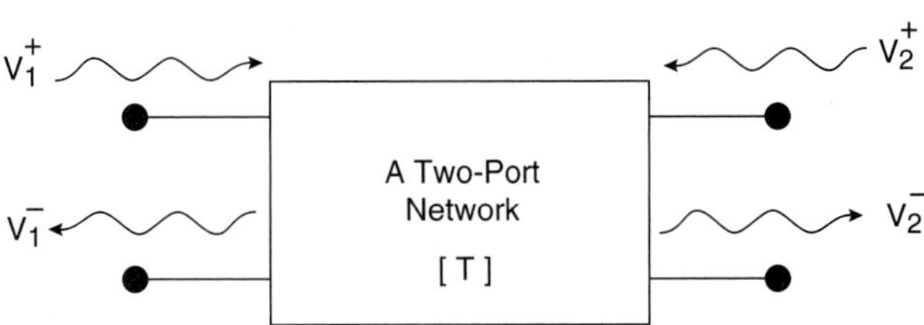

FIGURE 8.12 A two-port network.

$$\begin{bmatrix} V_1^+ \\ V_1^- \end{bmatrix} = \begin{bmatrix} T_{11} & T_{12} \\ T_{21} & T_{22} \end{bmatrix} \begin{bmatrix} V_2^- \\ V_2^+ \end{bmatrix} \tag{8.43}$$

The relationship between *S*- and *T*- parameters can be derived using the above basic definition as follows:

$$\begin{bmatrix} T_{11} & T_{12} \\ T_{21} & T_{22} \end{bmatrix} = \begin{bmatrix} \dfrac{1}{S_{21}} & -\dfrac{S_{22}}{S_{21}} \\ \dfrac{S_{11}}{S_{21}} & S_{12} - \dfrac{S_{11}S_{22}}{S_{21}} \end{bmatrix} \tag{8.44}$$

The reverse relationship expressing [*S*] in terms of [*T*] matrix can also be derived with the following result:

$$\begin{bmatrix} S_{11} & S_{12} \\ S_{21} & S_{22} \end{bmatrix} = \begin{bmatrix} \dfrac{T_{21}}{T_{11}} & T_{22} - \dfrac{T_{21}T_{12}}{T_{11}} \\ \dfrac{1}{T_{11}} & -\dfrac{T_{12}}{T_{11}} \end{bmatrix} \tag{8.45}$$

For a cascade connection of two-port networks, as shown in Figure 8.13, the overall *T*-matrix can be obtained as follows:

$$\begin{bmatrix} V_1^+ \\ V_1^- \end{bmatrix} = \begin{bmatrix} T_{11} & T_{12} \\ T_{21} & T_{22} \end{bmatrix} \begin{bmatrix} V_2^- \\ V_2^+ \end{bmatrix} \tag{8.46a}$$

$$\begin{bmatrix} V'^+_1 \\ V'^-_1 \end{bmatrix} = \begin{bmatrix} T'_{11} & T'_{12} \\ T'_{21} & T'_{22} \end{bmatrix} \begin{bmatrix} V'^-_2 \\ V'^+_2 \end{bmatrix} \tag{8.46b}$$

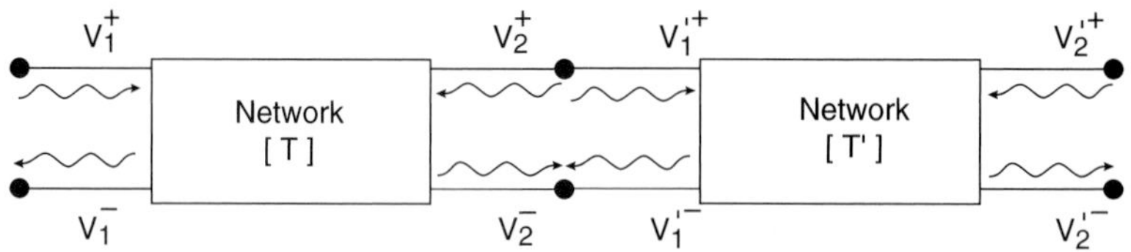

FIGURE 8.13 Cascade of two-port networks.

But we note that:

$$V_2^+ = V'^-_1, \tag{8.47a}$$

$$V_2^- = V'^+_1 \tag{8.47b}$$

Therefore, combining Equations 8.46 and 8.47, yields:

$$\begin{bmatrix} V_1^+ \\ V_1^- \end{bmatrix} = \begin{bmatrix} T_{11} & T_{12} \\ T_{21} & T_{22} \end{bmatrix} \begin{bmatrix} T'_{11} & T'_{12} \\ T'_{21} & T'_{22} \end{bmatrix} \begin{bmatrix} V'^-_2 \\ V'^+_2 \end{bmatrix} \tag{8.48}$$

Thus, the total T-matrix is the multiplication of the two T-matrices:

$$[T]_{tot} = [T][T'] \tag{8.49}$$

8.8 GENERALIZED SCATTERING PARAMETERS

The scattering matrix defined earlier was based on the assumption that all ports have the same characteristic impedances (usually $Z_o = 50\ \Omega$). Even though this is the case in many practical situations, however, there are cases where this may not apply and each port has a nonidentical characteristic impedance (see Figure 8.14). Thus, a need to generalize the scattering parameters arises.

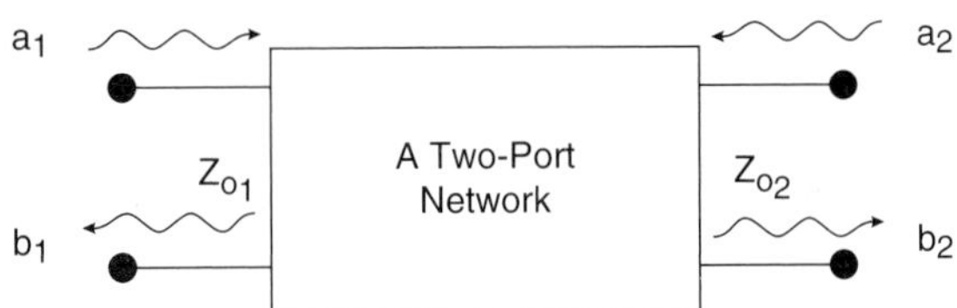

FIGURE 8.14 Normalized incident and reflected waves.

In this case, we have to modify our ordinary definition for the $[S]$ matrix to include the different characteristic impedances at each port. Taking each port's characteristic impedance into account, we need to define two normalized voltage waves as follows:

$$a_i = V_i^+ / \sqrt{Z_{oi}}\,, \quad i = 1,2 \tag{8.50}$$

$$b_i = V_i^- / \sqrt{Z_{oi}}\,, \quad i = 1,2 \tag{8.51}$$

where "i" is the port number, a_i represents the normalized incident voltage , b_i represents the normalized reflected voltage wave from the ith port and Z_{oi} is the characteristic impedance at the ith port (**NOTE:** *Z_{oi} is a real number for lossless lines*).

Thus, the total voltage and current at each port can now be written as:

$$V_i = V_i^+ + V_i^- = \sqrt{Z_{oi}}(a_i + b_i) \tag{8.52}$$

$$I_i = I_i^+ - I_i^- = V_i^+/Z_{oi} - V_i^-/Z_{oi} = (a_i - b_i)/\sqrt{Z_{oi}} \tag{8.53}$$

The average net power delivered to the ith port can now be expressed in terms of a_i and b_i with no further concern about different Z_{oi} at each port:

$$P_i = Re[V_i I_i^*]/2 = Re[|a_i|^2 - |b_i|^2 + (a_i^* b_i - a_i b_i^*)]/2 \tag{8.54}$$

Noting that $(a_i^* b_i - a_i b_i^*)$ term is purely imaginary yields the expression for the net real power:

$$P_i = (|a_i|^2 - |b_i|^2)/2 \tag{8.55}$$

This equation is meaningful because it is clearly showing that the net power delivered to each port is equal to the normalized incident power less the normalized reflected power.

The generalized [S] matrix can now be defined in terms of the normalized voltage waves as follows:

$$[b] = [S][a] \tag{8.56}$$

where each element of the generalized [S] matrix is now defined as:

$$S_{11} = \left.\frac{b_1}{a_1}\right|_{a_2=0} = \Gamma_{IN} \quad \text{reflection coefficient at port 1 with port 2 matched} \tag{8.57}$$

$$S_{12} = \left.\frac{b_1}{a_2}\right|_{a_1=0} \quad \text{transmission coefficient from port 2 to port 1 with port 1 matched} \tag{8.58}$$

$$S_{21} = \left.\frac{b_2}{a_1}\right|_{a_2=0} \quad \text{transmission coefficient from port 1 to port 2 with port 2 matched} \tag{8.59}$$

$$S_{22} = \left.\frac{b_2}{a_2}\right|_{a_1=0} = \Gamma_{OUT} \quad \text{reflection coefficient at port 2 with port 1 matched} \tag{8.60}$$

Clearly, this definition is very similar to the earlier one for the [S] matrix except that V_i^+ and V_i^- are replaced by a_i and b_i, respectively. Alternately, each element can be expressed as a general equation by:

$$S_{ij} = \left.\frac{b_i}{a_j}\right|_{a_k=0} \quad \text{for } i,j,k = 1,2 \text{ and } k \neq j \tag{8.61 a}$$

A Special Case. Consider a network, with identical characteristic impedances at all ports, having a known [S] matrix. If transmission lines of unequal characteristic impedances (Z_{oi}) are connected to each port, the new [S] matrix for the entire network with the help of Equation 8.61a, can now be written as:

$$(S_{ij})_{new} = \left.\frac{V_i^- \sqrt{Z_{o_j}}}{V_j^+ \sqrt{Z_{o_i}}}\right|_{V_k^+=0} \quad \text{for } i,j,k = 1,2 \text{ and } k \neq j \tag{8.61b}$$

we can simplify Equation 8.61b to yield:

$$(S_{ij})_{new} = (S_{ij})_{old} \sqrt{\frac{Z_{o_j}}{Z_{o_i}}} \quad \text{for } i,j = 1,2 \tag{8.61 c}$$

Equation 8.61c shows how the S-parameters of a network with equal characteristic impedances at each port can be converted to a network connected to transmission lines with unequal characteristic impedances.

In the next section, we will discuss the subject of signal flow graphs, whereby any complex circuit can be analyzed in terms of a simple diagram that can yield the relation between desired variables.

8.9 SIGNAL FLOW GRAPHS

Any linear system, or more specifically any linear electrical network, can be described by a set of simultaneous linear equations. Solutions to these equations can be obtained by the following methods:

- The elimination theory, which is a method of successive substitutions
- Cramer's rule, which is a method of solving by using deteminants
- Any of the topological techniques such as the "flow graph techniques" represented by the works of Mason

Although the algebraic manipulation methods, the first two methods listed, can be executed by a computer with relative ease and speed, they do not allow a pictorial analysis or perspective on the physical nature and the signal flows taking place inside the linear system. The signal flow graphs (SFGs), through the use of graphical diagrams, provide a physical insight into the cause-effect relationships between the variables of the system. This method of analysis enables the circuit analyst to gain an intuitive understanding about the system (or network) operation.

In the previous sections, we have seen how the incident, reflected, and transmitted waves are interrelated through a series of linear equations expressed concisely by the S-parameter matrix. This fact indicates that we are dealing with a linear system to which the signal flow graph can be applied directly.

In essence, a signal flow graph is an alternate and yet a simpler method to the "block diagram method" in representing a complicated linear system. The advantage of a signal flow graph method over a block diagram method is the availability of a flow-graph gain formula, which provides the relation between system variables without requiring any detailed procedure for manipulation or reduction of the flow graph.

In this section, we present a detailed discussion about the application and construction of signal flow graphs for analysis of any linear RF/MW network or system. Before we proceed into a detailed discussion of signal flow graphs, let us first define the term.

DEFINITION-SIGNAL FLOW GRAPH: *An abbreviated block diagram consisting of small circles (called nodes) representing the variables that are connected by several directed lines (called branches) representing one-way signal multipliers; an arrow on the line indicates direction of signal flow, a letter near the arrow indicates the multiplication factor.*

Having defined what a signal flow graph is, we will now discuss its main features and its construction as well as a reduction technique to generate ratios of any set of desired variables.

8.9.1 Main Features of a Signal Flow Graph

The main components of a signal flow graph are nodes and branches, defined as follows:

Node. Each port of a microwave network (e.g., the ith port) can be represented by two nodes:

- Node a_i, representing a wave entering port i (an independent variable)
- Node b_i, representing a wave reflected from port i (a dependent variable)

Branch. A branch is a directed path between an "a-node" (an independent variable) and a "b-node" (a dependent variable). A branch represents a signal flow from node a to b. The multiplication factor placed near the arrow is the associated S-parameter.

A signal flow graph is a convenient technique to represent and subsequently analyze the flow of waves in a microwave network. There are certain rules we need to follow in constructing one:

1. Each variable is shown as a node.
2. S-parameters are shown by branches.
3. Branches enter dependent variable nodes (reflected wave variables).
4. Branches emanate from independent variable nodes (incident wave variables).
5. A node is equal to the sum of the branches entering it. For example, consider Figure 8.15, where the dependent variable b_1 can be written as $b_1 = S_{11}a_1 + S_{12}a_2$.

FIGURE 8.15 An example of nodes and branches.

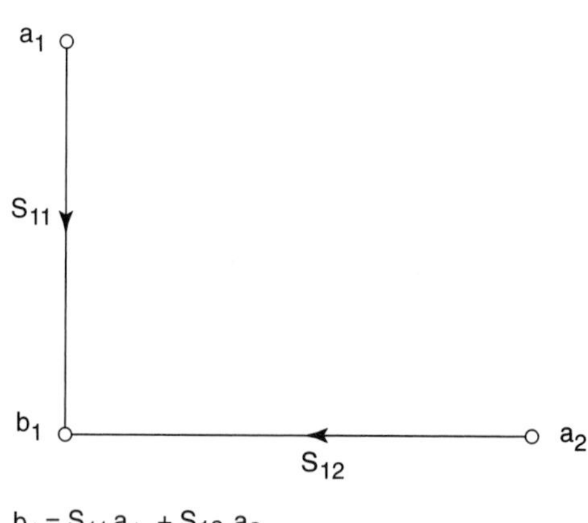

EXAMPLE 8.9

Draw the signal flow graph (SFG) for a linear two-port microwave network as shown in Figure 8.16.

Solution:

A two-port network at microwave frequencies can be characterized by S-parameters given by:

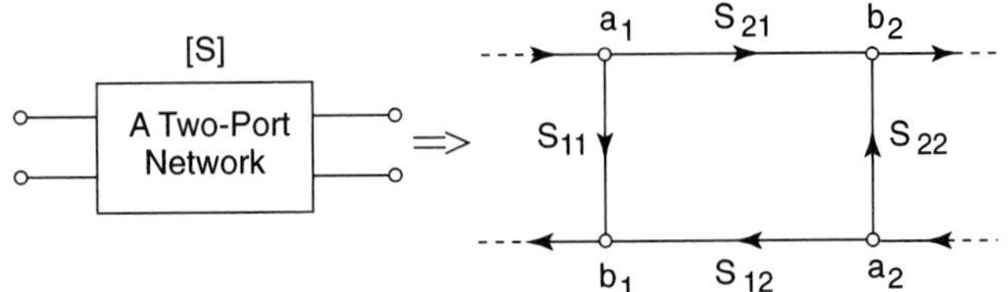

FIGURE 8.16 SFG for a two-port network.

$$b_1 = S_{11}a_1 + S_{12}a_2$$

$$b_2 = S_{21}a_1 + S_{22}a_2$$

Therefore, using the rules set forth (1–5), the signal flow graph for a two-port network consists of two a-nodes (a_1, a_2), two b-nodes (b_1, b_2), and four branches ($S_{11}, S_{12}, S_{21}, S_{22}$), as shown in Figure 8.16.

EXAMPLE 8.10

Find the signal flow graph (SFG) of a microwave amplifier shown in Figure 8.17.

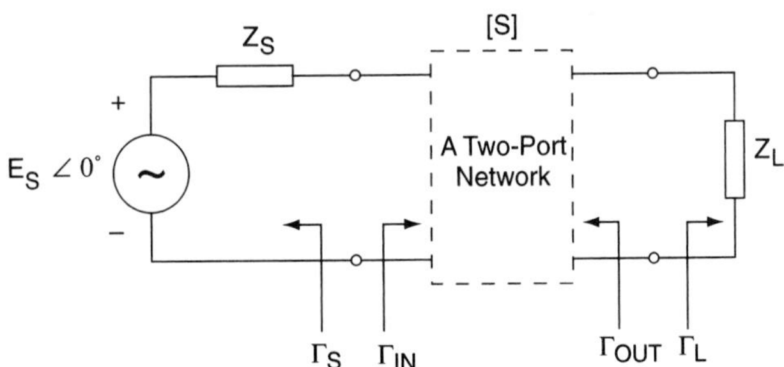

FIGURE 8.17 Diagram of a MW amplifier.

Solution:

In order to obtain the SFG for the microwave amplifier, we dissect or compartmentalize the problem into three separate areas, obtain SFG for each part, and then create a final SFG by combining these three SFGs into one (see Chapter 1, Section 1.5.5, *Solutions to Problems*):

a. Source SFG

The SFG of a signal generator with an internal impedance (Z_S) is obtained from Figure 8.18 as follows:

$$V_g = E_S + Z_S I_g \Rightarrow V_g^+ + V_g^- = E_S + Z_s\left(\frac{V_g^+}{Z_o} - \frac{V_g^-}{Z_o}\right)$$

$$\Rightarrow \frac{V_g^-}{\sqrt{Z_o}} = \frac{V_g^+}{\sqrt{Z_o}}\left(\frac{Z_s - Z_o}{Z_s + Z_o}\right) + E_s\frac{\sqrt{Z_o}}{Z_s + Z_o} \tag{8.62}$$

Equation 8.62 can be written as:

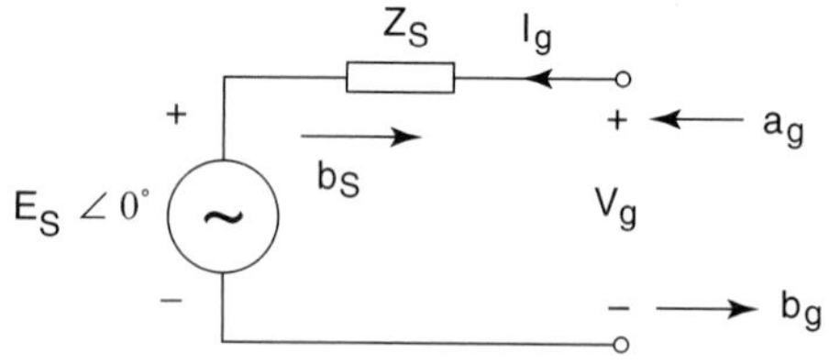

FIGURE 8.18 Schematic of a signal generator.

$$b_g = \Gamma_S a_g + b_S \tag{8.63}$$

where

$$b_g = \frac{V_g^-}{\sqrt{Z_o}}$$

$$a_g = \frac{V_g^+}{\sqrt{Z_o}}$$

$$\Gamma_s = \frac{Z_s - Z_o}{Z_s + Z_o} \tag{8.64}$$

$$b_s = E_S \frac{\sqrt{Z_o}}{Z_s + Z_o} \tag{8.65}$$

The source SFG is shown in Figure 8.19.

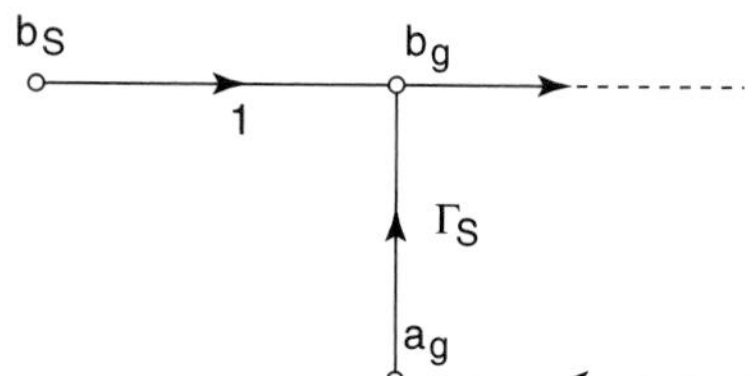

FIGURE 8.19 SFG of a signal generator.

b. Load SFG

The SFG of a load impedance (Z_L) is obtained from Figure 8.20 as follows:

$$V_L = Z_L I_L \Rightarrow V_L^+ + V_L^- = Z_L\left(\frac{V_L^+}{Z_o} - \frac{V_L^-}{Z_o}\right)$$

$$\Rightarrow \frac{V_L^-}{\sqrt{Z_o}} = \frac{V_L^+}{\sqrt{Z_o}}\left(\frac{Z_L - Z_o}{Z_L + Z_o}\right) \tag{8.66}$$

Equation 8.64 can be written as:

$$b_L = \Gamma_L a_L$$

where

$$b_L = \frac{V_L^-}{\sqrt{Z_o}}$$

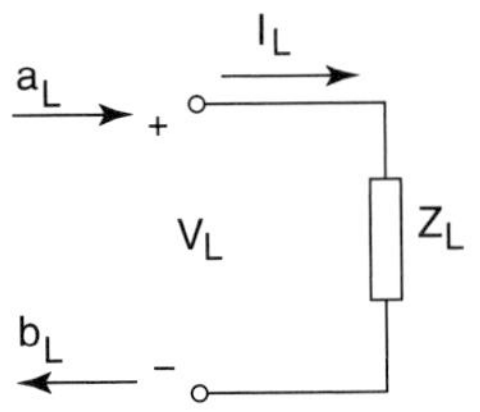

FIGURE 8.20 Schematic of a load.

$$a_L = \frac{V_L^+}{\sqrt{Z_o}}$$

$$\Gamma_L = \frac{Z_L - Z_o}{Z_L + Z_o} \tag{8.67}$$

The load SFG is shown in Figure 8.21.

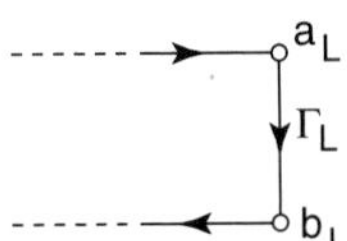

FIGURE 8.21 SFG of a load.

c. Linear two-port network SFG
The SFG of a linear two-port network has already been obtained in Example 8.9 and is shown in Figure 8.16.

d. Final SFG
Combine the SFG for parts (a), (b), and (c) to obtain the final SFG, as shown in Figure 8.22.

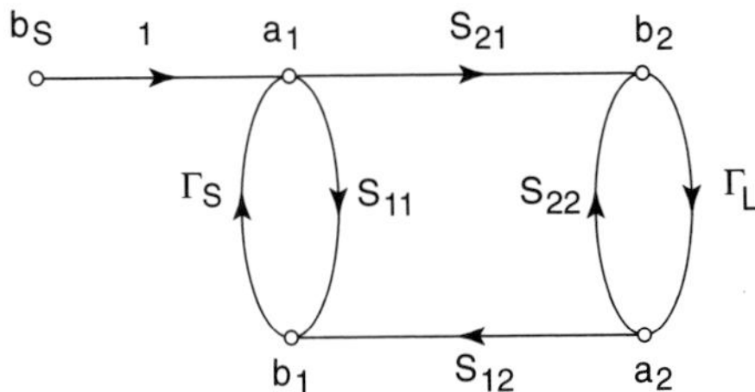

FIGURE 8.22 Final SFG of a two-port network.

8.9.2 Signal Flow Graph Reduction

Once a microwave network has been represented in terms of a signal flow graph, the wave amplitude ratio of any two variables can be obtained by using the following two techniques.

Mason's Rule (from Control System Theory). This method has been well documented in any "control system" text and will not be repeated here.

Signal Flow Graph Reduction Technique. This method is worth presentation and will be further discussed in this work. In its simplest form, it consists of reduction of a signal flow graph to a single branch using four basic decomposition rules. These rules are briefly summarized here, but each can easily be obtained by simple observation and basic application of the signal flow graph rules set forth earlier.

Rule #1-Series Rule: Two branches in series whose common node has only one incoming and outgoing wave may be combined into a single branch with a coefficient (or multiplication factor) equal to the product of the two coefficients (see Figure 8.23). That is,

$$S_c = S_a S_b \tag{8.68}$$

FIGURE 8.23 Series rule.

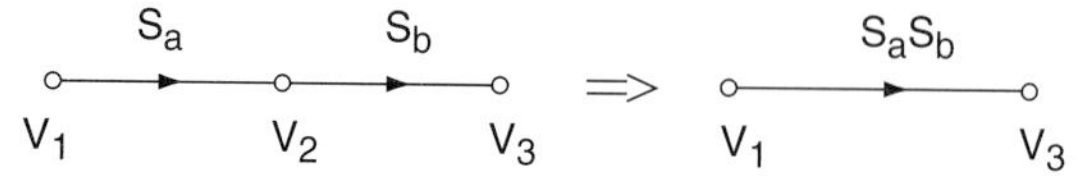

Rule #2-Parallel Rule: Two branches in parallel both going from one node to another may be combined into a single branch whose coefficient is the sum of the two coefficients (see Figure 8.24). That is,

$$S_c = S_a + S_b \tag{8.69}$$

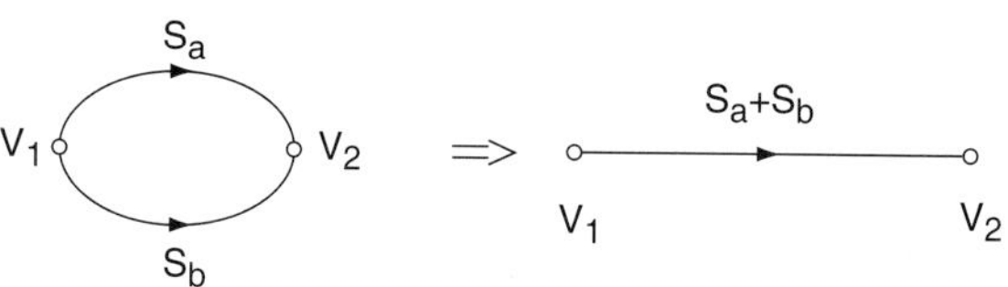

FIGURE 8.24 Parallel rule.

Rule #3-Self-Loop Rule: A branch beginning and ending on the same node (called a self-loop) with a coefficient S_ℓ (see Figure 8.25) can be eliminated by multiplying the coefficients of the branches (feeding that node) by $\frac{1}{1 - S_\ell}$.

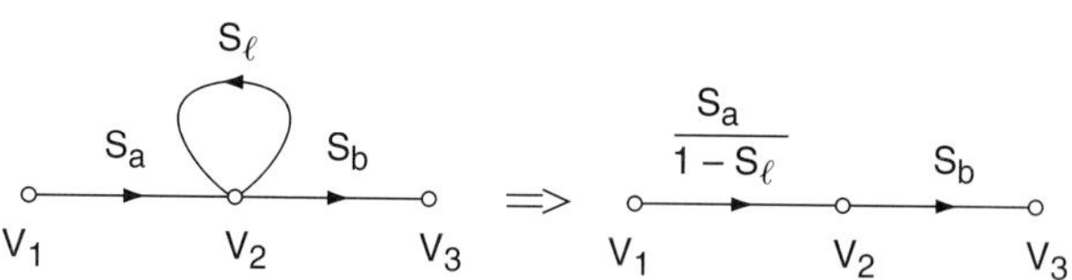

FIGURE 8.25 Self-loop rule.

Rule #4-Splitting Rule: Any node can be split into two separate nodes where each of the two nodes are connected only once to the incoming and the outgoing nodes with the same coefficient, as shown in Figure 8.26.

FIGURE 8.26 Splitting rule.

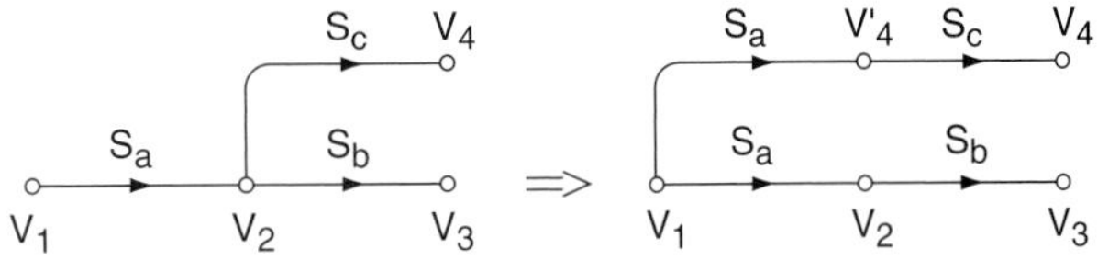

EXAMPLE 8.11

Consider a two-port network characterized by a scattering matrix [S] driven by a source with an internal impedance of (Z_S) and terminated in a load impedance (Z_L), as shown in Figure 8.27. Calculate:

a. *The input reflection coefficient (Γ_{IN}), and*

b. *The output reflection coefficient (Γ_{OUT}).*

These two coefficients are to be calculated in terms of [S], the source reflection coefficient (Γ_S), and the load reflection coefficient (Γ_L).

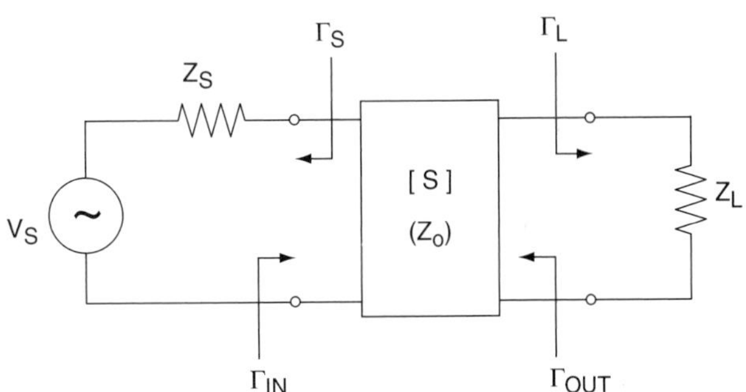

FIGURE 8.27 A terminated two-port network.

Solution:

The overall signal flow graph was shown earlier in Figure 8.22, from which we can derive the following for each of the two cases:

a. The input reflection coefficient ($\Gamma_{IN} = b_1/a_1$)

The signal flow graph (SFG) for this network is shown in Figure 8.28. Using the four decomposition rules stated previously, we can reduce the signal flow graph step by step as shown in Figure 8.29. Using Figure 8.29d, we can write Γ_{IN} as:

$$\Gamma_{IN} = S_{11} + \frac{S_{12}S_{21}\Gamma_L}{1 - S_{22}\Gamma_L} \tag{8.70}$$

b. The output reflection coefficient ($\Gamma_{OUT} = b_2/a_2$)

The signal flow graph for this part is shown in Figure 8.30. Very similar to part (a) for Γ_{IN}, we can reduce the signal flow graph step by step and write a similar equation for Γ_{OUT} as:

$$\Gamma_{OUT} = S_{22} + \frac{S_{12}S_{21}\Gamma_s}{1 - S_{11}\Gamma_s} \tag{8.71}$$

FIGURE 8.28 SFG for the input reflection coefficient calculation.

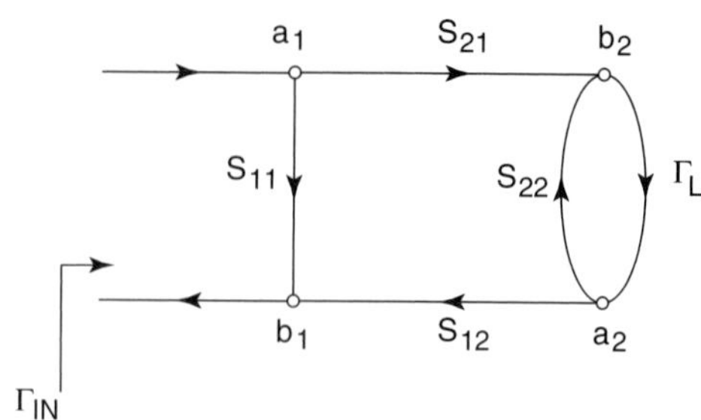

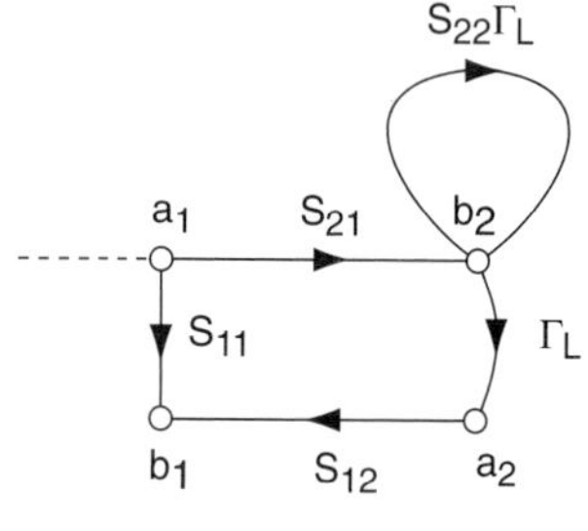

(a) Using Rule 4 on node a_2

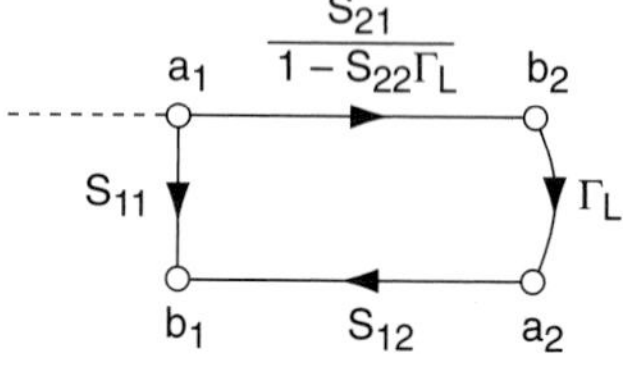

(b) Using Rule 3 for the self-loop

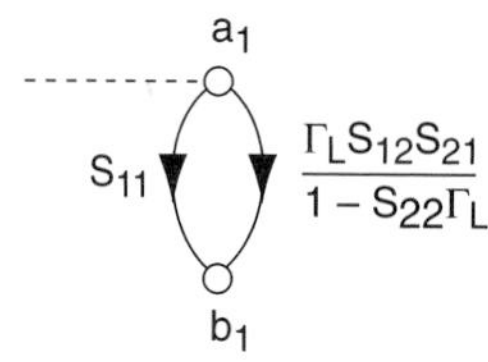

(c) Using Rule 1

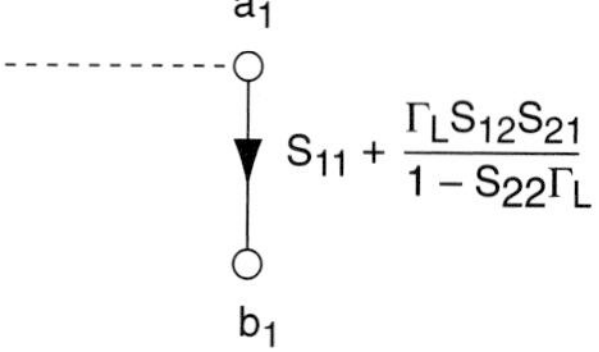

(d) Using Rule 2

FIGURE 8.29 Decomposition of the flow graph to find $\Gamma_{IN} = b_1/a_1$.

FIGURE 8.30 SFG for the output reflection coefficient calculations.

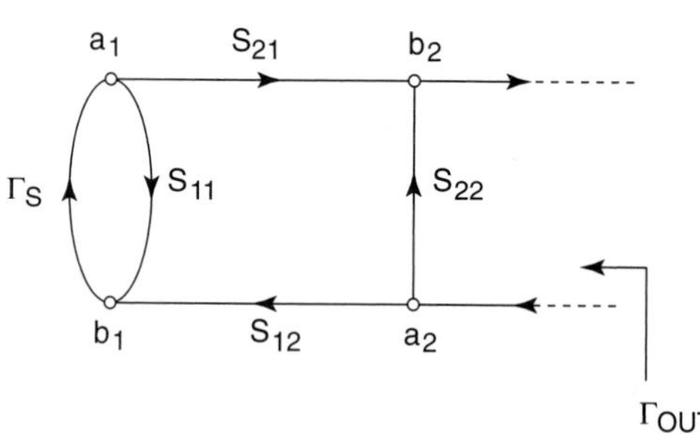

8.9.3 Applications of Signal Flow Graphs

Consider a two-port network (see Figure 8.17) and its corresponding signal flow graph, as shown earlier in Figure 8.22. We know that the square of the incident normalized wave represents the incident power while the reflected power is represented by the square of the reflected normalized wave. Using SFG, we can find the relationship between the various wave variables and therefore the several types of power as follows:

a. P_{IN} is the input power to the network given by:

$$P_{IN} = \frac{1}{2}|a_1|^2 - \frac{1}{2}|b_1|^2 = \frac{1}{2}|a_1|^2(1 - |\Gamma_{IN}|^2) \tag{8.72}$$

b. P_{AVS} is the power available from the source, defined as the input power (P_{IN}) delivered by the source to a conjugately matched input impedance given by:

$$P_{AVS} = P_{IN}\Big|_{\Gamma_{IN} = \Gamma_S^*} = \frac{1}{2}|b_g|^2 - \frac{1}{2}|a_g|^2 \tag{8.73}$$

where

$$a_g = \Gamma_S^* b_g$$

$$b_g = b_s + \Gamma_S a_g = \frac{b_S}{1 - |\Gamma_S|^2} \tag{8.74}$$

Thus, we have:

$$P_{AVS} = \frac{1}{2}|b_g|^2(1 - |\Gamma_S|^2) \Rightarrow P_{AVS} = \frac{\frac{1}{2}|b_S|^2}{1 - |\Gamma_S|^2} \tag{8.75}$$

c. P_L is power delivered to the load and is given by:

$$P_L = \frac{1}{2}|b_2|^2 - \frac{1}{2}|a_2|^2 = \frac{1}{2}|b_2|^2(1 - |\Gamma_L|^2) \tag{8.76}$$

d. P_{AVN} is power available from the network, defined as the power delivered to the load when the load is conjugately matched to the network and is given by:

$$P_{AVN} = P_L\Big|_{\Gamma_L = \Gamma_{OUT}^*} = \left[\frac{1}{2}|b_2|^2 - \frac{1}{2}|a_2|^2\right]_{\Gamma_L = \Gamma_{OUT}^*}$$

$$\Rightarrow P_{AVN} = \frac{1}{2}|b_2|^2(1 - |\Gamma_{OUT}|^2) \tag{8.77}$$

8.9.4 Power Gain Expressions

Using the results obtained in the previous section for various powers, we can now obtain several power gain expressions as follows:

Transducer power gain (G_T)

$$G_T \equiv \frac{P_L}{P_{AVS}} = \frac{|b_2|^2}{|b_S|^2}(1 - |\Gamma_S|^2)(1 - |\Gamma_L|^2) \tag{8.78}$$

Using the SFG technique, the ratio b_2/b_S is obtained to be:

$$\frac{b_2}{b_S} = \frac{S_{21}}{(1 - S_{11}\Gamma_S)(1 - S_{22}\Gamma_L) - S_{12}S_{21}\Gamma_S\Gamma_L} \tag{8.79}$$

Thus, G_T can be written as:

$$G_T = \frac{1-|\Gamma_S|^2}{|1-\Gamma_{IN}\Gamma_S|^2}|S_{21}|^2\frac{1-|\Gamma_L|^2}{|1-S_{22}\Gamma_L|^2} \tag{8.80}$$

Or, we can write an alternate form for G_T as:

$$G_T = \frac{1-|\Gamma_S|^2}{|1-S_{11}\Gamma_S|^2}|S_{21}|^2\frac{1-|\Gamma_L|^2}{|1-\Gamma_{OUT}\Gamma_L|^2} \tag{8.81}$$

Power gain (G_P)

$$G_P \equiv \frac{P_L}{P_{IN}} = \frac{|b_2/b_S|^2}{|a_1/b_S|^2}\left(\frac{1-|\Gamma_L|^2}{1-|\Gamma_{IN}|^2}\right) \tag{8.82}$$

Using the SFG technique, the ratio a_1/b_S is obtained to be:

$$\frac{a_1}{b_S} = \frac{1-S_{22}\Gamma_L}{1-(S_{11}\Gamma_S+S_{22}\Gamma_L+S_{12}S_{21}\Gamma_S\Gamma_L)+S_{11}S_{22}\Gamma_S\Gamma_L} \tag{8.83}$$

Thus, G_P can be written as:

$$G_P = \frac{1}{1-|\Gamma_{IN}|^2}|S_{21}|^2\frac{1-|\Gamma_L|^2}{|1-S_{22}\Gamma_L|^2} \tag{8.84}$$

Available power gain (G_A)

$$\Gamma_L = \Gamma_{OUT}^*$$

$$G_A \equiv \frac{P_{AVN}}{P_{AVS}} = \frac{|b_2|^2}{|b_S|^2}(1-|\Gamma_S|^2)(1-|\Gamma_{OUT}|^2) \tag{8.85}$$

Using Equation 8.79, the ratio b_2/b_S when $\Gamma_L = \Gamma_{OUT}^*$ becomes:

$$\frac{b_2}{b_S} = \frac{S_{21}}{(1-S_{11}\Gamma_S)(1-|\Gamma_{OUT}|^2)} \tag{8.86}$$

Thus G_A can be written as:

$$G_A = \frac{1-|\Gamma_S|^2}{|1-S_{11}\Gamma_S|^2}|S_{21}|^2\frac{1}{1-|\Gamma_{OUT}|^2} \tag{8.87}$$

NOTE: *So far we have discussed the various power gains that are mostly encountered in microwave amplifier design. In most audio and RF amplifier design, however, we use voltage gain. The voltage gain of an amplifier is defined to be the ratio of total output voltage to the total input voltage as follows:*

$$A_V = \frac{V_{OUT}}{V_{IN}} = \frac{a_2+b_2}{a_1+b_1} \tag{8.88}$$

Dividing Equation 8.88 by b_S, we obtain:

$$A_V = \frac{a_2/b_S + b_2/b_S}{a_1/b_S + b_1/b_S} \tag{8.89}$$

Using Equations 8.79 and 8.83 for b_2/b_S and a_1/b_S and similar derivations for b_1/b_S and a_2/b_S, we obtain A_V to be given by:

$$A_V = \frac{S_{21}(1+\Gamma_L)}{(1-S_{22}\Gamma_L)+S_{11}(1-S_{22}\Gamma_L)+S_{12}S_{21}\Gamma_L} \tag{8.90}$$

8.10 SUMMARY

Having defined the S-parameters and derived power gain expressions in this chapter, we will present important concepts about design of matching networks as well as stability of two-ports in the next chapters, which will lay the foundation for *RF/MW* active circuit design.

LIST OF SYMBOLS/ABBREVIATIONS

A symbol will not be repeated again, once it has been identified and defined in an earlier chapter, as long as its definition remains unchanged.

A, B, C, D	*ABCD* parameters
$h_{11}, h_{12}, h_{21}, h_{22}$	Hybrid or h-parameters
i_1	Current into port 1 of a network
i_2	Current into port 2 of a network
i_k	Current at branch k
SFG	Signal flow graph
$S_{11}, S_{12}, S_{21}, S_{22}$	Scattering parameters (or S-parameters)
$T_{11}, T_{12}, T_{21}, T_{22}$	Transmission parameters (or T-parameters)
v_1	Voltage at port 1 of a network
v_2	Voltage at port 2 of a network
v_k	Voltage at branch k
$Y_{11}, Y_{12}, Y_{21}, Y_{22}$	Admittance parameters (or Y-parameters)
$Z_{11}, Z_{12}, Z_{21}, Z_{22}$	Impedance parameters (or Z-parameters)

PROBLEMS

8.1 Determine the S-parameters of the circuit shown in Figure P8.1.

8.2 Find the scattering matrix and the transmission matrix of a lossless transmission line of length "ℓ" in a 50 Ω system when:

a. Characteristic impedance of the line is ($Z_o = 50\ \Omega$)

b. Characteristic impedance of the line is ($Z_o = 100\ \Omega$)

8.3 Find the generalized scattering matrix of a two-port consisting of a junction of two lossless transmission lines as shown in Figure P8.3.

8.4 Find the transmission matrix of the circuit shown in Figure P8.4.

FIGURE P8.1

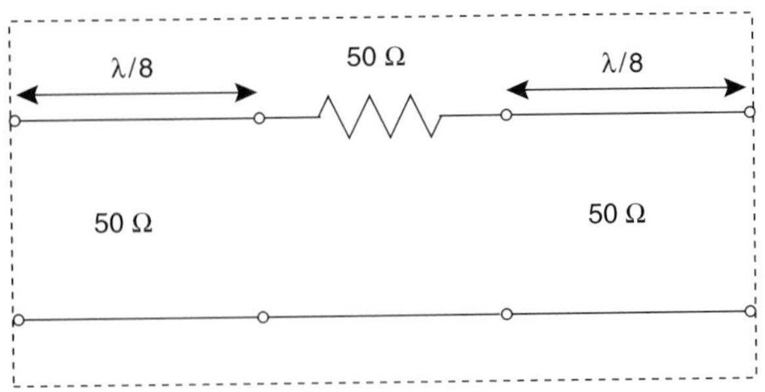

FIGURE P8.3

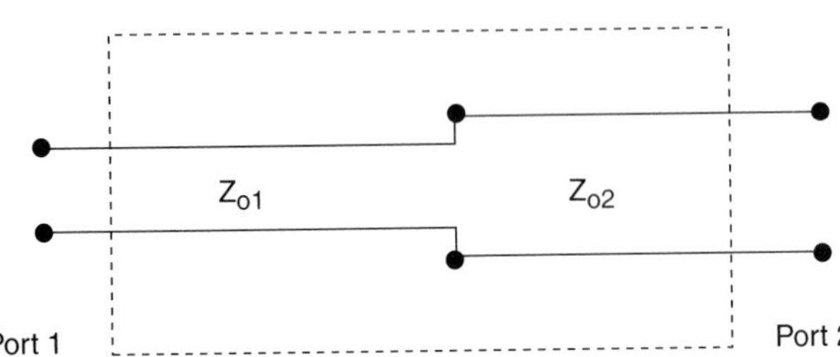

FIGURE P8.4

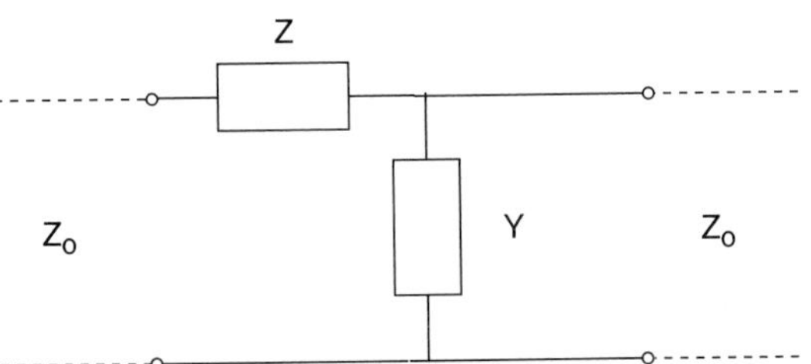

8.5 Derive the $[ABCD]$ matrix for the step-up transformer circuit shown in Figure P8.5.

FIGURE P8.5

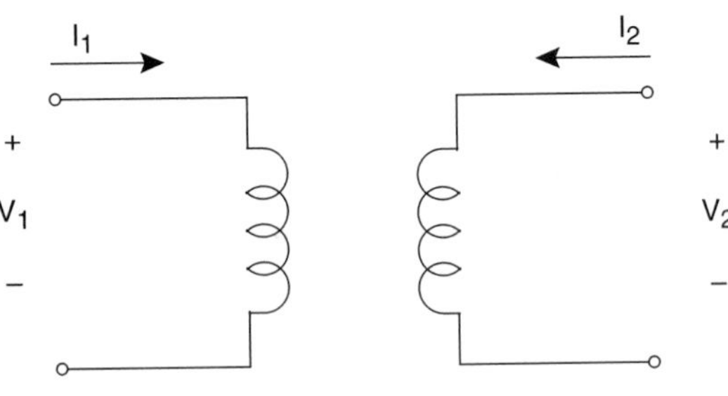

8.6 A two-port network has the scattering matrix as shown below. From these data:

a. Determine whether the network is reciprocal or lossless.

b. If the output terminals are shorted together, what will the input reflection coefficient be?

$$[S] = \begin{bmatrix} 0.2 & j0.9 \\ j0.9 & 0.4 \end{bmatrix}$$

REFERENCES

[8.1] Chang K. *Microwave and Optical Component.* Vols I, II. New York: John Wiley & Sons, 1989.

[8.2] Cheung, W. S. and F. H. Levien. *Microwave Made Simple.* Dedham: Artech House, 1985.

[8.3] Desor, C. A. and E. S. Kuh. *Basic Circuit Theory.* Tokyo: McGraw-Hill, 1969.

[8.4] Gonzalez, G. *Microwave Transistor Amplifiers, Analysis, and Design*, 2nd ed. Upper Saddle River: Prentice Hall, 1997.

[8.5] Gardiol, F. E. *Introduction to Microwaves.* Dedham: Artech House, 1984.

[8.6] Ishii, T. K. *Microwave Engineering.* 2nd ed., Orlando: Harcourt Brace Jovanovich Publishers, 1989.

[8.7] Laverahetta, T. *Practical Microwaves.* Indianapolis: Howard Sams, 1984.

[8.8] Pozar, D. M. *Microwave Engineering*, 2nd ed. New York: John Wiley & Sons, 1998.

[8.9] Saad, T. *Microwave Engineer's Handbook.* Vols I, II. Dedham: Artech House, 1988.

[8.10] Scott, A. W. *Understanding Microwaves.* New York: John Wiley & Sons, 1993.

[8.11] Vendelin, G. D. *Design of Amplifiers and Oscillators by the S-Parameter Method.* New York: John Wiley & Sons, 1981.

[8.12] Vendelin, G. D., A. M. Pavio, and Ulrich L. Rhode. *Microwave Circuit Design.* New York: John Wiley & Sons, 1990.

PART III

PASSIVE CIRCUIT DESIGN

CHAPTER 9

The Smith Chart

9.1 INTRODUCTION

One of the most valuable and pervasive graphical tools in all of microwave engineering is the Smith chart, originally developed in 1939 by P. Smith at the Bell Telephone Laboratories. This chart is the reflection coefficient-to-impedance/admittance converter or vice versa, and it can greatly simplify the analysis of complex design problems involving transmission lines or lumped elements.

Furthermore, the Smith chart provides valuable information about the circuit's performance when line lengths change or new elements are added to the circuit, particularly where obtaining the same amount of information through mathematical models and calculations would be very tedious and time consuming.

Over the years, it has proven itself to be a most useful tool and is thus employed frequently in all stages of circuit analysis or design whether done through manual methods or computer-aided-design (CAD) software techniques.

9.2 A VALUABLE GRAPHICAL AID: THE SMITH CHART

Considering the equation for the reflection coefficient (as given earlier in Chapter 8, *Circuit Representations of Two-Port RF/Microwave Networks*) we have:

$$\Gamma = \frac{Z - Z_o}{Z + Z_o} = \frac{Z_N - 1}{Z_N + 1} \tag{9.1}$$

where $Z_N = \dfrac{Z}{Z_o} = r + jx$ is the normalized impedance and Z_o is the characteristic impedance of the transmission line or a reference impedance value.

Based on Equation 9.1, the Smith chart can be derived mathematically, as discussed in the next section. This chart is a plot of Γ for different normalized resistance and reactance values, where the circuit is assumed to be passive, i.e., $Re(Z) \geq 0$. It can be shown that the loci of constant resistance values are circles centered on the horizontal (or real) axis while the loci of constant reactance values are circles centered on the vertical (or imaginary) axis offset by one unit.

9.3 DERIVATION OF SMITH CHART

The Smith chart is a plot of

$$\Gamma = \frac{Z_N - 1}{Z_N + 1} \tag{9.2}$$

in the Γ-plane as a function of r and x. Using Equation 9.2 and separating Γ in terms of its real part (U) and imaginary part (V) we obtain:

$$Z_N = r + jx$$

$$\Gamma = \frac{r + jx - 1}{r + jx + 1} = U + jV \tag{9.3}$$

$$U = \frac{r^2 - 1 + x^2}{(r+1)^2 + x^2} \tag{9.4}$$

$$V = \frac{2x}{(r+1)^2 + x^2} \tag{9.5}$$

At this juncture we note that by using Equations 9.4 and 9.5, we can obtain two families of circles that, when superimposed on each other, will make up the entire Smith chart. The procedure to obtain these two families of circles is described next.

1. **Constant-*r* circles:** The first family of circles is obtained by eliminating x from Equations 9.4 and 9.5, which gives:

$$\left(U - \frac{r}{r+1}\right)^2 + V^2 = \left(\frac{1}{r+1}\right)^2 \tag{9.6}$$

Equation 9.6 represents a family of circles with a center located at

$$(U_o, V_o) = \left(\frac{r}{r+1}, 0\right), \tag{9.7a}$$

and with a radius of

$$R = \left(\frac{1}{r+1}\right). \tag{9.7b}$$

From Equations 9.7, we can observe that all constant-r circles are centered on the real axis with a shrinking size as "r" is increased. In this regard, we note that the $r = 0$ circle is the outermost circle of the Smith chart while the $r = \infty$ circle is reduced to a point at (0,1). Figure 9.1 depicts this concept further.

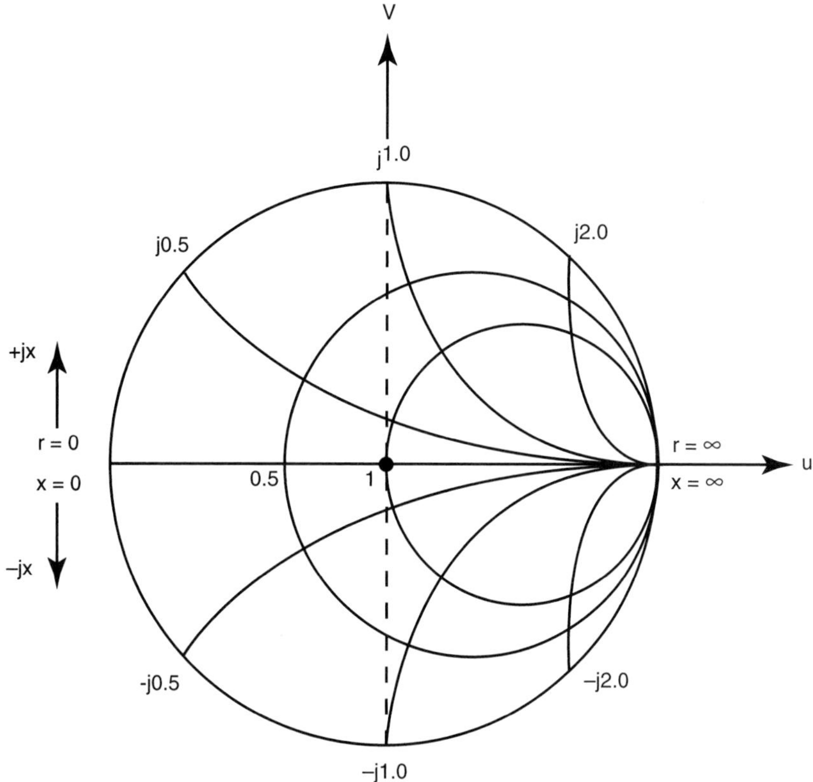

FIGURE 9.1 Construction of a standard Smith chart (for $r \geq 0, -\infty \leq x \leq \infty$).

2. **Constant-*x* circle:** The second family of circles is obtained by eliminating r from Equations 9.4 and 9.5, which gives:

$$(U-1)^2 + \left(V - \frac{1}{x}\right)^2 = \left(\frac{1}{x}\right)^2 \tag{9.8a}$$

Equation 9.8a represents a family of circles with a center located at

$$(U'_o, V'_o) = \left(1, \frac{1}{x}\right) \tag{9.8b}$$

and with a radius of

$$R' = \frac{1}{|x|} \tag{9.8c}$$

From Equations 9.8 we can observe that all constant-x circles are centered on a shifted line parallel to the imaginary axis (by +1 unit to the right), with a shrinking size as x increases. In this regard, we note that the $x = 0$ circle is the real axis of the Smith chart while the $x = \pm\infty$ circles are reduced to a point at (1,0). This is shown in Figure 9.1.

As described earlier, plotting the two families of circles as represented by Equations 9.6 and 9.8 for all values of (r,x) creates a circular chart commonly known as the Smith chart.

The standard Smith chart is a one-to-one correspondence between points in the normalized impedance (Z_N) plane [where $r = Re(Z_N) \geq 0$] and points in the reflection coefficient (Γ) plane. The upper half of the chart represents normalized impedance values with positive reactances $(x \geq 0)$ while the lower half corresponds to negative reactances $(x \leq 0)$.

NOTE: *The Smith chart could have also been developed based on normalized admittance (Y_N) as follows:*

$$Y_N = \frac{Y}{Y_o} = g + jb \tag{9.9a}$$

Where $Y_o = 1/Z_o$ is the normalized characteristic admittance or a reference admittance value. Thus, we can write Equation 9.2 as:

$$\Gamma = \frac{Z_N - 1}{Z_N + 1} = \frac{\frac{1}{Y_N} - 1}{\frac{1}{Y_N} + 1} = -\left(\frac{Y_N - 1}{Y_N + 1}\right) \tag{9.9b}$$

Now using the transformation:

$$\Gamma' = \left(\frac{Y_N - 1}{Y_N + 1}\right) \tag{9.10}$$

we obtain the same results as for impedance except that the transformation will be from Y_N-plane into Γ'-plane where

$$\Gamma' = -\Gamma = \Gamma e^{j180°} \tag{9.11}$$

Equation 9.11 indicates that Γ' and Γ are only 180° apart but have the same magnitude, which means that when dealing with admittances and impedances on the same chart we need to keep in mind the 180° phase adjustment every time we convert Z_N to Y_N or vice versa. Therefore, a Smith chart can be used as an impedance chart (Z-Smith chart) or equally as an admittance chart (Y-Smith chart).

SUMMARY: *In summary, using a Smith chart requires awareness and an understanding of the following transformations:*

$$Z_N \leftrightarrow \Gamma$$

$$Y_N \leftrightarrow \Gamma'$$

$$\Gamma' \leftrightarrow \Gamma e^{j180°}$$

The magic of the Smith chart lies in the fact that through the use of the previous transformation, a semi-infinite and an unbounded region (i.e., $0 \leq r \leq \infty, -\infty \leq x \leq +\infty$) is transformed into a finite and workable region (i.e., $0 \leq \Gamma \leq 1$), which creates easily understood graphical solutions to many complex microwave problems.

9.4 DESCRIPTION OF TWO TYPES OF SMITH CHARTS

As discussed in the previous sections, instead of plotting contours of constant reflection coefficient, contours of constant normalized resistance and reactance are plotted in the Γ-plane. A selected collection of these contours (which are circles), plotted in the Γ-plane, comprise the "entire Smith chart" (commonly known as the "compressed Smith chart"), which includes impedances with both positive and negative real parts. The compressed Smith chart is obtained when the entire impedance plane is mapped on a one-to-one basis onto the reflection coefficient plane, as shown in Figure 9.2.

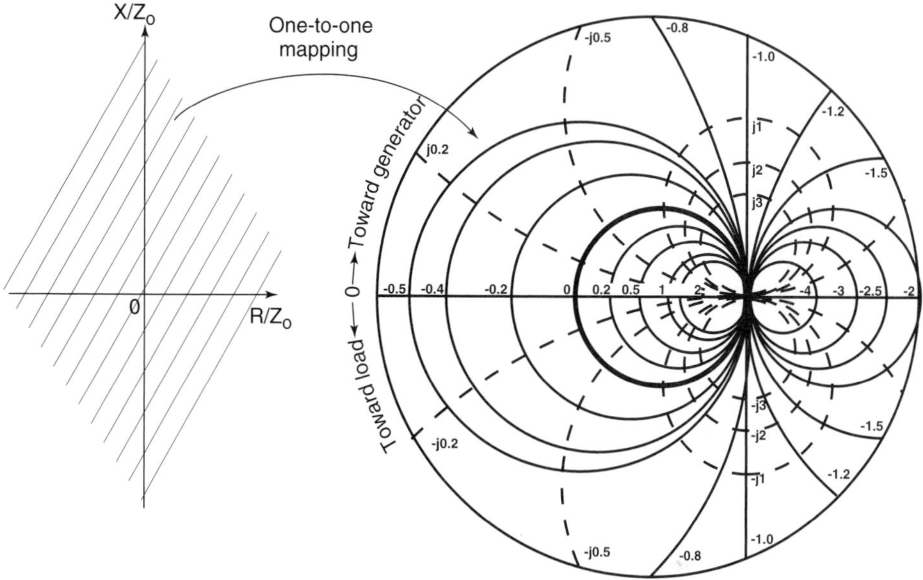

FIGURE 9.2 The whole impedance plane mapping into a compressed Smith chart.

The compressed Smith chart, even though it is very general and applies to both active and passive circuits, is yet impractical and is seldom used in design. Instead, a more useful part of this chart (called a standard Smith chart) is used in practice for all passive networks where $Re(Z) \geq 0$, which corresponds to mapping only the right half of the impedance plane into a circle in the reflection coefficient plane with radius $|\Gamma| \leq 1$, as shown in Figure 9.3.

The standard Smith chart represents a graphical display of impedance-to-reflection coefficient transformation, in which all values of impedance with $Re(Z) \geq 0$ (representing a semi-infinite region of the resistance-reactance rectangular plane) is mapped one-to-one into a circle with the radius of one unit in the reflection coefficient plane. A full blown-out version of the standard Smith chart is shown in Figure 9.4, where each circle is marked with its corresponding resistance or reactance value.

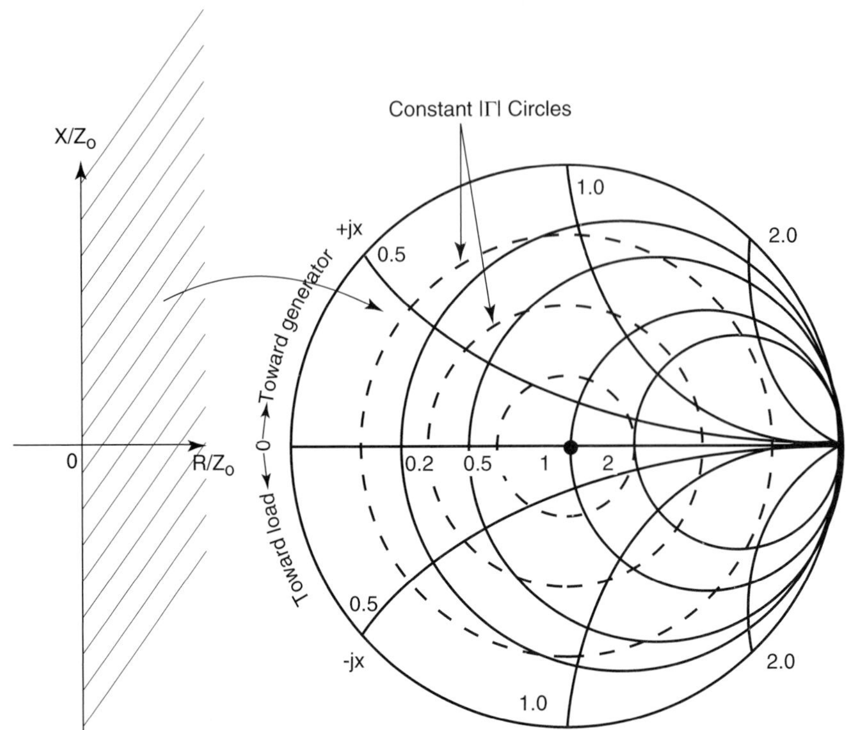

FIGURE 9.3 Positive half of impedance plane mapping into a standard Smith chart.

Thus, the Smith chart is made up of many circles either fully or partially enclosed within the outermost circle ($|\Gamma| = 1$) of the standard Smith chart.

The set of circles centered on the horizontal (or real) axis are circles of constant normalized resistance with values ranging from zero (extreme far left) to infinity (extreme far right) on the chart with each circle having a variable reactance.

On the other hand, the set of circles centered on the vertical axis, which is offset by one unit from the center, represents circles of constant normalized reactances with values ranging from $-\infty$ to $+\infty$ with each circle having a variable positive resistance. These are shown as partial circles starting from the right-hand side of the chart and going above the real axis (representing normalized positive reactances) and below (representing normalized negative reactances); the center real axis (horizontal line) represents the zero reactance circle with an infinite radius.

The markings for the positive and negative normalized reactances can be seen on the Smith chart close to the outermost circle.

The key to understanding the Smith chart is realizing that it is a polar plot of the reflection coefficient:

$$\Gamma = |\Gamma| e^{j\theta}, \quad 0 \le \theta \le 180° \tag{9.12}$$

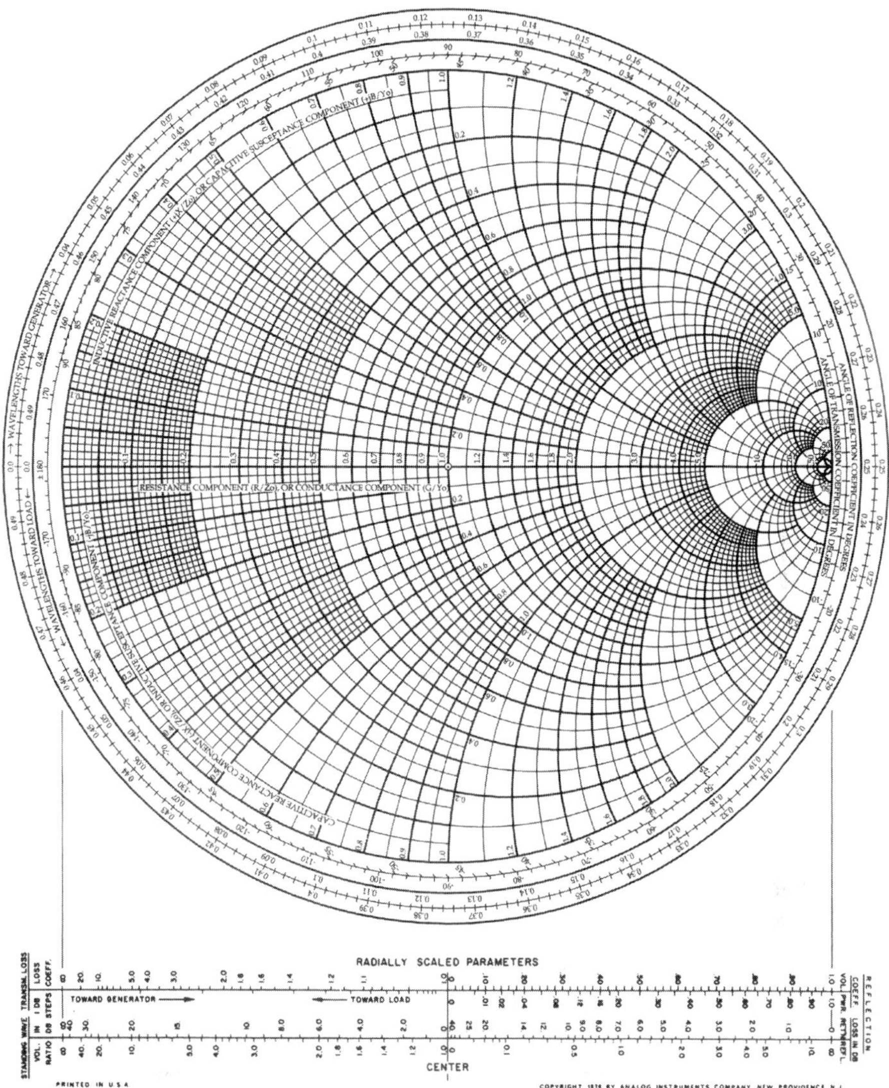

FIGURE 9.4 The standard Smith chart.
(Reproduced with permission of Analog Instruments Co., New Providence, NJ.)

with the reference of zero degrees at the right side of horizontal semi-axis. All passive networks ($|\Gamma| \leq 1$) have impedance values with $Re(Z) \geq 0$, which when normalized by the characteristic impedance of the transmission line (to which they are connected) can be represented uniquely on the Smith chart.

The real usefulness of the Smith chart lies in its ability to provide a one-to-one correspondence between reflection coefficient and input normalized impedance (or admittance) values.

Furthermore, moving a distance ℓ toward the load along a lossless transmission line corresponds to a change in the reflection coefficient by a factor of $e^{-2j\beta\ell}$, which corresponds to a counter-clockwise rotation of $2\beta\ell$ on the Smith chart, as shown in Figure 9.5.

FIGURE 9.5

(a) Traveling on a lossless transmission line

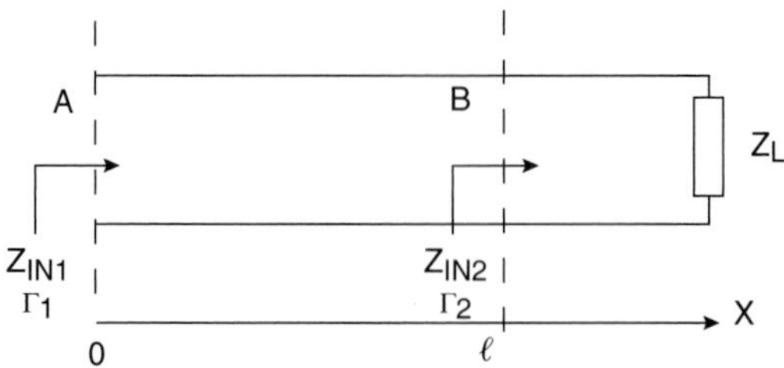

(b) Smith chart solution

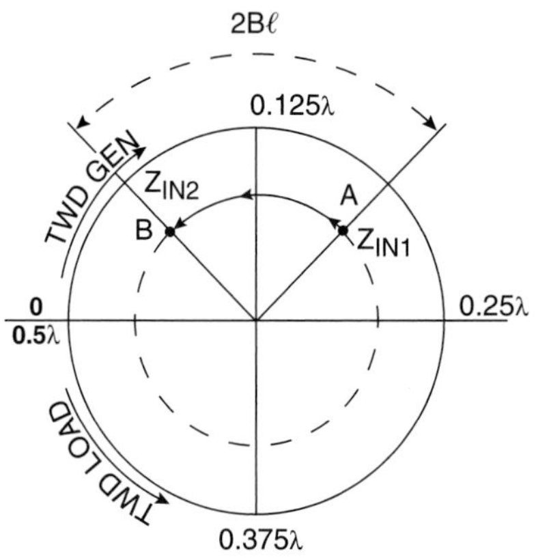

9.5 SMITH CHART'S CIRCULAR SCALES

Consider a standard Smith chart as shown in Figure 9.4. Any specific normalized impedance ($z = r + jx$) value can be uniquely located on this chart for $r \geq 0$. The r-values would be on the resistance circles and x-values on the partial circles for reactance. The positive reactance would be on the upper half of the chart whereas negative reactance values are plotted on the lower half of the chart.

Traveling on a transmission line toward the generator corresponds to moving clockwise on the Smith chart, whereas traveling toward the load corresponds to moving in a counter-clockwise direction as indicated by arrows on the left-hand side outer edge of the Smith chart.

The phase relationship and electrical length along a transmission line are shown on the outer edge of the Smith chart in terms of two secondary scales: One is graduated in fractional wavelength (ℓ/λ) and the other in degrees.

9.5.1 Wavelength Scale

Because the impedance value on a transmission line repeats itself every half wavelength, a complete revolution on the wavelength scale is equivalent to a half wavelength on the Smith chart.

9.5.2 Degree Scale

The degree scale goes through 180 degrees positive and 180 degrees negative with 0 degrees being on the right-hand semi-axis. This scale shows that in a complete revolution of the chart, the reflection coefficient's phase changes 360 degrees corresponding to a half wavelength on the wavelength scale.

These two scales (i.e., the wavelength and the degree scales) are important because they show that when a transmission line is terminated in a load impedance not equal to the line's characteristic impedance, the resulting impedance value on the line varies cyclically every half wavelength (this is, of course, due to the periodic nature of the standing wave pattern on the line). This fact is built into the Smith chart through these two scales and, thus, facilitates impedance calculations at various points along a line after the chart has been entered in for a specific impedance value.

9.6 SMITH CHART'S RADIAL SCALES

A number of radially marked scales, at the bottom of the Smith chart, are placed in such a manner that they can be radially set off and their values read off from the center of the Smith chart by using a pair of dividers or a compass.

These scales are described in the next sections.

9.6.1 Reflection

Starting with the scale on the right-hand side, the reflection scale is designed to show the ratio of the reflected wave to the incident wave and is further subdivided into four scales in the following manner:

1. **REFL. COEF** is the reflection coefficient and has the following two subscales:

   a. *VOL.* is the voltage reflection coefficient magnitude and is defined as:

$$|\Gamma| = |V^-/V^+| \tag{9.13}$$

   This scale starts from 0 at the center and ends at 1 at the outer rim of the chart.

   b. *PWR.* is the power reflection coefficient and is defined to be:

$$|\Gamma|^2 = |V^-/V^+|^2 \tag{9.14}$$

   Similar to VOL., this scale starts at 0 at the center and ends at 1 at the outer edge of the chart.

2. **Loss in DB** is the loss due to reflection and is expressed in dB with the following two subscales:

   a. *RETN.* is the return loss and is defined as the ratio of the incident power to the reflected power at any point on the transmission line, expressed in dB and is equal to

$$R_{loss}(\text{dB}) = 10\log_{10}(P_i/P_o) = 10\log_{10}(1/|\Gamma|^2) \Rightarrow R_{loss}(\text{dB}) = -20\log_{10}|\Gamma| \quad (9.15)$$

   This scale starts from 0 (corresponding to $|\Gamma| = 1$) at the outer edge of the chart and approaches infinity at the center of the chart (where $|\Gamma| = 0$), which indicates that the more perfect the load, the less reflection from the load and higher the return loss.

   b. *REFL.* (Reflected loss or mismatch loss) is the loss caused by reflection and is equal to the ratio of incident power to the difference between incident and reflected power expressed in decibels as follows:

$$M_{loss}(\text{dB}) = 10\log_{10}[P_i/(P_i - P_o)] = -10\log_{10}(1 - |\Gamma|^2) \quad (9.16)$$

   This scale starts from zero at the center and approaches infinity as $|\Gamma|$ approaches unity at the outer edge of the chart.

9.6.2 TRANSM. Loss

TRANSM. Loss is the transmission loss and is used primarily for lossy transmission lines and has two scales:

1. **LOSS COEF.** is the transmission loss coefficient and is used as a correction factor for the additional line losses created in a lossy transmission line due to high *VSWR*.

   A high *VSWR* on a line creates peaks of high current densities alternated with high voltage density peaks. Because resistive losses are proportional to the current value squared and dielectric losses are proportional to voltage value squared, the locale where these peaks of energy lie create additional losses on the line that are not accounted for through ordinary calculations. Thus, a correction factor is needed to provide a more accurate estimation of line losses when a high *VSWR* exists on the lossy transmission line. The "*LOSS COEF.*" scale provides the much needed correction factor when the *VSWR* on the line is greater than unity. The correction factor provided by this scale would increase the calculated line losses, which will affect the attenuation factor calculations.

   For example, when the *VSWR* is 1 (i.e., a matched case) the correction factor from this scale is read to be 1. On the other hand, when the *VSWR* of the line is increased to 4 (due to a load mismatch), then the correction factor is read off to be approximately 2.1, which means that the line losses have more than doubled due to this high *VSWR*.

2. **1 DB steps** is the transmission loss in 1 DB steps and is used to calculate *VSWR* on a lossy transmission line. This is shown on a graduated scale (with no numerical assignment) showing relative value in 1 DB steps, that is, each marking on the scale above or below a designated point shows a 1 DB increase or decrease in

attenuation value. Graphically, a lossy line can no longer be represented by a constant *VSWR* circle; it is instead represented by a spiral on the Smith chart due to the attenuation of both the incident wave's amplitude traveling "toward the load" and the reflected wave's amplitude propagating back to the generator. This power loss is shown in Figure 9.6. From Figure 9.6, it can be seen that:

$$|\Gamma_g|^2 = P_r/P_i \quad \text{(at the source end)} \tag{9.17}$$

$$|\Gamma_L|^2 = P_r'/P_i' \quad \text{(at the load end)} \tag{9.18}$$

$\Rightarrow |\Gamma_g| < |\Gamma_L|$, as long as the line connecting source and load remains lossy.

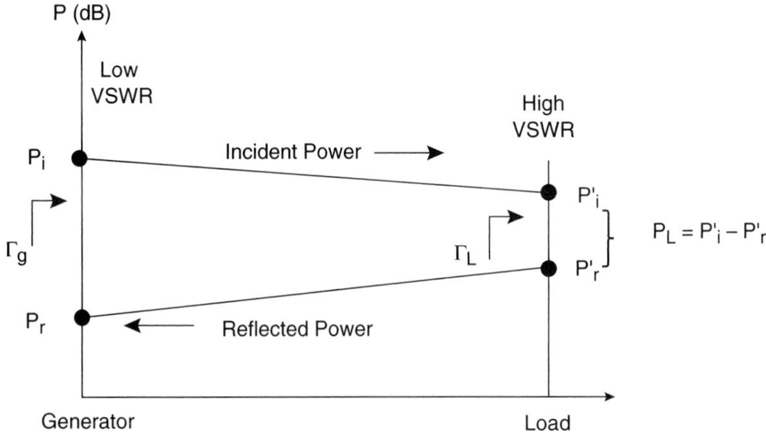

FIGURE 9.6 Power attenuation on a lossy transmission line.

It is important to note that as the measurement plane moves toward the generator and away from the load, the reflection coefficient becomes smaller and thus *VSWR* is reduced as illustrated in the next example.

EXAMPLE 9.1

Consider an unknown load connected to a lossy 50 Ω cable (with 2 dB of insertion loss) at one end and connected to a generator at the other. The VSWR at the generator end is measured to be 2.0. What is the VSWR at the load?

Solution:

We first plot the *VSWR* circles at the generator end by dropping a vertical line from the constant-*VSWR* circle to intersect the "1 DB Steps" scale at point *A* (see Figure 9.7). Now we add 2 dB correction to this value "toward load" as indicated on the scale to obtain point *B*. The radius related to point *B* is that of the load *VSWR* and is found by drawing a vertical line from point *B* to intersect the left-hand semi-axis on the Smith chart. By swinging this radius around, it is seen that the new *VSWR* circle has a *VSWR* = 3.2 at the load end.

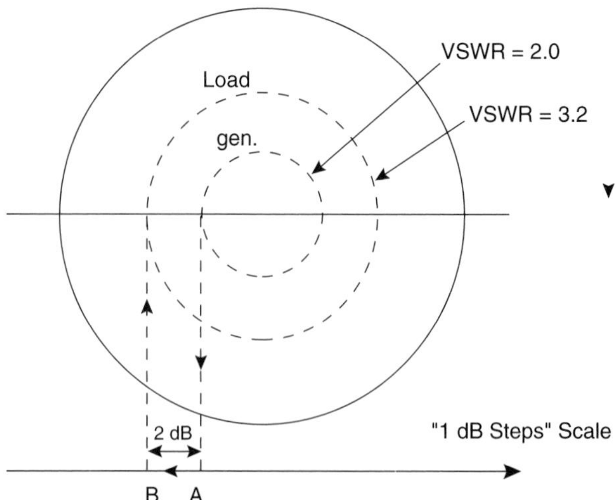

FIGURE 9.7 Use of "1 dB Steps" scale.

As can be seen from Example 9.1, a lossy line can improve the *VSWR* at the generator end at the expense of power loss, which may not always be desirable.

We also note from this example that moving away from the center (i.e., higher $|\Gamma|$) on this scale is labeled as "Toward load" while moving toward the center of the chart (i.e., lower $|\Gamma|$) is labeled as "Toward generator."

> ***Note:*** *Incidentally, it is interesting to note that the step size on "the transmission loss (in one dB steps)" scale are one-half of the corresponding step size values on the "Return loss in dB" scale. This factor of one-half is caused by the fact that the return loss scale indicates two-way power attenuation through a given piece of cable, whereas the transmission loss is defined merely as a one-way attenuation loss.*

9.6.3 Standing Wave

This scale shows the voltage standing wave ratio (*VSWR*) as follows:

1. **VOL. RATIO** (voltage ratio). This scale plots the *VSWR* as a ratio of maximum voltage to minimum voltage as given by the following equation:

$$VSWR = \frac{V_{max}}{V_{min}} = \frac{1+|\Gamma|}{1-|\Gamma|} \tag{9.19}$$

   The *VOL. RATIO* scale progresses from 1 at the center of the chart ($|\Gamma| = 0$) to infinity at the left-hand margin ($|\Gamma| = 1$).

2. **IN DB.** This scale expresses *VSWR* in dB by the relation:

$$(VSWR)_{\text{dB}} = 20 \log_{10} (VSWR)_{\text{ratio}} \tag{9.20}$$

   The maximum and minimum of the standing wave on a transmission line is a direct result of cancellation and addition of the incident and reflected wave phasors. The larger the reflected wave, the bigger will be the maximum and minimum of the voltage and current standing waves and the corresponding VSWR on the line.

EXAMPLE 9.2

To what value does VSWR = 2.0 on the "Voltage ratio" scale correspond in dB?

Solution:

Using the adjacent dB scale, we read a value of 6.0 dB on it.

9.7 THE NORMALIZED IMPEDANCE-ADMITTANCE (ZY) SMITH CHART

By superimposing two Smith charts, with one 180° rotated, we obtain a normalized impedance-admittance Smith chart (also known as a *ZY* Smith chart), as shown in Figure 9.8. The rotated represents the admittance, whereas the other chart represents impedance. The proof for 180° chart rotation to obtain admittance values is presented in Chapter 10, *Applications of the Smith Chart* (see application #4).

The *ZY* Smith chart therefore has two markings: one for the impedance chart and another for the admittance chart. Symbols $-X_S$ and $+X_S$ are used on the left-hand side for the impedance chart, and $-B_P$ and $+B_P$ are used for admittance chart, respectively. From these markings, we can see that positive reactances ($+X_S$) are on the upper half of the chart while negative reactances ($-X_S$) are on the lower half of the chart, respectively. This situation is reversed for the admittance chart, where positive susceptances ($+B_P$) are located on the lower half and the negative susceptances ($-B_P$) are on the upper half of the chart.

Each point on a *ZY* Smith chart represents the impedance and the corresponding admittance value simultaneously, whereby we can read off these values by a simple glance at the chart. This means that given an impedance (or admittance) value, its corresponding admittance (or impedance) value can readily be read off the chart without any resort to calculations. This is an important feature and a major improvement over a standard Smith chart because it greatly facilitates the circuit design process, particularly where complicated designs are desired. As will be seen in Chapter 11, *Design of Matching Networks*, the *ZY* Smith chart is an essential analytical tool and will be extensively used for RF/microwave circuit design applications

In the next chapter, we will discuss many applications of the Smith chart that are of great importance to the design of RF and microwave circuits.

LIST OF SYMBOLS/ABBREVIATIONS

A symbol/abbreviation will not be repeated again, once it has been identified and defined in an earlier chapter, as long as its definition remains unchanged.

M_{loss}	Reflected loss or mismatch loss
P_i	Power incident
P_r	Power reflected
PWR.	Power reflection coefficient
R	Radius of circle
REFL.	Reflection loss or mismatch loss

NAME	TITLE	DWG. NO.
SMITH CHART FORM ZY-01-N	ANALOG INSTRUMENTS COMPANY, NEW PROVIDENCE, N.J. 07974	DATE

NORMALIZED IMPEDANCE AND ADMITTANCE COORDINATES

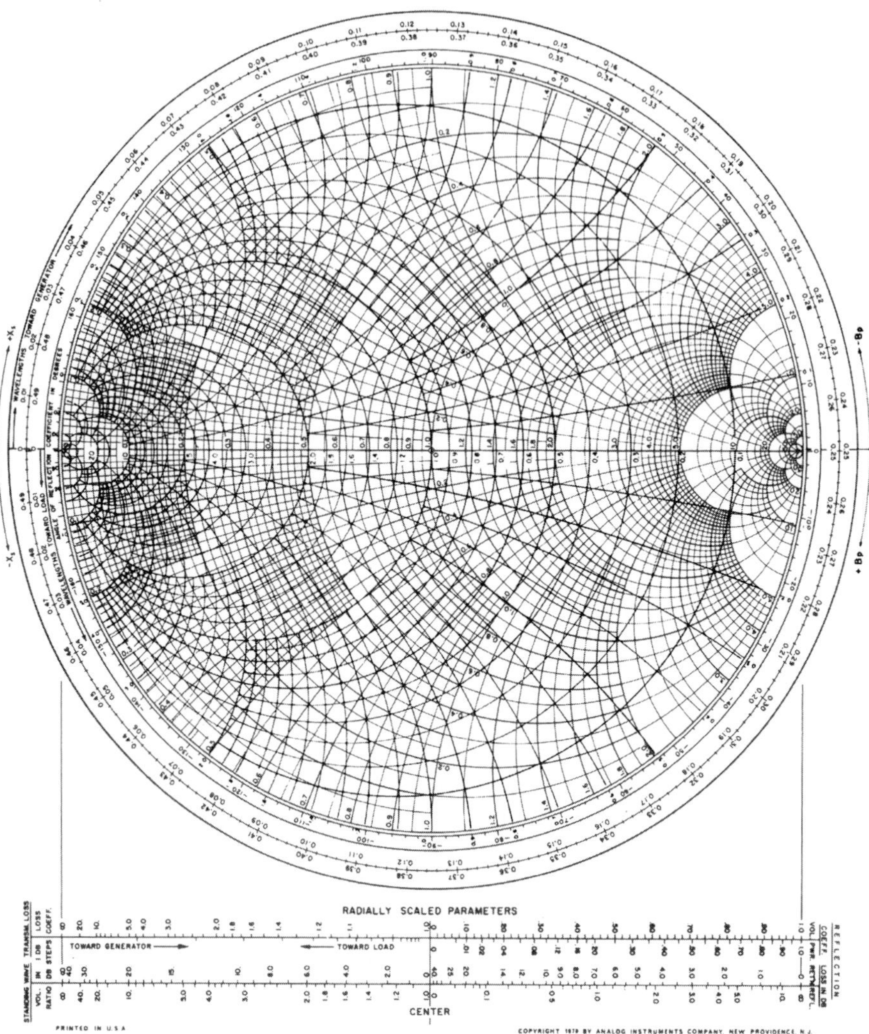

FIGURE 9.8 The normalized impedance and admittance coordinates Smith chart (ZY chart). (Reproduced with permission of Analog Instruments Co., New Providence, NJ.)

REFL. COEF.	Reflection coefficient
RETN.	Return loss
R_{loss}	Return loss
R_N	Normalized resistance
VOL.	Voltage reflection coefficient magnitude

VSWR	Voltage standing wave ratio
X_N	Normalized reactance
Z_o	Characteristic impedance
Z_N	Normalized impedance
Γ_g	Reflection coefficient at the generator/source
Γ_L	Reflection coefficient at the load

PROBLEMS

9.1 What is a standard Smith chart? What range of resistor and reactive values is mapped into a standard Smith chart?

9.2 What resistor values get mapped into a compressed Smith chart? Show by drawing a diagram.

9.3 A lossless transmission line is connected to a load $Z_L = 100 + j100\ \Omega$. Using a Smith chart:

- **a.** Determine the reflection coefficient at the load.
- **b.** Calculate the return loss.
- **c.** Find the *VSWR* on the line.
- **d.** Determine the reflection coefficient and the input impedance $\lambda/8$ away from the load.

9.4 Using a Smith chart, find Z_L for:

- **a.** $\Gamma = 0.6\, e^{j45°}$
- **b.** $\Gamma = -0.3$
- **c.** $\Gamma = 0.4 + j0.4$

9.5 VSWR on a lossless transmission line ($Z_o = 50\ \Omega$) is measured to be 3.0. Using a Smith chart, determine:

- **a.** The magnitude of the reflection coefficient in "ratio" and in "dB"
- **b.** The return loss in dB
- **c.** The mismatch loss in dB
- **d.** If the load is resistive ($R > Z_o$) and is located $3\lambda/8$ away from the source, determine the load impedance value, the input impedance of the transmission line, and the reflection coefficients at the load and at the source (see Figure P9.5).

FIGURE P9.5

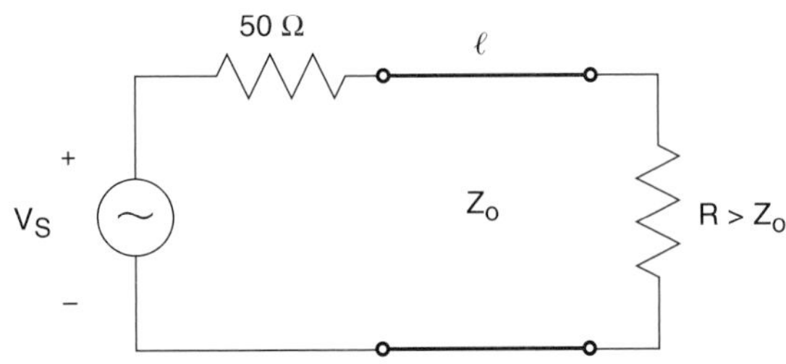

REFERENCES

[9.1] Cheng, D. K. *Field and Wave Electromagnetics*, 2nd ed., Reading: Addison Wesley, 1989.

[9.2] Cheung, W. S. and F. H. Levien. *Microwave Made Simple, Principles and Applications.* Norwood: Artech House, 1985.

[9.3] Ginzton, E. L. *Microwave Measurements*, New York: McGraw-Hill, 1957.

[9.4] Gonzalez, G. *Microwave Transistor Amplifiers, Analysis and Design*, 2nd ed. Upper Saddle River: Prentice Hall, 1997.

[9.5] Kosow, I. W. and Hewlett-Packard Engineering Staff, *Microwave Theory and Measurement*. Englewood Cliffs: Prentice Hall, 1962.

[9.6] Reich, H. J., F. O. Phillip, H. L. Krauss, and J. G. Skalnik, *Microwave Theory and Techniques,* New York: D. Van Norstrand Company, Inc., 1953.

[9.7] Schwarz, S. E. *Electromagnetics for Engineers*, Orlando: Saunders College Publishing, 1990.

[9.8] Smith, P. H. *Transmission-Line Calculator*, Electronics, 12, pp 29-31, Jan. 1939.

[9.9] Smith, P. H. *An Improved Transmission-Line Calculator*, Electronics, 17, pp 130, Jan. 1944.

CHAPTER 10

Applications of the Smith Chart

10.1 INTRODUCTION

The Smith chart applications in the analysis or design of RF and microwave circuits can be subdivided into three categories:

- Circuits containing primarily "distributed elements," particularly transmission lines (TLs)
- Circuits containing "lumped elements"
- Circuits containing "distributed and lumped elements" in combination

10.2 DISTRIBUTED CIRCUIT APPLICATIONS

The most common distributed circuit element is a transmission line (TL), and the Smith chart can be used effectively for calculation of values of its different parameters. Before we proceed into different Smith chart applications, it would serve us well, at the outset, to define the following notations that will be used throughout this book:

$$\text{Impedance: } Z = R + jX \ (\Omega) \tag{10.1}$$

$$\text{Admittance: } Y = G + jB \ (\text{S}) \tag{10.2}$$

The normalized values are given by:

$$(Z)_N = Z/Z_o = R/Z_o + jX/Z_o = r + jx \tag{10.3}$$

$$(Y)_N = Y/Y_o = G/Y_o + jB/Y_o = g + jb \tag{10.4}$$

where

$$r = R/Z_o, \tag{10.5}$$

$$x = X/Z_o, \tag{10.6}$$

$$g = G/Z_o, \tag{10.7}$$

$$b = B/Z_o \tag{10.8}$$

$$Y_o = 1/Z_o \tag{10.9}$$

and "Z_o" is the characteristic impedance of a transmission line or a reference impedance value.

10.2.1 Application #1

Input impedance (Z_{IN}) determination using a known load (Z_L).
The input impedance (Z_{IN}) at any point on a transmission line, a distance ℓ away from the load (Z_L), can be calculated by the following procedure:

1. Plot the normalized load impedance $[(Z_L)_N = Z_L/Z_o)]$ on the Smith chart.
2. Draw the constant *VSWR* circle that goes through $(Z_L)_N$.
3. Starting from $(Z_L)_N$, move "toward generator" on the constant *VSWR* circle a distance "ℓ/λ".
4. Read off the normalized input impedance value (Z_{IN}/Z_o) from the chart, as shown in Figure 10.1.

FIGURE 10.1 Smith chart plot (Application #1).

(a)

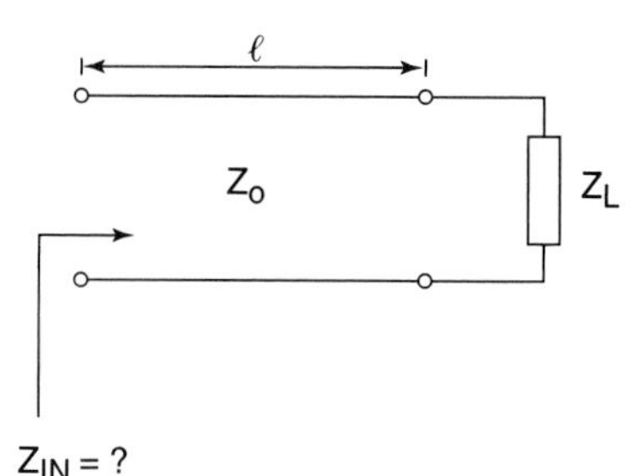

(b)

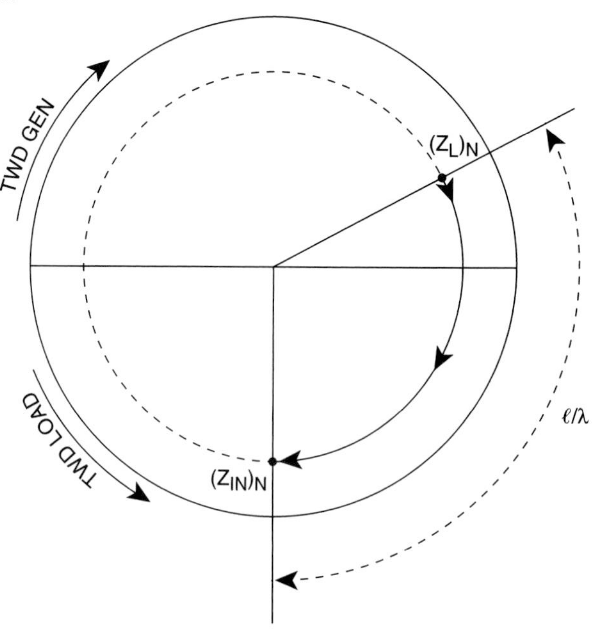

This process can be reversed easily when the input impedance (Z_{IN}) is known and the load impedance (Z_L) is unknown. In this case, starting from $(Z_{IN})_N$, we move "toward load" a distance ℓ/λ on the constant *VSWR* circle to arrive at $(Z_L)_N$.

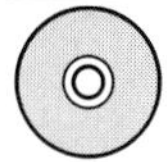

EXAMPLE 10.1

Find the input impedance of a transmission line ($Z_o = 50\ \Omega$) that has a length of $\lambda/8$ and is connected to a load impedance $Z_L = 50 + j50\ \Omega$.

Solution:

a. Locate $(Z_L)_N = Z_L/Z_o = 1 + j1$ on the Smith chart.

b. Draw the constant *VSWR* circle as shown in Figure 10.2.

c. Now move "toward generator" on the constant *VSWR* circle a distance of $\lambda/8$ (or 90°) to obtain:

$$(Z_{IN})_N = 2 - j1 \Rightarrow Z_{IN} = Z_o(Z_{IN})_N = 100 - j50\ \Omega.$$

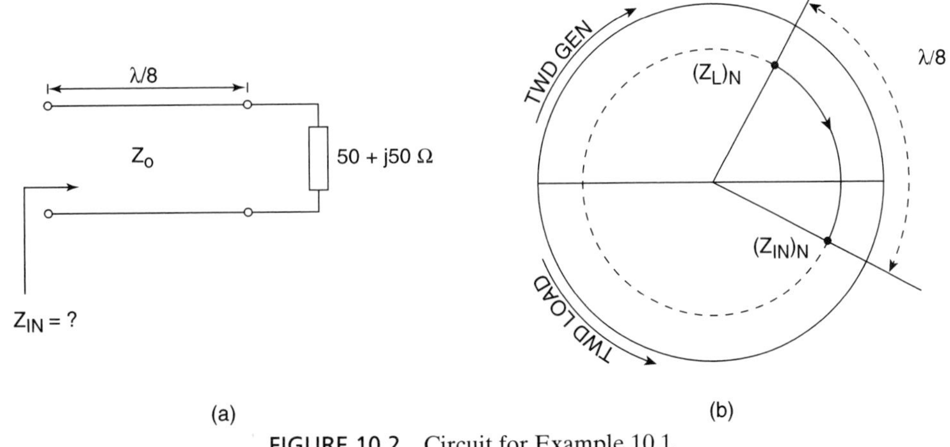

FIGURE 10.2 Circuit for Example 10.1.

10.2.2 Application #2

Input impedance determination using the input reflection coefficient ($|\Gamma_{IN}| \le 1$). When the reflection coefficient at any point on a transmission line is known, the input impedance at that point can be calculated as follows:

1. Locate $\Gamma_{IN} = |\Gamma_{IN}|e^{j\theta}$ on the Smith chart; the magnitude of $|\Gamma|$ can be read off the "Reflection coefficient voltage" radial scale at the bottom of the chart while θ is read off the circular scale (see Figure 10.3).
2. Normalized values of resistance and reactance (r,x) can be read off the Smith chart at point A, giving Z_{IN} as:

$$Z_{IN} = Z_o(r + jx)$$

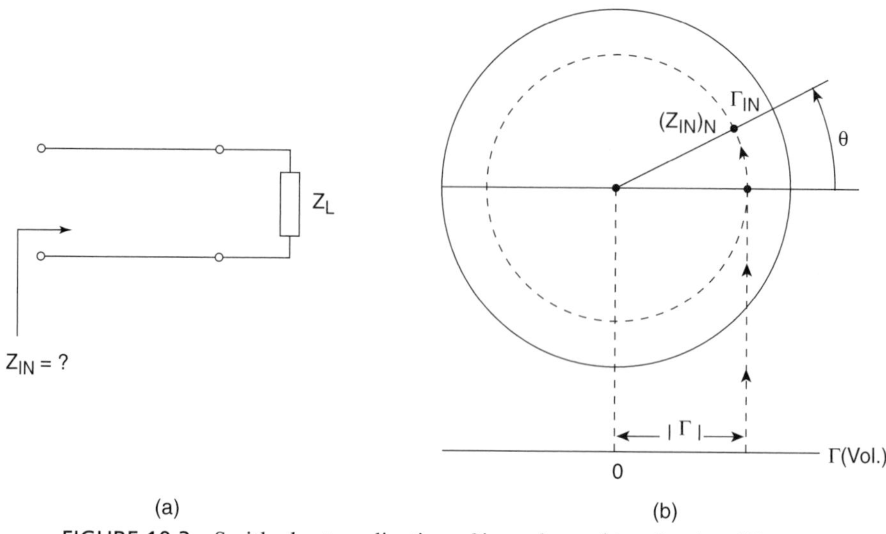

FIGURE 10.3 Smith chart application of impedance (Application #2).

NOTE 1: *If conversely, the input impedance (Z_{IN}) is known and the corresponding reflection coefficient is desired, the procedure would be reversed as follows:*

- **a.** *Plot the normalized input impedance $(Z_{IN})_N$ on the Smith chart and read off the angle θ on the circular scale.*
- **b.** *Draw the constant VSWR circle.*
- **c.** *The intersection of this circle with the right-hand horizontal axis is found and dropped off onto the "reflection coef." radial scale at the bottom, and the |Γ| value is read off, as shown in Figure 10.3.*

NOTE 2: *If the value of Z_{IN} at a distance ℓ from the current reflection coefficient location is sought, we need to use the procedure described in Application #1.*

10.2.3 Application #3

Impedance determination using reflection coefficient when |Γ| > 1.
When the magnitude of the reflection coefficient is greater than unity, the corresponding impedance has a negative resistance value and, thus, maps outside the standard Smith chart. In this case, another type of chart called a compressed Smith chart (as discussed earlier in Chapter 9, *The Smith Chart*) should be used. This chart includes both the standard Smith chart ($|\Gamma| \le 1$) and the ($|\Gamma| > 1$) region, which corresponds to the negative resistance region.

An alternate way of determining an impedance (Z) having ($|\Gamma| > 1$) is by using a standard Smith chart with the help of the following procedure:

1. Obtain the complex conjugate of the reflection coefficient at point A to obtain point B, ($\Gamma^* = |\Gamma| \angle -\theta$).
2. Plot $1/\Gamma^*$ on the standard Smith chart (see point C in Figure 10.4).

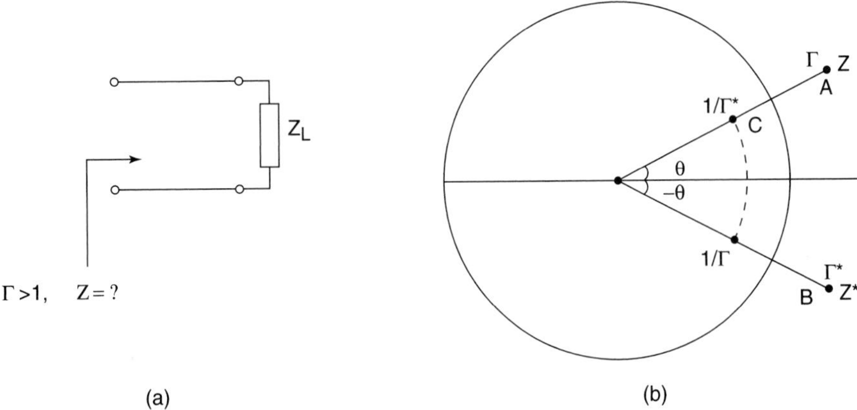

FIGURE 10.4 Smith chart for Application #3.

3. Read off the normalized impedance value ($r + jx$, corresponding to $1/\Gamma^*$) on the Smith chart.
4. The impedance (Z) value corresponding to Γ is obtained by negating r and keeping x intact, i.e.,

$$Z = Z_o(-r + jx)$$

This procedure can be proven as shown in the text that follows.

PROOF:
Assuming Z_o is a real number, the normalized impedance (Z/Z_o) corresponding to Γ is given by (see point A in Figure 10.4):

$$Z/Z_o = -r + jx, \ r > 0 \tag{10.10}$$

Where r and x are normalized values of resistance and reactance, respectively. Knowing that $\Gamma = |\Gamma| \angle \theta$, we can write:

$$\Gamma \leftrightarrow -r + jx$$

$$\Gamma = (Z - Z_o)/(Z + Z_o) = (-r + jx - 1)/(-r + jx + 1) \tag{10.11}$$

$$\Gamma^* = |\Gamma| \angle{-\theta} = (r + jx + 1)/(r + jx - 1) \tag{10.12}$$

Thus, we have:

$$1/\Gamma^* = \frac{1}{|\Gamma|} \angle \theta \tag{10.13a}$$

$$1/\Gamma^* = (r + jx - 1)/(r + jx + 1) = (Z'/Z_o - 1)/(Z'/Z_o + 1) \tag{10.13b}$$

where $Z'/Z_o = r + jx$ is the impedance corresponding to Γ^*.

$$1/\Gamma^* \leftrightarrow r + jx \tag{10.14}$$

From Equation 10.13a, we can see that $1/\Gamma^*$ (shown at point B in Figure 10.4) has the same angle as Γ, namely, they are on the same vector. Therefore, from Equations

10.11 and 10.14, we conclude that the impedance corresponding to Γ is simply obtained by reversing the sign of the real part of Z', i.e.,

$$Z = Z'|_{r \to -r} \tag{10.15}$$

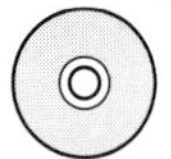

EXAMPLE 10.2

What is the impedance (Z_D) of a device having $\Gamma_D = 2.23 \angle 26.5°$? Assume $Z_o = 50\ \Omega$.

Solution:

a. We find $\Gamma_D^* = 2.23\angle{-26.5°}$.

b. Plot $1/\Gamma_D^* = 0.447\angle 26.5°$ on the Smith chart.
From the Smith chart, we obtain:

$$Z'_D = 50(2 + j1) = 100 + j50\ \Omega$$

c. Using Z'_D from step b, we can write Z_D as:

$$Z_D = -100 + j50\ \Omega$$

10.2.4 Application #4

Determination of admittance (*Y*) from impedance value (*Z*).
As discussed in Chapter 7, *Fundamental Concepts in Wave Propagation*, we know that the normalized input impedance $[Z_N(x)]$, and the normalized input admittance $[Y_N(x)]$ at any point on the line are given by:

$$Z_N(x) = [1 + \Gamma(x)]/[1 - \Gamma(x)] \tag{10.16}$$

and

$$Y_N(x) = 1/Z_N(x) = [1 - \Gamma(x)]/[1 + \Gamma(x)] \tag{10.17}$$

where

$$\Gamma(x) = \Gamma_L\, e^{j2\beta x} \tag{10.18}$$

is the reflection coefficient at any point (x) on the transmission line.

From the expression for $\Gamma(x)$, we note that for every phase change of $2\beta\ell = \pi$ (i.e., every $\ell = \lambda/4$), $\Gamma(x)$ changes sign, which leads to the inversion of the expressions given in Equations 10.16 and 10.17 causing $Z_N(x)$ to become $Y_N(x)$ and vice versa. This observation indicates that Y_N is located 180 degrees opposite to Z_N on the *VSWR* circle as shown below in Figure 10.5.

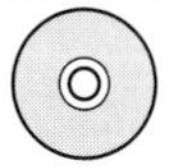

EXAMPLE 10.3

Find the admittance value for an impedance value of $Z = 50 + j50\ \Omega$, in a 50 Ω system.

Solution:

$$Z_o = 50\ \Omega \Rightarrow Y_o = 1/50 = 0.02\ \text{S}$$

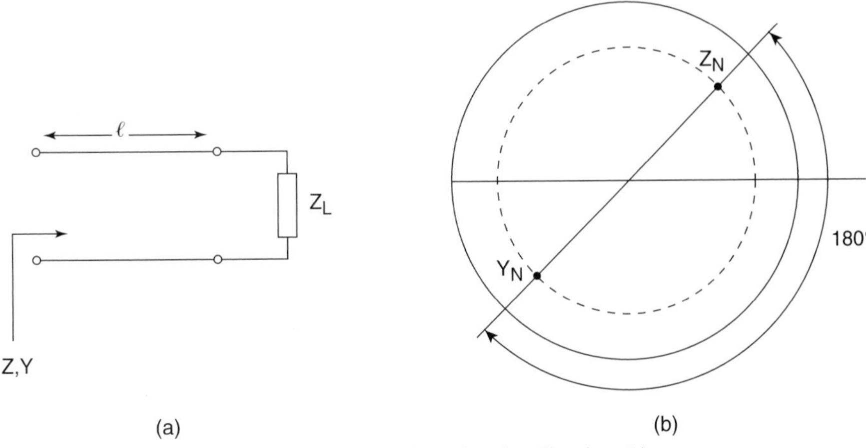

FIGURE 10.5 Smith chart for Application #4.

$$Z_N = Z/Z_o = 1 + j1$$

Using the standard Smith chart, Y_N can be read off at 180° away on the constant *VSWR* circle as:

$$Y_N = 0.5 - j0.5$$

$$Y = Y_o Y_N \Rightarrow Y = 0.01 - j0.01 \text{ S}$$

NOTE 1: *Z to Y conversion can also be obtained by rotating the Z-chart by 180° and superimposing it on the original chart, which will give a ZY-Smith chart. The Y-chart has negative susceptance on the upper half and positive susceptance on the lower half, exactly opposite of the Z-chart. Both values for Z and Y can be read off directly (without 180° shift) in the Z-Y chart as shown in Figure 10.6.*

FIGURE 10.6 The ZY-Smith chart.

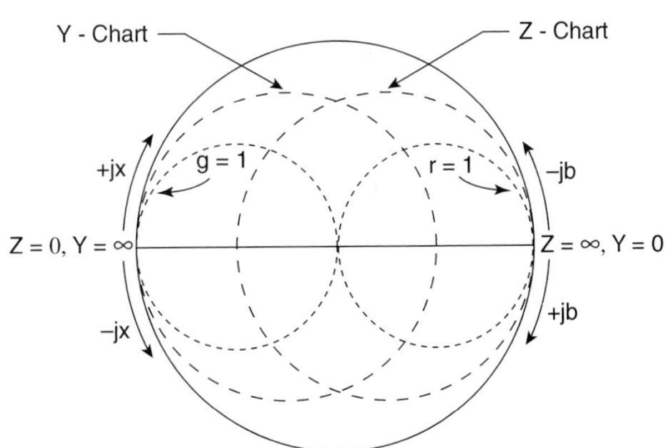

NOTE 2: *The standard Smith chart may be considered to be a Y- or Z-chart depending on the first time of entrance of values in it, being either admittance (Y) or impedance (Z).*

NOTE 3: *In the Z-Y chart, when working with* ***series elements****, the concept of impedance becomes important and we need to use the Z-****chart****. On the other hand, when working with* ***parallel (or shunt) elements****, the concept of admittance becomes paramount and therefore, we switch to the Y-****chart****.*

10.2.5 Application #5

Determination of value and location of Z_{max} and Z_{min} from a known load (Z_L).

Z_{max} and Z_{min} Value Determination. Given a known load $(Z_L)_N$, the *VSWR* circle can be drawn (see Figure 10.7). Furthermore, using the results for Γ and Z_{IN} from Chapter 7, we can write:

$$\Gamma(x) = \Gamma_L e^{j2\beta x}, \tag{10.19}$$

where $\Gamma_L = |\Gamma_L| e^{j\theta}$. Therefore, we can write:

$$\Gamma(x) = |\Gamma_L| e^{j\phi(x)}, \quad \phi(x) = 2\beta x + \theta \tag{10.20}$$

and

$$[Z_{IN}(x)]_N = Z_{IN}(x)/Z_o = [1 + \Gamma(x)]/[1 - \Gamma(x)] \tag{10.21}$$

From Equation 10.21, we note that maximum input impedance $(Z_{max})_N$ occurs when the numerator is maximum and denominator is minimum. By observation, this condition occurs when $\Gamma(x) = |\Gamma_L| e^{j\phi(x)}$ is a positive real number, i.e., $\phi(x) = 0$, which gives $(Z_{max})_N$ as:

$$(Z_{max})_N = [1 + |\Gamma_L|]/[1 - |\Gamma_L|] \tag{10.22}$$

This value can be read off the chart at the intersection of the *VSWR* circle with the right-hand horizontal axis (where $\phi = 0$), as shown in Figure 10.7.

From Equation 10.21, we can observe that minimum input impedance $(Z_{min})_N$, being the inverse of $(Z_{max})_N$, occurs when $\Gamma(x)$ is a negative real number (i.e., $\phi = 180°$):

$$(Z_{min})_N = 1/(Z_{max})_N = [1 - |\Gamma_L|]/[1 + |\Gamma_L|] \tag{10.23}$$

This value can be read off the chart where the *VSWR* circle intersects the left semi-axis (see Figure 10.7).

NOTE: *Because $(Z_{min})_N = 1/(Z_{max})_N = (Y_{max})_N$, the value and location of $(Z_{min})_N$ could have easily been found (using Application #4) by locating it 180° away from $(Z_{max})_N$ on the VSWR circle.*

Z_{max} and Z_{min} Location (i.e., Distance from Load) Determination.

Z_{max} Distance from Load Starting from the load on the *VSWR* circle, we now move "toward generator" to arrive at $\phi = 0°$ where $(Z_{max})_N$ is located. The distance

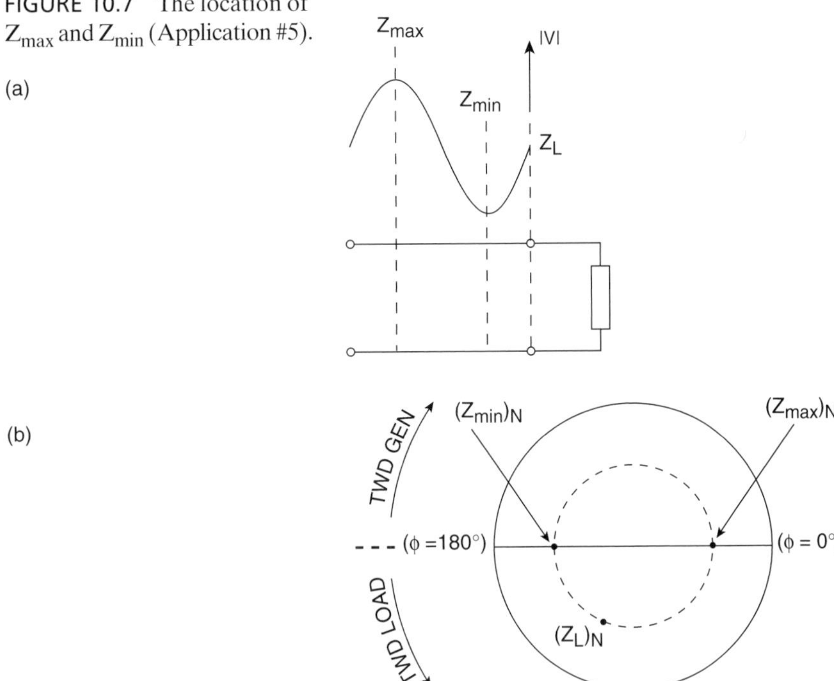

FIGURE 10.7 The location of Z_{max} and Z_{min} (Application #5).

ℓ_{max} can now be read off using the circular scale on the outer edge of the Smith chart, as shown in Figure 10.8.

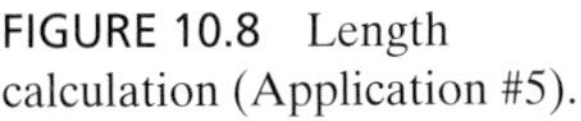

FIGURE 10.8 Length calculation (Application #5).

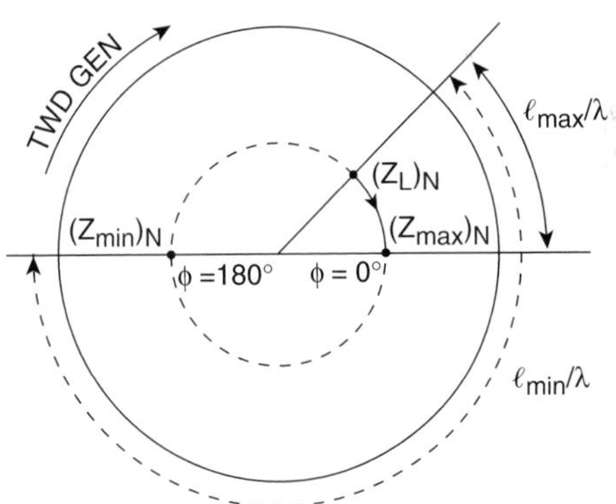

***Z_{min}* Distance from Load** Starting from the load, we now travel "toward generator" on the *VSWR* circle to arrive at $\phi = 180°$ where $(Z_{min})_N$ is located. The distance traveled is ℓ_{min}/λ, which can be read off on the outer circular scale (see Figure 10.8).

From the Smith chart, we can observe that the length difference between the locations of Z_{max} and Z_{min} is $\lambda/4$, i.e.,

$$|\ell_{max} - \ell_{min}| = \lambda/4 \tag{10.24}$$

This observation is further confirmed by our earlier discussion of the transmission line and Smith chart where we noted that the input impedance at any point on a line (e.g., Z_{max}) repeats itself every $\lambda/2$. Because Z_{min} is located one half of the distance between the two repeating maxima, thus the distance between Z_{max} and Z_{min} should be $\lambda/4$, as indicated by Equation 10.24 (see Figure 10.9).

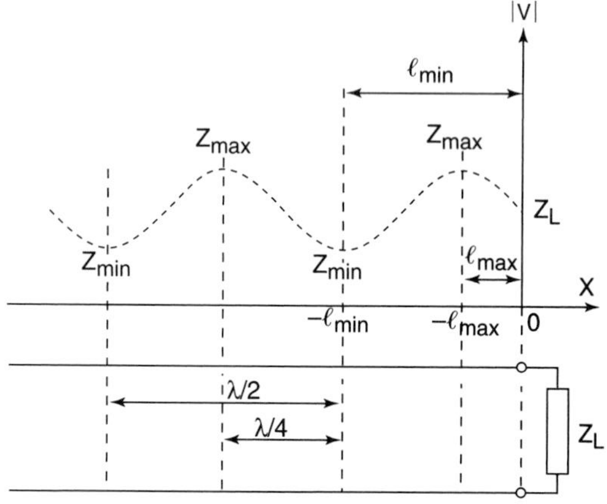

FIGURE 10.9 Voltage standing wave plot on a transmission line (Application #5).

For example, if ℓ_{max} is known and is nearest to the load, then ℓ_{min} is simply given by:

$$\ell_{min} = \ell_{max} + \lambda/4 \tag{10.25}$$

Without having to resort to the chart, and vice versa, if ℓ_{min} is known and is nearest to the load then:

$$\ell_{max} = \ell_{min} + \lambda/4 \tag{10.26}$$

A Special Case: Lines with Purely Resistive Loads. When a transmission line with a real characteristic impedance (Z_o) is terminated in a resistive load (i.e., has no reactive component, $Z_L = R_L$), then there are two possible cases.

Case I: $\boldsymbol{R_L > Z_o}$ In this case, the maximum impedance on the line equals the load value and is located at the load repeating every $\lambda/2$, that is:

$$Z_{max} = R_L, \tag{10.27}$$

$$\ell_{max} = n(\lambda/2), \quad n = 0,1,2, \ldots. \tag{10.28}$$

The minimum line impedance (Z_{min}) is located $\lambda/4$ away from the load and repeats itself every $\lambda/2$, i.e.,

$$\ell_{min} = \lambda/4 + n(\lambda/2), \quad n = 0,1,2, \ldots. \tag{10.29}$$

Furthermore, we can write:

$$(Z_{min})_N = 1/(Z_{max})_N \Rightarrow Z_{min}/Z_o = Z_o/Z_{max}$$

$$Z_{min} = Z_o^2/Z_{max} \tag{10.30}$$

or

$$Z_{min} = Z_o^2/R_L \tag{10.31}$$

These are shown in Figure 10.10.

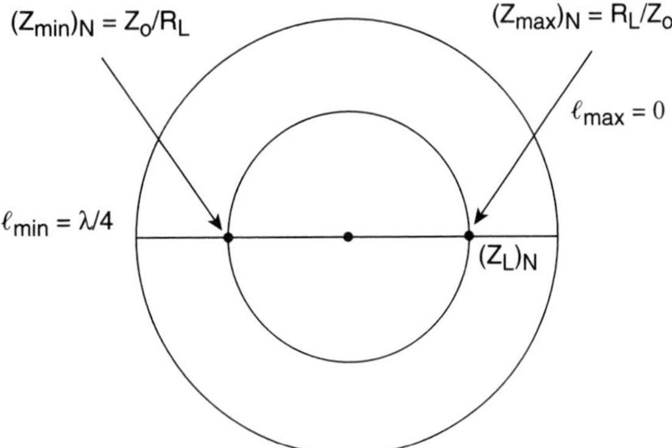

FIGURE 10.10 Plot of Z_{max} and Z_{min} for $Z_L = R_L$ $(R_L > Z_o)$ (Application #5).

EXAMPLE 10.4

A microwave signal is traveling on a transmission line that has $Z_o = 50\ \Omega$ and a load value of $Z_L = 100\ \Omega$. Find the values of Z_{max} and Z_{min} and their location on the transmission line.

Solution:

Because $Z_L > Z_o$, the maximum voltage and thus maximum impedance occurs at the load, i.e.,

$$Z_{max} = Z_L = 100\ \Omega$$

$$\ell_{max} = 0$$

The minimum impedance occurs λ/4 away from the load:

$$\ell_{min} = \lambda/4$$

$$Z_{min} = 50^2/100 = 25\ \Omega$$

This is shown in Figure 10.11.

***Case II:* $R_L < Z_o$** In this case, the minimum impedance on the line is located at the load and repeats every λ/2, i.e.,

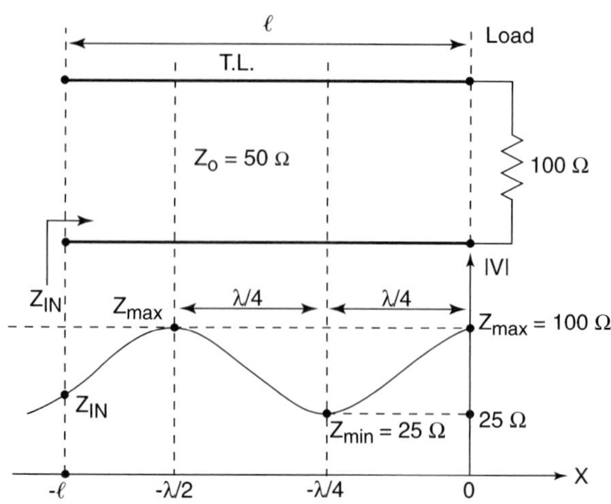

FIGURE 10.11 Circuit for Example 10.4.

$$Z_{min} = R_L, \tag{10.32}$$

$$\ell_{min} = n(\lambda/2), \quad n = 0,1,2, \ldots \tag{10.33}$$

The maximum line impedance is located λ/4 away from the load and also repeats every λ/2 as shown in Figure 10.12. Thus, using Equation 10.22 we can write:

$$(Z_{min})_N = 1/(Z_{max})_N$$

$$Z_{max} = Z_o^2/Z_{min} = Z_o^2/R_L, \tag{10.34}$$

$$\ell_{max} = \lambda/4 + n(\lambda/2), \quad n = 0,1,2, \ldots \tag{10.35}$$

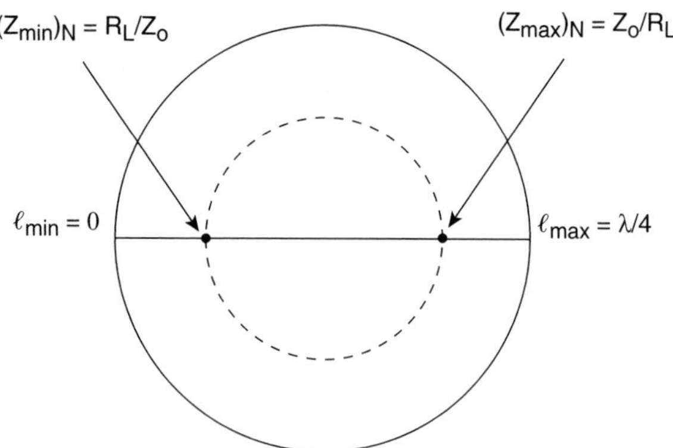

FIGURE 10.12 The location of Z_{max} and Z_{min} for $Z_L = R_L$ ($R_L < Z_o$) (Application #5).

EXAMPLE 10.5

A microwave signal at a frequency of $f = 1$ GHz, is traveling on a transmission line having $Z_o = 50\ \Omega$, and terminated in a load of $Z_L = 20\ \Omega$. Find the values of Z_{max} and Z_{min} and their location on the transmission line.

Solution:

Because $R_L < Z_o$, the minimum voltage or impedance on the line occurs at the load, i.e.,

$$Z_{min} = 20\ \Omega$$

$$\ell_{min} = 0$$

From Equations 10.34 and 10.35, the value and location of Z_{max} is given by:

$$Z_{max} = Z_o^2/R_L = 50^2/20 = 125\ \Omega$$

$$\lambda = c/f \ \Rightarrow \lambda(\text{cm}) = 30/f(\text{GHz}) = 30\ \text{cm}$$

The first maximum occurs at $\ell_{max} = \lambda/4 = 7.5$ cm away from the load, as shown in Figure 10.13. The standing wave pattern is plotted in Figure 10.14.

FIGURE 10.13 Circuit for Example 10.5.

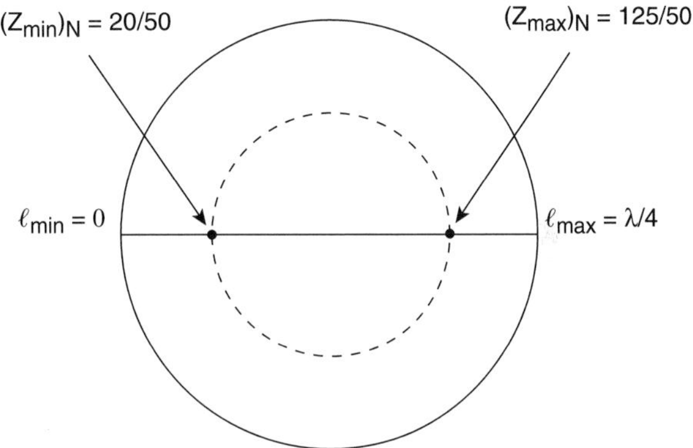

FIGURE 10.14 Plot of the standing wave pattern.

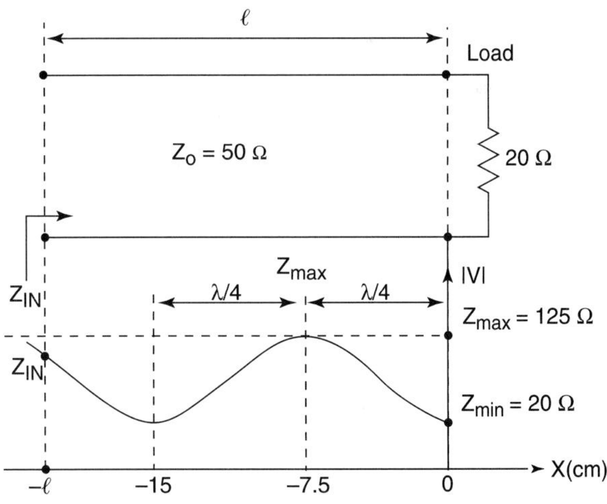

10.2.6 Application #6

Determination of VSWR using a known load.
There are two methods to find the *VSWR* on a transmission line with a given load (Z_L or Γ_L) as follows:

Method #1. Using $|\Gamma_L|$ as the radius, draw the constant *VSWR* circle; from the intersection of this circle with the left-hand horizontal axis, drop a vertical line onto the *VSWR* scale on the bottom of the chart to find the *VSWR*, as shown in Figure 10.15.

FIGURE 10.15
Determination of VSWR using the radial scale (Application #6).

(a)

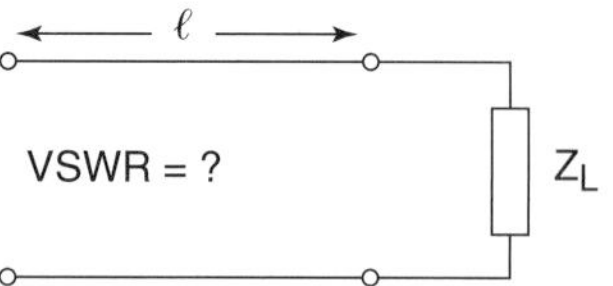

(b)

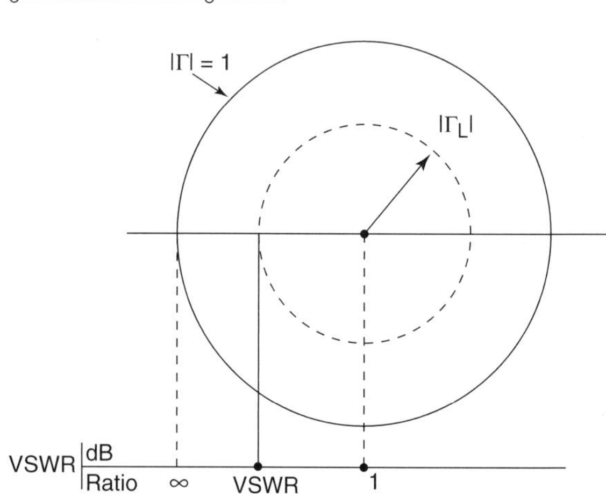

Method #2. For a lossless transmission line, we know that $|\Gamma| = |\Gamma_L|$ anywhere on the line, which is the radius of the *VSWR* circle. Therefore, *VSWR* can be calculated by:

$$VSWR = \frac{1 + |\Gamma_L|}{1 - |\Gamma_L|} \tag{10.36}$$

where

$$|\Gamma_L| = \frac{Z_L - Z_o}{Z_L + Z_o} \tag{10.37a}$$

From Application #5, we have:

$$(Z_{max})_N = 1/(Z_{min})_N = \frac{1 + |\Gamma_L|}{1 - |\Gamma_L|} = VSWR \tag{10.37b}$$

Thus,

$$Z_{max} = Z_o(VSWR), \tag{10.38a}$$

and

$$Z_{min} = Z_o / VSWR \tag{10.38b}$$

The *VSWR* value can be read off from the *VSWR* circle intersection with the horizontal semi-axis at $\theta = 0°$ (where Z_{max} is located), as shown in Figure 10.16.

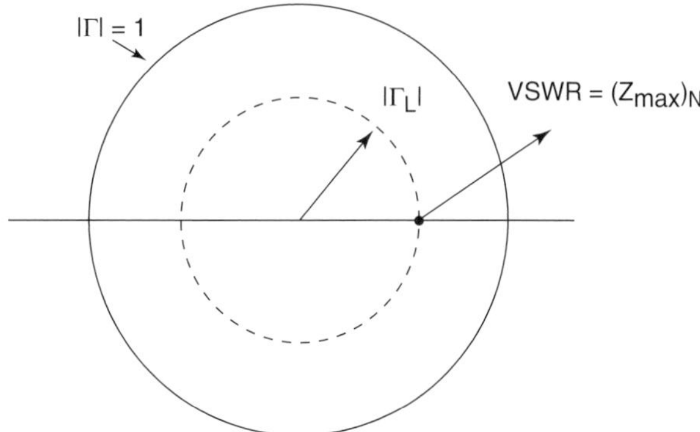

FIGURE 10.16 Determination of VSWR using Method #2 (Application #6).

10.2.7 Application #7

Plot of standing wave pattern using a known load.
As discussed in Chapter 7, when the incident wave encounters a discontinuity of any kind (such as a load) that is different from the characteristic impedance of the propagating media, a portion (or all) of the wave will be reflected. The reflected wave when combined with the incident wave will create a standing wave pattern in voltage and current.

Voltage and current waves simultaneously coexist, and each have their own standing wave patterns with peaks and valleys occurring at different points along the line, as shown in Figure 10.17.

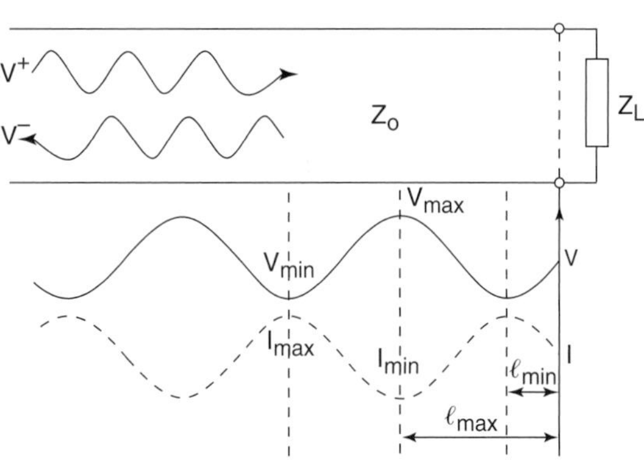

FIGURE 10.17 Standing waves on a transmission line (Application #7).

Use of the Smith chart will help us determine the exact standing wave pattern for voltage and current for a known load with a relatively good degree of accuracy.

To determine the standing wave pattern let's consider an incident voltage wave (V^+) causing an incident current wave (I^+) traveling toward the load Z_L on a lossless transmission line (Z_o). The reflected voltage and current waves (V^-, I^-) will interact with the incident voltage and current waves (V^+, I^+) to create the standing wave pattern.

To get an exact pattern determination in terms of its peak and valley magnitude and location, the following procedure can be used:

1. Locate $(Z_L)_N = Z_L/Z_o$ on the Smith chart, draw the constant $VSWR$ circle and determine the $VSWR$ on the line.
2. Using Application #5, determine the location and value of Z_{max} (ℓ_{max} away from the load). The location of Z_{max} corresponds to the location of the peak of that voltage standing wave pattern (V_{max}) and the valley of the current standing wave pattern (I_{min}) because

$$Z_{max} = V_{max}/I_{min} \tag{10.39}$$

Furthermore, to calculate the magnitude of the maximum voltage (V_{max}) on the line, we note that:

$$V_{max} = |V^+| + |V^-| = |V^+|(1 + |V^+|/|V^-|) = |V^+|(1 + |\Gamma_L|) \tag{10.40}$$

From Equation 10.36, we can write:

$$|\Gamma_L| = \frac{VSWR - 1}{VSWR + 1} \tag{10.41}$$

Substituting Equation 10.41 in Equation 10.40, we get:

$$V_{max} = |V^+|\left(\frac{2VSWR}{VSWR + 1}\right) \tag{10.42}$$

Similarly, I_{min} is given by:

$$I_{min} = V_{max}/Z_{max} = |I^+| - |I^-| = |V^+|/Z_o - |\Gamma_L||V^+|/Z_o = \frac{|V^+|}{Z_o}(1 - |\Gamma_L|)$$

Thus,

$$I_{min} = \frac{|V^+|}{Z_o}\left(\frac{2}{VSWR + 1}\right) \tag{10.43}$$

Thus, the location and magnitude of V_{max} and I_{min} can easily be determined once the load value (Z_L or Γ_L) and the incident voltage value ($|V^+|$) are known.

3. Similarly, the location and value of Z_{min} (ℓ_{min} away from the load) can be determined from the Smith chart. The location of Z_{min} corresponds to the valley of the voltage standing wave pattern (V_{min}) and the peak of the current standing wave pattern (I_{max}) because

$$Z_{min} = V_{min}/I_{max} \tag{10.44}$$

To calculate the value of V_{min} and I_{max} in terms of the magnitude of the incident voltage $|V^+|$ and the load, we note that:

$$V_{min} = |V^+| - |V^-| = |V^+|(1 - |V^+|/|V^-|) = |V^+|(1 - |\Gamma_L|) \tag{10.45}$$

Using Equation 10.41 in Equation 10.45, we get:

$$V_{min} = |V^+|\left(\frac{2}{VSWR + 1}\right) \tag{10.46}$$

And similarly, I_{max} is given by:

$$I_{max} = V_{min}/Z_{min} = |I^+| + |I^-| = |V^+|/Z_o + |\Gamma_L|\,|V^+|/Z_o$$
$$= |V^+|(1 + |\Gamma_L|)/Z_o = \frac{|V^+|}{Z_o}\left(\frac{2VSWR}{VSWR + 1}\right) \tag{10.47}$$

NOTE: *Comparing Equations 10.42 and 10.47, we can see that:*

$$I_{max} = V_{max}/Z_o$$

And similarly, from Equations 10.43 and 10.46, we can write:

$$I_{min} = V_{min}/Z_o$$

This would give an alternate way to find I_{max} and I_{min}.

Knowing the value and location of V_{max}, V_{min}, I_{max} and I_{min}, the patterns for voltage and current can now be plotted easily.

Figure 10.18 shows the standing wave pattern for voltage and current on a transmission line.

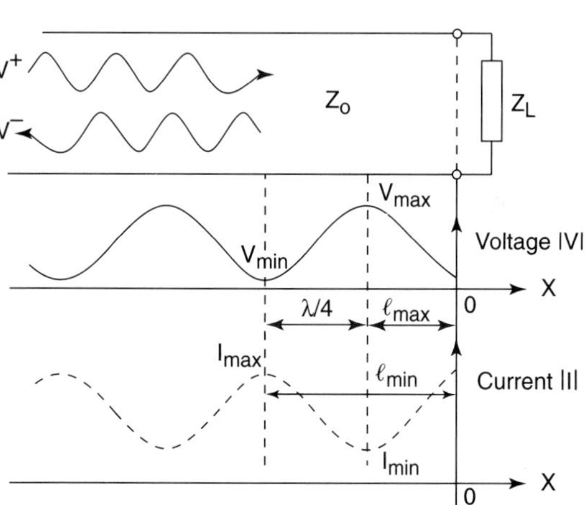

FIGURE 10.18 Standing wave pattern (Application #7) for net voltage and net current.

The following example may further help to illustrate the concept of standing waves.

EXAMPLE 10.6

Determine the standing wave pattern on a transmission line ($Z_o = 50\ \Omega$) terminated in $Z_L = 100 + j100\ \Omega$ with an incident voltage of $V^+ = 1\angle 0°$ as shown in Figure 10.19.

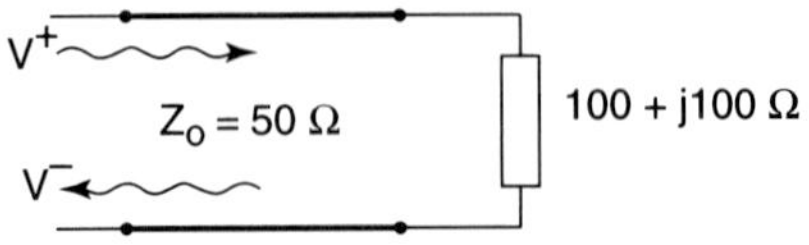

FIGURE 10.19 Circuit for Example 10.6.

Solution:

a. Locate $(Z_L)_N = Z_L/Z_o = 2 + j2$ on the Smith chart and draw the constant *VSWR* circle as shown in Figure 10.20. From this figure, we can read off [$VSWR = 4.4$ at $(Z_{max})_N$] and calculate the following:

$$VSWR = (Z_{max})_N = 4.4$$

$$Z_{max} = 4.4 \text{ x } 50 = 220\ \Omega$$

$$\ell_{max} = 0.292 - 0.250 = 0.042\lambda$$

Thus, from Equations 10.39 to 10.43, V_{max} and I_{min} are given by:

$$V_{max} = [2 \text{ x } 4.4/(1 + 4.4)] = 1.63 \text{ V}$$

$$I_{min} = V_{max}/Z_{max} = 1.63/220 = 7.4 \text{ mA}$$

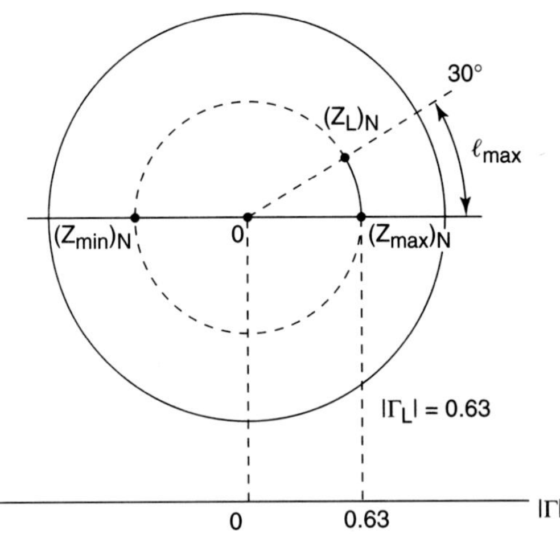

FIGURE 10.20 Solution for Example 10.6.

b. Similarly, from the *VSWR* circle, we can write:

$$(Z_{min})_N = 1/(Z_{max})_N = 1/4.4$$

$$Z_{min} = 50(1/4.4) = 11.36\ \Omega$$

$$\ell_{min} = 0.042\lambda + \lambda/4 = 0.292\lambda$$

Thus, from (10.44) to (10.47), V_{min} and I_{max} are given by:

$$V_{min} = [2/(1 + 4.4)] = 0.37\ \text{V}$$

$$I_{max} = V_{min}/Z_{min} = 0.37/11.36 = 32.6\ \text{mA}$$

The final standing wave pattern is plotted in Figure 10.21.

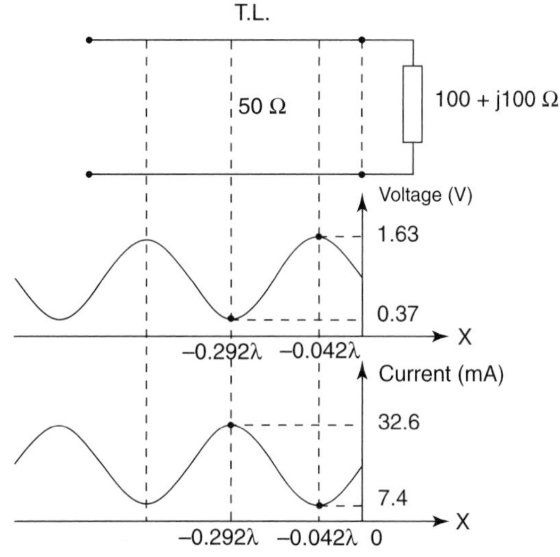

FIGURE 10.21 Final standing wave pattern for Example 10.6.

NOTE: *An alternate method would be to calculate V_{max} and V_{min} using the VSWR value and then find I_{max} and I_{min} as follows:*

$$VSWR = 4.4$$

$$V_{max} = 2 \times 4.4/(1 + 4.4) = 1.63\ \text{v}$$

$$V_{min} = 2/(1 + 4.4) = 0.37\ \text{v}$$

$$I_{max} = V_{max}/Z_o = 1.63/50 = 32.6\ \text{mA}$$

$$I_{min} = V_{min}/Z_o = 0.37/50 = 7.4\ \text{mA}$$

These are the same values that were obtained earlier.

10.2.8 Application #8

Input impedance determination using single stubs.

DEFINITION-STUB: *A short section of a transmission line (usually terminated in either an open or a short) often connected in parallel and sometimes in series with a feed transmission line in order to transform the load to a desired value.*

In general, the stub can have any general termination (Z'_L); however, in practice as explained earlier, Z'_L is either a short or an open circuit, as shown in Figures 10.22, 10.23, and 10.24.

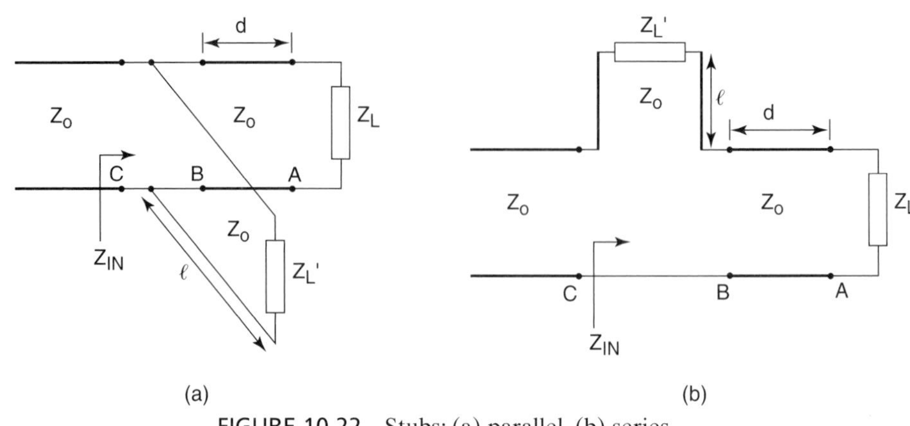

FIGURE 10.22 Stubs: (a) parallel, (b) series.

There are two cases that will be considered separately:

- Parallel stubs
- Series stubs

Parallel (or Shunt) Stubs. Consider the stub located a distance d away from a load (Z_L), as shown in Figure 10.22. We would like to determine the input impedance of the combination.

Before we proceed to find the input impedance, we need to determine the stub's susceptance. Because the stub is connected in parallel, we use the Smith chart as a Y-chart. The stub has a length (ℓ) that can be used to determine its input admittance (or susceptance). If the stub is terminated in a short, we use the Y-chart and start from $Y = \infty$; see point A in Figure 10.23c and travel ℓ toward the generator to arrive at point B. We read off the stub's susceptance from the chart. In a similar fashion, an open stub's susceptance can be found except we should start at $Y = 0$ on the opposite side (see Figure 10.23).

To find the input impedance, the following steps are carried out:

1. Locate Z_L on the Smith chart (use a ZY chart) at point A in Figure 10.25.
2. Draw the constant $VSWR$ circle.
3. Travel a distance (d) toward the generator on the $VSWR$ circle to arrive at point B.
4. Now because we are adding the parallel stub, we must switch to the Y-chart and travel on a constant conductance circle an amount equal to the susceptance of the stub to arrive at point C, as shown in Figure 10.25.
5. To find the input impedance, we switch back to the Z-chart and read off the normalized values (r, x) at point C corresponding to $(Z_{IN})_N$. The total input impedance is given by:

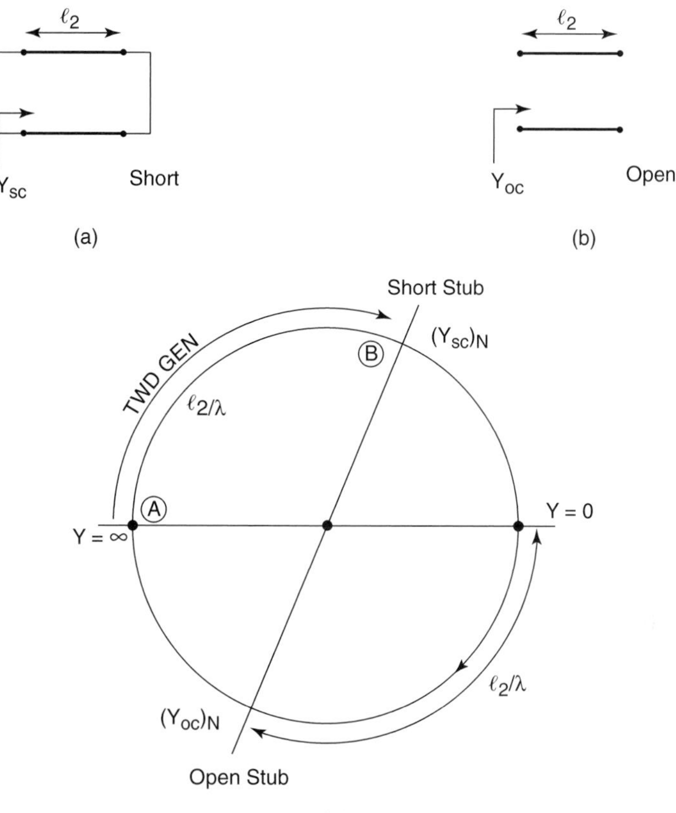

FIGURE 10.23 (a) Short stub, (b) open stub, and (c) Smith chart plot.

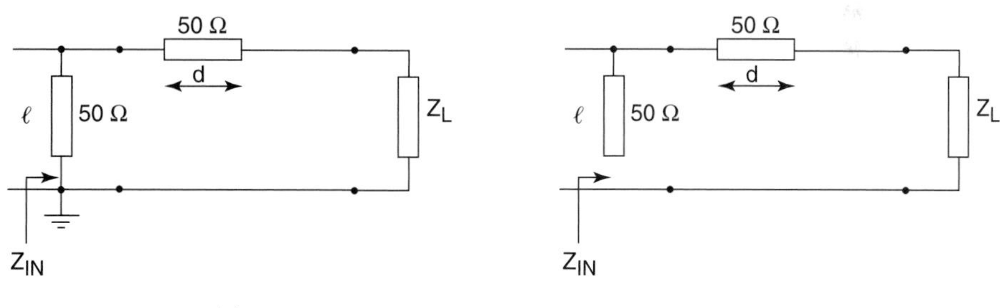

FIGURE 10.24 Shorthand schematic: (a) short shunt stub, (b) open shunt stub.

$$Z_{IN} = Z_o(Z_{IN})_N$$

Series Stubs. Consider a series stub located a distance (d) away from the load (Z_L), as shown in Figure 10.26. Similar to the parallel stub case, we need to know the

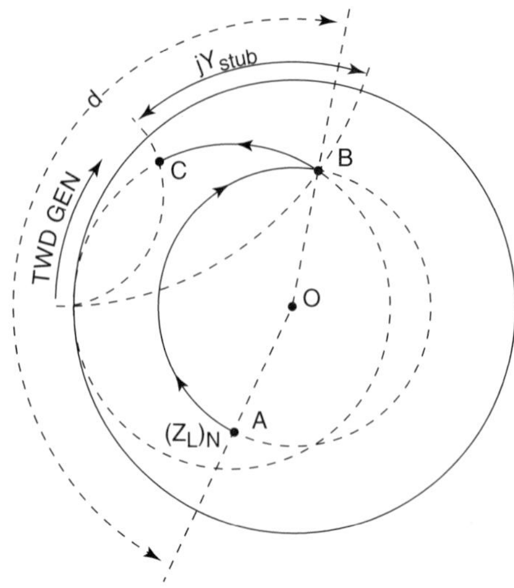

FIGURE 10.25 Smith chart solution for a shunt stub (Application #8).

series stub's reactance (jX) based on its electrical length ($\beta\ell$). Because the stub is in series, we use the Smith chart as a Z-chart. If the stub is terminated in a short, start from $Z = 0$, point A in Figure 10.27c, and travel a distance of ℓ/λ "toward generator" to arrive at point B. Read off the normalized stub's reactance ($jx = jX/Z_o$) from the chart, as shown in Figure 10.27b. Similarly, an open stub's reactance can be determined by following the previous procedure except by starting from $Z = \infty$ on the chart, as shown in Figure 10.27c.

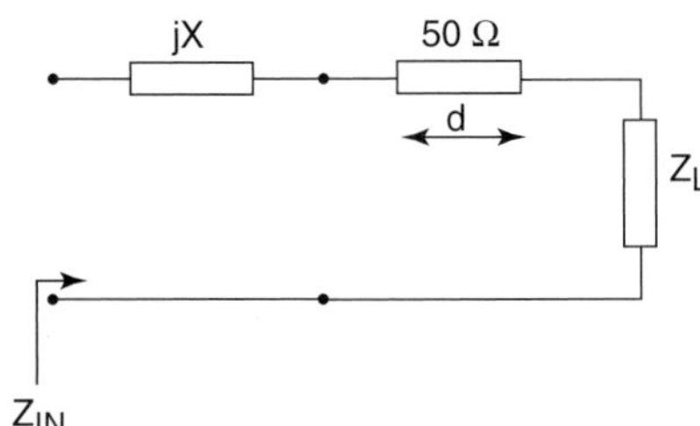

FIGURE 10.26 A series stub in circuit.

To find the input impedance, the following steps are carried out:

1. **1.** Locate $(Z_L)_N$ on the Smith chart at point A, as shown in Figure 10.28 (use a Z-chart).
2. **2.** Draw the constant $VSWR$ circle.
3. **3.** From $(Z_L)_N$, travel a distance (d) toward the generator on the $VSWR$ circle to arrive at point B.
4. **4.** Now, because we are adding the series stub, we travel on a constant resistance circle an amount equal to the reactance of the stub, jx, to arrive at point C.
5. **5.** The input impedance is read off at point C in Figure 10.28.

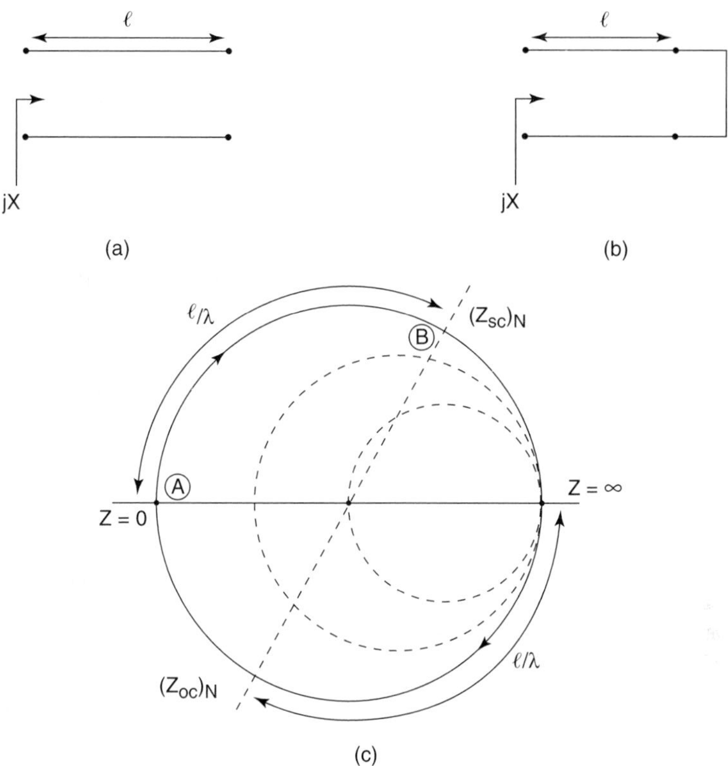

FIGURE 10.27 Reactance of a series stub: (a) open stub, (b) short stub, and (c) Smith chart solution.

FIGURE 10.28 Smith chart solution for a series stub (Application #8).

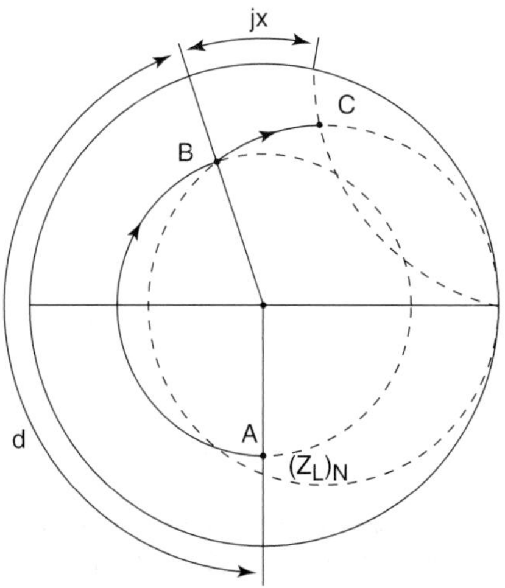

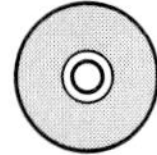

EXAMPLE 10.7

Consider a transmission line ($Z_o = 50\ \Omega$) terminated in a load $Z_L = 15 + j10\ \Omega$, as shown in Figure 10.29. Calculate the input impedance of the line where the shunt open stub is located a distance of $d = 0.044\lambda$ from the load and has a length of $\ell = 0.147\lambda$.

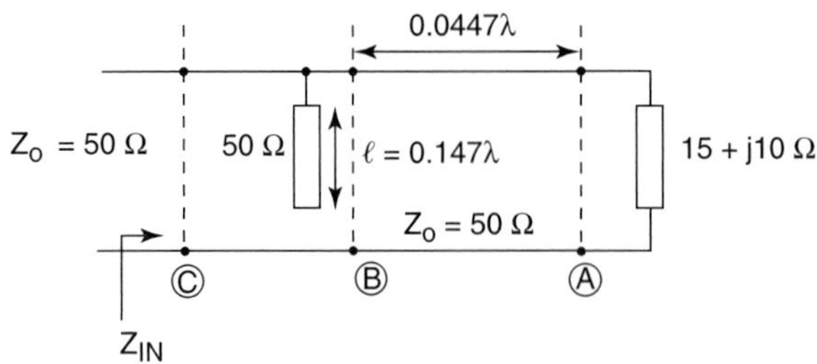

FIGURE 10.29 Circuit for Example 10.7.

Solution:

a. The susceptance of the open stub is first calculated by moving on a Smith chart from $Y = 0$ and moving a distance of 0.147λ toward the generator to arrive at $(Y_{oc})_N = j1.33$, as shown in Figure 10.30. Next, the input impedance is found by the following procedure:

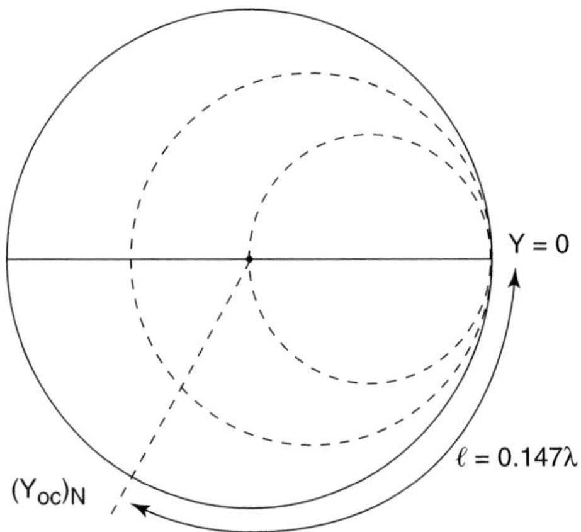

FIGURE 10.30 Shunt open stub.

b. Locate $(Z_L)_N = (15 + j10)/50 = 0.3 + j\,0.2$ on the Smith chart (see point A in Figure 10.31).

c. Draw the constant $VSWR$ circle.

d. From Z_L, travel a distance of 0.044λ to arrive at point B. The admittance is read off to be:

$$(Y_B)_N = 1 - j1.33 \text{ (point } B \text{ in Figure 10.31)}$$

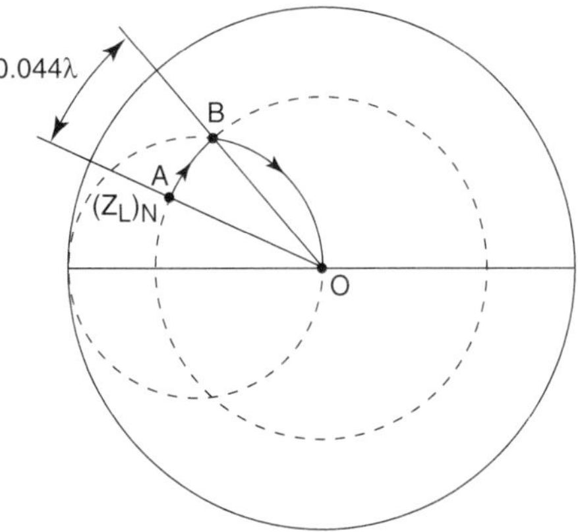

FIGURE 10.31 Smith chart solution.

e. Adding an open shunt stub of length $\ell = 0.147$ with $(Y_{oc})_N = j1.33$ gives:

$$(Y_{IN})_N = (Y_B)_N + (Y_{oc})_N = (1 - j1.33) + j1.33 = 1$$

$$(Z_{IN})_N = 1/(Y_{IN})_N = 1 \Rightarrow Z_{IN} = Z_o = 50\ \Omega$$

Adding the shunt stub on the Smith chart results in arriving at point *O*, which is obtained by moving on $r = 1$ constant resistance circle by $-j1.33$.

NOTE: *Use of Application #8 in the design of circuits to bring about reflectionless loads is widely explored in the next chapter of this book, where matching circuits are treated in depth.*

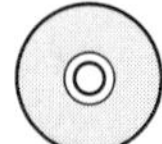

EXAMPLE 10.8

Consider a transmission line (Z_o = 50 Ω) with Z_L = 100 Ω, as shown in Figure 10.32. Calculate the input impedance of the line where the shorted series stub is located a distance of d = λ/4 from the load and has a length ℓ = λ/8.

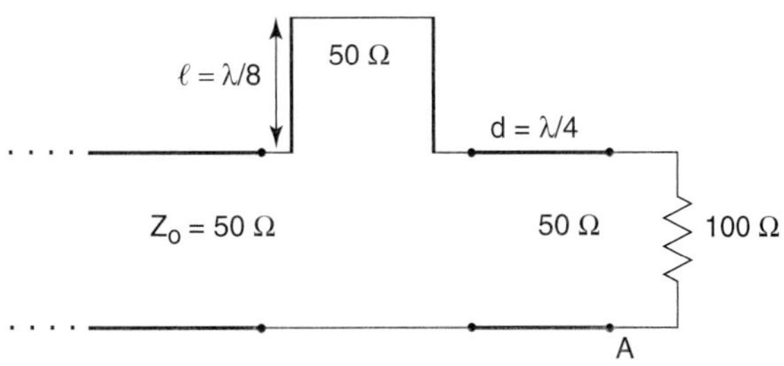

FIGURE 10.32 Circuit for Example 10.8.

Solution:

a. The reactance of the series shorted stub is first calculated by moving on a Smith chart from $Z = 0$ a distance of 0.125λ toward generator to arrive at $(Z_{SC})_N = j1$, as shown in Figure 10.33. Next, to find the input impedance we perform the following steps.

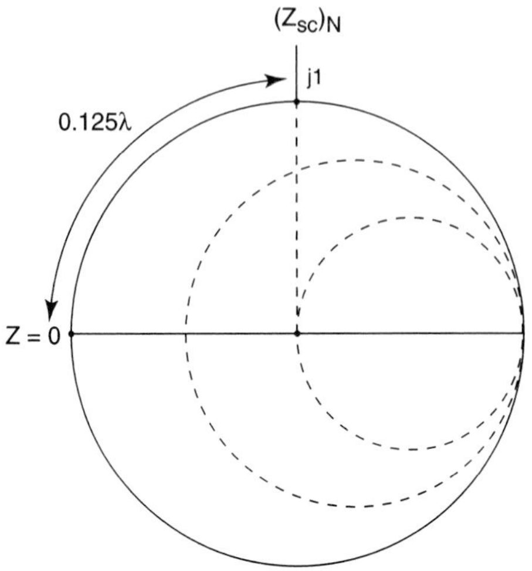

FIGURE 10.33 Short series stub.

b. Locate $(Z_L)_N = 100/50 = 2$ on the Smith chart (see point A in Figure 10.34).

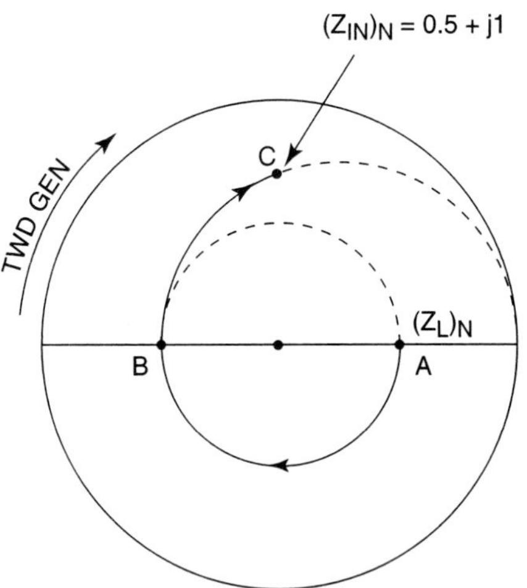

FIGURE 10.34 Smith chart solution.

c. Draw the constant $VSWR$ circle. From Z_L, travel a distance of 0.25λ to arrive at point B. The impedance is read off to be:

$$(Z_B)_N = 0.5 \text{ (at point } B)$$

NOTE: *Because the load is resistive and has a value more than Z_o, the $(Z_L)_N$ value and location correspond to $(Z_{max})_N$ (at point A) and Z_B corresponds to $(Z_{min})_N$ (for more details, see Application #5).*

d. From point *B*, move toward generator on a constant resistance circle to $0.5+j1$ (point *C* in Figure 10.34), which corresponds to adding a series stub of length $\ell = 0.125\lambda$ or $(Z_{SC})_N = j1$, giving:

$$(Z_{IN})_N = (Z_B)_N + (Z_{SC})_N = 0.5 + j1$$

$$Z_{IN} = Z_o(Z_{IN})_N = 25 + j50\ \Omega$$

10.3 LUMPED ELEMENT CIRCUIT APPLICATIONS

Lumped elements are usually employed in the design of microwave circuits. These elements are mostly lossless reactive elements (such as inductors or capacitors) and are added either in series or in parallel in the circuit.

10.3.1 Application #9

Input impedance determination for a series lumped element.
Consider the circuit shown in Figure 10.35 where a load (Z_L) is in series with a series element (Z_S). The lumped element can be reactive (lossless), resistive (lossy), or a combination of both. In this application, we consider a very general lumped element consisting of both resistive and reactive components.

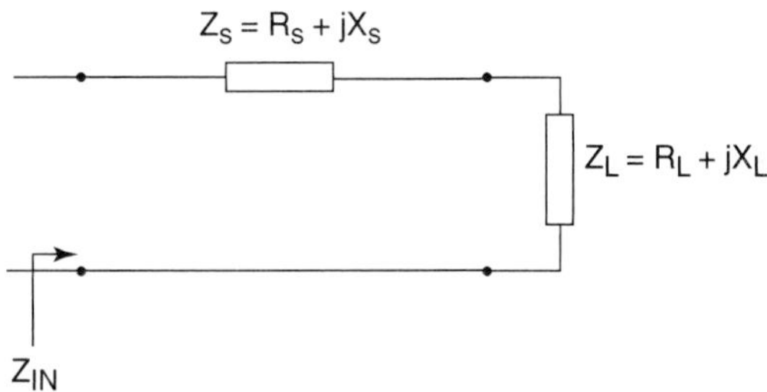

FIGURE 10.35 Circuit for series lumped element (Application #9).

Because the lumped element is in series with the load, we need to consider only the *Z*-chart markings of the *ZY*-Smith chart (or only a *Z*-chart), in order to determine Z_{IN}.

We know mathematically that:

$$Z_{IN} = Z_L + Z_S$$

Thus,

$$(Z_{IN})_N = (r_L + r_S) + j(x_L + x_S) \tag{10.48}$$

The purpose of this application is to show how to achieve this result graphically where the exact steps are delineated below:

1. Locate $(Z_L)_N$ on the Smith chart (see point *A* in Figure 10.36).
2. Moving on the constant resistance circle that passes through Z_L, add a reactance of jx_S to arrive at point *B*.
3. Now moving on a constant reactance circle that passes through point *B*, add a resistance of r_S to arrive at point *C*.
4. The input impedance value is read off at point *C*, using the *Z*-chart markings.

Alternate Procedure

Point *C* could have equally been reached by the following steps (see Figure 10.36):

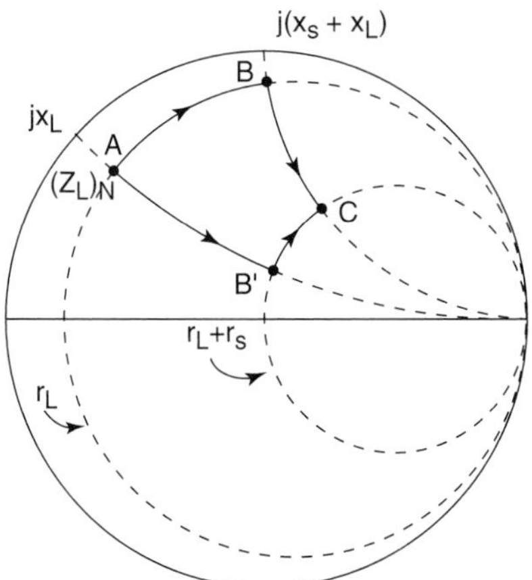

FIGURE 10.36 Graphical solution (Application #9).

- Move on a constant reactance circle (that passes through *A*) and add the resistance of r_S to arrive at point *B*' (see Figure 10.36).
- Now moving on a constant resistance circle (that passes through point *B*), add the reactance of jx_S to arrive at point *C*.

10.3.2 Application #10

Input admittance determination for a shunt lumped element.

Consider the circuit shown in Figure 10.37, where a load (Y_L) is in parallel with a shunt element (Y_P). In general, the lumped element is considered to have both resistive and reactive components (similar to Application #9).

Because the lumped element is in parallel with the load, only the *Y*-chart markings of the *ZY*-Smith chart need be considered.

The total admittance is given mathematically by:

$$Y_{IN} = Y_L + Y_P$$

$$(Y_{IN})_N = (g_L + g_P) + j(b_L + b_P) \tag{10.49}$$

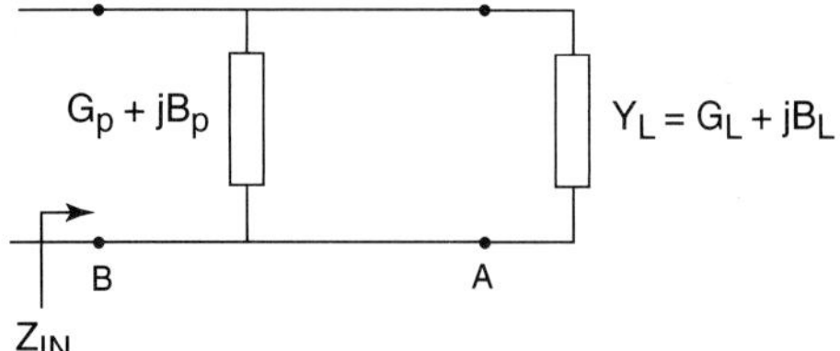

FIGURE 10.37 Circuit for shunt lumped element (Application #10).

Similar to Application #9, we now present the procedure to determine $(Y_{IN})_N$ graphically:

1. Locate $(Y_L)_N$ on the Y-chart at point A in Figure 10.38.
2. Move on the constant conductance circle that passes through $(Y_L)_N$ and add a susceptance of jb_P to arrive at point B.
3. Move on the constant susceptance circle (passing through B) by adding a conductance of g_P to arrive at point C.
4. The input admittance is read off at point C using the Y-chart markings.

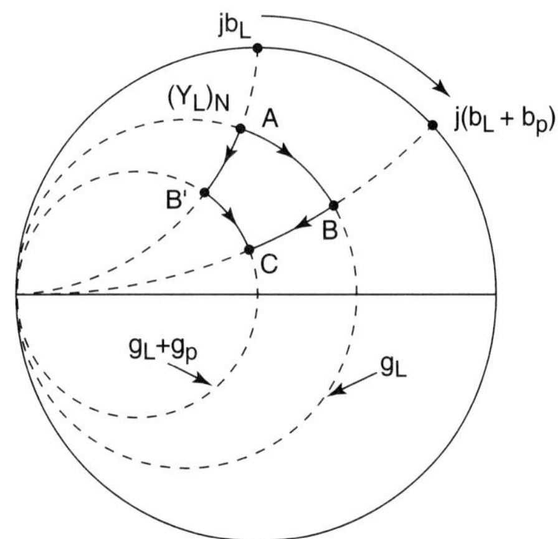

FIGURE 10.38 Smith chart solution (Application #10).

Alternate Procedure

Similar to Application #9, the input admittance could have equally been determined by the following:

- Move on a constant susceptance circle and add g_P to arrive at point B', as shown in Figure 10.38.
- Now add jb_P on a constant conductance circle to arrive at point C.

10.3.3 Application #11

Input impedance of single shunt/series reactive elements.

This is a special case of Applications #9 and #10 where the series or the shunt elements are lossless (i.e., purely reactive). In this case, there are four possible combinations (see Figure 10.39):

- Series L
- Series C
- Shunt L
- Shunt C

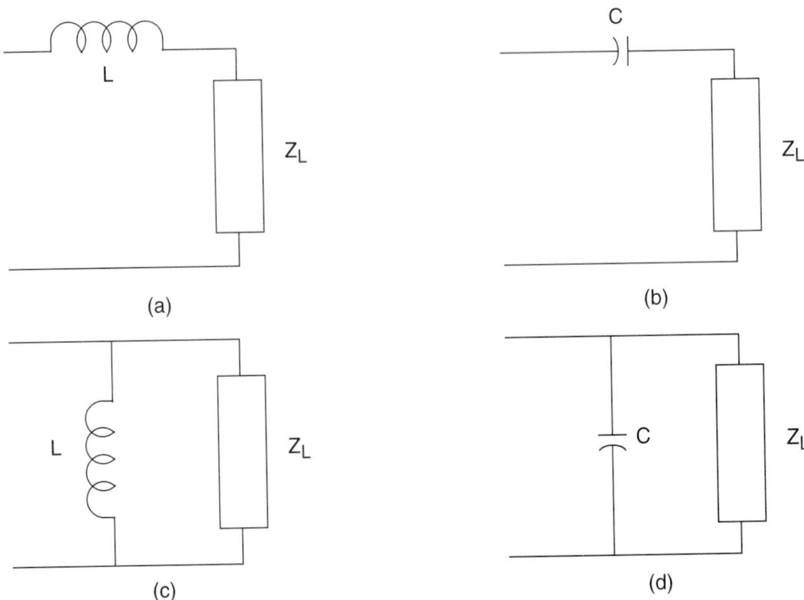

FIGURE 10.39 Four possible combinations (Application #11): (a) series L, (b) series C, (c) shunt L, and (d) shunt C.

This case has already been discussed in Chapter 6, *RF Electronics Concepts*, under the heading, "*L*-network Matching," which is now revisited and treated with the help of the Smith chart.

To find the input impedance, we first calculate the normalized series reactance ($jx = jX/Z_o$) or normalized shunt susceptance value ($jb = jB/Y_o$) of the lumped element before entering the Smith chart. Next, we locate $(Z_L)_N$ on the chart as point A (see Figure 10.40). Now starting from point A, the following steps are applied:

1. To add a series L: on a constant resistance circle, move up by $jx_S = j\omega L/Z_o$.
2. To add a series C: on a constant resistance circle, move down by $jx_S = -j/\omega CZ_o$.
3. To add a shunt L: on a constant conductance circle, move up by $jb_P = -j/\omega LY_o$.
4. To add a shunt C: on a constant conductance circle, move down by $jb_P = j\omega C/Y_o$.

These are all shown in Figure 10.40.

Rule of Thumb. On close observation of these four cases, it appears that for the majority of load values, adding a series (or shunt) inductor would move point A upward on the constant resistance (or conductance) circle while adding a series (or shunt) capacitance would move point A downward.

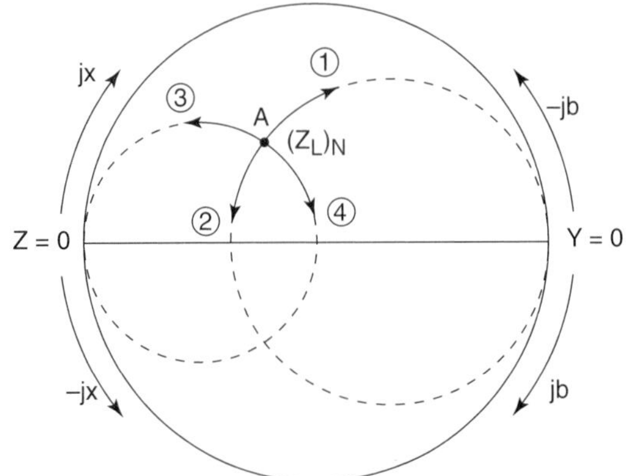

FIGURE 10.40 A graphical representation of 4 elements in the Smith chart (Application #11).

It should be noted, however, that the preceding is a good rule of thumb to follow when dealing with purely reactive elements, but it should never be generalized outside the scope of this discussion. This rule of thumb is limited but workable and will never actually replace the reasoning and the understanding that goes into making it.

EXAMPLE 10.9

Calculate the total input admittance of a combination of a load $Z_L = 50 + j50$ Ω with a shunt inductor of $L = 8$ nH at $f_o = 1$ GHz as shown in Figure 10.41. Assume a 50 Ω system.

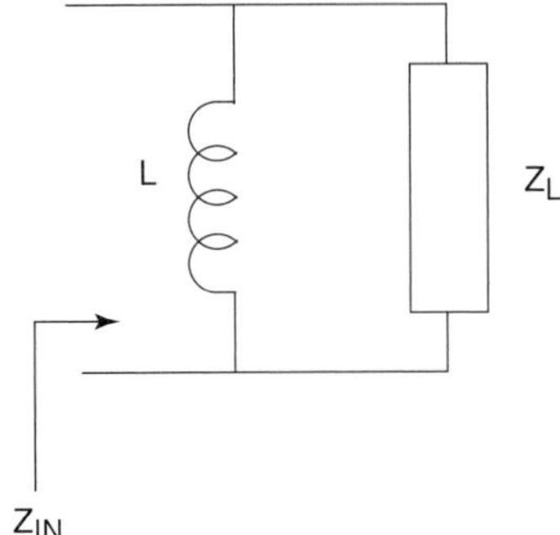

FIGURE 10.41 Circuit for Example 10.9.

Solution:

$$Z_o = 50\ \Omega \Rightarrow Y_o = 0.02\ \text{S}$$

a. We first find the susceptance of the shunt inductor:

$$jB_P = -j/(\omega_o L) = -j\,0.02\ \text{S} \Rightarrow jb_P = jB_P/Y_o = -j1$$

b. Locate $(Z_L)_N = Z_L/Z_o = 1 + j1$ on the Smith chart at point A shown in Figure 10.42.

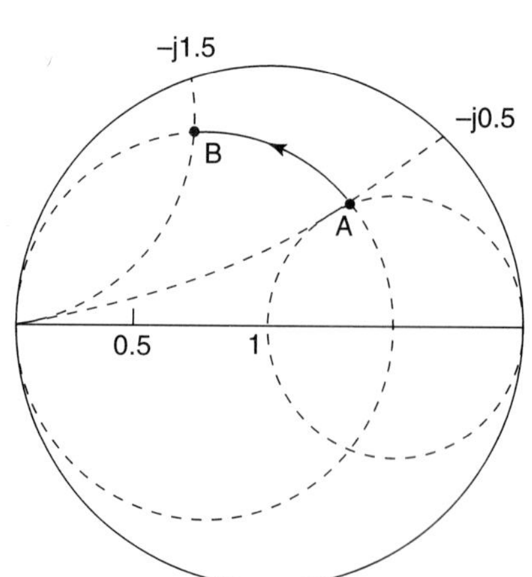

FIGURE 10.42 Smith chart solution for Example 10.9.

c. Because this is a shunt inductor, we need to move upwards from point A on a constant conductance circle by $-j1$ to arrive at point B.

d. The normalized input admittance is read off at point B as:

$$(Y_{IN})_N = 0.5 - j1.5$$

$$Y_{IN} = Y_o(Y_{IN})_N = 0.01 - j0.03 \text{ S}$$

Or, from the chart we have $(Z_{IN})_N = 0.2 + j0.6$, which gives:

$$Z_{IN} = 10 + j30 \ \Omega$$

10.3.4 Application #12

Input impedance determination for any combination of series and shunt reactive elements. In this application, we will consider the case where there are several series and shunt elements in combination with the load (as shown in Figure 10.43). Application #11 can be used repeatedly to arrive at the total input impedance as described in the following steps:

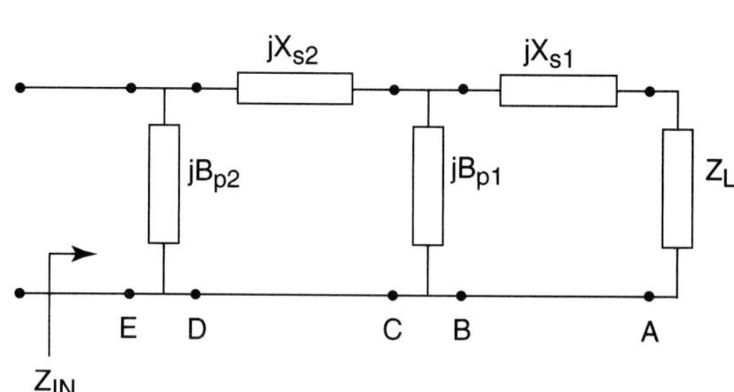

FIGURE 10.43 Combination of series and shunt inductive elements (Application #12).

1. Because the first element adjacent to the load is connected in series, we start with $(Z_L)_N$ and locate it on the Z-chart (see point A in Figure 10.44).

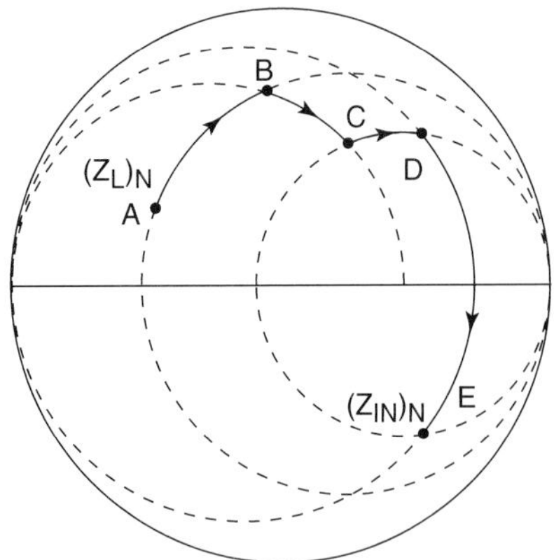

FIGURE 10.44 The Smith chart solution (Application #12).

2. On the constant resistance circle passing through $(Z_L)_N$, a reactance of $jx_{S1} = jX_{S1}/Z_o$ is added to arrive at point *B*.
3. Now switching to the *Y*-chart, we move on the constant conductance circle and add a susceptance of $jb_P = jB_{P1}/Y_o$ to arrive at point *C*.
4. Because the next element is in series, we switch back to the *Z*-chart and move on a constant resistance circle by adding a reactance of $jx_{S2} = jX_{S2}/Z_o$ to arrive at point *D*.
5. The final element is in parallel, so we switch to the *Y*-chart and add a susceptance of $jb_{P2} = jB_{p2}/Y_o$ to arrive at point *E*.
6. The total impedance is now read off on the *Z*-chart at point *E*, as shown in Figure 10.44.

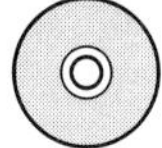

EXAMPLE 10.10

Find the input impedance at f = 100 MHz for the circuit shown in Figure 10.45.

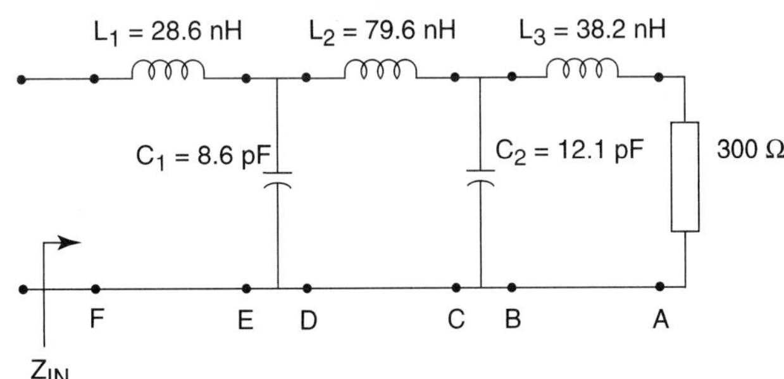

FIGURE 10.45 Circuit for Example 10.10.

Solution:

First, we choose the normalizing factor arbitrarily to be:

$$Z_o = 50\ \Omega,$$

and

$$Y_o = 1/50 = 0.02 \text{ S}.$$

Then, we normalize all impedance and admittance values:

$$jx_{S1} = (jX_1)_N = j\omega L_1/Z_o = j0.36$$

$$jb_{P1} = (jB_1)_N = j\omega C_1/Z_o = j0.27$$

$$jx_{S2} = (jX_2)_N = j\omega L_2/Z_o = j1.0$$

$$jb_{P2} = (jB_2)_N = j\omega C_2/Z_o = j0.38$$

$$jx_{S3} = (jX_3)_N = j\omega L_3/Z_o = j0.48$$

$$(Z_L)_N = 300/50 = 6$$

a. Locate $(Z_L)_N$ on the Smith chart (point *A* in Figure 10.46).

b. Because the first element (L_3) adjacent to the load is a series inductor, we move upward from point *A* on a constant resistance circle by a reactance of $j0.36$ to arrive at point *B*.

c. Now switch to constant conductance circle, and add the next shunt element by moving downward by $j0.27$ to arrive at point *C*.

d. For the next series inductor, switch to the constant resistance circle and move upward by $j1.0$ to arrive at point *D*.

e. Next, for the shunt capacitor, switch to a constant conductance circle and move downward by $j0.38$ to arrive at point *E*.

f. Finally, for the series inductor, switch to the constant resistance circle and move upward by $j0.48$ to arrive at point *F*.

g. Now we read off the value of the normalized input impedance at point *F* as:

$$(Z_{IN})_N = Z_{IN}/Z_o = 0.5 - j0.7 \Rightarrow Z_{IN} = 25 - j35\ \Omega$$

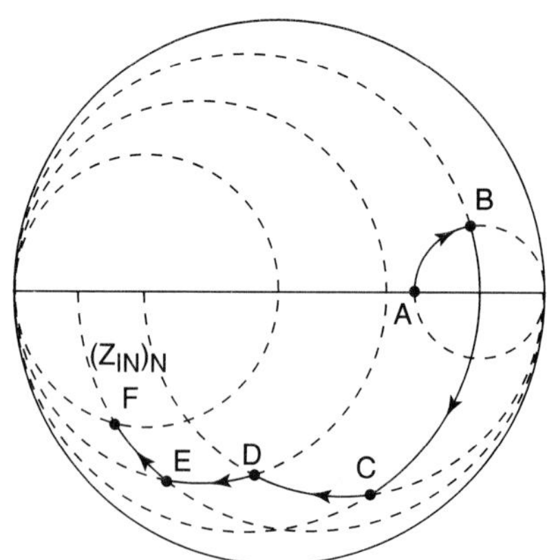

FIGURE 10.46 Smith chart solution for Example 10.10.

10.3.5 Application #13

Frequency response of "*RLC*" circuits.
The Smith chart can be used effectively to plot out the course of input impedance variation over a frequency range $(f_2 - f_1)$ as frequency is increased from f_1 to f_2. Vice versa, from the frequency response plot of a complex circuit in the Smith chart, we can develop an equivalent circuit model.

Of all possible combinations of three elements (R, L, and C), there are four non-trivial combinations that are of interest and, therefore, are discussed here.

Case 1: Series RC + Shunt L Combination. This combination is shown in Figure 10.47.

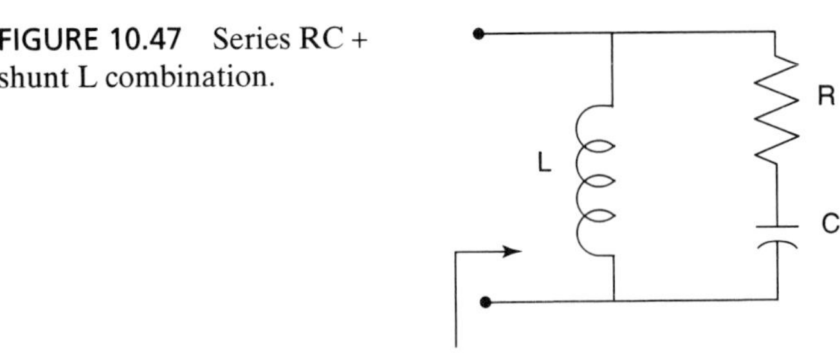

FIGURE 10.47 Series RC + shunt L combination.

In the "series RC + shunt L" combination, we see that series RC, as a capacitive load, is located on the lower half of the Smith chart at $f = f_1$ (point A on the Smith chart in Figure 10.48). As frequency is increased (from f_1), the capacitive reactance magnitude decreases and point A moves on a constant resistance circle toward the real axis to arrive at point B where $f = f_2$.

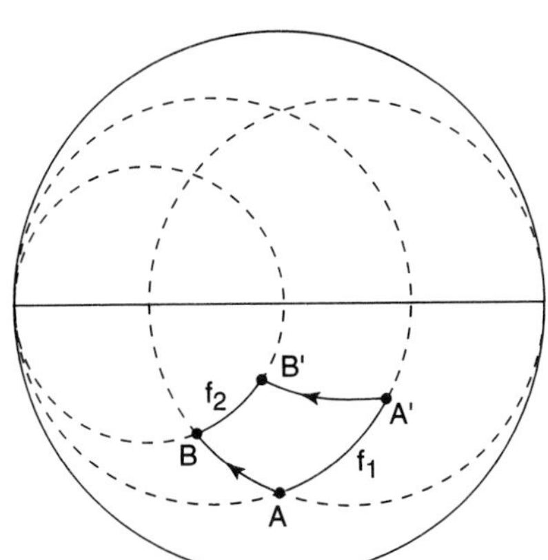

FIGURE 10.48 Smith chart solution (Case 1).

Now adding a shunt inductor (with a susceptance of $1/\omega L$) would move points A and B more toward the real axis (to points A' and B') as frequency is increased from f_1 to f_2.

These points move on the constant conductance circles that pass through points A and B to arrive at points A' and B'. Connecting points A' and B' would map the course of the frequency response for this RLC combination in the $(f_2 - f_1)$ frequency range (see Figure 10.48).

Case 2: Series RL + Shunt C Combination. This combination is shown in Figure 10.49.

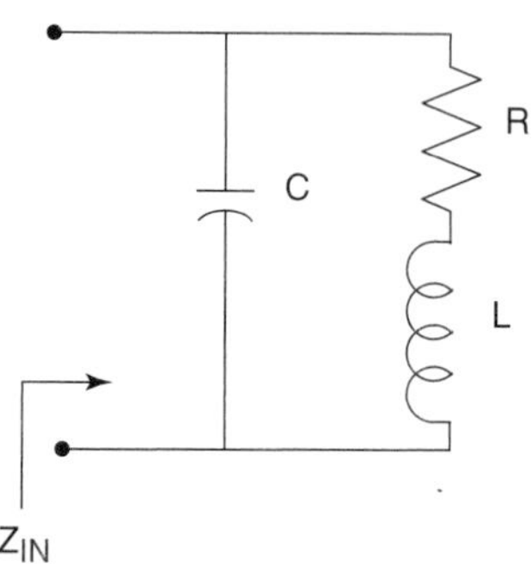

FIGURE 10.49 Series RL + shunt C combination.

In the "series RL + shunt C" combination, the "series RL," as an inductive load, is located on the upper half of the ZY chart, as shown at point A in Figure 10.50.

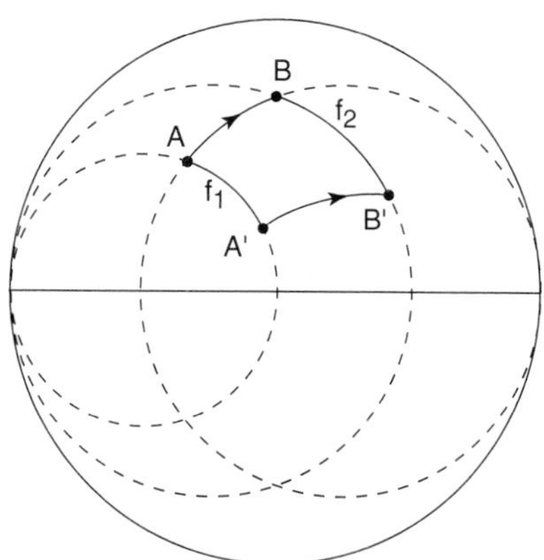

FIGURE 10.50 Smith chart solution (Case 2).

As frequency is increased from f_1 to f_2, the inductive reactance magnitude (ωL) increases and point A at $f = f_1$ moves on a constant resistance circle away from the real axis to point B at $f = f_2$. Adding a shunt capacitor (with a susceptance value of ωC) would move points A (for $f = f_1$) and B (for $f = f_2$) on constant conductance circles to points A' and B', as shown in Figure 10.50. Similar to Case 1, connecting points A' and B' would yield the input impedance (or admittance) variation in the frequency range $(f_2 - f_1)$.

Case 3: Shunt RC + Series L Combination. For the "shunt *RC* + series *L*" combination, we need to consider the *Y*-chart (see Figure 10.51). We are interested in its performance in the frequency range $(f_2 - f_1)$. Thus, we first plot the admittance of the parallel *RC* combination at $f = f_1$ at point *A*, as shown in Figure 10.52 (located on the lower half of the *Y*-chart because it is a capacitive load).

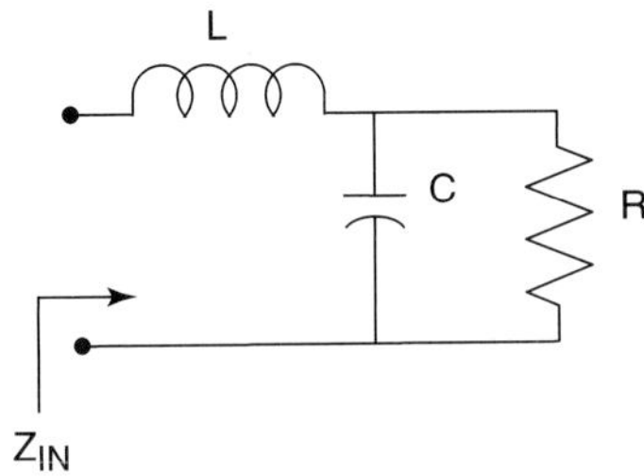

FIGURE 10.51 Shunt RC + series L combination.

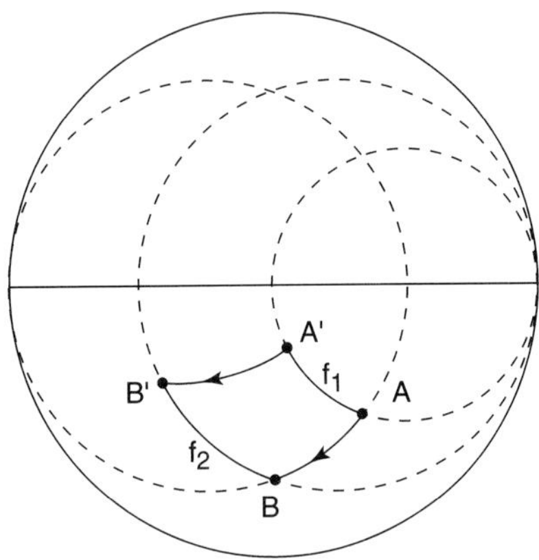

FIGURE 10.52 Smith chart solution (Case 3).

As frequency increases from f_1 to f_2, the capacitive susceptance magnitude (ωC) increases and point *A* (at $f = f_1$) moves on a constant conductance circle to point *B* (at $f = f_2$). Adding a series *L* (having a reactance ωL) will move points *A* and *B* on constant resistance circles to points *A'* and *B'*. Connecting points *A'* and *B'* provides the plot of input impedance frequency response on the Smith chart as shown in Figure 10.51.

Case 4: Shunt RL + Series C Combination. For the "shunt *RL* + series *C*" combination, we need to consider the *Y*-chart (see Figure 10.53). We are interested in its performance in the frequency range $(f_2 - f_1)$. Thus, we first plot the admittance of the parallel *RL* combination at $f = f_1$ at point *A* as shown in Figure 10.54 (located on the upper half of the *Y*-chart because it is an inductive load).

As frequency increases from f_1 toward f_2, point *A* would move on a constant conductance circle toward the real axis to point *B*. This occurs because the magnitude of the inductive susceptance ($1/\omega L$) decreases with frequency.

Now adding a series *C* (with a reactance of $-1/\omega C$) will move points *A* and *B* on constant resistance circles to points *A'* and *B'*. Connecting *A'* to *B'* provides the plot of

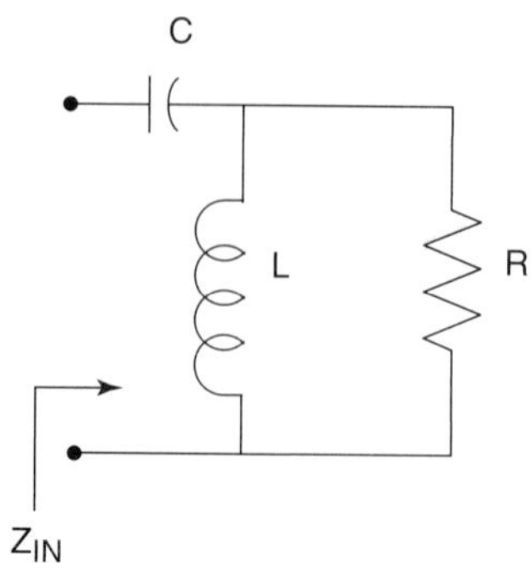

FIGURE 10.53 Shunt RL + series C combination.

the input impedance frequency response of this *RCL* combination on the Smith chart (see Figure 10.54).

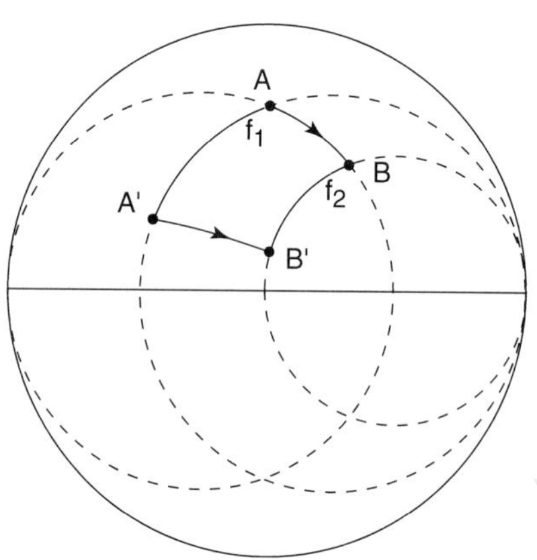

FIGURE 10.54 Smith chart solution (Case 4).

OBSERVATION: *By looking at the Smith chart plots (Cases 1 through 4), we note that the input impedance moves clockwise in the Smith chart as the frequency is increased. This observation is actually based on a much deeper concept, Foster's reactance theorem. This theorem is briefly discussed in the next section.*

10.4 FOSTER'S REACTANCE THEOREM

Foster's reactance theorem can be stated as follows:

For a passive lossless one-port network, the reactance and susceptance are strictly and monotonically increasing functions of the frequency.

This theorem, in essence, states that:

The slope of the reactance function $X(\omega)$ or the susceptance function $B(\omega)$, is always positive, i.e.,

$$\partial X(\omega)/\partial\omega > 0 \text{ for } 0 < \omega < \infty$$

$$\partial B(\omega)/\partial\omega > 0 \text{ for } 0 < \omega < \infty$$

This positive-slope condition means that the impedance or admittance frequency response of a passive lossless one-port network moves in a clockwise direction in the Smith chart as frequency is increased.

This is a general concept and applies to both lumped and distributed elements, as illustrated in the next two examples.

EXAMPLE 10.11 (Lumped Circuit)

Plot the frequency response (of the input admittance) for a parallel LC lumped circuit, as shown in Figure 10.55.

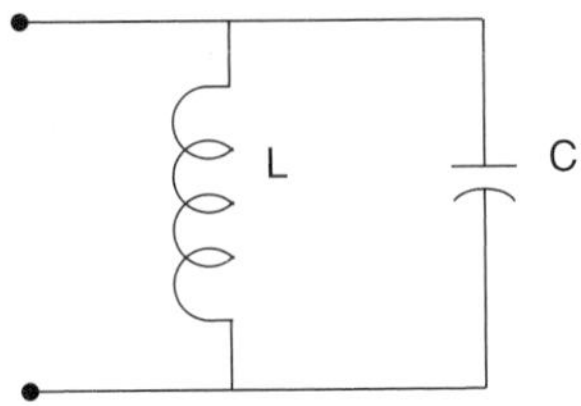

FIGURE 10.55 Circuit for Example 10.11.

Solution:

$$Y_{IN} = jB = j(\omega C - 1/\omega L)$$

$$\partial B(\omega)/\partial\omega = C + 1/\omega^2 L > 0$$

The input admittance (or susceptance) as a function of ω is sketched in Figure 10.56.

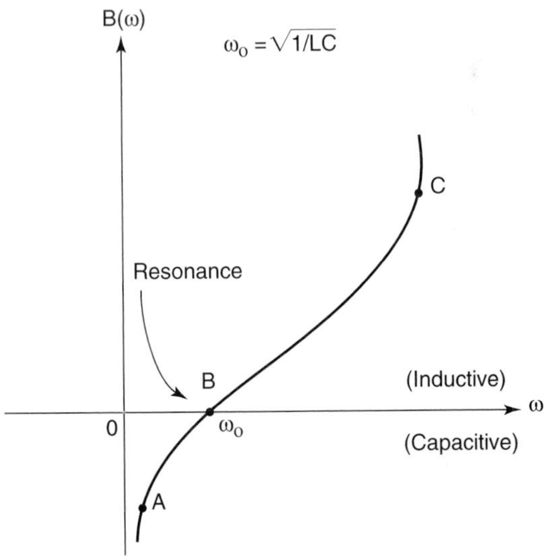

FIGURE 10.56 Frequency response of circuit in Example 10.11.

Setting $B(\omega)$ equal to zero yields the resonant frequency (ω_o) as:

$$\omega_o = 1/\sqrt{LC}$$

Furthermore, as the frequency increases, the impedance goes from point A, through point B (resonance), and then to point C. This can be plotted on a Smith chart on the outermost circle ($r = 0$), as shown in Figure 10.57.

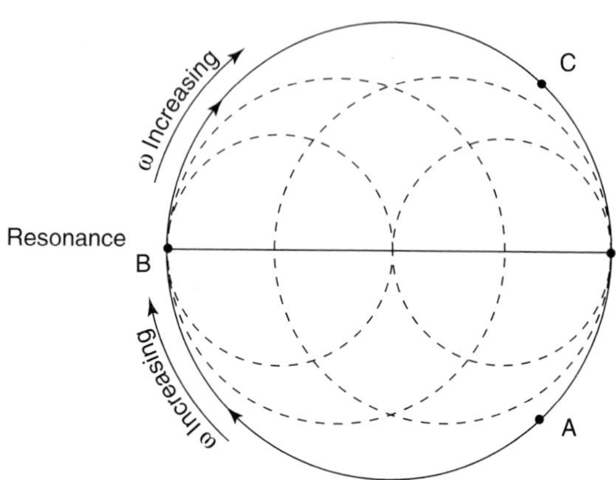

FIGURE 10.57 Smith chart plot of frequency response.

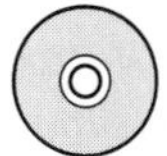

EXAMPLE 10.12 (Distributed Circuit)

Plot the frequency response of the input impedance of a shorted transmission line, as shown in Figure 10.58.

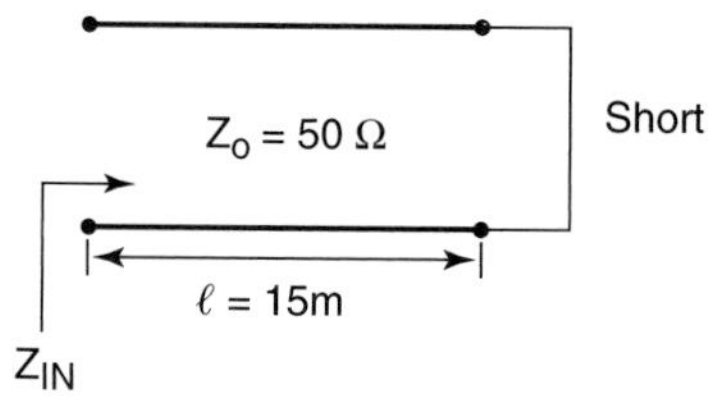

FIGURE 10.58 Circuit for Example 10.12.

Solution:

$$Z_{IN} = jZ_o \tan\beta\ell$$

where

$$\beta = 2\pi/\lambda$$

$$\lambda = c/f = 300/f\,(\text{MHz}) \quad (\text{m})$$

Therefore, we can write:

$$\beta\ell = (2\pi/\lambda) \times 15 = (\pi/10)f, \ (f \text{ in MHz}) \tag{10.50}$$

In order to get resonance at a frequency $f = f_o$, Z_{IN} must approach infinity (∞). But because Z_{IN} is proportional to $\tan\beta\ell$, we can write:

$$\tan\beta\ell = \infty \Rightarrow \beta\ell = (2n+1)\pi/2, \quad n = 0,1,2,3,\ldots \tag{10.51}$$

Substituting for $\beta\ell$ from Equation 10.51 in Equation 10.50, we have:

$$(\pi/10)f_o = (2n+1)\pi/2 \Rightarrow f_o = 5(2n+1)\ \text{MHz}, \ n = 0,1,2,3,\ldots$$

Thus, there are an infinite number of frequencies at which the circuit resonates. The resonant frequencies occur at: $f = 5,15,25,\ldots$(MHz).

These resonant frequencies are shown in Figure 10.59 in the frequency domain and on the Smith chart, as shown in Figure 10.60.

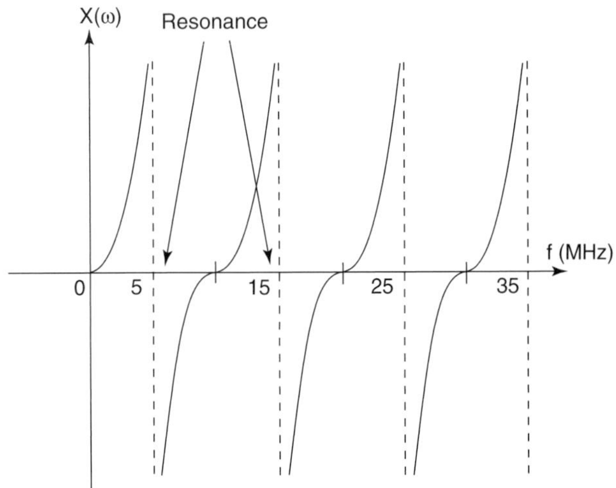

FIGURE 10.59 Frequency plot of Example 10.12.

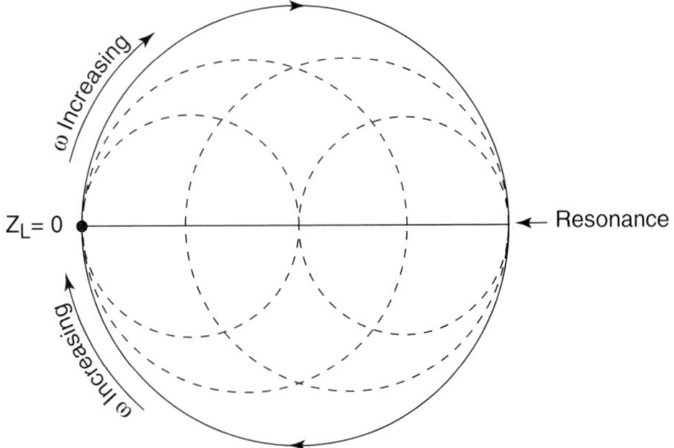

FIGURE 10.60 Frequency response plot on Smith chart.

NOTE: *This example shows that there are an infinite number of resonances that can occur for a "distributed circuit," which is in contrast with "lumped circuits" that have a finite number of resonances.*

10.4.1 Application #14

Input impedance (or admittance) determination for a combination of distributed and lossless lumped elements.

This final application deals with circuits having distributed elements (such as transmission lines) and lossless lumped elements (such as capacitors and inductors).

To obtain the input impedance (or admittance) we use the following two rules:

1. When dealing with distributed elements, for ease and convenience we start from the load end. Then we travel on a constant *VSWR* circle a length (ℓ) toward the generator.
2. When dealing with lossless lumped elements, we also start from the load end but move on a constant resistance (or conductance) circle depending on whether the lumped element is in series (or shunt) with the rest of the circuit.

The overall procedure is the same as delineated in the previous applications. The example below will illustrate this concept further.

EXAMPLE 10.13

In the circuit shown below (see Figure 10.61), determine the input impedance at f = 10 GHz.

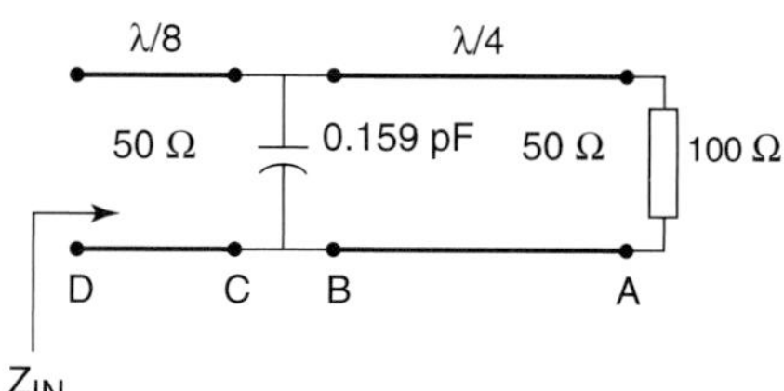

FIGURE 10.61 Circuit for Example 10.13.

Solution:

To find Z_{IN}, we perform the following steps:

a. Locate $(Z_L)_N = 100/50 = 2$ on the Smith chart (see Figure 10.62).

b. Because the first element adjacent to the load is a series transmission line, we draw the constant *VSWR* circle.

c. Starting from $(Z_L)_N$, at point *A*, we move on this circle a length of λ/4 "toward generator" to arrive at point *B*.

d. Now because the next element is a shunt capacitor, we switch to the *Y*-chart and move on the constant conductance circle to arrive at point *C*. The shunt capacitor has a susceptance of:

$$jB = j\omega C = j2\pi \times 10^{10} \times 0.159 \times 10^{-12} = j\,0.01\text{ S}$$

e. The next element is a series transmission line, so we switch back to the *Z*-chart and draw the constant *VSWR* circle that passes through *C*.

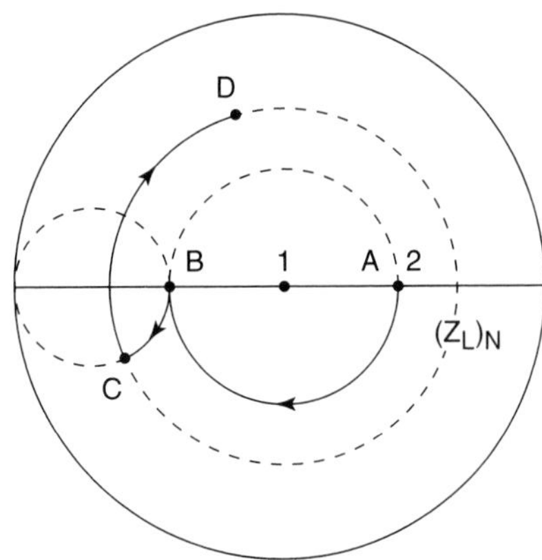

FIGURE 10.62 Smith chart solution for Example 10.13.

f. Now from point C we move a distance of $\lambda/8$ "toward generator" to arrive at point D, as shown in Figure 10.62.

g. The value of the input impedance is read off at point D as:

$$(Z_{IN})_N = 0.4 + j0.55 \Rightarrow Z_{IN} = 20 + j27.5\ \Omega$$

LIST OF SYMBOLS/ABBREVIATIONS

A symbol will not be repeated again, once it has been identified and defined in an earlier chapter, as long as its definition remains unchanged.

b	Normalized susceptance, $b = B/Z_o$
b_L	Normalized susceptance at the load
b_P	Normalized susceptance for a parallel element
g	Normalized conductance, $g = G/Z_o$
g_L	Load normalized conductance
g_P	Shunt normalized conductance
I_{max}	Maximum current on a transmission line
I_{min}	Minimum current on a transmission line
ℓ_{max}	location of Z_{max} on the transmission line
ℓ_{min}	location of Z_{min} on the transmission line
r	Normalized resistance, $r = R/Z_o$
x	Normalized reactance, $x = X/Z_o$
V_{max}	Maximum voltage on a transmission line
V_{min}	Minimum voltage on a transmission line
Y_o	Characteristic admittance
Z_o	Characteristic impedance
Z_D	Device impedance
Z_{IN}	Input impedance
$(Z_{IN})_N$	Normalized input impedance

Z_{max} Maximum impedance, corresponding to the location of the peak of the voltage and the valley of the current in a standing wave pattern on a transmission line

Z_{min} Minimum impedance, corresponding to the location of the valley of voltage and the peak of the current in a standing wave pattern on a transmission line

Γ_D Device reflection coefficient

PROBLEMS

10.1 The normalized impedance of an unknown device is measured (from 1 to 2 GHz) to have a frequency response as plotted in Figure P10.1. Determine an equivalent circuit for the unknown device with all element values correctly calculated.

FIGURE P10.1

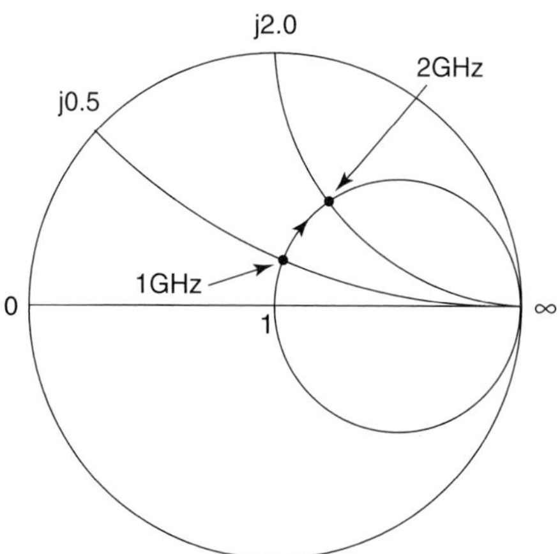

10.2 Using a Smith chart, determine Z_{IN} for the circuit shown in Figure P10.2.

FIGURE P10.2

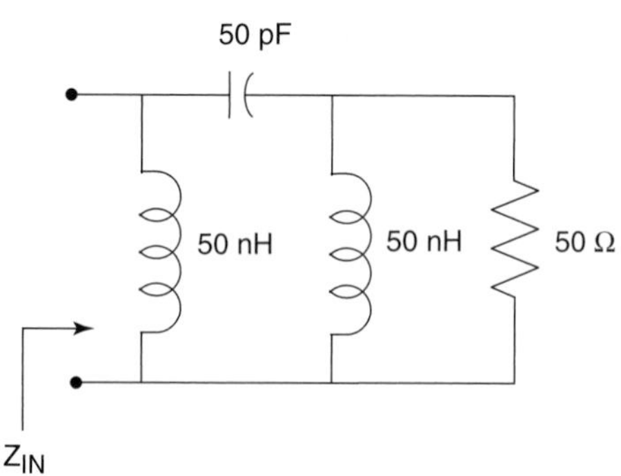

10.3 A lossless coaxial line (Z_o = 50 Ω) is terminated in a 100 Ω load. If the incident voltage wave has an rms magnitude of 10 V, determine:

a. The reflection coefficient and the *VSWR* on the line

b. The magnitude and location of V_{MAX}, V_{MIN}, Z_{MAX}, and Z_{MIN} on the line

c. The magnitude and location of I_{MAX} and I_{MIN} on the line

d. The power absorbed by the load

e. The voltage standing wave pattern on the line for both voltage and current

10.4 A lossless transmission line (Z_o = 50 Ω) is terminated in a load (Z_L = 100 + j100 Ω). A single shorted stub (ℓ = λ/8) is inserted λ/4 away from the load, as shown in Figure P10.4. Using a Smith chart, determine the line's input impedance (Z_{IN}).

FIGURE P10.4

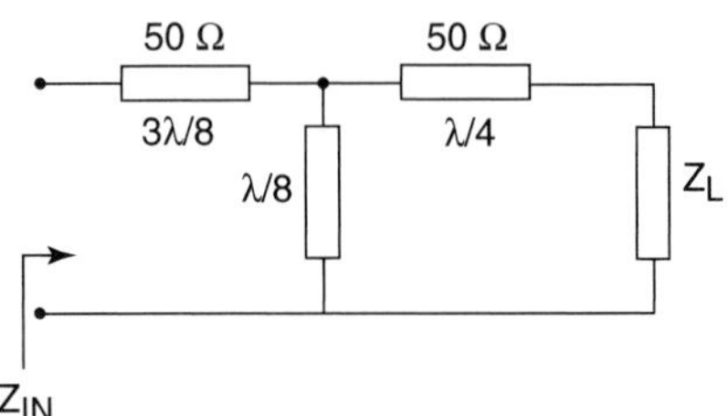

10.5 A lossless transmission line (Z_o=75 Ω) is terminated in an unknown load. Determine the load if the *VSWR* on the line is found to be 2 and the adjacent voltage maxima are at x = –15 cm and –35 cm where the load is located at x = 0 (see Figure P10.5).

FIGURE P10.5

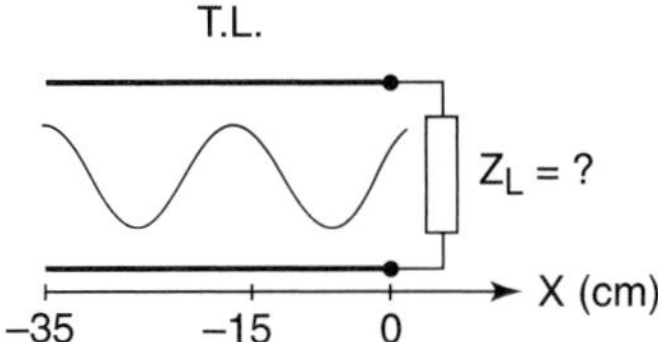

10.6 Determine the input impedance of a transmission line at a distance of 2 cm from the load impedance if the wavelength on the transmission line is found to be 16 cm. What is the *VSWR* on the line? Assume Z_L = 40 + j50 Ω and Z_o= 50 Ω.

10.7 Determine the input impedance of the lumped element network shown in Figure P10.7 (all values are in Ω).

10.8 Find the input impedance of the transmission line circuit in a 50 Ω system (as shown in Figure P10.8) for Z_L = 25 + j25 Ω, d_1 = 3λ/8, d_2 = λ/4 and ℓ= λ/8. What is the *VSWR* at the input terminals?

FIGURE P10.7

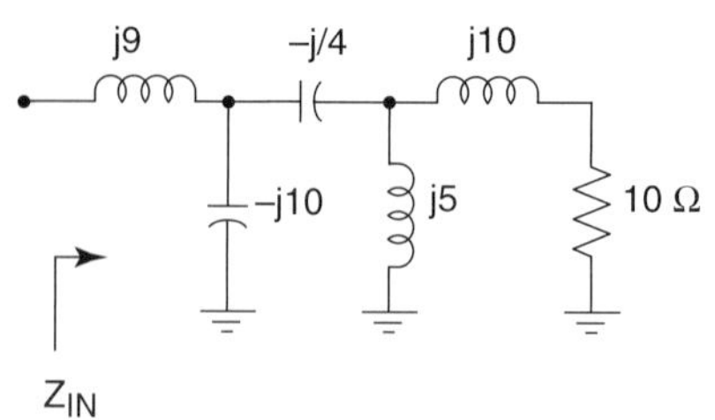

FIGURE P10.8

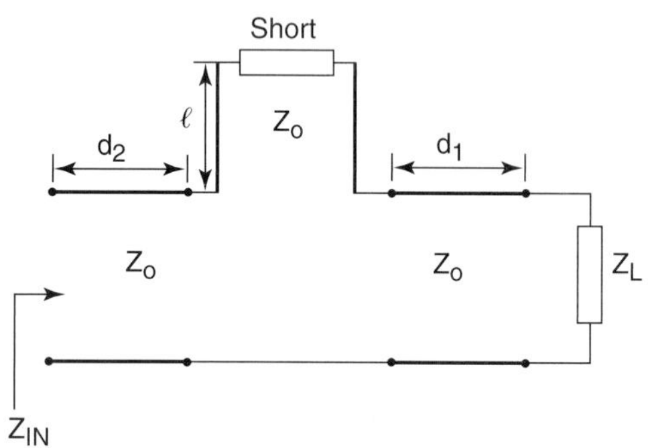

10.9 Find the input impedance of a double-stub shunt tuner as shown in Figure P10.9. Assume that the stubs are short circuited and $\ell_1 = 0.23\lambda$, $\ell_2 = 0.1\lambda$, $d = \lambda/8$ and $Z_o = 50\ \Omega$. What is the reflection coefficient at the input terminals?

FIGURE P10.9

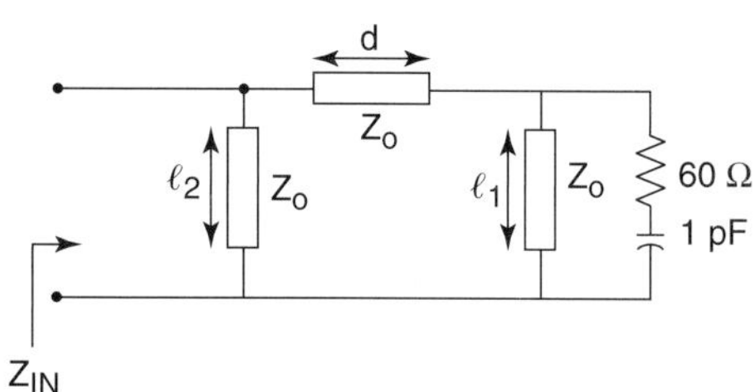

REFERENCES

[10.1] Anderson, E. M. *Electric Transmission Line Fundamentals.* Reston: Prentice Hall, 1985.

[10.2] Cheng, D. K. *Field and Wave Electromagnetics*, 2nd ed., Reading: Addison Wesley, 1989.

[10.3] Cheung, W. S. and F. H. Levien. *Microwave Made Simple, Principles and Applications.* Norwood: Artech House, 1985.

[10.4] Ginzton, E. L. *Microwave Measurements*, New York: McGraw-Hill, 1957.

[10.5] Gonzalez, G. *Microwave Transistor Amplifiers, Analysis and Design*, 2nd ed. Upper Saddle River: Prentice Hall, 1997.

[10.6] Kosow, I. W. and Hewlett-Packard Engineering Staff. *Microwave Theory and Measurement*. Englewood Cliffs: Prentice Hall, 1962.

[10.7] Liao, S. Y. *Microwave Circuit Analysis and Amplifier Design*. Upper Saddle River: Prentice Hall, 1987.

[10.8] Pozar, D. M. *Microwave Engineering*, 2nd ed. New York: John Wiley & Sons, 1998.

[10.9] Reich, H. J., F. O. Phillip, H. L. Krauss, and J. G. Skalnik. *Microwave Theory and Techniques,* New York: D. Van Norstrand Company, Inc., 1953.

[10.10] Schwarz, S. E. *Electromagnetics for Engineers.* New York: Saunders College Publishing, 1990.

[10.11] Smith, P. H. *Transmission-Line Calculator*, Electronics, 12, pp. 29-31, Jan. 1939.

[10.12] Smith, P. H. *An Improved Transmission-Line Calculator*, Electronics, 17, pp. 130, Jan. 1944.

CHAPTER 11

Design of Matching Networks

11.1 INTRODUCTION

Having studied the Smith chart in full detail and seen the ease and simplicity that it brings to the analysis of distributed or lumped element circuits, we now turn to the design of matching networks.

Applications #1 through #14 in Chapter 10, *Applications of the Smith Chart*, have in reality set the stage for most of the possible ways a Smith chart could be used as an essential tool in RF/microwave circuit analysis and, more importantly, in network design.

Many of these applications will be cited as reference throughout the rest of this chapter to simplify and speed up the process of the design of a matching network, which is an essential part of any modern active circuit.

11.2 DEFINITION OF IMPEDANCE MATCHING

At the outset of this section, we will define an important term:

DEFINITION-MATCHING: *Connecting two circuits (source and load) together via a coupling device or network in such a way that the maximum transfer of energy occurs between the two circuits.*

This is one of the most important design concepts in amplifier and oscillator design as shown in Figure 11.1.

The concept of impedance matching (also referred to as "tuning") is the third step in the overall design process (see Chapter 5, *Introduction to Radio Frequency and Microwave Concepts and Applications*, Figure 5.1). Impedance matching is a very important concept at RF/microwave frequencies because it allows the following:

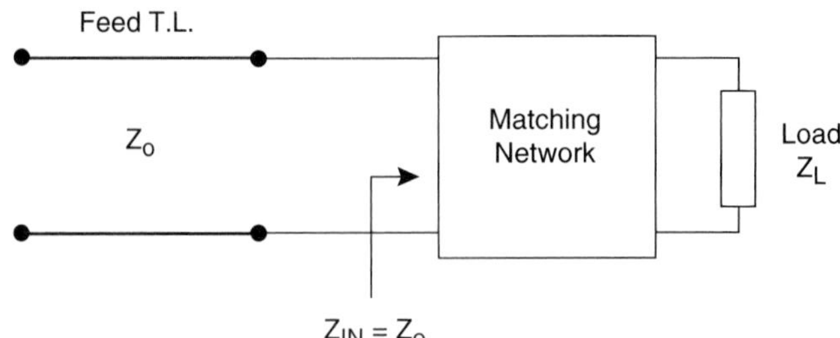

FIGURE 11.1 The concept of matching.

- Maximum power transfer to occur from source to load
- Signal-to-noise ratio to be improved due to an increase in the signal level

These are the primary reasons to employ tuning in practically all RF/microwave active circuit design. To get a conceptual understanding of why a matching network is needed in a circuit in general, we can visualize an active circuit in which a load impedance is different from the transmission line characteristic impedance causing power reflections back to the source. To alleviate this problem and bring about zero power reflection from the load (i.e., maximum power transfer), a matching network needs to be inserted between the transmission line and the load.

Ideally, the matching network is lossless to prevent further loss of power to the load. It acts as an intermediary circuit between the two nonidentical impedances in such a way that the feeding transmission line sees a perfect match (eliminating all possible reflections) while the multiple reflections existing between the load and the matching network will be unseen by the source.

11.3 SELECTION OF A MATCHING NETWORK

Selection of a lossless matching network is always possible as long as the load impedance is not purely imaginary and has, in fact, a non-zero real part.

There are many considerations in selecting a matching network including the following:

Simplicity. The simplest design is usually highly preferable because simpler matching networks have fewer elements, require less work to manufacture, are cheaper, are less lossy, and are more reliable compared to a more complicated and involved design.

Bandwidth. Any matching network can provide zero reflection at a single frequency; however, to achieve impedance matching over a frequency band more complex designs need to be used. Thus, there is a trade-off between design simplicity and matching bandwidth and eventually the network price, as shown in Figure 11.2.

Feasibility of manufacturing. To manufacture a certain design, we need to consider first the type of transmission line technology with which the matching net-

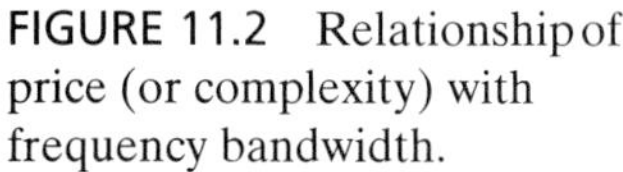

FIGURE 11.2 Relationship of price (or complexity) with frequency bandwidth.

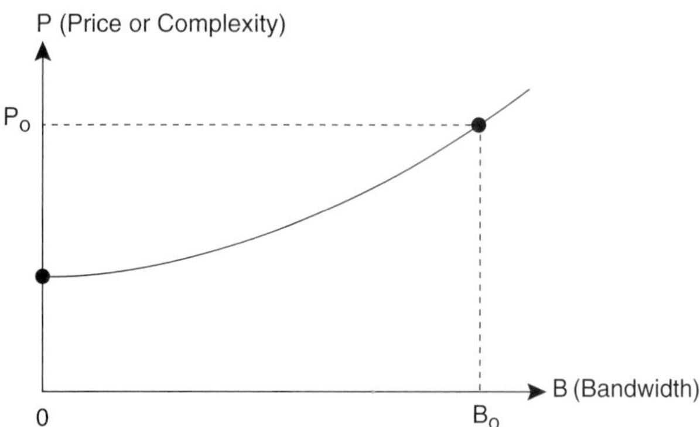

work will be implemented. This means that before the matching circuit is designed we need to know beforehand whether it is a microstrip, coaxial line or waveguide technology, so that the matching circuit will be designed properly to integrate most efficiently with the rest of the network. For example, in microstrip line technology, due to its planar configuration, the use of quarter-wave transformers, stubs, and chip lumped elements for matching is feasible. On the other hand, in waveguide technology, implementing tuning stubs for matching purposes is more predominant than lumped elements or $\lambda/4$ multisection transformers.

Ease of tunability. Variable loads require variable tuning. Thus, the matching network design and implementation should account for this. Implementing such an adjustable matching network may require a more complex design or even switching to a different type of transmission line technology to accommodate such a requirement.

These four considerations form the backbone of all design criteria. For the sake of clarity and ease, we will focus on only the first consideration—simplicity—for the rest of this work and leave the other three considerations to more advanced texts.

NOTE: *There are cases where the matching network has two or more solutions for the same load impedance. The preference of one design over the other would greatly depend on bringing the other three considerations into view, which will place the ensuing discussions outside the scope of this work.*

11.4 THE GOAL OF IMPEDANCE MATCHING

The most important design tool in amplifier and oscillator design is the concept of impedance matching. **The goal of impedance matching** in all of its different forms can be summed up into one issue:

Making the input impedance of the load in combination with the added matching network theoretically equal to the characteristic impedance of the feeding transmission line, thus allowing the maximum amount of power to transfer to the load.

The next section will delineate conditions under which maximum power transfer does take place.

11.4.1 Maximum Average Power Transfer

Consider the circuit (shown in Figure 11.3), which is a problem of great practical importance. In this circuit, source impedance (Z_S) is a known and fixed value and V_S is the phasor representation of the source sinusoidal voltage at angular frequency (ω):

$$V_S = Re(|V_S|e^{j\omega t}) \tag{11.1}$$

The problem is to select the load impedance (Z_L) such that the maximum average power (P_{av}), at steady state, is obtained from the source and fed to the load. This problem can be easily solved with the help of the maximum power transfer theorem.

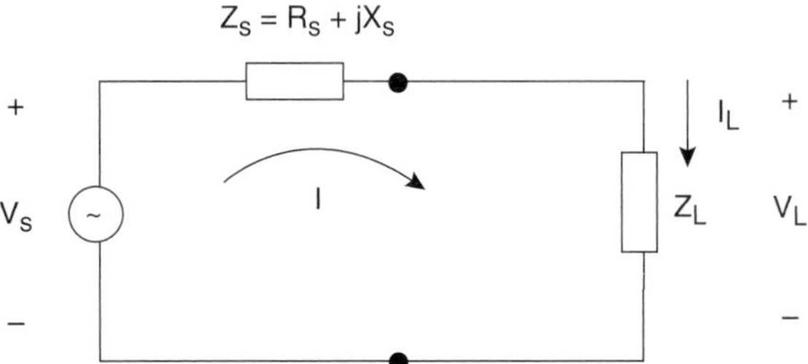

FIGURE 11.3 A general circuit.

Maximum Power Transfer Theorem. *Consider the general circuit having a known source impedance and an unknown load, as shown in Figure 11.3. The maximum power transfer theorem states that the maximum power that can be delivered to a load is feasible only when the load has an optimum impedance value $(Z_L)_{opt}$ equal to the complex conjugate of the source impedance value (Z_S); that is,*

$$(Z_L)_{opt} = Z_S^* \tag{11.2}$$

From Equation 11.2, we can see that:

$$R_L = R_S$$

$$X_L = -X_S$$

Considering the maximum power transfer as the cornerstone of matching, we can make the following conclusive observation about the goal of matching:

The goal of matching is adding a matching network to a load (Z_L) in such a way that the input impedance of the total combination will be located at the center of the Smith chart (assuming a real source impedance).

The proof for maximum power transfer theorem is presented below.

PROOF:

The average power delivered to the load can be written as:

$$V_L = Z_L I_L$$

$$I_L = \frac{V_S}{Z_S + Z_L}$$

$$P_{av} = \frac{1}{2}Re(V_L I_L^*) = \frac{1}{2}Re(Z_L|I_L|^2) = |V_S|^2 \frac{Re(Z_L)}{2|Z_S + Z_L|^2} \qquad (11.3)$$

Where V_S, $Z_S = R_S + jX_S$ and $Z_L = R_L + jX_L$ are the source voltage phasor, source impedance, and load impedance, respectively. Substitution in Equation 11.3 gives:

$$P_{av} = |V_S|^2 \frac{R_L}{2[(R_S + R_L)^2 + (X_S + X_L)^2]} \qquad (11.4)$$

In Equation 11.4, V_S, R_S, and X_S are given; R_L and X_L are to be chosen such that their value will maximize P_{av}. The reactance X_L is found by differentiating P_{av} with respect to X_L and setting it to zero, which yields:

$$\frac{\partial P_{av}}{\partial X_L} = 0$$

$$X_L = -X_S \qquad (11.5)$$

With this choice, the term $(X_L + X_S)^2$ in the denominator of P_{av} becomes zero, which minimizes the denominator and maximizes the expression with respect to X_L. Thus, Equation 11.4 can now be written as:

$$P_{av} = |V_S|^2 \frac{R_L}{2(R_S + R_L)^2} \qquad (11.6)$$

Now to determine optimum R_L, we set the partial derivative of P_{av} with respect to R_L equal to zero, i.e.,

$$\frac{\partial P_{av}}{\partial R_L} = 0$$

Upon differentiation and setting it equal to zero, we obtain:

$$R_L = R_S \quad \text{Q.E.D.} \qquad (11.7)$$

Using $Z_L = Z_S^*$ (referred to as a conjugately matched load), from Equation 11.4 we obtain the maximum power delivered to the load $(P_{av})_{max}$ as:

$$(P_{av})_{max} = \frac{|V_S|^2}{8R_S} \qquad (11.8)$$

Furthermore, under these conditions the power produced by the source is given by:

$$P_S = \frac{1}{2}Re(V_S I_L^*) = \frac{1}{2}Re\left(\frac{|V_S|^2}{Z_S + Z_L}\right) = \frac{|V_S|^2}{4R_S} \qquad (11.9)$$

Thus, we have:

$$P_S = 2(P_{av})_{max} \qquad (11.10)$$

From Equation 11.10, we can observe that the efficiency of a conjugately matched load is 50 percent:

$$\frac{(P_{av})_{max}}{P_S} = 0.5 = 50\% \text{ for } Z_L = Z_S^* \qquad (11.11)$$

NOTE 1: *For RF and microwave engineers this fact (i.e., 50 percent efficiency) is of much significance because the energy in the incoming electromagnetic waves would have been lost if it were not absorbed by a conjugately matched load (which is the first or "front-end stage" of a receiver).*

NOTE 2: *For power engineers and electric power companies, this situation is never allowed to occur; in fact, the reverse is desired. This is because power engineers and power companies are extremely interested in efficiency and want to deliver as much of the average power as possible to the load (the customer). Thus, huge power generators are never conjugately matched.*

NOTE 3: *The maximum power transfer theorem assumes that the source impedance (Z_S) is a fixed and known quantity while the load impedance is a variable and unknown quantity (Z_L), whose value is varied to be the complex conjugate of Z_S, to achieve a maximum power transfer. Under this condition, due to the complex conjugate condition, the total resistance in the circuit is given by:*

$$Z_{tot} = Z_S + Z_L = 2R_S = 2R_L$$

NOTE 4: *If the reverse is true, namely the load impedance (Z_L) is a known and fixed quantity and the source impedance (Z_S) a variable quantity, the requirement that source and load impedances be complex conjugates of each other is no longer valid and does not apply in this case. Furthermore, to obtain maximum power transfer from the source to the load for this case, it can easily be observed that we need to have minimum loss in the source. Thus, we can write the following for the source:*

$$R_S = 0$$

$$X_S = -X_L$$

Under this condition, because the reactances cancel out, the total resistance in the circuit is given by:

$$Z_{tot} = Z_S + Z_L = R_L$$

11.5 DESIGN OF MATCHING CIRCUITS USING LUMPED ELEMENTS

Considering the size of most modern RF/microwave circuits, actual *discrete* lumped element resistors, capacitors and inductors are used in the design process at low RF/microwave frequencies (up to around 1–2 GHz) or at higher frequencies (up to 60 GHz) if the circuit size is much smaller than the wavelength ($\ell < \lambda/10$).

Although microwave integrated circuit (MIC) technology has pushed the frequency limitation of lumped elements into the high microwave range, there are a large number of circuits whose size has become comparable with the signal wavelength at higher frequency ranges where using lumped elements would become completely impractical. Thus, one of the biggest limitations of the use of lumped elements is in circuits whose size has become comparable with the signal wavelength.

Furthermore, if the length (ℓ) of the lumped component is below ($\lambda/10$) as mentioned previously, then lumped components can be used in hybrid or "monolithic" MICs at frequencies up to 60 GHz. At these high frequencies, electrical elements can be realized via several methods. These methods for each component can be summarized as follows:

Capacitors

- A single-gap capacitor ($C < 0.5$ pF)
- An interdigital gap capacitor in a microstrip line ($C < 0.5$ pF)
- A short or open transmission line stub ($C < 0.1$ pF)
- A chip capacitor
- A metal-insulator-metal (MIM) capacitor ($C < 25$ pF)

These are shown in Figure 11.4.

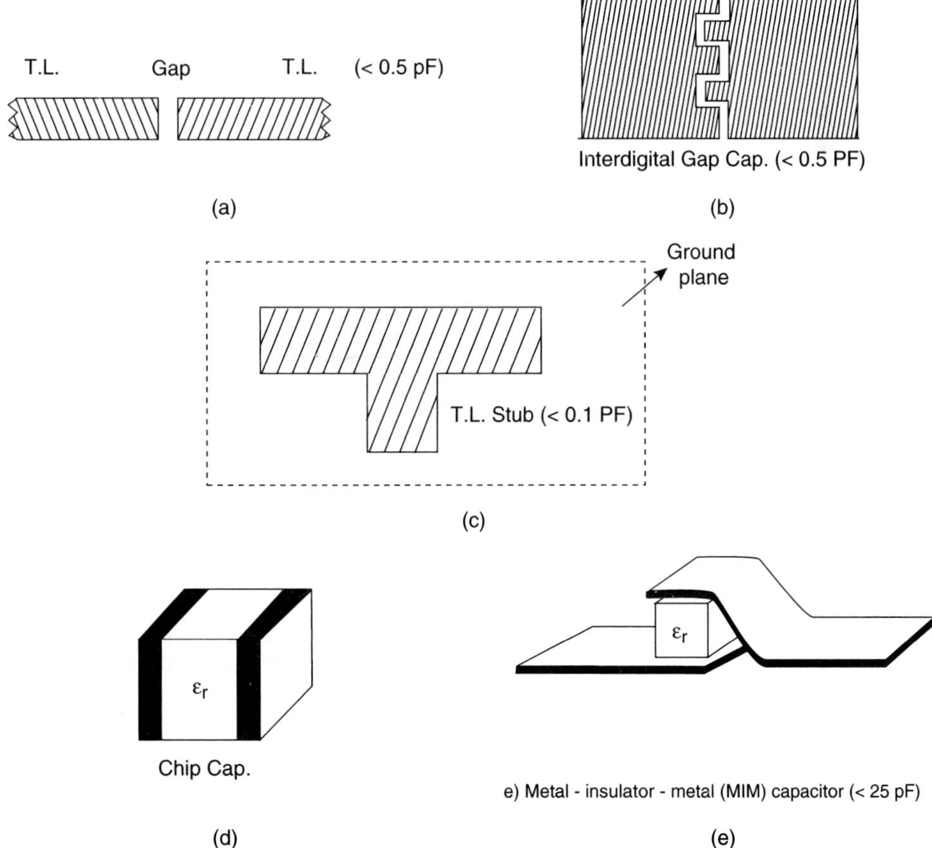

FIGURE 11.4 Several types of capacitors: (a) single-gap capacitor, (b) interdigital capacitor, (c) transmission line stub capacitor, (d) chip capacitor, (e) metal-insulator-metal (MIM) capacitor.

Resistors

- Thin film technology using NiChrome or doped semiconductor material
- Chip resistor

These are shown in Figures 11.5a and b.

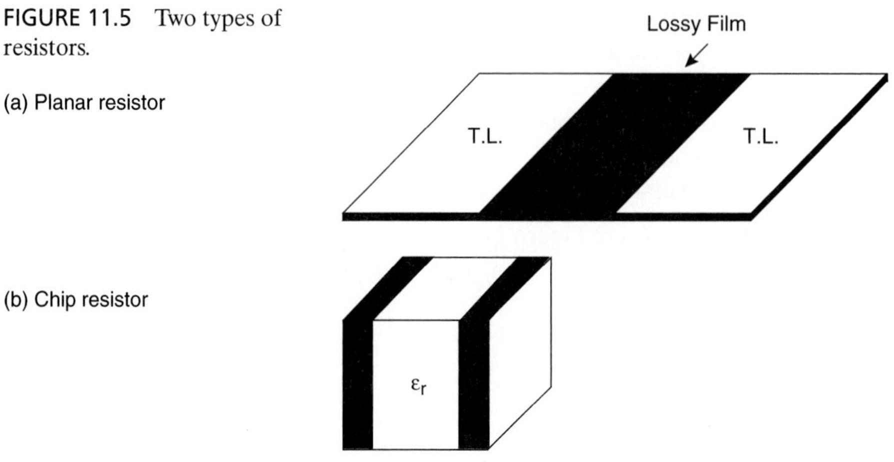

FIGURE 11.5 Two types of resistors.

(a) Planar resistor

(b) Chip resistor

Inductors

- A loop of a transmission line
- A short length of a transmission line
- A spiral inductor using an air bridge

These are shown in Figures 11.6a, b, and c.

11.5.1 Matching Network Design Using L-Sections

As already discussed in Chapter 6, *RF Electronics Concepts*, the simplest type of matching network is the L-section, consisting of two reactive elements that match a load to a transmission line. The actual configuration is of the form of an inverted L. The two possibilities are shown in Figure 11.7.

Considering the fact that either of the two reactive elements can be an inductor or a capacitor, circuit configurations L1 and L2 provide a total of eight different possibilities for a given load. The location of the load on the Smith chart determines the useful configuration, as discussed in the next section.

11.5.2 Design Based on the Load Location

Depending on the location of the normalized load impedance on the Smith chart, one or both of the two configurations L1 and L2 may become practical.

The location of the load becomes crucial in the choice of circuit configuration for the purpose of matching. The load location can have three distinct possibilities.

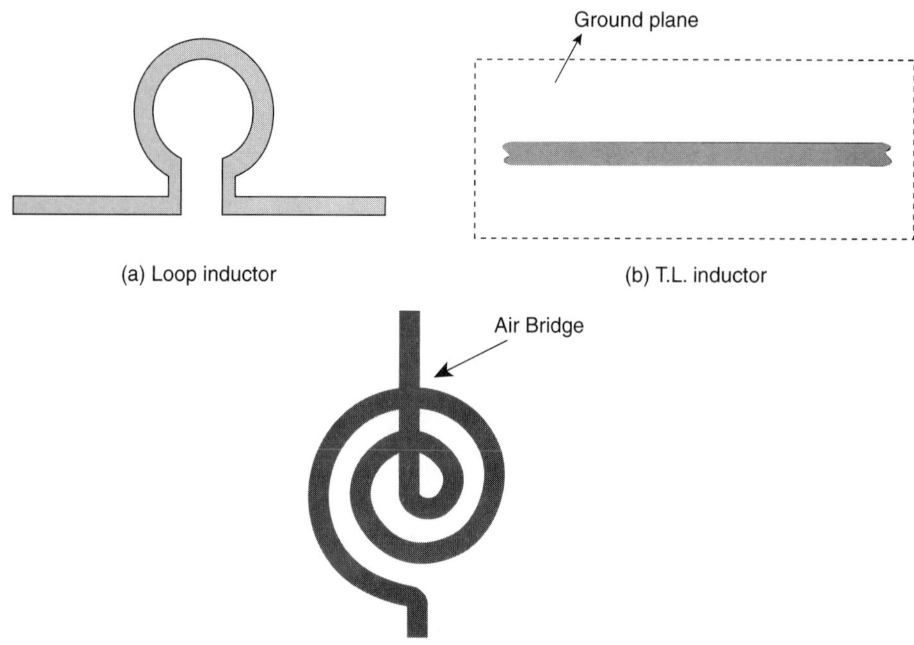

FIGURE 11.6 Three types of inductors.

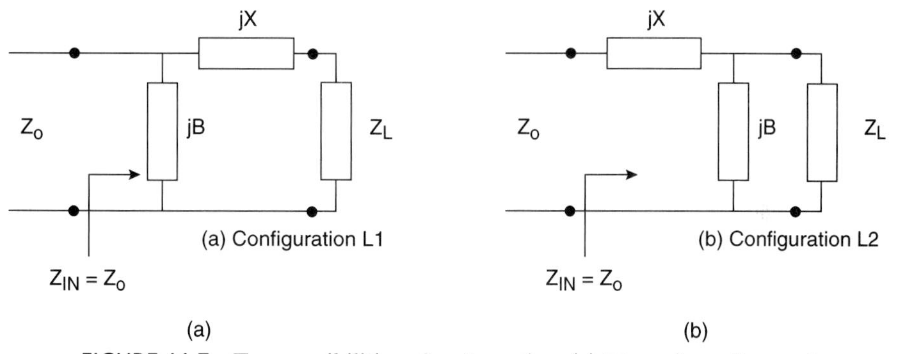

FIGURE 11.7 Two possibilities of an L-section: (a) L1 configuration and (b) L2 configuration.

Case I: The load is located inside the $(1 + jx)$ circle (resistance unity circle). In this case, we can see from Figure 11.8 that the first element has to be a shunt element; thus, configuration L1 is the only practical one. Using configuration L1, two possible solutions exist:

- Solution 1: shunt L and series C
- Solution 2: shunt C and series L

These are shown in Figure 11.9.

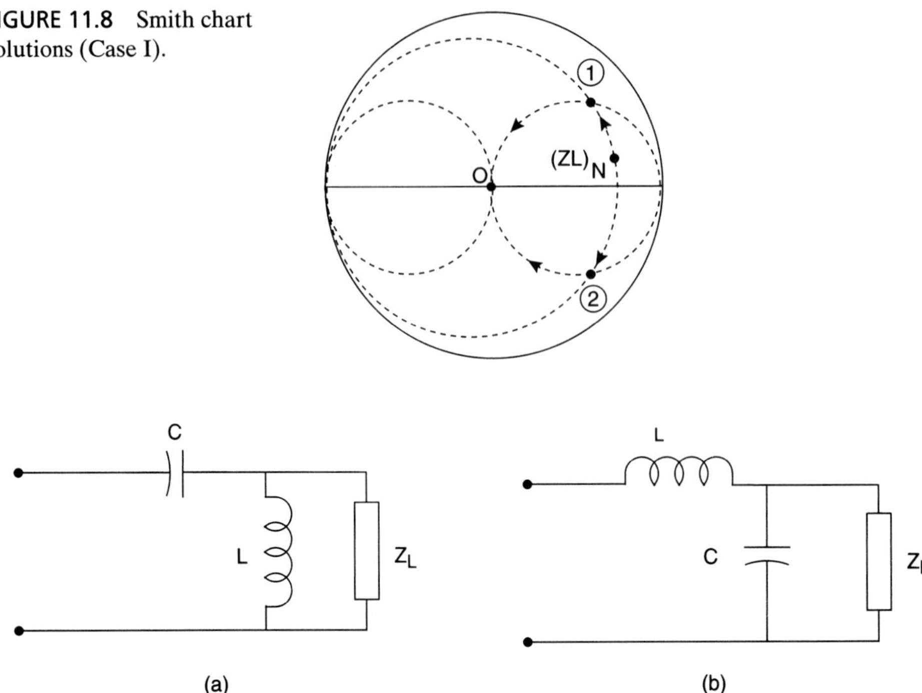

FIGURE 11.8 Smith chart solutions (Case I).

FIGURE 11.9 Two solutions for Case I: (a) Solution 1 and (b) Solution 2.

Case II: The load is located inside the (1 + *jb*) circle (conductance unity circle). In this case, from Figure 11.10 we can see that the first element has to be a series element; thus, only configuration L1 becomes useful. Similar to Case 1, two solutions are possible, as shown in Figure 11.10:

- Solution 1: series L and shunt C
- Solution 2: series C and shunt L

The circuits for these two solutions are shown in Figure 11.11.

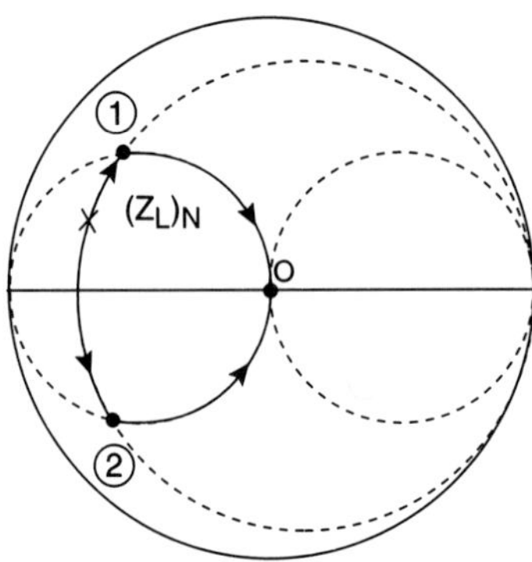

FIGURE 11.10 Smith chart solutions (Case II).

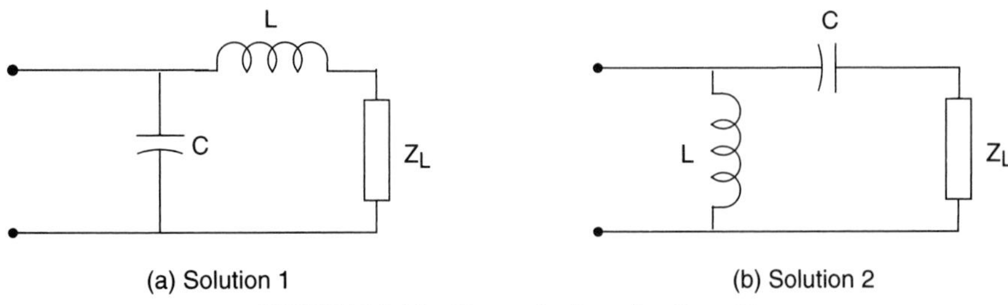

FIGURE 11.11 Two solutions for Case II.

Case III: The load is located outside the (1 + *jx*) and (1+ *jb*) circle. In this case, there are four solutions possible, as shown in Figure 11.12. These four possibilities are described next.

FIGURE 11.12 Smith chart solutions for Case.

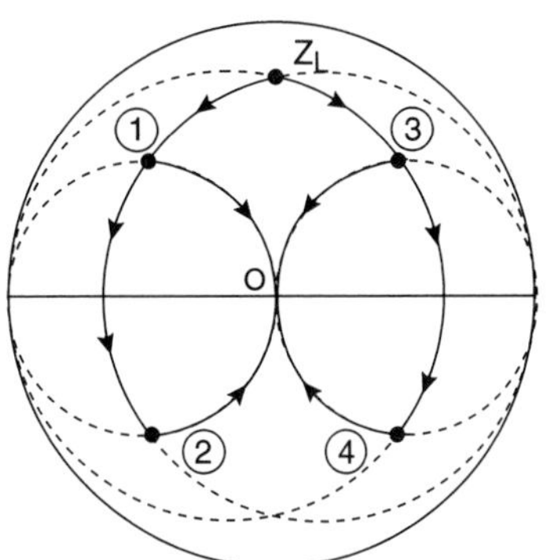

Solutions 1 and 2: Both require a series element first, which makes configuration L1 the only practical one. Thus the solutions are these:

- Solution 1: series *C* and shunt *C*
- Solution 2: series *C* and shunt *L*

The circuits for the two solutions are shown in Figure 11.13.

Solutions 3 and 4: Both require a shunt element inserted first, which makes configuration L2 useful. These two solutions are as follows:

- Solution 3: shunt *C* and series *C*
- Solution 4: shunt *C* and series *L*

These solutions are shown in Figure 11.14.

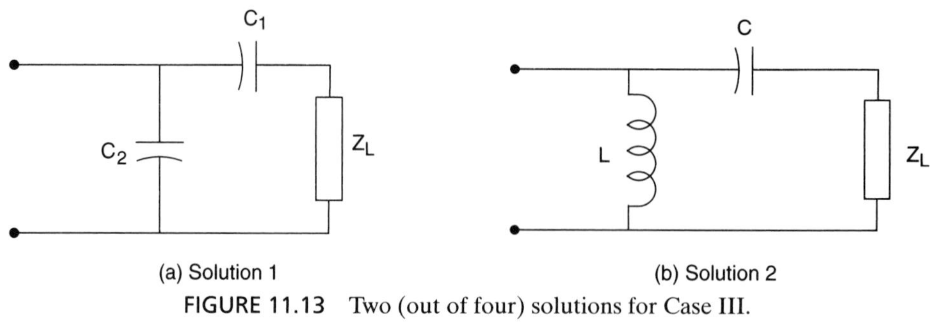

(a) Solution 1 (b) Solution 2

FIGURE 11.13 Two (out of four) solutions for Case III.

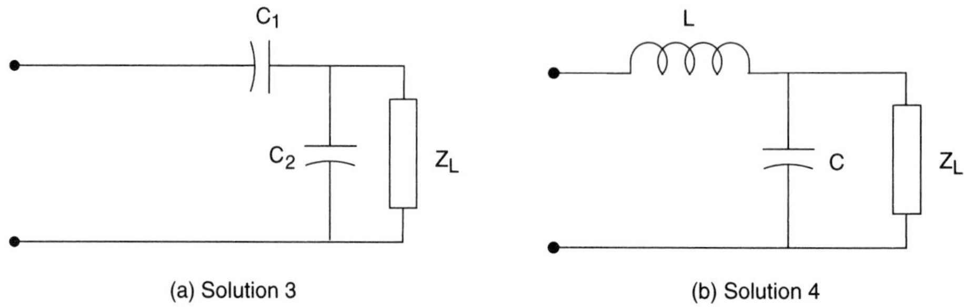

(a) Solution 3 (b) Solution 4

FIGURE 11.14 Solutions 3 and 4.

11.5.3 Design Flexibility

Considering all three cases, it appears that Case III has the highest flexibility because when the load is located outside the unity circle, we have the highest number of design choices to suit the designer's needs. If the load falls inside any of the unity circles, we may be able to add a reactive element to the load in such a way as to bring the combined load to the outside of the two unity circles and then take advantage of the matching possibilities available at the outside of these two unity circles.

EXAMPLE 11.1

Consider a load $(Z_L)_N$ located inside the $(1 + jb)$ circle. Discuss the matching possibilities for this load.

Solution:

This load obviously has two matching possibilities (1 and 2) as discussed earlier in Case II.

Let's add a series L to take the load Z_L to Z'_L outside the $(1 + jb)$ circle, as shown in Figure 11.15.

Now because $(Z'_L)_N$ is outside, we have two additional possibilities (3 and 4) that may be more suitable for our design needs.

Selecting solution 3, we can see that the final matching circuit will have the following three elements, as shown in Figure 11.16.

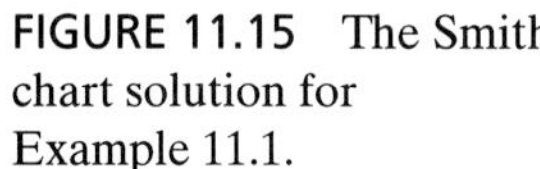

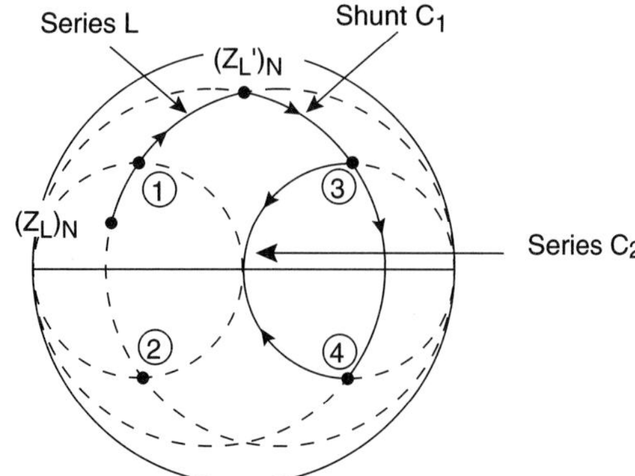

FIGURE 11.15 The Smith chart solution for Example 11.1.

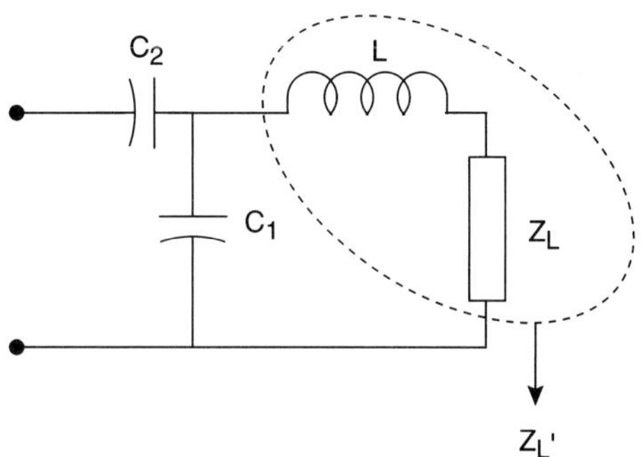

FIGURE 11.16 Matching solutions.

As discussed in Section 11.3, *Selection of a Matching Network*, four main considerations seem to govern the matching circuit design process: simplicity, bandwidth, feasibility of manufacturing, and ease of tunability. These four criteria heavily influence our decision regarding the choice of the matching circuit's design. Therefore, even though from the simplicity point of view Solution 3 seems to be more complex than a simple L-design, there are instances where the other three considerations would become paramount and thus make this design a valuable one.

11.5.4 Design Rules for Matching Networks—Lumped Elements

Based on the discussion presented in the previous two sections, certain rules, if followed, simplify and even speed up the matching circuit design process. These rules can be summarized as follows:

Rule #1. Use a *ZY* Smith chart at all times.

Rule #2. Always start off from the load end and travel "toward generator" (to prevent uncertainty and confusion about the starting point).

Rule #3. Always move on a constant-*R* or constant-*G* circle in such a way as to arrive eventually at the center of the Smith chart.

Rule #4. Each motion along a constant-*R* or constant-*G* circle gives the value of a reactive element.

RULE #5. Moving on a constant-R circle yields series reactive elements, whereas moving on a constant-G circle yields shunt reactive elements.

RULE #6. The direction of travel (or motion) on a constant-R or constant-G circle determines the type of element to be used: a capacitor or an inductor.

Rules 5 and 6 lead to the following additional two rules.

RULE #7. When the motion is upward, it usually corresponds to a series or a shunt inductor.

RULE #8. When the motion is downward, it usually corresponds to a series or a shunt capacitor.

NOTE: *These rules are merely a guideline to be followed in the matching circuit design process. They will never replace the theoretical and practical understandings that go into making them. The understandings contained in this work are the essentials from which all of these rules have been derived.*

EXAMPLE 11.2

Given the circuit shown in Figure 11.17, design a lumped element matching network at 1 GHz that would transform $Z_L = 10 + j10\ \Omega$ into a 50 Ω transmission line.

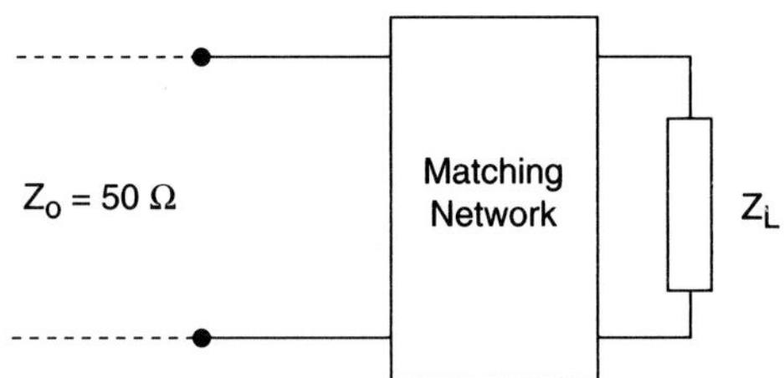

FIGURE 11.17 Circuit for Example 11.2.

Solution:

a. $Z_o = 50\ \Omega, Y_o = 0.02$ S

$$(Z_L)_N = (10 + j10)/50 = 0.2 + j0.2$$

b. Locate $(Z_L)_N$ on the Smith chart, as shown in Figure 11.18.

c. Because $(Z_L)_N$ is inside the unity conductance circle, this would correspond to Case II and has two possible solutions.

Solution 1:

Start from load on a constant-R circle and move up from point A to B. This yields the series L:

$$j\omega L = j(0.4 - 0.2) \times 50 \Rightarrow L = 1.59 \text{ nH}$$

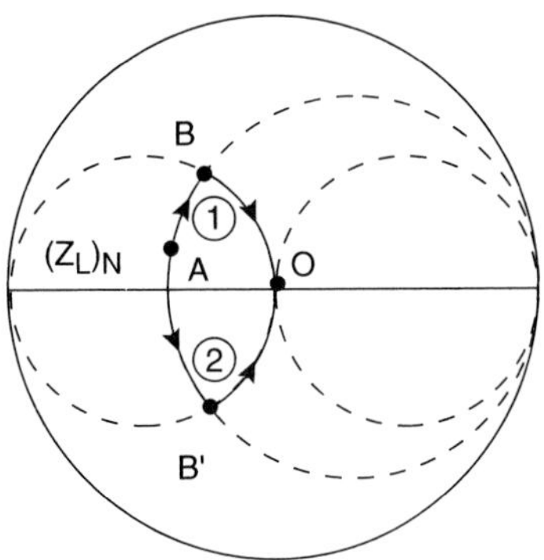

FIGURE 11.18 Smith chart solution for Example 11.2.

Now starting from point B, a motion downward on the unity conductance circle yields a shunt C:

$$j\omega C = j2.0 \times 0.02 \Rightarrow C = 6.37 \text{ pF}$$

The final circuit schematic is shown in Figure 11.19.

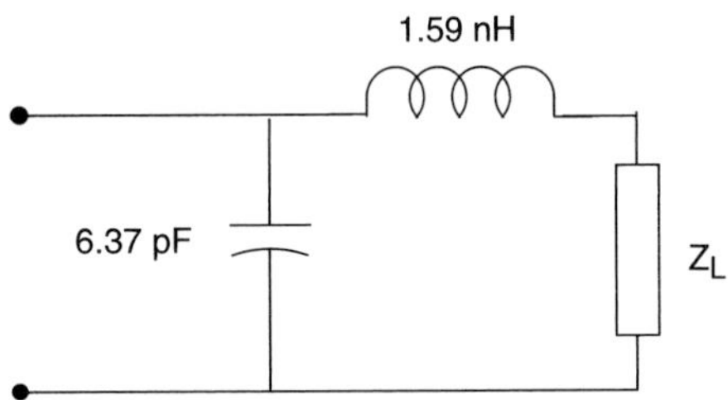

FIGURE 11.19 Circuit for Solution 1.

Solution 2:

Starting from load, move downward on a constant-R circle (series C) to point B' and then upward (shunt L) to arrive at O as follows:

$$\text{Series } C \Rightarrow 1/j\omega C = -j0.6 \text{ x } 50 \Rightarrow C = 5.3 \text{ pF}$$

$$\text{Shunt } L \Rightarrow 1/j\omega L = -j2 \text{ x } 0.02 \Rightarrow L = 3.98 \text{ nH}$$

The schematic for Solution 2 is shown in Figure 11.20.

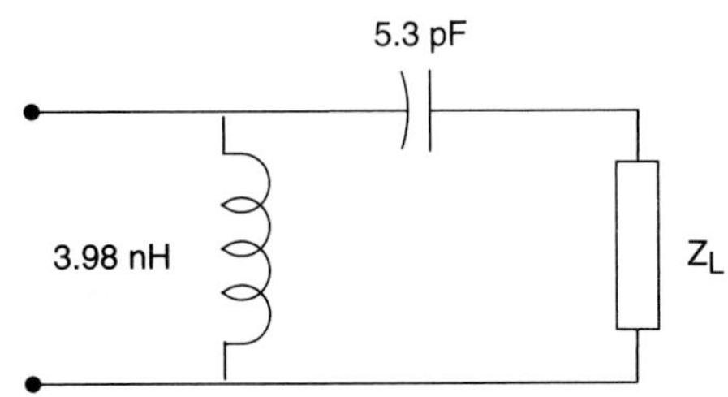

FIGURE 11.20 Circuit for Solution 2.

11.6 MATCHING NETWORK DESIGN USING DISTRIBUTED ELEMENTS

At higher frequencies where the component or circuit size is comparable with wavelength, distributed components may be used to match the load to the transmission line.

The most common technique in this type of design is the use of a single open-circuited or short-circuited length of transmission line (called a stub) connected either in parallel or in series with the transmission feed line at a certain distance from the load as shown in Figure 11.21.

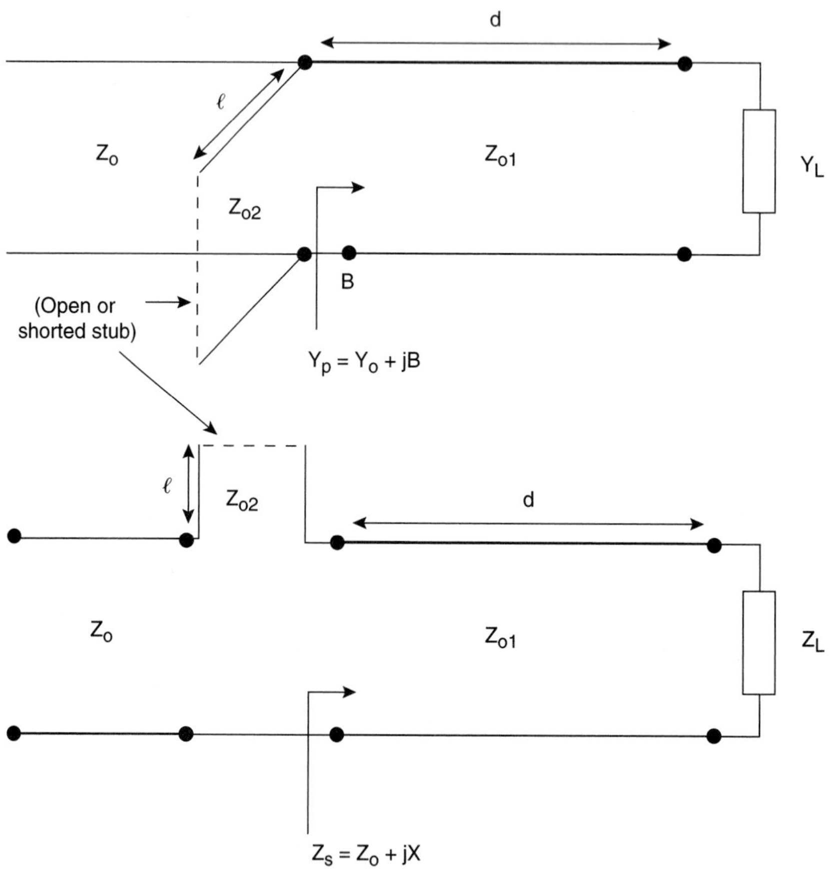

FIGURE 11.21 A stub located a distance d from the load; (a) shunt stub and (b) series stub.

Rather than using a two-wire transmission line schematic, alternate microstrip line schematic for short- and open-stubs can be drawn more effectively as shown in Figures 11.22a, b, c, and d.

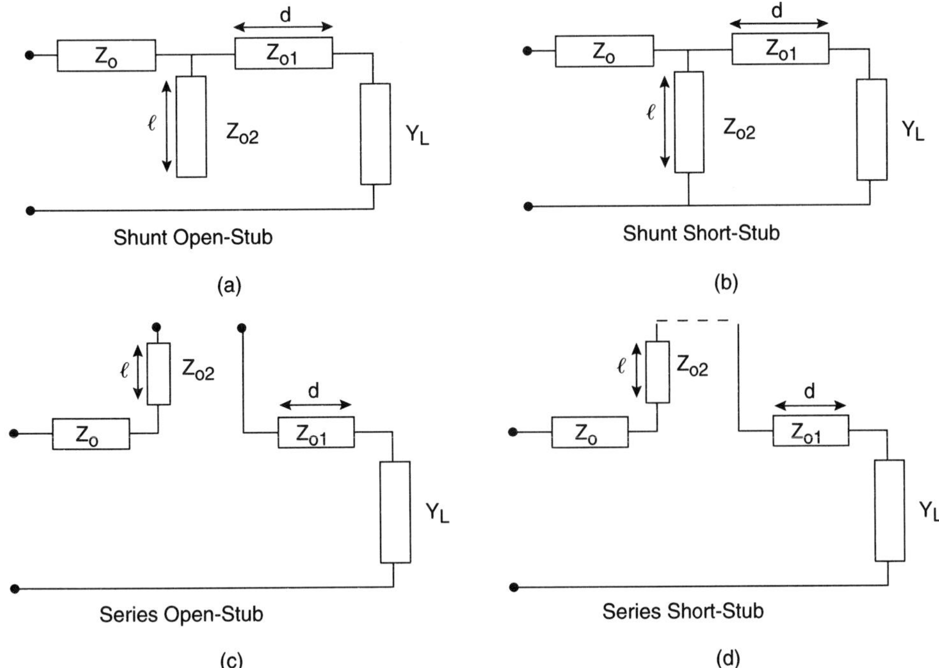

FIGURE 11.22 Microstrip-line schematic for shunt stubs (a, b) and series stubs (c, d).

Such a matching network is easy to build using microstrip or stripline technology. In single-stub matching networks, the two variable parameters are the distance d (from load to stub) and the length ℓ (stub's length), which provides the value of stub susceptance or reactance.

Selection of distance d is crucial for both shunt and series stub:

1. For the shunt stub case, d should be chosen such that the input admittance Y_P (seen looking into the line before adding the stub) is of the form $Y_P = Y_o + jB$ with the stub susceptance selected as $(-jB)$ resulting in a matched condition.
2. For the series stub case, d should be chosen such that the impedance Z_S (seen looking into the line before addition of the stub) is of the form $Z_S = Z_o + jX$ with the stub reactance selected as $-jX$, resulting in a matched condition.

11.6.1 Choice of Short- or Open-Circuited Stubs

With a $\lambda/4$ difference in length between the two, a short or open transmission line with proper length can provide any value of reactance or susceptance needed for the design. Structural considerations behind the choice of a short- versus an open-stub are as follows:

Open stubs. For microstrip and stripline technology use of open stubs is preferred. Use of short stubs requires a via-hole through the substrate to the ground plane, which adds extra work and can be eliminated through the use of open circuits.

Short stubs. For coaxial line or waveguide as a transmission line media, use of short stubs is preferred because the open stubs may radiate, causing power losses and thus making the stub no longer a purely reactive element.

11.6.2 Stub Realization Using Microstrip Lines

Series transmission lines and shunt stubs (short or open) can easily be realized using design steps for microstrip line technology (as outlined in Chapter 7, *Fundamental Concepts in Wave Propagation*). Given a dielectric constant (ε_r), a height (h), and a certain characteristic impedance value (Z_o), the width of the microstrip line (W) can be calculated. For example, a series transmission line and an open/short shunt stub problem (see Figures 11.22a and b) can be solved using a Smith chart with its solution and stub length calculation as shown in Figures 11.23 and 11.24.

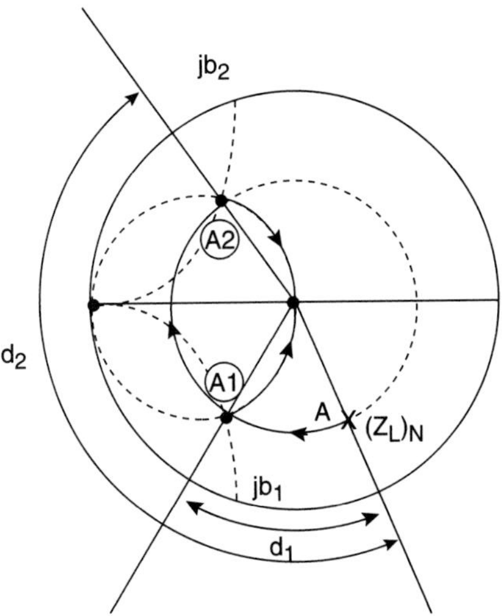

FIGURE 11.23 Smith chart solution.

To minimize the microstrip transition interaction and improve the input *VSWR*, many designers use a balanced approach for shunt stubs rather than a single stub. Using the balanced shunt stubs technique, two stubs of the same length (as the single stub) but twice the characteristic impedance are placed in parallel. That is:

$$\ell'_2 = \ell_2 \tag{11.12a}$$

$$Z'_{o2} = 2Z_{o2} \tag{11.12b}$$

The reason we use twice the characteristic impedance for example for each open shunt stub is due to the fact that each half of the balanced stub (Y_{stub}) must provide half the total admittance (Y_{tot}), as shown below:

$$Y_{stub} = jY'_{o2}\tan\beta\ell'_2 = \frac{1}{2}Y_{tot} \tag{11.13}$$

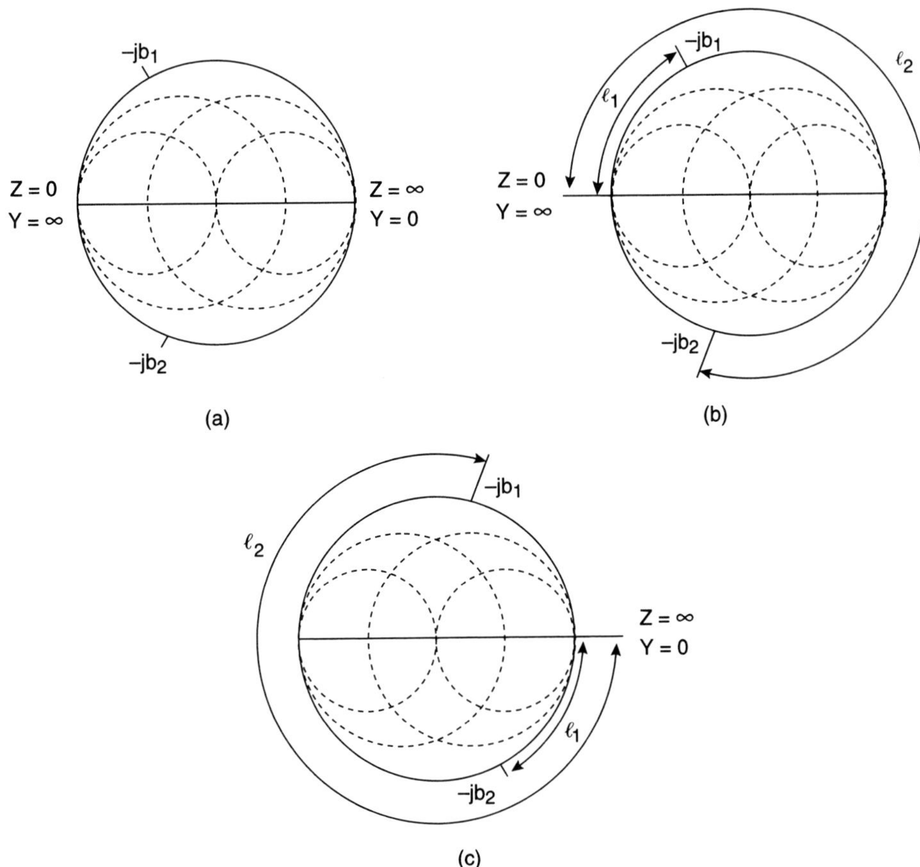

FIGURE 11.24 (a) Plotting stub value on the Smith chart; (b) short stub lengths; (c) open stub lengths.

where

$$Y_{tot} = jY_{o2} \tan\beta\ell_2 \tag{11.14}$$

Thus, substituting Equation 11.14 in 11.13, we have:

$$Y'_{o2} \tan\beta\ell'_2 = \frac{1}{2} Y_{o2} \tan\beta\ell_2 \tag{11.15}$$

Choosing $\ell'_2 = \ell_2$ yields:

$$Y'_{o2} = \frac{1}{2} Y_{o2} \tag{11.16}$$

or

$$Z'_{o2} = 2Z_{o2} \tag{11.17}$$

A similar discussion applies to a short shunt stub with the same conclusions.

11.6.3 Design Steps for Single Stub Matching (Using the Same Characteristic Impedance)

The Smith chart can be used effectively to find the distance d of the stub to the load and the length ℓ of the stub to create the proper value of susceptance or reactance (for more details see Application #8, Chapter 10).

There are two circuit configurations—parallel stub design and series stub design—where the stub is either in parallel or in series. Each case needs to be treated separately.

Parallel Stub Design. This is shown in Figure 11.23. The design process has the following steps:

STEP 1. Plot $(Y_L)_N$ on the ZY chart. (Please note that a single Y-chart could also be used as well, but the load impedance has to be inverted first.)

STEP 2. Draw the appropriate $VSWR$ circle that goes through $(Z_L)_N$.

STEP 3. On the $VSWR$ circle, move (toward generator) to intersect the $(1 + jb)$ conductance unity circle at two solutions located at points $A1$ and $A2$, as shown in Figure 11.23:

- Solution 1: $Y_1 = 1 + jb_1$ (distance d_1)
- Solution 2: $Y_2 = 1 + jb_2$ (distance d_2)

STEP 4. Now add a shunt susceptance of either jb_1 (Solution 1) or jb_2 (Solution 2) to arrive at the center of the chart.

STEP 5. To determine lengths ℓ_1 and ℓ_2, we first locate $-jb_1$ and $-jb_2$ on the Smith chart, as shown in Figure 11.24a.

Then starting from $Z = \infty$ (for open stubs) or $Z = 0$ (for short stubs), we travel along the outer edge of the chart "toward generator" to arrive at $-jb_1$ (for open) or $-jb_2$ (for short) stubs. The lengths can be read off on the circular scale on the outer edge of the chart. These are shown in Figures 11.24b and c.

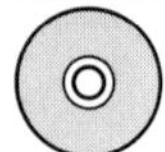

EXAMPLE 11.3 (Parallel Stub Design)

Design a matching network using a single shunt open stub as a tuning element to match a load impedance $Z_L = 15 + j10\ \Omega$ to a 50 Ω transmission line (see Figure 11.25).

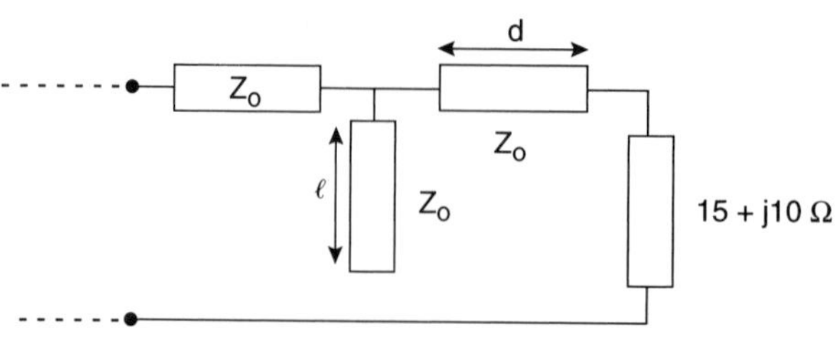

FIGURE 11.25 Circuit for Example 11.3.

Solution:

a. Plot $(Z_L)_N = (15 + j10)/50 = 0.3 + j0.2$ on the ZY-chart.

b. Draw the $VSWR$ circle through $(Z_L)_N$ as shown in Figure 11.26.

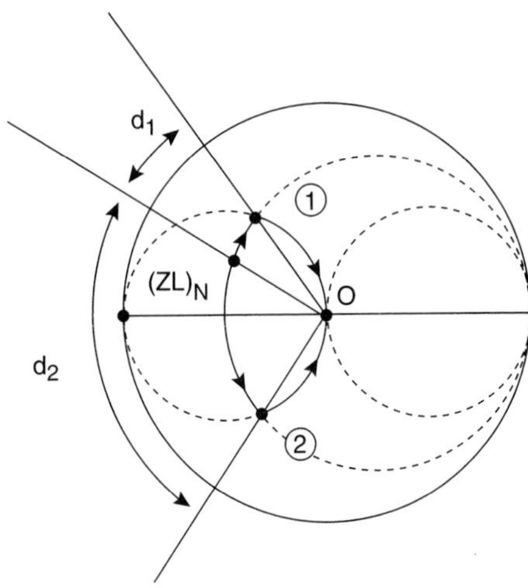

FIGURE 11.26 Smith chart solution.

c. Move toward generator to meet the $(1 + jb)$ circle at two points, giving two solutions:

$$(Y_1)_N = 1 - j1.33 \text{ with } d_1 = 0.044\lambda \text{ and } jb_1 = j1.33$$

$$(Y_2)_N = 1 + j1.33 \text{ with } d_2 = 0.387\lambda \text{ and } jb_2 = -j1.33$$

d. From Figure 11.27, we can read off ℓ_1 and ℓ_2 as:

$$\ell_1 = 0.147\lambda$$

$$\ell_2 = 0.353\lambda$$

The two possible design schematics are shown in Figures 11.28a and b.

NOTE: *Design #1 is more desirable because it is usually preferred to keep the matching stub as close as possible to the load (i.e., smaller d) in order to improve:*

- *The sensitivity of the matching network to frequency, thus providing a larger bandwidth*
- *The standing wave losses occurring on the line between the stub and the load due to a high VSWR*

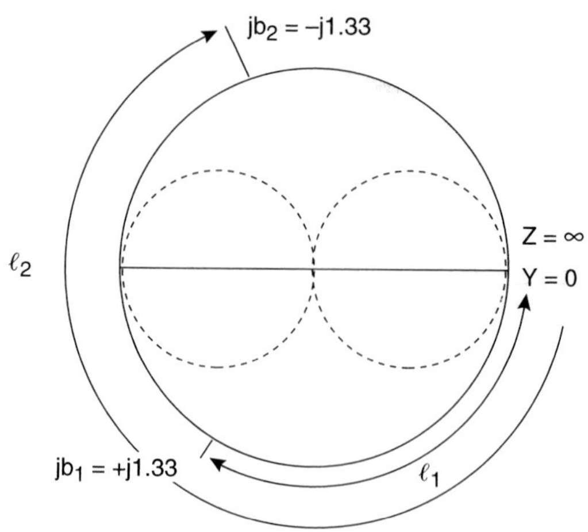

FIGURE 11.27 Stub lengths.

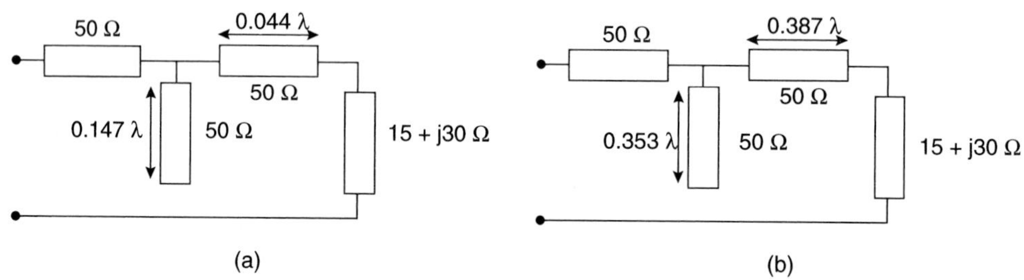

FIGURE 11.28 Two possible circuit designs for Example 11.3.

Series Stub Design. From Figure 11.29, the design steps are as follows:

STEP 1. Plot $(Z_L)_N$ on the ZY-chart.

STEP 2. Draw the appropriate $VSWR$ circle.

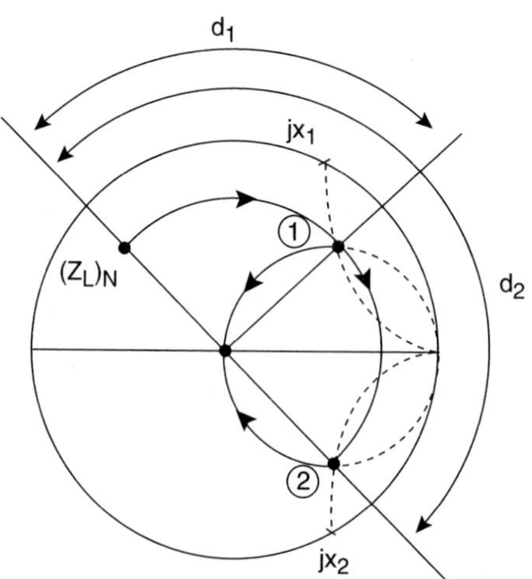

FIGURE 11.29 Smith chart solution.

STEP 3. Move "toward generator" to intersect the $(1 + jx)$, unity resistance circle, at two points, (1) and (2), as shown in Figure 11.29:

- Solution 1: $Z_1 = 1 + jx_1$ (distance d_1), $Z_{1stub} = -jx_1$
- Solution 2: $Z_2 = 1 + jx_2$ (distance d_2), $Z_{2stub} = -jx_2$

STEP 4. Now add a series reactance of $-jx_1$ for Solution 1 (or $-jx_2$ for Solution 2) to cancel the existing reactance.

STEP 5. Stub length ℓ_1 (or ℓ_2) is now calculated by first locating reactance $-jx_1$ (or $-jx_2$) on the Smith chart; see Figures 11.30a and b.

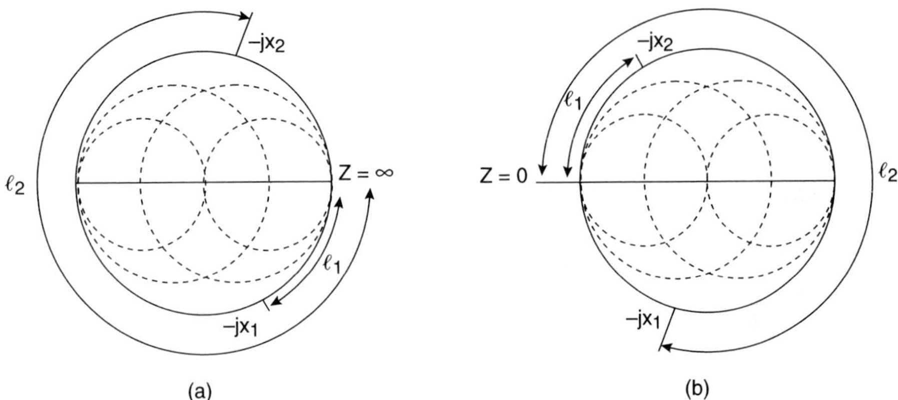

FIGURE 11.30 Stub lengths for (a) open stubs and (b) short stubs.

For a series open stub, start from $Z = \infty$ and travel on the outer edge of the chart "toward generator" to arrive at $-jx_1$ for Solution 1 (or $-jx_2$ for Solution 2). On the other hand, for a series short stub, repeat the previous procedure except start from $Z = 0$, as shown in Figures 11.30a and b.

EXAMPLE 11.4 (Series Stub Design)

Using a single series open stub, design a matching network that will transform a load impedance $Z_L = 100 + j80\ \Omega$ to a $50\ \Omega$ feed transmission line, as shown in Figure 11.31.

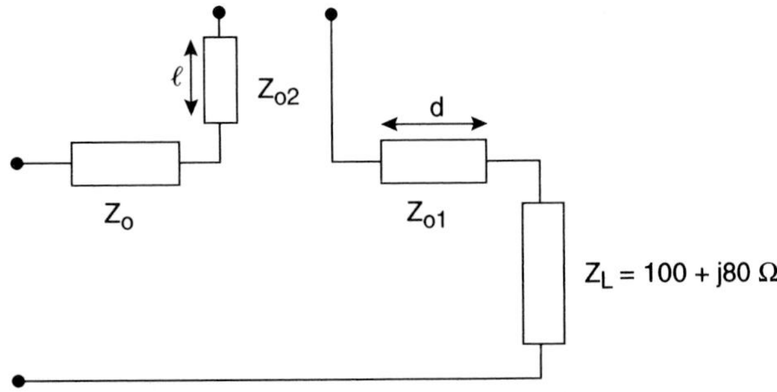

FIGURE 11.31 Circuit for Example 11.4.

Solution:

a. Plot $(Z_L)_N = (100 + j80)/50 = 2 + j1.6$ on the Smith chart.
b. Draw the *VSWR* circle (see Figure 11.32).

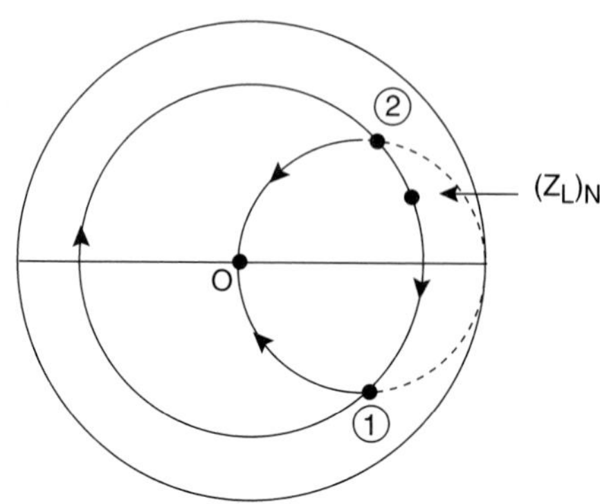

FIGURE 11.32 Smith chart solution.

c. Move "toward generator" to intersect $(1 + jx)$ circle at two points (1) and (2) giving:

$$Z_1 = 1 - j1.33 \text{ with } d_1 = 0.120\lambda \text{ and } Z_{1stub} = j1.33$$

$$Z_2 = 1 + j1.33 \text{ with } d_2 = 0.463\lambda \text{ and } Z_{2stub} = -j1.33$$

d. Lengths ℓ_1 and ℓ_2 can be read off from Figure 11.33 as:

$$\ell_1 = 0.397\lambda$$

$$\ell_2 = 0.103\lambda$$

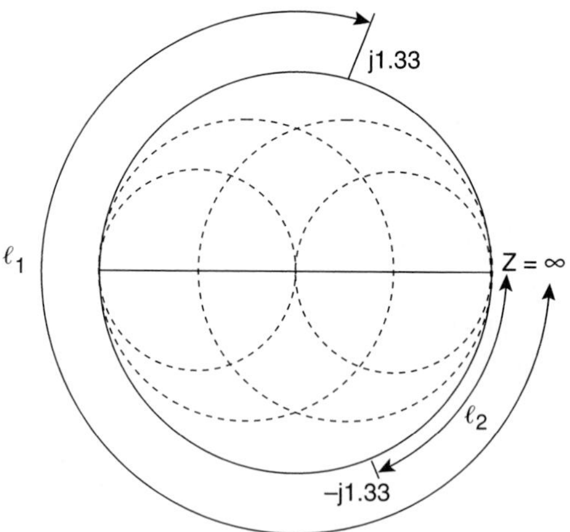

FIGURE 11.33 Stub lengths.

The two possible design schematics are shown in Figures 11.34a and b.

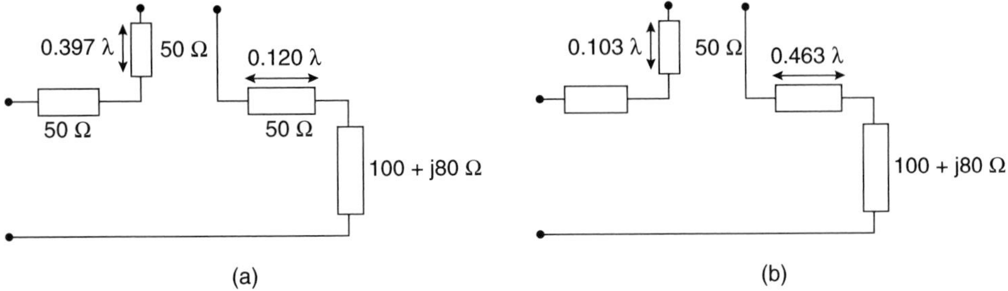

FIGURE 11.34 Final design schematics: (a) Design #1, (b) Design #2.

11.6.4 Design of Single Stub Matching (Using a Different Characteristic Impedance)

As can be seen from the previous two examples, the stub, the feed transmission line, and the line between the load and the stub all have the same characteristic impedance (Z_o).

If a characteristic impedance other than Z_o is desired, the location and length of the stub will now be different. The following procedure can be used effectively to solve for this situation:

We add a stub at the load such that it will reduce the load to a purely resistive value, as shown in Figure 11.35. Two types of stubs (each with the possibility of being short- or open-circuited) should be considered: the shunt stub and the series stub.

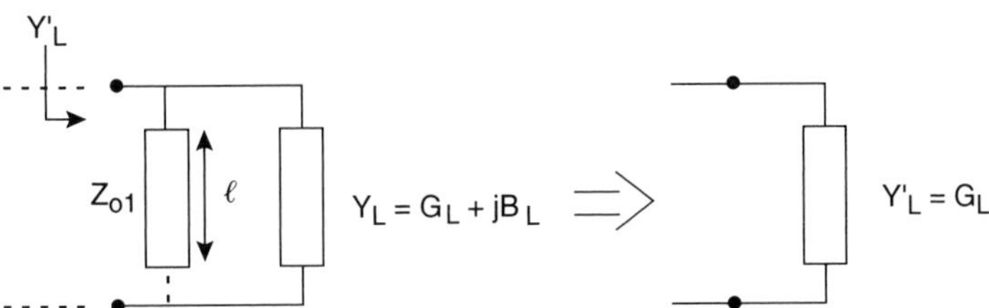

FIGURE 11.35 Conversion of the load to a purely resistive load.

a) Shunt Stub.

Load: $Y_L = G_L + jB_L$
Stub: $-jB_L$
Result: $Y'_L = G_L$

The new load (Y'_L) is purely resistive. To determine the length of the stub, we have to specify the stub termination as being short or open.

Short Shunt Stub(Z_{o1})

$$Y_{SC} = -jY_{o1}/\tan\beta\ell = -jB_L \Rightarrow Y_{o1} = B_L\tan\beta\ell$$

$$Z_{o1} = 1/Y_{o1} = 1/B_L\tan\beta\ell$$

By proper choice of stub length (ℓ), any positive value of stub characteristic impedance can be realized, for example:

$$\left.\begin{array}{l}\ell = \lambda/8 \Rightarrow Z_{o1} = 1/B_L \ (for\, B_L > 0) \\ \ell = 3\lambda/8 \Rightarrow Z_{o1} = -1/B_L \ (for\, B_L < 0)\end{array}\right\} \Rightarrow Z_{o1} = 1/|B_L|$$

Open Shunt Stub(Z_{o1})

$$Y_{OC} = jY_{o1}\tan\beta\ell = -jB_L \ \Rightarrow\ Y_{o1} = -B_L/\tan\beta\ell$$

$$Z_{o1} = 1/Y_{o1} = -\tan\beta\ell/B_L$$

By proper choice of stub length (ℓ), any positive value of stub characteristic impedance can be realized, for example:

$$\left.\begin{array}{l}\ell = \lambda/8 \Rightarrow Z_{o1} = -1/B_L \ (for\, B_L < 0) \\ \ell = 3\lambda/8 \Rightarrow Z_{o1} = 1/B_L \ (for\, B_L > 0)\end{array}\right\} \Rightarrow Z_{o1} = 1/|B_L|$$

b) Series Stub.

Load: $Z_L = R_L + jX_L$
Stub: $-jX_L$
Result: $Z'_L = R_L$

The new load (Z'_L) is purely resistive. To determine the length of the stub, we have to specify the stub termination as being short or open.

Short Series Stub(Z_{o1})

$$Z_{SC} = +jZ_{o1}\tan\beta\ell = -jX_L \ \Rightarrow\ Z_{o1} = -X_L/\tan\beta\ell$$

By proper choice of stub length (ℓ), any positive value of stub characteristic impedance can be realized, for example:

$$\left.\begin{array}{l}\ell = \lambda/8 \Rightarrow Z_{o1} = -X_L \ (for\, X_L < 0) \\ \ell = 3\lambda/8 \Rightarrow Z_{o1} = +X_L (for\, X_L > 0)\end{array}\right\} \Rightarrow Z_{o1} = |X_L|$$

Open Series Stub(Z_{o1})

$$Z_{OC} = -jZ_{o1}/\tan\beta\ell = -jX_L \ \Rightarrow\ Z_{o1} = +X_L\tan\beta\ell$$

By proper choice of stub length (ℓ), any positive value of stub characteristic impedance can be realized, for example:

$$\left.\begin{array}{l}\ell = \lambda/8 \Rightarrow Z_{o1} = +X_L \ (for\, X_L > 0) \\ \ell = 3\lambda/8 \Rightarrow Z_{o1} = -X_L \ (for\, X_L < 0)\end{array}\right\} \Rightarrow Z_{o1} = |X_L|$$

Considering that the new load (Y'_L or Z'_L) is now purely resistive (G_L or R_L), a simple quarter-wave ($\lambda/4$) transformer (Z'_o) can be used efficiently to transform (Y'_L or Z'_L) to Z_o. The $\lambda/4$ transformer has a characteristic impedance of:

Shunt stub: $Z'_o = (Z_o/Y'_L)^{1/2} = (Z_o/G_L)^{1/2}$

or

Series stub: $Z'_o = (Z'_L Z_o)^{1/2} = (R_L Z_o)^{1/2}$

where Z_o is the characteristic impedance of the feed line (assumed to be lossless) and is a positive real number. This method is depicted for a shunt stub on the Smith chart (as shown in Figure 11.36). Starting from point *A*, we travel on a constant conductance circle to point *B* (located on the purely resistive axis). A quarter-wave transformer would then take it from point *B* to *O*, the center of the chart. The final circuit schematic for the matching network is shown in Figure 11.37.

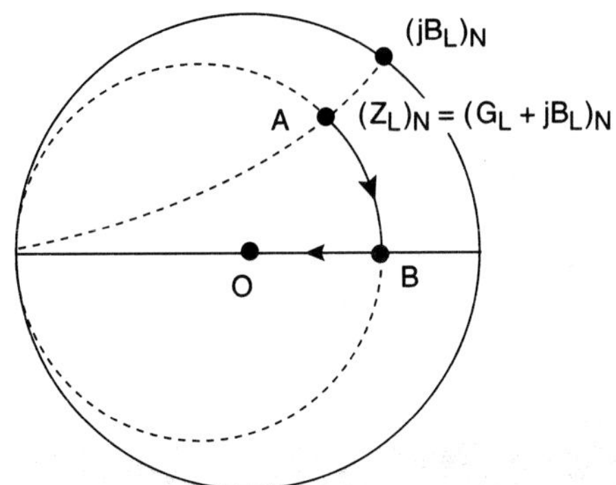

FIGURE 11.36 Smith chart solution.

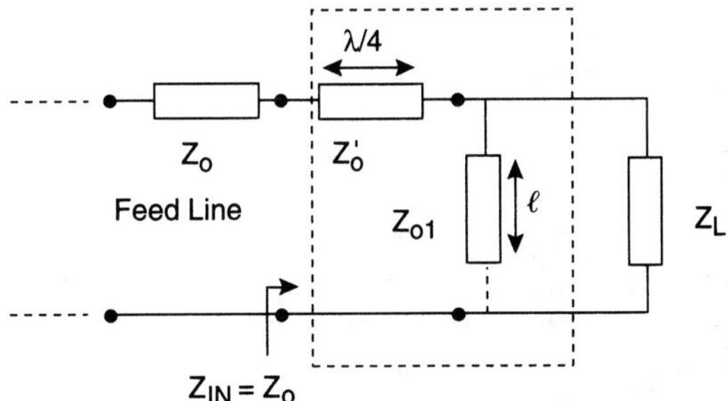

FIGURE 11.37 Final circuit schematic.

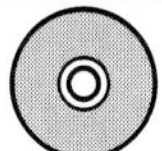

EXAMPLE 11.5

Design a matching network using a quarter-wave transformer that would transform a 50 Ω load to $\Gamma_{IN} = 0.68\angle 97°$ as shown in Figure 11.38.

Solution:

To design the matching network, first we need to transform the load to the resistive part of Z_{IN} and then add a reactance equal to the reactance of Z_{IN} as follows:

a. Plot $\Gamma_{IN} = 0.68\angle 97°$ on the Smith chart as shown in Figure 11.38 and read off $(Y_{IN})_N$:

$$(Y_{IN})_N = (0.4 - j1.05) \Rightarrow Y_{IN} = Y_o(Y_{IN})_N = 0.008 - j0.021 \text{ S}$$

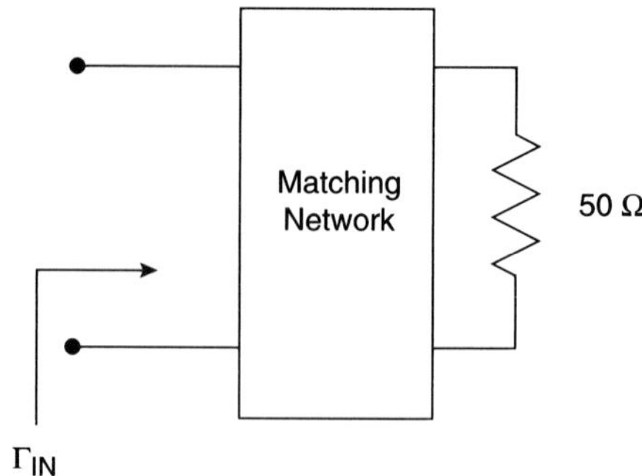

FIGURE 11.38 Circuit for Example 11.5.

b. Starting from the center of the chart (i.e., load at point O), a quarter-wave transformer (Z_{o1}) is now added to transform 50 Ω to ($R_{IN} = 1/0.008 = 125\ \Omega$) (point A in Figure 11.39):

$$Z_{o1} = (50/0.008)^{1/2} = 79\ \Omega$$

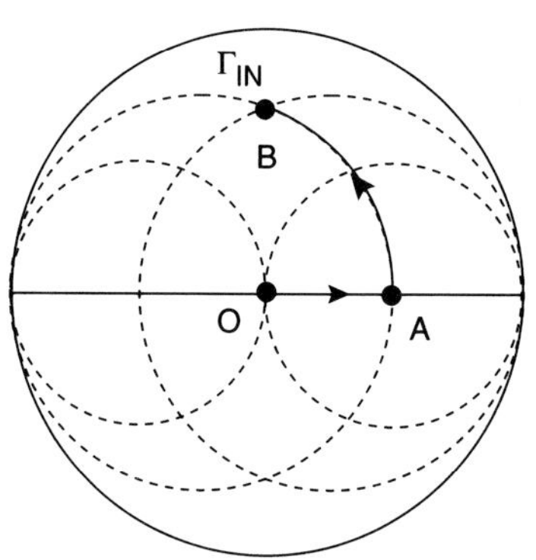

FIGURE 11.39 Smith chart solution.

c. The reactive part ($B_{IN} = -j\,0.021$ S) can now be synthesized by using an open-circuited shunt stub of length $\ell = 3\lambda/8$ to arrive at point B. The open-circuited stub has:

$$Y_{OC} = jY_{o2}\tan\beta\ell = -j0.021\ \text{S}$$

The characteristic impedance is given by:

$$\ell = 3\lambda/8 \Rightarrow \tan\beta(3\lambda/8) = -1$$

Thus,

$$Z_{o2} = 1/0.021 = 47.6\ \Omega$$

d. The final circuit schematic is shown in Figure 11.40.

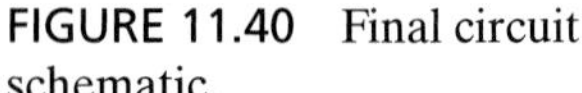

FIGURE 11.40 Final circuit schematic.

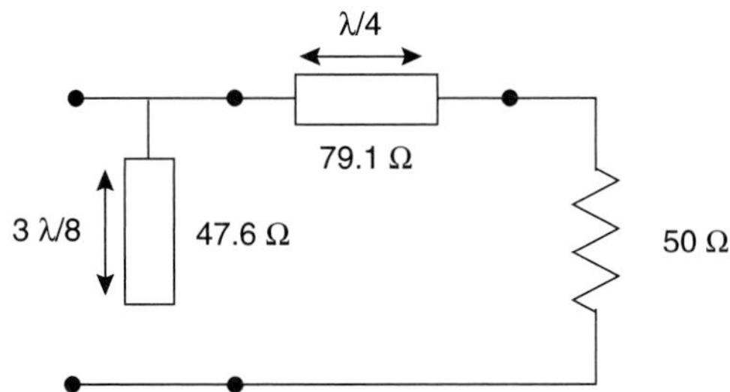

11.6.5 Generalized Impedance Matching Network for Non-Zero Reflection Coefficients

When the input and the output reflection coefficients are both non-zero, then the matching network would transform the load impedance to a point (other than the center of the chart) corresponding to the desired input reflection coefficient. This case is a very generalized concept of a matching network and is in contrast with the case where either the input or the output reflection coefficient was at the center of the chart. This type of matching network could occur, for example, in an intermediate matching stage of a two-stage amplifier. Figure 11.41 shows the generalized concept of a matching network along with the plot of the input and output reflection coefficients in the Smith chart.

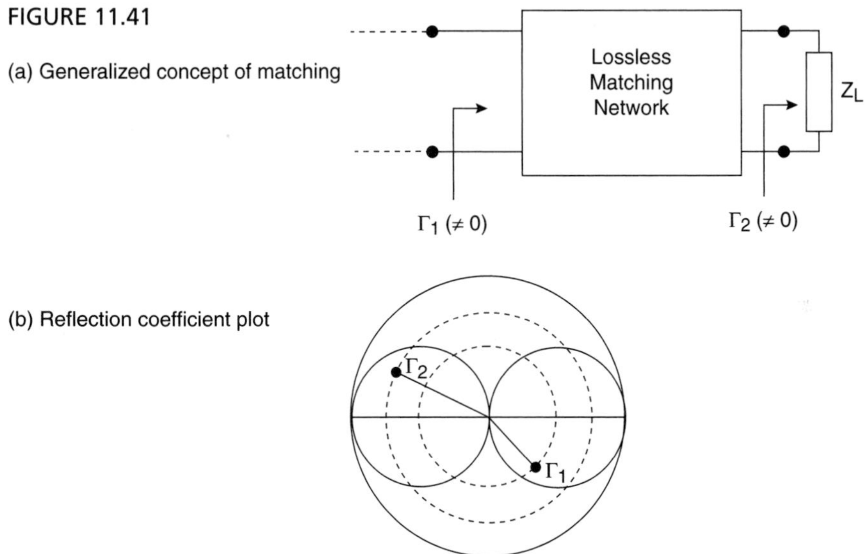

FIGURE 11.41

(a) Generalized concept of matching

(b) Reflection coefficient plot

There are a number of possible solutions that would lead to the desired matching network.

Solution #1. Assuming the input and output reflection coefficients to be Γ_1 and Γ_2, respectively, an obvious solution is to convert the load (Γ_2) to 50 Ω first and then match Γ_1 to 50 Ω, which is located at the center of the chart, as shown in Figure 11.42a.

To realize the matching circuit, we start from the load at point A and travel "toward generator" on the constant *VSWR* circle to arrive at point B. Adding a shunt stub at this point will move this point to the center of the Smith chart ($\Gamma'_2 = 0$), which is the new load at point O as shown.

Now starting from point O, add a series stub (to arrive at point C) and then a series transmission line to end up at Γ_1, which is the input reflection coefficient. The resulting distributed circuit is shown in Figure 11.42b.

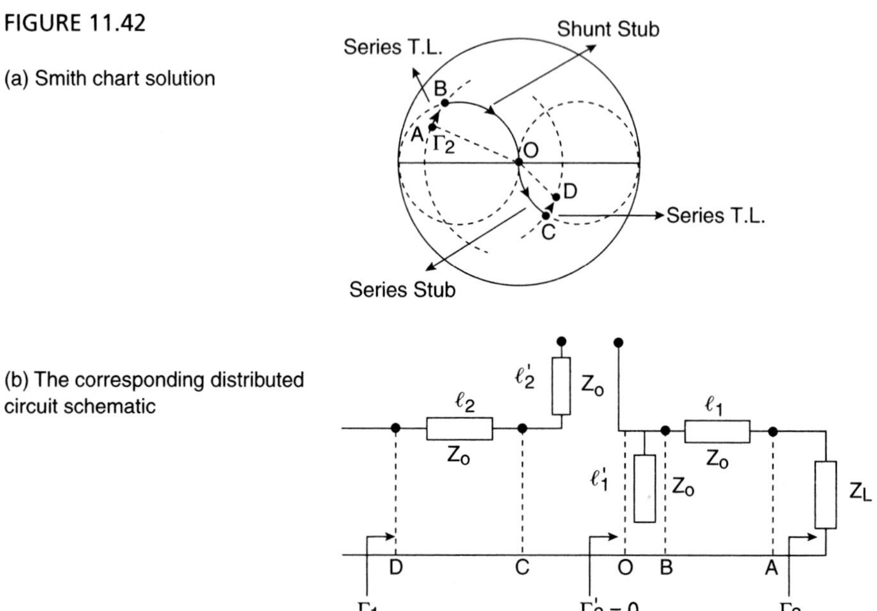

FIGURE 11.42

(a) Smith chart solution

(b) The corresponding distributed circuit schematic

NOTE 1: *As described earlier, it is best to follow the matching philosophy of moving from the load impedance toward the center of the chart, which means we move from the output (load) and progress backward to the input end.*

NOTE 2: *The matching network described above could have been alternately realized with lumped elements using the constant conductance or resistance circle (see Section 11.5, Design of Matching Circuits Using Lumped Elements), rather than the distributed element design.*

Solution #2. First, we draw the constant *VSWR* circle for (Γ_1) as shown in Figure 11.43. Then, starting from (Γ_2) at point A, we travel on a constant conductance circle to intersect the (Γ_1) *VSWR* circle at point B. This corresponds to a shunt stub element.

Next, we travel on the constant *VSWR* circle "toward generator" (clockwise) to arrive at point C. This would correspond to a series transmission line of length ℓ, as shown in the schematic in Figure 11.44.

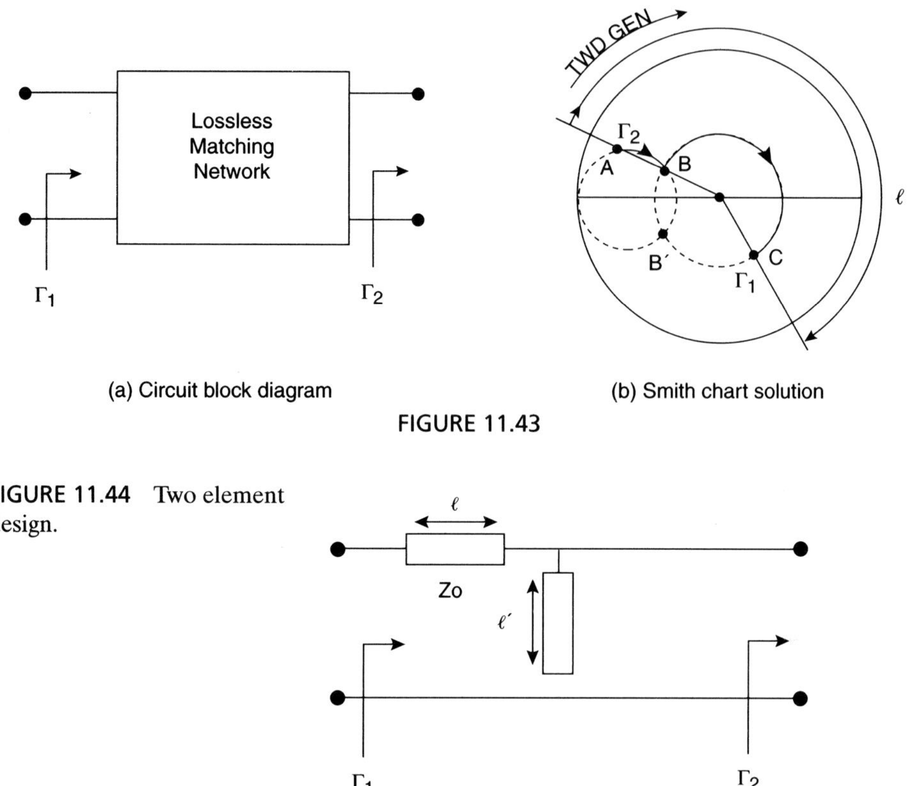

(a) Circuit block diagram (b) Smith chart solution

FIGURE 11.43

FIGURE 11.44 Two element design.

NOTE 1: *There is a second solution that uses the second intersection of the constant-G circle with the constant VSWR circle at point B' leading to a similar design procedure as described previously.*

NOTE 2: *Solution #2 provides only two elements (compared to four elements given by Solution #1), which usually results in the best bandwidth for the design of an interstage circuit.*

POINT OF INTEREST *In some cases, it is more convenient to work with the equivalent problem of conjugate reflection coefficients, which functionally yield an equivalent circuit as shown in Figures 11.45 and 11.46.*

In Figure 11.47, because we are matching Γ^*_1 to Γ^*_2, we have to start from Γ^*_1 (as the load) and progress backward to Γ^*_2 at the other end. Traveling from point A to B gives a series transmission line, followed by going from B to C (producing a shunt stub), which is identical to what was obtained earlier.

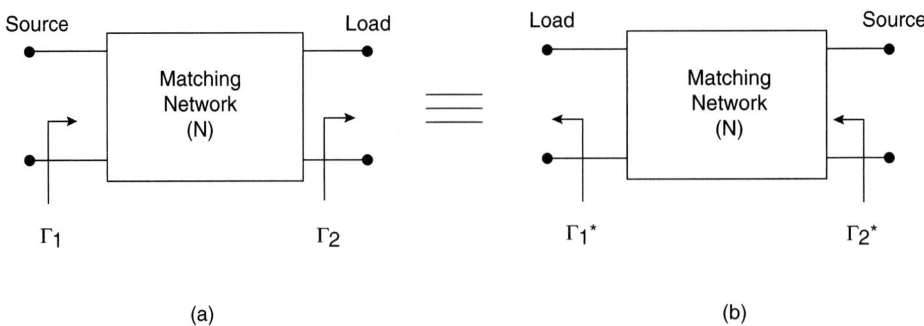

FIGURE 11.45 An equivalent representation of reflection coefficients at the terminals.

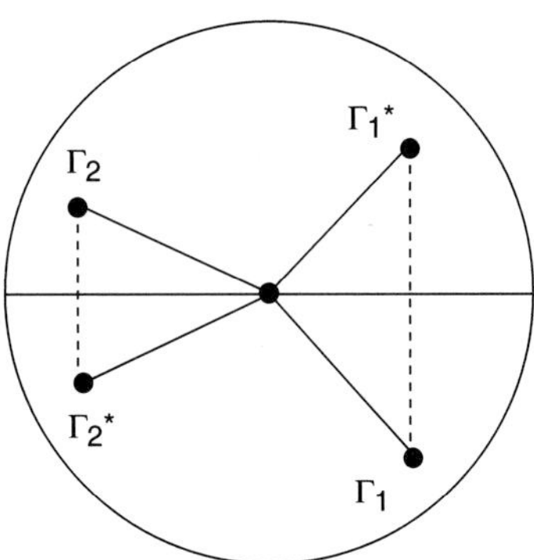

FIGURE 11.46 Plot of conjugate reflection coefficients in the Smith chart.

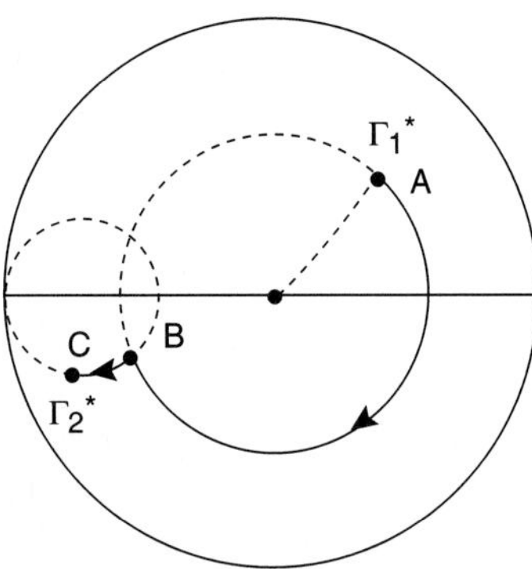

FIGURE 11.47 Smith chart solution for conjugate reflection coefficients.

LIST OF SYMBOLS/ABBREVIATIONS

A symbol will not be repeated again once it has been identified and defined in an earlier chapter, as long as its definition remains unchanged.

P_{av}	Average power to the load
$(P_{av})_{max}$	Maximum average power to the load
$(Z_L)_{opt}$	Optimum load impedance for maximum power transfer

PROBLEMS

11.1 Design a single stub matching circuit that transforms a load ($Z_L = 30 + j50\ \Omega$) to a transmission line as follows (see Figure P11.1):

a. Assume $Z_{o1} = Z_{o2} = 100\ \Omega$

b. Assume $Z_{o1} = 100\ \Omega$ and $Z_{o2} = 200\ \Omega$

FIGURE P11.1

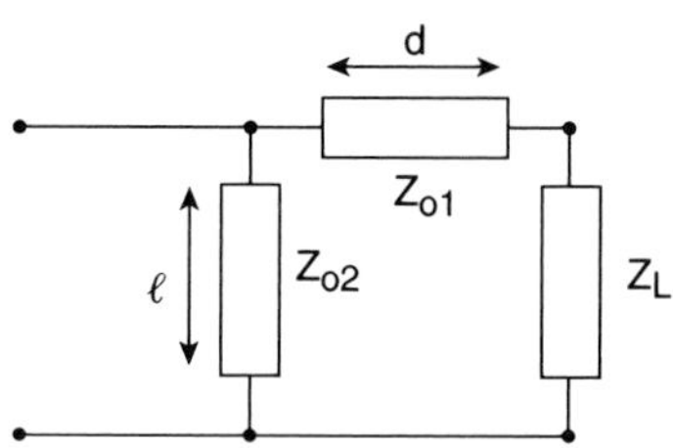

11.2 Design a lumped matching network to match the load $Y_L = (4 - j6) \times 10^{-3}$ S to a transmission line ($Z_o = 100\ \Omega$) as shown in Figure P11.2. Find the element values at 10 GHz.

FIGURE P11.2

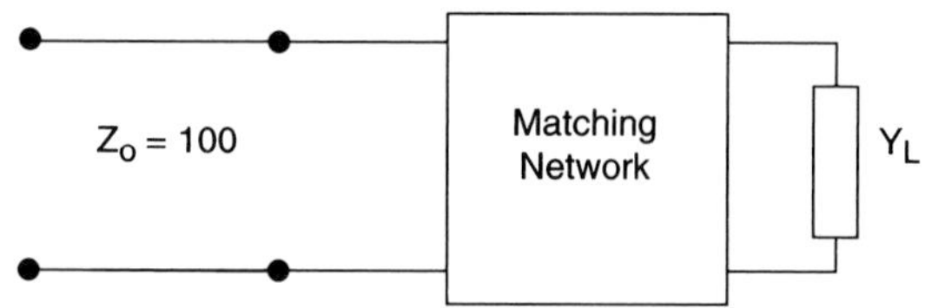

11.3 Design the matching network shown in Figure P11.3 to match the load, $Z_L = 100 + j100\ \Omega$ to a 50 Ω transmission line.

FIGURE P11.3

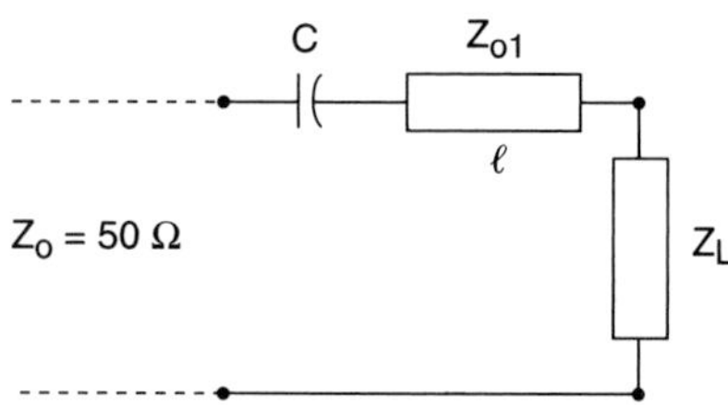

11.4 Design a matching network to transform a load impedance ($Z_L = 50 + j50\ \Omega$) to the input impedance ($Z_{IN} = 25 - j25\ \Omega$) as shown in Figure P11.4 for the following two cases:

a. Lumped element design

b. Distributed element design

FIGURE P11.4

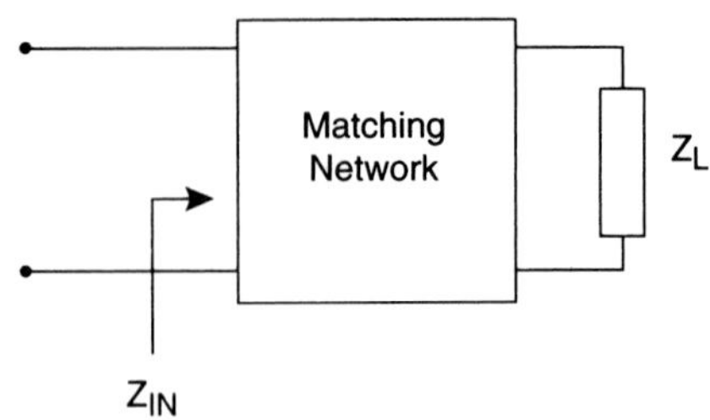

11.5 Design a matching network that will match a 50 Ω load to an input reflection coefficient of $\Gamma_{IN} = 0.5\angle\ 150°$ in a 50 Ω system as shown in Figure P11.5. The matching network should use a quarter-wave transformer.

FIGURE P11.5

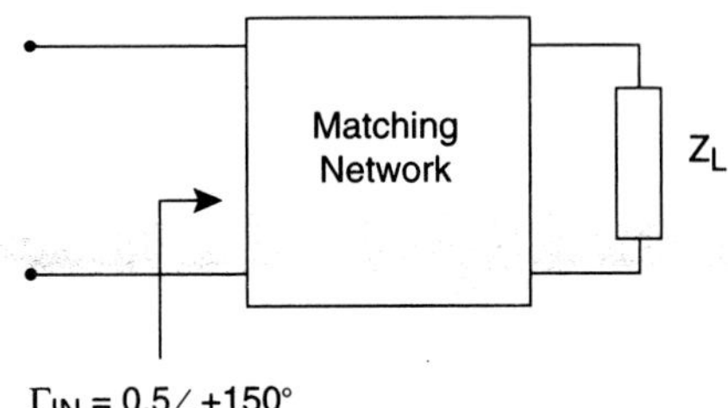

11.6 In the circuit shown below, a load $Z_L = 90 + j60\ \Omega$ is to be matched to a line as shown in Figure P11.6. Determine lumped element Z_1, Z_{o1}, and ℓ.

FIGURE P11.6

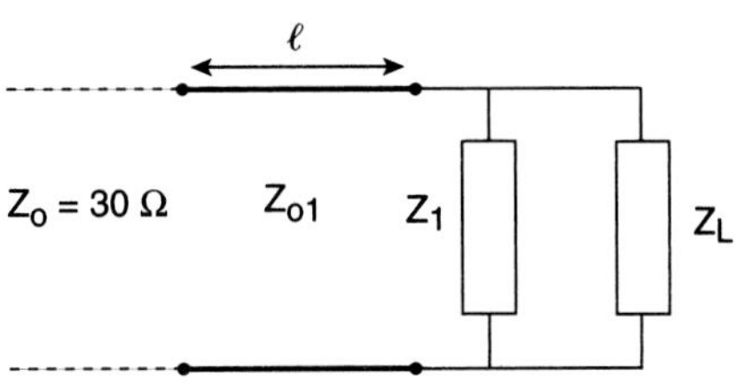

11.7 A certain microwave device has $Z_d = 50 - j50\ \Omega$. Design a matching network to match the device impedance to a 25 Ω system as shown in Figure P11.7 for:

a. Lumped element circuit design

b. Distributed element circuit design

FIGURE P11.7

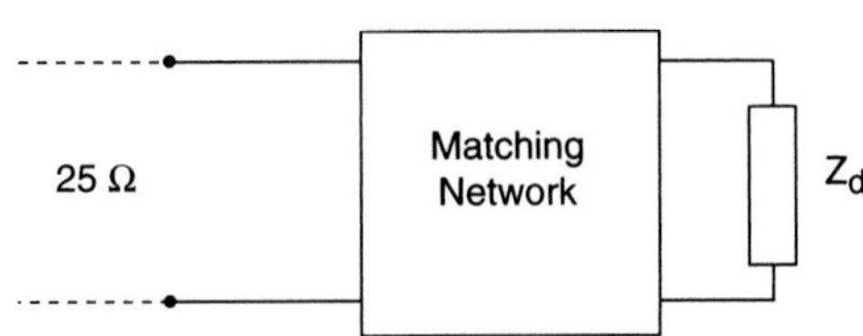

11.8 A lossless transmission line ($Z_o = 50\ \Omega$) is to be matched to a load, $Z_L = 5.5 - j10.5\ \Omega$, by means of a short-circuited stub ($Z_{o1} = 100\ \Omega$) as shown in Figure P11.8. Determine the position and length of the stub.

FIGURE P11.8

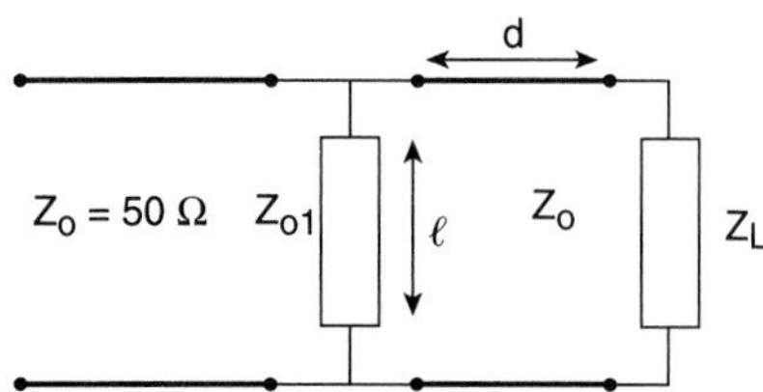

REFERENCES

[11.1] Anderson, E. M. *Electric Transmission Line Fundamentals.* Upper Saddle River: Prentice Hall, 1985.

[11.2] Cheng, D. K. *Fundamentals of Engineering Electromagnetics.* Reading: Addison Wesley, 1993.

[11.3] Cheung, W. S. and F. H. Levien. *Microwave Made Simple, Principles and Applications.* Norwood: Artech House, 1985.

[11.4] Gonzalez, G. *Microwave Transistor Amplifiers, Analysis and Design,* 2nd ed. Upper Saddle River: Prentice Hall, 1997.

[11.5] Liao, S. Y. *Microwave Circuit Analysis and Amplifier Design.* Upper Saddle River: Prentice Hall, 1987.

[11.6] Pozar, D. M. *Microwave Engineering,* 2nd ed. New York: John Wiley & Sons, 1998.

[11.7] Schwarz, S. E. *Electromagnetics for Engineers.* Orlando: Sanders College Publishing, 1990.

PART IV

BASIC CONSIDERATIONS IN ACTIVE NETWORKS

CHAPTER 12

Stability Considerations in Active Networks

12.1 INTRODUCTION

In any amplifier design, one important consideration is the stability of the circuit under different source or load conditions. This term needs to be defined at this point:

DEFINITION-STABILITY: *Is the ability of an amplifier to maintain effectiveness in its nominal operating characteristics in spite of large changes in the environment such as physical temperature, signal frequency, source or load conditions, etc.*

In this chapter, the stability requirements for an amplifier circuit or in general a two-port network with known *S*-parameters are discussed (see Figure 12.1). In this text, we limit our stability considerations primarily to source and load conditions, and develop exact criteria for unconditional stability as well as conditional stability (also called potentially unstable condition).

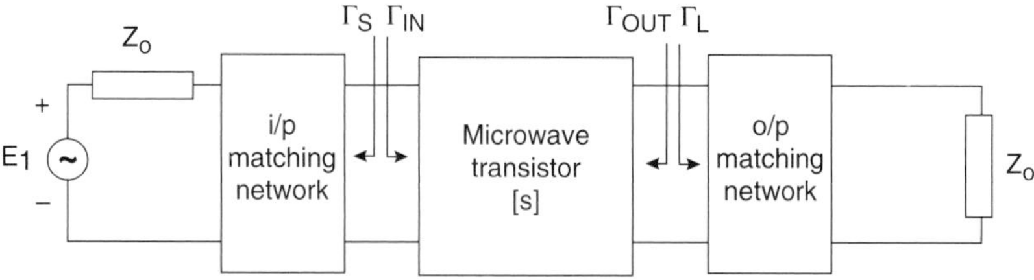

FIGURE 12.1 A single stage amplifier circuit.

When an amplifier becomes unstable (i.e., not able to maintain its nominal characteristics), it no longer acts as an amplifier but as an oscillator.

From Figure 12.1, we can observe that because Γ_{IN} and Γ_{OUT} depend on the load and source matching networks, respectively, the stability of the amplifier circuit therefore depends on Γ_L and Γ_S.

It should be noted that because the S-parameters of a two-port network are frequency dependent, the stability condition of a circuit depends on the frequency of operation. Thus, it is possible to have a circuit functioning well as an amplifier at an intended frequency while possibly oscillating for out-of-band frequencies.

12.1.1 Analysis

In a two-port network as shown in Figure 12.2, when the input impedance or output impedance presents a negative resistance, we can write:

$$Re(Z_{IN}) < 0 \tag{12.1}$$

or

$$Re(Z_{OUT}) < 0 \tag{12.2}$$

This negative resistance condition at the input (or the output) port means that the reflected signal from the input (or output) port has higher power than the incident signal thus making $|\Gamma_{IN}|$ or $|\Gamma_{OUT}|$ larger than unity, i.e.,

$$|\Gamma_{IN}| > 1 \tag{12.3}$$

or

$$|\Gamma_{OUT}| > 1 \tag{12.4}$$

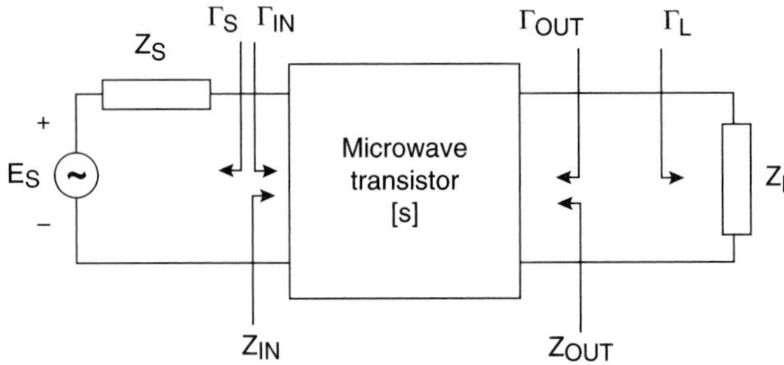

FIGURE 12.2 General two-port network.

Thus, considering a passive source or load impedance where $|\Gamma_S| < 1$ and $|\Gamma_L| < 1$, two types of "stability conditions" can be defined as follows:

- **Unconditional stability:** A network is said to be "Unconditionally stable" in a frequency range if, and only if:

$$|\Gamma_{IN}| < 1, \tag{12.5a}$$

and

$$|\Gamma_{OUT}| < 1 \text{ for all } |\Gamma_S| < 1 \text{ and } |\Gamma_L| < 1 \tag{12.5b}$$

- **Conditional stability:** A network is said to be "conditionally stable" or "potentially unstable" in a frequency range if:

$$|\Gamma_{IN}| < 1 \tag{12.6a}$$

and

$$|\Gamma_{OUT}| < 1 \tag{12.6b}$$

only for a limited range of values of passive source and load impedances (or $|\Gamma_S|$ and $|\Gamma_L|$), but not for all values.

Using signal flow graphs from Chapter 8, Γ_{IN} and Γ_{OUT} from Equations 8.70 and 8.71 are given by:

$$\Gamma_{IN} = S_{11} + \frac{S_{12}S_{21}\Gamma_L}{1 - S_{22}\Gamma_L} \tag{12.7}$$

$$\Gamma_{OUT} = S_{22} + \frac{S_{12}S_{21}\Gamma_S}{1 - S_{11}\Gamma_S} \tag{12.8}$$

A Special Case: Unilateral Transistor. If the transistor is unilateral, i.e., if $S_{12} = 0$, then the equations for unconditional stability simplify to the following:

$$|S_{11}| < 1 \tag{12.9a}$$

$$|S_{22}| < 1 \tag{12.9b}$$

If the above condition cannot be met by the transistor at all times, then the amplifier circuit is considered to be conditionally stable (for a certain range of Γ_S and Γ_L), which will be discussed in more depth shortly.

12.2 STABILITY CIRCLES

As discussed earlier, the unconditional stability for a general two-port network having passive source/load impedances (as shown in Figure 12.3) requires the conditions as given by Equations 12.5a and b. The boundaries between stable and potentially unstable regions of Γ_S and Γ_L, however, are determined by replacing the inequality signs in Equations 12.5 with equality signs, giving:

$$|\Gamma_{IN}| = \left| S_{11} + \frac{S_{12}S_{21}\Gamma_L}{1 - S_{22}\Gamma_L} \right| = 1 \tag{12.10a}$$

$$|\Gamma_{OUT}| = \left| S_{22} + \frac{S_{12}S_{21}\Gamma_S}{1 - S_{11}\Gamma_S} \right| = 1 \tag{12.10b}$$

when

$$|\Gamma_S| < 1 \tag{12.10c}$$

and

$$|\Gamma_L| < 1 \tag{12.10d}$$

12.3 GRAPHICAL SOLUTION OF STABILITY CRITERIA

Considering the Γ_L and Γ_S planes, the loci of points for which $|\Gamma_{IN}| = 1$ and $|\Gamma_{OUT}| = 1$ are found to be two circles called:

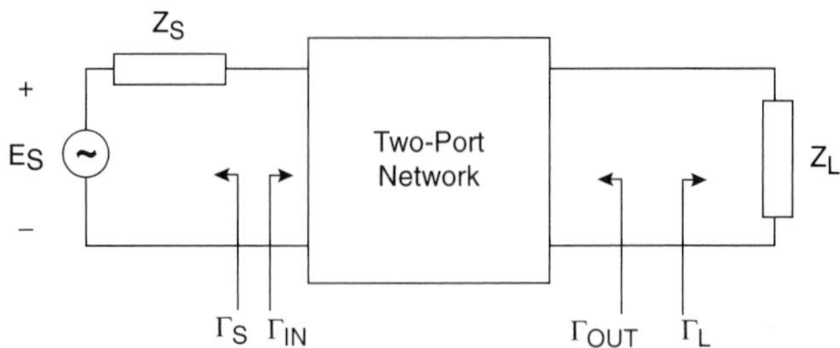

FIGURE 12.3 Stability of a two-port network.

- The input stability circle
- The output stability circle

These two stability circles define the boundaries between "stable" and "unstable" regions for different values of Γ_S and Γ_L. For passive matching networks, these values lie inside the standard Smith chart ($|\Gamma_S| \leq 1$, $|\Gamma_L| \leq 1$). Thus, the intersection of the two stability circles with the standard Smith chart provides the stable and unstable regions.

The general equation of a circle in the Γ plane can be written as:

$$|\Gamma - C| = R \tag{12.11}$$

where C is a complex number representing the center of the circle and R is a real positive number representing the circle's radius.

The equation for the output stability circle ($|\Gamma_{IN}| = 1$) drawn in the Γ_L plane can be written as (see Equation 12.11 and Figure 12.4):

$$|\Gamma_L - C_L| = R_L \tag{12.12}$$

where:

$$C_L = \frac{(S_{22} - \Delta S_{11}^*)^*}{D_L} \tag{12.13a}$$

$$R_L = |S_{12}S_{21}/D_L| \tag{12.13b}$$

$$D_L = |S_{22}|^2 - |\Delta|^2 \tag{12.13c}$$

$$\Delta = S_{11}S_{22} - S_{12}S_{21} \tag{12.13d}$$

Similarly, the input stability circle ($|\Gamma_{OUT}| = 1$) drawn in the Γ_S-plane (see Figure 12.5) is obtained by interchanging S_{11} for S_{22} in Equations 12.12–12.13 and is given by:

$$|\Gamma_S - C_S| = R_S \tag{12.14}$$

where:

$$C_S = \frac{(S_{11} - \Delta S_{22}^*)^*}{D_S} \tag{12.15a}$$

$$R_S = |S_{12}S_{21}/D_S| \tag{12.15b}$$

$$D_S = |S_{11}|^2 - |\Delta|^2 \tag{12.15c}$$

$$\Delta = S_{11}S_{22} - S_{12}S_{21} \tag{12.15d}$$

Figures 12.4 and 12.5 illustrate the graphical plot of the input and output stability circles where the circles divide their respective planes (Γ_S-plane or Γ_L-plane) into two regions:

- The first region is the stable region and is characterized by $|\Gamma_{OUT}| < 1$ for input stability circle in the Γ_S-plane (or $|\Gamma_{IN}| < 1$ for output stability circle in the Γ_L-plane).
- The second region is the unstable region and is characterized by $|\Gamma_{OUT}| > 1$ for input stability circle in the Γ_S-plane (or $|\Gamma_{IN}| > 1$ for output stability circle in the Γ_L-plane).

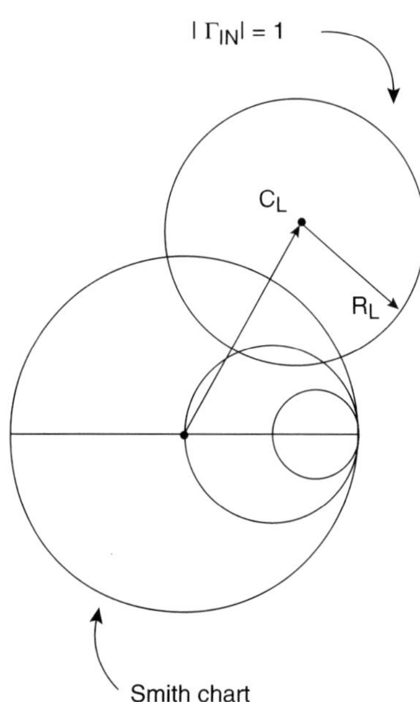

FIGURE 12.4 Output stability circle in the Γ_L plane.

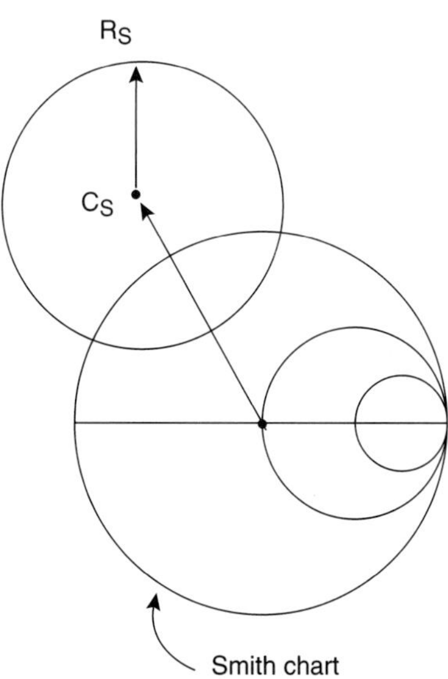

FIGURE 12.5 Input stability circle in the Γ_S plane.

We now need a method to determine which region corresponds to the stable region so that it can be subsequently used for amplifier design.

OBSERVATION: *Because the S-parameters are frequency dependent, the size and position of the input and output stability circles would change as frequency is varied. This occurs because the center and radius of each circle are expressed in terms of the S-parameters.*

12.3.1 Output Stability Circle

The output stability circles are plotted in the Γ_L plane, as shown in Figure 12.4. If we set $\Gamma_L = 0$ (i.e., $Z_L = Z_o$, the center of the chart), then from Equation 12.7 we have:

$$|\Gamma_{IN}| = |S_{11}| \tag{12.16}$$

Now if the $|S_{11}|$ of the device is less than unity (i.e., $|S_{11}| < 1$), then that region of the output stability circle that includes the center of the Smith chart is the "stable region," and the second region would be the unstable area. For example, if the stability circle does not include the center of the Smith chart, as shown in Figure 12.4, then Γ_L values located outside of the stability circle are in the stable region and Γ_L values inside the stability circle are in the unstable region.

Vice versa, if $|S_{11}|$ of the device is more than unity (i.e., $|S_{11}| > 1$), then the center of the Smith chart is in the unstable region. For example, in Figure 12.4, the stable region would be inside the stability circle and the outside would be the unstable region.

12.3.2 Input Stability Circle

Similar to the output stability circles, the input stability circles are plotted in the Γ_S plane. If we set $\Gamma_S = 0$ (i.e., the center of the Smith chart), then from Equation 12.8 we have:

$$|\Gamma_{OUT}| = |S_{22}| \tag{12.17}$$

In this case, if $|S_{22}|$ of the device is less than unity (i.e., $|S_{22}| < 1$), then one region of the circle containing the center of the Smith chart is the stable region and the other region is unstable. For example, the input stability circle, as shown in Figure 12.5, does not include the center of the Smith chart. Thus, the outside of the circle is stable and the inside unstable. Vice versa, for $|S_{22}| > 1$ the outside of the input stability circle in Figure 12.5 is unstable and the inside stable.

12.3.3 Special Case: Unconditional Stability

For $|S_{11}| < 1$ and $|S_{22}| < 1$, the amplifier circuit will be unconditionally stable when either of the following two conditions hold true:

- Both stability circles fall completely outside the Smith chart
- Both stability circles completely enclose the smith chart

Therefore, for all passive sources and load impedances, the unconditional stability criteria (for the two preceding cases) can be concisely stated in mathematical form as:

$$||C_L| - R_L| > 1 \text{ for } |S_{11}| < 1 \tag{12.18a}$$

$$||C_S| - R_S| > 1 \text{ for } |S_{22}| < 1 \tag{12.18b}$$

Equations 12.18a and b in essence state that the distance between the center of the Smith chart and the center of the stability circle when subtracted by the radius of the stability circle in absolute value form (i.e., neglecting the sign) must be larger than the Smith chart's radius. An example of the input stability circle in the Γ_S plane for unconditional stability is shown in Figure 12.6.

FIGURE 12.6 Two possibilities of input stability circle relative to the Smith chart: (a) completely outside, (b) completely enclosing the Smith chart.

(a) $|S_{22}| < 1$

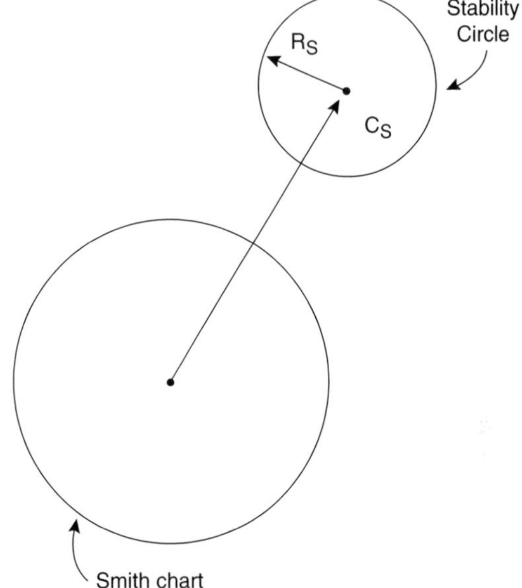

(b) $|S_{22}| < 1$

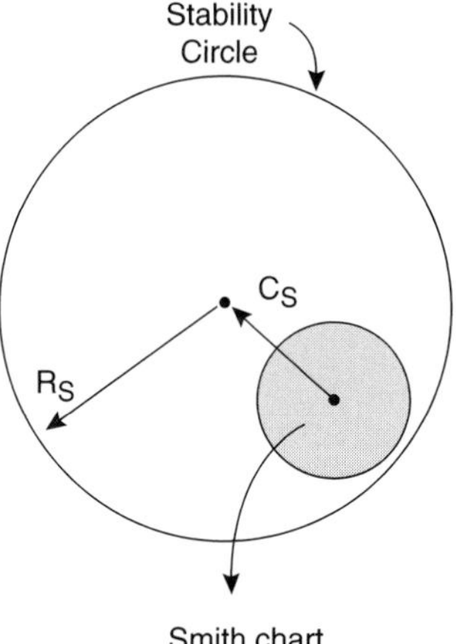

NOTE: *If $|S_{11}| > 1$ or $|S_{22}| > 1$, then the amplifier cannot be unconditionally stable because at the center of the Smith chart (where $\Gamma_S = 0$, $\Gamma_L = 0$), we have $|\Gamma_{IN}| = |S_{22}| > 1$ and $|\Gamma_{OUT}| = |S_{11}| > 1$, which obviously is an unstable region.*

12.4 ANALYTICAL SOLUTION OF STABILITY CRITERIA

The necessary and sufficient conditions for an amplifier to be unconditionally stable can be mathematically derived from Equations 12.10a, b, c and d. The result of this derivation is a set of mathematical conditions that can be concisely referred to as "two- or three-parameter test criteria," as follows.

Let's define the determinant (Δ) of the S-matrix and factors K and B_1 as:

$$\Delta = S_{11}S_{22} - S_{12}S_{21}, \tag{12.19}$$

$$K = \frac{1 - |S_{11}|^2 - |S_{22}|^2 + |\Delta|^2}{2|S_{12}S_{21}|} \tag{12.20}$$

and

$$B_1 = 1 + |S_{11}|^2 - |S_{22}|^2 - |\Delta|^2 \tag{12.21}$$

Based on these definitions, a two-port network will be unconditionally stable if, and only if, either one of the following mathematically equivalent criteria are satisfied.

Criterion 1: Three-Parameter Test Criterion.

$$K > 1, \tag{12.22}$$

and

$$\frac{1 - |S_{11}|^2}{|S_{12}S_{21}|} > 1, \tag{12.23}$$

and

$$\frac{1 - |S_{22}|^2}{|S_{12}S_{21}|} > 1 \tag{12.24}$$

We can shrink the three-parameter test into a two-parameter test as follows.

Criterion 2: Two-Parameter Test Criterion (K-Δ Test).

$$K > 1 \tag{12.25}$$

$$|\Delta| < 1 \tag{12.26}$$

This is often referred to as the "K-Δ Test."

Criterion 3: Two-Parameter Test Criterion (K-B_1 Test).

$$K > 1 \tag{12.27}$$

$$B_1 > 0 \tag{12.28}$$

This may also be called the "K-B_1 Test."

These three criteria are mathematically equivalent, and if a device satisfies any one of the three criteria, the other two are automatically satisfied.

Thus, a two-port network will be unconditionally stable if and only if any one of these three criteria are satisfied.

NOTE 1: *The two-parameter test criteria (Criteria 2 and 3) are more popular and more often used than Criterion 1, primarily due to their simplicity and ease of calculation.*

NOTE 2: *For a unilateral transistor, we have:*

$$S_{12} = 0 \quad \Rightarrow \quad K = \infty > 1 \tag{12.29}$$

and

$$|\Delta| = |S_{11}S_{22}| \tag{12.30}$$

Because K > 1 has already been satisfied, in order to satisfy the condition for unconditional stability we desire $|\Delta| < 1$*, which requires:*

$$|S_{11}| < 1 \tag{12.31}$$

$$|S_{22}| < 1 \tag{12.32}$$

for all passive values of Z_S *and* Z_L.

This conclusion is in agreement with the earlier discussion.

EXAMPLE 12.1

Determine the stability of a GaAs FET that has the following S-parameters at 2 GHz in a 50 Ω system both graphically and mathematically:

$$S_{11} = 0.89\angle{-60^\circ}$$

$$S_{21} = 3.1\angle{123^\circ}$$

$$S_{12} = 0.02\angle{62^\circ}$$

$$S_{22} = 0.78\angle{-27^\circ}$$

Solution:

a. *Graphical method.* We calculate the following values:

$$C_L = 1.36\angle{47^\circ}, \; R_L = 0.5$$

$$C_S = 1.13\angle{68^\circ}, \; R_S = 0.2$$

Input and output stability circles are plotted in Figure 12.7. From this figure, we can see that the GaAs FET is "potentially unstable" and the center of the Smith chart ($\Gamma_s = 0$, $\Gamma_L = 0$), being outside of the circles, represents stable regions because at this point we have:

$$|\Gamma_{IN}| = |S_{11}| = 0.89 < 1$$

$$|\Gamma_{OUT}| = |S_{22}| = 0.78 < 1$$

b. *Mathematical method.* We calculate the following values:

$$K = 0.66$$

$$\Delta = 0.7\angle{-83^\circ}$$

Because $K < 1$, the transistor is potentially unstable, which is in agreement with the graphical method.

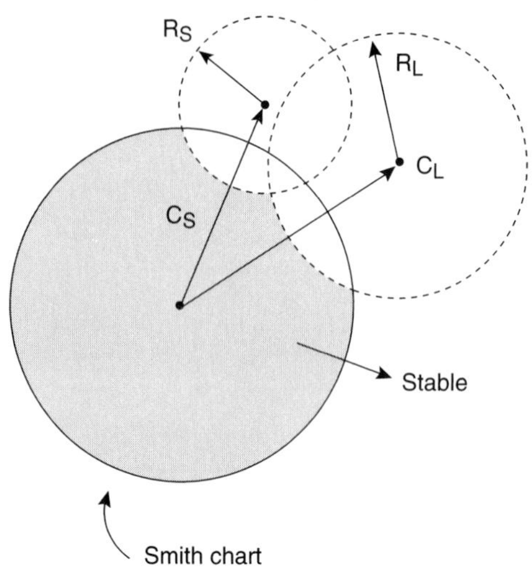

FIGURE 12.7 Input and output stability circles. ($|S_{11}| < 1, |S_{22}| < 1$).

Single-Parameter (or μ-Parameter) Test Considering the more popular two-parameter test, we can see that the two-parameter test (in particular the K-Δ test) provides a set of mathematical conditions on two parameters for unconditional stability, and it indicates only whether a device is stable or not. Due to the fact that certain constraints are imposed on the two parameters, the "two-parameter" test cannot be used to show the degree of stability of one device relative to other similar devices.

To determine the unconditional stability of a device as well as its degree of stability relative to other devices, a new criterion has been derived that combines the K–Δ parameters into a single-parameter test; this is often referred to as the "**μ-parameter test**." The parameter μ is defined as:

$$\mu = \frac{1 - |S_{11}|^2}{|S_{22} - S_{11}^*\Delta| + |S_{21}S_{12}|} \tag{12.33a}$$

For unconditional stability, the following must be satisfied:

$$\mu > 1 \tag{12.33b}$$

Furthermore, if device A has a parameter μ_A that is greater than μ_B corresponding to device B, i.e.,

$$\mu_A > \mu_B \tag{12.34}$$

then device A is said to be more stable than device B.

Equation 12.34 indicates that a device with a larger value of μ is more desirable for an amplifier design because it implies a greater degree of stability.

12.5 POTENTIALLY UNSTABLE CASE

Sometimes when Γ_s and Γ_L are chosen such that:

$$|\Gamma_{IN}| > 1 \text{ or } |\Gamma_{OUT}| > 1 \tag{12.35}$$

then the amplifier circuit becomes potentially unstable. In these situations, the device could be made unconditionally stable if the total input and output loop resistance is made to be positive, i.e.,

$$Re(Z_S + Z_{IN}) > 0 \tag{12.36}$$

$$Re(Z_L + Z_{OUT}) > 0 \tag{12.37}$$

To achieve a positive loop resistance and thus make a potentially unstable transistor into a conditionally stable one, two methods normally are employed:

- Resistively loading the transistor
- Adding negative feedback

Use of these techniques brings about a reduction in the gain, an increase in the noise figure, and a degradation of the amplifier power output.

These two techniques are useful in broadband, potentially unstable amplifiers where the wide frequency range increases the probability of instability. First, the resistive loading is used to stabilize the transistor and then negative feedback is used to provide a relatively constant gain with a low input and output *VSWR*.

In narrowband amplifiers, use of these techniques is not recommended; instead careful selection of Γ_s and Γ_L in the early stages of the design is necessary to ensure a stable amplifier.

EXAMPLE 12.2

A BJT has the following S-parameters:

$$S_{11} = 0.65\angle{-95°}$$

$$S_{21} = 5.0\angle 115°$$

$$S_{12} = 0.035\angle 40°$$

$$S_{22} = 0.8\angle{-35°}$$

Is this transistor unconditionally stable? If not, use resistive loading to make the transistor conditionally stable. What are the resistor values?

Solution:

Simple calculations give us:

$$K = 0.547$$

$$\Delta = 0.504\angle{-110.4°}$$

Because $K < 1$, the transistor is potentially unstable!

To draw the output stability circles, we find the output stability circle C_L and R_L from Equation 12.13 as follows:

$$C_L = \frac{(S_{22} - \Delta S_{11}^*)^*}{D_L},$$

$$R_L = |S_{12}S_{21}/D_L|$$

$$D_L = |S_{22}|^2 - |\Delta|^2$$

$$\Delta = S_{11}S_{22} - S_{12}S_{21}$$

$$C_L = 1.3\angle 48°, R_L = 0.45$$

Similarly, for the input stability circle, we calculate C_S and R_S from Equation 12.15 as follows:

$$C_S = \frac{(S_{11} - \Delta S_{22}^*)^*}{D_S}$$

$$R_S = |S_{12}S_{21}/D_S|$$

$$D_S = |S_{11}|^2 - |\Delta|^2$$

$$C_S = 1.79\angle 122°, R_S = 1.04$$

These two circles are drawn in Figure 12.8.

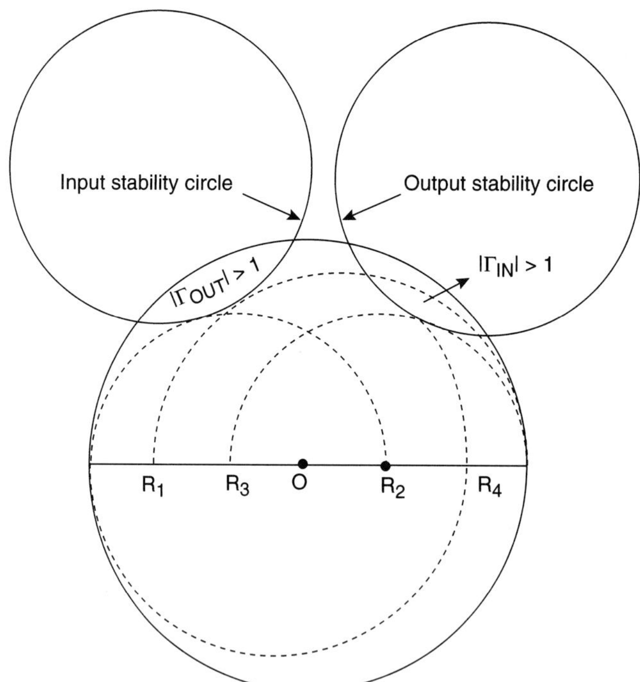

FIGURE 12.8 Input and output stability circles.

Four types of resistive loading are possible to improve stability, as shown in Figure 12.9.

Using the input stability circle, it can be seen that a series resistor of $R_1 = 9\ \Omega$ (see Figure 12.9a) or a shunt resistor of $R_2 = 71\ \Omega$ (see Figure 12.9b) at the input of the transistor will restore the stability.

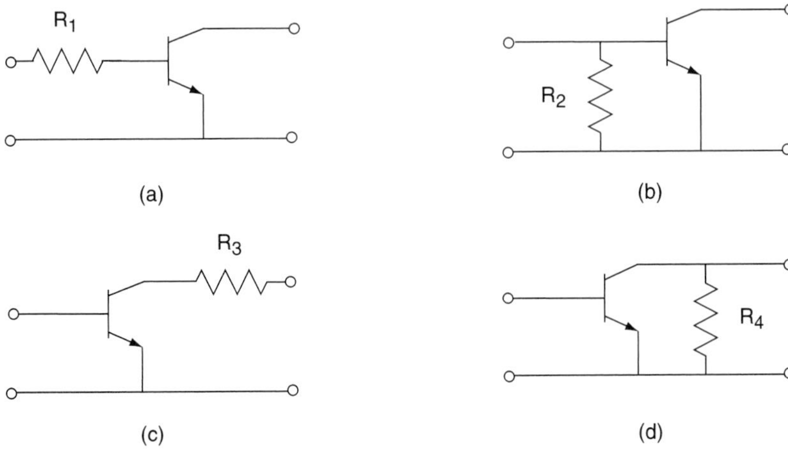

FIGURE 12.9 Four types of resistive loading.

On the other hand, using the output stability circle, we can see that a series resistor of $R_3 = 43\ \Omega$ (see Figure 12.9c) or a shunt resistor of $R_4 = 500\ \Omega$ (see Figure 12.9d) at the output of the transistor will ensure stability

---

It should be noted that in most cases, stabilizing either the input or the output port will restore stability to the transistor. Thus, any one of the four types of resistive loading should be sufficient to create a stable amplifier.

POINT OF CAUTION: *Use of resistive loading at the input of the transistor (see Figures 12.9a and b) is not recommended due to an increase in the input loss, which translates into a higher noise figure at the output of the amplifier. Any resistive loading is preferred to take place at the output of the amplifier to minimize the increase in the amplifier's noise figure. This effect will be studied in more depth in Chapter 14, Noise Considerations in Active Networks.*

LIST OF SYMBOLS/ABBREVIATIONS

A symbol/abbreviation will not be repeated again once it has been identified and defined in an earlier chapter, as long as its definition remains unchanged.

- B_1 A factor used in stability calculations (K-B_1 test)
- K Stability factor used to evaluate the stability of a two-port network (K-Δ test)
- Δ Determinant of the S-matrix used to evaluate stability (K-Δ test)
- μ A parameter used to evaluate the stability of a network; the test is called the μ-parameter test

PROBLEMS

12.1 In each of the stability circle drawings shown in Figure P12.1, clearly indicate the possible locations for a stable source reflection coefficient.

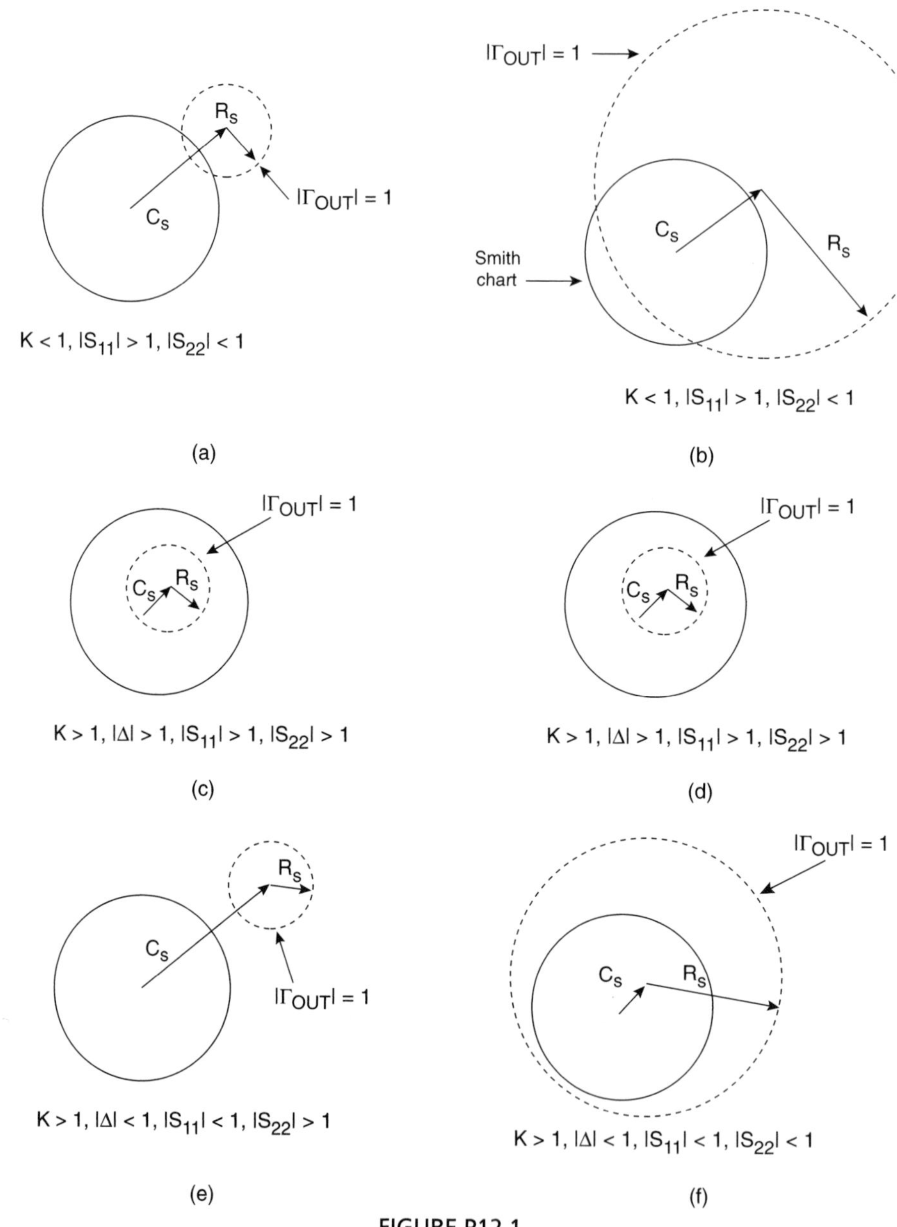

FIGURE P12.1

12.2 Output stability circles are shown in Figure P12.2. Determine the stable region for the load reflection coefficient.

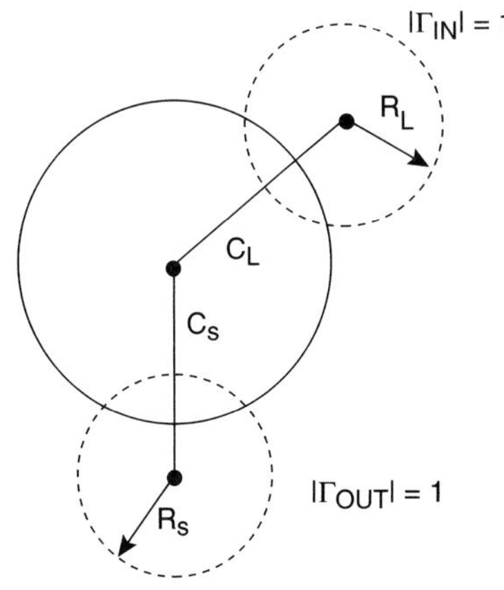

(a)

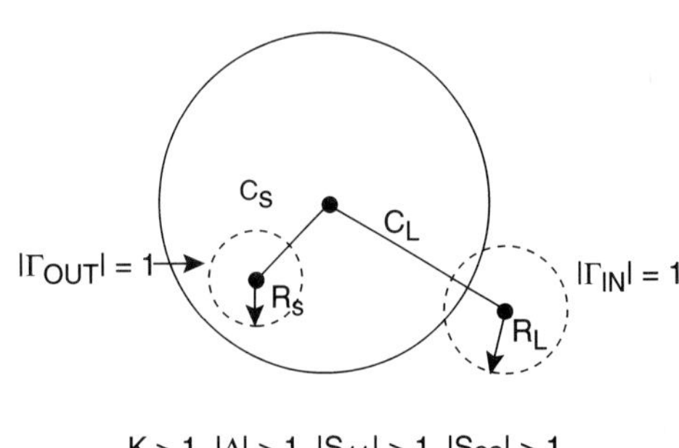

(b)

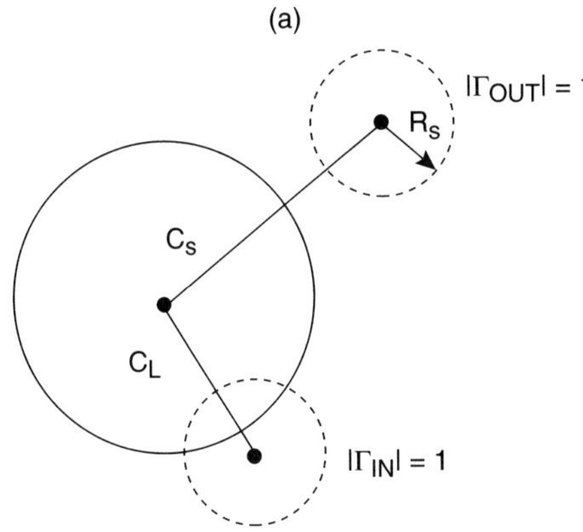

(c)

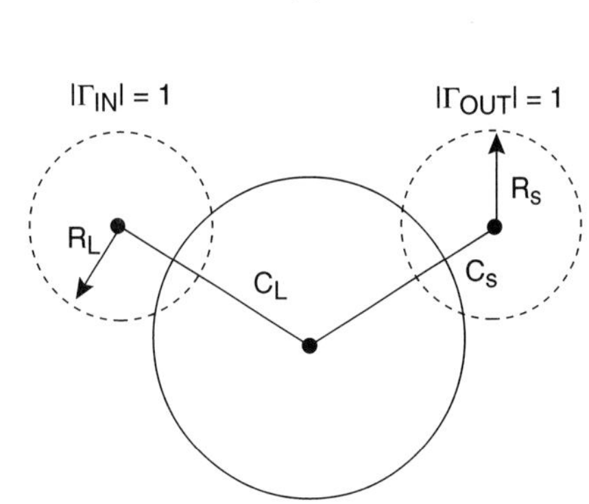

(d)

FIGURE P12.2

12.3 The scattering parameters for three different transistors are given below. Determine the stability in each case, and in a potentially unstable case, draw the input and output stability circles:

a.
$$S = \begin{bmatrix} 0.67\angle{-67^\circ} & 0.075\angle 6.2^\circ \\ 1.74\angle 36.4^\circ & 0.6\angle{-92.6^\circ} \end{bmatrix}$$

b.
$$S = \begin{bmatrix} 0.385\angle{-55^\circ} & 0.045\angle 90^\circ \\ 2.7\angle 78^\circ & 0.89\angle{-26.5^\circ} \end{bmatrix}$$

c.
$$S = \begin{bmatrix} 0.7\angle{-50^\circ} & 0.27\angle 75^\circ \\ 5\angle 120^\circ & 0.6\angle 80^\circ \end{bmatrix}$$

12.4 The S-parameters of a GaAs FET at a certain Q-point are given in the following table:

f(GHz)	S_{11}	S_{12}	S_{21}	S_{22}
4	$0.9\angle{-67^\circ}$	$0.076\angle 43^\circ$	$2.3\angle{-118^\circ}$	$0.68\angle{-39^\circ}$
6	$0.84\angle{-97^\circ}$	$0.112\angle 24^\circ$	$2.06\angle 87^\circ$	$0.6\angle{-58^\circ}$
8	$0.73\angle{-140^\circ}$	$0.135\angle{-5^\circ}$	$2.04\angle 53^\circ$	$0.47\angle{-85^\circ}$
10	$0.67\angle{-178^\circ}$	$0.146\angle{-27^\circ}$	$1.81\angle 18^\circ$	$0.42\angle{-120^\circ}$
12	$0.63\angle 115^\circ$	$0.133\angle{-66^\circ}$	$1.42\angle{-38^\circ}$	$0.36\angle{-172^\circ}$

Draw the input stability as well as the output stability circles (at each frequency) in a Smith chart. Indicate the unstable regions.

12.5 The S-parameters of several two-port networks are given by:

a.
$$S = \begin{bmatrix} 0.7\angle 0^\circ & 0.7\angle 180^\circ \\ 0.7\angle 0^\circ & 0.7\angle 0^\circ \end{bmatrix}$$

b.
$$S = \begin{bmatrix} 0.7 & 1.7 \\ 1.7 & 0.7 \end{bmatrix}$$

c.
$$S = \begin{bmatrix} 1 & 0.7 \\ 0.7 & 1 \end{bmatrix}$$

Determine K and $|\Delta|$. Draw the input and output stability circles for each case as well.

12.6 **a.** Show that in the limit as S_{12} approaches zero, the center and radius of the stability circles are given by $C_S \approx 1/S_{22}$, $R_S \approx 0$, $C_L \approx 1/S_{11}$, and $R_L \approx 0$.

b. The S-parameters of a two-port network are given by:

$$S = \begin{bmatrix} 2\angle 90^\circ & 0 \\ 2 & 0.1\angle 45^\circ \end{bmatrix}$$

Draw the stability circles and show the unstable regions.

12.7 Show how resistive loading can stabilize a resistor whose S-parameters at f=750 MHz are:

$$S = \begin{bmatrix} 0.69\angle 78^\circ & 0.033\angle 41.4^\circ \\ 5.67\angle 123^\circ & 0.84\angle 25^\circ \end{bmatrix}$$

Consider all four types of resistive loading for this problem.

12.8 A microwave GaAs FET has the following S-parameters measured at

$$V_{DS} = 3\text{ V and } I_D = 10\text{ mA at 4 GHz:}$$

$$S = \begin{bmatrix} 0.89\angle{-50^\circ} & 0.06\angle 66^\circ \\ 3.26\angle 141^\circ & 0.58\angle{-24^\circ} \end{bmatrix}$$

a. Calculate the delta factor (Δ).
b. The stability factor (K).
c. Find the center and radius of the input stability circle and plot the circle.
d. Determine the center and radius of the output stability circle and plot the circle.

12.9 Considering the diagram shown in Figure P12.9, indicate the possible locations for a stable source impedance and stable load impedance. The solid circle is the Smith chart.

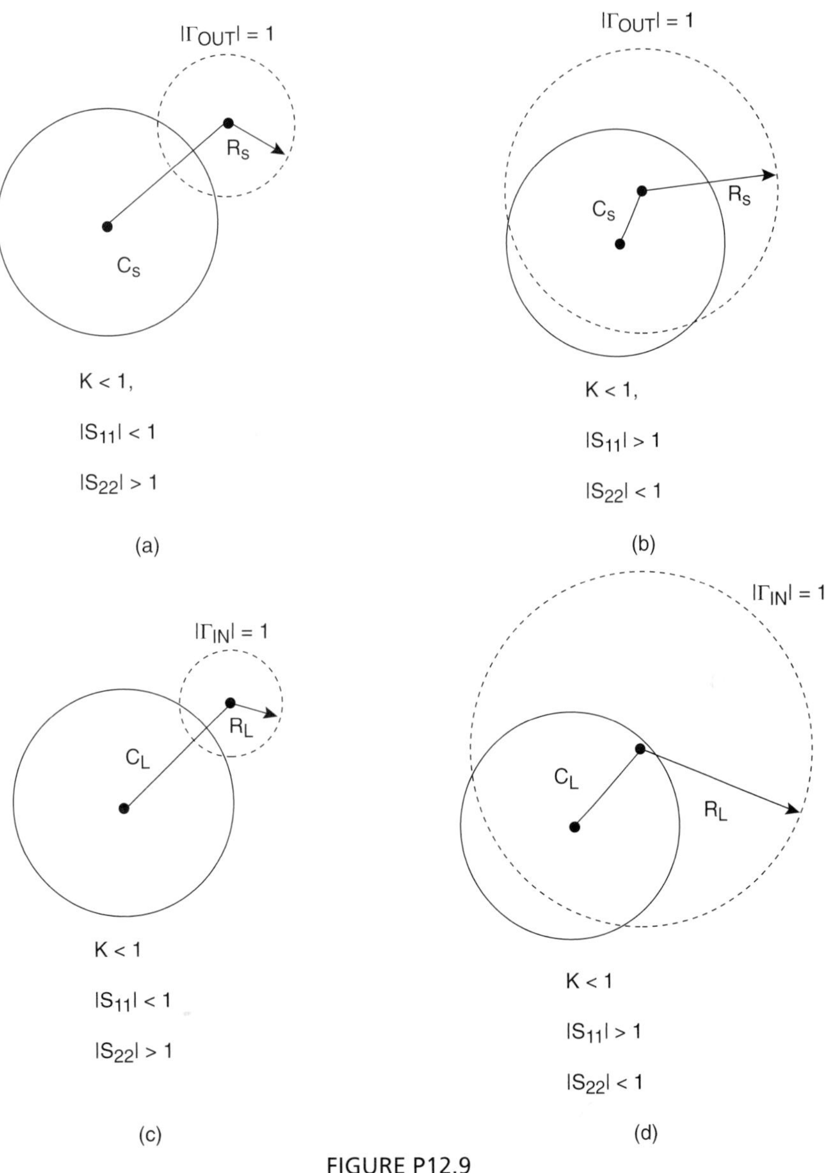

FIGURE P12.9

12.10 A microwave transistor has the following S-parameters:

$$S = \begin{bmatrix} 0.8\angle{-170^\circ} & 0.1\angle 80^\circ \\ 5.1\angle 70^\circ & 0.62\angle{-40^\circ} \end{bmatrix}$$

Determine the stability, and plot the stability circles if the device is potentially unstable.

REFERENCES

[12.1] Cheung, W. S. and F. H. Levien. *Microwave Made Simple, Principles and Applications.* Norwood: Artech House, 1985.

[12.2] Gonzalez, G. *Microwave Transistor Amplifiers, Analysis and Design*, 2nd ed. Upper Saddle River: Prentice Hall, 1997.

[12.3] Liao, S. Y. *Microwave Circuit Analysis and Amplifier Design.* Upper Saddle River: Prentice Hall, 1987.

[12.4] Pozar, D. M. *Microwave Engineering*, 2nd ed. New York: John Wiley & Sons, 1998.

[12.5] Schwarz, S. E. *Electromagnetics for Engineers.* Orlando: Saunders College Publishing, 1990.

[12.6] Vendelin, George D. *Design of Amplifiers and Oscillators by the S-Parameter Method.* New York: John Wiley & Sons, 1982.

[12.7] Vendelin, George D., Anthony M. Pavio, and Ulrich L. Rhode. *Microwave Circuit Design, Using Linear and Non-Linear Techniques.* New York: John Wiley & Sons, 1990.

[12.8] Woods, D. *Reappraisal of the Unconditional Stability Criteria.* IEEE Transactions on Circuits and Systems, Feb. 1976.

CHAPTER 13

Gain Considerations in Amplifiers

13.1 INTRODUCTION

Gain consideration in an amplifier plays an important role in the design process. As discussed earlier, the primary consideration in an amplifier is its stability, with its power gain following very closely as second in importance.

13.2 POWER GAIN CONCEPTS

Consider the single stage microwave transistor amplifier with the transistor straddled by two matching networks on either side, as shown in Figure 13.1.

Several power gain concepts are commonly used in the amplifier design process. Each one of these power gains has a specific name with a specific definition. Therefore at the outset, let's first define the various power levels existing in the circuit, as shown in Figure 13.2.

$P_{IN} \equiv$ Power input to the transistor or to the input matching network.

$P_{AVS} \equiv$ Power available from the source under matched condition. This is a special case of P_{IN} when $\Gamma_{IN} = \Gamma^*_S$.

$P_L \equiv$ Power delivered to the load or the output matching network.

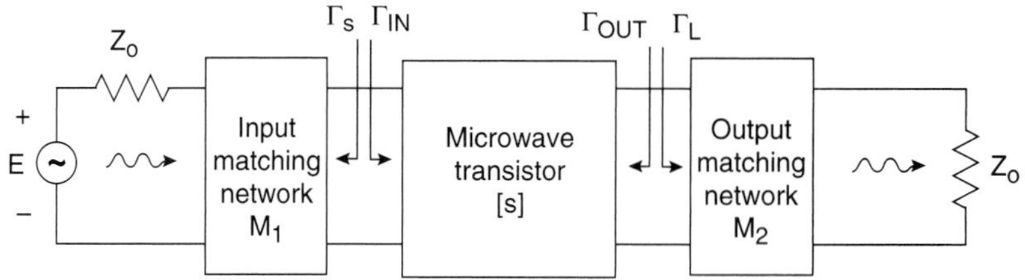

FIGURE 13.1 A general block diagram for a transistor amplifier.

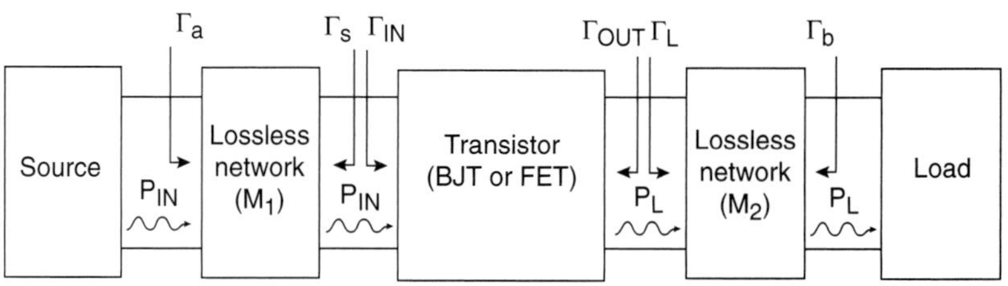

FIGURE 13.2 A transistor amplifier.

P_{AVN} ≡ Power available from the transistor under matched condition; a special case of P_L when $\Gamma_L = \Gamma^*_{OUT}$.

Based on these definitions of power, we can now define the following power gain equations:

$G_T \equiv \dfrac{P_L}{P_{AVS}}$ Transducer power gain

$G_P \equiv \dfrac{P_L}{P_{IN}}$ Operating power gain (also called power gain)

$G_A \equiv \dfrac{P_{AVN}}{P_{AVS}}$ Available power gain

Using signal flow graphs (or Mason's rule), the three power ratios as defined above, were already derived in Chapter 8, and the results are stated as follows:

a. Transducer gain (G_T)

$$G_T = \frac{1-|\Gamma_S|^2}{|1-\Gamma_{IN}\Gamma_S|^2}|S_{21}|^2\frac{1-|\Gamma_L|^2}{|1-S_{22}\Gamma_L|^2} \tag{13.1}$$

Equation 13.1 can be written as:

$$G_T = G_S.G_o.G_L \tag{13.2}$$

where

$$G_S = \frac{1-|\Gamma_S|^2}{|1-\Gamma_{IN}\Gamma_S|^2} \tag{13.3}$$

$$G_o = |S_{21}|^2, \tag{13.4}$$

$$G_L = \frac{1 - |\Gamma_L|^2}{|1 - S_{22}\Gamma_L|^2} \tag{13.5}$$

From Equation 13.2, we may attribute G_o to the gain of the transistor while G_S and G_L are attributable to the effective gains of the input and output matching networks.

Alternately, Equation 13.1 can also be written in terms of Γ_{OUT} (rather than Γ_{IN}) as:

$$G_T = \frac{1 - |\Gamma_S|^2}{|1 - S_{11}\Gamma_S|^2} |S_{21}|^2 \frac{1 - |\Gamma_L|^2}{|1 - \Gamma_{OUT}\Gamma_L|^2} \tag{13.6}$$

b. Operating power gain (G_P)

$$G_P = \frac{1}{|1 - \Gamma_{IN}|^2} |S_{21}|^2 \frac{1 - |\Gamma_L|^2}{|1 - S_{22}\Gamma_L|^2} \tag{13.7}$$

c. Available power gain (G_A)

$$G_A = \frac{1 - |\Gamma_S|^2}{|1 - S_{11}\Gamma_S|^2} |S_{21}|^2 \frac{1}{1 - |\Gamma_{OUT}|^2} \tag{13.8}$$

where

$$\Gamma_{IN} = S_{11} + \frac{S_{12}S_{21}\Gamma_L}{1 - S_{22}\Gamma_L} \tag{13.9}$$

$$\Gamma_{OUT} = S_{22} + \frac{S_{12}S_{21}\Gamma_S}{1 - S_{11}\Gamma_S} \tag{13.10}$$

NOTE: *From Equation 13.2, the terms (G_S and G_L) represent input and output matching networks' degree of matching to the transistor at its input or its output. The matching networks are made up of passive components and have no inherent gain; thus, they are incapable of generating power. Nevertheless, because input and output matching networks are capable of increasing the degree of matching in the circuit as the signal flows through, they can be considered to have a positive gain in a relative manner. Thus, we can write Equation 13.2 in dB as:*

$$G_T(\text{dB}) = G_S(\text{dB}) + G_o(\text{dB}) + G_L(\text{dB}) \tag{13.11}$$

13.3 A SPECIAL CASE: UNILATERAL TRANSISTOR

If the transistor is unilateral, i.e., $S_{12} = 0$, then G_S and G_L gain blocks as well as Γ_{IN} and Γ_{OUT} simplify into:

$$G_{TU} = G_{SU}.G_o.G_{LU} \tag{13.12}$$

where

$$\Gamma_{IN} = S_{11} \tag{13.13}$$

$$\Gamma_{OUT} = S_{22} \tag{13.14}$$

$$G_{SU} = \frac{1-|\Gamma_S|^2}{|1-S_{11}\Gamma_S|^2} \tag{13.15}$$

$$G_{LU} = \frac{1-|\Gamma_L|^2}{|1-S_{22}\Gamma_L|^2} \tag{13.16}$$

13.4 THE MISMATCH FACTOR

The source mismatch factor (M_S), can be defined as:

$$M_S \equiv \frac{P_{IN}}{P_{AVS}} \tag{13.17a}$$

This ratio is always less than or at best equal to unity, (i.e., $M_S \leq 1$) because P_{IN} is always less than or equal to P_{AVS}.

By observation, we can see that the equation for M_S can be obtained simply by taking the ratio of G_T/G_P, which yields:

$$M_S = \frac{(1-|\Gamma_S|^2)(1-|\Gamma_{IN}|^2)}{|1-\Gamma_S\Gamma_{IN}|^2} \tag{13.17b}$$

Source mismatch factor (M_S) is used to quantify the portion of P_{AVS} that is delivered to the input of the transistor.

If the input port is matched (i.e., $\Gamma_{IN} = \Gamma^*_S$) making $P_{IN} = P_{AVS}$, then we can see that $M_S = 1$. This means that all of the available power from the source is delivered to the transistor and no mismatch exists at the input port, i.e.,

$$P_{IN} = P_{AVS}\big|_{\Gamma_{IN}=\Gamma_S^*} \Rightarrow M_S = 1 \tag{13.18}$$

Similarly, the load mismatch factor (M_L), can be defined as:

$$M_L \equiv \frac{P_L}{P_{AVN}} \tag{13.19a}$$

This ratio is always less than or at best equal to unity, (i.e., $M_L \leq 1$) because P_L is always less than or equal to P_{AVN}.

By observation, we can see that the equation for M_L can be obtained simply by taking the ratio of G_T/G_A, which yields:

$$M_L = \frac{(1-|\Gamma_L|^2)(1-|\Gamma_{OUT}|^2)}{|1-\Gamma_{OUT}\Gamma_L|^2} \tag{13.19b}$$

The load mismatch factor (M_L) is used to quantify the portion of P_{AVN} that is delivered to the load.

If the output port is matched (i.e., $\Gamma_{OUT} = \Gamma^*_L$) making $P_L = P_{AVN}$, then we can see that $M_L = 1$. This means that all of the available power from the transistor is delivered to the load and no mismatch exists at the output port, i.e.,

$$P_{OUT} = P_{AVN}\big|_{\Gamma_{OUT}=\Gamma_L^*} \Rightarrow M_L = 1 \tag{13.20}$$

NOTE: *The "mismatch factor" is also called "mismatch loss," which (in dB) signifies the amount of power loss due to mismatch. From Equations 13:17 and 13.19, we can write the following:*

$$M_S(\text{dB}) = P_{IN}(\text{dBm}) - P_{AVS}(\text{dBm}), \quad M_S < 0 \tag{13.21a}$$

$$M_L(\text{dB}) = P_L(\text{dBm}) - P_{AVN}(\text{dBm}), \quad M_L < 0 \tag{13.21b}$$

13.4.1 Constancy of the Mismatch Factor

For a "lossless network," because the output power equals the input power, it can be shown mathematically that the mismatch factor always remains constant.

For example, the mismatch factor (M_S) at the input of the lossless matching network (M_1), where the source is connected, has the same value as at its output where the transistor input is connected, i.e.,

At input of M_1: $P_{IN} = M_S P_{AVS}$ (power into M_1)
At output of M_1: $P'_{IN} = M'_S P_{AVS}$ (power into the transistor)
Lossless network: $P'_{IN} = P_{IN} \Rightarrow M_S = M'_S$

Similarly, the mismatch factor (M_L) remains unchanged at the input and output of the lossless matching network (M_2), i.e.,

At output of M_2: $P_L = M_L P_{AVN}$ (power into the load)
At input of M_2: $P_{OUT} = M'_L P_{AVN}$ (power into M_2)
Lossless network: $P_{OUT} = P_L \Rightarrow M_L = M'_L$

In summary, we can state the following:

For a lossless network, the mismatch factor is an "invariant quantity."

13.5 INPUT AND OUTPUT VSWR

In many cases, the microwave amplifier's specification is in terms of the input *VSWR* and the output *VSWR*. Therefore, we would like to obtain a relationship between the mismatch factor and *VSWR*.

13.5.1 Input-Port VSWR

From Figure 13.2, we can express the input power (P_{IN}) entering the input port of the matching network (M_1), in terms of the input reflection coefficient (Γ_a) as follows:

$$P_{IN} = P_{AVS}(1 - |\Gamma_a|^2) \text{ (cf., } P_{IN} = M_S P_{AVS}) \tag{13.22a}$$

Thus,

$$M_S = 1 - |\Gamma_a|^2 \tag{13.22b}$$

where

$$\Gamma_a = \frac{Z_a - Z_o}{Z_a + Z_o} \tag{13.23}$$

From Equation 13.22b, we can write:

$$|\Gamma_a| = \sqrt{1 - M_S} \tag{13.24}$$

Therefore, at the input of the lossless matching network (M_1), the input *VSWR* is given by:

$$(VSWR)_{IN} = \frac{1 + |\Gamma_a|}{1 - |\Gamma_a|} = \frac{1 + \sqrt{1 - M_S}}{1 - \sqrt{1 - M_S}} \tag{13.25}$$

Thus, $(VSWR)_{IN}$ can be calculated simply by knowing the source mismatch factor (M_S).

13.5.2 Output-Port VSWR

Similarly, from Figure 13.2, we can express the output power (P_L) exiting the output port of the matching network (M_2). This power can be expressed in terms of the ouput reflection coefficient (Γ_b) as follows:

$$P_L = P_{AVN}(1 - |\Gamma_b|^2) \text{ (cf., } P_L = M_L P_{AVN}) \tag{13.26a}$$

Thus,

$$M_L = 1 - |\Gamma_b|^2 \tag{13.26b}$$

where

$$\Gamma_b = \frac{Z_b - Z_o}{Z_b + Z_o} \tag{13.27}$$

From Equation 13.26b, we can write:

$$|\Gamma_b| = \sqrt{1 - M_L} \tag{13.28}$$

Therefore, at the output of the lossless matching network (M_2), the output *VSWR* is given by:

$$(VSWR)_{OUT} = \frac{1 + |\Gamma_b|}{1 - |\Gamma_b|} = \frac{1 + \sqrt{1 - M_L}}{1 - \sqrt{1 - M_L}} \tag{13.29}$$

Thus, $(VSWR)_{OUT}$ can be calculated simply by knowing the load mismatch factor (M_L).

NOTE: *In all of the preceding equations the normalizing factor, Z_o (usually the source impedance: $Z_S = Z_o$), is assumed to be a real positive number. In cases where this assumption cannot be upheld (i.e., when the source impedance is a complex number), the preceding equations no longer apply and have to be modified appropriately to account for this fact. Such a modification is outside the scope of this work and will not be discussed here.*

EXAMPLE 13.1

Given the amplifier circuit shown in Figure 13.3, having:

$$\Gamma_S = 0.5\angle 120°,$$

$$\Gamma_L = 0.4\angle 90°,$$

$$S_{11} = 0.6\angle -160°, S_{12} = 0.045\angle 16°,$$
$$S_{21} = 2.5\angle 30°, S_{22} = 0.5\angle -90°$$

determine:

a. *G_T, G_A, and G_P*
b. *The power levels P_L, P_{IN}, P_{AVS}, and P_{AVN}*
c. *The mismatch loss (in dB) at the input and output of the transistor*
d. *The input and output reflection coefficient magnitude ($|\Gamma_a|$, $|\Gamma_b|$)*
e. *$(VSWR)_{IN}$ and $(VSWR)_{OUT}$*

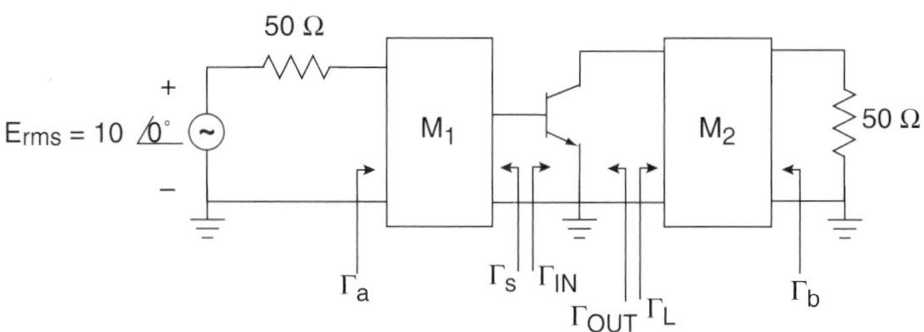

FIGURE 13.3 Figure for Example 13.1.

Solution:

a. The following are calculated to be:

$$\Gamma_{IN} = 0.627\angle -165°$$
$$\Gamma_{OUT} = 0.47\angle -98°$$
$$G_T = 9.4 = 9.75 \text{ dB}$$
$$G_P = 13.5 = 11.3 \text{ dB}$$
$$G_A = 9.6 = 9.8 \text{ dB}$$

b. To find the power levels, we note that:

$$G_T = P_L/P_{AVS} = 9.4$$
$$G_P = P_L/P_{IN} = 13.5$$
$$\Rightarrow P_{AVS}/P_{IN} = G_P/G_T = 13.5/9.4 = 1.43$$

P_{AVS} is the available power from the source (or the input power under matched conditions) as shown in Figure 13.4 (see Equation 11.8):

$$P_{AVS} = \frac{[\sqrt{2}E_{rms}]^2}{8Z_o} = 0.5 \text{ W}$$

Thus,

$$P_{IN} = P_{AVS}/1.43 = 0.35 \text{ W}$$

$$P_L = G_P P_{IN} = 4.72 \text{ W}$$

$$P_{AVN} = G_A P_{AVS} = 4.8 \text{ W}$$

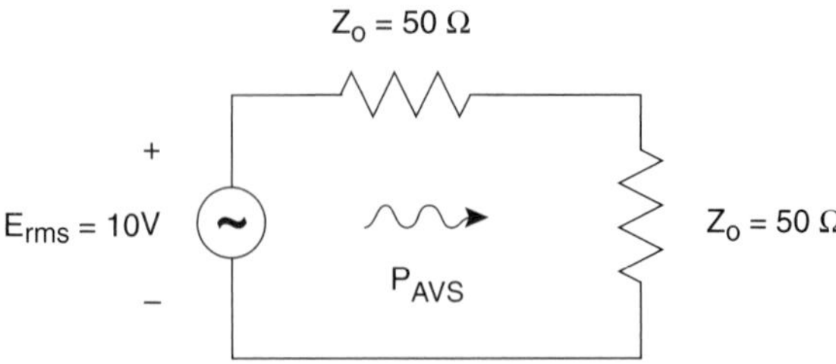

FIGURE 13.4 P_{AVS} in Example 13.1.

c. The mismatch factor (being an invariant quantity) at the input of the transistor has the same value as at the input of the matching network (M_1):

$$M_S = P_{IN}/P_{AVS} = 1/1.43 = 0.70 = -1.55 \text{ dB}$$

Similarly, the mismatch factor at the output of the transistor is the same value as at the load:

$$M_L = P_L/P_{AVN} = 4.72/4.8 = 0.99 = -0.055 \text{ dB}$$

d.

$$|\Gamma_a| = (1 - M_S)^{1/2} = (1 - 0.7)^{1/2} = 0.55$$

$$|\Gamma_b| = (1 - M_L)^{1/2} = (1 - 0.99)^{1/2} = 0.1$$

e.

$$(VSWR)_{IN} = \frac{1 + 0.55}{1 - 0.55} = 3.44$$

$$(VSWR)_{OUT} = \frac{1 + 0.1}{1 - 0.1} = 1.22$$

13.6 MAXIMUM GAIN DESIGN

From Equation 13.4, we can observe that because G_o is fixed for any given transistor, the overall gain of the amplifier is controlled by the gain blocks G_S and G_L corresponding to the input and output matching networks, respectively.

Therefore, in order to obtain the maximum possible gain from the amplifier circuit, we must maximize G_S and G_L values, which effectively implies that the input and output matching sections must provide a conjugate match at the transistor's input and output port. Furthermore, under this conjugate matched condition at the input and the output of the transistor, maximum power will be transferred into the input port and out of the output port of the transistor as shown in Figure 13.5.

Based on conjugate impedance matching concept, maximum power transfer from the input matching network to the transistor and from the transistor to the output matching network will occur when:

$$\Gamma_{IN} = \Gamma^*_S \tag{13.30}$$

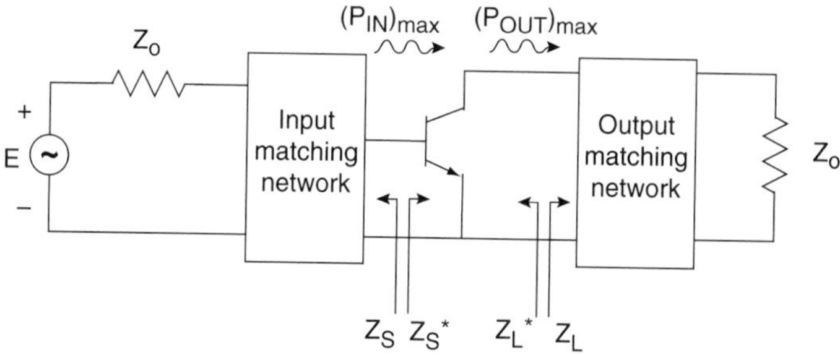

FIGURE 13.5 Maximum gain under conjugate matched conditions.

$$\Gamma_{OUT} = \Gamma^*_L \tag{13.31}$$

NOTE: *Due to the inherent mismatch between the transistor and the matching networks (M_1, M_2) and the fact that the conjugate match and maximum power transfer will occur theoretically only at one particular frequency, this type of circuit design is considered to be a "narrowband design."*

13.7 UNILATERAL CASE (MAXIMUM GAIN)

When the transistor is unilateral (i.e., $S_{12} = 0$) then as stated earlier in Section 13.3, we have:

$$\Gamma_{IN} = S_{11} \tag{13.32}$$

$$\Gamma_{OUT} = S_{22} \tag{13.33}$$

$$G_{TU} = G_S . G_o . G_L \tag{13.34}$$

where

$$G_S = \frac{1 - |\Gamma_S|^2}{|1 - \Gamma_{IN}\Gamma_S|^2} \tag{13.35}$$

$$G_o = |S_{21}|^2, \tag{13.36}$$

$$G_L = \frac{1 - |\Gamma_L|^2}{|1 - S_{22}\Gamma_L|^2} \tag{13.37}$$

Under conjugately matched (maximum gain) conditions:

$$\Gamma_S = S^*_{11} \tag{13.38}$$

$$\Gamma_L = S^*_{22} \tag{13.39}$$

$$G_{TU,max} = G_{S,max} G_o G_{L,max} \tag{13.40}$$

where

$$G_{S,max} = \frac{1}{1 - |S_{11}|^2} \tag{13.41}$$

$$G_o = |S_{21}|^2, \tag{13.42}$$

$$G_{L,max} = \frac{1}{1-|S_{22}|^2} \tag{13.43}$$

Thus, we can write Equation 13.40 as:

$$G_{TU,max} = G_{S,max}G_oG_{L,max} = \frac{1}{1-|S_{11}|^2}|S_{21}|^2\frac{1}{1-|S_{22}|^2}$$

NOTE: *Each gain block (G_S or G_L) is bound at the lower end by a gain of zero and at the upper end by the maximum gain ($G_{S,max}$ or $G_{L,max}$), as follows:*

$$0 \le G_S \le G_{S,max} \tag{13.44}$$

$$0 \le G_L \le G_{L,max} \tag{13.45}$$

We can normalize these two equations to obtain:

$$0 \le g_S \le 1 \tag{13.46}$$

$$0 \le g_L \le 1 \tag{13.47}$$

where the normalized gain factors (g_S, g_L) are defined as:

$$g_S = \frac{G_S}{G_{S,max}} = \frac{1-|\Gamma_S|^2}{|1-S_{11}\Gamma_S|^2}(1-|S_{11}|^2) \tag{13.48}$$

$$g_L = \frac{G_L}{G_{L,max}} = \frac{1-|\Gamma_L|^2}{|1-S_{22}\Gamma_L|^2}(1-|S_{22}|^2) \tag{13.49}$$

13.8 CONSTANT GAIN CIRCLES (UNILATERAL CASE)

Considering Equations 13.35 and 13.37, it can be shown (see Appendix M, *Derivation of the Constant Gain and Noise Figure Circles*) that the values of Γ_S and Γ_L that produce a constant gain (or normalized gain) lie in a circle in the Smith chart. These circles are called constant G_S and G_L circles, respectively.

To obtain the equations for these circles, we start with Equations 13.48 and 13.49. It is shown that the values of Γ_S (or Γ_L) that produce a constant value of g_S (or g_L) lie in a circle described by the following equations:

$$|\Gamma_S - C_{gS}| = R_{gS}, \tag{13.50a}$$

$$|\Gamma_L - C_{gL}| = R_{gL} \tag{13.50b}$$

where the center and radius (C_S, R_S) and (C_L, R_L) for each of the two circles are given by:

$$C_{gS} = \frac{g_S S_{11}^*}{1-|S_{11}|^2(1-g_S)} \tag{13.51a}$$

$$R_{gS} = \frac{\sqrt{(1-g_S)}(1-|S_{11}|^2)}{1-|S_{11}|^2(1-g_S)} \tag{13.51b}$$

and

$$C_{gL} = \frac{g_L S_{22}^*}{1-|S_{22}|^2(1-g_L)} \tag{13.52a}$$

$$R_{gL} = \frac{\sqrt{(1-g_L)}(1-|S_{22}|^2)}{1-|S_{22}|^2(1-g_L)} \tag{13.52b}$$

The two equations in Equation 13.50 represent equations of two families of circles where the centers of each family of circles lie along the straight line given by the angle of S_{11}^* and S_{22}^* as shown in Figures 13.6a and b.

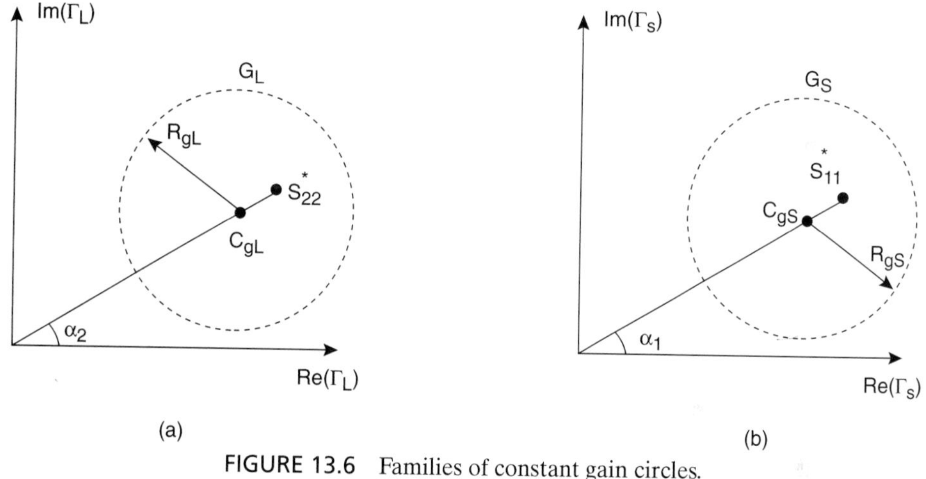

FIGURE 13.6 Families of constant gain circles.

13.8.1 Observations

a. From Equations 13.51 and 13.52, we note that when g_S or $g_L = 1$ (i.e., maximum gain condition), we have:

$$R_{gS} = 0, \tag{13.53a}$$

$$R_{gL} = 0, \tag{13.53b}$$

$$C_{gS} = S_{11}^* \tag{13.53c}$$

and

$$C_{gL} = S_{22}^* \tag{13.53d}$$

This indicates that the maximum gain occurs only at one point located at S_{11}^* and S_{22}^* in the Γ_S or Γ_L plane, respectively. This observation is in agreement with our earlier results as expressed by Equations 13.38 and 13.39.

b. The 0-dB circles (i.e., $G_S = 1$, $G_L = 1$ circles) will always pass through the origin (i.e., $\Gamma_S = 0$, $\Gamma_L = 0$ points). This can be shown by noting that g_S from Equation 13.48 can be written as:

$$G_S = 1 \text{ when } \Gamma_S = 0 \Rightarrow g_S = 1 - |S_{11}|^2$$

$$|C_{gS}| = R_{gS} = \frac{|S_{11}|}{1 + |S_{11}|^2} \tag{13.54}$$

Similarly, for g_L:

$$G_L = 1 \text{ when } \Gamma_L = 0 \Rightarrow g_L = 1 - |S_{22}|^2$$

$$|C_{gL}| = R_{gL} = \frac{|S_{22}|}{1 + |S_{22}|^2}$$

which shows that the radius and distance from the origin to the center of the 0-dB constant G_S or G_L circle are identical and proves our observation.

c. At the outer edge of the Smith chart:

$$|\Gamma_S| = 1 \Rightarrow G_S = 0 = -\infty \text{ dB}$$

$$|\Gamma_L| = 1 \Rightarrow G_L = 0 = -\infty \text{ dB}$$

Because this gain value is impossible to achieve, the gain circles never intersect the outer edge of the Smith chart.

d. For a particular gain value, there are an infinite number of points on the constant gain circle that provide the same gain. Thus, the choices of Γ_S and Γ_L along the constant gain circles are not unique, but in order to minimize mismatch loss and maximize bandwidth it is best to choose points close to the center of the Smith chart. This is true only for cases where noise is not of importance. In fact, for low-noise amplifier design (as we will see in the next chapter) we need to use mismatch at the input matching network in order to obtain minimum noise from the amplifier.

13.9 UNILATERAL FIGURE OF MERIT

We already noticed that under the unilateral assumption, power gain analysis greatly simplifies. In most cases $S_{12} \neq 0$. If we still wish to use the unilateral assumption and the simplified unilateral gain equations for the amplifier design (when $S_{12} \neq 0$), we need to determine the error involved in our analysis.

The error involved lies in the magnitude ratio of G_T/G_{TU}, which is obtained by dividing Equation 13.1 by Equation 13.12:

$$\frac{G_T}{G_{TU}} = \frac{1}{|1 - X|^2} \tag{13.55}$$

where

$$X = \frac{S_{12}S_{21}\Gamma_S\Gamma_L}{(1 - S_{11}\Gamma_S)(1 - S_{22}\Gamma_L)} \tag{13.56}$$

It can be shown that ratio of G_T/G_{TU} is bounded by:

$$\frac{1}{(1+|X|)^2} < \frac{G_T}{G_{TU}} < \frac{1}{(1-|X|)^2} \tag{13.57}$$

When $\Gamma_S = S_{11}^*$ and $\Gamma_L = S_{22}^*$, G_{TU} achieves its maximum value, $G_{TU,max}$. The maximum error introduced using the unilateral assumption (i.e., using G_{TU} instead of G_T) is bounded by:

$$\frac{1}{|1+U|^2} < \frac{G_T}{G_{TU,max}} < \frac{1}{|1-U|^2} \tag{13.58}$$

where

$$U = \frac{|S_{12}||S_{21}||S_{11}||S_{22}|}{(1-|S_{11}|^2)(1-|S_{22}|^2)} \tag{13.59}$$

U is defined to be the "*unilateral figure of merit,*" which varies with frequency due to its S-parameter dependence. Thus, U needs to be calculated at each frequency in order to obtain the limits of the error involved due to unilateral assumption. From Table 13.1, which lists various values of U versus G_T/G_{TU}, we can determine if the calculated value of U gives a tolerable error value for G_T/G_{TU}.

TABLE 13.1 Tabulation of Values of "U" versus $G_T/G_{TU,max}$ (in Ratio and dB).

U		$R = G_T/G_{TU}$	
(ratio)	(dB)	(ratio)	(dB)
0.010	−20.0	0.980 < R < 1.020	−0.086 < R < 0.087
0.020	−17.0	0.961 < R < 1.041	−0.170 < R < 0.180
0.030	−15.2	0.943 < R < 1.063	−0.26 < R < 0.26
0.040	−14.0	0.925 < R < 1.085	−0.34 < R < 0.36
0.050	−13.0	0.907 < R < 1.108	−0.42 < R < 0.45
0.060	−12.2	0.890 < R < 1.132	−0.51 < R < 0.54
0.070	−11.5	0.873 < R < 1.156	−0.59 < R < 0.63
0.080	−11.0	0.857 < R < 1.181	−0.67 < R < 0.72
0.090	−10.5	0.842 < R < 1.208	−0.75 < R < 0.82
0.10	−10.0	0.826 < R < 1.235	−0.83 < R < 0.92
0.11	−9.6	0.812 < R < 1.262	−0.91 < R < 1.01
0.12	−9.2	0.797 < R < 1.291	−0.98 < R < 1.11
0.13	−8.9	0.783 < R < 1.321	−1.06 < R < 1.21
0.14	−8.5	0.769 < R < 1.352	−1.13 < R < 1.31
0.15	−8.2	0.756 < R < 1.384	−1.21 < R < 1.41

Usually an error of a few tenths of a dB in the $G_T/G_{TU,max}$ ratio is justifiable when using the unilateral assumption. The following example illustrates this point further.

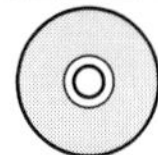

EXAMPLE 13.2

Assume $U = 0.05$ at 1 GHz for a microwave amplifier where $S_{12} \neq 0$.

a. *Find the maximum error if we use unilateral gain equations for this transistor.*

b. *If the transistor is used in an amplifier design with a gain of 15 dB, can unilateral assumption be used in this case?*

Solution:

a. From Table 13.1, we can see that for $U = 0.05$, $G_T/G_{TU,max}$ is bounded by:

$$0.907 < G_T/G_{TU,max} < 1.108$$

Or, in dB, we have:

$$-0.42 \text{ dB} < G_T/G_{TU,max} < 0.45 \text{ dB}$$

b. Since the error due to unilateral assumption is bound between −0.42 dB and +0.45 dB, it is small enough (compared to 15 dB) to justify the unilateral assumption.

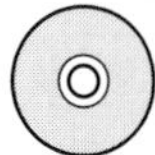

EXAMPLE 13.3

Calculate the maximum error range for the transducer gain value if we use unilateral gain equations for a transistor that has the following S-parameters:

$$S_{11} = 0.6\angle -160°,\ S_{12} = 0.045\angle 16°,$$

$$S_{21} = 2.5\angle 30°,\ S_{22} = 0.5\angle -90°$$

Solution:

From Equation 13.59, we have:

$$U = \frac{|S_{12}||S_{21}||S_{11}||S_{22}|}{(1-|S_{11}|^2)(1-|S_{22}|^2)} = \frac{0.045 \times 2.5 \times 0.6 \times 0.5}{(1-0.6^2)(1-0.5^2)} = 0.070$$

The lower limit is:

$$\frac{1}{(1+U)^2} = \frac{1}{(1+0.070)^2} = 0.873 = -0.59 \text{ dB}$$

The upper limit is:

$$\frac{1}{(1-U)^2} = \frac{1}{(1-0.070)^2} = 1.156 = 0.63 \text{ dB}$$

Thus, the error range for G_T is given by:

$$-0.59 < G_T < 0.63 \text{ dB}$$

13.10 BILATERAL CASE

When $S_{12} \neq 0$ and the unilateral figure of merit causes an unjustifiably high error in the gain equations, we are faced with the bilateral case where S_{12} can no longer be ignored.

We know that from Equations 13.30 and 13.31 the maximum power transfer occurs when:

$$\Gamma_{IN} = \Gamma_S^* = S_{11} + \frac{S_{12}S_{21}\Gamma_L}{1 - S_{22}\Gamma_L} \tag{13.60}$$

$$\Gamma_{OUT} = \Gamma_L^* = S_{22} + \frac{S_{12}S_{21}\Gamma_S}{1 - S_{11}\Gamma_S} \tag{13.61}$$

Under these conditions, the overall maximum gain using lossless matching networks is given by:

$$G_T = \frac{1}{1 - |\Gamma_S|^2}|S_{21}|^2 \frac{1 - |\Gamma_L|^2}{|1 - S_{22}\Gamma_L|^2} \tag{13.62}$$

From Equation 13.60, we note that for a bilateral transistor Γ_S depends on Γ_L, and vice versa, from Equation 13.61, Γ_L depends on Γ_S. This means that these two equations are cross-coupled and must be solved simultaneously to obtain the simultaneous conjugate match values of Γ_S and Γ_L.

Solving Equations 13.60 and 13.61 simultaneously, we obtain the simultaneous conjugate match values of Γ_S and Γ_L (referred to as Γ_{MS} and Γ_{ML}) as:

$$\Gamma_{MS} = \frac{B_1 \pm \sqrt{B_1^2 - 4|C_1|^2}}{2C_1} \tag{13.63a}$$

$$\Gamma_{ML} = \frac{B_2 \pm \sqrt{B_2^2 - 4|C_2|^2}}{2C_2} \tag{13.63b}$$

where

$$B_1 = 1 + |S_{11}|^2 - |S_{22}|^2 - |\Delta|^2 \tag{13.64a}$$

$$B_2 = 1 + |S_{22}|^2 - |S_{11}|^2 - |\Delta|^2 \tag{13.64b}$$

$$C_1 = S_{11} - \Delta S_{22}^* \tag{13.65a}$$

$$C_2 = S_{22} - \Delta S_{11}^* \tag{13.65b}$$

NOTE: *It can be shown that for an unconditionally stable two-port network ($K > 1$, $|\Delta| < 1$), the solutions (from Equations 13.63) with a minus sign (–) should be considered in order to obtain meaningful values for Γ_{MS} and Γ_{ML} (i.e., $|\Gamma_{MS}| < 1$ and $|\Gamma_{ML}| < 1$). Furthermore, under simultaneous conjugate matched conditions we have: $G_T = G_A = G_P$.*

Under simultaneous conjugate matched conditions, $G_{T,max}$ from Equation 13.62 is obtained to be:

$$\Gamma_S = \Gamma_{IN}^* = \Gamma_{MS} \tag{13.66a}$$

$$\Gamma_L = \Gamma_{OUT}^* = \Gamma_{ML} \tag{13.66b}$$

$$G_{T,max} = \frac{1}{1 - |\Gamma_{MS}|^2}|S_{21}|^2 \frac{1 - |\Gamma_{ML}|^2}{|1 - S_{22}\Gamma_{ML}|^2} \tag{13.67}$$

Substituting for Γ_{MS} and Γ_{ML} from Equations 13.63a and b into Equation 13.67, we obtain:

$$G_{T,max} = \frac{|S_{21}|}{|S_{12}|}(K - \sqrt{K^2 - 1}) \tag{13.68}$$

where K was defined earlier in Chapter 12 as one of the stability factors. It can be shown easily that:

$$G_{T,max} = G_{A,max} = G_{P,max}$$

When $K = 1$, we obtain the maximum stable gain (G_{MSG}) from Equation 13.68 as:

$$G_{MSG} = G_{T,max}|_{K=1} = |S_{21}|/|S_{12}|$$

G_{MSG} is a figure of merit showing the maximum value that $G_{T,max}$ can achieve. Thus, by looking at a transistor's forward (S_{21}) and reverse (S_{12}) transmission coefficients, we can decide if the transistor is useful in providing the needed gain for a particular amplifier design.

13.11 SUMMARY

Before embarking on the task of designing a functional amplifier we need to consider one more aspect, namely "noise figure," which for a sensitive receiver plays a very important role in the design process. This topic will be considered in detail in the next chapter and will consummate all of the major considerations that accompany the design process of an amplifier. This is done so that when we start discussing the actual design process for several types of amplifiers in Chapter 15, where each type of amplifier accentuates one or more of these major considerations (e.g., stability, gain, noise figure, etc.), a full knowledge of the subject has already been fully experienced and explored by the reader.

LIST OF SYMBOLS/ABBREVIATIONS

A symbol/abbreviation will not be repeated again once it has been identified and defined in an earlier chapter, as long as its definition remains unchanged.

G_A	Available power gain
G_o	Transistor power gain

G_L	Output matching network power gain
G_{LU}	Output matching network power gain for unilateral case
G_{MSG}	Maximum stable gain
G_P	Operating power gain
G_S	Input matching network transducer power gain
G_{SU}	Input matching network transducer power gain for unilateral transistor
G_T	Transducer power gain
$G_{T,max}$	Maximum transducer power gain
G_{TU}	Transducer power gain for special case-unilateral transistor
M_L	Load mismatch factor
M_S	Source mismatch factor
P_{AVN}	Power available from the transistor under matched condition
P_{AVS}	Power available from the source under matched condition
P_{IN}	Input power to the transistor (or to the input matching network)
P_L	Output power to the load (or to the output matching network)
P_{OUT}	Output power from network under consideration
Γ_{ML}	Reflection coefficient for conjugate match value of Γ_L
Γ_{MS}	Reflection coefficient for conjugate match value of Γ_S

PROBLEMS

13.1 The S-parameters of a transistor are:

$$S = \begin{bmatrix} 0.7\angle 30^\circ & 0 \\ 4\angle 90^\circ & 0.5\angle 0^\circ \end{bmatrix}$$

The transistor is used in the amplifier shown in Figure P13.1, where the output matching network produces $\Gamma_L = 0.5\angle 90^\circ$. Determine the values of G_T, G_P, and G_A.

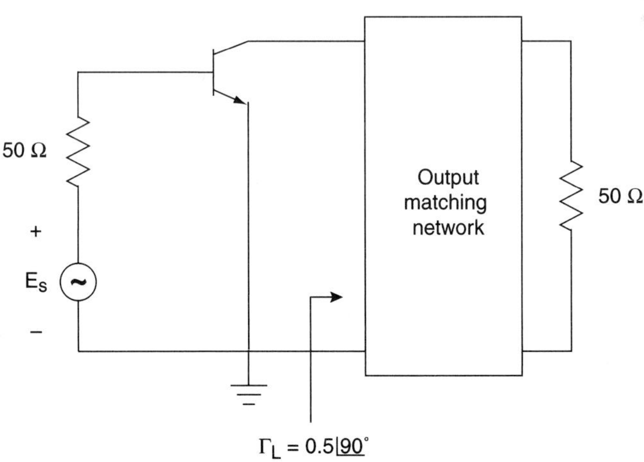

FIGURE P13.1

13.2 **a.** Determine G_T, G_P, and G_A in a microwave amplifier as shown in Figure P13.2, if $\Gamma_S = 0.49\angle{-150°}$, $\Gamma_L = 0.56\angle 90°$ and the S-parameters of the transistor are:

$$S = \begin{bmatrix} 0.54\angle 165° & 0.09\angle 20° \\ 2\angle 30° & 0.5\angle{-80°} \end{bmatrix}$$

b. Calculate P_{AVS}, P_{IN}, P_{AVN}, and P_L if $E_1 = 10\angle 30°$, $Z_1 = 50\ \Omega$, and $Z_2 = 50\ \Omega$.

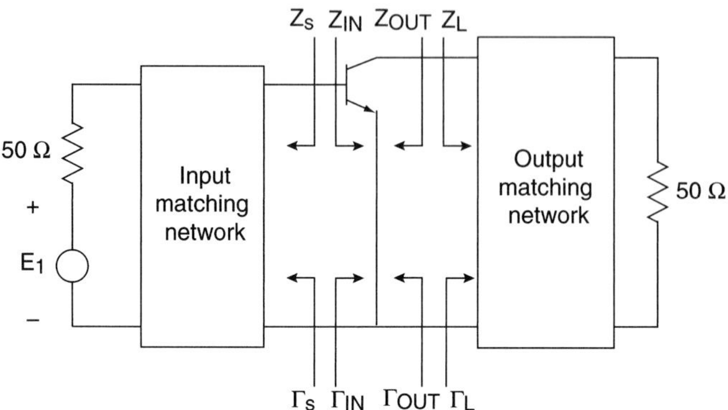

FIGURE P13.2

13.3 Prove that the maximum unilateral transducer power gain is obtained when $\Gamma_S = S_{11}^*$ and $\Gamma_L = S_{22}^*$.

13.4 Design a microwave transistor amplifier for $G_{TU,max}$ using a BJT whose S-parameters in a 50 Ω system at $V_{CE} = 10$ V, $I_C = 10$ mA, and $f = 1$ GHz are given by:

$$S = \begin{bmatrix} 0.7\angle{-160°} & 0 \\ 5\angle 85° & 0.5\angle{-20°} \end{bmatrix}$$

13.5 A microwave amplifier is to be designed for $G_{TU,max}$ using a transistor with:

$$S = \begin{bmatrix} 0.5\angle 140° & 0 \\ 5\angle 45° & 0.6\angle{-95°} \end{bmatrix}$$

The S-parameters were measured in a 50 Ω system at $f = 900$ MHz, $V_{CE} = 15$ V and $I_C = 20$ mA.

a. Determine $G_{TU,max}$.
b. Draw the constant gain circle for $G_L = 1$ dB.
c. If the S-parameters at 1 GHz are:

$$S = \begin{bmatrix} 0.48\angle 137° & 0 \\ 4,6\angle 48° & 0.57\angle{-99°} \end{bmatrix}$$

Calculate the gain G_T at 1 GHz if $\Gamma_S = 0.49\angle{-150°}$ and $\Gamma_L = 0.56\angle 90°$.

13.6 A GaAs FET amplifier has the following S-parameters for the active device:

$$S = \begin{bmatrix} 0.5\angle 180^\circ & 0 \\ 4\angle 90^\circ & 0.5\angle -45^\circ \end{bmatrix}$$

a. Is the amplifier stable?
b. What is the maximum gain in dB?
c. What is the input impedance Z_{IN} in a 50 Ω system?
d. What is the load impedance Z_L for the maximum-gain case?

13.7 An amplifier is operating at 10 GHz using a GaAs FET with the following S-parameters:

$$S = \begin{bmatrix} 0.45\angle -45^\circ & 0 \\ 5\angle 30^\circ & 0.8\angle -160^\circ \end{bmatrix}$$

Design an amplifier for maximum gain using a 50 Ω input and output transmission lines as shown in Figure P13.7.

a. Is the amplifier stable?
b. What is the maximum gain in dB?
c. Design the input matching network for maximum gain using series C_1 and shunt L_1 elements to match the 50 Ω line to Z_S.
d. What are the values of Γ_{IN} and its corresponding Z_{IN}?

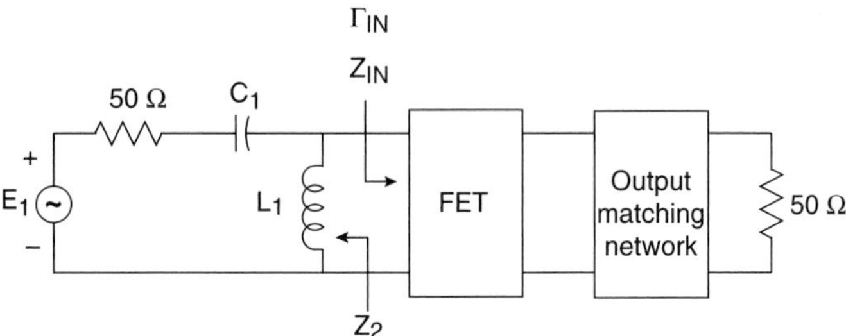

FIGURE P13.7

13.8 An FET device has the following S-parameters at 3 GHz:

$$S = \begin{bmatrix} 0.3\angle -60^\circ & 0 \\ 2\angle 45^\circ & 0.8\angle -30^\circ \end{bmatrix}$$

Design an amplifier for maximum gain using this transistor and 50 Ω input and output transmission lines:

a. Check the stability.
b. What is the maximum gain in dB?
c. Design an input-matching network for maximum gain using series L and shunt C elements to match a 50 Ω line to Z_S.

13.9 The FET amplifier shown in Figure P13.9 has the following S-parameters in a 50 Ω system:

$$S = \begin{bmatrix} 0.5\angle 180^\circ & 0 \\ 3.0\angle 90^\circ & 0.5\angle -90^\circ \end{bmatrix}$$

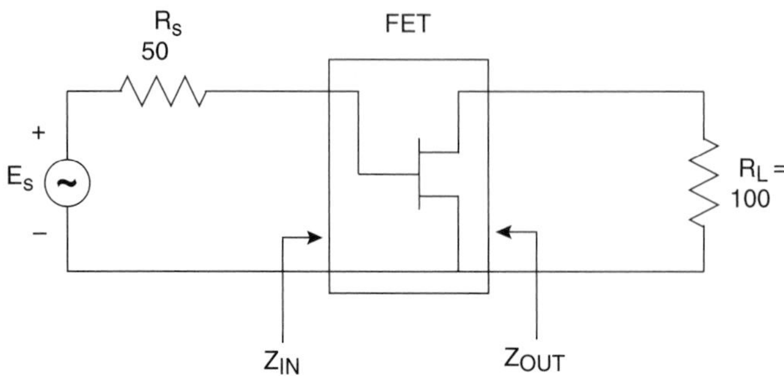

FIGURE P13.9

The circuit is terminated by $R_S = 50\ \Omega$ and $R_L = 100\ \Omega$. Find:

a. Z_{IN} and Z_{OUT}.
b. The unilateral power gain in dB.
c. If matching networks are used at the input and output ports such that maximum power transfer occurs, find the maximum unilateral power gain in dB.

13.10 Consider the microwave network shown in Figure P13.10, consisting of a 50 Ω source, a 50 Ω, 3 dB matched attenuator, and a 50 Ω load.

a. Compute the available power gain, the transducer power gain, and the actual power gain.
b. How do these gains change if the load is changed to 25 Ω? How do these gains change if the source impedance is changed to 25 Ω?

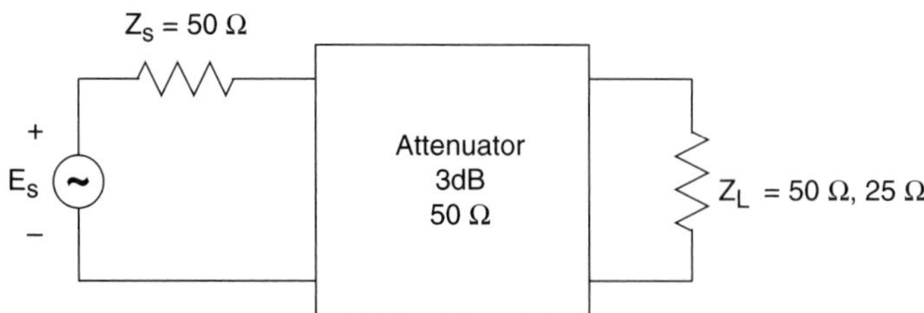

FIGURE P13.10

13.11 Use the new μ-parameter test to determine which of the following devices are unconditionally stable and, of these, which has the greatest stability:

Device	S_{11}	S_{12}	S_{21}	S_{22}
A	$0.34\angle -170°$	$0.06\angle 70°$	$4.3\angle 80°$	$0.45\angle -250°$
B	$0.75\angle -60°$	$0.2\angle 70°$	$5.0\angle 90°$	$0.51\angle 60°$
C	$0.65\angle -140°$	$0.04\angle 60°$	$2.4\angle 50°$	$0.70\angle -65°$

13.12 Show that for a unilateral device where $S_{12} = 0$, the μ-parameter test implies that $|S_{11}| < 1$ and $|S_{22}| < 1$ (for unconditional stability).

REFERENCES

[13.1] Bahl, I. and P. Bhartia. *Microwave Solid State Circuit Design*. New York: Wiley Interscience, 1988.

[13.2] Gonzalez, G. *Microwave Transistor Amplifiers, Analysis and Design*, 2nd ed. Upper Saddle River: Prentice Hall, 1997.

[13.3] Ha, T. T. *Solid State Microwave Amplifier Design*. New York: John Wiley & Sons, 1987.

[13.4] Liao, S. Y. *Microwave Circuit Analysis and Amplifier Design*. Upper Saddle River: Prentice Hall, 1987.

[13.5] Pozar, D. M. *Microwave Engineering*, 2nd ed. New York: John Wiley & Sons, 1998.

[13.6] Vendelin, George D. *Design of Amplifiers and Oscillators by the S-Parameter Method.* New York: John Wiley & Sons, 1982.

[13.7] Vendelin, George D., Anthony M. Pavio, and Ulrich L. Rhode. *Microwave Circuit Design, Using Linear and Non-Linear Techniques.* New York: John Wiley & Sons, 1990.

CHAPTER 14

Noise Considerations in Active Networks

14.1 INTRODUCTION

Having done a stability check and having met the gain requirements of an amplifier, we shall consider noise as our next milestone in active circuit design considerations. In an RF/microwave amplifier, the existence of the noise signal plays an important role in the overall design procedure, and its impact needs to be grasped before a meaningful design process can be developed.

Noise power results from random processes that exist in nature. These random processes can be classified in several important classes, each generating a certain type of noise, which will be characterized shortly. Some of the most important types of random processes are the following:

- Thermal vibrations of atoms, electrons, and molecules in a component at any temperature above 0° K
- Flow of charges (electrons and/or holes) in a wire or a device
- Emission of charges (electrons or ions) from a surface such as the cathode of a diode or an electron tube, etc.
- Wave propagation through atmosphere or any other gas

14.2 IMPORTANCE OF NOISE

Noise is passed into a microwave component or system from an external source, or it is generated within the unit itself. Regardless of the manner of entrance of the noise signal, the noise level of a system greatly affects the performance of the system by setting the minimum detectable signal in the presence of noise. Therefore, it is often desirable to reduce the influence of external noise signals and minimize the generation of noise signals within the unit to achieve the best performance.

14.3 NOISE DEFINITION

Because noise considerations have important consequences, we need to define noise first:

DEFINITION-ELECTRICAL NOISE (OR NOISE): *Any unwanted electrical disturbance or spurious signal. These unwanted signals are random in nature and are generated either internally in the electronic components or externally through impinging electromagnetic radiation.*

Because noise signals are totally random and uncorrelated in time, they are best analyzed through statistical methods. Their statistical properties can be briefly summarized as follows:

a. The "mean value" of the noise signal is zero, i.e.,

$$\overline{V}_n = \lim_{T \to \infty} \frac{1}{T} \int_{t_1}^{t_1 + T} V_n(t) dt = 0 \tag{14.1}$$

where $\overline{V}_n$ is the noise mean value, $V_n(t)$ is the instantaneous noise voltage, t_1 is any arbitrary point in time, and T is any arbitrary period of time, ideally a large one approaching infinity.

b. The "mean-square value" of the noise signal is a constant value, i.e.,

$$\overline{V_n^2} = \lim_{T \to \infty} \frac{1}{T} \int_{t_1}^{t_1 + T} [V_n(t)]^2 dt = \text{Constant} \tag{14.2}$$

c. The "root-mean-square" (rms) of a noise signal is given by:

$$V_{n,\,rms} = \sqrt{\overline{V_n^2}} \tag{14.3}$$

or

$$V_{n,\,rms}^2 = \overline{V_n^2} \tag{14.4}$$

The concept of "root-mean-square value" (rms value) of noise, as given by Equation 14.3, is based on the fact that the "mean-square value," $\overline{V_n^2}$, is proportional to the "noise power." Thus, if we take the square root of Equation 14.2, we obtain the rms value of the noise voltage, which is the "effective value" of the noise voltage.

14.4 SOURCES OF NOISE

There are several types of noise that need to be defined:

DEFINITION-THERMAL NOISE (OR JOHNSON NOISE OR NYQUIST NOISE): *The most basic type of noise, which is caused by thermal vibration of bound charges and thermal agitation of electrons in a conductive material. This is common to all passive or active devices.*

DEFINITION-SHOT NOISE (OR SCHOTTKY NOISE): *Caused by random passage of discrete charge carriers (causing a current I, due to motion of electrons or holes) in a solid-state device while crossing a junction or other discontinuities. It is commonly found in a semiconductor device (e.g., in a pn junction of a diode or a transistor) and is proportional to* $(I)^{1/2}$.

DEFINITION-FLICKER NOISE (ALSO CALLED 1/*f* NOISE): *Small vibrations of a current due to the following factors:*

a. *Random injection or recombination of charge carriers at an interface, such as at a metal and semiconductor interface (in semiconductor devices)*

b. *Random changes in cathode emissions of electric charges such as at a cathode-air interface (in a thermionic tube)*

Flicker noise is a frequency-dependent noise, which distorts the signal by adding more noise to the lower part of the signal band than the upper part. It exists at lower frequencies, almost from DC extending down to approximately 500 kHz to 1 MHz at a rate of –10 dB per decade.

14.5 THERMAL NOISE ANALYSIS

To analyze noise, let's consider the circuit shown in Figure 14.1a where a noisy resistor is connected to the input port of a two-port network. Focusing primarily on thermal noise, we note that the available noise power (i.e., maximum power available under matched conditions) from any arbitrary resistor has been shown by Nyquist to be:

$$P_N = kTB \tag{14.5}$$

where

k = Boltzmann's constant (= 1.374×10^{-23} J/K)
T = The resistor's physical temperature (in Kelvin)
B = The two-port network's bandwidth (i.e., $B = f_H - f_L$)

Because the noise power does not depend on the center frequency of operation but only on the bandwidth, it is called "white noise," as shown in Figure 14.1b.
A few observations about noise power (P_N) are worth considering:

- As bandwidth (B) is reduced, so is the noise power, which means narrower bandwidth amplifiers are less noisy.
- As temperature (T) is reduced, the noise power is also lessened, which means cooler devices or amplifiers generate less noise power.

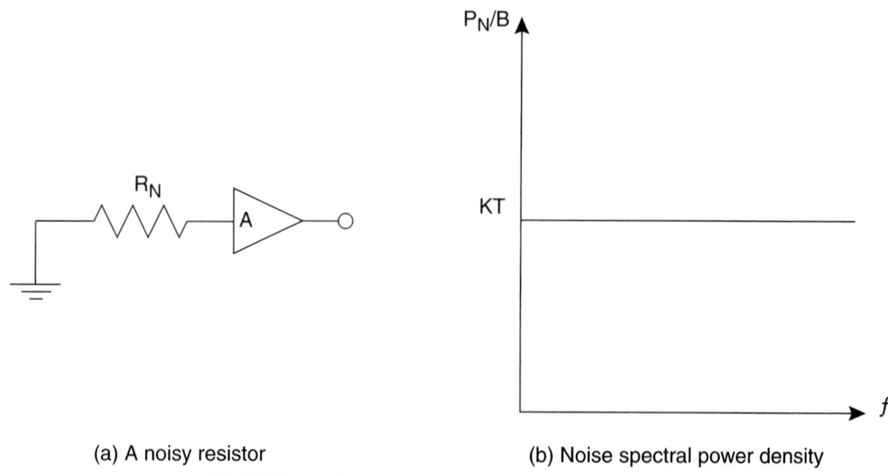

(a) A noisy resistor (b) Noise spectral power density

FIGURE 14.1 White noise in an amplifier.

- Increasing bandwidth to infinity causes an infinite noise power (called ultraviolet catastrophe), which is incorrect because Equation 14.5 for noise power is valid only up to approximately 1000 GHz.

14.6 NOISE MODEL OF A NOISY RESISTOR

A noisy resistor (R_N) at a temperature (T) can be modeled by an ideal noiseless resistor (R_{No}) at 0° K in conjunction with a noise voltage source ($V_{n,rms}$), as shown Figure 14.2.

FIGURE 14.2 Model of a noisy resistor.

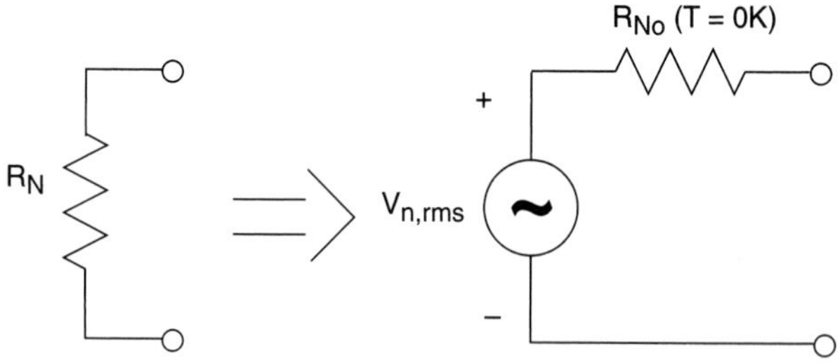

If we assume that the resistor value is independent of temperature then $R_{No} = R_N$.

From this model, the available noise power to the load (under matched condition) is given by (see Figure 14.3):

$$P_N = \frac{V_{n,\,rms}^2}{4R_N} \tag{14.6}$$

Equation 14.6 provides the noise power available from a noisy resistor, which equals Equation 14.5 for any arbitrary resistor. Thus:

FIGURE 14.3 Available noise power.

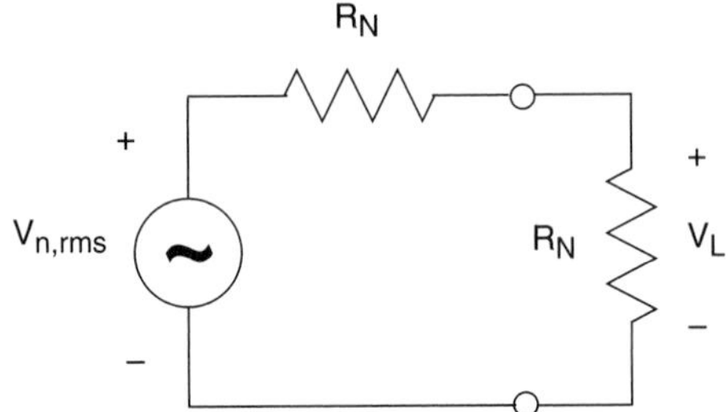

$$P_N = kTB \qquad (14.7\text{a})$$

$$V_{n,\,rms} = 2\sqrt{P_N R_N} = 2\sqrt{kTBR_N} \qquad (14.7\text{b})$$

From Equation 14.7b, we can observe that the noise voltage is proportional to $R_N^{1/2}$. Thus, higher-valued resistors have higher noise voltage even though they provide the same noise power level as the lower-valued resistors.

EXAMPLE 14.1

Calculate the noise power (in dBm) and rms noise voltage at T = 290 ° K for:

a. $R_N = 1\ \Omega, B = 1$ Hz
b. $R_N = 2\ \text{M}\Omega, B = 5$ kHz

Solution:

a. The noise power is given by:

$$k = 1.374 \times 10^{-23}\ \text{J/° K}$$

$$B = 1\ Hz$$

$$P_N = kTB = kT = 1.374 \times 10^{-23} \times 290 \times 1 = 3.985 \times 10^{-21}\ \text{W}$$

Or, in dBm, we have:

$$P_N(\text{dBm}) = 10\log(3.985 \times 10^{-21}/10^{-3}) = -174\ \text{dBm}$$

This is the power per unit Hz. The corresponding noise voltage for a 1 Ω resistor is given by:

$$V_{n,\,rms} = 2\sqrt{P_N R_N} = 2\sqrt{3.958 \times 10^{-21} \times 1} = 12.6 \times 10^{-11}\ \text{V} = 12.6 \times 10^{-5}\ \mu\text{V}$$

b. For a 5 kHz bandwidth, we have

$$P_N = kTB = 3.985 \times 10^{-21} \times 5000 = 19.925 \times 10^{-18}\ \text{W} = 19.925 \times 10^{-15}\ \text{mW}$$

Or, in dBm:

$$P_N\ (\text{dBm}) = -137\ \text{dBm}$$

(alternately, $\text{P}_\text{N} = 10\log(\text{kT}) + 10\log\text{B} = -174 + 10\log 5000 = -137\ \text{dBm}$)

The corresponding noise voltage for a 2 MΩ resistor is given by

$$V_{n,\,rms} = 2\sqrt{P_N R_N} = 2\sqrt{19.925 \times 10^{-18} \times 2 \times 10^{6}} = 12.6 \times 10^{-6}\ \text{V} = 12.6\ \mu\text{V}$$

14.7 EQUIVALENT NOISE TEMPERATURE

Any type of noise, in general, has a power spectrum that can be plotted in the frequency domain. If the noise power spectrum is not a strong function of frequency (i.e., it is white noise), then it can be modeled as an equivalent thermal noise source characterized by an "equivalent noise temperature" (T_e).

To define the equivalent noise temperature (T_e), we consider an arbitrary white noise source with an available power (P_S) having a noiseless source resistance (R_S), as shown in Figure 14.4a. This white noise source can be replaced by a noisy resistor with an equivalent noise temperature (T_e) defined by:

$$T_e = \frac{P_S}{kB} \tag{14.8}$$

where B is the bandwidth of the system or the component under consideration.

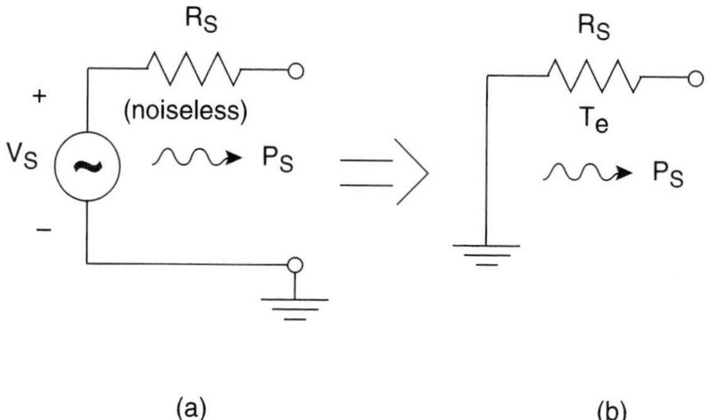

FIGURE 14.4 (a) An arbitrary white noise source, (b) equivalent circuit.

EXAMPLE 14.2

Consider a noisy amplifier with available power gain (G_A) and bandwidth (B) connected to a source and load resistance (R) both at $T = T_S$, as shown in Figure 14.5. Determine the overall noise temperature of the combination and the total output noise power if the amplifier alone creates an output noise power of P_n.

Solution:

To simplify the analysis, let's first assume that the source resistor is at $T = 0°$ K. This means that no noise enters the amplifier, i.e., $P_{Ni} = 0$.

The noisy amplifier can be modeled by a noiseless amplifier with an input resistor at an equivalent noise temperature of:

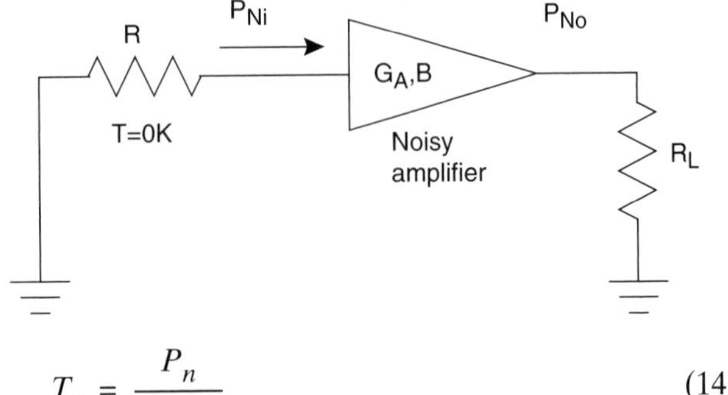

FIGURE 14.5 A noisy amplifier.

$$T_e = \frac{P_n}{G_A kB} \tag{14.9}$$

T_e is called the equivalent noise temperature of the amplifier "referred to the input," as shown in Figure 14.6.

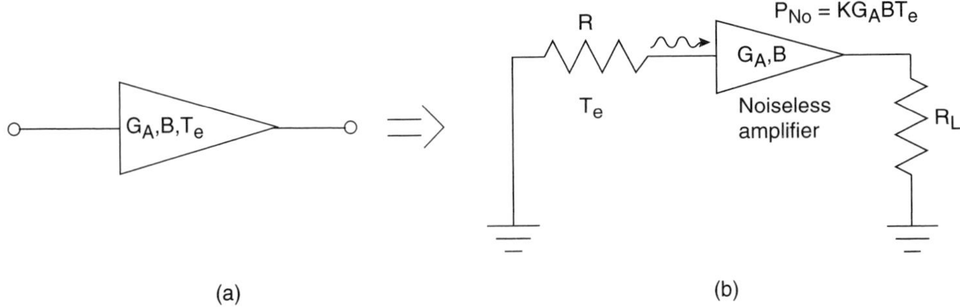

FIGURE 14.6 Equivalent models of a noisy amplifier.

Because source resistor (R) is at a physical temperature other than zero ($T = T_S$), then as a result the combined equivalent noise temperature (T'_e) is the addition of the two noise temperatures:

$$T'_e = T_e + T_S \tag{14.10}$$

Assuming the noise power at the input terminals of the amplifier is P_{Ni} ($= kT_SB$), the total output noise power due to the amplified input thermal noise power will be (G_AP_{Ni}), which adds to the amplifier's generated noise power (P_n) linearly by using the superposition theorem (see Figure 14.7), i.e.,

$$P_{No,tot} = G_AP_{Ni} + P_n = G_AkB(T_S + T_e)$$

$$P_{No,tot} = G_AkBT'_e \tag{14.11}$$

NOTE: *It is important to note that from Equation 14.11, the "equivalent noise temperature" (T'_e) is defined by "referring" the total output noise power to the input port. Thus, the same noise power is delivered to the load by driving a "noiseless amplifier" with a resistor at an equivalent temperature ($T'_e = T_e + T_S$).*

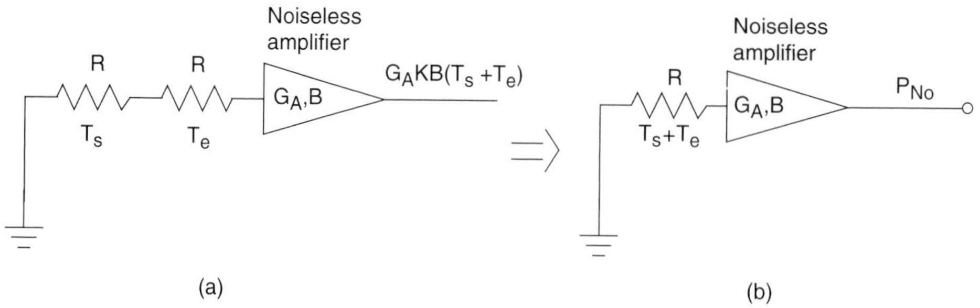

Noiseless amplifier

R R G_A, B $G_A kB(T_s + T_e)$

T_s T_e

Noiseless amplifier

R G_A, B P_{No}

$T_s + T_e$

(a) (b)

FIGURE 14.7 Total output noise power and its equivalent circuit.

14.7.1 A Measurement Application: Y-Factor Method

The concept of equivalent noise temperature is commonly used in the measurement of noise temperature of an unknown amplifier using the "*Y*-factor method." In this method, the physical temperature of a matched resistor is changed to two distinct and known values:

- One temperature (T_1) is at boiling water ($T_1 = 100°C$) or at room temperature ($T_1 = 290°\ K$)
- The second temperature (T_2) is obtained by using either a noise source (hotter source than room temperature) or a load immersed in liquid nitrogen at $T = 77°$ K (a colder source than room temperature), as shown in Figure 14.8.

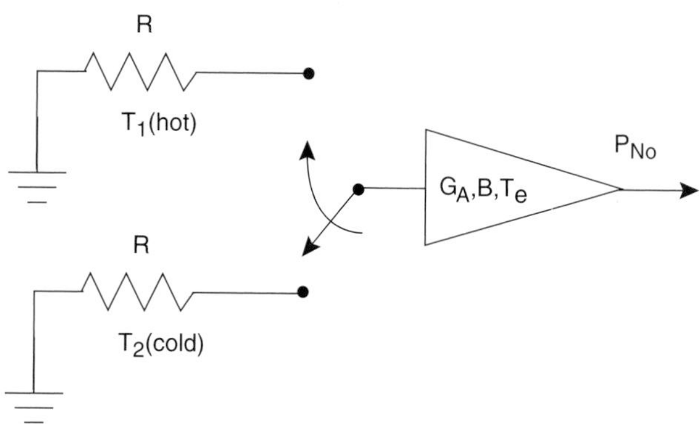

FIGURE 14.8 Y-factor method.

The amplifier's unknown noise temperature (T_e) can be obtained as follows:

$$P_{No,1} = G_A kB(T_1 + T_e) \tag{14.12}$$

$$P_{No,2} = G_A kB(T_2 + T_e) \tag{14.13}$$

Now define:

$$Y \equiv \frac{P_{No,1}}{P_{No,2}}$$

Thus, we can write:

$$Y = \frac{T_1 + T_e}{T_2 + T_e} \tag{14.14}$$

or

$$T_e = \frac{T_1 - YT_2}{Y - 1} \tag{14.15}$$

From a measurement of T_1, T_2 and Y, the unknown amplifier's noise temperature (T_e) can be found.

POINT OF CAUTION: *To obtain an accurate value for Y, the two temperatures ideally must be far apart; otherwise, $Y \approx 1$ and the denominator of Equation 14.15 will create relatively inaccurate results.*

NOTE: *A noise source "hotter" than room temperature, as used in the Y-factor measurement, would be a solid-state noise source (such as an IMPATT diode) or a noise tube. Such active sources, providing a calibrated and specific noise power output in a particular frequency range, are most commonly characterized by their "excess noise ratio" values versus frequency. The term excess noise ratio or ENR is defined as:*

$$ENR(\text{dB}) = 10 \log_{10}\left(\frac{P_N - P_o}{P_o}\right) = 10 \log_{10}\left(\frac{T_N - T_o}{T_o}\right) \tag{14.16}$$

where P_N and T_N are the noise power and equivalent noise temperature of the active noise generator, and P_o and T_o are the noise power and temperature of a room-temperature passive source (e.g., a matched load), respectively.

14.8 DEFINITIONS OF NOISE FIGURE

As discussed earlier, a noisy amplifier can be characterized by an equivalent noise temperature (T_e). An alternate method to characterize a noisy amplifier is through the concept of noise figure, which we need to define first.

DEFINITION-NOISE FIGURE: *The ratio of the total available noise power at the output, $(P_o)_{tot}$, to the output available noise power $(P_o)_i$ due to thermal noise coming only from the input resistor at the standard room temperature ($T_o = 290°\,K$).*

To formulate an equation for noise figure (F), let us transfer the noise generated inside the amplifier (P_n) to its input terminals and model it as a "noiseless" amplifier that is connected to a noisy resistor (R) at noise temperature (T_e) in series to another resistor (R) at $T = T_o$, both connected at the input terminals of the "noiseless" amplifier, as shown in Figure 14.9.

From this configuration, we can write:

$$P_n = G_A k T_e B \tag{14.17a}$$

$$(P_o)_i = G_A P_{Ni} = G_A k B T_o \tag{14.17b}$$

$$(P_o)_{tot} = P_{No} = P_n + (P_o)_i \tag{14.18}$$

FIGURE 14.9 A noisy amplifer.

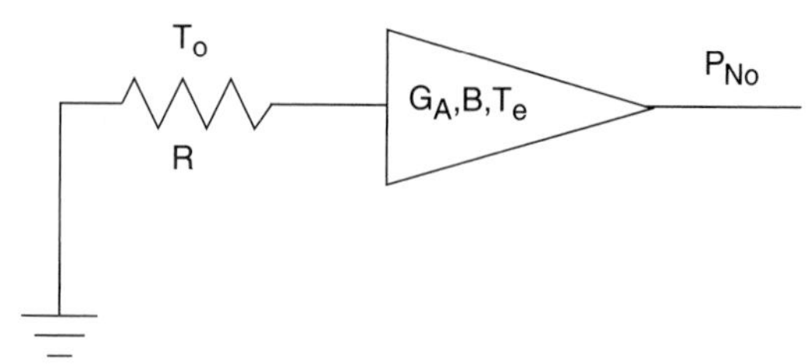

$$F = \frac{(P_o)_{tot}}{(P_o)_i} = \frac{(P_o)_i + P_n}{(P_o)_i} = 1 + \frac{P_n}{G_A P_{Ni}} \tag{14.19a}$$

or

$$F = 1 + \frac{T_e}{T_o} \tag{14.19b}$$

Or, in dB, we can write:

$$F = 10 \log_{10}\left(1 + \frac{T_e}{T_o}\right) \tag{14.20}$$

From Equations 14.19, we can see that F is bounded by:

$$1 \le F \le \infty \tag{14.21}$$

The lower boundary ($F = 1$) is the best-case scenario and is the noise figure of an ideal noiseless amplifier where $T_e = 0$.

From Equation 14.19b, we can write:

$$T_e = (F - 1)T_o \tag{14.22}$$

NOTE 1: *Temperature (T_e) is the equivalent noise temperature of the amplifier referred to the input.*

NOTE 2: *Either F or T_e can interchangeably be used to describe the noise properties of a two-port network. For small noise figure values (i.e., when $F \approx 1$), use of T_e becomes preferable.*

POINT OF CAUTION: *It is interesting to note that the noise figure is defined with reference to a matched input termination at room temperature (T_o= 290° K). Therefore, if the physical temperature of the amplifier changes to some value other than T_o, we still use the room temperature (T_o = 290° K) to find the noise figure value.*

14.8.1 Alternate Definition of Noise Figure

From Equations 14.17 and 14.18, we can write:

$$P_{No} = G_A P_{Ni} + P_n \tag{14.23}$$

$$(P_o)_i = G_A P_{Ni} \tag{14.24}$$

where $P_n = G_A k T_e B$ is the generated noise power inside the amplifier. The noise figure can now be written as:

$$F = \frac{P_{No}}{(P_o)_i} = \frac{P_{No}}{G_A P_{Ni}} \tag{14.25}$$

The available power gain (G_A) by definition is given by:

$$G_A = \frac{P_{So}}{P_{Si}}$$

where P_{So} and P_{Si} are the available signal power at the output and the input, respectively. Thus, Equation 14.25 can now be written as:

$$F = \frac{P_{Si}/P_{Ni}}{P_{So}/P_{No}} = \frac{(SNR)_i}{(SNR)_o} \tag{14.26}$$

where $(SNR)_i$ and $(SNR)_o$ are the available signal-to-noise ratio at the input and output ports, respectively.

Equation 14.26 indicates that the noise figure can also be defined in terms of the ratio of the available signal-to-noise power ratio at the input to the available signal-to-noise power ratio at the output.

14.8.2 Noise Figure of a Lossy Two-Port Network

This is an important case, where the two-port network considered earlier is a lossy passive component, such as an attenuator or a lossy transmission line, as shown in Figure 14.10.

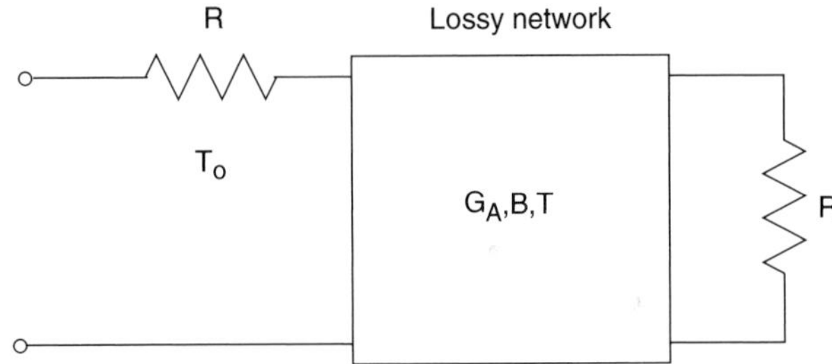

FIGURE 14.10 A lossy two-port network.

A lossy network has a gain $\left(G_A = \frac{P_o}{P_i}\right)$ less than unity, which can be expressed in terms of the loss factor or attenuation (L) as:

$$G_A = \frac{1}{L} \quad (G_A < 1) \tag{14.27}$$

Because the gain of a lossy network is less than unity it follows that the loss or attenuation factor (L) is more than unity (i.e., $L = P_i/P_o > 1$) for any lossy network or component.

Expressing the attenuation factor (L) in dB gives the following:

$$L(\text{dB}) = 10\log_{10}\left(\frac{P_i}{P_o}\right) \tag{14.28}$$

For example, if the lossy component attenuates the input power by ten times, then we can write:

$$G_A = \frac{P_o}{P_i} = 0.1 \Rightarrow L = 1/G_A = 10 \text{ dB}$$

If the lossy network is held at a temperature (T), the total available output noise power according to Equation 14.5 is given by:

$$P_{No} = kTB \tag{14.29}$$

On the other hand, from Equation 14.23 the available output noise power is also given by the addition of the input noise power and the generated noise inside the circuit (P_n):

$$P_{No} = G_A kTB + P_n = KTB/L + P_n \tag{14.30}$$

where P_n is the noise generated inside the two-port network. Equating Equations 14.29 and 14.30, we obtain P_n as:

$$P_n = \left(\frac{L-1}{L}\right)kTB \tag{14.31 a}$$

NOTE: *If we refer the noise generated inside the amplifier (P_n) to the input side $(P_n)_i$ from Equation 14.31a, we have:*

$$(P_n)_i = P_n/G_A = LP_n = (L-1)kTB \tag{14.31b}$$

Using Equation 14.31b, we can now define the equivalent noise temperature (T_e) of a lossy two-port network, referred to the input terminals, as:

$$T_e = \frac{(p_n)_i}{kB} \Rightarrow T_e = (L-1)T \tag{14.32}$$

Thus, the noise figure of a lossy network is given by:

$$F = 1 + \frac{T_e}{T_o} = 1 + (L-1)\frac{T}{T_o} \tag{14.33}$$

A Special Case. For a lossy network at room temperature, i.e., $T = T_o$, Equation 14.33 gives:

$$F = L \tag{14.34}$$

Equation 14.34 indicates that the noise figure of a lossy network at room temperature equals the attenuation factor (L).

For example, if $G_A = 1/5$ then $L = 1/G_A = 5$, giving $F = 5$ or 7 dB for $T = T_o = 290° \text{ K}$.

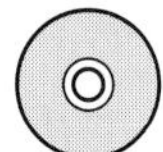

EXAMPLE 14.3

A wideband amplifier (2–4 GHz) has a gain of 10 dB, an output power of 10 dBm and a noise figure of 4 dB at room temperature. Find the output noise power in dBm.

Solution:

$$B = 2 \text{ GHz}$$

$$G_A = 10 \text{ dB}$$

$$F = 4 \text{ dB}$$

$$F = P_{No}/G_A P_{Ni} = P_{No}/G_A k T_o B$$

Thus,

$$P_{No} = F G_A k T_o B$$

$$10 \log_{10} P_{No} = P_{No}(\text{dBm}) = F(\text{dB}) + G_A(\text{dB}) + 10 \log_{10}(kT_o) + 10 \log_{10}(B)$$
$$= 4 + 10 - 174 + 10 \log_{10}(2 \times 10^9) = -67 \text{ dBm}$$

14.9 NOISE FIGURE OF CASCADED NETWORKS

A microwave system usually consists of several stages or networks connected in cascade where each adds noise to the system, thus degrading the overall signal-to-noise ratio.

If the noise figure (or noise temperature) of each stage is known, the overall noise figure (or noise temperature) can be determined.

14.9.1 Cascade of Two Stages

To analyze a two-stage amplifier, let's consider a cascade of two amplifiers each with its own gain, noise temperature, or noise figure, as shown in Figure 14.11. The noise power of each stage is given as follows:

$$P_{No1} = G_{A1} k B (T_o + T_{e1}) \tag{14.35}$$

$$P_{No2} = G_{A2} P_{No1} + G_{A2} k T_{e2} B \tag{14.36}$$

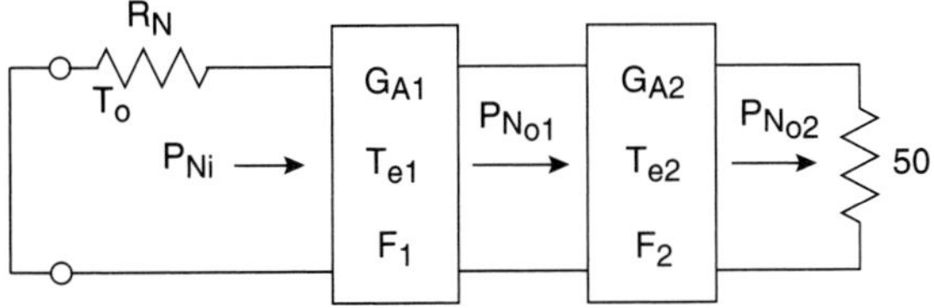

FIGURE 14.11 Cascade of two stages.

Combining Equations 14.35 and 14.36, we get:

$$P_{No2} = G_{A1} G_{A2} k B (T_o + T_{e1} + T_{e2}/G_{A1}) \tag{14.37}$$

The two-stage amplifier as a whole has a total gain of $G_A = G_{A1} G_{A2}$, an overall equivalent noise temperature (T_e), and a total output noise power (P_{No}) given by:

$$P_{No} = G_A k B (T_o + T_e) \tag{14.38}$$

Comparing Equation 14.38 to 14.37, we have:

$$T_e = T_{e1} + T_{e2}/G_{A1} \tag{14.39}$$

The overall noise figure (F) for the two-stage amplifier is found by using Equation 14.39:

$$F = 1 + T_e/T_o = 1 + (T_{e1} + T_{e2}/G_{A1})/T_o \tag{14.40}$$

By noting that:

$$F_1 = 1 + T_{e1}/T_o, \tag{14.41}$$

$$F_2 = 1 + T_{e2}/T_o \tag{14.42}$$

Equation 14.40 can be written as:

$$F = F_1 + \frac{F_2 - 1}{G_{A1}} \tag{14.43}$$

Equations 14.39 and 14.43 show that the first-stage noise figure F_1 (or noise temperature, T_{e1}) and gain (G_{A1}) have a large influence on the overall noise figure (or noise temperature). This is because the second-stage noise figure, F_2 (or noise temperature, T_{e2}) is reduced by the gain of the first stage (G_{A1}).

Thus, the key to low overall noise figure is a primary focus on the first stage by reducing its noise and increasing its gain. Later stages have a greatly reduced effect on the overall noise figure.

Noise Measure. In order to determine systematically the order or sequence in which two similar amplifiers need be connected to produce the lowest possible noise figure, we first must define a quantity called "noise measure" as:

$$M = \frac{F - 1}{1 - 1/G_A} \tag{14.44}$$

If amplifier #1 (AMP1) has a noise measure (M_1) and amplifier #2 (AMP2) a noise measure (M_2), then there are two possible cases that need to be addressed (in order to obtain the lowest possible noise figure from the cascade), as follows:

Case I: $M_1 < M_2$ Then AMP1 should precede AMP2 because $F_{12} < F_{21}$.

Case II: $M_2 < M_1$ Then AMP2 should precede AMP1 because $F_{21} < F_{12}$.

where F_{12} is the overall noise figure of the two-stage amplifier when AMP1 precedes AMP2, and vice versa, F_{21} is for the case when AMP2 precedes AMP1.

NOTE: *It can easily be shown mathematically that, for example:*

$$\text{If } M_1 < M_2 \text{ then } F_{12} < F_{21} \tag{14.45}$$

where

$$F_{12} = F_1 + \frac{F_2 - 1}{G_{A1}} \tag{14.46}$$

$$F_{21} = F_2 + \frac{F_1 - 1}{G_{A2}} \tag{14.47}$$

and vice versa: if $M_2 < M_1$ then $F_{21} < F_{12}$.

EXAMPLE 14.4

An antenna is connected to an amplifier via a transmission line that has an attenuation of 3 dB (see Figure 14.12). The amplifier has the following specifications:

$$G_A = 20 \text{ dB}$$

$$B = 200 \text{ MHz}$$

$$T_e = 145 \text{ K}$$

Calculate the overall noise figure and gain of the cascade at 300 K.

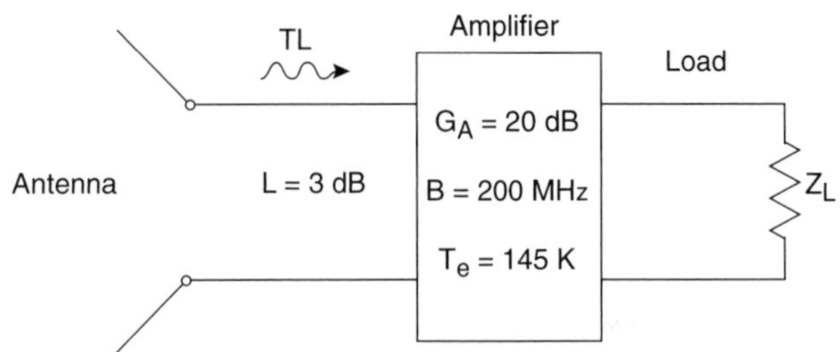

FIGURE 14.12 Circuit for Example 14.4.

Solution:

a. For the transmission line, we have:

$$\text{L} = 1/G_{TL} \Rightarrow L(\text{dB}) = -G_{TL}(\text{dB})$$

Thus,

$$L = 3 \text{ dB} = 2 \Rightarrow G_{TL} = -3 \text{ dB} = 1/2$$

$$F_{TL} = 1 + (L-1)T/T_o = 1 + (2-1)300/290 = 2.03 = 3.1 \text{ dB}$$

b. For the amplifier, we have:

$$F_{AMP} = 1 + T_e/T_o = 1.5 = 1.8 \text{ dB}$$

c. The overall noise figure and gain are calculated to be:

$$F_{tot} = F_{TL} + (F_{AMP} - 1)/G_{TL} = 2.03 + (1.5-1)/0.5 = 3.03 = 4.8 \text{ dB}$$

$$G_{tot} = G_{TL} + G_{AMP} = -3 + 20 = 17 \text{ dB}$$

Therefore, we can see that due to the addition of a lossy transmission line in front of the amplifier, we have three deleterious effects: (1) the overall noise figure has increased (from 1.8 dB to 4.8 dB); (2) the second-stage noise contribution has been intensified because the transmission line has a gain less than unity ($G_{TL} < 1$); and (3) the overall gain dropped by 3 dB, which represents the third side effect.

14.9.2 Cascade of *n* Stages

For a cascade of n amplifiers (see Figure 14.13), the overall noise figure is the generalization of equations for equivalent noise temperature ($T_{e,cas}$) and noise figure (F_{cas}) of a two-stage cascade, as follows:

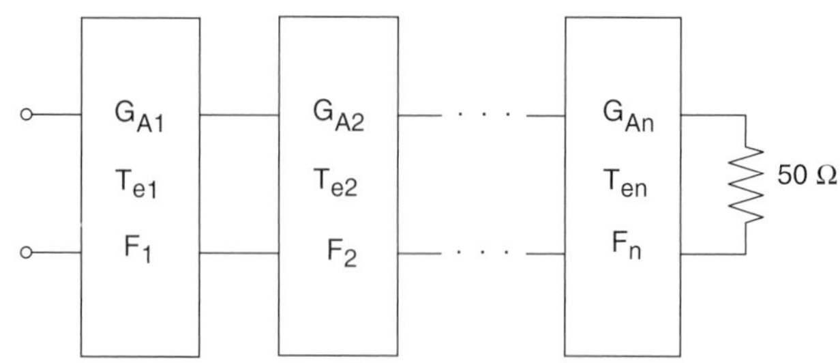

FIGURE 14.13 Cascade of n amplifier stages.

$$T_{e,cas} = T_{e1} + \frac{T_{e2}}{G_{A1}} + \frac{T_{e3}}{G_{A1}G_{A2}} + \ldots + \frac{T_{en}}{G_{A1}G_{A2}\ldots G_{An-1}} \tag{14.48a}$$

$$F_{cas} = F_1 + \frac{F_2 - 1}{G_{A1}} + \frac{F_3 - 1}{G_{A1}G_{A2}} + \ldots + \frac{F_n - 1}{G_{A1}G_{A2}\ldots G_{An-1}} \tag{14.48b}$$

Special Case: Identical Stages. If all stages are identical, i.e.,

$$G_{A1} = G_{A2} = \ldots. = G_{An} = G_A \tag{14.49a}$$

$$T_{e1} = T_{e2} = \ldots.. = T_{en} = T_e \tag{14.49b}$$

$$F_1 = F_2 = \ldots... = F_n = F \tag{14.49c}$$

Then, Equations 14.48a and b would greatly simplify as follows:

$$T_{e,cas} = T_e(1 + X + X^2 + \ldots. + X^{n-1}), \tag{14.50a}$$

$$F_{cas} = (F-1)(1 + X + X^2 + \ldots. + X^{n-1}) + 1 \tag{14.50b}$$

where

$$X = \frac{1}{G_A}$$

Using the following identity for the geometric series:

$$1 + X + X^2 + \ldots.. + X^{n-1} = (1 - X^n)/(1 - X), \quad |X| < 1 \tag{14.51}$$

we can write Equations 14.50a and b as:

$$T_{e,cas} = T_e\left(\frac{1 - (1/G_A)^n}{1 - 1/G_A}\right) \tag{14.52a}$$

$$F_{cas} = (F-1)\left(\frac{1 - (1/G_A)^n}{1 - 1/G_A}\right) + 1 \tag{14.52b}$$

An Infinite Chain of Identical Amplifiers. If n is very large (i.e., $n \to \infty$) then:

$$\lim_{n \to \infty} (X)^n = 0, \qquad |X| < 1 \tag{14.53a}$$

And the geometric series identity in Equation 14.51 further simplifies into:

$$1 + X + X^2 + \ldots. + X^{n-1} + \ldots. = \frac{1}{1-X}, \quad |X| < 1 \tag{14.53b}$$

Using Equation 14.53b, we can see that Equations 14.52a and b for an infinite chain of amplifiers become:

$$T_{e,cas} = T_e\left(\frac{1}{1-1/G_A}\right) \tag{14.54a}$$

$$F_{cas} = (F-1)\left(\frac{1}{1-1/G_A}\right) + 1 \tag{14.54b}$$

In terms of noise measure, M, defined earlier as:

$$M = \frac{F-1}{1-1/G_A} \tag{14.55}$$

we can write Equations 14.54 as:

$$T_{e,cas} = T_e\left(\frac{M}{F-1}\right) \tag{14.56a}$$

$$F_{cas} = M + 1 \tag{14.56b}$$

NOTE 1: *For a "minimum-noise amplifier," where each stage operates at minimum noise figure (i.e., $F_1 = F_2 =. \ldots = F_n = F_{min}$), we have:*

$$M_{min} = \frac{F_{min}-1}{1-1/G_A} \tag{14.57}$$

We can write Equations 14.56 as:

$$T_{e,cas} = T_{e,min}\left(\frac{M_{min}}{F_{min}-1}\right) \tag{14.58a}$$

$$F_{cas} = M_{min} + 1 \tag{14.58b}$$

NOTE 2: *If the gain of each stage is very large (i.e., $G_A \to \infty$), then Equations 14.56 become:*

$$G_A \to \infty \Rightarrow M = F - 1 \tag{14.59}$$

$$T_{e,cas} = T_e \tag{14.60a}$$

$$F_{cas} = F \tag{14.60b}$$

This result indicates that a large cascade of very high-gain amplifiers will result only in the degradation of the signal by the first stage, and the effect of all the many stages is null and void as far as the added noise is concerned.

This result is in agreement with the conclusion we made earlier, in which it became apparent that the first stage's gain and noise figure value dominate and greatly affect the overall noise figure of the cascade.

14.10 CONSTANT NOISE FIGURE CIRCLES

It can be shown that the noise figure of a two-port amplifier is given by:

$$F = F_{min} + \frac{r_n}{g_S}|Y_S - Y_{opt}|^2 \tag{14.61}$$

where

$r_n = \frac{R_n}{Z_o}$ the equivalent normalized noise resistance of the two-port

$Y_S = g_S + jb_S$ the normalized source admittance corresponding to Γ_S as defined in Chapter 13, *Gain Considerations in Amplifiers*

$Y_{opt} = g_{opt} + jb_{opt}$ the normalized source admittance for minimum noise figure, (i.e., at $\Gamma_S = \Gamma_{opt} \Rightarrow F = F_{min}$)

Because Y_S and Y_{opt} are related to Γ_S and Γ_{opt} by the relations:

$$Y_S = \frac{1 - \Gamma_S}{1 + \Gamma_S} \tag{14.62}$$

$$Y_{opt} = \frac{1 - \Gamma_{opt}}{1 + \Gamma_{opt}} \tag{14.63}$$

Using Γ_S and Γ_{opt} instead of Y_S and Y_{opt} in Equation 14.61, we can write:

$$F = F_{min} + \frac{4r_n|\Gamma_S - \Gamma_{opt}|^2}{(1 - |\Gamma_S|^2)|1 + \Gamma_{opt}|^2} \tag{14.64}$$

We now define a parameter called the noise figure parameter (N):

$$N = \frac{|\Gamma_S - \Gamma_{opt}|^2}{1 - |\Gamma_S|^2} \tag{14.65}$$

Thus, Equation 14.64 can be written as:

$$F = F_{min} + \frac{4r_nN}{|1 + \Gamma_{opt}|^2} \tag{14.66}$$

Parameters r_n, Γ_{opt}, and F_{min} are called the "noise parameters" of the transistor and are usually provided in the data sheets by the manufacturer.

NOTE 1: *Using Equation 14.66, we can write Equation 14.65 as:*

$$N = \frac{|\Gamma_S - \Gamma_{opt}|^2}{1 - |\Gamma_S|^2} = \frac{F - F_{min}}{4r_n}|1 + \Gamma_{opt}|^2 \tag{14.67}$$

From Equation 14.67, we can see that for a fixed (F), the parameter (N) is a positive real number (because $F \geq F_{min}$).

NOTE 2: *Noise parameters may also be determined experimentally by the following procedure:*

a. *Vary Γ_S until a minimum noise figure occurs. This is recorded as F_{min}.*

b. *Now, using a vector network analyzer, measure Γ_S, which provides the value for Γ_{opt}.*

c. *We find r_n, by setting Γ_S to zero and then measure the noise figure F_o at this point. By using Equation 14.64 and the value of Γ_{opt} from the second step, we can obtain r_n as:*

$$r_n = \Delta F\left(\frac{|1+\Gamma_{opt}|^2}{4|\Gamma_{opt}|^2}\right) \tag{14.68}$$

where

$$\Delta F = F_o - F_{min}$$

14.10.1 Analysis

By using Equation 14.67 and through rearranging terms and further mathematical manipulation of Equation 14.64, we obtain an equation of a circle in the Γ_S-plane as:

$$|\Gamma_S - C_F| = R_F \tag{14.69}$$

where C_F and R_F are the center and radius of noise figure circles given by:

$$C_F = \frac{\Gamma_{opt}}{N+1} \tag{14.70}$$

$$R_F = \frac{\sqrt{N^2 + N(1-|\Gamma_{opt}|^2)}}{1+N} \tag{14.71}$$

Equation 14.69 represents a family of noise figure circles with the noise figure (F) value as a parameter.

NOTE: *For derivation of constant noise figure circles please see Appendix M, Derivation of the Constant Gain and Noise Figure Circles.*

When $F = F_{min}$, then:

$$\Gamma_S = \Gamma_{opt} \Rightarrow N = 0 \tag{14.72}$$

$$C_F = \Gamma_{opt} \tag{14.73}$$

$$R_F = 0 \tag{14.74}$$

Equations 14.73 and 14.74 indicate that F_{min} is a point uniquely located at Γ_{opt}. Furthermore, from Equation 14.67, we can see that because N is a positive real number, then all noise figure circles have centers located along Γ_{opt} vector in the Γ_S plane, as shown in Figure 14.14 (see Equation 14.70).

EXERCISE 14.1

Derive Equation 14.69 from Equations 14.64 through 14.67.

HINT: Use the following identity:

$$|\Gamma_S - \Gamma_{opt}|^2 = (\Gamma_S - \Gamma_{opt})(\Gamma_S - \Gamma_{opt})^*$$

to write (14.65) as:

FIGURE 14.14 Family of noise figure circles all located on the Γ_{opt} vector.

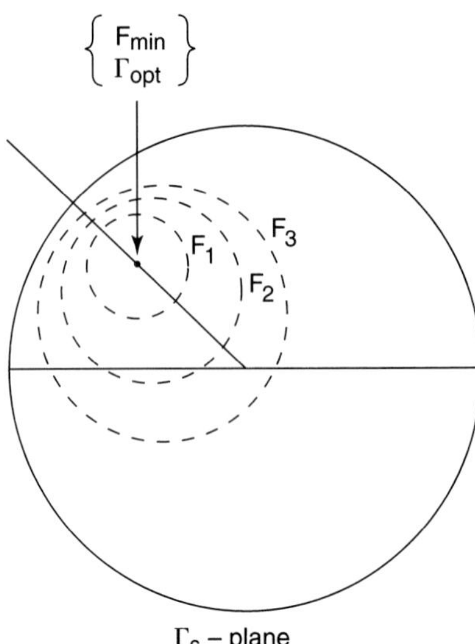

$$|\Gamma_S|^2 - \frac{(\Gamma_S\Gamma_{opt}^* + \Gamma_S^*\Gamma_{opt})}{1+N} + \frac{|\Gamma_{opt}|^2}{1+N} = \frac{N}{1+N}$$

Now add $|\Gamma_{opt}|^2/(N+1)^2$ to both sides of the above equation to obtain the desired relation for the constant noise circle (center and radius) as given by Equations 14.70 and 14.71 (see Appendix M).

LIST OF SYMBOLS/ABBREVIATIONS

A symbol/abbreviation will not be repeated again once it has been identified and defined in an earlier chapter, as long as its definition remains unchanged.

B	Bandwidth
F	Noise figure
k	Boltzmann's constant
M	Noise measure
N	Noise figure parameter
P_N	Thermal noise power
P_{NI}	Input noise power
$P_{No,tot}$	Total output noise power
r_n	Normalized noise resistance of a two-port
R_n	Noise resistance of a two-port
R_N	Noisy resistor
R_{No}	Noiseless resistor
SNR	Signal-to-noise ratio
T	Temperature

T_e	Equivalent noise temperature
T_o	Standard room temperature (290° K)
T_S	Source resistor's temperature
$V_{n,rms}$	Root-mean-square (rms) of noise
$\overline{V_n^2}$	The mean-square value of noise
Y	*Y*-factor used in noise measurement
Y_{opt}	Normalized source admittance for minimum noise figure at which $F = F_{min}$
Γ_{opt}	The optimum source reflection coefficient at which $F = F_{min}$

PROBLEMS

14.1 The *Y*-factor method is to be used to measure the equivalent noise temperature of a component. A hot load of $T_1 = 300°$ K and a cold load of $T_2=77°$ K will be used. If the noise temperature of the amplifier is $T_e = 250°$ K, what will be the ratio of power meter readings at the output of the component for the two loads?

14.2 A transmission line has a noise figure $F = 1$ dB at a temperature $T_o = 290°$ K. Calculate and plot the noise figure of this line as its physical temperature ranges from $T = 0°$ K to 1000° K.

14.3 Assume that measurement error introduces an uncertainty of ΔY into the measurement of Y in a *Y*-factor measurement. Derive an expression for the normalized error of the equivalent noise temperature $(\Delta T_e/T_e)$ in terms of $(\Delta Y/Y)$ and the temperatures T_1, T_2, and T_e. Minimize this result with respect to T_e to obtain an expression in terms of T_1 and T_2 that will result in minimum error.

14.4 An amplifier with a bandwidth of 1 GHz has a gain of 15 dB and a noise temperature of 250° K. If it is used as a preamplifier in a cascade, preceding a microwave amplifier of 20 dB gain and 5 dB noise figure, determine the overall noise temperature.

14.5 An amplifier with a gain of 12 dB, a bandwidth of 150 MHz, and a noise figure of 4 dB feeds a receiver with a noise temperature of 900 K. Find the noise figure of the overall system.

14.6 Consider the microwave system shown in Figure P14.6, where the bandwidth is 1 GHz centered at 20 GHz, and the physical temperature of the system is $T = 300°$ K. What is the equivalent noise temperature of the source? What is the noise figure of the amplifier in dB? When the noisy source is connected to the system what is the total noise power output of the amplifier in dBm?

14.7 Consider the wireless local area network (WLAN) receiver front-end shown in Figure P14.7, where the bandwidth of the bandpass filter is 100 MHz centered at 2.4 GHz. If the system is at room temperature, find the noise figure of the overall system. What is the resulting signal-to-noise ratio at the output if the input signal power level is –90 dBm? Can the components be rearranged to give a better noise figure?

14.8 A two-way power divider has one output port terminated in a matched load, as shown in Figure P14.8. Find the equivalent noise temperature of the resulting two-port network if the divider is an equal-split two-way resistive divider.

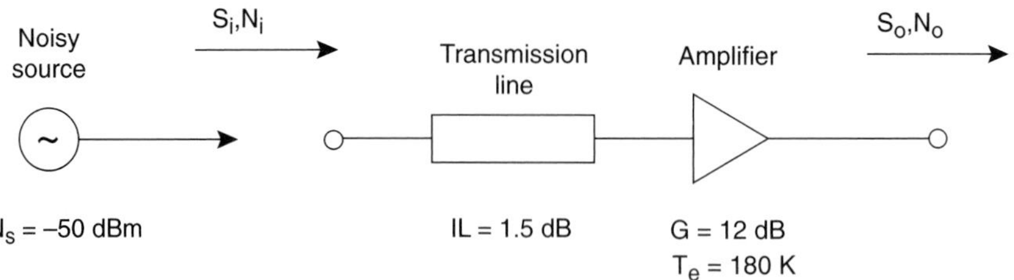

FIGURE P14.6

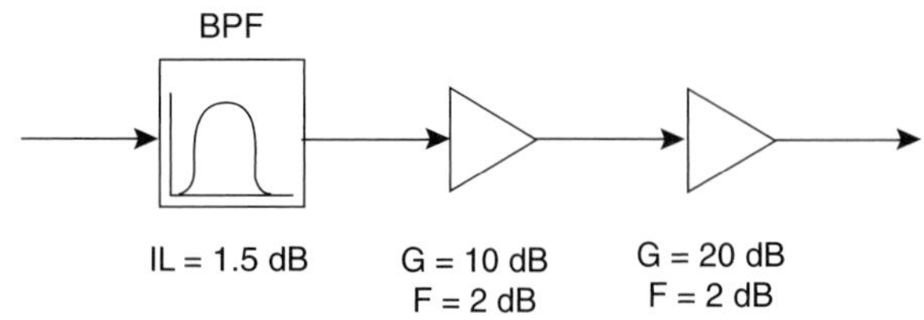

FIGURE P14.7

FIGURE P14.8

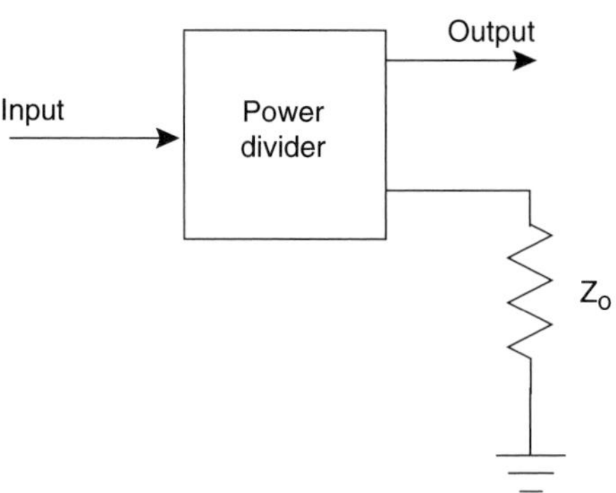

14.9 For a two-stage cascaded network with gain values of G_1 and G_2 and noise figures of F_1 and F_2, as shown in Figure P14.9, the input noise power is $N_i = kTB$. The output noise powers are N_1 and N_2 at the output of the first and second stages. Which of the following expressions are correct?

FIGURE P14.9

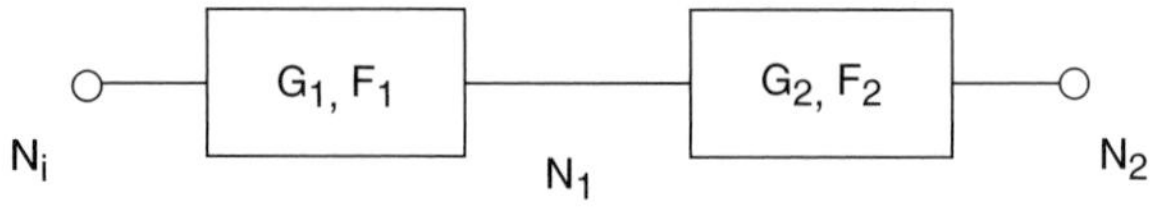

G_1, F_1 G_2, F_2

N_i N_1 N_2

a. $F_1 = N_1/G_1N_i$
b. $F_2 = N_2/G_2N_1$
c. $F_2 = N_2/G_1G_2F_1N_i$

14.10 A receiver has the block diagram shown in Figure P14.10. Calculate:

a. The total gain (or loss) in dB

b. The overall noise figure in dB

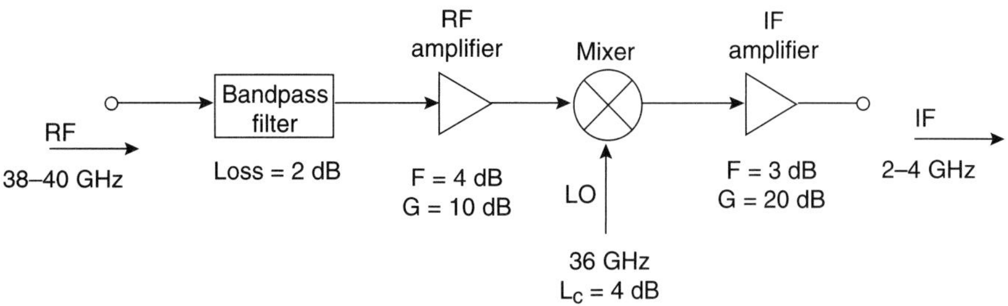

FIGURE P14.10

14.11 Two satellite receiver systems have the following specifications for their components:

RF amplifier: $F = 5$ dB, $G = 10$ dB

Mixer: $L_c = 5$ dB

IF amplifier: $F = 2$ dB, $G = 15$ dB

Bandpass filter: $IL = 2$ dB

Compare the two systems in terms of the overall gain and noise figure values (see Figure P14.11).

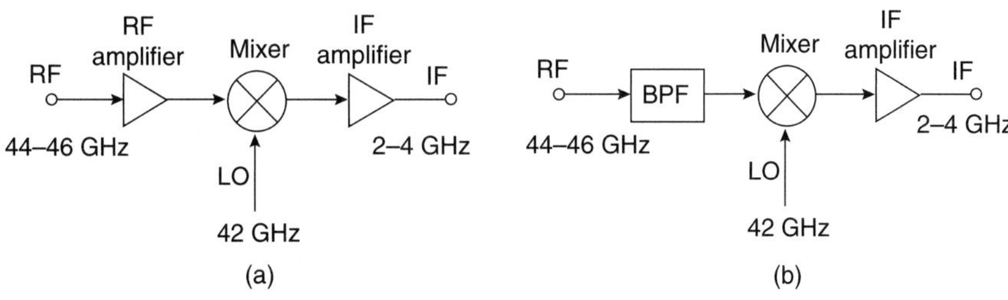

FIGURE P14.11

14.12 Calculate the overall noise figure and gain for the receiver system shown in Figure P14.12.

14.13 The S-parameters and the noise parameters of a GaAs FET at 10 GHz in a 50 Ω system are:

$$\begin{bmatrix} 0.6\angle{-170^\circ} & 0.05\angle 16^\circ \\ 2\angle 30^\circ & 0.5\angle{-95^\circ} \end{bmatrix}$$

$$F_{min} = 2.5 \text{ dB}$$

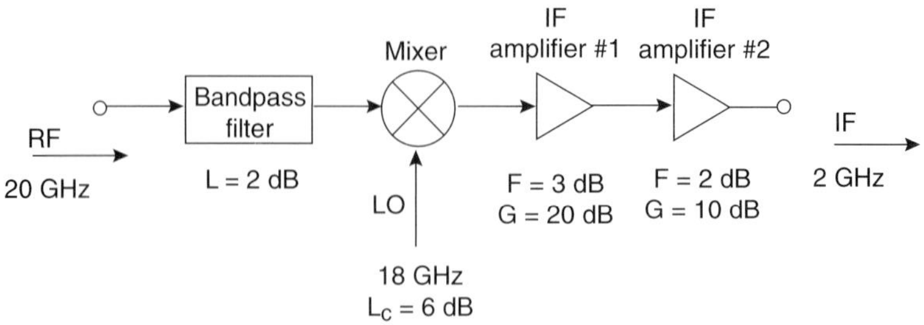

FIGURE P14.12

$$\Gamma_{opt} = 0.5\angle 145°$$

$$R_n = 5\ \Omega$$

a. Is the transistor unconditionally stable?

b. Determine $G_{A,max}$.

c. Determine the noise figure if the transistor is used in an amplifier designed for maximum available gain ($G_{A,max}$).

14.14 Consider the low noise block (LNB) shown in Figure P14.14. Calculate the total noise figure and the available gain of this LNB.

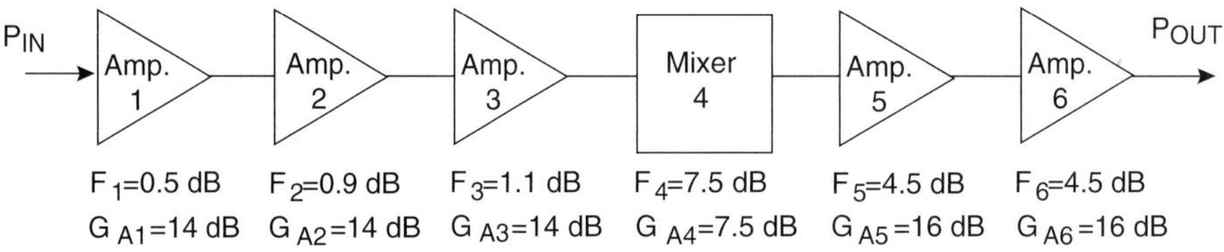

FIGURE P14.14

REFERENCES

[14.1] Ambrozy, A. *Electronic Noise.* New York: McGraw-Hill, 1982.

[14.2] Cappy, A. Noise Modeling and Measurement Techniques. *IEEE Transactions on Microwave Theory and Technique*, Vol. MTT-36, January 1988, pp. 1–10.

[14.3] Friis, H. T. Noise Figure of Radio Receivers. *Proceedings of IRE,* Vol. 32, July 1944, pp. 419–22.

[14.4] Fukui, H. Available Power Gain, Noise Figure, and Noise Measure of Two-Ports and Their Graphical Representation. *IEEE Transactions on Circuit Theory,* June 1966, pp. 137–42.

[14.5] Gonzalez, G. *Microwave Transistor Amplifiers, Analysis and Design*, 2nd ed. Upper Saddle River: Prentice Hall, 1997.

[14.6] Haus, H. A. and R. B. Adler. *Circuit Theory of Linear Noisy Networks.* Cambridge: MIT Press, 1959 and New York: John Wiley & Sons, 1959.

[14.7] Pozar, D. M. *Microwave Engineering,* 2nd ed. New York: John Wiley & Sons, 1998.

[14.8] Radmanesh, M. M., and J. M. Cadwallader. Millimeter-Wave Noise Sources at V-Band (50 to 75 GHz), *Microwave Journal,* Vol. 36, No. 2, pp. 128–134, Sept. 1993.

[14.9] Radmanesh, M. M. and J. M. Cadwallader. Solid State Noise Sources at mm-Waves: Theory and Experiment. *Microwave Journal,* Vol. 34, No. 10, pp. 125–133, Oct. 1991.

[14.10] Vendelin, George D., Anthony M. Pavio, and Ulrich L. Rhode. *Microwave Circuit Design, Using Linear and Non-Linear Techniques.* New York: John Wiley & Sons, 1990.

PART V

ACTIVE NETWORKS: LINEAR AND NONLINEAR DESIGN

CHAPTER 15

RF/Microwave Amplifiers I: Small-Signal Design

15.1 INTRODUCTION

One of the most basic concepts in microwave circuit design is amplification. In the past, microwave tubes and microwave diodes (biased in the negative resistance region) were commonly used; however, nowadays use of microwave transistors (BJT or FET) has become very popular.

Transistor amplifiers are built rugged and are reliable for low-power to medium-power applications. In this chapter, we will consider the design of small-signal, narrow-band, high-gain, low-noise, broadband, and finally multistage amplifiers. The design methods presented are based on the *S*-parameters of the transistor, which forms the heart of the amplifier circuit. We will relegate the design of high-power amplifiers to the next chapter when we consider large-signal amplifiers.

15.2 TYPES OF AMPLIFIERS

Having considered basic concepts in stability, gain, and noise of a two-port amplifier in the previous three chapters, we are now ready to embark on the task of designing a functional amplifier.

In general, any type of amplifier requires an optimization of the gain, noise figure, power, or bandwidth. Therefore, each type of design will be an interplay of constant-gain, noise figure circles or power contours, depending on the design requirements.

15.2.1 Classes of Amplifiers Based on Operating Point

An amplifier usually operates in one of the following classes:

- Class A amplifier
- Class B amplifier
- Class AB amplifier
- Class C amplifier

Class A amplifier: In this mode, each transistor in the amplifier operates in its active region for the entire signal cycle.

Class B amplifier: In this mode of operation, each transistor is in its active region for approximately half of the signal cycle.

Class AB amplifier: In this mode, an amplifier operates in class A for small signals and in class B for large signals.

Class C amplifier: In this mode of operation, each transistor is in its active region for significantly less than half of the signal cycle.

15.2.2 Classes of Amplifiers Based on Signal Level

Even though this classification describes the operation of the transistor amplifiers under different operating points, most microwave amplifiers are usually further classified according to the signal level, i.e., small-signal mode or large-signal mode. The method of analysis of an amplifier under each signal condition is drastically different. Each mode of operation offers its own unique style of analysis as follows:

Small-signal analysis is a method of analysis of an active circuit in which it is assumed that the signals deviate from (or fluctuate to either side of) the steady bias levels by such a small amount that only a small part of the operating characteristic of the device is covered and thus the operation is always linear.

Large-signal analysis is a method of analysis of an active circuit under high-amplitude signals that traverse such a large part of the operating characteristic of a device that nonlinear portions of the characteristic are usually encountered, causing nonlinear operation of the device.

In this chapter, the primary focus will be on small-signal amplifiers, and the operating point most suitable for this purpose would be class A amplifiers, which will be studied here. Large-signal amplifiers will be studied in the next chapter.

15.3 SMALL-SIGNAL AMPLIFIERS

To design small-signal amplifiers properly, we need to determine the biasing values for the transistor first and then use great care to design and connect the DC circuitry correctly to the RF portion of the amplifier. Thus, generally speaking, the task of building an amplifier (or for that matter any microwave circuit) consists of two separate steps:

- DC circuit design
- RF/MW circuit design

These two circuits are then integrated and packaged seamlessly into one complete unit, as discussed next.

15.3.1 Amplifier's DC-Bias Circuit Design

To operate an amplifier under small-signal conditions, linear operation is required. This demands that the amplifier operate in class A mode. Under this condition, the DC Q-point of the amplifier must be chosen approximately midrange in the $I_C - V_{CE}$ (for BJT) or $I_D - V_{DS}$ (for FET) characteristic curves, ensuring active mode (BJT) or saturation mode (FET) operation of the transistor, as shown in Figure 15.1.

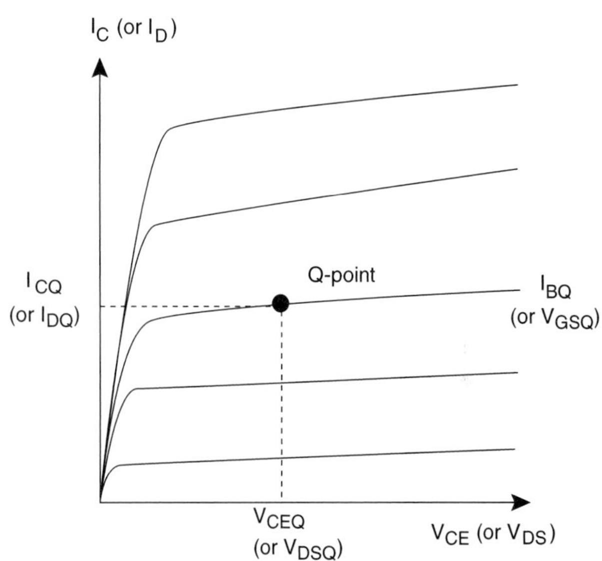

FIGURE 15.1 General transistor characteristic curves for BJT (or FET).

Furthermore, it is essential that the amplifier's bias circuitry be connected to the RF circuit in such a way that it will create minimum interaction and leakage for the RF/microwave signals (which are traveling from the input to the output). In other words, the DC circuitry should be completely isolated from the path of RF/microwave signals so that none of the signals leak or couple improperly to the DC source or, worse yet, to ground. To successfully achieve such a task, we use several schemes that can be briefly stated as follows:

a. Connect an inductor, commonly referred to as an RF choke (RFC), between the DC source and the RF/MW circuitry. One simple method to implement an RF choke in practice is through the use of a "ferrite bead."

b. Connect a quarter-wave transformer between the DC source and the RF/MW circuitry. The characteristic impedance (Z_o) of the transformer should preferably be very high (i.e., $Z_o >> 1$) because it will create a high impedance path for any traveling RF/MW signal.

c. Connect a high-value capacitor (as a load) to the quarter-wave transformer to effectively short out any residual RF/MW signal that would leak into the DC circuitry. The use of a high-value capacitor as a load for the quarter-wave transformer effectively creates an open circuit at its input end, where it is connected to the RF/MW circuitry.

Combining all of the above three schemes into any RF/MW design simultaneously would guarantee a high degree of isolation between the DC and RF/MW circuitry and proper operation of the amplifier (see Figure 15.2). Balanced (or symmetrical) stubs may also be used for better input *VSWR* as shown in Figure 15.3. This arrangement will minimize transition interactions between shunt and series TLs.

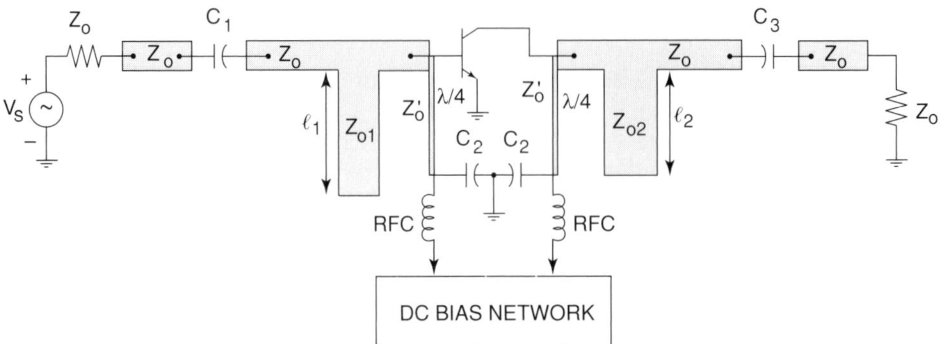

FIGURE 15.2 Connection of the DC bias to the RF circuitry (unbalanced shunt stubs).

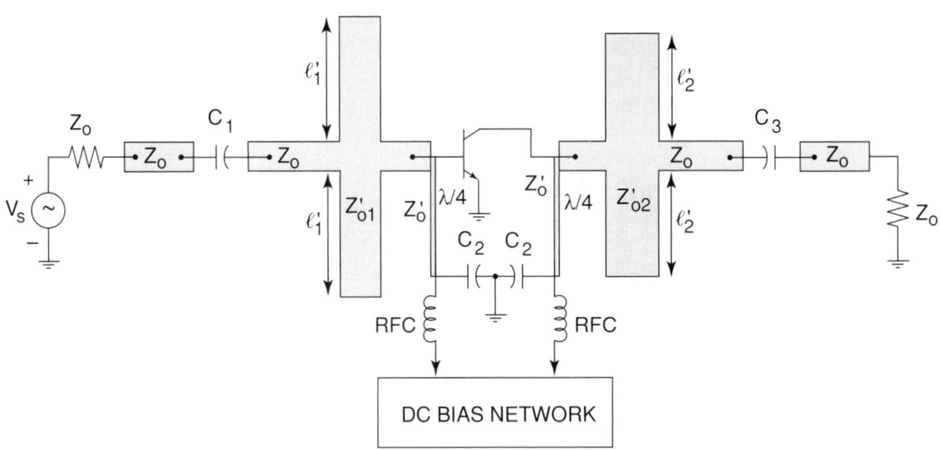

FIGURE 15.3 Use of balanced shunt stubs for better VSWR.

Design of balanced shunt stubs was discussed in Chapter 11 (see Section 11.6). We will now summarize the two possible cases for the design of balanced shunt stubs:

a) case I - Same characteristic impedances

$$Z'_{o1} = Z_{o1} = Z_o$$

$$Z'_{o2} = Z_{02} = Z_o$$

We now obtain the stub lengths by reading off ℓ'_1 and ℓ'_2 from the Smith chart for $(Y_{stub})_{bal}$ such that:

$$(Y_{stub})_{bal} = \frac{1}{2}(Y_{stub})_{unbal}$$

b) case II - Different characteristic impedances

$$\ell'_1 = \ell_1,\ \ell'_2 = \ell_2$$
$$Z'_{o1} = 2Z_{o1}$$
$$Z'_{o2} = 2Z_{o2}$$

15.3.2 Amplifier's RF/MW Circuit Design

To design the RF/microwave circuits, we need to follow these steps:

STEP 1. Based on the microwave amplifier's specifications, select an appropriate device, e.g., if gain (G) is given, choose a transistor with typical $|S_{21}/S_{12}| > G$ that is in the desired frequency range. Or, if the noise figure (F_o) is given, make sure that it is greater than the F_{min} of the selected transistor, i.e., $F_o > F_{min}$.
STEP 2. Bias the transistor in midrange of the $I_C - V_{CE}$ curves (BJT) or $I_D - V_{DS}$ curves (FET).
STEP 3. Measure the S-parameters of the transistors at the selected Q-point.
STEP 4. Check stability conditions ($K > 1$, $|\Delta| < 1$) at a particular frequency. If the condition is not met, draw the input and output stability circles and determine the stable regions.
STEP 5. If $S_{12} = 0$, then use unilateral design formulas. If $S_{12} \neq 0$, compute the unilateral figure of merit (U) and find the error range, and if small enough, then use unilateral assumption. Otherwise, use bilateral analysis and design formulas.
STEP 6. The input and output matching networks should now be designed based on any one of the following requirements:

- Narrowband amplifier (NBA)
- High-gain amplifier design (HGA)
- Maximum-gain amplifier design (MGA) (This is a special case of HGA.)
- Low-noise amplifier design (LNA)
- Minimum-noise amplifier design (MNA) (This is a special case of LNA.)
- Broadband amplifier design (BBA)
- Multistage amplifier design (MSA)

We will discuss the design techniques and criteria for each one of the categories in the next section. A summary of all of the steps taken in the design of RF/microwave amplifiers is shown in Figure 15.4.

15.4 DESIGN OF DIFFERENT TYPES OF AMPLIFIERS

Now we will discuss the exact steps required for the design of each of the seven types of small-signal amplifiers introduced in the previous section. Each type of amplifier optimizes a particular amplifier characteristic (such as gain, noise figure, etc.) that needs to be dealt with properly in the design process.

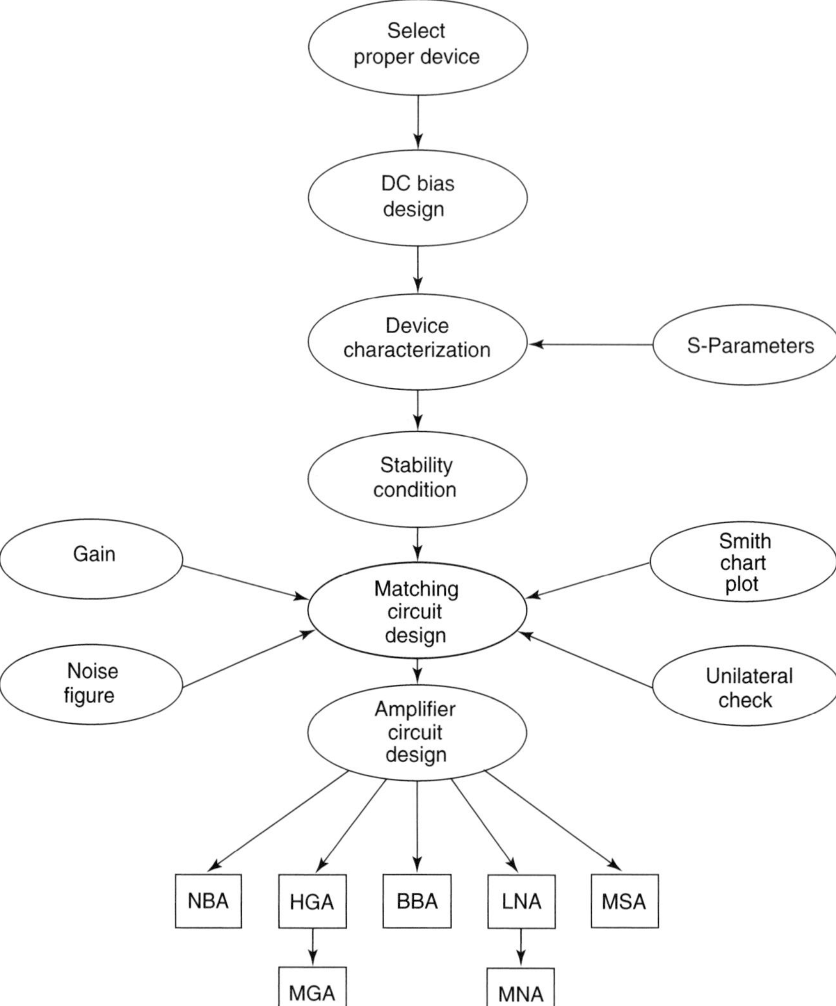

FIGURE 15.4 Overall view of design steps.

15.4.1 Narrowband Amplifier (NBA) Design

By definition, a narrowband amplifier is an amplifier in which the amplification takes place over a bandwidth that is 10 percent (or less) of the center frequency of operation. For example, a typical 1 GHz narrowband amplifier operates from 950 MHz to 1050 MHz.

Because the small-signal amplifiers normally have a narrow bandwidth of less than 10 percent of the center frequency, they may be referred to as "narrow-band amplifiers" as well. This means that the design considerations and techniques that will be discussed for small-signal amplifiers also apply to the design of narrowband amplifiers where the *S*-parameters, the noise figure, and other parameters are measured at the *center of the frequency bandwidth*.

15.4.2 High-Gain Amplifier (HGA) Design

High-gain amplifier design requires a specific gain that is not equal to the maximum gain available from the amplifier. Therefore, for high-gain operation, we need to draw input and output constant gain circles and select Γ_S and Γ_L, which are appropriate points on the respective constant-gain circles located in stable regions. Once Γ_S and Γ_L are selected, the rest of the design reduces to the design of appropriate input and output matching networks. The following example will illustrate the HGA design process.

EXAMPLE 15.1

Design a high-gain amplifier for a power gain of 15 dB at a frequency of 3 GHz, if the selected bipolar transistor has the following S-parameters (at $V_{CE} = 4$ V and $I_C = 5$ mA):

$$[S] = \begin{bmatrix} 0.7\angle{-155^\circ} & 0 \\ 4\angle 180^\circ & 0.51\angle{-20^\circ} \end{bmatrix}$$

Solution:

1. Using the selected transistor, we need to check the stability condition first:

$$S_{12} = 0 \Rightarrow K = \infty$$

$$|\Delta| = 0.357 < 1$$

Therefore, the transistor is unconditionally stable.

2. Because $S_{12} = 0$, this is a unilateral design.

$$G_T = G_S G_o G_L$$

$$G_o = |S_{21}|^2 = 16 = 12 \text{ dB}$$

Because the transistor can provide a gain of 12 dB, we need to obtain the remaining 3dB from the input and output matching networks as follows:

$$G_{S,max} = 1/(1 - |S_{11}|^2) = 1.96 = 2.9 \text{ dB}$$

$$G_{L,max} = 1/(1 - |S_{22}|^2) = 1.35 = 1.3\text{dB}$$

From $G_{S,max}$ and $G_{L,max}$ values, we assign the following gain values to the input and output matching networks:

$$G_S = 2 \text{ dB}$$

$$G_L = 1 \text{ dB}$$

3. Input 2 dB constant gain circle and output 1 dB constant gain circle are now plotted:

a.

$$G_S = 2 \text{ dB} = 1.59$$

$$g_S = 1.59/1.96 = 0.81$$

$$C_{gS} = 0.63\angle 155^\circ$$

$$R_{gS} = 0.25$$

b.

$$G_L = 1 \text{ dB} = 1.26$$

$$g_L = 1.26/1.35 = 0.93$$

$$C_{gL} = 0.48\angle 20°$$

$$R_{gL} = 0.20$$

The 2 dB and 1 dB constant gain circles are plotted in the Smith chart as shown in Figure 15.5.

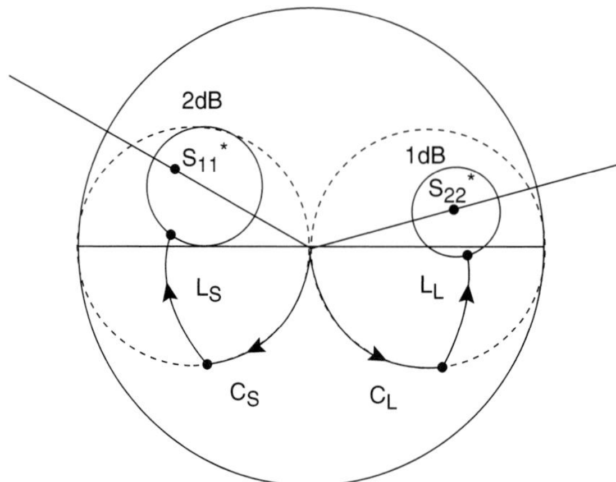

FIGURE 15.5 Smith chart solution for Example 15.1.

4. Because the transistor is unconditionally stable, Γ_S and Γ_L can be chosen anywhere on the constant gain circles. One possible design is given by:

   a. Input matching network:
   Shunt capacitance: $C_S = 2.31$ pF
   Series inductor: $L_S = 1.14$ nH

   b. Output matching network:
   Series capacitance: $C_L = 0.73$ pF
   Shunt inductor: $L_L = 6.17$ nH

   c. The final schematic for the high-gain amplifier is shown in Figure 15.6.

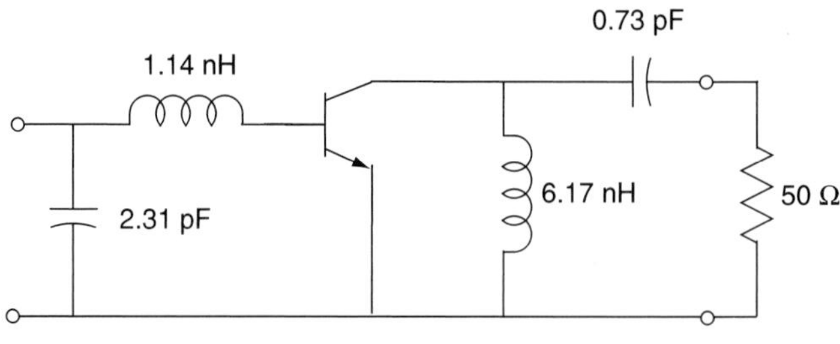

FIGURE 15.6 Example 15.1 design.

15.4.3 Maximum-Gain Amplifier (MGA) Design

Maximum-gain amplifier (MGA) design is a special case of high-gain amplifier (HGA) design where the input and output gain circles are reduced to single points, and thus the design process is reduced to a single design choice.

Assuming that the first five steps of the design are successfully completed, we now turn to step 6 and require the following conditions to be met for this type of amplifier design:

$$\Gamma_S = \Gamma_{IN}^*$$
$$\Gamma_L = \Gamma_{OUT}^*$$

We need to consider two cases: unilateral design and bilateral design.

Unilateral Design. If the unilateral assumption holds valid, i.e.,

$$-\varepsilon \leq \frac{G_T}{G_{TU}} \leq \varepsilon \tag{15.1}$$

where ε is the maximum tolerable error, then we can write:

$$\Gamma_S = S_{11}^* \tag{15.2a}$$
$$\Gamma_L = S_{22}^* \tag{15.2b}$$

This condition provides the maximum transducer gain ($G_{TU,max}$):

$$G_{TU,max} = G_{S,max}\, G_o\, G_{L,max} \tag{15.3}$$

where

$$G_{S,max} = \frac{1}{1-|S_{11}|^2} \tag{15.4a}$$
$$G_{L,max} = \frac{1}{1-|S_{22}|^2} \tag{15.4b}$$

Bilateral Design. In this case, Γ_S and Γ_L should be chosen equal to Γ_{MS} and Γ_{ML} as given in Chapter 13, *Gain Considerations in Amplifiers*, by:

$$\Gamma_{MS} = \frac{B_1 \pm \sqrt{B_1^2 - 4|C_1|^2}}{2C_1} \tag{15.5a}$$
$$\Gamma_{ML} = \frac{B_2 \pm \sqrt{B_2^2 - 4|C_2|^2}}{2C_2} \tag{15.5b}$$

where

$$B_1 = 1 + |S_{11}|^2 - |S_{22}|^2 - |\Delta| \tag{15.6a}$$
$$B_2 = 1 + |S_{22}|^2 - |S_{11}|^2 - |\Delta| \tag{15.6b}$$
$$C_1 = S_{11} - \Delta S_{22}^* \tag{15.7a}$$
$$C_2 = S_{22} - \Delta S_{11}^* \tag{15.7b}$$

The following example illustrates this concept further.

EXAMPLE 15.2

A GaAs MESFET is measured to have the following S-parameters for a midrange Q-point where $V_{DS} = 5$ V, $I_D = 10$ mA at 10 GHz with:

$$S_{11} = 0.55\angle{-150°}$$
$$S_{12} = 0.04\angle 20°$$
$$S_{21} = 2.82\angle 180°$$
$$S_{22} = 0.45\angle{-30°}$$

Using this transistor, design a microwave amplifier for maximum power gain at 10 GHz.

Solution:

1. Check stability condition:

$$|\Delta| = |S_{11}S_{22} - S_{12}S_{21}| = 0.16 < 1$$
$$K = 2.28 > 1$$

Therefore, the device is unconditionally stable at 10 GHz.

2. Because $S_{12} \neq 0$, we need to check the unilateral figure of merit:

$$U = 0.05$$
$$1/(1 + 0.05)^2 < G_T/G_{TU,max} < 1/(1 - 0.05)^2$$
$$-0.41 \text{ dB} < G_T/G_{TU,max} < 0.45 \text{ dB}$$

If we consider approximately ± 0.5 dB error as acceptable, then we will use unilateral assumption, and at the end we will verify this assumption when we find the maximum gain value.

3.

$$G_{S,max} = 1.43 = 1.55 \text{ dB}$$
$$G_o = |S_{21}|^2 = 7.95 = 9 \text{ dB}$$
$$G_{L,max} = 1.25 = 0.97 \text{ dB}$$
$$G_{TU,max} = 1.55 + 9 + 0.97 = 11.55 \text{ dB}$$

4. **Verification:** Compared to a gain of 11.55 dB, an error of ±0.5 dB is justifiable!
5. Now we design the input and output matching network using a 50 Ω system. From Figure 15.7, we can see that:

Input matching network: $\Gamma_S = S_{11}^* = 0.55\angle 150°$

First element—a shunt capacitor:

$$j\omega C = j1.66/50 \Rightarrow C_S = 0.59 \text{ pF}$$

Second element—a series inductor:

$$j\omega_L = j0.71 \times 50 \Rightarrow L_S = 0.63 \text{ nH}$$

FIGURE 15.7 Smith chart solution for Example 15.2.

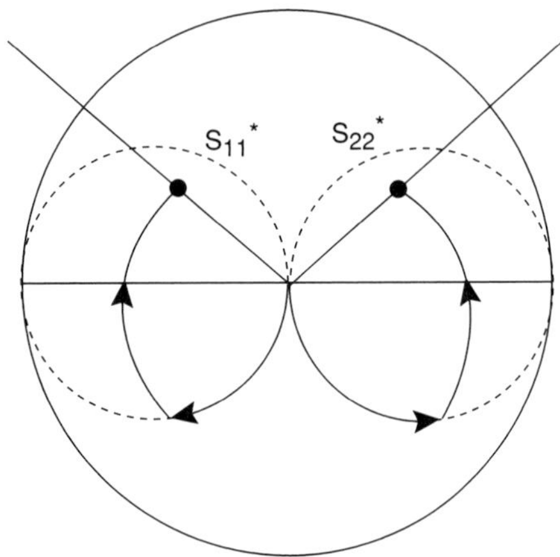

Output matching network: $\Gamma_L = S_{22}^* = 0.45\angle 30°$

First element—a series capacitor:

$$1/j\omega C = -j1.1 \times 50 \Rightarrow C_L = 0.29 \text{ pF}$$

Second element—a series inductor:

$$1/j\omega L = -j0.73/50 \Rightarrow L_L = 1.21 \text{ nH}$$

The final design is shown in Figure 15.8.

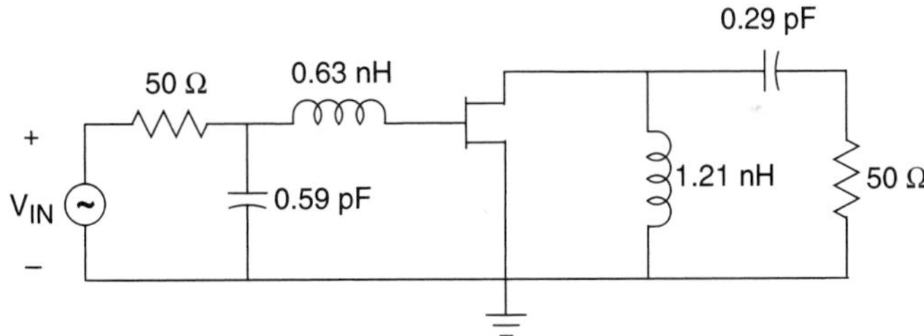

FIGURE 15.8 Final circuit for Example 15.2.

15.4.4 Low–Noise Amplifier (LNA) Design

In this type of amplifier design, the objective is not to exceed a specified noise figure value while achieving the highest possible gain. Because maximum power gain and minimum noise figure cannot be achieved simultaneously, in most cases we need to trade off one for the other to obtain our design objective.

Given a design requirement with a specific noise figure and an exact gain value, we need to carry out the following design procedure (assuming that we have successfully performed the first five design steps as outlined previously):

Step 1. Compute the allocated gain values for the input and the output matching networks.

Step 2. Plot the source constant gain circle and the constant noise figure circle on the same Smith chart.

Step 3. Choose a source constant gain circle to intercept the desired constant noise figure circle.

Step 4. Using a value of Γ_S (selected anywhere on the constant gain circle, such that it is between the intercept points and inside the constant noise figure circle), the input matching network can be designed.

Step 5. Plot the load constant gain circle with the allocated gain of step 1, and choose a Γ_L on this circle; finalize the amplifier circuit with the design of output matching network.

EXAMPLE 15.3

Design a low-noise amplifier (LNA) for a noise figure of 3.5 dB and a power gain of 16 dB. A bipolar transistor has been selected and is biased at midrange for class A amplifier design: $V_{CE} = 4$ V, $I_C = 30$ mA. The S-parameters and the noise parameters at 1 GHz are as follows:

$$[S] = \begin{bmatrix} 0.7\angle -155^\circ & 0 \\ 5.0\angle 180^\circ & 0.51\angle -20^\circ \end{bmatrix}$$

$$F_{min} = 3 \text{ dB}$$

$$\Gamma_{opt} = 0.45\angle 180^\circ$$

$$R_n = 4\ \Omega$$

Solution:

$$S_{12} = 0 \Rightarrow K = \infty$$

$$|\Delta| = 0.36 < 1$$

Therefore, the amplifier is unconditionally stable.

1. This is a unilateral amplifier design:

$$G_{T,max} = G_{S,max} G_o G_{L,max}$$

$$G_{S,max} = 1/(1 - |S_{11}|)^2 = 1.96 = 3 \text{ dB}$$

$$G_o = |S_{21}|^2 = 25 = 14 \text{ dB}$$

$$G_{L,max} = 1/(1 - |S_{22}|)^2 = 1.35 = 1.30 \text{ dB}$$

$$G_{T,max} = 3 + 14 + 1.3 = 18.3 \text{ dB}$$

Because the maximum gain is more than the required gain, the transistor will serve our purpose splendidly!

2. We know the transistor provides 14 dB of gain; therefore, to obtain the required gain of 16 dB, we allocate the gain for each stage as follows:

$$G_o = 14 \text{ dB (transistor)}$$

$$G_S = 1.22 \text{ dB (input matching network)}$$

$$G_L = 0.78 \text{ dB (output matching network)}$$

3. Plot the input 1.22 dB gain circle:

$$G_S = 1.22 \text{ dB} = 1.32$$

$$g_S = G_S / G_{S,max} = 1.32 / 1.96 = 0.67$$

$$C_{gS} = 0.56\angle 155°$$

$$R_{gS} = 0.35$$

The 1.22 dB gain circle is plotted in Figure 15.9.

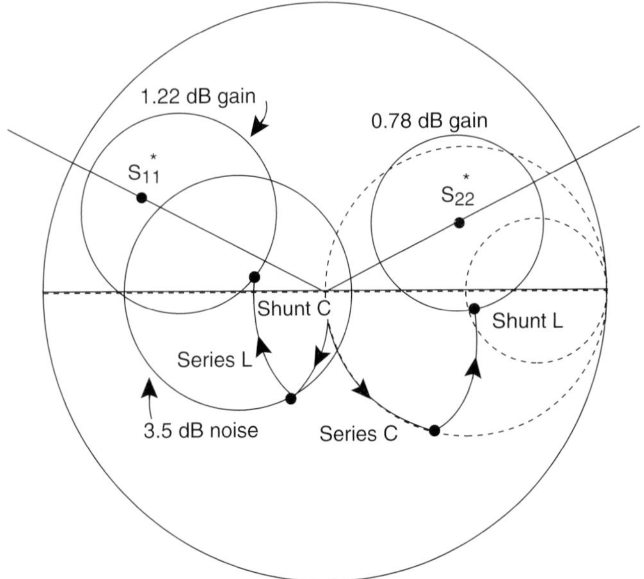

FIGURE 15.9 Graphical solution for Example 15.3.

4. The 3.5 dB noise figure circle is plotted (as shown in Figure 15.9) with:

$$F = 3.5 \text{ dB} = 2.24$$

$$N = 0.23$$

$$C_F = 0.37\angle 180°$$

$$R_F = 0.39$$

5. The input matching network is obtained by choosing any point located both on the 1.22 dB gain circle and inside (or on) the 3.5 dB noise figure circle. One such point is shown in Figure 15.9, giving the following element values:
   Shunt capacitor: $jB = j0.75/50 = j15 \times 10^{-3}$ S $\Rightarrow C_s = 2.39$ pF
   Series inductor: $jX = j0.44 \times 50 = j22.0\ \Omega \Rightarrow L_s = 3.5$ nH
6. Plot the output 0.78 dB gain circle:

$$G_L = 0.78 \text{ dB} = 1.20$$

$$g_L = G_L/G_{L,max} = 1.20/1.35 = 0.89$$

$$C_{gL} = S^*_{22} = 0.47\angle 20°$$

$$R_{gL} = 0.25$$

   The 0.78 dB gain circle is plotted in Figure 15.9.
7. The output matching network is obtained by choosing an arbitrary point on the 0.78 dB gain circle giving the following element values:
   Series capacitor: $jX = j1.4 \times 50 = j70\ \Omega \Rightarrow C_L = 2.27$ pF
   Shunt inductor: $jB = -j0.4/50 = -j8.0 \times 10^{-3}$ S $\Rightarrow L_L = 19.9$ nH
8. The final amplifier schematic is drawn in Figure 15.10.

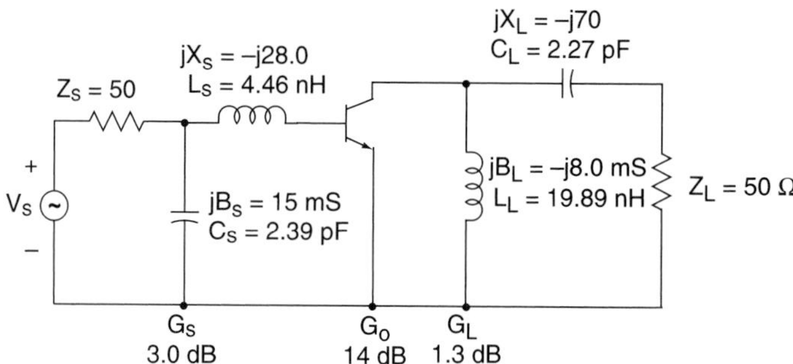

FIGURE 15.10 Designed matching network for the amplifier of Example 15.3.

15.4.5 Minimum-Noise Amplifier (MNA) Design

Minimum-noise amplifier (MNA) design is a special case of low-noise amplifier (LNA) design where the noise figure circles are reduced to a single point (Γ_{opt}) and thus the design process is reduced to a single design choice.

In this type of amplifier, after performing the first five steps of the design, we are ready to design the input and output matching networks.

From Chapter 14, *Noise Considerations in Active Networks*, we know that to achieve minimum noise, we need to select:

$$\Gamma_S = \Gamma_{opt} \tag{15.8}$$

and for best *VSWR* at the output, choose:

$$\Gamma_L = \Gamma_{OUT}^* = \left(S_{22} + \frac{S_{12}S_{21}\Gamma_{opt}}{1 - S_{11}\Gamma_{opt}}\right)^* \tag{15.9}$$

Once Γ_S and Γ_L are known, the input and output matching networks can be easily designed.

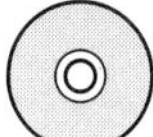

EXAMPLE 15.4

Design a minimum noise amplifier (see Figure 15.11) using a MESFET with its S-parameters measured at a Q-point in midrange of $I_D - V_{DS}$ characteristic curves (i.e., $V_{DS} = 8$ V, $I_D = 5$ mA) at 12 GHz:

$$[S] = \begin{bmatrix} 0.61\angle 37^\circ & 0.144\angle -89^\circ \\ 2.34\angle -84^\circ & 0.15\angle 46^\circ \end{bmatrix}$$

$$F_{min} = 1.2 \text{ dB}$$

$$\Gamma_{opt} = 0.47\angle -65^\circ$$

$$R_n = 40\ \Omega$$

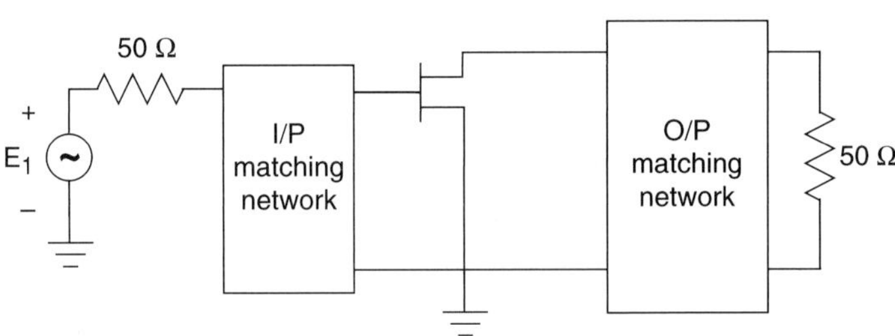

FIGURE 15.11 Circuit diagram for Example 15.4.

Solution:

1. Because the transistor has already been selected and measured, we check the stability condition as the next step:

$$K = 1.10 > 1$$

$$|\Delta| = 0.37 < 1$$

Thus, the transistor is unconditionally unstable.

2. For minimum noise design, we choose:

$$\Gamma_S = \Gamma_{opt} = 0.47\angle -65^\circ$$

3. Now we choose $\Gamma_L = \Gamma^*_{OUT}$ for maximum power transfer:

$$\Gamma_L = \Gamma_{OUT}^* = \left(S_{22} + \frac{S_{12}S_{21}\Gamma_{opt}}{1 - S_{11}\Gamma_{opt}}\right)^* = 0.303\angle -85^\circ$$

4. Using the selected Γ_S and Γ_L, we obtain:

$$G_T = 9.12 \text{ dB}$$

5. Using established techniques, the input and output matching network can be realized. The final circuit is shown in Figure 15.12.

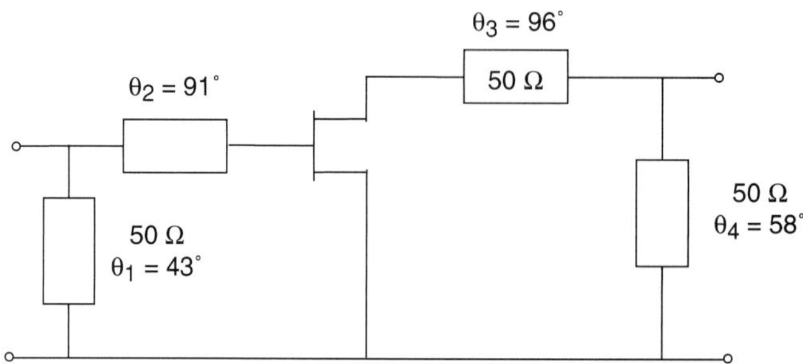

FIGURE 15.12 Final design for Example 15.4.

15.4.6 Broadband Amplifier (BBA) Design

Before we design what a broadband amplifier is, we need to define a few terms:

DEFINITION-BROADBAND: *The bandwidth, in dB, spanning approximately 6 dB per octave or more.*

DEFINITION-BANDWIDTH (IN DB): *Is the common logarithm (base 10) of the highest to the lowest frequency ratio, i.e.,* $20\log_{10}(f_H/f_L)$, *where* f_H *and* f_L *are the highest and lowest frequencies in the band, respectively.*

DEFINITION-OCTAVE: *The interval between any two frequencies having a ratio of 2 to 1. In general, an octave is the region between any given frequency (f) and either twice that frequency (2f) or a half of that frequency (f/2). In other words, the ratio from one frequency to the next is 2:1. Therefore, two octaves exist in the range of frequencies from f/2 to 2f. For example, from f = 200 MHz to 400 MHz is an octave as is from 100 MHz to 200 MHz. Thus, from 100 MHz to 400 MHz is 2 octaves.*

As another example, 2–4 GHz, 4–8 GHz, and 8–16 GHz each provide one octave of bandwidth; the bandwidth for each case, respectively, is 2 GHz, 4 GHz, and 8 GHz. As can be seen from this example, doubling the frequency corresponds to an increase of 6 dB, i.e., 6 dB/octave.

DEFINITION-DECADE: *A frequency band whose upper limit is 10 times the lower limit (10:1 ratio). In general, we can see that a decade is the region between any given frequency (f) and either ten times that frequency (10f) or a tenth of that frequency (f/10). In other words, the ratio from one frequency to the next is 10:1. Therefore, two decades exist in the range of frequencies from f/10 to 10f. For example, from f=200 MHz to 2 GHz is a decade as well as from 20 MHz to 200 MHz. Thus, from 20 MHz to 2 GHz is 2 decades. Furthermore, we can see that there is 20 dB per decade.*

DEFINITION-BROADBAND AMPLIFIER: *An amplifier having essentially a flat response (i.e., no resonant element) over a wide range of frequencies—one to several octaves or decades of bandwidth. Examples are a video amplifier or an instrument amplifier operating from DC to 10 MHz.*

The design of a broadband amplifier is the same as the design of a high-gain amplifier but over a broader frequency range. This design requirement (i.e., multi-octave bandwidth) introduces new complications to the design process that must be dealt with properly.

The complications encountered stem from variation of the device parameters and the matching network's properties with frequency. These variations can be categorized briefly as follows:

a. The variation of transistor's $|S_{21}|$ and $|S_{12}|$ with frequency is shown in Figure 15.13. As can be seen from Figure 15.13, the forward transmission coefficient magnitude, $|S_{21}|$, decreases with frequency at the rate of 6 dB/octave while the reverse transmission coefficient magnitude, $|S_{12}|$, increases with frequency at the same rate.

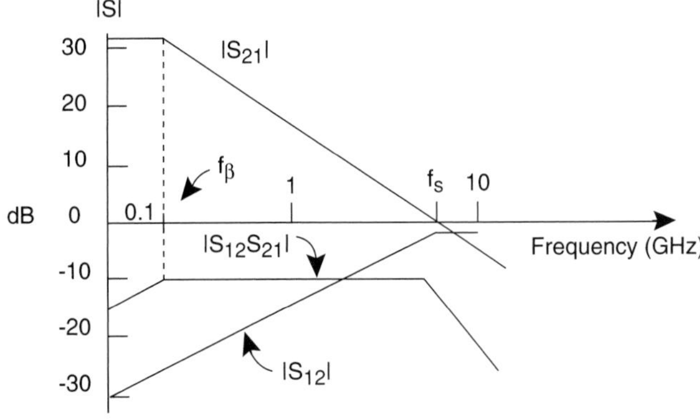

FIGURE 15.13 A typical frequency behavior of $|S_{12}|$, $|S_{21}|$, and $|S_{12}S_{21}|$.

b. The stability of the circuit as given by the equation for the K-factor (see Chapter 12, *Stability Considerations in Active Networks*), depends on the product $|S_{12}S_{21}|$, where its variation with frequency is very important. The stability of the amplifier is dependent on the flat region of $|S_{12}S_{21}|$ curve (see Figure 15.13).

c. Over a broad frequency range, the variation of the transistor's input and output reflection coefficients (S_{11}, S_{22}) becomes very important since they affect all of the power-gain equations.

d. The variation of the noise figure (F) with frequency is another important factor in broadband amplifier design that should be considered.

e. The variation of the $VSWR$ of the matching networks with frequency should also be taken into consideration.

The design of a broadband microwave amplifier requires proper handling of the problem of power gain roll-off with frequency. In general, there are four techniques for the design of broadband amplifiers:

- Compensated matching design technique
- Network synthesis design technique
- Balanced amplifier design technique
- Negative feedback design technique

I. Compensated Matching Design Technique.

The technique of "compensated matching" is a method of compensating for the forward gain, $|S_{21}|$, variation with frequency by properly mismatching the input and output matching networks. The design can be performed analytically with the help of the Smith chart.

Through the use of a good computer-aided design software package, the amplifier circuit can be simulated and the design process can be greatly enhanced by either of the following:

- Optimizing the circuit elements using the values obtained by the analytical method (as a starting point), or
- Designing the whole circuit from beginning to end using the device and circuit models stored in the software library

Example 15.5 illustrates the concept of compensated matching design technique.

EXAMPLE 15.5

Using lumped element L-networks, design a broadband amplifier operating in the frequency range of 300-700 MHz with a transducer power gain of 10 dB. The S-parameters of a BJT are given in Table 15.1.

TABLE 15.1 S-parameter Sets at Three Different Frequencies Spanning the Band

f(MHz)	S_{11}	S_{12}	S_{21}	S_{22}
300	$0.3\angle -45°$	0	$4.47\angle 40°$	$0.86\angle -5°$
450	$0.27\angle -70°$	0	$3.16\angle 35°$	$0.855\angle -14°$
700	$0.2\angle -95°$	0	$2.0\angle 30°$	$0.85\angle -22°$

Solution:

Based on Table 15.1, we create Table 15.2 for compensation of $|S_{21}|$ variation with frequency in order to obtain a flat gain of 10 dB:

TABLE 15.2 $|S_{21}|^2$ Values at the Three Frequencies

f(MHz)	$|S_{21}|^2$ (dB)	Compensation (dB)
300	13	−3
450	10	0
700	6	+4

From the transistor *S*-parameters, the input and the output matching network's maximum gain are calculated, and the results are summarized in Table 15.3.

TABLE 15.3 Maximum Gain Values at the Three Frequencies

f(MHz)	$G_{S,max}$ (dB)	$G_{L,max}$ (dB)
300	0.409	5.84
450	0.329	5.70
700	0.177	5.56

From Table 15.3, we can see that there is little to be gained by matching the source because the maximum possible gain is less than 0.5 dB in the desired frequency range. So, the input matching network can be a direct connection from the source resistance to the base of the transistor.

Therefore, we will focus primarily on the design of the output matching network because it can provide up to approximately 5.5–5.8 dB of gain.

Next, we plot the constant gain circles for $G_L = -3$ dB (at 300 MHz), $G_L = 0$ dB (at 450 MHz), and $G_L = 4$ dB (at 700 MHz), as shown in Figure 15.14.

The output matching network should transform a 50 Ω load to a point somewhere on each of the three constant gain circles as frequency varies from 300 MHz to 700 MHz.

The *L*-matching network consists of a shunt inductor and a series inductor, as shown in Figure 15.15. The inductor element values are found from Figure 15.14.

Using the lowest frequency ($f = 300$ MHz), the value of L_1 can be determined as:

$$j\omega L_1 = 50(-j1.2) \Rightarrow L_1 = 22.1 \text{ nH}$$

Using the highest frequency ($f = 700$ MHz), the value of L_2 can be determined as:

$$j\omega L_2 = 50(j3.2 - j0.4) \Rightarrow L_2 = 31.8 \text{ nH}$$

The maximum input *VSWR* is at 300 MHz and is given by:

$$VSWR = (1 + 0.3)/(1 - 0.3) = 1.86$$

The relatively high input *VSWR* is a disadvantage of the compensated matching technique.

As noted in Example 15.5, the design of compensated matching networks (in order to obtain gain flatness) requires impedance mismatching of input and output matching networks, which degrades the input and output *VSWR*.

II. Network Synthesis Design Technique.
Passive network synthesis methods (such as Chebyshev or Butterworth response methods, etc.) for the design of networks using lumped elements are well developed and can be used analytically or by CAD software. This type of matching technique is well beyond the scope of this book and will not be covered here.

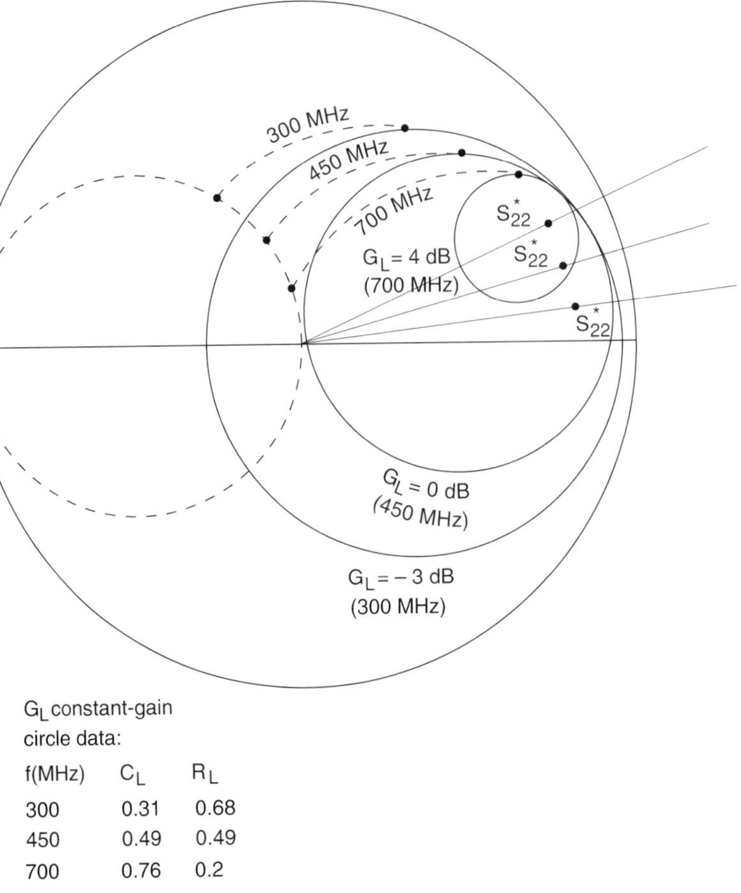

FIGURE 15.14 Smith chart plots for Example 15.5.

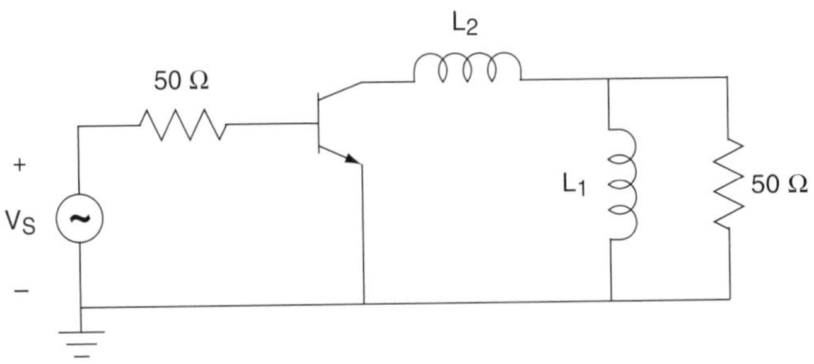

FIGURE 15.15 Figure for Example 15.5.

III. Balanced Amplifier Design Technique.

Balanced amplifier technique is a practical and popular method of obtaining a broadband amplifier, a flat power gain versus frequency, and a good input and output *VSWR* for microwave solid-state amplifiers (unlike the compensated matching technique).

Figure 15.16 compares two amplifiers: one in a series configuration and the other in a balanced configuration. A balanced amplifier consists of two hybrid couplers (90°), with one at the input and another at the output. From Figure 15.16b, we can see that the input power is split into two parts; therefore this type of amplifiers can also be called a two-way combined amplifier. The 3 dB input coupler divides the input power equally between Ports 2 and 3 but with a 90° phase shift, and the output coupler recombines the amplified output signals in phase, from the transistor amplifiers.

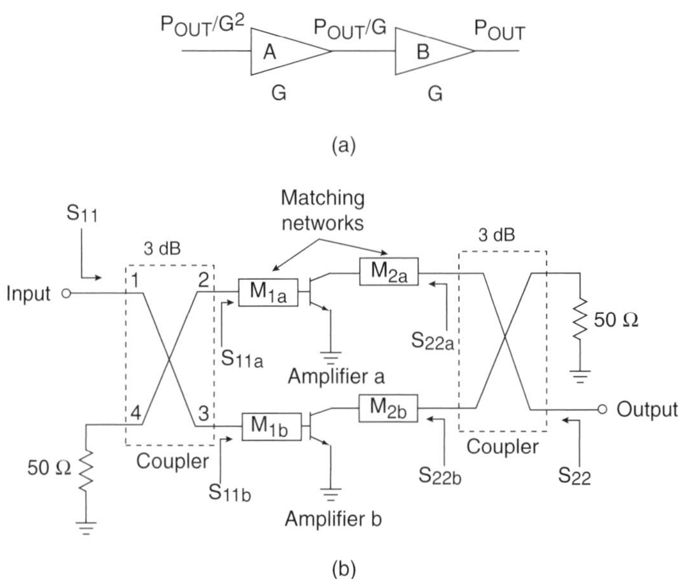

FIGURE 15.16 (a) A series configuration of two amplifiers; (b) A balanced amplifier configuration.

The hybrid coupler in microwave planar integrated circuits using microstrip technology can be realized by a 3 dB Lange coupler, a 3 dB branch-line coupler or a 3 dB Wilkinson coupler as shown in Figure 15.17. A simple comparison between the three types of couplers reveals that the Lange, the branch line, and the Wilkinson couplers are all limited by the frequency sensitivity of the λ/4 lines and the bandwidth is about 50 percent of the center frequency. The Lange coupler occupies less space than the branch line or the Wilkinson couplers, and thus it lends itself to very compact designs.

The reflection coefficients for the total structure along with power gain and the *VSWR* can be obtained as follows:

$$S_{11} = e^{-j\pi}(S_{11a} - S_{11b})/2 \tag{15.10a}$$

$$S_{21} = e^{-j\pi/2}(S_{21a} + S_{21b})/2 \tag{15.10b}$$

$$S_{12} = e^{-j\pi/2}(S_{12a} + S_{12b})/2 \tag{15.10c}$$

$$S_{22} = e^{-j\pi}(S_{22a} - S_{22b})/2 \tag{15.10d}$$

FIGURE 15.17 Microstrip implementation of a (a) Lange coupler (3 dB), (b) 90° branch-line coupler, (c) Wilkinson coupler.

(a)

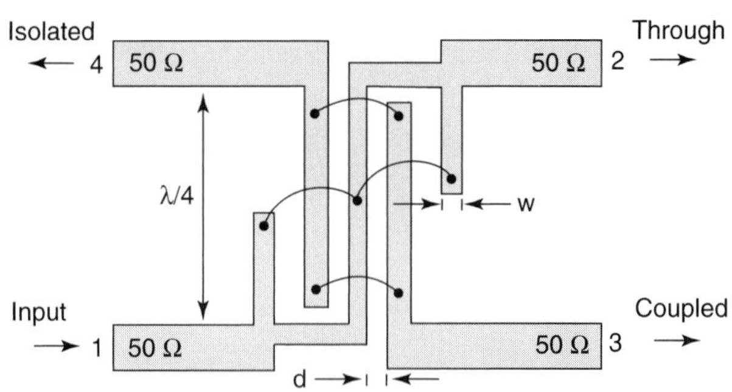

(b)

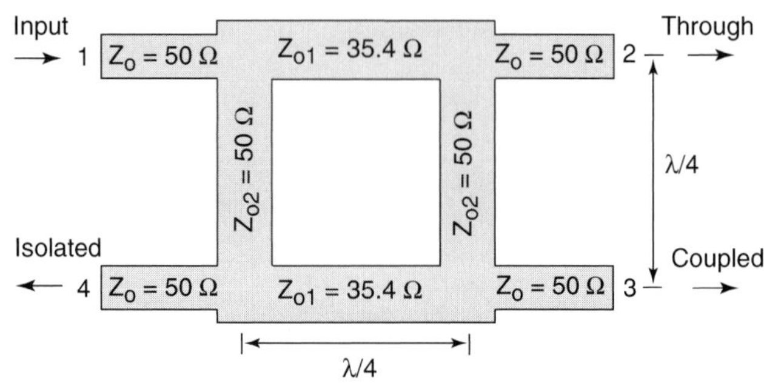

(c)

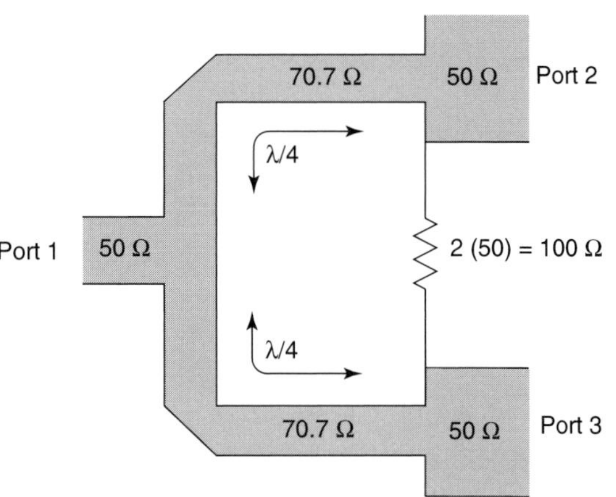

The forward power gain $(G_T)_F$, the reverse power gain $(G_T)_R$, and the input and output $VSWR$ are given by:

$$(G_T)_F = |S_{21}|^2 = |S_{21a} + S_{21b}|^2/4 \tag{15.11a}$$

$$(G_T)_R = |S_{12}|^2 = |S_{12a} + S_{12b}|^2/4 \tag{15.11b}$$

$$(VSWR)_{IN} = \frac{1 + |S_{11}|}{1 - |S_{11}|} \tag{15.12a}$$

$$(VSWR)_{OUT} = \frac{1 + |S_{22}|}{1 - |S_{22}|} \tag{15.12b}$$

A Special Case: Identical Transistors The balanced amplifier design configuration allows the input and output amplifier ports to be mismatched, and yet the combined amplifier will appear matched if the two parallel amplifiers are balanced (i.e., identical). For the case of identical amplifiers, we have:

$$S_{11} = 0 \tag{15.13a}$$

$$S_{22} = 0 \tag{15.13b}$$

$$(G_T)_F = |S_{21}|^2 = |S_{21a}|^2 \tag{15.13c}$$

$$(G_T)_R = |S_{12}|^2 = |S_{12a}|^2 \tag{15.13d}$$

$$(VSWR)_{IN} = 1 \tag{15.13e}$$

$$(VSWR)_{OUT} = 1 \tag{15.13f}$$

Therefore, for two identical parallel amplifiers in balanced amplifier configuration: (a) The overall gain equals the gain of one side of the coupler and (b) the input as well as the output ports are completely matched.

The bandwidth of the combination is limited by the bandwidth of the coupler, which is about two octaves.

The balanced amplifier configuration has several advantages:

- They are highly stable and have excellent input and output *VSWR* for identical amplifiers even if each individual amplifier has a high *VSWR*.
- The output power of the combination is twice that of each single amplifier.
- They are easy to cascade together because the couplers isolate each unit from the adjacent ones and thus prevent loading effects.
- They will still operate even if one individual amplifier fails (this is called graceful degradation). The gain, however, will be reduced by 6 dB.

The advantages listed here outweigh greatly a few disadvantages (such as more DC power and its larger size), and they make the balanced amplifier technique the ideal choice for broadband amplifier design.

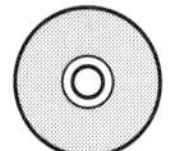

EXAMPLE 15.6

Two GaAs MESFET amplifiers are used in a balanced amplifier configuration, each with the S-parameters listed in Table 15.4.

a. *Calculate the reflection coefficients and VSWRs at the input and output port of the amplifier.*

b. *Calculate the overall power gain in dB.*

c. *Calculate the output power capability (in the linear range) compared with two FETs in series.*

d. *Determine the power gain if one of the individual amplifiers fails.*

TABLE 15.4 S-Parameters for the Two GaAs FETs

Amplifier A	Amplifier B
$S_{11a} = 0.75\angle 40°$	$S_{11b} = 0.82\angle 30°$
$S_{12a} = 0.15\angle -30°$	$S_{12b} = 0.17\angle -32°$
$S_{21a} = 7.0\angle 180°$	$S_{21b} = 7.5\angle 170°$
$S_{22a} = 0.6\angle 20°$	$S_{22b} = 0.75\angle 40°$

Solution:

a.

$$S_{11} = e^{-j\pi}(0.75\angle 40° - 0.82\angle 30°)/2 = 0.078\angle -28°$$

$$S_{22} = e^{-j\pi}(0.6\angle 20° - 0.75\angle 40°)/2 = 0.139\angle 88°$$

$$(VSWR)_{IN} = \frac{1+|S_{11}|}{1-|S_{11}|} = \frac{1+0.078}{1-0.078} = 1.17$$

$$(VSWR)_{OUT} = \frac{1+|S_{22}|}{1-|S_{22}|} = \frac{1+0.139}{1-0.139} = 1.32$$

b. $(G_T)_F = |S_{21}|^2 = |S_{21a} + S_{21b}|^2/4 = |7.0\angle 180° + 7.50\angle 170°|/4 = 52.2 = 17.2\text{ dB}$

c. Two identical FET amplifiers in series will have the overall output power of only one amplifier. On the other hand, the same two amplifiers in balanced amplifier configuration (i.e., in parallel) will give twice as much power, as shown in Figure 15.16.

$$(P_{OUT})_{balamp} = 2\,(p_{OUT})_{series}$$

$$\text{or } (P_{OUT})_{balamp}\,(\text{dB}) = (p_{OUT})_{series}\,(\text{dB}) + 3\text{ dB}$$

d. If one amplifier (let us say amplifier B) fails, the reduced gain is given by:

$$(G_T)_{F,\,reduced} = |S_{21a}|^2/4 = 12.25 = 10.9\text{ dB}$$

Comparing the reduced gain with the full gain of (17.2 dB), we can see that:

$$(G_T)_{F,\,reduced} - (G_T)_F = 17.2 - 10.9 = 6.3\text{ (dB)}$$

Thus, we can write approximately:

$$(G_T)_{F,\,reduced} \approx (G_T)_F - 6\text{ (dB)}$$

IV. Negative Feedback Design Technique.

In this design technique, a negative feedback (from the output to the input) provides a flat gain response and reduces input and output *VSWR*. The main application of this technique is for amplifiers with very wide bandwidth design requirements (more than a decade) where the compensated matching design technique or the balanced amplifier design technique (both based on the concept of matching) fail.

Negative feedback is a very powerful technique when bandwidths greater than two decades with very small gain variations (tenths of dB) are required. The only drawback is the use of resistors in the feedback path, which will degrade the noise figure and reduce the maximum power gain available from a transistor.

a) Analysis The most general configuration for a negative feedback amplifier consists of a series element (R_1) and a shunt element (R_2), as shown in Figures 15.18a and b, for a BJT or a MESFET device. Using small signal models for the BJT and the MESFET, the equivalent circuits are obtained as shown in Figures 15.19a and b.

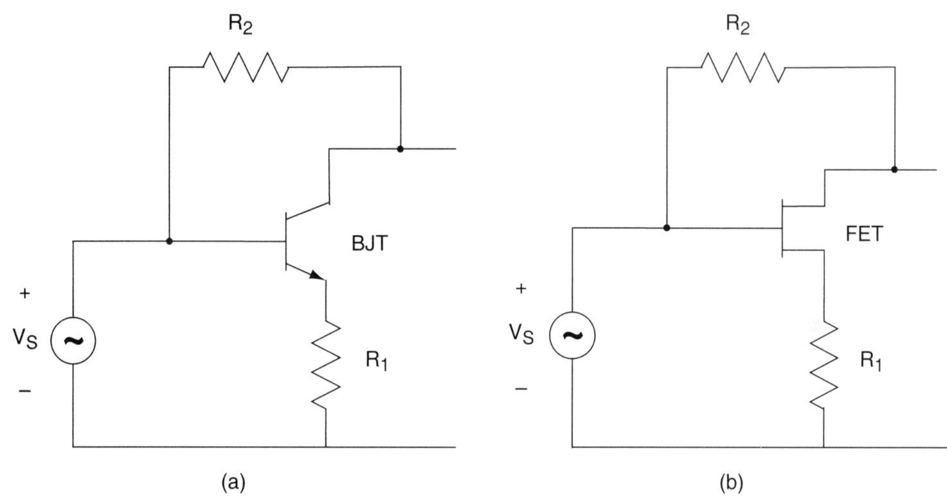

FIGURE 15.18 A general configuration for a negative feedback amplifier (a) BJT, (b) FET.

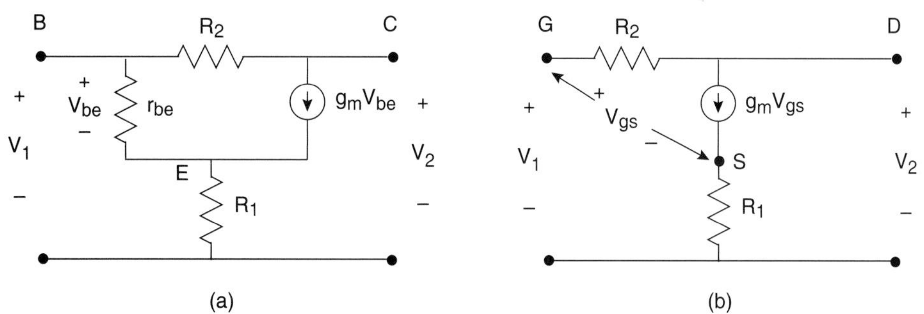

FIGURE 15.19 Negative feedback amplifier models for (a) BJT, (b) MESFET.

Considering Figure 15.19b, the admittance matrix $[Y]$ for the MESFET negative feedback amplifier is obtained to be:

$$\begin{bmatrix} i_1 \\ i_2 \end{bmatrix} = \begin{bmatrix} Y \end{bmatrix} \begin{bmatrix} v_1 \\ v_2 \end{bmatrix} \tag{15.14}$$

where

$$[Y] = \begin{bmatrix} \frac{1}{R_2} & -\frac{1}{R_2} \\ \frac{g_m}{1+g_m R_1} - \frac{1}{R_2} & \frac{1}{R_2} \end{bmatrix} \tag{15.15}$$

NOTE: *A similar admittance matrix can be written for the BJT circuit (see Figure 15.19a) if we assume* $r_{be} + \beta R_1 >> R_2$.

Converting the *Y*-parameters from Equation 15.15 into *S*-parameters gives:

$$S_{11} = S_{22} = \frac{1}{D}\left[1 - \frac{g_m Z_o^2}{R_2(1+g_m R_1)}\right] \tag{15.16}$$

$$S_{21} = \frac{1}{D}\left[\frac{-2g_m Z_o}{1+g_m R_1} + \frac{2Z_o}{R_2}\right] \tag{15.17}$$

$$S_{12} = \frac{2Z_o}{DR_2} \tag{15.18}$$

where Z_o is the characteristic impedance of the connecting transmission lines, and

$$D = 1 + \frac{2Z_o}{R_2} + \frac{g_m Z_o^2}{R_2(1+g_m R_1)}$$

b) A Special Case: Matched-Ports Condition From Equation 15.16, we can see that in order to obtain no reflection at the input or the output port (i.e., $VSWR = 1$), we need to set S_{11} and S_{22} equal to zero, which gives:

$$S_{11} = S_{22} = 0$$

$$1 + g_m R_1 = g_m Z_o^2 / R_2 \tag{15.19a}$$

$$R_1 = \frac{Z_o^2}{R_2} - \frac{1}{g_m} \tag{15.19b}$$

Substituting for R_1 from Equation 15.19b into 15.17, we have

$$S_{21} = \frac{Z_o - R_2}{Z_o} \Rightarrow R_2 = Z_o(1 - S_{21}) \tag{15.19c}$$

$$S_{12} = \frac{Z_o}{R_2 + Z_o} \tag{15.19d}$$

Equation 15.19b imposes a fixed condition that the two feedback resistors, R_1 and R_2, must satisfy in order to have no reflection loss at the input or the output ports.

In fact, not every BJT or MESFET is suitable for shunt-series feedback amplifier configuration and only those having:

$$g_m > (g_m)_{min} \tag{15.19e}$$

should be considered. The value of $(g_m)_{min}$ is obtained by noting that the value of R_1 is always a positive number (i.e., $R_1 \geq 0$); therefore, for the worst case (i.e., when $R_1 = 0$), from Equation 15.19b, we can write:

$$\frac{Z_o^2}{R_2} - \frac{1}{(g_m)_{min}} = 0 \Rightarrow (g_m)_{min} = R_2/Z_o^2 \tag{15.20}$$

Substituting for R_2 from Equation 15.19c in Equation 15.20, we obtain a simplified equation for $(g_m)_{min}$ under matched-ports condition as:

$$(g_m)_{min} = \frac{1 - S_{21}}{Z_o} \tag{15.21}$$

Once the transistor has met $(g_m)_{min}$ condition, from Equation 15.19c we can see that the forward gain for the overall configuration (S_{21}) depends only on the feedback resistor (R_2) (and not on the device parameters), thus allowing gain flattening to occur over a wide range of frequencies as long as the device behaves linearly.

NOTE 1: *Under matched-ports condition (i.e., $S_{11} = S_{22} = 0$), we can see from Equation 15.19b that when g_m is large (i.e., $1/g_m \approx 0$), the following condition must be met in order to obtain minimum input and output VSWR:*

$$R_1R_2 \approx Z^2{}_o \tag{15.22}$$

Because BJTs ordinarily have a high value of g_m, they are quite useful in negative feedback amplifiers. On the other hand, one must be cautious in using GaAs FETs, which usually have low values of g_m; thus, they may not lend themselves easily to the design of this type of amplifiers.

NOTE 2: *For low-frequency transistor amplifiers, the overall gain (S_{21}) is negative, i.e.,*

$$S_{21} = -|S_{21}| \tag{15.23}$$

Thus, given a desired constant value for the overall gain (S_{21}), the required parameters for the device (g_m), and the feedback resistors (R_1,R_2) can easily be determined. Examples 15.7 and 15.8 further illustrate this concept.

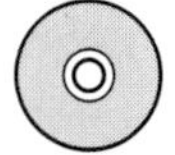

EXAMPLE 15.7

Design a completely matched GaAs MESFET shunt feedback (i.e., $R_1 = 0$) amplifier for a 10 dB overall gain in a 50 Ω system operating at 50 MHz, as shown in Figure 15.20. Specify the FET and the shunt element.

Solution:

$$10\log|S_{21}|^2 = 10 \text{ dB} \Rightarrow |S_{21}| = 10^{1/2} = 3.16$$

Because the amplifier operates at low frequencies,

$$S_{21} = -3.16$$

To have the matched-ports condition satisfied, we need to have:

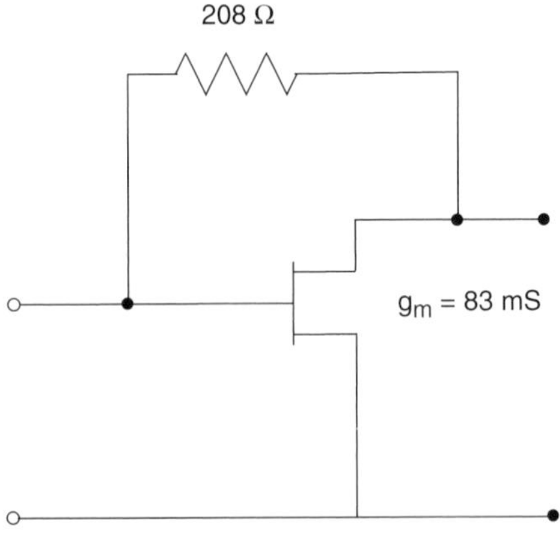

FIGURE 15.20 Figure for Example 15.7.

a. A shunt resistor:

$$S_{21} = \frac{Z_o - R_2}{Z_o} \Rightarrow R_2 = Z_o(1 - S_{21})$$

$$R_2 = 50(1 + 3.16) = 208\ \Omega$$

b. We need a MESFET having at least a value of g_m given by:

$$(g_m)_{min} = R_2/Z_o^2 = 208/50^2 = 83\ \text{mS}$$

Therefore,

$$g_m \geq 83\ \text{mS}$$

The final design is shown in Figure 15.20.

EXAMPLE 15.8

Design a series-shunt negative feedback MESFET amplifier such that:

a. *It has a 20 dB flat gain from 1 MHz to 10 MHz*

b. $(VSWR)_{IN} = (VSWR)_{OUT} = 1$

Assuming that the FET has $V_t = 2$ V and is biased to have $I_D = 0.5$ mA at $V_{GS} = 4.4$ V, is this FET suitable for negative feedback configuration? If so, determine the series and shunt elements.

Solution:

$$10 \log_{10}|S_{21}|^2 = 20\ \text{dB}$$

$$|S_{21}| = 10$$

Because the transistor is operating at low frequencies, we have:

$$S_{21} = -10$$

We need to determine $(g_m)_{min}$ first:

$$(g_m)_{min} = (1 + 10)/50 = 0.22 \text{ S}$$

But the FET's actual g_m is given by (see Chapter 4, *DC and Low-Frequency Circuits Concepts*):

$$g_m = 2I_D/(V_{GS} - V_t) = 2 \times 0.5/(4.4 - 2) = 0.416 \text{ S}$$

Because $g_m > (g_m)_{min}$, thus the MESFET is suitable for shunt-series negative feedback configuration. The shunt resistor (R_2) and series resistor (R_1) are given by:

$$S_{21} = \frac{Z_o - R_2}{Z_o} \Rightarrow R_2 = Z_o(1 - S_{21})$$

$$R_2 = 50(1 + 10) = 550\ \Omega$$

$$R_1 = \frac{Z_o^2}{R_2} - \frac{1}{g_m}$$

$$R_1 = (50^2/550) - 1/0.416 \approx 2\ \Omega$$

The final MESFET amplifier configuration is shown in Figure 15.21.

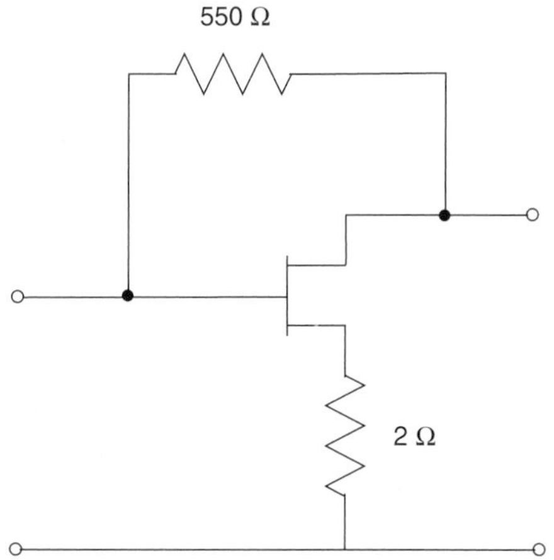

FIGURE 15.21 Figure for Example 15.8.

c) High-Frequency Considerations (Negative Feedback Case) At low frequencies, the phase of S_{21} is around 180°. As the frequency is increased, the phase of S_{21} starts to approach 90°, which creates an undesirable positive feedback condition. This means that a portion of the output signal will add in phase with the input signal, causing an increase in gain and consequent amplifier instability. To alleviate this problem, the amount of feedback can be reduced as the frequency increases and as the phase of S_{21} approaches 90°. A typical method to achieve the reduction in feedback amount is by adding an inductor in series with the shunt feedback element. Example 15.9 illustrates this point further.

EXAMPLE 15.9

Find the value of an inductor that would provide negative feedback above $f = 600$ MHz for the amplifier shown in Figure 15.22. Assume the phase of S_{21} approaches 90° around 600 MHz.

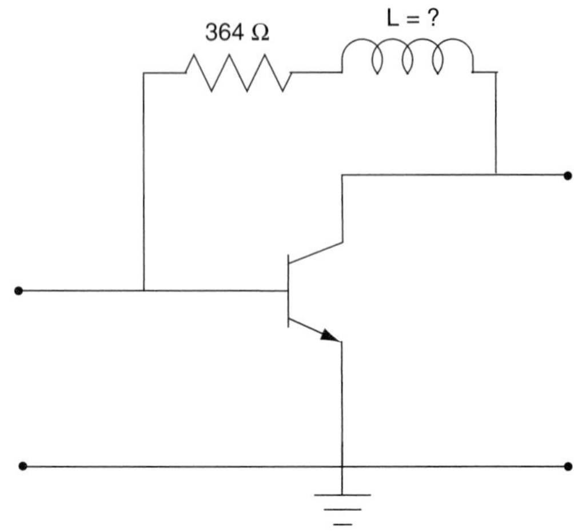

FIGURE 15.22 Circuit for Example 15.9.

Solution:

To keep the phase of S_{21} above 90°, the value of L is selected such that at $f = 600$ MHz we have:

$$\omega L = R_2 \Rightarrow \mathrm{L} = \mathrm{R}_2/\omega = 364/2\pi \times 600 \times 10^6$$

$$L = 96.6 \text{ nH}$$

The effect of the inductor is insignificant below 600 MHz; however, as the frequency increases above 600 MHz, the inductor's reactance increases proportionately and reduces the amount of feedback automatically.

15.5 MULTISTAGE SMALL-SIGNAL AMPLIFIER DESIGN

Practical transistor amplifiers usually consist of a number of stages connected in cascade, forming a multistage amplifier. Depending on the type of small-signal amplifier, whether it is a maximum-gain amplifier (MGA) or minimum-noise amplifier (MNA), each stage has a certain function and must satisfy certain conditions, which will be discussed shortly. In this section, two of the most important type of multistage amplifiers (MSA), namely MGA and MNA, will be discussed.

Consider a general n-stage amplifier configuration, as shown in Figure 15.23. To have a stable amplifier, the stability of the individual stages as well as overall stability must be checked.

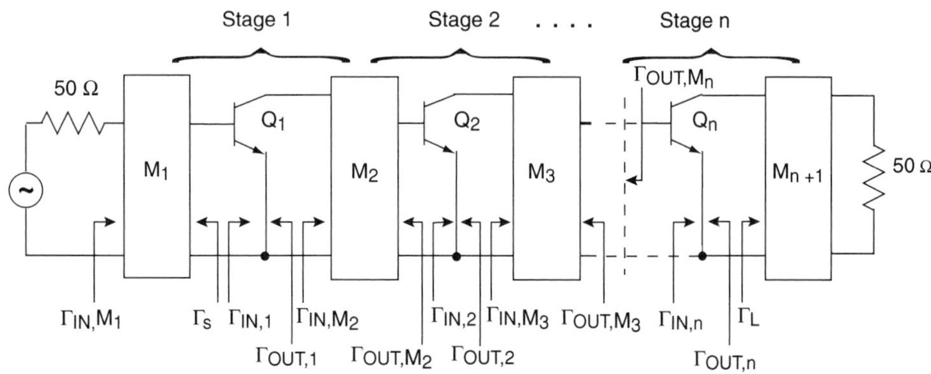

FIGURE 15.23 N-stage amplifier configuration.

The two types of amplifiers considered in this section are maximum-gain multistage amplifiers and low-noise multistage amplifiers.

Each type has its own specific criteria for the optimization of its circuit parameters, as discussed next.

15.5.1 Maximum-Gain Multistage Amplifier Design

In this type of amplifier, the goal is to obtain an overall maximum possible gain, which means that each of the stages should operate at maximum gain. Thus, all ports should be conjugately matched simultaneously, i.e.,

$$\Gamma_S = (\Gamma_{IN,1})^* \tag{15.24a}$$

$$\Gamma_{IN,M2} = (\Gamma_{OUT,1})^* \tag{15.24b}$$

$$\Gamma_{OUT,M2} = (\Gamma_{IN,2})^* \tag{15.24c}$$

$$\vdots$$

$$\vdots$$

$$\Gamma_{OUT,Mn} = (\Gamma_{IN,n})^* \tag{15.24d}$$

$$\Gamma_L = (\Gamma_{OUT,n})^* \tag{15.24e}$$

A Special Case: Unilateral Design. For a unilateral design, the preceding equations simplify:

$$\Gamma_S = (S_{11,1})^* \tag{15.25a}$$

$$\Gamma_{IN,M2} = (S_{22,1})^* \tag{15.25b}$$

$$\Gamma_{OUT,M2} = (S_{11,2})^* \tag{15.25c}$$

$$\vdots$$

$$\vdots$$

$$\Gamma_{OUT,Mn} = (S_{11,n})^* \tag{15.25d}$$

$$\Gamma_L = (S_{22,n})^* \tag{15.25e}$$

15.5.2 Minimum-Noise Multistage Amplifier Design

As discussed earlier, minimum-noise amplifiers (MNAs) are a special case of low-noise amplifiers (LNAs), wherein the goal of the design process is to have an overall minimum noise figure, which means that each stage should operate at the minimum noise figure (F_{min}) condition.

Having the requirement of $F = F_{min}$ means that the optimum noise source reflection coefficient for each transistor (e.g., $\Gamma_{opt,n}$ for the nth stage transistor) should equal the reflection coefficient of the output of each matching network (i.e., $M_1, M_2, \ldots ,M_n$) excluding the M_n+1 network (which is terminated in a 50Ω load). Furthermore, the input of each matching network (except the first stage, M_1) should be conjugately matched to the output of each transistor to maximize gain and minimize *VSWR*. These conditions can be written mathematically as:

$$\Gamma_S = \Gamma_{opt,1} \tag{15.26a}$$

$$\Gamma_{IN,M2} = (\Gamma_{OUT,1})^* \tag{15.26b}$$

$$\Gamma_{OUT,M2} = \Gamma_{opt,2} \tag{15.26c}$$

$$\vdots$$

$$\Gamma_{OUT,Mn} = \Gamma_{opt,n} \tag{15.26d}$$

$$\Gamma_L = (\Gamma_{OUT,n})^* \tag{15.26e}$$

Noise Figure Analysis of an *n*-Stage Amplifier. For a cascade of n amplifiers, the overall noise figure is the generalization of equations for equivalent noise temperature ($T_{e,cas}$) and noise figure (F_{cas}) (see Chapter 14) as follows:

$$T_{e,cas} = T_{e1} + \frac{T_{e2}}{G_{A1}} + \frac{T_{e3}}{G_{A1}G_{A2}} + \ldots + \frac{T_{en}}{G_{A1}G_{A2}\ldots G_{An-1}} \tag{15.27}$$

$$F_{cas} = F_1 + \frac{F_2 - 1}{G_{A1}} + \frac{F_3 - 1}{G_{A1}G_{A2}} + \ldots + \frac{F_n - 1}{G_{A1}G_{A2}\ldots G_{An-1}} \tag{15.28}$$

a) Special Case: Identical Stages If all stages are identical, i.e.,

$$G_{A1} = G_{A2} = \ldots. = G_{An} = G_A \tag{15.29}$$

$$T_{e1} = T_{e2} = \ldots.. = T_{en} = T_e \tag{15.30}$$

$$F_1 = F_2 = \ldots\ldots = F_n = F \tag{15.31}$$

Then Equations 15.27 and 15.28 would greatly simplify as follows:

$$T_{e,cas} = T_e\left(\frac{1 - (1/G_A)^n}{1 - 1/G_A}\right) \tag{15.32a}$$

$$F_{cas} = (F-1)\left(\frac{1-(1/G_A)^n}{1-1/G_A}\right)+1 \tag{15.32b}$$

NOTE:

$$1 + x + x^2 + \ldots + x^n = \frac{1-x^n}{1-x},\ |x| < 1 \tag{15.33}$$

b) An Infinite Chain of Identical Amplifiers If n is very large (i.e., $n \to \infty$), then:

$$T_{e,cas} = T_e\left(\frac{1}{1-1/G_A}\right) \tag{15.34}$$

$$F_{cas} = (F-1)\left(\frac{1}{1-1/G_A}\right)+1 \tag{15.35}$$

In terms of noise measure, M, of one of the amplifiers (defined earlier in Chapter 14):

$$M = \frac{F-1}{1-1/G_A} \tag{15.36}$$

we can write (15.34) and (15.35) as

$$T_{e,cas} = T_e\left(\frac{M}{F-1}\right) \tag{15.37}$$

$$F_{cas} = M + 1 \tag{15.38}$$

NOTE 1: *For a minimum-noise amplifier, where each stage operates at minimum noise figure (i.e., $F_1 = F_2 = \ldots. = F_n = F_{min}$), we have:*

$$M_{min} = \frac{F_{min}-1}{1-1/G_A} \tag{15.39}$$

Thus, we can write Equations 15.37 and 15.38 as:

$$T_{e,cas} = T_{e,min}\left(\frac{M_{min}}{F_{min}-1}\right) \tag{15.40}$$

$$F_{cas} = M_{min} + 1 \tag{15.41}$$

NOTE 2: *If the gain of each stage is very large (i.e., $G_A \to \infty$), then Equations 15.36 through 15.38 become:*

$$G_A \to \infty \Rightarrow M = F - 1 \tag{15.42}$$

$$T_{e,cas} = T_e \tag{15.43}$$

$$F_{cas} = F \tag{15.44}$$

This result indicates that a large cascade of very high-gain amplifiers will be equivalent to only the degradation of the signal by the first stage. This effectively means that the effect of added noise by all the following stages is null and void. This result is expected because from Chapter 14 we know that the first stage dominates the overall noise figure (or noise temperature) of the cascade.

LIST OF SYMBOLS/ABBREVIATIONS

A symbol/abbreviation will not be repeated again once it has been identified and defined in an earlier chapter, as long as its definition remains unchanged.

BBA	Broadband amplifier
F_{min}	Minimum noise figure
F_{cas}	Noise figure for cascaded components
HGA	High-gain amplifier
GaAs	Gallium Arsenide
$G_{L,MAX}$	Maximum output matching network power gain
$G_{S,MAX}$	Maximum input matching network power gain
LNA	Low-noise amplifier
MESFET	Metal semiconductor field effect transistor
M_{min}	Minimum noise measure
MGA	Maximum gain amplifier
MNA	Minimum noise amplifier
MSA	Multistage amplifier
$T_{e,cas}$	Equivalent noise temperature for cascaded components
$\Gamma_{IN,M1}$	Reflection coefficient for input of stage one of a multistage amplifier
$\Gamma_{OUT,M2}$	Reflection coefficient for output of stage two of a multistage amplifier

PROBLEMS

15.1 Design a microwave transistor amplifier for maximum gain using a BJT whose S-parameters in a 50 Ω system are:

$$S = \begin{bmatrix} 0.7\angle{-160^\circ} & 0 \\ 5\angle 85^\circ & 0.5\angle{-20^\circ} \end{bmatrix}$$

15.2 The S-parameters of a GaAs FET in a 50 Ω system are given by:

$$S = \begin{bmatrix} 2.3\angle{-135^\circ} & 0 \\ 4\angle 60^\circ & 0.8\angle{-60^\circ} \end{bmatrix}$$

a. Determine the unstable region in the Smith chart.

b. Design the input matching network for $G_S = 4$ dB with the greatest degree of stability.

c. Draw the complete RF amplifier schematic.

15.3 Design a broadband microwave GaAs FET amplifier to have a unilateral gain of 12 dB from 150 to 400. The S-parameters of the transistor as a function of frequency are as follows:

TABLE 15.5 Microwave GaAs FET S-Parameters

f(MHz)	S_{11}	S_{21}	S_{22}
150	$0.31\angle{-36^\circ}$	5	$0.91\angle{-6^\circ}$
250	$0.29\angle{-55^\circ}$	4	$0.86\angle{-15^\circ}$
400	$0.25\angle{-76^\circ}$	2.8	$0.81\angle{-26^\circ}$

15.4 Design a broadband microwave BJT amplifier with a gain of 10 dB from 1 to 2 GHz with a noise figure of less than 4 dB. Assuming $S_{12} \approx 0$, the remaining scattering and noise parameters of the BJT are as follows:

TABLE 15.6 Microwave BJT *S*-Parameters

f(MHz)	S_{11}	S_{21}	S_{22}	Γ_{opt}	$R_n(\Omega)$	Fmin(dB)
1000	$0.64\angle{-98°}$	$5\angle 113°$	$0.79\angle{-30°}$	$0.48\angle 23°$	23.3	1.45
1500	$0.60\angle{-127°}$	$3.9\angle 87°$	$0.76\angle{-35°}$	$0.45\angle 61°$	15.6	1.49
2000	$0.59\angle{-149°}$	$3.15\angle 71°$	$0.75\angle{-43°}$	$0.41\angle 88°$	15.7	1.61

15.5 The source and load impedance in the balanced amplifier shown in Figure P15.5 are 50 Ω. The *S*-parameters of each of the amplifiers A and B are identical and are given by:

$$S = \begin{bmatrix} 0.5\angle 160° & 0.08\angle 60° \\ 3.4\angle 70° & 0.4\angle{-45°} \end{bmatrix}$$

a. Determine S_{11}, S_{22}, the input *VSWR*, the output *VSWR*, and the power gain for the balanced amplifier.

b. If the values of the *S*-parameters for only amplifier B (both phase and magnitude) are 5 percent higher, determine the new S_{11}, S_{22}, the input *VSWR*, the output *VSWR*, and the power gain for the new balanced amplifier.

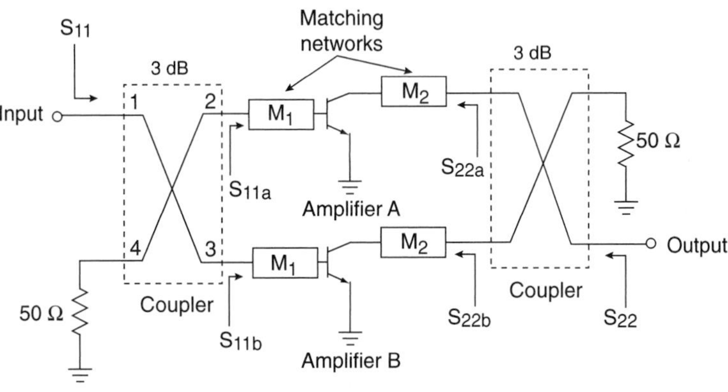

FIGURE P15.5

15.6 An amplifier is operating at 10 GHz using an FET device with the following *S*-parameters:

$$S = \begin{bmatrix} 0.5\angle{-45°} & 0 \\ 5\angle 30° & 0.8\angle{-160°} \end{bmatrix}$$

Design the amplifier for maximum gain using a 50 Ω input and output and transmission lines as shown in Figure P15.6.

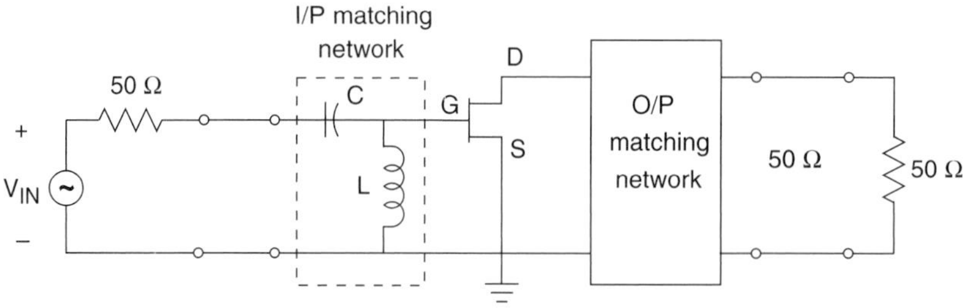

FIGURE P15.6

a. Is the amplifier stable?

b. What is the maximum gain in dB?

c. Design the input matching network to work for maximum gain using series capacitor and shunt inductor to match the 50 Ω line to Z_S.

d. What are the values of Γ_{IN} and Z_{IN}?

15.7 A GaAs FET device has the following S-parameters at 3 GHz:

$$S = \begin{bmatrix} 0.3\angle-60^\circ & 0 \\ 2\angle 45^\circ & 0.8\angle-30^\circ \end{bmatrix}$$

Design an amplifier for maximum gain using this transistor and 50 Ω input and output transmission lines. The input matching network should use lumped elements, as shown in Figure P15.7.

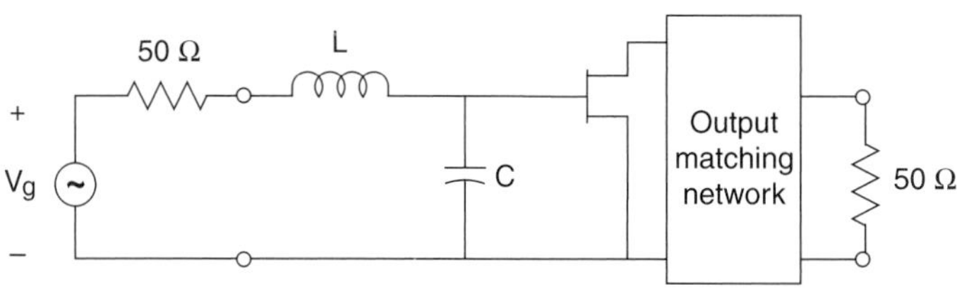

FIGURE P15.7

15.8 Design an FET amplifier for a minimum noise figure and maximum possible gain. Use open stub and quarter-wavelength transformers for the matching networks. The S-parameters of the device at 3 GHz over a 100 MHz bandwidth at a low-noise bias point ($V_{DS} = 5$ V, $I_D = 10$ mA) are given by:

$$S = \begin{bmatrix} 0.9\angle-60^\circ & 0 \\ 2\angle 90^\circ & 0.5\angle-45^\circ \end{bmatrix}$$

$$F_{min} = 3 \text{ dB}$$

$$\Gamma_{opt} = 0.5\angle-135^\circ$$

$$R_n = 4\ \Omega$$

Show the final DC and RF schematic with the input and output matching networks (use $Z_o = 50\ \Omega$).

15.9 Design an amplifier with maximum unilateral gain at 6.0 GHz using a GaAs FET transistor with the following S-parameters:

$$S = \begin{bmatrix} 0.61\angle{-170^\circ} & 0 \\ 2.24\angle 32^\circ & 0.72\angle{-83^\circ} \end{bmatrix}$$

Design L-section matching networks using lumped elements (use $Z_o = 50\ \Omega$.)

15.10 A GaAs FET has the following S-parameters in a 50 Ω system:

$$S = \begin{bmatrix} 0.7\angle{-170^\circ} & 0.02\angle 60^\circ \\ 3.5\angle 60^\circ & 0.8\angle{-70^\circ} \end{bmatrix}$$

$$F_{min} = 2 \text{ dB}$$

$$\Gamma_{opt} = 0.7\angle 120^\circ$$

$$R_n = 15\ \Omega$$

Design an amplifier with minimum noise figure and maximum possible gain. Use transmission lines in the matching networks.

15.11 If the RF output voltage (v_o) of an FET amplifier is represented by:

$$v_o = C_1 v_i + C_2 v_i^2 + C_3 v_i^3$$

where the input voltage is: $v_i = |v_i| \cos \omega t$

show that the voltage gain of the amplifier, G in dB, is given by:

$$G = 20 \log_{10}(C_1 + 3C_3|v_i|^2/4)$$

15.12 Design a two-stage low-noise amplifier for the minimum possible noise at 5 GHz. Calculate the total amplifier noise figure and the total amplifier transducer power gain. Use transmission lines for the matching networks M_1, M_2, and M_3. The two transistors are identical each with the following specifications in a 50 Ω system ($f = 5$ GHz):

$$S = \begin{bmatrix} 0.8\angle{-70^\circ} & 0 \\ 1.5\angle 60^\circ & 0.8\angle{-60^\circ} \end{bmatrix}$$

$$F_{min} = 2 \text{ dB}$$

$$\Gamma_{opt} = 0.65\angle 60^\circ$$

$$R_n = 25\ \Omega$$

15.13 By using the feedback technique, a broadband microwave amplifier with a gain of 10 dB (in a 50 Ω system) is to be designed. Calculate:

a. The required minimum transconductance ($g_{m,min}$)

b. The required shunt feedback resistor (R_2)

c. The transducer power gain

REFERENCES

[15.1] Anderson, R. W. *S-Parameter Techniques for Faster, More Accurate Network Design*. Hewlett-Packard Application Note 95–1. Feb. 1967.

[15.2] Bahl, I. and P. Bhartia. *Microwave Solid State Circuit Design*. New York: Wiley Interscience, 1988.

[15.3] Gonzalez, G. *Microwave Transistor Amplifiers, Analysis and Design*, 2nd ed. Upper Saddle River: Prentice Hall, 1997.

[15.4] Carson, R. S. *High-Frequency Amplifiers*. New York: Wiley Interscience, 1975.

[15.5] Chang, K. *Microwave Solid-State Circuits and Applications*. New York: John Wiley & Sons, 1994.

[15.6] Froehner, W. H. Quick Amplifier Design with Scattering Parameters, *Electronics*, October 1967.

[15.7] Ha, T. T. *Solid-State Microwave Amplifier Design*. New York: John Wiley & Sons, 1987.

[15.8] Liao, S. Y. *Microwave Circuit Analysis and Amplifier Design*. Upper Saddle River: Prentice Hall, 1987.

[15.9] *Microwave Transistor Bias Configurations*. Hewlett-Packard Application Note 944–1. April 1975.

[15.10] Pozar, D. M. *Microwave Engineering*, 2nd ed. New York: John Wiley & Sons, 1998.

[15.11] Vendelin, George D. *Design of Amplifiers and Oscillators by the S-Parameter Method*. New York: John Wiley & Sons, 1982.

[15.12] Vendelin, George D., Anthony M. Pavio, and Ulrich L. Rhode. *Microwave Circuit Design, Using Linear and Non-Linear Techniques*. New York: John Wiley & Sons, 1990.

CHAPTER 16

RF/Microwave Amplifiers II: Large-Signal Design

16.1 INTRODUCTION

Having considered small-signal amplifiers, we are now prepared to study large-signal amplifiers. These amplifiers are also called high-power amplifiers (HPAs) in some texts. This type of amplifier is used in many microwave applications such as radar, communication, surveillance electronic transmitter systems and so on.

In this chapter, we will consider the design of various large-signal (or high-power) amplifiers.

The design methods presented are based on three possible characterization methods of the high-power transistor:

- The modified *S*-parameters method
- The source and load reflection coefficients method
- Power contours plot method

These design methods will be explored later in this chapter.

16.2 HIGH-POWER AMPLIFIERS

Having considered the basic concepts in stability, gain, and noise of a two-port network, particularly small-signal amplifiers in Chapter 15, *RF/Microwave Amplifiers I: Small-Signal Design*, we are now prepared to take on the task of designing a large-signal amplifier.

As discussed in Chapter 15, the design of a small-signal amplifier is based on the small-signal *S*-parameters of the device, which are normally supplied by the device manufacturer or are easily measurable on a network analyzer. As will be seen shortly, this by itself will not be adequate for the design of high-power amplifiers and needs to be modified.

In general, any type of amplifier design requires an optimization of the gain, noise figure, power, or bandwidth. Therefore, each type of design will be an interplay of constant gain circles, noise figure circles, or power contours depending on the design requirements. But because the design of a large-signal amplifier focuses primarily on the amount of power available at the output, some of the small-signal amplifier considerations (such as noise figure) will have no significance or value in the design process.

As discussed in Chapter 15, an amplifier usually operates in one of the following classes:

- Class A amplifier
- Class B amplifier
- Class AB amplifier
- Class C amplifier

The description for each class is mentioned here once again for the ease and continuity that it brings to the design process:

Class A amplifier: In this mode, each transistor in the amplifier operates in its active region for the entire signal cycle.

Class B amplifier: In this mode of operation, each transistor is in its active region for approximately half of the signal cycle.

Class AB amplifier: In this mode, an amplifier operates in class A for small signals and in class B for large signals.

Class C amplifier: In this mode of operation, each transistor is in its active region for significantly less than half of the signal cycle.

In the course of the design of a high-power amplifier, we will be using a method of analysis called "large-signal analysis." Before we proceed any further, let's define what is meant by large-signal analysis:

DEFINITION-LARGE-SIGNAL ANALYSIS: *A method of analysis of an active circuit under high-amplitude signals that traverse such a large part of the operating characteristics of a device that nonlinear portions of the characteristic are usually encountered, causing nonlinear operation of the device.*

A large-signal amplifier can be designed in class A mode (behaving relatively linear), where a modified version of small-signal *S*-parameters could be used. A large-signal amplifier can also be designed to operate in classes AB, B, and C. Due to nonlinearities involved for class AB, B, and C modes, we need to have the high-power characteristics of a transistor available before proceeding with the design process.

16.3 LARGE-SIGNAL AMPLIFIER DESIGN

As discussed in the previous chapter, small-signal amplifier design is based on the assumption that the RF signals are small enough to assume that the active device operates in a linear fashion. For this reason the small-signal S-parameters enable us to design these amplifiers rather efficiently. As the signal magnitude is increased to a large value, however, the small-signal S-parameters will not be useful in power amplifier design depending on the class of operation of the amplifier as explained below.

16.3.1 Design Method 1: Class A Amplifiers

For class A operation and under large signals, all of the small-signal S-parameters remain almost unchanged except for S_{21}, which is reduced as the power level and frequency are increased. The reduction in S_{21} (forward transmission gain) is primarily caused by power saturation at the output of the amplifier and power-gain roll-off as frequency is increased.

This leads to one possible design technique in which large-signal amplifiers are designed using the same small-signal design technique involving the S-parameters except for replacing S_{21} with its new value under large-signal conditions. Example 16.1 illustrates this point further.

EXAMPLE 16.1

A certain MESFET is biased for large-signal class A operation with the following small-signal S-parameters at 5 GHz:

$$S_{11} = 0.55\angle{-150^\circ}$$

$$S_{12} = 0.04\angle 20^\circ$$

$$S_{21} = 3.5\angle 170^\circ$$

$$S_{22} = 0.45\angle{-30^\circ}$$

The large-signal forward transmission coefficient, S_{21}, is measured to be $S_{21} = 2.8\angle 180^\circ$. Design a large-signal class A amplifier with maximum transducer gain in a 50 Ω system. Assume ±0.5 dB error in gain is small enough to justify simplifications. What is the high-power amplifier gain?

Solution:

We will use the new large-signal S_{21} in all of our calculations that follow:

The amplifier's stability is first checked:

$$|\Delta| = 0.16 < 1$$

$$K = 2.28 > 1$$

Therefore, the MESFET is unconditionally stable.

Now we check for unilateral assumption:

$$\frac{1}{(1+U)^2} < \frac{G_T}{G_{TU,max}} < \frac{1}{(1-U)^2}$$

where

$$U = \frac{|S_{12}||S_{21}||S_{11}||S_{22}|}{(1-|S_{11}|^2)(1-|S_{22}|^2)} = 0.05$$

$$-0.41 \text{ dB} < G_T/G_{TU,max} < 0.45 \text{ dB}$$

Because the maximum errors involved are within our tolerance range of ±0.5 dB, we can utilize the unilateral amplifier design technique.

The maximum gain is calculated to be:

$$G_{TU,max} = 14.2 = 11.5 \text{ dB}$$

To determine the input and output matching networks, we plot S^*_{11} and S^*_{22} as shown in Figure 16.1. Using standard Smith chart techniques, we obtain (see Chapter 11):

Input matching network:

$$C_S = 0.59 \text{ nF}$$

$$L_S = 0.63 \text{ nH}$$

Output matching network:

$$C_L = 0.29 \text{ nF}$$

$$L_L = 1.2 \text{ nH}$$

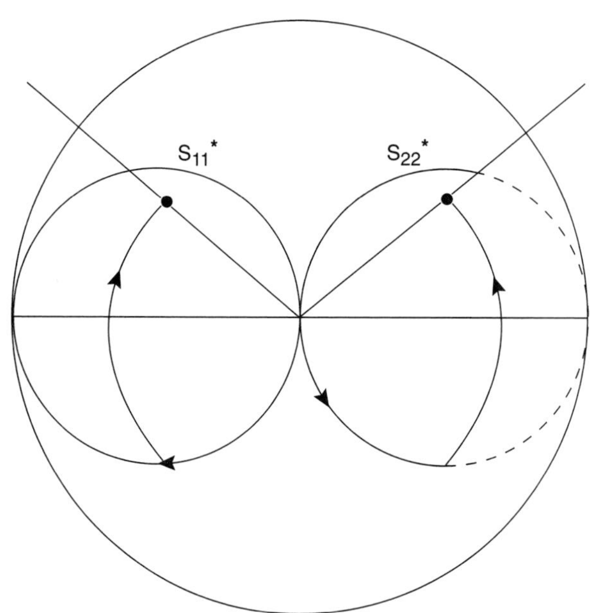

FIGURE 16.1 Smith chart plot.

The final design schematic is shown in Figure 16.2.

16.3.2 Design Method 2: Class A Amplifiers

There are amplifiers operating in class A (which is linear operation) where the use of small-signal *S*-parameters becomes fruitless. For design purposes, the measurement of large-signal *S*-parameters is also often not feasible due to the inability of test equipment to handle high power.

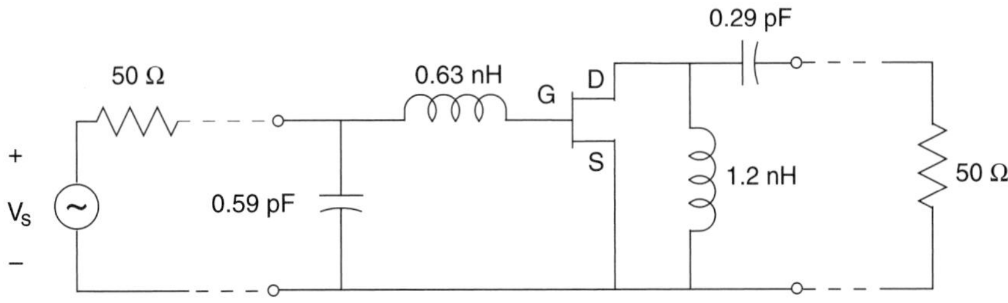

FIGURE 16.2 Circuit for Example 16.1.

For this reason, manufacturers often specify another set of data that brings about a whole new design methodology and gives us the second design method.

Therefore, for class A operation, rather than providing the small-signal *S*-parameters along with the large-signal S_{21} parameter, manufacturers may provide an alternate set of data that specifies the source reflection coefficient (Γ_{SP}), the load reflection coefficient (Γ_{LP}), and its output power when the transistor is operated at its 1 dB gain compression point ($P_{1\text{dB}}$). Thus, the design method 2 is based on the use of Γ_{SP}, Γ_{LP}, and $\text{P}_{1\text{dB}}$.

Design method 2 can be broken down into the following concepts.

a) Operation at 1 dB Gain Compression Point. Before we proceed any further, we need to define an important term:

DEFINITION-1 DB GAIN COMPRESSION POINT: *The point (on the P_{OUT} versus P_{IN} plot) at which the power gain of the transistor, due to nonlinearities, is reduced by 1 dB from its small-signal linear power gain value, i.e.,*

$$G_{1\text{dB}} = G_o(\text{dB}) - 1\text{ dB} \tag{16.1}$$

where G_o(dB) is the small-signal linear power gain in dB.

If we designate the input power at the 1-dB gain compression point as $P_{IN,1\text{dB}}$ and the output power as $P_{1\text{dB}}$, as shown in Figure 16.3, then we can write:

$$G_{1\text{dB}} = P_{1\text{dB}}/P_{IN,1\text{dB}} \tag{16.2}$$

or,

$$P_{1\text{dB}}(\text{dBm}) = P_{IN,1\text{dB}}(\text{dBm}) + G_{1\text{dB}}\,(\text{dB}). \tag{16.3}$$

NOTE: *The relationship between P_{OUT} and P_{IN} is given by:*

$$G_P = P_{OUT}/P_{IN}$$

b) Operation at the Minimum Signal Level. From Figure 16.3, we can see that for minimum input detectable signal ($P_{i,mds}$), the corresponding output power ($P_{o,mds}$) must be above the noise power level in order to be detected. The difference in the output power (in dB) in the linear portion of the power plot that is limited at one end by ($P_{o,mds}$) and at the other by $P_{1\text{dB}}$ is called the dynamic range (*DR*) of the power amplifier, i.e.,

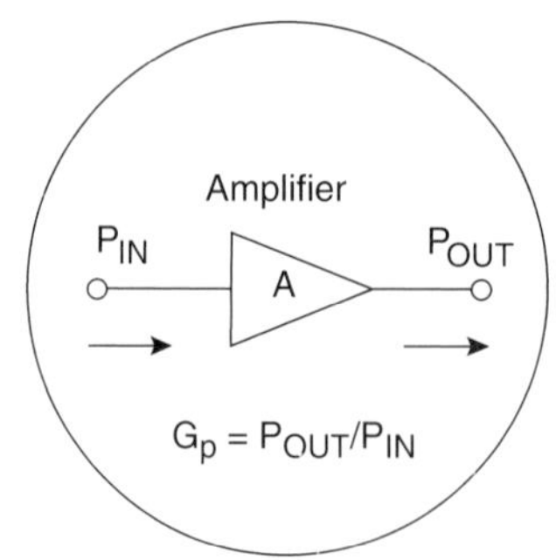

(a) High-power amplifier

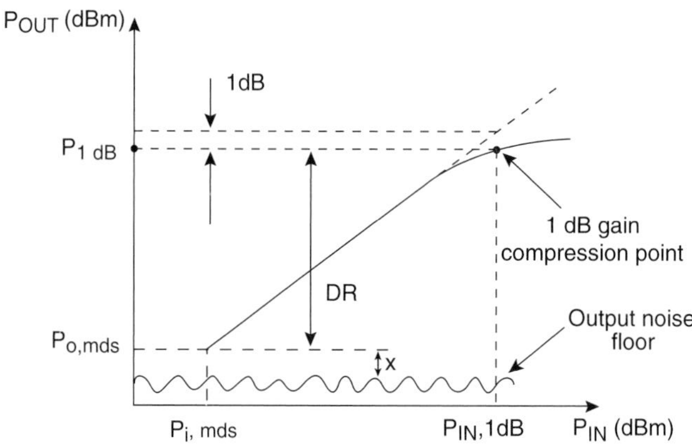

(b) The definition of 1-dB gain compress

FIGURE 16.3

$$DR = P_{1\mathrm{dB}} - P_{o,mds}\ (\mathrm{dB}) \tag{16.4}$$

The dynamic range (expressed in dB) is a very important characteristic of a power amplifier as it identifies its range of operation as well as its output power for a given input when operated in class A mode.

c) Analysis. Let's consider an amplifier with available power gain (G_A) and noise figure (F), as shown in Figure 16.4. To find $P_{o,mds}$, which is the lower end of the linear portion of the output power plots, we need to determine the minimum acceptable input signal ($P_{i,mds}$) because:

$$P_{o,mds} = P_{i,mds}\, G_A \tag{16.5}$$

or

$$P_{o,mds}(\mathrm{dBm}) = P_{i,mds}(\mathrm{dBm}) + G_A(\mathrm{dB}) \tag{16.6}$$

At the amplifier's input, the **input thermal noise floor** (P_{NI}) at $T_o = 290°$ K (see Chapter 14, *Noise Considerations in Active Networks*), is given by:

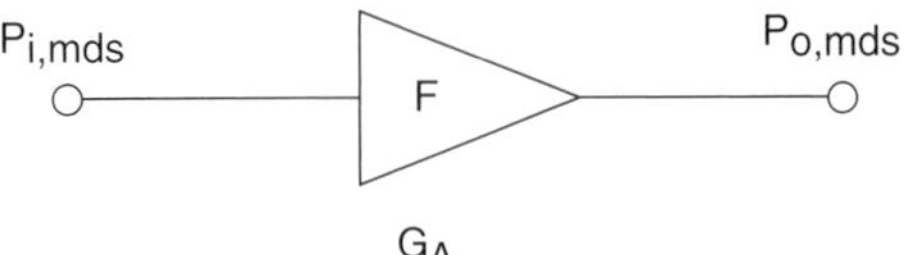

FIGURE 16.4 An amplifier with gain (G_A) and noise figure (F).

$$P_{NI} = kT_oB = -174 \text{ dBm/Hz at } T_o = 290° \text{ K} \tag{16.7}$$

At the amplifier's output, the **output thermal noise floor** (P_{No}) at $T_o = 290°$ K is given by:

$$P_{No} = kT_oBG_AF \tag{16.8}$$

$$P_{No}(\text{dBm}) = -174 \text{ dBm} + 10\log_{10}B + G_A(\text{dB}) + F(\text{dB}) \tag{16.9}$$

In order to detect the output signal, we assume that $P_{o,mds}$ is X dB (usually ≈3 dB) above the output thermal noise floor (P_{No}) and thus is given by:

$$P_{o,mds}(\text{dBm}) = P_{No}(\text{dBm}) + X(\text{dB})$$

$$= -174 \text{ dBm} + 10\log_{10}B + G_A(\text{dB}) + F(\text{dB}) + X(\text{dB}) \tag{16.10}$$

NOTE: *We can see that minimum detectable input noise power ($P_{i,mds}$), is $G_A(dB)$ less than ($P_{o,mds}$), i.e.,*

$$P_{i,mds}(\text{dBm}) = P_{o,mds}(\text{dBm}) - G_A(\text{dB}) \tag{16.11}$$

d) Γ_{SP}, Γ_{LP}, and P_{1dB} Set of Data. Consider Figure 16.5, where a power transistor is under large-signal condition. A power transistor, as a two-port, can be characterized in terms of the large-signal source reflection (Γ_{SP}), load reflection (Γ_{LP}) coefficients, and the output power at the 1-dB gain compression point (P_{1dB}) where each of these three parameters are a function of frequency and DC bias conditions. These parameters are normally measured under large-signal conditions and plotted on the Smith chart. A typical example is shown in Figure 16.6 where Γ_{SP} and Γ_{LP} are plotted from 4 to 12 GHz.

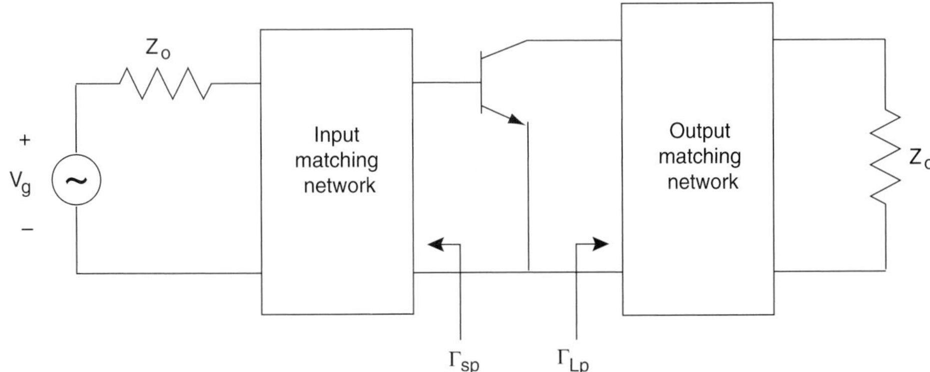

FIGURE 16.5 A power transistor.

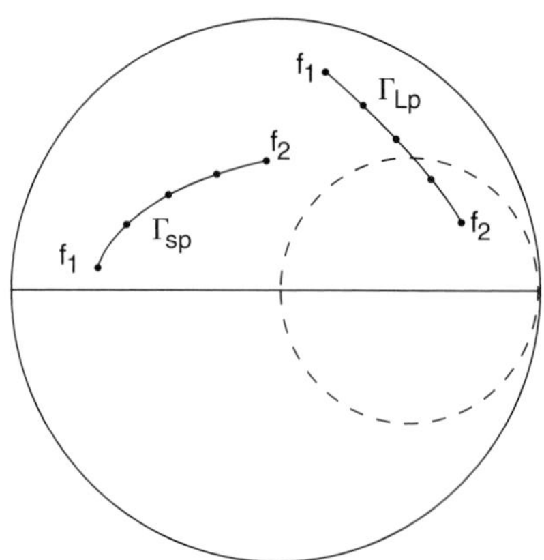

FIGURE 16.6 Γ_{SP} and Γ_{LP} plots in Smith chart ($f_1 = 4$ GHz, $f_2 = 12$ GHz).

e) Power Contours. An alternate set of large-signal data can be provided for a power transistor by plotting the output power in the load reflection coefficient plane (Γ_{LP}). Figure 16.7 shows a typical plot of output power contours in Γ_{LP}-plane with the maximum output power being $P_{max} = P_{1\text{dB}} = 19$ dBm. These power contours are not circles, as expected, due to nonlinear operation of the transistor under large-signal conditions.

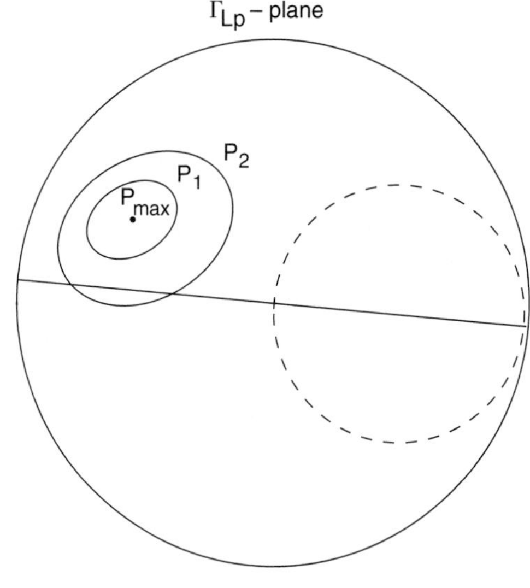

FIGURE 16.7 Plot of output power (P_{OUT}) contours.

f) Experimental Setup. A typical measuring system for measurement of large-signal parameters (Γ_{SP}, Γ_{LP}) is shown in Figure 16.8. To measure Γ_{SP} and Γ_{LP}, the following steps are carried out:

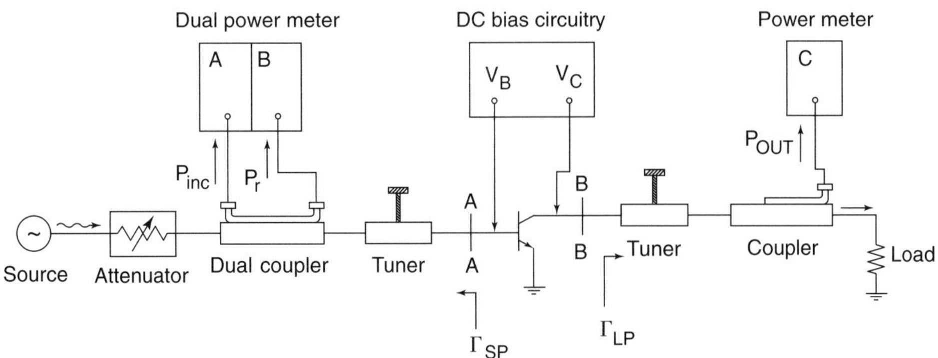

FIGURE 16.8 Experimental system for measurement of Γ_{SP} and Γ_{LP}.

Step 1. Place the transistor between planes A-A and B-B.
Step 2. Output tuning stubs are adjusted until power meter (C) measures a given power level (P_{OUT}) at the output.
Step 3. Input tuning stubs are adjusted for zero reflected power where the reading is taken at power meter (B).
Step 4. Power meter (A) reads the incident power (P_{inc}), thus:

$$G_P = P_{OUT}/P_{inc} \tag{16.12}$$

Step 5. Because the input tuning stubs are adjusted for zero reflected power in step 3, the input port is conjugately matched (i.e., $\Gamma_{SP} = \Gamma^*_{IN}$). The output impedance (Z_{LP} corresponding to Γ_{LP}) is the impedance looking into plane B-B, which produces the output power level (P_{OUT}).
Step 6. The transistor is now disconnected from the setup, and impedances at reference planes A-A and B-B are measured with a vector automatic network analyzer (VANA).
Step 7. These two measurements produce Γ_{SP} and Γ_{LP} for a given P_{OUT} and gain (G_P) at any given frequency.

Design Procedure (Method 2): Class A Operation. To design a class A power amplifier using the large-signal device parameters, the following steps are performed:

Step 1. For a given output power level, locate that particular power level contour on the Smith chart (if not available exactly, interpolate and draw an approximate contour between two known power contours).
Step 2. Select $\Gamma_L = \Gamma_{LP}$ on the contour (determined in step 1) at an appropriate point (i.e., in a stable region).
Step 3. Based on Γ_{LP} (obtained from step 2), calculate Γ_{IN}.
Step 4. For maximum power transfer, set $\Gamma_{SP} = \Gamma^*_{IN}$.
Step 5. Realize a proper matching network for Γ_{SP} and Γ_{LP} using any of the matching techniques discussed earlier in Chapter 11.

NOTE: *In this amplifier design and as always, the transistor should first be checked for stability before proceeding with the design process.*

The following examples will bring about a clearer understanding of the large-signal amplifier design process.

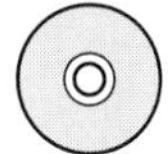

EXAMPLE 16.2

A power amplifier uses a GaAs FET transistor that has the following large-signal S-parameters at 3 GHz in a 50 Ω system:

$$S_{11} = 0.62\angle 140°$$

$$S_{12} = 0.06\angle -10°$$

$$S_{21} = 2.58\angle 20°$$

$$S_{22} = 0.53\angle -120°$$

$$P_{1dB} = 30 \text{ dBm}$$

Design a class A power amplifier for maximum output power.

Assume ±0.5 dB error in gain is allowable in our design.

Solution:

Check transistor stability:

$$K = 1.18 > 1$$

$$|\Delta| = 0.17 < 1$$

Therefore, the transistor is unconditionally stable!
Because $S_{12} \neq 0$, the unilateral figure needs to be found:

$$U = 0.115$$

$$-0.95 \text{ dB} < G_T/G_{TU,max} < 1.06 \text{ dB}$$

The error is larger than ±0.5 dB, so we use the bilateral design procedure.

For maximum power transfer, both input and output ports of the transistor must be conjugately matched:

$$G_A = G_{TU,max} = \left|\frac{S_{21}}{S_{12}}\right|(K - \sqrt{K^2 - 1}) = 22.36 = 13.5 \text{ dB}$$

$$G_{1dB} = G_A - 1 = 13.5 - 1 = 12.5 \text{ dB}$$

Γ_{MS} and Γ_{ML} are found to be:

$$B_1 = 1.07$$

$$C_1 = 0.53\angle 138.5° \Rightarrow \Gamma_{MS} = 0.83\angle -138.5°$$

Similarly,

$$B_2 = 0.86$$

$$C_2 = 0.42\angle{-122°} \Rightarrow \Gamma_{ML} = 0.79\angle 122°$$

The input and output matching networks are now realized.

Input matching network:

$$\Gamma_{MS} \Rightarrow Z_{MS} = 50(0.14 - j0.37)\Omega$$

$$Y_{MS} = 0.018 + j0.047 \text{ S}$$

For better *VSWR*, we use two parallel open stubs ($\ell = \lambda/8$), each having an admittance of $Y_{o1} = j0.047/2 = j0.0235$ S with a corresponding characteristic impedance of:

$$\ell = \lambda/8 \Rightarrow \beta\ell = \pi/4$$

$$Z_{OC} = -jZ_{o1}\cot\beta\ell$$

$$Y_{OC} = j0.0235 = j\tan\beta\ell/Z_{o1}$$

$$Z_{o1} = \tan(\pi/4)/0.0235 = 42.6\ \Omega$$

Now we use a $\lambda/4$ series transmission line (i.e., a $\lambda/4$ transformer) with a characteristic impedance (Z_{o2}) of:

$$Z_{o2} = [50(1/0.018)]^{1/2} = 52.7\ \Omega$$

Similarly, for the output matching network:

$$\Gamma_{ML} \Rightarrow Z_{ML} = 50(0.24 + j0.58)\Omega$$

$$Y_{ML} = 0.014 - j0.031 \text{ S}$$

Similar to the input section, we use two parallel open stubs ($\ell = 3\lambda/8$) each having an admittance of $Y_{o1} = j0.031/2 = j0.0155$ S or with a corresponding characteristic impedance of:

$$Z'_{o1} = \tan(\pi/4)/0.0155 = 64.5\ \Omega$$

Now we use a $\lambda/4$ series transmission line (i.e., a $\lambda/4$ transformer) with a characteristic impedance (Z'_{o2}) of:

$$Z'_{o2} = [50(1/0.014)]^{1/2} = 59.8\ \Omega$$

The final design schematic is shown in Figure 16.9.

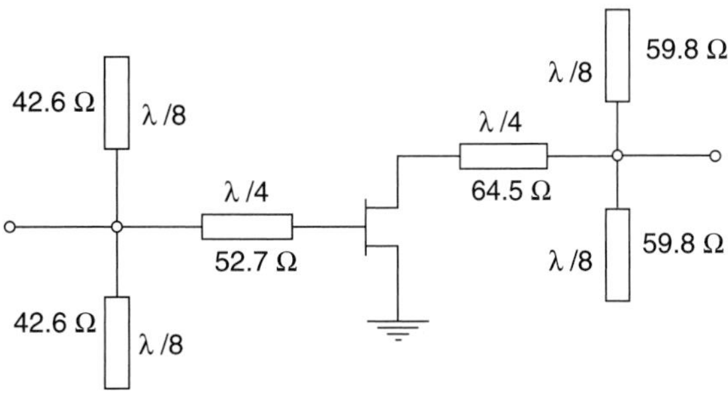

FIGURE 16.9 Final design schematic for Example 16.2.

NOTE: *The required input power to obtain P_{1dB} = 30 dBm is given by:*

$$P_{IN,1\mathrm{dB}} = P_{1\mathrm{dB}} - G_{1\mathrm{dB}} = 30 - 12.5 = 17.5 \text{ dBm}$$

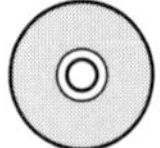

EXAMPLE 16.3

A high-power broadband amplifier with a bandwidth of 8 GHz, a gain of 20 dB, and a noise figure of 7 dB can produce a power of 30 dBm at 1-dB gain compression point. What is its dynamic range (DR)? (Assume T = 290° K and X = 3 dB.)

Solution:

$$P_{i,mds} = -174 \text{ dBm} + 10\log_{10} B + F(\text{dB}) + X(\text{dB})$$

$$P_{i,mds} = -174 \text{ dBm} + 10\log_{10}(8 \times 10^9) + 7 + 3 = -65 \text{ dBm}$$

$$P_{o,mds}(\text{dBm}) = P_{i,mds}(\text{dBm}) + G_A$$

$$P_{o,mds} = -65 + 20 = -45 \text{ dBm}$$

$$DR = 30 - (-45) = 75 \text{ dB}$$

The dynamic range is shown in Figure 16.10.

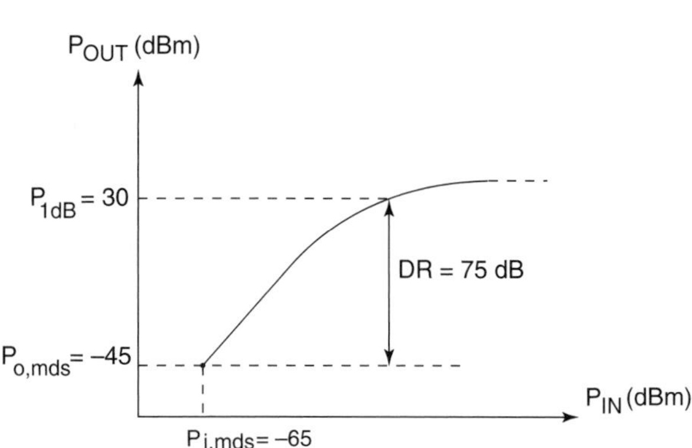

FIGURE 16.10 Dynamic range for Example 16.3.

16.3.3 Classes AB, B, and C Amplifier Design

While most amplifiers at microwave frequencies are class A, at frequencies below 1 GHz such as the RF range (f ≈ 300 kHz to 1 GHz) and below, there are amplifiers operating in classes AB, B, and C (which is nonlinear operation), where the use of small-signal *S*-parameters becomes fruitless. For this reason, and as discussed earlier, manufacturers often specify another set of data (Γ_{SP}, Γ_{LP}, and $P_{1\mathrm{dB}}$) that brings about design method 2, already discussed for class A operation.

Characterizing a power transistor for operation in classes AB, B, and C consists of Smith chart plots of source and load reflection coefficients (Γ_{SP}, Γ_{LP}) or (Z_{SP}, Z_{LP}) and output power (P_{OUT}) as a function of frequency (very similar to the class A transistor data). The design procedure is identical to design method 2 described for class A operation except for the output matching network, which must also provide the required Q needed for resonance, in order to suppress the harmonics. Further discussion on this topic is relegated to more advanced texts.

16.4 MICROWAVE POWER COMBINING/DIVIDING TECHNIQUES

In special cases when more power is required such as in a radar transmitter or in space communication, a single microwave power amplifier cannot supply the required amount of microwave power, and power combining techniques should be used.

To achieve a high output power, several individual power amplifiers are combined by means of "power combiners" to produce the needed power. Conversely, when the input power is high enough to be beyond the power handling capabilities of a single amplifier. We need to use power dividers to bring the power down to the acceptable range. The balanced amplifier discussed in Chapter 15 uses two Lange couplers and sets a very good example of power dividing/combining technique in that it first divides and then combines the amplified power to produce the desired level of signal power.

There are two general categories of combiners or dividers (as shown in Figure 16.11) that need to be discussed.

16.4.1 N-way in Multiple Stages

N-way in multiple stages power combiners (or dividers) are coupling devices that combine (or divide) the output of N amplifiers in several stages, as shown in Figure 16.12.

This category of combiners/dividers, due to its flexibility and overall simplicity of structure, has a greater popularity and is more widely used. This category of combiners/dividers is further subdivided into the following:

a. Binary structures (also called tree or corporate type): In this type of structure, the number of power sources for combining (or being divided into) are powers of 2 (i.e., 2, 4, 8, etc.). Examples include Lange couplers, 90°-hybrid branch line, Wikinson combiner, and rat-race ring (see Figure 16.13). These couplers will be examined briefly later in this chapter.

b. Nonbinary structures (also called chain or serial type): In this type of structure, the number of power sources for combining (or being divided into) are not multiples of 2. Examples include microstrip line, coaxial line, and waveguide couplers. A general schematic is shown in Figure 16.14. For an N-stage combining structure, each successive combiner adds $1/N$ of its output power to the total output power.

16.4.2 N-way in a Single Stage

N-way in a single stage power combiners (or dividers) combine or divide the output of an amplifier in one single stage, as shown in Figures 16.15 and 16.16. This category is further subdivided into two as follows:

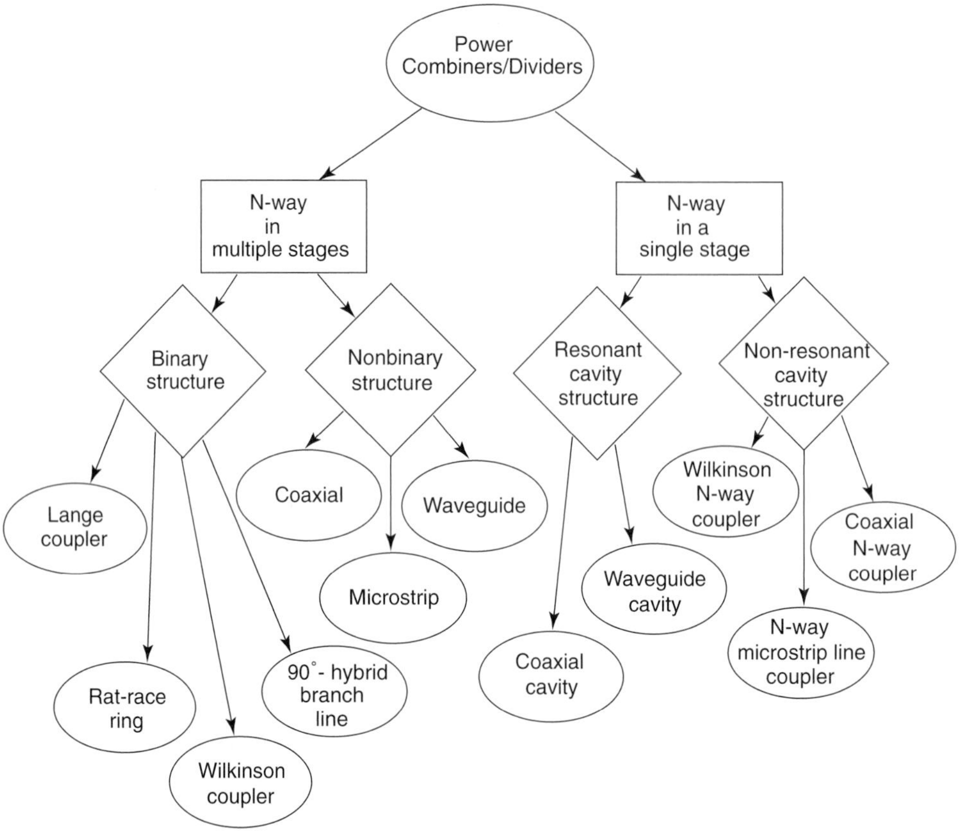

Power Combiners/Dividers
N-way in multiple stages
N-way in a single stage
Binary structure
Nonbinary structure
Resonant cavity structure
Non-resonant cavity structure
Lange coupler
Rat-race ring
Wilkinson coupler
90°- hybrid branch line
Coaxial
Microstrip
Waveguide
Coaxial cavity
Waveguide cavity
Wilkinson N-way coupler
N-way microstrip line coupler
Coaxial N-way coupler

FIGURE 16.11 Variety of microwave power combiners/dividers.

P1
P2
P3
P4
1
2
3
4
N-3
N-2
N-1
N
Power combiner #1
Power combiner #2
Power combiner #N/2-1
Power combiner #N/2
Power combiner
Power combiner
Power combiner
Po = Σ Pi, i=1 to N

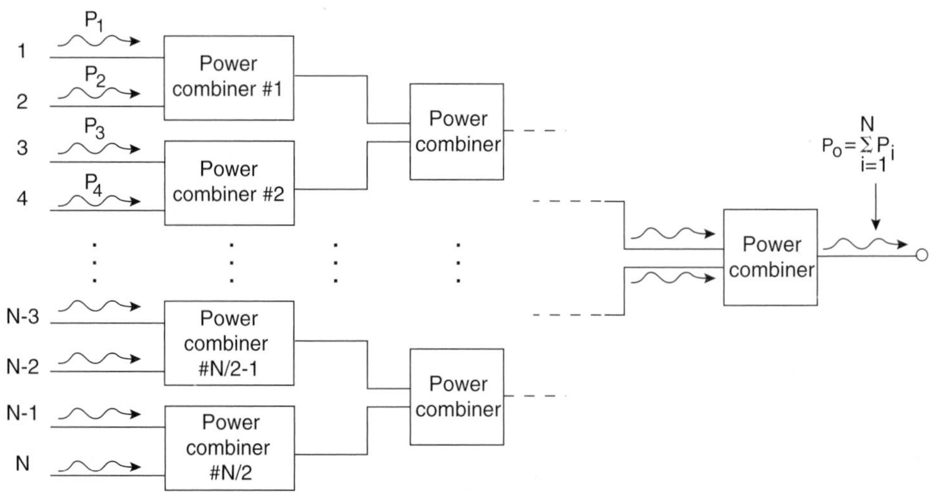

FIGURE 16.12 N-way in multiple stages combiner (N is given by powers of 2).

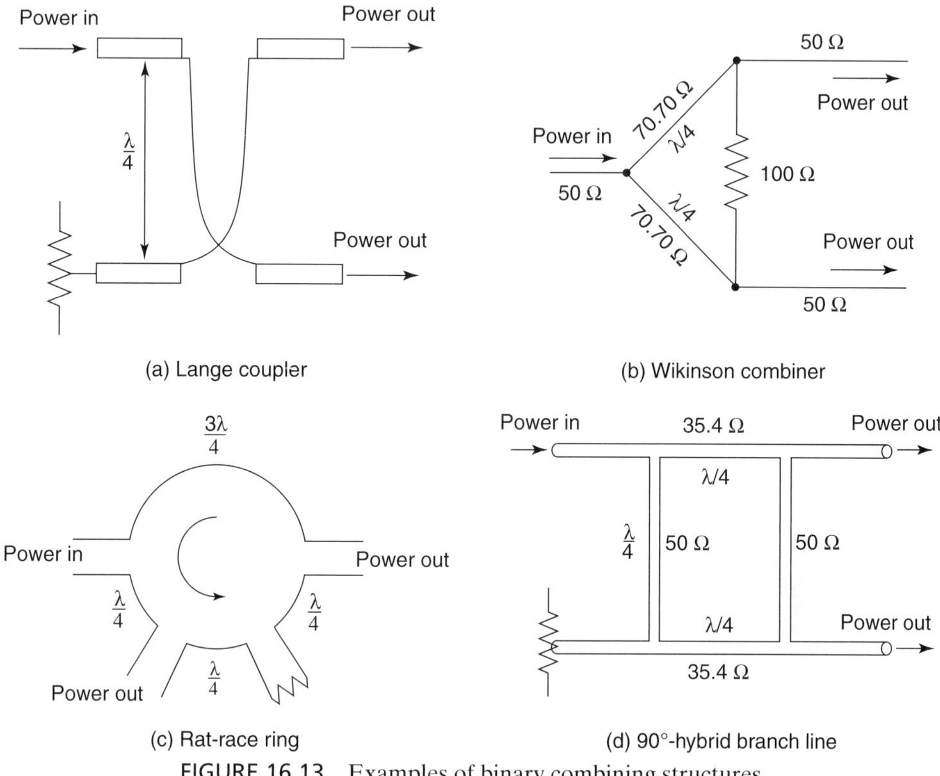

FIGURE 16.13 Examples of binary combining structures.

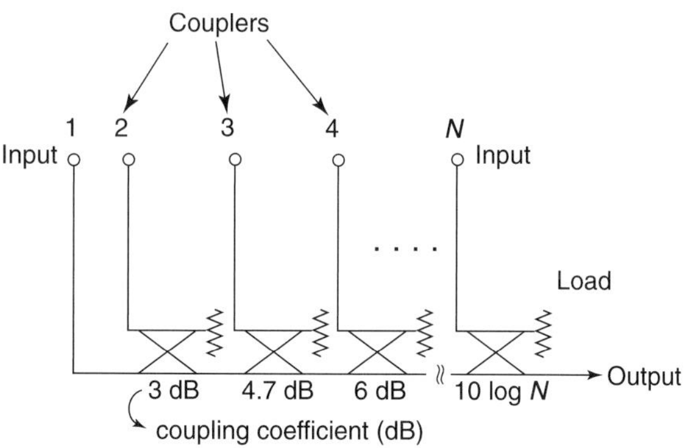

FIGURE 16.14 A general non-binary structure (also called a serial combiner).

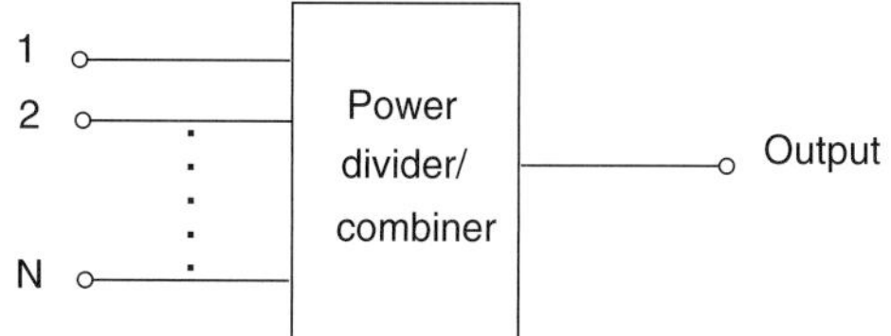

FIGURE 16.15 N-way in a single-stage divider/combiner.

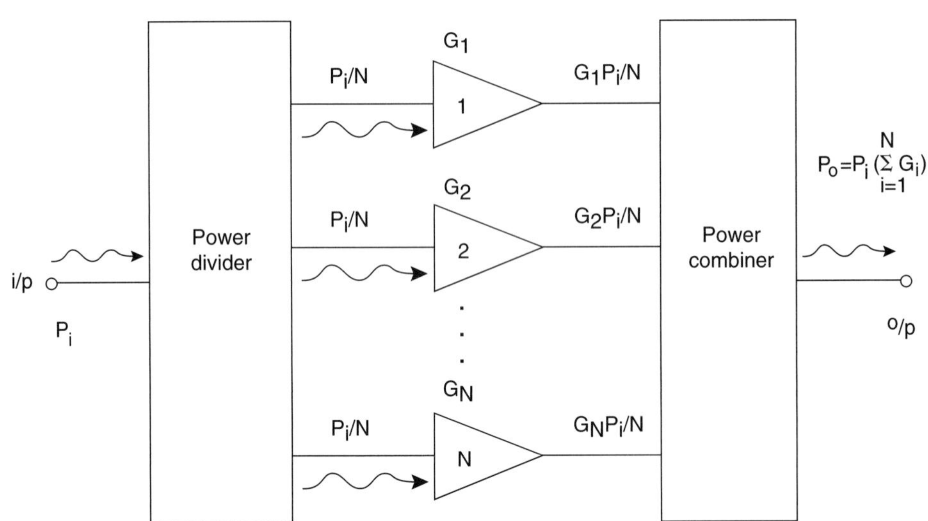

FIGURE 16.16 N-way in a single stage used in a power amplifier.

a. Resonant cavity structure: Examples include waveguide and coaxial resonant cavity structures (see Figure 16.17).

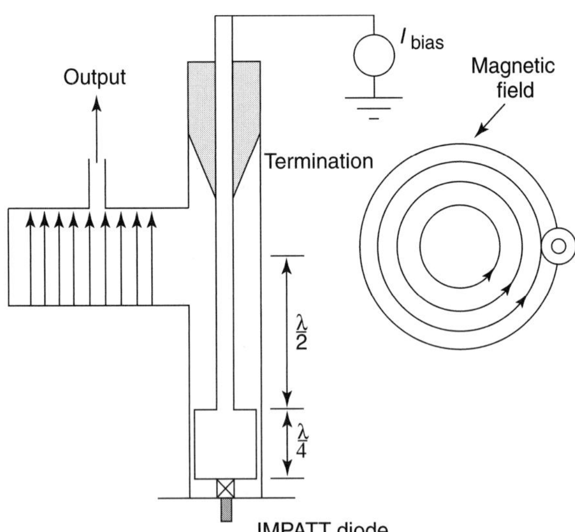

FIGURE 16.17 A coaxial-cavity resonator. (Reprinted by permission of the IEEE, Inc.)

b. Nonresonant cavity structure: Examples include Rucker 32-way combiner, Wilkinson N-way combiner, and 12-way microstrip coupler (see Figure 16.18).

In this book, we will focus primarily on binary structures in the first category and leave the treatment of all others to more advanced texts.

FIGURE 16.18 Examples of nonresonant couplers.

(a) Wilkinson N-way combiner

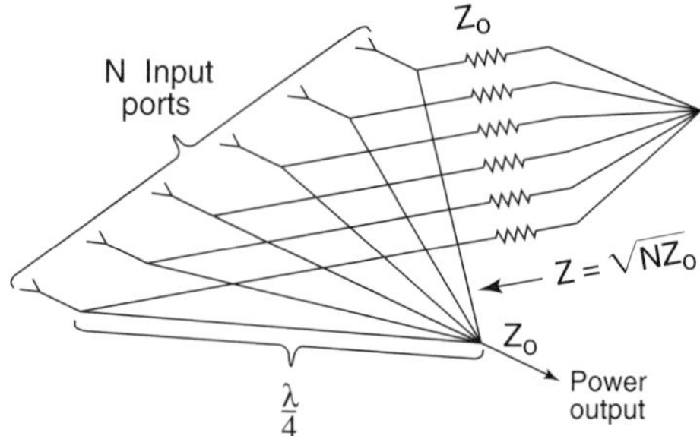

(b) 12-way microstrip coupler

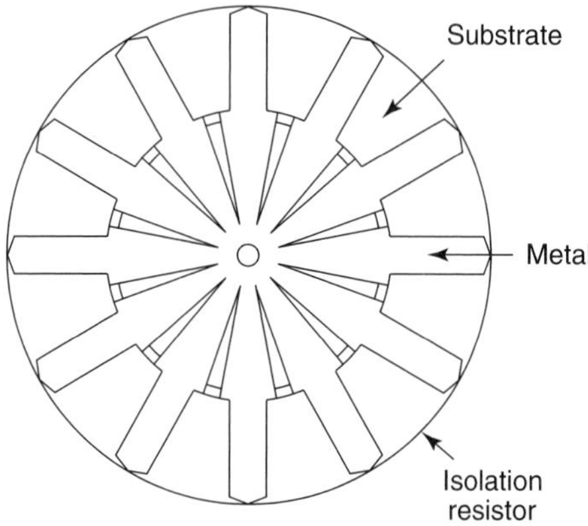

(c) Equivalent circuit for Figure 16.18b

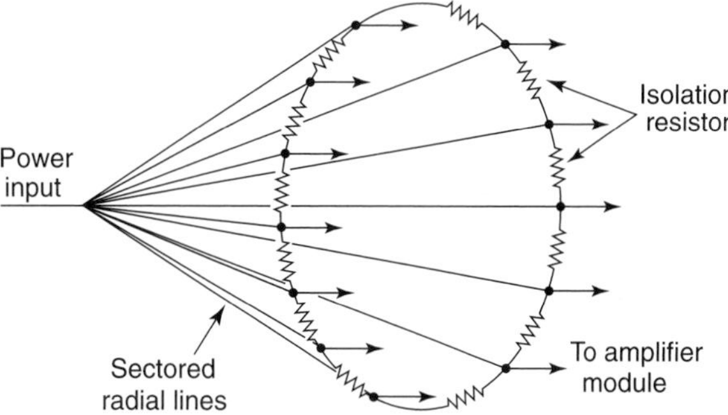

16.4.3 Binary Combiners/Dividers

The configuration of a binary power combiner is shown in Figure 16.12. From this figure we notice that if we reverse the power flow, the power combiner will be turned into a power divider. We further note that in a binary structure, the number of power sources (N) for combining or division are in powers of 2 (see Figure 16.12). For example, combining eight sources of power in a binary combiner (i.e., 8 to 1) requires three stages of power combining as follows:

STAGE 1: Four coupling devices, power doubles.
STAGE 2: Two coupling devices, power quadruples.
STAGE 3: One coupling device, power octuples.

In other words, the number of coupling devices in each stage is a binary multiple of the succeeding stage. Based on this observation, we can easily see that, for example, a 32 to 1 binary combiner requires 5 stages, where each stage (starting from input) progressively has diminishing numbers of coupling devices, i.e., 16, 8, 4, 2, down to 1 at the output. Thus, by observation we can see that:

$$N = 2^S \tag{16.13}$$

where

N = Number of inputs (or power sources)

S = Number of stages

From Figure 16.12, we can observe that each coupling device must have the following properties:

- The output port of each coupling device must be matched to its input port to reduce the mismatch loss.
- The two inputs of each coupling device must be isolated from one another to prevent cross coupling and power leakage, which would otherwise increase the overall loss of the binary structure.

Aside from the losses that occur due to port mismatch or poor isolation, the overall loss grows when the number of coupling devices is increased. This is due to the insertion loss of each coupling device, which causes power loss as the signal travels through.

To calculate the overall efficiency of a binary coupler, we can do the following simple analysis.

Let:

L (dB) = loss per coupling device in dB

S = Number of stages

$$\text{Total loss in } S \text{ stages} = P_i(\text{dB}) - P_o(\text{dB}) = SL \text{ (dB)} \tag{16.14}$$

Rewriting Equation 16.14 in ratio format, we have:

$$P_i/P_o = 10^{SL/10}$$

$$\text{Efficiency} = \eta = \frac{P_o}{P_i} = 10^{-SL/10} \tag{16.15}$$

where P_i and P_o are the total input power and total output power, respectively. Example 16.4 will elucidate these points further.

EXAMPLE 16.4

In a 64 to 1 binary combiner, if the insertion loss of each coupling device is 1 dB, find:

a. *The combining efficiency of the combiner*

b. *The maximum output power if each input source can deliver a maximum of 10 dBm power*

c. *The combining efficiency if the loss in each coupling device is reduced to 0.1 dB. (Ignore losses due to mismatch or isolation.)*

Solution:

a.

$$N = 64 = 2^S \Rightarrow S = 6$$

$$\eta = 10^{-6(1)/10} = 0.25 = 25\%$$

This means that 75 percent of the input power is lost in the combiner.

b. To find the maximum power, two methods are possible:

Method #1

The output power is combined in six stages, where the power doubles in each stage (i.e., 2 × or 3dB). Considering that the loss in each stage is 1 dB, we can write:

$$P_o = 10\ (\text{dBm}) + 6 \times 3(\text{dB}) - 6 \times 1(\text{dB}) = 22\ \text{dBm}$$

Method #2

Because we know the number of input sources and the combining efficiency, we can calculate the total output power easily by finding the total input power first:

$$(P_i)_{source} = 10\ \text{dBm} = 10\ \text{mW}$$

$$P_i = N(P_i)_{source} = 64 \times 10\ \text{mW} = 640\ \text{mW}$$

$$P_o = \eta P_i = 0.25 \times 640 = 160\ \text{mW} = 22\ \text{dBm}$$

c. Reducing the loss per coupling device to 0.1 dB, we have:

$$L = 0.1\ \text{dB}$$

$$\eta = 10^{-6(0.1)/10} = 0.87 = 87\%$$

This dramatic increase in efficiency from 25 percent to 87 percent clearly shows that the proper selection of a low-loss coupling device is essential for an efficient operation of a combiner/divider.

Types of Coupling Devices for Binary Structure. As mentioned earlier, types of coupling devices that can be used in N-way multiple stages power combiner/dividers include the following:

- Lange couplers (a quadrature hybrid)
- 90° branch-line hybrid couplers (a quadrature hybrid)
- Wilkinson couplers
- Rat-race rings or hybrid ring (a 180° hybrid)

These four types of couplers are shown in Figure 16.13. We will now briefly describe the structure and properties of each.

Lange Couplers A Lange coupler is an interdigitated microstrip four-port coupler consisting of four parallel microstrip lines with alternate lines tied together. Figure 15.17a (or Figure 16.13a) shows a schematic diagram of a Lange coupler. A 3 dB Lange coupler is designed such that when an incident wave enters port 1, equal powers are sent to ports 2 and 3 (that are 90° out of phase) but none to the opposite port (port 4). A Lange coupler is an example of a larger class of 3 dB couplers called "quadrature (90°) hybrid." Other examples of quadrature hybrids include 90° branch-line hybrid, coupled line couplers, etc.

The scattering matrix for an ideal 3 dB Lange coupler is given by:

$$[S] = \frac{-1}{\sqrt{2}}\begin{bmatrix} 0 & j & 1 & 0 \\ j & 0 & 0 & 1 \\ 1 & 0 & 0 & j \\ 0 & 1 & j & 0 \end{bmatrix} \tag{16.16}$$

90° Branch-Line Hybrid The 90° branch-line hybrid is a four-port network, as shown in Figure 15.17b (or Figure 16.13d). In this type of coupler, with all ports matched, the power entering port 1 is equally divided between ports 2 and 3, with a 90°-phase shift between these outputs. Furthermore, an input signal incident at port 1 (or port 4) couples to port 2 and 3 but not to port 4 (or port 1). Therefore, ports 1 and 4 are designed to be isolated from each other. As mentioned earlier, a 90° branch-line hybrid is another example of a larger class of 3 dB couplers called "quadrature (90°) hybrid." The scattering parameter matrix for an ideal 90° branch-line hybrid can be written as:

$$[S] = \frac{-1}{\sqrt{2}}\begin{bmatrix} 0 & j & 1 & 0 \\ j & 0 & 0 & 1 \\ 1 & 0 & 0 & j \\ 0 & 1 & j & 0 \end{bmatrix} \tag{16.17}$$

From Figure 16.13d, we can observe that the 90° branch-line hybrid has a high degree of symmetry because any port can be used as the input port with the output ports at the opposite side, and the isolated port will be the remaining port on the same side as the input port. This degree of symmetry can also be observed from the scattering matrix given previously.

There are two pairs of quarter-wave transmission line (TL) sections in a 90° branch-line hybrid where each pair has a different characteristic impedance than the input or output transmission lines. The design equations for a general 90° branch-line hybrid coupler with a coupling factor (C in dB) are given by:

$$C(\text{dB}) = -10\log_{10}\left[1 - \left(\frac{Z_{o1}}{Z_o}\right)^2\right] \tag{16.18}$$

and,

$$Z_{o2} = \frac{Z_{o1}}{\sqrt{1 - \left(\frac{Z_{o1}}{Z_o}\right)^2}} \tag{16.19}$$

EXAMPLE 16.5

Calculate a 3 dB 90° branch-line hybrid coupler's impedance values for 50 Ω input and output transmission lines.

$$C = 3 = -10 \log_{10}\left[1 - \left(\frac{Z_{o1}}{Z_o}\right)^2\right] \Rightarrow Z_{o1}/Z_o = 1/(\sqrt{2})$$

Thus,

$$Z_{o1} = 50/\sqrt{2} = 35.36\ \Omega$$

and

$$Z_{o2} = \frac{35.36}{\sqrt{1 - \left(\frac{1}{\sqrt{2}}\right)^2}} = \sqrt{2}(35.36) = 50\ \Omega$$

Wilkinson Couplers A Wilkinson 3 dB coupler is a three-port network that divides the input power (at port 1) equally between ports 2 and 3 (assuming ports 2 and 3 are matched). In this type of coupler, the output signals at ports 2 and 3 are not only equal in magnitude but also have the same phase. A schematic diagram for a 3 dB Wilkinson power divider is shown in Figure 15.17c (or Figure 16.13b). Reversing the direction of flow (i.e., input at ports 2 and 3 and output at port 1) will make this divider act like a power combiner.

From Figure 16.13b, we can see that there is a $\lambda/4$ transmission line between the input and each output port (λ is the TL wavelength at the center frequency). The characteristic impedance of this TL section needs to be calculated properly. Assuming that the input and output transmission lines have a characteristic impedance (Z_o) and both are matched, we can calculate the characteristic impedance of the $\lambda/4$ TL section $(Z_o)_{\lambda/4}$ by the following procedure.

In order for the signal entering port 1 (see a matched line), it must see an input impedance of $Z_{IN} = Z_o$. There are two matched TL sections where each should produce an input impedance of $2Z_o$, so that when it is combined with its identical parallel arm, it would yield the desired input impedance of Z_o. Thus, each parallel TL section has $2Z_o$ as the input impedance, which leads to calculation of $(Z_o)_{\lambda/4}$:

$$(Z_o)_{\lambda/4} = \sqrt{Z_o \cdot 2Z_o} = Z_o\sqrt{2}$$

For example, if $Z_o = 50\ \Omega$, then $(Z_o)_{\lambda/4} = 50\sqrt{2} = 70.7\ \Omega$.

In this analysis we have assumed that both parallel arms are matched; however, in practice a perfect match seldom occurs so we add a resistor with a value of $2Z_o$ between the two parallel arms to absorb any reflected power. When a mismatch occurs at one of the output ports, the reflected signal will now split equally between the TL section and the resistor. The two signals will travel to the other output port with a 180° phase shift and cancellation occurs.

The scattering matrix of the Wilkinson coupler is given by:

$$[S] = \frac{-j}{\sqrt{2}}\begin{bmatrix} 0 & 1 & 1 \\ 1 & 0 & 0 \\ 1 & 0 & 0 \end{bmatrix} \tag{16.20}$$

Rat-Race Hybrid Ring (or Hybrid Ring) A rat-race ring circuit is a four-port network consisting of an annular ring $3\lambda/2$ in circumference (λ is the wavelength at the center frequency of operation), as shown in Figure 16.19 (or Figure 16.13c). There are four arms connected at 60° intervals of angular rotation (but $\lambda/4$ or 90° of electrical length apart) by means of either series or parallel junctions.

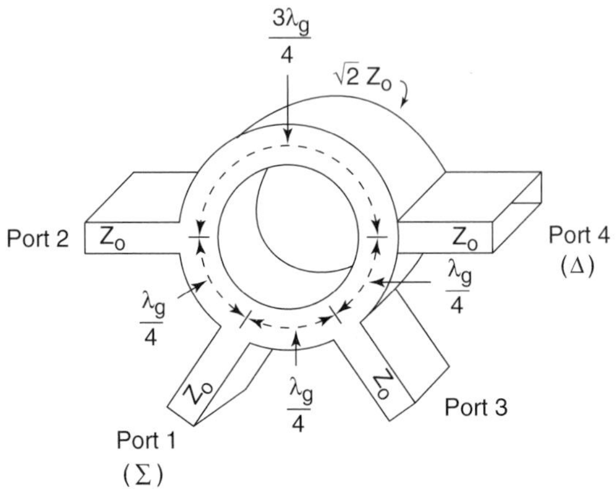

FIGURE 16.19 A rat-race hybrid ring.

When a signal at the center frequency is fed into port 1, it will split equally into ports 2 and 3, with 180° phase difference between the two signals due to electrical lengths between the two ports. No signal will appear at port 4, however, because the phase difference between two traveling waves (i.e., one traveling clockwise and the other counter-clockwise) is 180°, thus causing cancellation at this port. If now a signal is fed into port 4, it will split equally between ports 2 and 3 and none will reach port 1. Therefore, ports 1 and 4 are isolated. Similarly, using the same line of reasoning we can observe that ports 2 and 3 are also isolated. Obviously, from the discussion presented so far, a rat-race hybrid ring (like all other couplers presented earlier) is frequency sensitive and will not support a bandwidth more than 10 to 20 percent of the center frequency.

To obtain a low *VSWR* at the ports, the characteristic impedance of the annular-ring transmission line (Z'_o) can be shown to be:

$$Z'_o = \sqrt{2}\, Z_o$$

where Z_o is the characteristic impedance of each arm. For example, for a rat-race ring with 50 Ω arms, we need to design for a 70.7 Ω annular ring in order to minimize the *VSWR* at the ports.

The rat-race hybrid ring is considered to be an example of a general class of couplers called "180° hybrid." The scattering matrix of an ideal hybrid ring can be shown to be:

$$[S] = \frac{-j}{\sqrt{2}}\begin{bmatrix} 0 & 1 & 1 & 0 \\ 1 & 0 & 0 & -1 \\ 1 & 0 & 0 & 1 \\ 0 & -1 & 1 & 0 \end{bmatrix} \tag{16.21}$$

Other examples of a 180° hybrid other than hybrid ring are tapered coupled-line hybrid (which is a planar circuit), hybrid waveguide junction (or magic-T), etc.

The following examples further illustrate the topic of couplers.

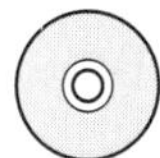

EXAMPLE 16.6

Consider two power BJT amplifiers used in a circuit configuration, as shown in Figure 16.20 having the following specifications:

Amplifier	G_o(dB)	G_{1dB}(dB)	P_{1dB}(dBm)
AMP1	8	7	27
AMP2	10	9	22

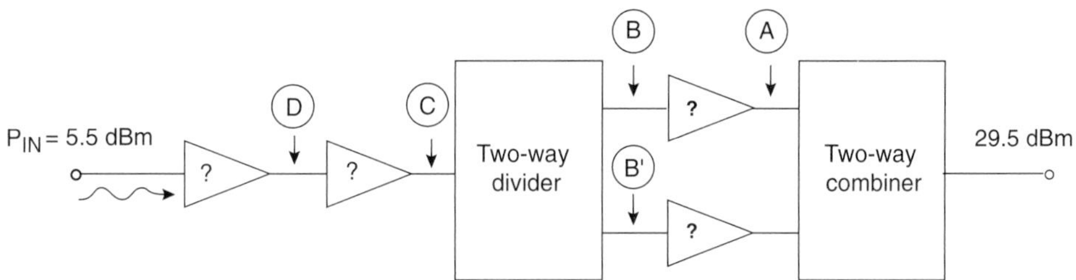

FIGURE 16.20 Circuit for Example 16.6.

Assuming the operating frequency is 1 GHz and the input power 5.5 dBm, specify the correct BJT amplifiers that must be used in each stage to obtain 29.5 dBm output power. Assume that each two-way divider/combiner has 0.5 dB insertion loss.

Solution:

Starting from the output and working backward, we have the following power and gain at each stage:

Point A: $P_A = 29.5 + 0.5 - 3 = 27$ dBm

Thus, we need to use AMP1 for both parallel arms because AMP1 has $P_{1dB} = 27$ dBm and $G_{1dB} = 8 - 1 = 7$ dB.

Point B: $P_B = P'_B = 27 - 7 = 20$ dB

Point C: $P_C = 20 + 0.5 + 3 = 23.5$ dBm

Thus, we need to use AMP1 at point C because 23.5 dBm is above the P_{1dB} (= 22 dBm) of AMP2. AMP1 operates linearly with a gain of $G_o = 8$ dB.

Point D: Because AMP1 operates linearly thus,

$$P_D = 23.5 - 8 = 15.5 \text{ dBm}$$

Either of the two amplifiers can be used at the input, but because the input power is 5.5 dBm, we need to choose a higher-gain amplifier. This condition requires us to choose to place AMP2 with $G_o = 10$ dB at the input, as shown in Figure 16.21.

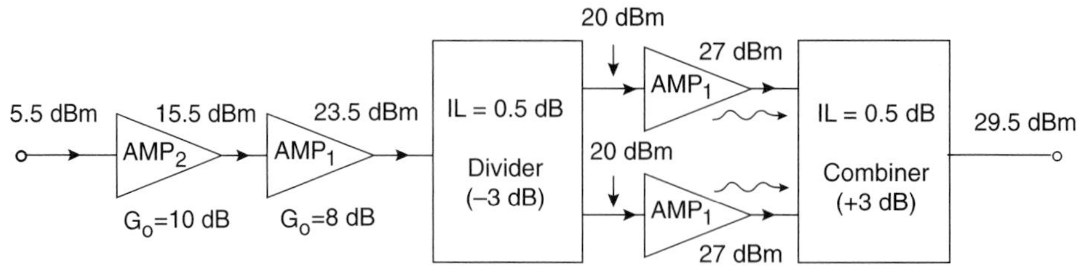

FIGURE 16.21 Solution for Example 16.6.

16.5 SIGNAL DISTORTION DUE TO INTERMODULATION PRODUCTS

Operating an amplifier under large-signal conditions causes distortions in the output signal. This distortion is primarily caused by deviation from linear operation, which causes new frequencies to appear at the output port, usually referred to as "intermodulation products."

DEFINITION-INTERMODULATION PRODUCTS: *The additional frequencies at the output of a nonlinear amplifier (or in general any nonlinear network) when two or more sinusoidal signals are applied at the input.*

To illustrate this concept, let's consider the following input signal consisting of two frequencies, each with unity amplitude:

$$V_i(t) = \cos(2\pi f_1 t) + \cos(2\pi f_2 t) \tag{16.22}$$

If $V_i(t)$ is applied to a nonlinear amplifier (see Figure 16.22) with the output/input voltage characteristic of:

$$V_o(t) = AV_i(t) + BV_i^2(t) + CV_i^3(t) \tag{16.23}$$

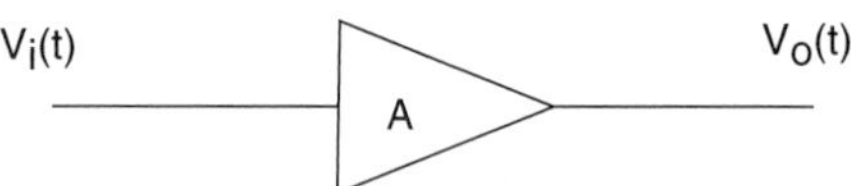

FIGURE 16.22 Nonlinear amplifier.

Then, the output signal $V_o(t)$ will contain not only the original frequencies f_1 and f_2, but also the following intermodulation products: DC, $2f_1, 2f_2, 3f_1, 3f_2, f_1 \pm f_2, 2f_1 \pm f_2$, and $2f_2 \pm f_1$.

We may classify these intermodulation products as follows:

Second harmonics: $2f_1, 2f_2$ (caused by V_i^2 term).

Third harmonics: $3f_1, 3f_2$ (caused by V_i^3 term).
Second-order intermodulation products: $f_1 \pm f_2$ (caused by V_i^2 term).
Third-order intermodulation products: $2f_1 \pm f_2$, $2f_2 \pm f_1$ (caused by V_i^3 term).

These are plotted on the frequency scale along with the original frequencies, as shown in Figure 16.23. From this figure, we can see that all additional frequencies can be filtered out except the intermodulation products $2f_1 - f_2$ and $2f_2 - f_1$, which are very close to f_1 and f_2 and fall within the amplifier bandwidth and cannot be filtered out. Thus, they are capable of signal distortions at the output.

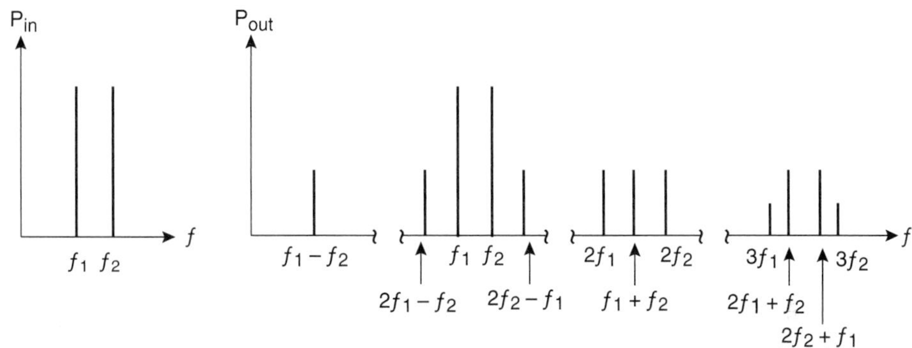

FIGURE 16.23 Input and output power spectrum.

Third-order two-tone intermodulation products $(2f_1 - f_2)$ and $(2f_2 - f_1)$ have special importance because they set the upper limit on the dynamic range or bandwidth of the amplifier.

A measure of the second- or third-order intermodulation distortion is given by two theoretical intercept points, as shown in Figure 16.24. As can be seen from Figure 16.24, the third-order product has a lower intercept point than the second-order product and thus is more significant in distortion analysis.

If the third-order product output power is measured versus the input power, then the third-order intercept (*TOI*) point can be theoretically obtained, as shown in Figure 16.25. The higher the value of power at *TOI* (P_{TOI} or P_{IP}), the larger the dynamic range of the amplifier will be.

The power at the third-order intercept point can be theoretically and experimentally obtained to be approximately given by:

$$P_{IP}(\text{dBm}) \approx P_{1\text{dB}}(\text{dBm}) + 10(\text{dB}) \tag{16.24}$$

Furthermore, the difference between the two curves $(P_{f1} - P_{2f1-f2})$ is a variable quantity and is maximum at $P_{o,mds}$ and zero at P_{IP}. It can be shown that:

$$P_{f_1} - P_{2f_1 - f_2} = \frac{2}{3}(P_{IP} - P_{2f_1 - f_2}) \ (\text{dBm}) \tag{16.25}$$

We now define the "spurious free dynamic range" (DR_f) to be the difference between two powers $P_{f_1} - P_{2f_1 - f_2}$ when the third-order intermodulation product is equal to the minimum detectable signal. That is:

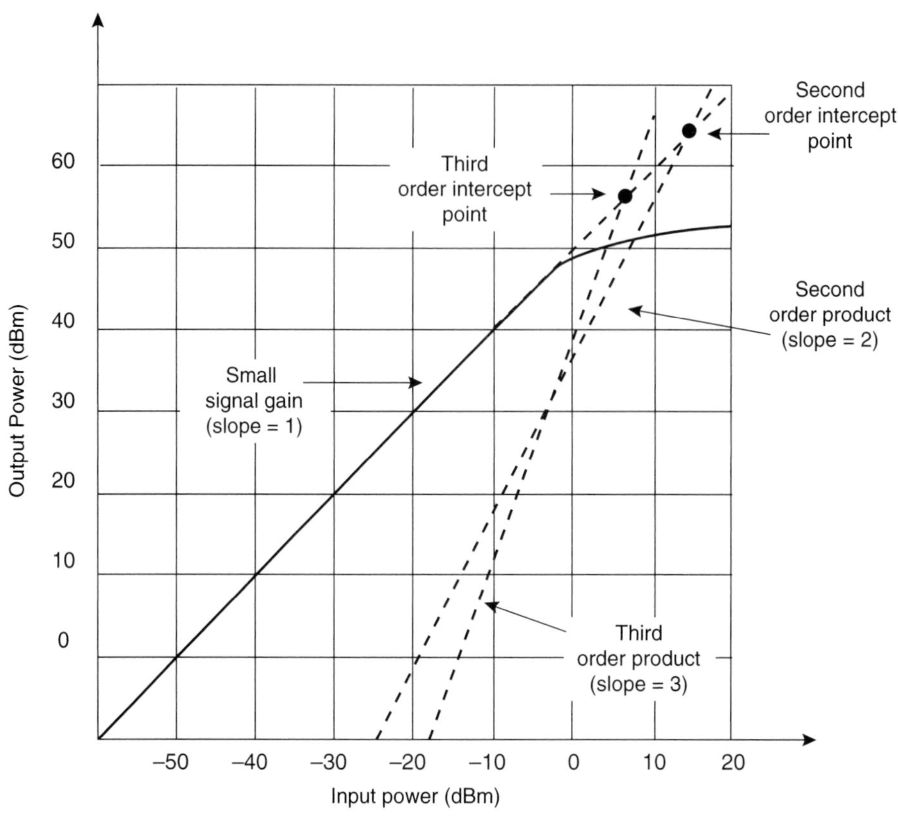

FIGURE 16.24 Second and third order intercept points.

$$DR_f = (P_{f_1} - P_{2f_1 - f_2}) \quad \text{when} \quad (P_{2f_1 - f_2} = P_{o,mds})$$

Thus, we can write DR_f as:

$$DR_f = (P_{f_1} - P_{o,mds}) = \frac{2}{3}(P_{IP} - P_{o,mds}) \quad \text{(dBm)} \tag{16.26}$$

where from Equation 16.10,

$$P_{o,mds}(\text{dBm}) = -174 \text{ dBm} + 10\log_{10}\text{B} + G_A(\text{dB}) + F(\text{dB}) + X(\text{dB})$$

EXAMPLE 16.7

Calculate dynamic range (DR) and spurious free dynamic range (DR_f) for a microwave high-power/broadband amplifier that has a gain of 20 dB, a noise figure of 5 dB, a bandwidth of 250 MHz and can deliver a power of P_{1dB} = 30 dBm (assume X = 3 dB).

Solution:

$$P_{o,mds}(\text{dBm}) = -174 \text{ dBm} + 10\log_{10}B + G_A(\text{dB}) + F(\text{dB}) + X(\text{dB})$$

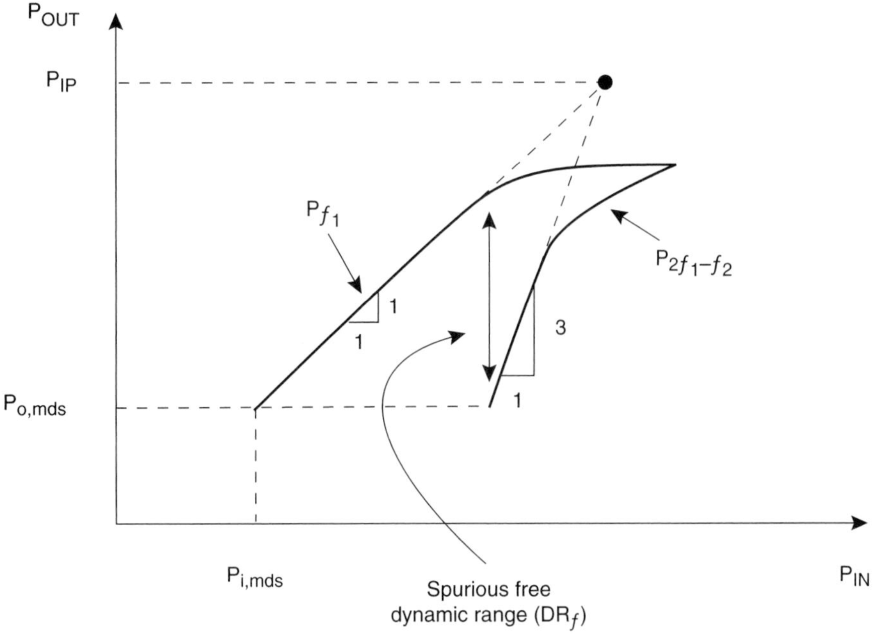

FIGURE 16.25 Third-order intercept point.

$$= -174 + 10\log_{10}(250 \times 10^6) + 5 + 20 + 3 = -62\text{ dBm}$$

$$DR = 30 - (-62) = 92\text{ dB}$$

$$DR_f = (2/3)(30 + 10 + 62) = 68\text{ dB}$$

Two-Tone Measurement Technique. Third-order intercept power (P_{IP}) is a figure of merit for intermodulation product suppression. A high intercept point is a good indicator and signifies a high suppression of undesired intermodulation products. An experimental method for finding P_{IP} is by the use of a technique called "two-tone measurement technique."

In this technique, two signals of close but different frequencies that have equal magnitude are applied to the input of the amplifier, as shown in Figure 16.23. Using a spectrum analyzer, the outputs are examined (see Figure 16.23), and from a simple measurement of the difference in power between the main output (P_{f_1} in dBm) and the third-order intermodulation product ($P_{2f_1-f_2}$ in dBm), we can obtain P_{IP} (in dBm). To find P_{IP}, first let's define:

$$\Delta = P_{f_1} - P_{2f_1-f_2}\ \text{(dB)} \tag{16.27a}$$

Then, substituting for $P_{2f_1-f_2}$ from Equation 16.27a in 16.25, we can write:

$$\Delta = \frac{2}{3}(P_{IP} - P_{2f_1 - f_2}) = \frac{2}{3}[P_{IP} - (P_{f_1} - \Delta)] \quad \text{(dB)} \tag{16.27b}$$

By rearranging terms in Equation 16.27b we obtain $(P_{IP})_3$ for the third-order harmonic as :

$$(P_{IP})_3 = P_{f_1} + \frac{\Delta}{2} \quad \text{(dBm)} \tag{16.28a}$$

Thus, by halving the difference (in dB) between the main output and one of the third-order intermodulation products and adding it to the main output, we can obtain the third-order intercept point (in dBm) as in dicated by Equation 16.28a.

EXAMPLE 16.8

If through measurement we find that $P_{f_1} = 8$ dBm and $\Delta = 40$ dB for the third-order intermodulation product, what is the power at the third-order intercept point?

$(P_{IP})_3$ can easily be calculated to be:

$$(P_{IP})_3 = 8 + 40/2 = 28 \text{ dBm}$$

NOTE: *It can be shown that in general, for the nth order intermodulation product ($n \neq 1$), Equation 16.28a can be generalized as:*

$$(P_{IP})_n = P_{f_1} + \frac{\Delta}{n-1} \quad \text{(dBm)} \tag{16.28b}$$

where Δ is the difference between the fundamental harmonic power and the undesired n^{th} intermodulation product power.

16.6 MULTISTAGE AMPLIFIERS: LARGE-SIGNAL DESIGN

As discussed in the last chapter, most practical transistor amplifiers usually consist of a number of stages connected in cascade forming a multistage amplifier. In a high-power amplifier, each stage should be designed for operation at maximum power such that the maximum power transfer condition is met. In the next several sections, we will present a detailed analysis of a multistage high-power amplifier.

16.6.1 Analysis

Consider a general N-stage amplifier configuration, as shown in Figure 16.26. To have a stable amplifier, the stability of the individual stages as well as overall stability must be checked.

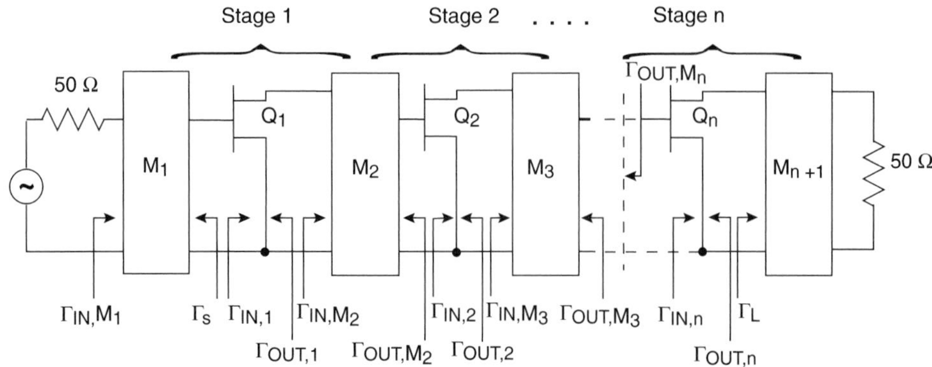

FIGURE 16.26 *N*-stage FET amplifier configuration.

In this type of amplifier, the goal is to produce the overall highest possible power. Thus, each stage must operate at or close to its 1-dB gain compression point under large-signal conditions. This means that using power contours we need to select Γ_{LP} at the point where $P_{OUT} = P_{max}$, (which is at the output port of each transistor) and then use conjugate matched condition for the input port to minimize *VSWR* and create the maximum power transfer condition, i.e.,

$$\Gamma_S = (\Gamma_{IN,1})^* \tag{16.29a}$$

$$\Gamma_{IN,M2} = \Gamma_{LP,1} \tag{16.29b}$$

$$\Gamma_{OUT,M2} = (\Gamma_{IN,2})^* \tag{16.29c}$$

$$\vdots$$

$$\Gamma_{OUT,Mn} = (\Gamma_{IN,n})^* \tag{16.29d}$$

$$\Gamma_L = \Gamma_{LP,n} \tag{16.29e}$$

where $\Gamma_{LP,1}$, $\Gamma_{LP,2}....\Gamma_{LP,n}$ represent points on the power contours where P_{max} occurs for transistors $Q_1, Q_2, ...Q_n$, respectively.

16.6.2 Overall Third-Order Intercept Point Power

High-power amplifiers are designed not only to obtain large amounts of output power but also to have a high third-order intercept point (*TOI*). If each individual stage has a known power value at *TOI* (P_{TOI}), then assuming in-phase addition (of the P_{TOI} of each stage), overall P_{TOI} of the multistage power amplifier is given by:

$$\frac{1}{P_{TOI}} = \frac{1}{P_{TOI,n}} + \frac{1}{G_{Pn}P_{TOI,n-1}} + \ldots + \frac{1}{G_{Pn}G_{Pn-1}\ldots G_{P2}P_{TOI,1}} \tag{16.30}$$

where G_P is the power gain.

If all stages are identical, i.e.,

$$G_{Pk} = G_P,\ k = 1,2,\ldots,n \tag{16.31a}$$

$$P_{TOI,k} = P,\ k = 1,2,\ldots,n \tag{16.31b}$$

Then, Equation 16.22 reduces to:

$$\frac{1}{P_{TOI}} = \frac{1}{P}\left(1 + \frac{1}{G_P} + \frac{1}{G_P^2} + \ldots + \frac{1}{G_P^{n-1}}\right) \tag{16.32}$$

Using Equation 15.33, the geometric series identity, Equation 16.32 becomes:

$$\frac{1}{P_{TOI}} = \frac{1}{P}\left(\frac{1 - 1/G_P^n}{1 - 1/G_P}\right) \tag{16.33}$$

For an infinite chain of amplifier stages (i.e., $n \to \infty$), we can write Equation 16.32 as:

$$P_{TOI} = P(1 - 1/G_P) \tag{16.34}$$

In practice, n may be large (but is not infinite), thus Equation 16.34 gives a best case scenario for the amplifier's overall power at the third-order intercept point (P_{TOI}), which is a power amplifier's figure of merit—very similar to noise measure (M), which is a figure of merit for an LNA as discussed earlier.

16.6.3 Dynamic Range Considerations

As discussed earlier, the dynamic range of an amplifier is bound at the lower end by noise considerations ($P_{o,mds}$) and at the upper end by 1-dB gain compression point ($P_{1\text{dB}}$). Thus, for an n-stage amplifier, we can write:

a) Lower Limit of Dynamic Range.

$$(P_{o,mds})_{cas} = KT(\text{dBm}) + 10\log_{10}B + 10\log_{10}F_{cas} + G_{A,tot}(\text{dB}) + X(\text{dBm}) \tag{16.35}$$

where (see Equation 15.28):

$$G_{A,tot} = G_{A1}G_{A2}\ldots G_{An}, \tag{16.36}$$

$$F_{cas} = F_1 + \frac{F_2 - 1}{G_{A1}} + \frac{F_3 - 1}{G_{A1}G_{A2}} + \ldots + \frac{F_n - 1}{G_{A1}G_{A2}\ldots G_{An-1}} \tag{16.37}$$

Special Case: Identical Amplifiers For identical amplifiers and n very large ($n \to \infty$), Equations 16.36 and 16.37 simplify as:

$$G_{Ak} = G_A, \quad k = 1,2,\ldots,n \tag{16.38a}$$

$$F_k = F, \quad k = 1,2,\ldots,n \tag{16.38b}$$

$$G_{A,tot} = (G_A)^n \tag{16.39}$$

$$F_{cas} = 1 + \frac{F - 1}{1 - 1/G_A} = 1 + M \tag{16.40}$$

$$(P_{o,mds})_{cas} = KT(\text{dBm}) + 10\log_{10}B + 10\log_{10}(1 + M) + nG_A(\text{dB}) + X(\text{dB})\ (\text{dBm}) \tag{16.41}$$

where (see Equation 15.36):

$$M = \frac{F-1}{1-1/G_A} \tag{16.42}$$

NOTE: *$(P_{o,mds})_1$ for the first stage is given by:*

$$(P_{o,mds})_1(\text{dBm}) = KT(\text{dBm}) + 10\log_{10}B + 10\log_{10}F + G_A(\text{dB}) + X\,(\text{dB}) \tag{16.43}$$

Combining Equations 16.41 and 16.43 we can see that:

$$(P_{o,mds})_{cas}\,(\text{dBm}) = (P_{o,mds})_1(\text{dBm}) + \Delta P_{o,n} \tag{16.44a}$$

where

$$\Delta P_{o,n} = 10\log_{10}\left(\frac{1+M}{F}\right) + (n-1)G_A \quad (\text{dB}) \tag{16.44b}$$

Because $\Delta P_{o,n}$ is always positive, thus we can write:

$$(P_{o,mds})_{cas} \geq (P_{o,mds})_1 \tag{16.45}$$

Equation 16.45 shows an important consideration where the output minimum detectable signal for the whole cascade $(P_{o,mds})_{cas}$ is determined by and depends greatly on the minimum detectable signal of the first stage of the cascade, $(P_{o,mds})_1$. Thus, it is important to have the first stage operate at the lowest possible output noise level.

b) Upper Limit of Dynamic Range. At the upper limit, the total output power at 1-dB gain compression point ($P_{1\text{dB},cas}$) can be shown to be of similar form to Equation 16.30:

$$\frac{1}{P_{1\text{dB},cas}} = \frac{1}{P_{1\text{dB},n}} + \frac{1}{G_{Pn}P_{1\text{dB},n-1}} + \dots + \frac{1}{G_{Pn}G_{Pn-1}\dots G_{P2}P_{1\text{dB},1}} \tag{16.46}$$

If all stages are identical, i.e.,

$$G_{Pk} = G_P,\ k = 1,2,\dots,n \tag{16.47a}$$

$$P_{1\text{dB},k} = P_{1\text{dB}},\ k = 1,2,\dots,n \tag{16.47b}$$

then Equation 16.46 reduces to:

$$\frac{1}{P_{1\text{dB},cas}} = \frac{1}{P_{1\text{dB}}}\left(1 + \frac{1}{G_P} + \frac{1}{G_P^2} + \dots + \frac{1}{G_P^{n-1}}\right) \tag{16.48}$$

Using Equation 15.33, the geometric series identity, Equation 16.48 becomes:

$$\frac{1}{P_{1\text{dB},cas}} = \frac{1}{P_{1\text{dB}}}\left(\frac{1-1/G_P^n}{1-1/G_P}\right) \tag{16.49}$$

For an infinite and idenfical chain of amplifier stages (i.e., $n\to\infty$) and very similar to Equation 16.34, we can write Equation 16.49 as:

$$P_{1\text{dB},cas} = P_{1\text{dB}}(1 - 1/G_P) \tag{16.50}$$

NOTE: *Because $G_P \geq 1$ and $P_{1dB,n} = P_{1dB}$, we can see from Equation 16.50 that at all times:*

$$P_{1\text{dB},cas} \leq P_{1\text{dB},n} \tag{16.51}$$

That is, the signal at the output of the cascade is below or at the 1 dB gain compression point of the last-stage amplifier.

As can be seen from Equation 16.51, the power output at 1 dB gain compression point for the whole cascade ($P_{1\text{dB},cas}$) is limited by the last stage 1 dB gain compression point power capability ($P_{1\text{dB},n}$). Thus, we can conclude the following:

CONCLUSION: *It is crucial to have the last stage of the cascade be designed such that it has the highest power handling capability.*

16.6.4 Wide Dynamic Range Multi-stage Amplifier Design

For an n-stage amplifier, the actual dynamic range (DR) is given by:

$$DR = P_{1\text{dB},cas}(\text{dBm}) - (P_{o,mds})_{cas}(\text{dBm}) \tag{16.52}$$

As noted from the previous section, however, for a wide dynamic range design we need to have the following considerations firmly in place:

a. From Equation 16.45 it can be concluded that the first stage sets the lower limit. Therefore, ideally we would like to have:

$$(P_{o,mds})_{cas} \approx (P_{o,mds})_1 \tag{16.53}$$

b. From Equation 16.51, we can observe that the upper limit is determined by the 1 dB gain compression point of the last stage and ideally, we would like to have:

$$P_{1\text{dB},cas} \approx P_{1\text{dB},n} \tag{16.54}$$

Thus, the maximum dynamic range or the **best estimate** of dynamic range (DR_{max}) that we can hope for, can be written as:

$$DR_{max} = P_{1\text{dB},n}(\text{dBm}) - (P_{o,mds})_1(\text{dBm}) \tag{16.55a}$$

Thus, for n identical amplifiers ($n \to \infty$):

$$DR_{max} - DR = \Delta P_{o,n} + P_{1\text{dB}}/G_P \tag{16.55b}$$

POINT OF CAUTION: *From Equations 16.39 and 16.44 we can observe that increasing the number of stages (n) and/or available gain (G_A) of each stage will increase the overall gain but will reduce the effective dynamic range by increasing $(P_{o,mds})_{cas}$. Thus, there is a trade-off between the overall gain and the dynamic range of a multistage amplifier.*

EXAMPLE 16.9

Design a 10 Watt (i.e., 40 dBm) power amplifier at 1GHz as shown in Figure 16.27. The input signal is given to be 1mW (i.e., 0dBm). The following table sets forth the amplifier choices that are available in our design:

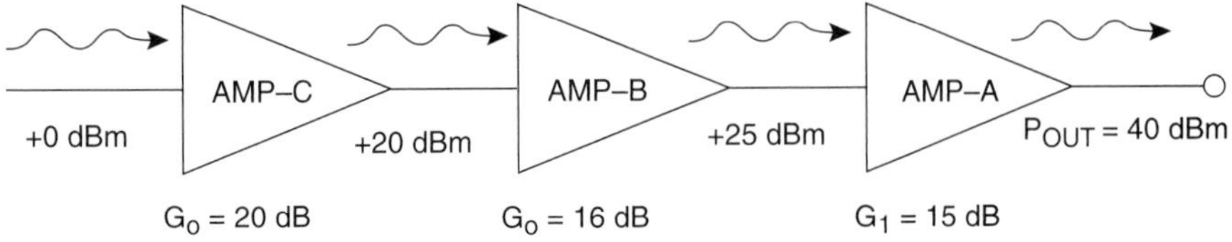

FIGURE 16.27 Circuit for Example 16.9.

Amplifier	G_o(dB)	P_{1dB}(dBm)	G_{1dB}(dB)	F(dB)
AMP-A	16	40	15	3
AMP-B	16	38	15	2
AMP-C	20	32	19	2

1. *In what sequence should the three amplifiers be cascaded in order to get 10 W power output in a linear range?*
2. *Using results from part (a) what is the best estimate for the dynamic range at 1 GHz for 100 MHz bandwidth at room temperature ($T = 290°K$). Assume $X = 3$ dB?*

Solution:

1. The correct cascade sequence of amplifiers is established as follows:
   a. Because the output is 40 dBm, we need AMP-A at the output stage because it would be able to provide the desired amount of power ($P_{1dB,A}$ = 40 dBm). However, because AMP-A is operating at 1 dB gain compression point, its linear gain ($G_{o,A}$) is reduced by 1 dB. Thus, a gain of $G_{1dB,A}$ = 15 dB should be used in the calculations instead.
   b. AMP-C has the lowest 1 dB gain compression point power and the highest gain (the high gain at the input of the cascade will reduce the second-stage noise contribution). Therefore, it must be placed as the first stage.
   c. AMP-B has a relatively higher value for P_{1dB} and a low noise figure value and, thus, will be placed in the middle, as shown in Figure 16.27, where the power levels as well as the gain values at each stage are also shown.
2. The best estimate for the dynamic range is given by:

$$DR_{max} = P_{1dB,A} - (P_{o,mds})_C$$

$$(P_{o,mds})_C = -174 + 10\log_{10}(10^8) + 3 + 2 + 20 = -69 \text{ dBm}$$

Thus,

$$DR_{max} = 40 - (-69) = 109 \text{ dB}$$

EXERCISE 16.1

Calculate the actual dynamic range for the three-stage amplifier of Example 16.9 and compare it with the "best estimate" DR_{max} calculated in this example.

LIST OF SYMBOLS/ABBREVIATIONS

A symbol will not be repeated again once it has been identified and defined in an earlier chapter, as long as its definition remains unchanged.

DR	Dynamic range (expressed in dB)
DR_{max}	Best estimate dynamic range
f_1	First harmonic
f_2	Second harmonic
$G_{1\text{dB}}$	1 dB gain compression point
$G_o(\text{dB})$	Small-signal linear power gain
HPA	High-power amplifier
$P_{1\text{dB}}$	Output power at the 1 dB gain compression point
P_{f1}	Power curve for f_1
P_{2f1-f2}	Power curve for $2f_1 - f_2$
$P_{IN,1\text{dB}}$	Input power at the 1 dB gain compression point
$P_{i,mds}$	Minimum detectable signal input power
P_{inc}	Incident power
P_{IP}	Power at the third-order intercept point
P_{OUT}	Output power
P_{NI}	Input thermal noise floor power
P_{No}	Output thermal noise floor power
$P_{o,mds}$	Output power level for minimum detectable signal
P_{TOI}	Output power at the third-order intercept point (same as P_{IP})
TOI	Third-order intercept point
Γ_{LP}	Load reflection coefficient for large-signal condition
Γ_{SP}	Source reflection coefficient for large-signal condition

PROBLEMS

16.1 **a.** Show that the stability factor (K) for a balanced amplifier using a 3 dB Lange coupler is given by:

$$K = (1 + |S_{21}S_{12}|^2)/2\,|S_{21}S_{12}|$$

Furthermore, show that $K \geq 1$ for all values and the minimum occurs when

$$|S_{21}S_{12}| = 1$$

b. The S-parameters of a transistor at 3 GHz are given by:

$$S = \begin{bmatrix} -0.7 - j0.35 & -0.009 - j0.015 \\ -10.9 + j7.9 & -0.31 - j0.46 \end{bmatrix}$$

Find and compare the stability factors of the transistor by itself and when used in a balanced amplifier configuration.

16.2 The specifications for two power GaAs FETs at 4 GHz are as follows:

Device	P_{1dB}(dBm)	G_{1dB}(dB)	G_P(dB)
FET#1	25	6	7
FET#2	20	8	9

Design the amplifier configuration (as shown in Figure P16.2) by using an appropriate device such that at 1-dB gain compression point the output power will be 27.7 dBm for an input power of 6.3 dBm. Assume that each two-way combiner/divider has an insertion loss of 0.3 dB.

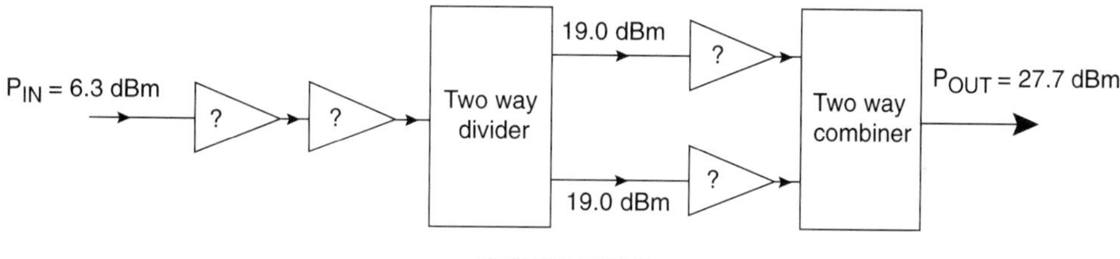

FIGURE P16.2

16.3 Design a power amplifier for maximum power using a BJT having the following S-parameters and power characteristics at 4 GHz:

$$S = \begin{bmatrix} 0.32\angle{-145°} & 0.08\angle{-98°} \\ 1.38\angle{-113°} & 0.8\angle{-177°} \end{bmatrix}$$

$$\Gamma_{LP} = 0.1\angle 0°$$

$$P_{1dB} = 27.5 \text{ dBm}$$

$$G_{1dB} = 7 \text{ dB}$$

16.4 A 4 GHz microwave amplifier consists of four stages as shown in Figure P16.4. Determine the input power (P_{IN}) and output power (P_{OUT}) in dBm.

16.5 Calculate the input minimum detectable signal for a receiver with a bandwidth of 10 GHz and a noise figure of 10 dB operating at $T = 100°$ K.

16.6 The receiver shown in Figure P16.6 is designed for a signal frequency range of 10 to 11 GHz. Calculate:

a. The overall gain or loss of the system in dB

b. The overall noise figure in dB

c. The input and output minimum detectable signals for the receiver at $T = 290°$ K

16.7 For the receiver system shown in Figure P16.7, calculate:

a. The total system gain

b. The total noise figure (in dB)

c. The minimum detectable signal

d. The output power (P_{OUT}) for an input power of 0.1 mW

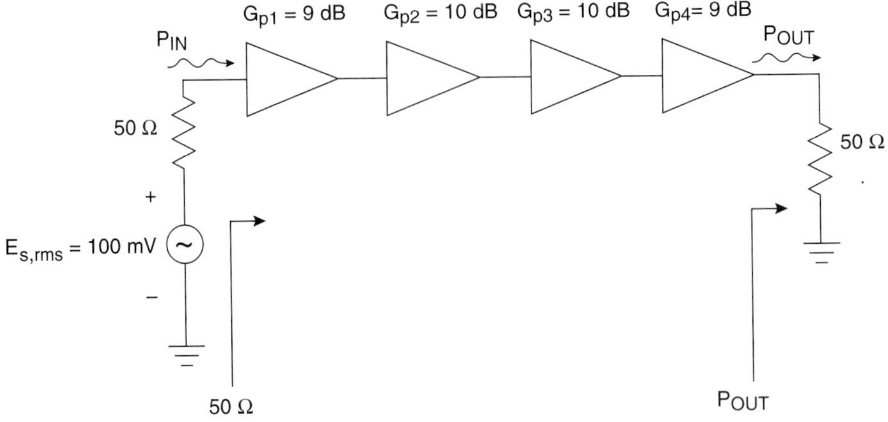

FIGURE P16.4

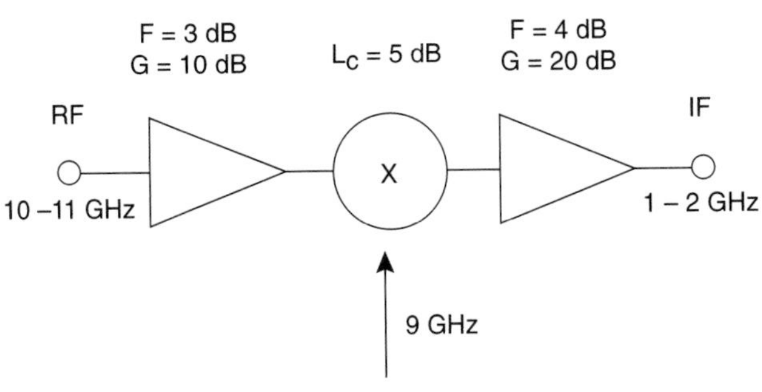

FIGURE P16.6

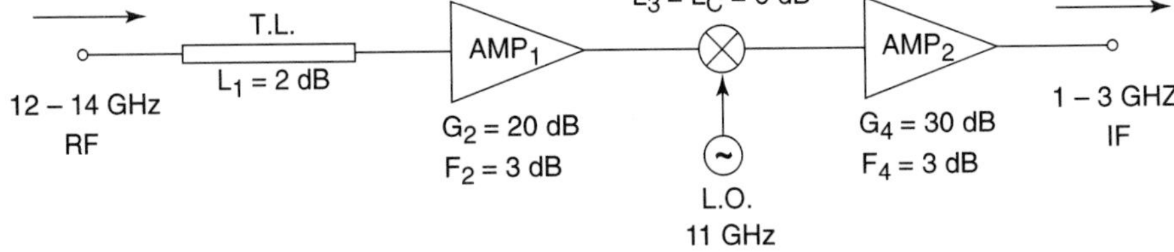

FIGURE P16.7

16.8 A three-stage low noise amplifier has a noise figure of 3 dB and a total gain of 30 dB over a 1 GHz bandwidth at room temperature. Calculate the dynamic range and the spurious-free dynamic range if the amplifier has a 1 dB compression point power ($P_{1\text{dB}}$) of 15 dBm.

16.9 Design the input and output matching networks of a power amplifier using a microwave power transistor with the following large-signal parameters at 3 GHz (in a 50 Ω system):

$$S = \begin{bmatrix} 0.67\angle 140° & 0.04\angle -15° \\ 2.32\angle 30° & 0.46\angle -125° \end{bmatrix}$$

$$P_{1\text{dB}} = 30 \text{ dBm}$$

$$G_{1\text{dB}} = 10 \text{ dB}$$

REFERENCES

[16.1] Anderson, R. W. *S-Parameter Techniques for Faster, More Accurate Network Design.* Hewlett-Packard Application Note 95-1, February 1967.

[16.2] Bahl, I. and P. Bhartia. *Microwave Solid State Circuit Design*. New York: Wiley Interscience, 1988.

[16.3] Carson, R. S. *High Frequency Amplifiers.* New York: Wiley Interscience, 1975.

[16.4] Fukui, H. "Available Power Gain, Noise Figure, and Noise Measure of Two-Ports and their Graphical Representation." *IEEE Transactions on Circuit Theory,* June 1966.

[16.5] Gonzalez, G. *Microwave Transistor Amplifiers, Analysis and Design,* 2nd ed. Upper Saddle River: Prentice Hall, 1997.

[16.6] Ha, T. T. *Solid State Microwave Amplifier Design*. New York: John Wiley & Sons, 1987.

[16.7] Kurokawa, K. Design Theory of Balanced Transistor Amplifiers. *Bell System Technical Journal,* Vol. 44, October 1965, pp. 1675-98.

[16.8] Liao, S. Y. *Microwave Circuit Analysis and Amplifier Design*. Upper Saddle River: Prentice Hall, 1987.

[16.9] Presser, A. "Interdigitated Microstrip Coupler Design." *IEEE Transactions on Microwave Theory and Techniques,* October 1978.

[16.10] Pozar, D. M. *Microwave Engineering,* 2nd ed. New York: John Wiley & Sons, 1998.

[16.11] Russel, K. J. Microwave Power Combining Techniques. *IEEE Transactions on Microwave Theory and Technique,* Vol. MTT-27, May 1979, pp. 472-78.

[16.12] Vendelin, George D. *Design of Amplifiers and Oscillators by the S-Parameter Method.* New York: John Wiley & Sons, 1982.

[16.13] Vendelin, George D., Anthony M. Pavio, and Ulrich L. Rhode. *Microwave Circuit Design, Using Linear and Non-Linear Techniques.* New York: John Wiley & Sons, 1990.

[16.14] Vizmuller, P. *RF Design Guide, System, Circuits, and Equations.* Norwood: Artech House, 1995.

CHAPTER 17

RF/Microwave Oscillator Design

17.1 INTRODUCTION

In this chapter, we will study the design of transistor oscillators at microwave frequencies. An RF/microwave oscillator is designed to convert DC power to microwave power and thus forms one of the most basic and yet essential parts of any microwave system.

At the heart of any modern oscillator, we find an active solid-state device (e.g., a tunnel diode, a transistor, etc.) at work straddled by two passive networks to produce the desired microwave signal. Because the device is producing RF power with no input microwave signal, it must have a negative resistance. Examples of two-terminal negative resistance devices include tunnel diodes, IMPATT diodes, and GUNN diodes; three-terminal devices include bipolar junction transistors (BJTs) and field effect transistors (FETs), particularly MESFETs at microwave frequencies.

Use of two-terminal devices leads to the design of one-port negative resistance oscillators. On the other hand, use of three-terminal devices usually provides two-port oscillators where the device is operated in an unstable region in contrast to an amplifier circuit where a stable operating point is required. At this point, let's define an important term:

DEFINITION-NEGATIVE RESISTANCE (NR) DEVICE: *A device in which the voltage and current are 180° out of phase with respect to each other.*

This definition indicates that an increase in voltage across an NR device will produce a decrease in current and vice versa, leading to the product of the current and voltage being negative, which corresponds to the concept of power generation (see Figure 17.1).

Therefore, the concept of a negative resistance (NR) device is directly related to the concept of power generation, which is an essential part of any workable oscillator.

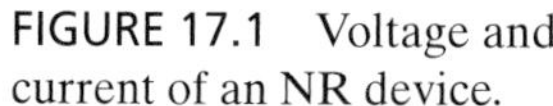

FIGURE 17.1 Voltage and current of an NR device.

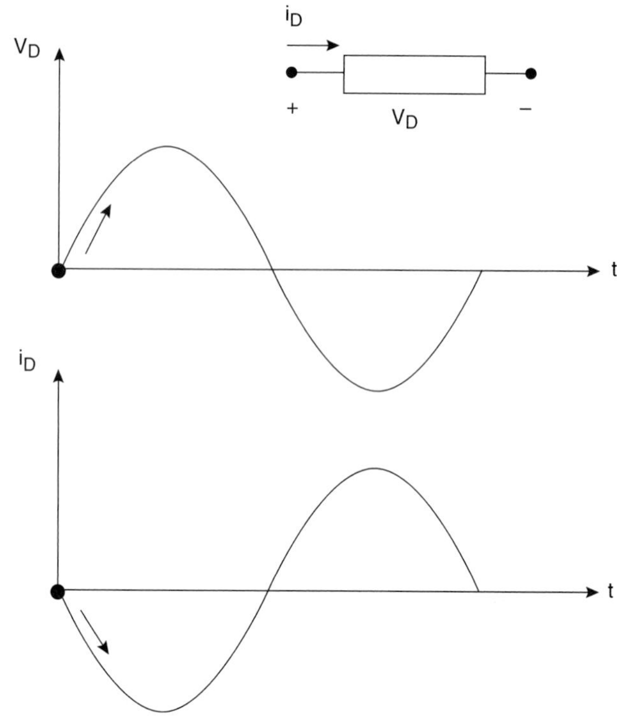

17.2 OSCILLATOR VERSUS AMPLIFIER DESIGN

The design of a microwave oscillator is very similar to the design of a microwave amplifier, in that the biasing circuits and the device selection procedures are quite similar.

The main difference between the two is the fact that an amplifier requires a microwave signal at its input while an oscillator needs none. This major difference between the two leads to the following considerations:

a. An amplifier can be designed using a normal Smith chart; however, because power is generated in an oscillator, the reflection coefficients are greater than unity (i.e., $\Gamma_{IN} > 1$, $\Gamma_{OUT} > 1$), thus a compressed Smith chart is needed in the oscillator design process.

b. Because $\Gamma_{IN} > 1$ and $\Gamma_{OUT} > 1$, the input and output matching network design requirements for an oscillator will be different from those of an amplifier (where $\Gamma_{IN} < 1$ and $\Gamma_{OUT} < 1$).

c. The input and output matching networks used earlier for an amplifier design are now called the generator-tuning network and the terminating (or load matching) network, respectively, as shown in Figure 17.2.

d. The generator-tuning network determines the oscillation frequency while the terminating network provides the proper loading function such that oscillation does take place (even though the actual location of the load can be at either end).

e. At startup, the oscillation is triggered by random noise spikes or transients after which it will reach a stable oscillation state very quickly (see Figure 17.3).

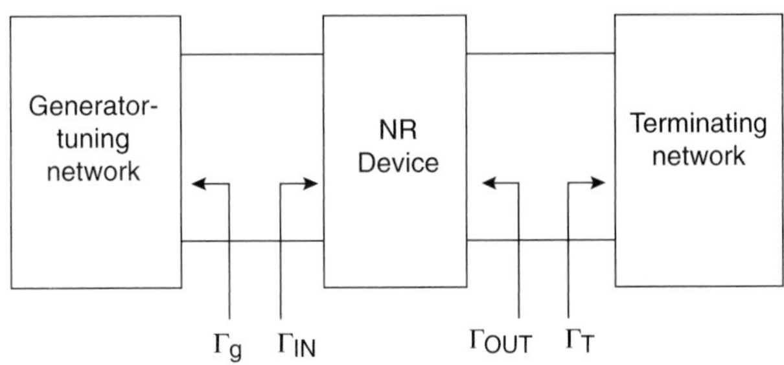

FIGURE 17.2 A two-port oscillator block diagram.

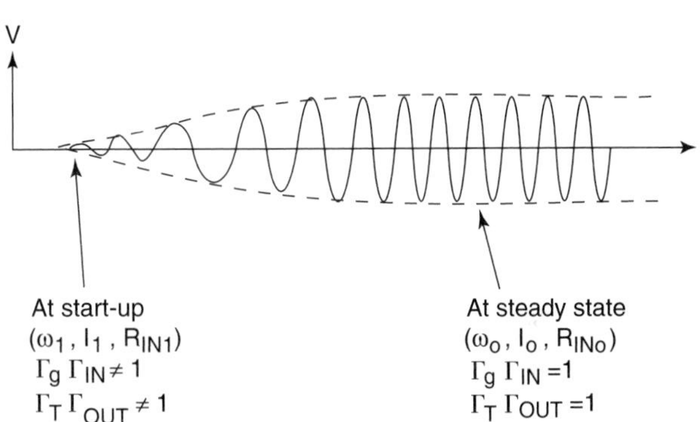

FIGURE 17.3 Oscillation buildup from startup to steady state.

This process requires the active device to be nonlinear unlike a class A type amplifier, which needs linear device operation. Because of this nonlinear device operation, the complete analysis of oscillator operation becomes relatively complicated.

17.3 OSCILLATION CONDITIONS

In general, an oscillator may be realized as a one-port or a two-port. As it turns out, a one-port is a special case of a two-port oscillator network. Thus, we will first consider and analyze oscillation conditions for a general two-port oscillator and then will introduce a one-port as a special case.

17.3.1 Two-Port NR Oscillators

Consider a general two-port oscillator circuit, as shown in Figure 17.2. Three conditions need to be satisfied at "steady state" for an oscillation to occur:

Condition #1: Unstable device

$$K < 1 \tag{17.1}$$

Condition #2: Oscillating input port

$$\Gamma_{IN}\Gamma_g = 1 \tag{17.2}$$

Condition #3: Oscillating output port

$$\Gamma_{OUT}\Gamma_T = 1 \tag{17.3}$$

Where K, Γ_{IN}, and Γ_{OUT} are given by (see Chapter 12, *Stability Considerations in Active Networks*, for details):

$$K = \frac{1-|S_{11}|^2-|S_{22}|^2+|\Delta|^2}{2|S_{12}S_{21}|}, \tag{17.4}$$

$$\Delta = S_{11}S_{22}-S_{12}S_{21}, \tag{17.5}$$

$$|\Gamma_{IN}| = \left|S_{11}+\frac{S_{12}S_{21}\Gamma_T}{1-S_{22}\Gamma_T}\right| = 1, \tag{17.6}$$

and

$$|\Gamma_{OUT}| = \left|S_{22}+\frac{S_{12}S_{21}\Gamma_g}{1-S_{11}\Gamma_g}\right| = 1 \tag{17.7}$$

Condition #1 indicates that the NR device itself (operating in the unstable region) is in the oscillation mode.

Conditions #2 and #3 are dependent on one another, that is to say, condition #2 is satisfied if and only if condition #3 is satisfied and vice versa. The dependency of conditions #2 and #3 on each other indicates that if one port is oscillating then the other port must also be oscillating.

Furthermore, because the generator-turning network and the load-matching network are passive, therefore, we have:

$$|\Gamma_g| < 1 \tag{17.8}$$

$$|\Gamma_T| < 1 \tag{17.9}$$

From Equations 17.8 and 17.9 we can observe that in order to have conditions #2 and #3 met, as stated in Equations 17.2 and 17.3, we need to have:

$$|\Gamma_{IN}| > 1 \tag{17.10}$$

$$|\Gamma_{OUT}| > 1 \tag{17.11}$$

The necessary conditions as given by Equations 17.10 and 17.11 confirm condition #1 (which requires the device to be unstable) and indicate that an oscillator design requires the use of a compressed Smith chart.

17.3.2 A Special Case: One-Port NR Oscillator

An appropriate load termination at the output port can change a two-port transistor network into a one-port NR oscillator network (series or shunt configurations), as shown in Figure 17.4. Additionally, a one-port oscillator can also be obtained through the use of microwave diodes (e.g., IMPATT diode, GUNN diode, etc.). The following theoretical analysis applies to a one-port oscillator regardless of whether it was obtained through proper termination (of the output port) of a two-port transistor oscillator or obtained through the use of a microwave diode.

One-Port OscillatorAnalysis. The steady-state condition of oscillation for a two-port oscillator simplifies into the following condition for a one-port oscillator:

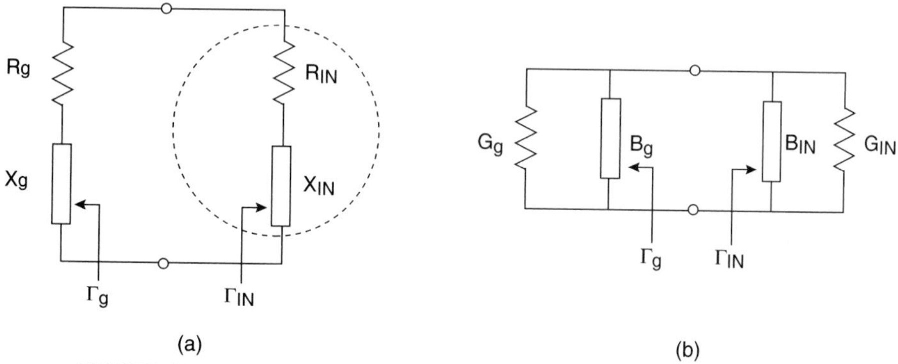

FIGURE 17.4 A one-port oscillator: (a) series, (b) shunt configuration.

$$Z_g + Z_{IN} = 0 \quad \text{(Series configuration)} \tag{17.12a}$$

or

$$Y_g + Y_{IN} = 0 \quad \text{(Shunt configuration)} \tag{17.12b}$$

Equating the real and imaginary parts in Equation 17.12, the oscillation condition for the series or shunt configuration becomes:

- Series configuration (steady state)

$$R_g + R_{IN} = 0 \tag{17.13a}$$

$$X_g + X_{IN} = 0 \tag{17.13b}$$

- Shunt configuration (steady state)

$$G_g + G_{IN} = 0 \tag{17.14a}$$

$$B_g + B_{IN} = 0 \tag{17.14b}$$

NOTE: *Equations 17.12 is stating the same mathematical fact as Equation 17.2, which can easily be proven by noting that:*

$$\Gamma_{IN} = \frac{Z_{IN} - Z_o}{Z_{IN} + Z_o} \tag{17.15a}$$

$$\Gamma_g = \frac{Z_g - Z_o}{Z_g + Z_o} \tag{17.15b}$$

EXERCISE 17.1

Derive Equations 17.12 from Equation 17.2.

The following example provides further insight into the design of one-port oscillators.

EXAMPLE 17.1

Design a one-port oscillator for maximum power output if the device has a negative resistance varying linearly with the current amplitude (I_o) at its terminal, as shown in Figure 17.5. Assume the device's reactance is negligible.

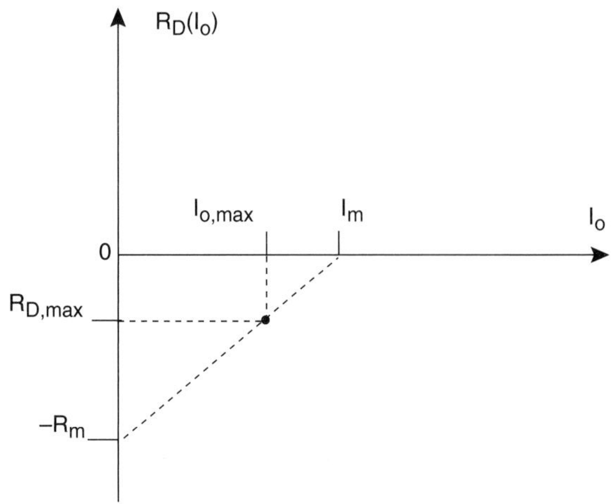

FIGURE 17.5 Negative resistance variation with current.

Solution:

Because the device has no reactance, its impedance is purely resistive and has a linear equation given by (see Figure 17.5):

$$R_D(I_o) = -R_m(1 - I_o/I_m)$$

The one-port oscillator schematic is shown in Figure 17.6. From this figure, we can see that the voltage phasor across the device is given by:

$$V = R_D I = R_D\, I_o \angle\theta$$

where $I = I_o\angle\theta$ is the current through the device.

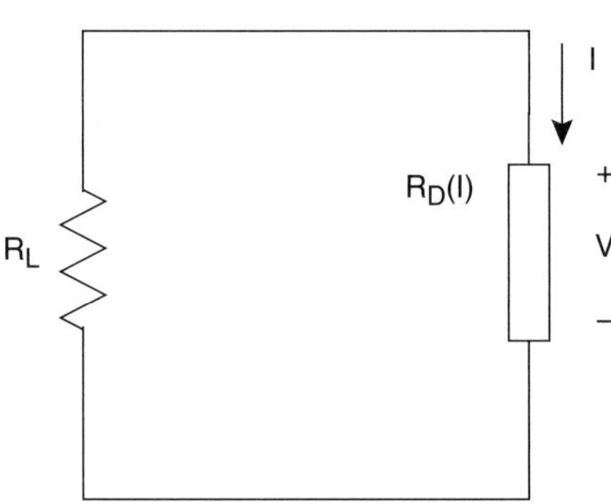

FIGURE 17.6 Circuit for Example 17.1.

The power generated by the device (for $I_o < I_m$) is given by:

$$P_D = \frac{1}{2}Re[VI^*] = \frac{1}{2}R_D I_o^2 = -\frac{R_m I_o^2}{2}\left(1-\frac{I_o}{I_m}\right)$$

The maximum power output by the device occurs when

$$\frac{dP_D}{dI_o} = 0$$

Upon differentiation of the equation for P_D, we obtain the following equation for the current ($I_{o,max}$) under maximum power condition:

$$R_m\left(2I_{o,max} - \frac{3I_{o,max}^2}{I_m}\right) = 0$$

Solving the above equation for $I_{o,max}$, we obtain:

$$I_{o,max} = \frac{2}{3}I_m$$

Using this value of $I_{o,max}$ (obtained under maximum power condition), we find $R_{D,max}$ (shown as a point in Figure 17.4) and P_{max} as follows:

$$R_{D,max} = -\frac{1}{3}R_m$$

$$P_{max} = \frac{2}{27}R_m I_m^2$$

Thus, at steady state, from Equation 17.13a the load resistance (R_L) providing the maximum power, is given by:

$$R_L = -R_{D,max} = \frac{1}{3}R_m = \frac{1}{3}R_D(0)$$

This value of R_L results in a net negative resistance (R_{net}) at startup and a successful oscillation buildup because:

$$I_o = 0 \quad \text{(at start-up)}$$

$$R_D(0) = -R_m$$

$$R_{net} = R_D(0) + R_L = -R_m + \frac{R_m}{3} = -\frac{2}{3}R_m \tag{17.16}$$

This value of load resistance ($R_L = R_m/3$) has generally shown to be a good and reliable rule of thumb and thus is used in the practical design of one-port oscillators, as will be seen in the rest of this chapter.

17.3.3 Condition of Stable Oscillation

The process of oscillation depends on the nonlinear behavior of Z_{IN} because initially it is necessary for the overall circuit to be unstable at a certain frequency (ω). The one-port network is unstable at ω if the net resistance (R_{net}) of the network is negative, that is,

$$R_{net} = R_{IN}(I,\omega) + R_g < 0 \tag{17.17}$$

Any transient excitation due to noise in the circuit will initiate an oscillation at the frequency (ω) for which the net reactance of the network (X_{net}) is equal to zero, that is:

$$X_{net} = X_{IN}(I,\omega) + X_g = 0 \tag{17.18}$$

A growing sinusoidal current (I) will flow through the oscillator and the oscillation will continue to build up as long as Equation 17.17 is valid (i.e., the net resistance remains negative), as shown in Figure 17.3.

The current amplitude eventually will reach a steady state value (I_o) at a final frequency of oscillation (ω_o), which occurs when the loop impedance is zero, as given by Equations 17.12:

$$R_{IN}(I_o,\omega_o) + R_g = 0 \tag{17.19a}$$

$$X_{IN}(I_o,\omega_o) + X_g(\omega_o) = 0 \tag{17.19b}$$

NOTE: *The final frequency of oscillation (ω_o), obtained from Equations 17.19, generally is different from the startup frequency (ω) and is* ***not stable*** *because R_{IN} and X_{IN} are current dependent, which means that:*

$$R_{IN}(I_o,\omega_o) \neq R_{IN}(I'_o,\omega_o)$$

$$X_{IN}(I_o,\omega_o) \neq X_{IN}(I'_o,\omega_o) \tag{17.20}$$

where I'_o is an arbitrary current. This fact further connotes that the conditions of oscillation, as set forth by Equation 17.19, are not sufficient to guarantee a stable state of operation. Therefore, it is necessary to find another condition to guarantee a stable oscillation.

For a stable oscillation, any perturbation or deviation in current (δI) or frequency ($\delta\omega$) will be damped out, allowing the oscillator to return to its original state. Following this line of reasoning, Kurokawa has shown that for a stable oscillation, we need:

$$\left.\frac{\partial R_{IN}(I,\omega)}{\partial I}\right|_{I=I_o} \left.\frac{dX_g(\omega)}{d\omega}\right|_{\omega=\omega_o} - \left.\frac{\partial X_{IN}(I,\omega)}{\partial I}\right|_{I=I_o} \left.\frac{dR_g(\omega)}{d\omega}\right|_{\omega=\omega_o} > 0 \tag{17.21}$$

In practice, the load resistance is a constant value, that is:

$$\frac{dR_g(\omega)}{d\omega} = 0 \tag{17.22}$$

Thus, Equation 17.21 simplifies as follows:

$$\left.\frac{\partial R_{IN}(I,\omega)}{\partial I}\right|_{I=I_o} \left.\frac{dX_g(\omega)}{d\omega}\right|_{\omega=\omega_o} > 0 \tag{17.23}$$

Thus, we can see that the necessary and sufficient condition for stable oscillation, as given by Equation 17.23, can always be satisfied if we choose:

$$\left.\frac{dX_g(\omega)}{d\omega}\right|_{\omega=\omega_o} >> 0 \tag{17.24}$$

Equation 17.24 can be satisfied simply by using a high-Q tuning circuit (such as a cavity or a dielectric resonator circuit) for the generator-tuning network, which will result in maximum oscillator stability.

A complete design of an oscillator, however, requires more advanced considerations such as phase noise, maximum power output, and selection of the Q-point for stable operation, which will be relegated to more advanced texts.

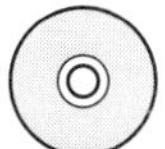

EXAMPLE 17.2

Design a one-port oscillator using a tunnel diode with $\Gamma_{IN} = 1.25\angle 40°$ at 8 GHz in a 50 Ω system (see Figure 17.7).

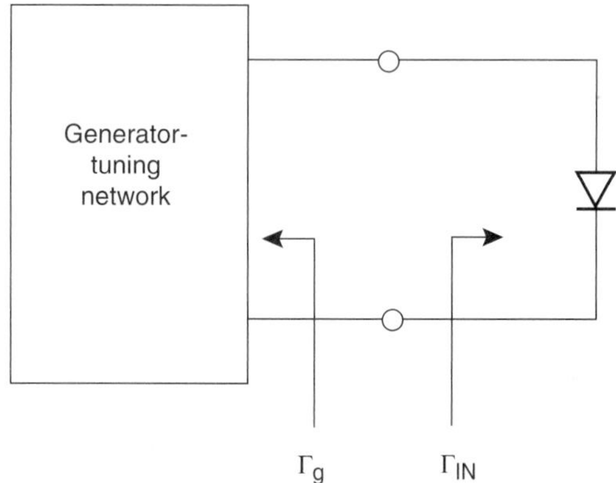

FIGURE 17.7 Circuit for Example 17.2.

Solution:

The following design steps are in order:

1. Plot $1/\Gamma^*_{IN}$ on the Smith chart and read off Z_{IN} as (see Chapter 10, Section 10.2.3, *Application #3*):

$$1/\Gamma^*_{IN} = 0.8\angle 40° \Rightarrow Z_{IN} = -43 + j124\ \Omega$$

2. Using the results from Example 17.1, we can write:

$$R_g = -R_{IN}/3 = 43/3 = 14.3\ \Omega$$

$$X_g = -X_{IN} = -124\ \Omega$$

The value of R_g as given here will ensure a successful oscillation buildup right from the start!

Thus, $Z_g = 14.3 - j124\ \Omega$

Several circuit designs could be implemented. One possible design would be using a shunt configuration, which requires us to work in admittances. Thus, we plot $(Z_g)_N = 0.29 - j2.48$ on the Smith chart (see Figure 17.8a); the admittance is read off to be:

$$(Y_g)_N = g_g + jb_g = (0.05 + j0.4)$$

$$Y_g = G_g + jB_g = (0.05 + j0.4) \times 0.02 = 0.001 + j0.008\ \text{S}$$

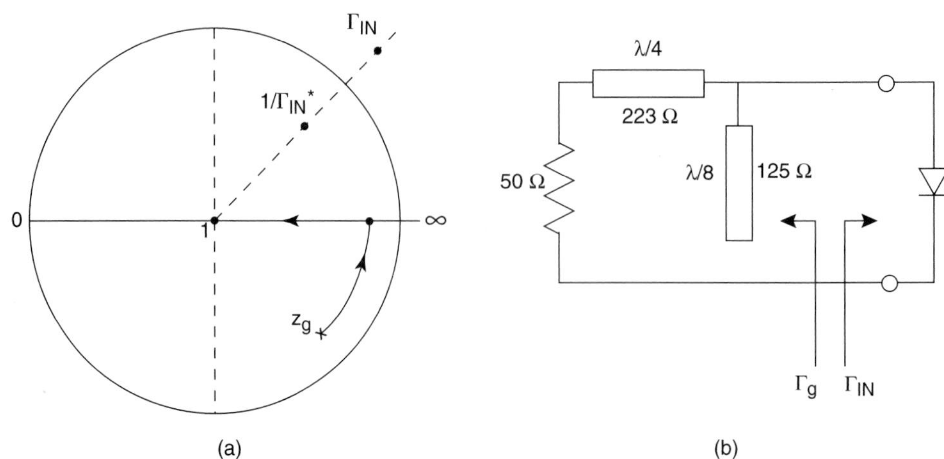

FIGURE 17.8 Solution for Example 17.2: (a) Smith chart plot, (b) final circuit schematic.

3. To design the matching network and thus realize the generator-tuning network (Z_g), we employ the technique from Chapter 11, *Design of Matching Networks* (see Section 11.6) as follows:

a. A series λ/4 transformer with characteristic impedance of:

$$(Z_o)_{\lambda/4} = (50/0.001)^{1/2} = 223\ \Omega$$

will transform a 50 Ω generator load to 14.3 Ω.

b. We need to create $B_g = 0.008$ S using an open shunt stub ($\ell = \lambda/8$) with a characteristic impedance of:

$$(Z_o)_{stub} = 1/B_g = 1/0.008 = 125\ \Omega$$

The final circuit is realized and shown in Figure 17.8b.

17.4 DESIGN OF TRANSISTOR OSCILLATORS

Transistor oscillator design (as shown in Figure 17.9) is actually a two-port circuit design with the transistor operating in the unstable region. In practice a two-port oscillator can be effectively reduced to a one-port oscillator design by terminating a potentially unstable transistor with an appropriate impedance in the unstable region.

Unlike an amplifier (where the device is required either to have a high degree of stability or to be unconditionally stable), we desire the device to be highly unstable. To achieve this, we normally desire the transistor to be in common base (or common emitter) configuration (for BJT) and in common gate (or common source) configuration (for FET), often with positive feedback to further increase the degree of instability of the device.

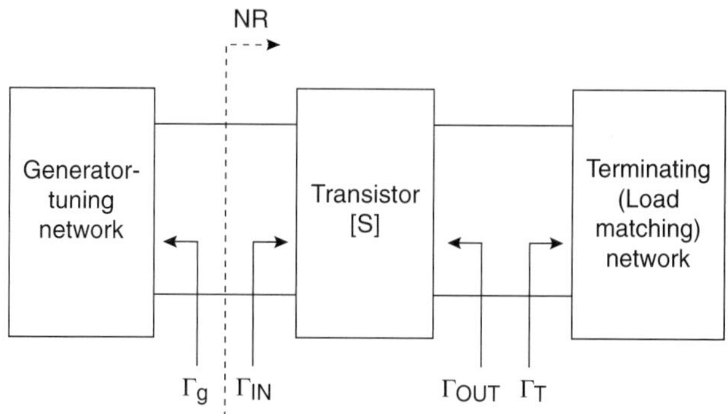

FIGURE 17.9 A two-port oscillator.

17.4.1 Design Procedure for Transistor Oscillators

Based on the previous discussion, the following design procedure for a transistor oscillator is set forth:

STEP 1. Select a potentially unstable transistor at the desired frequency of oscillation.

STEP 2. Select an appropriate transistor circuit configuration:

- For BJTs: Common base or common emitter
- For FETs: Common gate or common source

STEP 3. Draw the output stability circle in the Γ_T-plane.

STEP 4. Select an appropriate value for Γ_T in the unstable region to produce a negative resistance (preferably as large as possible) at the input of the transistor, yielding:

$$|\Gamma_{IN}| > 1 \Rightarrow Z_{IN} < 0 \tag{17.25}$$

STEP 5. Select the generator-tuning network impedance (Z_g) as if the circuit were a one-port oscillator, by choosing its value such that:

$$R_g + R_{IN} < 0 \tag{17.26}$$

In practice, to satisfy the oscillation startup condition of Equation 17.26, a typical value for R_g is chosen as follows (see Example 17.1)

$$R_g = |R_{IN}|/3, \quad R_{IN} < 0 \tag{17.27}$$

The reactive part of Z_g is chosen to resonate the circuit:

$$X_g = -X_{IN} \tag{17.28}$$

STEP 6. Design the generator-tuning and the terminating networks with lumped or distributed elements using techniques described in Chapter 10, *Applications of the Smith Chart.*

This design procedure is based on practical considerations and its popularity, to a great extent, is due to a successful and workable circuit that it creates. The following example illustrates this design procedure.

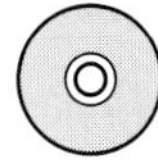

EXAMPLE 17.3

Design a transistor oscillator at 10 GHz using a GaAs FET in a common gate configuration, as shown in Figure 17.10. An inductor ($L_G = 5\ nH$) is placed in series with the gate to increase the positive feedback and thus create further instability. The S-parameters of the transistor have been converted from the common source into the common gate configuration and are given by:

$$S = \begin{bmatrix} 2.18\angle{-35°} & 1.26\angle 18° \\ 2.75\angle 96° & 0.52\angle 155° \end{bmatrix}$$

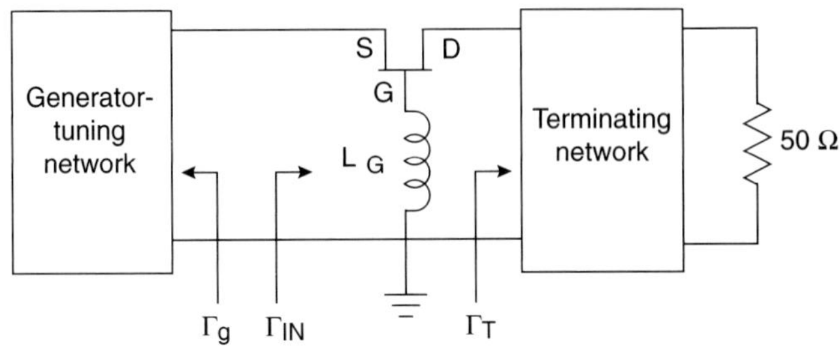

FIGURE 17.10 Circuit for Example 17.3.

Solution:

a. Steps 1 and 2 of the design procedure are already done, therefore now we draw the output stability circle in the Γ_T-plane with the center and radius given by:

$$C_T = 1.08\angle 33°$$

$$R_T = 0.665$$

Because $|S_{11}| = 2.18$ (>1), the stable region is inside the output stability circle (shaded) and the unshaded area is unstable, as shown in Figure 17.11.

b. There are infinite choices for selection of Γ_T, but the best value is obtained when it maximizes the $|\Gamma_{IN}|$ value. Through trial and error, Γ_T is selected as:

$$\Gamma_T = 0.59\angle{-104°} \quad (\text{or } Z_T = 20 - j35\ \Omega)$$

This selection of Γ_T leads to:

$$\Gamma_{IN} = 3.96\angle{-2.4°}$$

or

$$Z_{IN} = -84 - j1.9\ \Omega$$

c. In this step, we determine the impedance value of the generator-tuning network as follows:

$$R_g = |R_{IN}|/3 = 28\ \Omega$$

$$X_g = -X_{IN} = 1.9\ \Omega$$

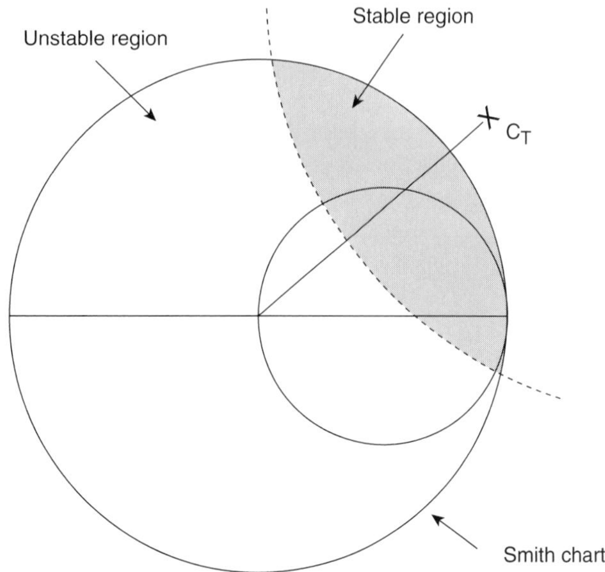

FIGURE 17.11 The output stability circle.

The value of $R_g = 28\ \Omega$ should ensure an adequate negative resistance at startup for a successful oscillation buildup. Thus:

$$Z_g = 28 + j1.9\ \Omega$$

Based on the values of Z_g and Z_T, the generator-tuning and terminating networks can readily be realized. Two possible solutions are these:

Solution #1: One easy and fast solution is using a 28 Ω resistor with a series 50 Ω transmission line having a length of $\ell_1 = 0.006\ \lambda$, as shown in Figure 17.12. Due to such a small length of series transmission line ($\ell_1 = 0.006\ \lambda$), this solution seems impractical, unless as an approximate solution, we completely ignore the 50 Ω series transmission line.

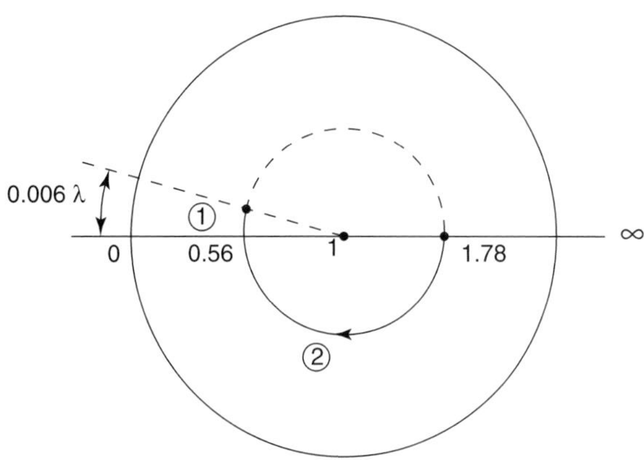

FIGURE 17.12 Smith chart solutions for Example 17.3.

Solution #2: Another reasonable solution is use of a higher resistance (see Figure 17.12):

$$(R_L)_N = 1.78 \Rightarrow R_L = 1.78 \times 50 = 89\ \Omega$$

and a series 50 Ω transmission line having a length of:

$$\ell_2 = 0.256\ \lambda$$

A similar technique is used to provide the terminating network element values.

The final circuit schematic is shown in Figure 17.13.

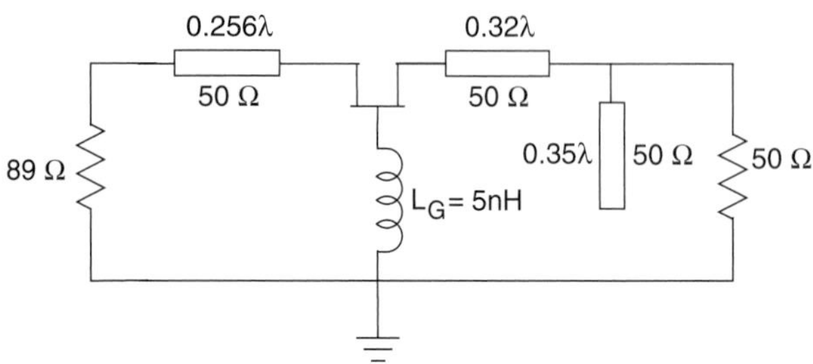

FIGURE 17.13 Solution for Example 17.3.

The next exercise will consider the design of BJT oscillator.

EXERCISE 17.2

Consider a common-base transistor oscillator circuit. The *S*-parameters for the bipolar transistor have been measured at the bias point ($V_{CE} = 15$ V, $I_C = 25$ mA) and are shown below in Table 17.1. A feedback inductor (L_B) is added in the base of the transistor to produce even more instability as shown in Figure 17.14. The new *S*-parameters at the same *Q*-point are also shown in Table 17.1:

TABLE 17.1

L_B	0.0 nH	0.5 nH
S_{11}	0.94∠174°	1.04∠173°
S_{12}	1.90∠–28°	2.00∠–30°
S_{21}	0.013∠98°	0.043∠153°
S_{22}	1.01∠–17°	1.05∠–18°
K	–0.09	–0.83

Determine the generator-tuning network.

FIGURE 17.14 Figure for Exercise 17.2.

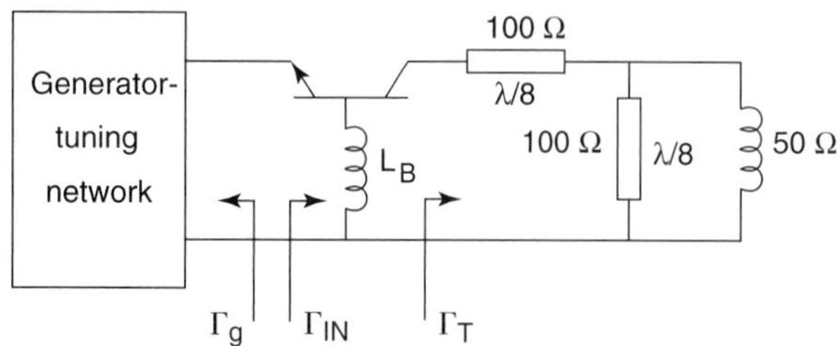

HINT: *Using the given terminating network, first show that $\Gamma_T = 0.62\angle 30°$. Then, using Γ_T, show that $\Gamma_{IN} = 1.18\angle 173°$ and consequently*

$$\Gamma_g = 1/\Gamma_{IN} = 0.84\angle -173°$$

Now use Γ_g to realize the generator-tuning network.

17.5 GENERATOR-TUNING NETWORKS

The generator-tuning networks can be roughly classified into two basic categories:

- Fixed-frequency oscillators: Lumped-element circuits and dielectric resonator (DR) circuits
- Frequency-tunable oscillators: YIG-tuned circuits, varactor-tuned circuits, dielectric resonator (DR) circuits, and cavity-tuned circuits

These are described in detail in the following sections.

17.5.1 Fixed-Frequency Oscillators

Fixed-frequency oscillators find useful applications as local oscillators in communication and radar systems. The circuit configurations in this category can be divided into two major classes:

I. Lumped Elements. This class of circuits can be further subdivided into:

- RF and low microwave frequencies
- High microwave frequencies

Each subclass is described next.

a) RF and Low Microwave Frequencies At RF and low microwave frequencies, lumped-element tuning networks are used. The four conventional oscillator tuning network configurations are as follows:

- The Armstrong circuit adopts a tapped inductor for the tuning function.
- The Clapp circuit employs an inductor in series with a capacitor as the tuning device.
- The Colpitts circuit utilizes a capacitor voltage divider for tuning purposes.
- The Hartley circuit uses a tapped inductor for the tuning mechanism.

The 3 basic transistor topologies for a BJT (common base, common emitter, and common collector) or for a FET (common gate, common source, and common drain) apply to each one of these 4 types of circuits, thus making a total of 12 possible circuits for each type of transistor. It should be noted that the most effective and easily tunable topology for low-power applications is the common base for BJTs and common gate configuration for FETs.

Figure 17.15 shows typical circuits for each type of the four conventional oscillator tuning network configurations in common mode topology using an FET. The design procedure for these four types of oscillators can be found in standard texts on RF oscillator circuit design.

NOTE: *For efficient tuning, the Armstrong or Hartley oscillator requires a high-Q inductor, which is difficult to build. This requirement makes the Colpitts or Clapp oscillator a preferred choice.*

b) High Microwave Frequencies At higher microwave frequencies, we can design the oscillator utilizing two techniques:

Use of series and shunt feedback resonant circuits: Considering that each type of transistor provides three possible circuit configurations for each type of feedback (series or shunt), therefore, we can see that each transistor provides a total of six oscillator circuit configurations, as shown in Figure 17.16. The lumped elements in each of the six configurations can be calculated from the device's Z or Y parameters. If the S-parameters are provided, a simple conversion would provide the desired Z or Y parameters.

Use of two-port NR oscillator design technique: In this technique, use of S-parameters to characterize the transistor becomes necessary. The transistor is treated as a negative-resistance device and the oscillator can be designed using the design technique already outlined in Section 17.4. In this technique, we select a transistor that is potentially unstable at the desired frequency of oscillation. Then, the transistor is terminated by using an appropriate load value in the unstable region such that it produces the largest possible negative resistance at the input of the transistor. The final step involves choosing the generator-tuning network such that it satisfies the oscillation startup condition as given by Equation 17.15.

All three transistor topologies can be used to create a relatively low-power microwave oscillator. The common base for BJTs (or common gate for FETs) is preferred due to the ease of tuning that it provides. Using this configuration, we may want to add positive feedback (e.g., an inductor in the base) to increase the negative resistance of the transistor even further (see Figure 17.17).

NOTE: *For higher-power oscillators, use of common emitter for BJTs (or common source for FETs) topology is recommended whereas common collector (or common drain) topology is unpopular due to difficulties encountered in its implementation.*

II. Dielectric Resonator (DR) Circuits. The transistor dielectric resonator oscillator (DRO or sometimes TDRO) is a high-Q circuit that can be used for applications

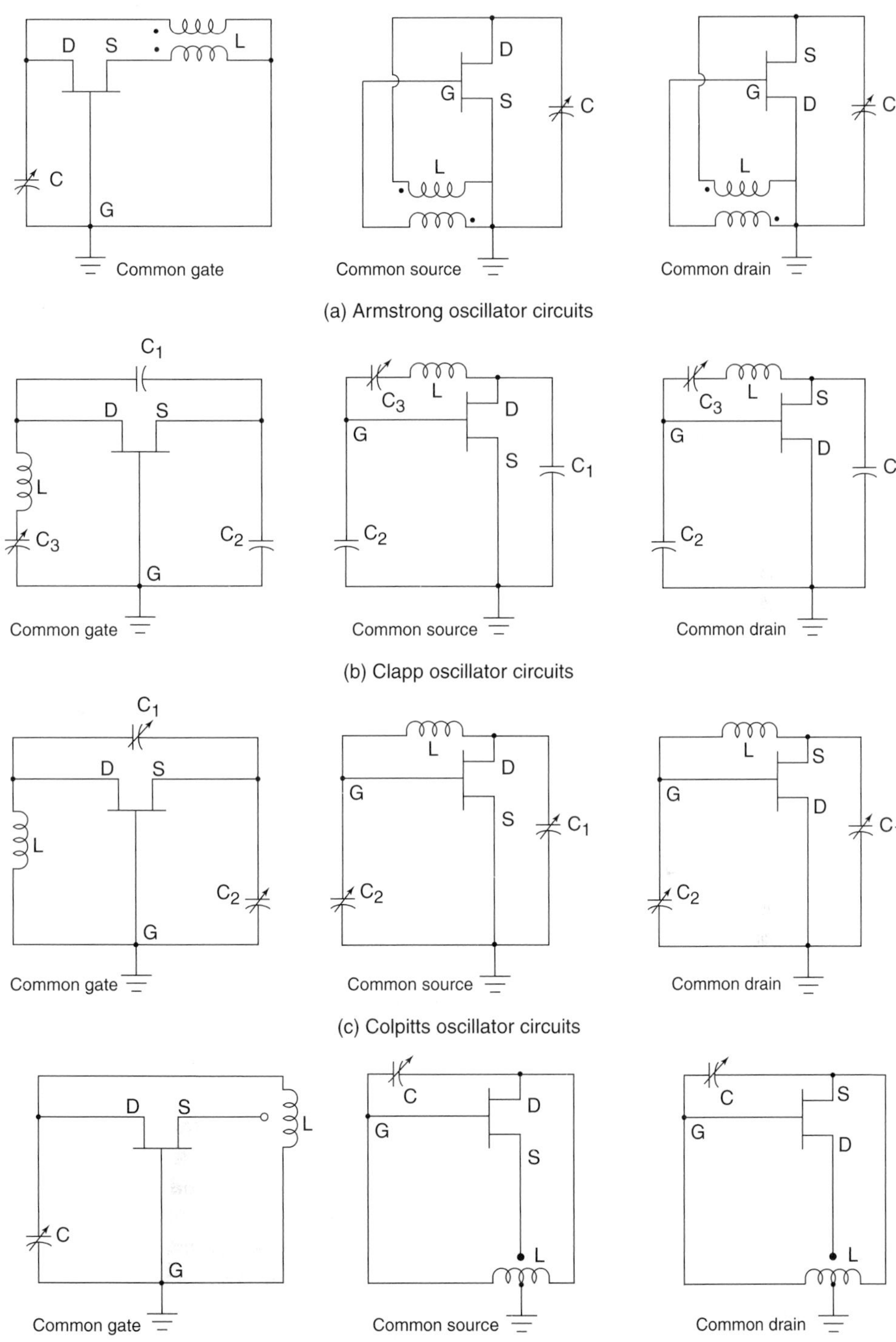

FIGURE 17.15 RF and low microwave oscillator circuit configurations using an FET device.

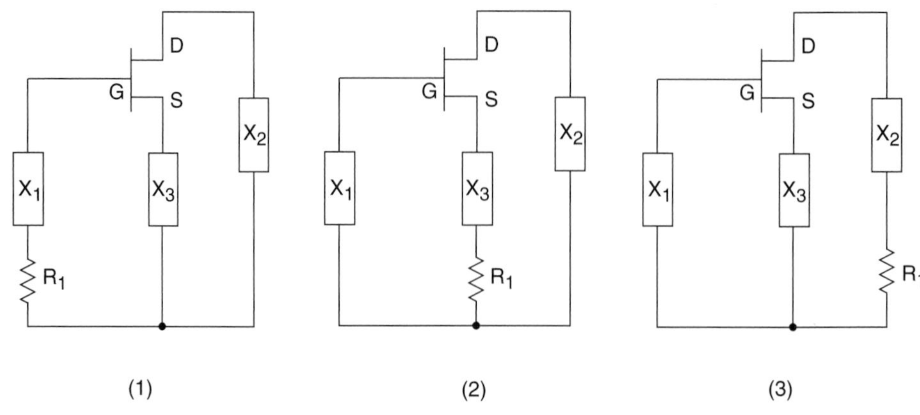

(a) Three series oscillator circuit configurations

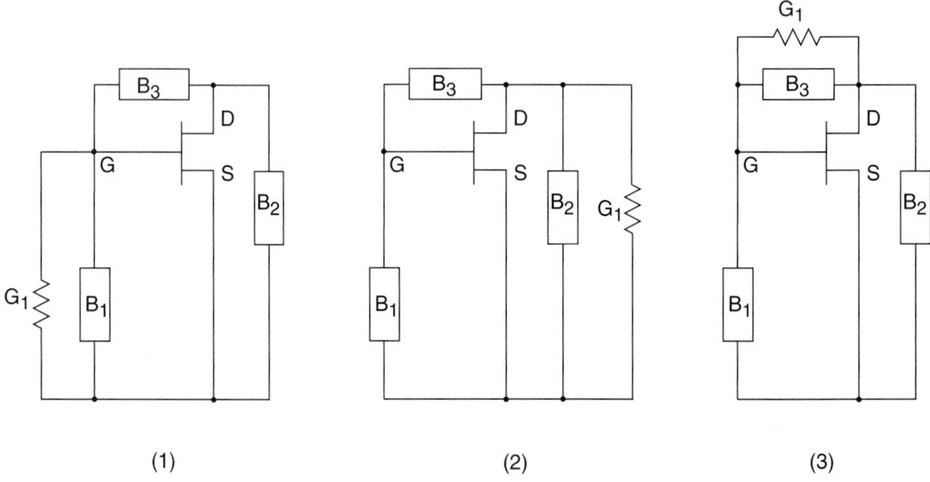

(b) Three shunt oscillator circuit configurations

FIGURE 17.16 High microwave-frequency oscillator circuit configuration using an FET.

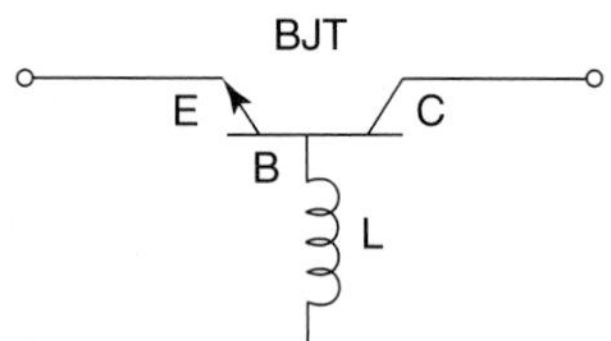

FIGURE 17.17 A BJT with external feedback to increase the region of instability.

requiring high-quality oscillators having a fixed frequency or being tunable over a narrow band of frequencies (from 0.1% to 2% of f_o). Currently, DROs are commonly used as low-noise and temperature-stable fixed-frequency oscillators. Typical DRO circuits are shown in Figure 17.18.

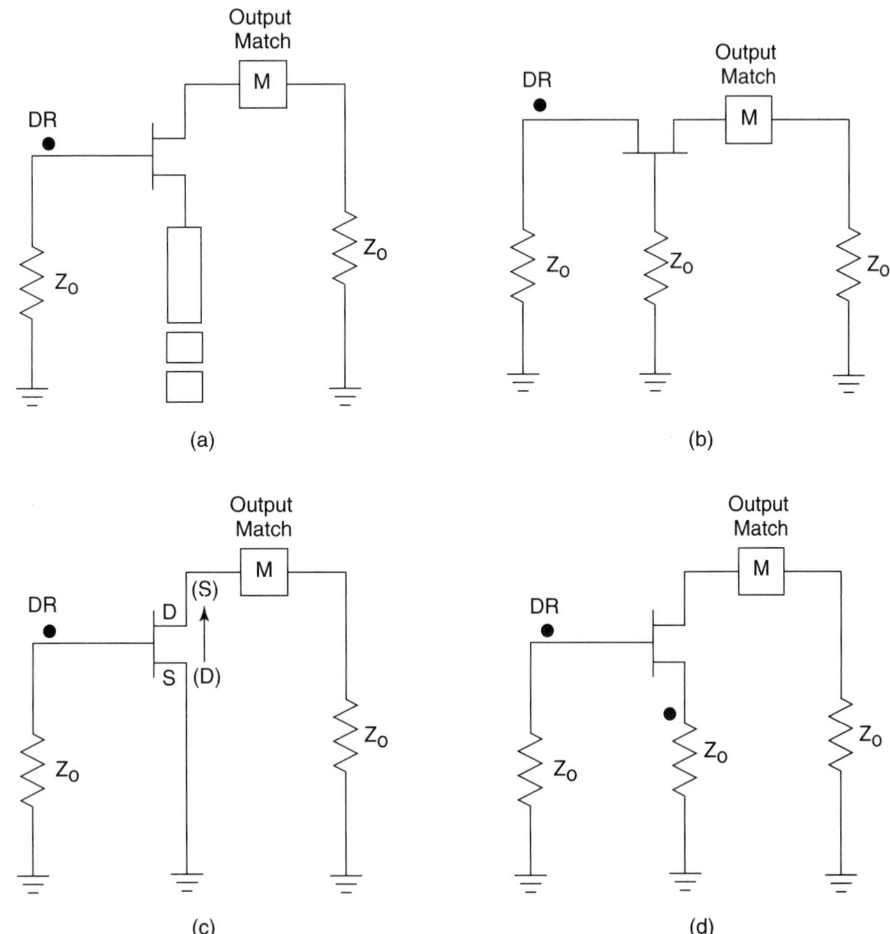

FIGURE 17.18 Typical DRO circuits in series feedback configuration.

One of the main advantages of a DRO is that its size is considerably smaller than the size of a hollow resonant cavity operating at the same frequency. The reason is that the relative dielectric constant is substantially larger than unity (ε_r for air). Typically, ε_r is between 20 and 80. As the frequency decreases below 500 MHz, the resonator dimensions become unmanageably large and thus impractical for any miniature-sized application.

Furthermore, aside from their considerable reduction in size, DROs have high Q, high temperature stability, and excellent integrability in microwave integrated circuit (MIC) technology, which, as a result, allows them to be used efficiently as stable MIC sources. Popular dielectric resonators consist of a high-Q ceramic resonator that is properly sized and placed for microwave integrated circuit (MIC) applications. DROs use BJTs (up to 15 GHz) and GaAs FETs (up to 35 GHz), with a typical output power level in the order of 10 to 30 mW. With special designs the operating frequency can be extended up to 100 GHz. We relegate the design of this category of circuits to more advanced texts.

17.5.2 Frequency-Tunable Oscillators

Tunable and wideband tunable oscillators are commonly employed in synthesizers, instrumentation, and electronic warfare (EW) systems where the need for a wide range of frequencies is imperative. This category of oscillators subdivides further into the following types of circuits.

I. YIG-Tuned Oscillator (YTO) Circuits. YIG is an abbreviation for Yttrium-Iron-Garnet ($Y_3Fe_5O_{12}$), which is a magnetic crystal material with a frequency of oscillation directly proportional to the applied bias magnetic field. Use of a YIG crystal with a proper bias provides a high-Q resonance circuit in which tuning over a frequency range can be accomplished. Tuning is achieved by positioning the YIG crystal in a magnetic field (provided by a solenoid) and then, by varying the amount of DC current through the solenoid, we change the magnetic field applied to the YIG sphere. Its frequency of oscillation will change accordingly, and therefore wideband tuning in a microwave oscillator circuit can be achieved.

A typical YIG-tuned oscillator circuit can be obtained by placing a YIG crystal in the source leg (for FETs) or emitter leg (for BJTs), as shown in Figure 17.19. A YIG resonator circuit can be modeled by a parallel RLC resonant circuit where the G_o, L_o, and C_o element values depend on, among other things, the applied DC magnetic field (see Figure 17.20).

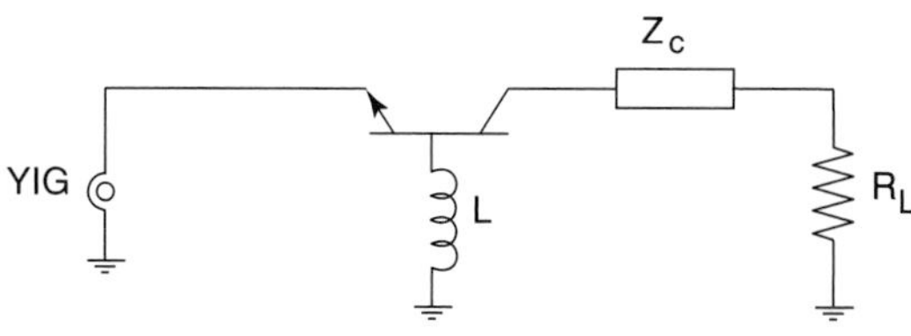

FIGURE 17.19 YIG-tuned oscillator circuit.

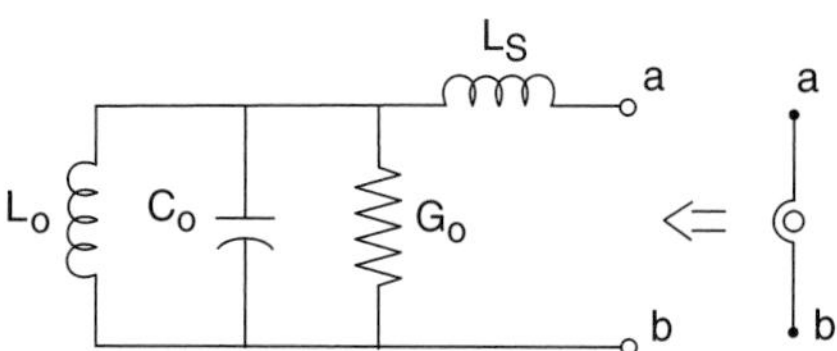

FIGURE 17.20 Equivalent network of a YIG sphere in a YIG-tuned oscillator (YTO).

II. Varactor-Tuned Oscillator (VTO) Circuits. A varactor-tuned oscillator uses a varactor diode to achieve the desired frequency tuning, as shown in Figure 17.21. A varactor diode is a two-terminal semiconductor device that utilizes the voltage-sensitive property of a pn junction. In these diodes, unlike regular diodes, the pn junction capacitance under reverse-bias conditions is accentuated by proper choice of diode profiles (e.g., use of hyper-abrupt junctions).

Assuming a one-sided junction (e.g., n^+p for microwave frequencies), the junction capacitance due to a diode's junction capacitance (C_j) can be shown to be a function of the applied bias voltage as follows:

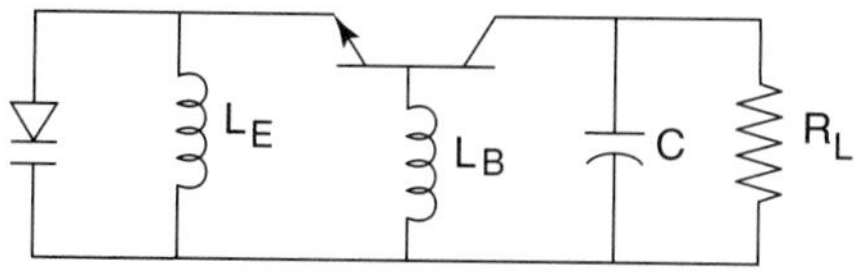

FIGURE 17.21 Varactor-tuned BJT oscillator.

$$C_j = \frac{C_{jo}}{\left(1 - \frac{V_a}{V_{bi}}\right)^S} \tag{17.29}$$

where

C_{jo} = The junction capacitance at zero bias voltage

V_a = The applied bias voltage across the junction (e.g., for reverse bias $V_a = -V_r$)

V_{bi} = The built-in voltage (or contact potential)

and

$$S = \frac{1}{m + 2}, \tag{17.30}$$

where

"m" is the power parameter in the doping distribution function (N) of the lighter side, defined as:

$$N(x) = N_B x^m$$

and N_B is an arbitrary constant established by the doping profile.

Equation 17.29 is plotted in Figure 17.22. From this figure, we can see that the capacitor value becomes very high for $V_a \approx V_{bi}$, which is impractical. A varactor is usually operated in the reverse bias region that is $V_a = -V_r < 0$.

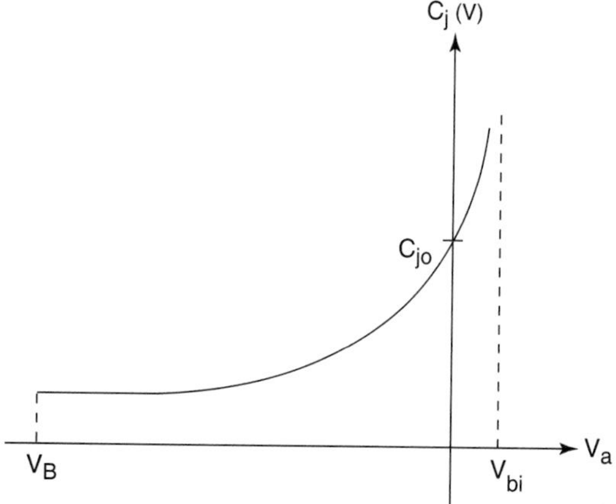

FIGURE 17.22 Varactor characteristic (C_j vs. V_a) with V_B as the breakdown voltage.

From Equation 17.29, we can see that the larger the S, the larger will be the junction capacitance with reverse bias voltage. For example, for an abrupt junction ($m = 0$) under reverse bias, we have (see Figure 17.22):

$$m = 0 \Rightarrow S = 1/2$$

$$C_j = \frac{C_{jo}}{\sqrt{1 + \frac{V_r}{V_{bi}}}} \qquad (17.31)$$

An important parameter in characterizing a varactor is the sensitivity (γ) defined by:

$$\gamma \equiv -\frac{dC_j}{C_j}\frac{V'_r}{dV'_r} = -\left[d(\log_{10} C_j)\cdot\frac{1}{d(\log_{10} V'_r)}\right] = -\frac{V'_r}{C_j}\frac{dC_j}{dV'_r} = S \qquad (17.32)$$

where

$$V'_r = V_r + V_{bi}$$

From Equation 17.32 we can see that the larger the S, the larger will be the capacitance variation with the biasing voltage. Therefore, for an efficient and sensitive varactor, we use a high value for sensitivity (S). For example, choosing $S = 1, 2,$ or 3 corresponds to doping profile parameters $m = -1, -3/2,$ or $-5/2$, respectively, as shown in Figure 17.23. These are called hyper-abrupt junctions, which give rise to the largest capacitance variation.

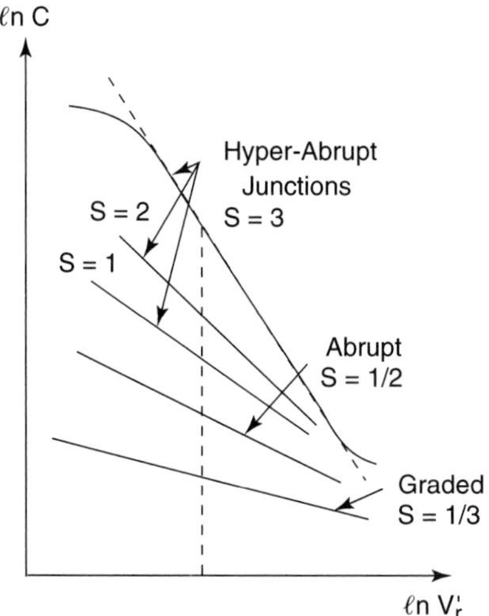

FIGURE 17.23 Capacitance versus voltage variation for diodes with various doping profiles.

A varactor-tuned oscillator may use a varactor diode made of silicon or GaAs. The difference between the two is that GaAs varactors provide higher Q than silicon, due to lower resistivity of GaAs for a given doping level. Compared to YIG oscillators, which provide a decade or more of bandwidth, varactor-tuned oscillators generally provide a limited tuning range of about an octave but a higher tuning speed.

Due to their capability in the control of frequency (within a certain range) by means of an applied bias voltage, varactor-tuned oscillators are also called voltage-controlled oscillators (VCOs).

Dielectric Resonator (DR) Circuits. Even though the transistor dielectric resonator oscillator (DRO) is commonly used for applications requiring high-quality fixed frequency oscillators, however, the oscillation frequency can be tuned over a narrow frequency range using a number of approaches. These approaches broadly fall into two categories:

Mechanical tuning: This technique consists of adjusting a screw above the DR (up or down) because the resonant frequency is highly sensitive to the proximity of the ground plane. Lowering the screw would increase the resonant frequency. A mechanical frequency tuning range of approximately ±2 percent of the operating frequency can be achieved without any noticeable change in output power, FM noise, and overall temperature stability (see Figure 17.24).

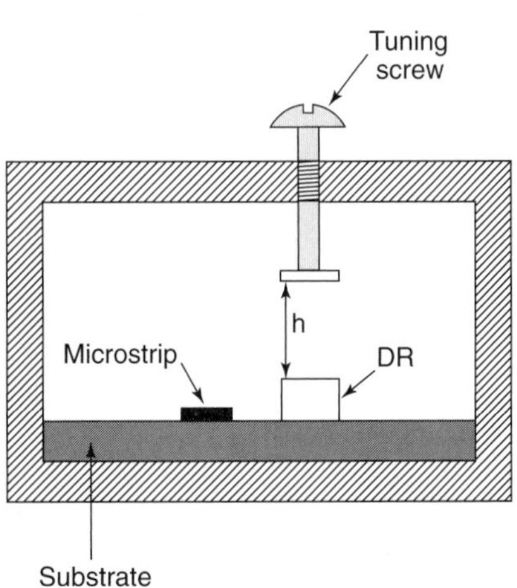

FIGURE 17.24 A mechanical tuning arrangement for DROs.

Electrical tuning: In this category a number of practical approaches can be employed utilizing varactor-tuning, bias-tuning, and optical-tuning techniques. Use of any of the three techniques alters one of the main properties of the DR, thus causing a change in frequency to take place. For example, in the case of varactor tuning, the frequency of a DR, coupled electromagnetically to a microstrip line (which is resonant and has a varactor in it), is varied by a change in the varactor's capacitance. Because the fields are coupled, a change in the frequency of the resonant varactor-microstrip structure will shift the frequency of operation in the DRO circuit.

In the case of bias tuning, the Q-point of the transistor is slightly altered, causing a shift in the frequency of operation. This will affect the power output of the oscillator, which needs to be taken into consideration (about ±0.05 percent of the operating frequency).

Use of an optical beam (coming from a laser or LED through an optical fiber) can alter the electrical conductivity of a photosensitive material directly placed on a DR. The change in conductivity will perturb the EM field around the DR, thus causing a shift in the frequency operation (around 0.1 percent of center frequency).

Cavity-Tuned Circuits. In these types of oscillators, the tuning mechanism consists of a cavity-tuned circuit. The microwave cavity used can be either a coaxial cavity or a waveguide cavity. The frequency tuning is obtained usually from a manual adjustment of a sliding plunger or an adjustable screw. These types of oscillators can operate either as fixed-frequency or as tunable oscillators. Figure 17.25 shows an example of a full-height waveguide rectangular cavity.

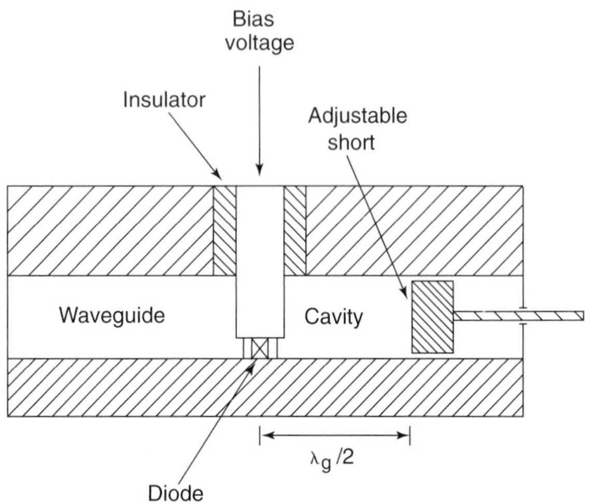

FIGURE 17.25 A full-height rectangular waveguide cavity.

A full discussion of this subject is beyond the scope of this work, and its mention here is only for completeness of presented ideas.

LIST OF SYMBOLS/ABBREVIATIONS

A symbol/abbreviation will not be repeated again once it has been identified and defined in an earlier chapter, as long as its definition remains unchanged.

B_g	Susceptance of the generator/source (shunt configuration)
C_j	Junction Capacitance of a pn junction
C_{jo}	Junction Capacitance of a pn junction at zero bias
G_g	Conductance of the generator-tuning network (shunt configuration)
I_o	Oscillator current
$I_{o,max}$	Oscillator current at maximum power
$N(x)$	The doping profile
N_B	An arbitrary constant in the doping profile: $N(x) = N_B x^m$
NR	Negative resistance
$R_{D,max}$	Negative resistance of an NR device that generates maximum power in an oscillator design
R_g	Resistance of the generator-tuning network (series configuration)
S	Sensitivity parameter in a varactor
X_g	Reactance of the generator-tuning network (series configuration)

Y_g Admittance of the generator-tuning network (shunt configuration)
Z_g Impedance of the generator-tuning network (series configuration)
$(Z_o)_{\lambda/4}$ Characteristic impedance of a series $\lambda/4$ transformer
γ Sensitivity
Γ_g Reflection coefficient at the generator-tuning network
Γ_L Reflection coefficient at the load
Γ_T Reflection coefficient at the terminating network

PROBLEMS

17.1 Design a 10 GHz oscillator using a common-gate GaAs FET with the following S-parameters (at $V_{DS} = 6$ V, $I_{DS} = 150$ mA):

$$S = \begin{bmatrix} 0.85\angle{-36^\circ} & 0.22\angle{-36^\circ} \\ 0.53\angle 96^\circ & 1.13\angle 171^\circ \end{bmatrix}$$

Show the RF schematic as well as the DC bias network.

17.2 It has been shown by Johnson that the output power of an oscillator can be approximated by:

$$P_{OUT} = P_{sat}(1 - e^{G_o P_{IN}/P_{sat}})$$

where

P_{sat} = Saturated output power of the amplifier
P_{IN} = Input power
G_o = Small signal power gain

a. Knowing that $P_{osc} = P_{OUT} - P_{IN}$, and the maximum oscillator power occurs at $\partial P_{osc}/\partial P_{IN} = 0$, show that the maximum oscillator power is given by:

$$(P_{osc})_{max} = P_{sat}(1 - 1/G_o - \ln G_o/G_o)$$

b. If $G_o = 7.5$ dB and $P_{sat} = 1$ W, find $(P_{osc})_{max}$.

17.3 Given below are the S-parameters for the following two transistors at 2 GHz. Determine which one would be more suitable to be used for an oscillator design at 2 GHz:

$$S_A = \begin{bmatrix} 0.48\angle 25^\circ & 0 \\ 5.0\angle 30^\circ & 0.3\angle{-120^\circ} \end{bmatrix} \qquad S_B = \begin{bmatrix} 0.8\angle 90^\circ & 0 \\ 4.0\angle 65^\circ & 2.0\angle 180^\circ \end{bmatrix}$$

17.4 Prove that:

$$\text{If } \Gamma_{IN}\Gamma_{OUT} = 1 \Rightarrow \text{Then } X_{OUT} + X_L = 0 \text{ and } R_{OUT} + R_L = 0$$

17.5 A one-port oscillator at 8 GHz is to be designed using a microwave diode having (50 Ω system):

$$\Gamma_D = 1.75\angle 40^\circ$$

a. Determine the diode's impedance.
b. Find the proper load value required for oscillation.
c. Draw the complete circuit diagram (AC and DC) for the designed oscillator.

17.6 Design a two-port oscillator at 10 GHz using a GaAs FET in the common source configuration with the following S-parameters:

$$S = \begin{bmatrix} 0.79\angle{-140^\circ} & 0.10\angle 50^\circ \\ 2.75\angle 130^\circ & 0.75\angle{-40^\circ} \end{bmatrix}$$

17.7 Design an oscillator using a common-collector bipolar transistor at $f = 4$ GHz, $V_{CE} = 8$ V and $I_C = 40$ mA with the following S-parameters:

$$S = \begin{bmatrix} 0.91\angle{-135^\circ} & 0.67\angle 30^\circ \\ 1.41\angle 130^\circ & 0.60\angle 90^\circ \end{bmatrix}$$

Draw the final RF schematic for the design using lumped elements.

REFERENCES

[17.1] Anderson, R. W. *S-Parameter Techniques for Faster, More Accurate Network Design.* Hewlett-Packard Application Note 95–1, February 1967.

[17.2] Bahl, I. and P. Bhartia. *Microwave Solid State Circuit Design.* New York: Wiley Interscience, 1988.

[17.3] Elmi, N. and M. M. Radmanesh. "A Novel Design of Low Noise, Highly Stable GaAs Dielectric Resonator Oscillators." *Microwave Journal*, Vol. 39, No. 11, pp. 104–12. November 1996.

[17.4] Gilmore, R. J., and F. J. Rosenbaum. "An Analytic Approach to Optimum Oscillator Design Using S-Parameters." *IEEE Transactions on Microwave Theory and Technique*, Vol. MTT-31, August 1983, pp. 633–9.

[17.5] Gonzalez, G. *Microwave Transistor Amplifiers, Analysis and Design*, 2nd ed. Upper Saddle River: Prentice Hall, 1997.

[17.6] Johnson, K. M. "Large-Signal GaAs MESFET Oscillator Design." *IEEE Transactions on Microwave Theory and Techniques*, Vol. MTT-27, March 1979.

[17.7] Khanna, A. P. S. and J. Obregon. "Microwave Oscillator Analysis." *IEEE Transactions on Microwave Theory and Technique*, Vol. MTT-29, June 1981, pp. 606–7.

[17.8] Kurokawa, K. "Some Basic Characteristics of Broadband Resistive Oscillator Circuits." *Bell System Technical Journal*, Vol. 48, July 1969.

[17.9] Liao, S. Y. *Microwave Circuit Analysis and Amplifier Design.* Upper Saddle River: Prentice Hall, 1987.

[17.10] Pozar, D. M. *Microwave Engineering*, 2nd ed. New York: John Wiley & Sons, 1998.

[17.11] Radmanesh, M. M., C. M. Chu, and G. I. Haddad. *Magnetostatic Wave Propagation in a Finite YIG-Loaded Rectangulor Waveguide.* IEEE Transactions on Microwave Theory and Technique (MTT), Vol. 34, No. 12, pp. 1377–82, Dec. 1986.

[17.12] Radmanesh, M. M., C. M. Chu, and G. I. Haddad. *Numerical Analysis and Computer Simulation of Magnetostatic Waves in a Waveguide,* IEEE Transactions on Microwave Theory and Technique (MTT), Vol. 41, No. 1, pp. 89–96, Jan. 1993.

[17.13] Vendelin, George D. *Design of Amplifiers and Oscillators by the S-Parameter Method.* New York: John Wiley & Sons, 1982.

[17.14] Vendelin, George D., Anthony M. Pavio, and Ulrich L. Rhode. *Microwave Circuit Design, Using Linear and Non-Linear Techniques.* New York: John Wiley & Sons, 1990.

CHAPTER 18

RF/Microwave Frequency Conversion I: Rectifier and Detector Design

18.1 INTRODUCTION

Detectors make use of the nonlinear characteristics of a solid-state device to bring about frequency conversion of an input signal. The more nonlinear the device's *I-V* characteristic curves are, the more efficient the detection process will be, that is, a higher percentage of the signal power at the input frequency will be converted to signal power at the output frequency.

The three basic types of frequency conversion circuits are these:

- A rectifier
- A detector
- A mixer

Each is described briefly as follows:

A **rectifier** is a circuit that converts the RF signal into a zero frequency signal (i.e., a DC signal) with time and frequency-domain signals as shown in Figure 18.1. The rectifier is used for automatic gain control (AGC) circuits or power monitor circuits, etc.

A **detector** (also called a demodulator) is a circuit that demodulates a modulated carrier wave by discarding the carrier wave and outputting the modulating signal as shown in Figure 18.2. Detectors are used in circuits such as the AM demodulator circuit and are discussed in depth later in this chapter.

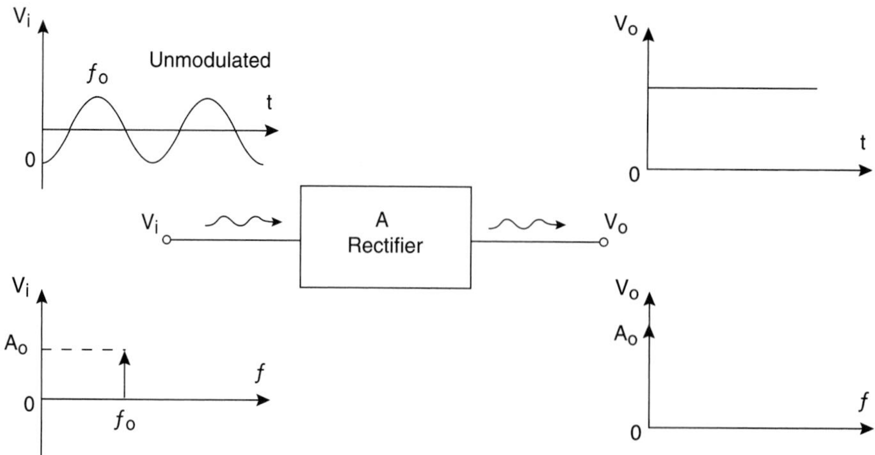

FIGURE 18.1 Response of a rectifier in time and frequency domains.

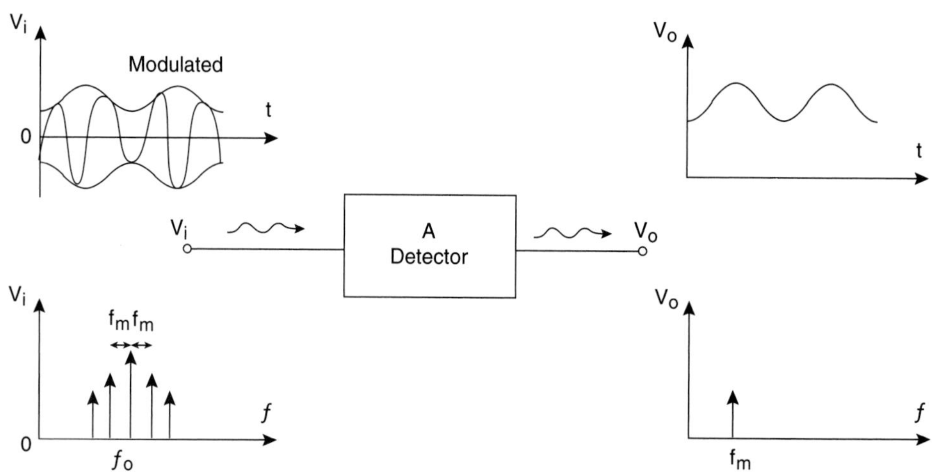

FIGURE 18.2 Response of a detector in time and frequency domains.

NOTE: *A rectifier is a special case of a detector where proper filtering is used at the output to reject all frequencies except for the DC component. For this reason we will consider detectors as a general subject, knowing well that rectifiers are also included in the discussion.*

A **mixer** (also called a converter) is a circuit that converts an input signal to a higher-frequency signal (called an up-converter) or to a lower-frequency signal (called a down-converter) while ideally preserving all of the original signal characteristics (such as sidebands, waveshape, etc.). This frequency conversion can be readily obtained by mixing the input signal with another signal (called the "local oscillator" signal, or "LO" for short) as shown in Figure 18.3.

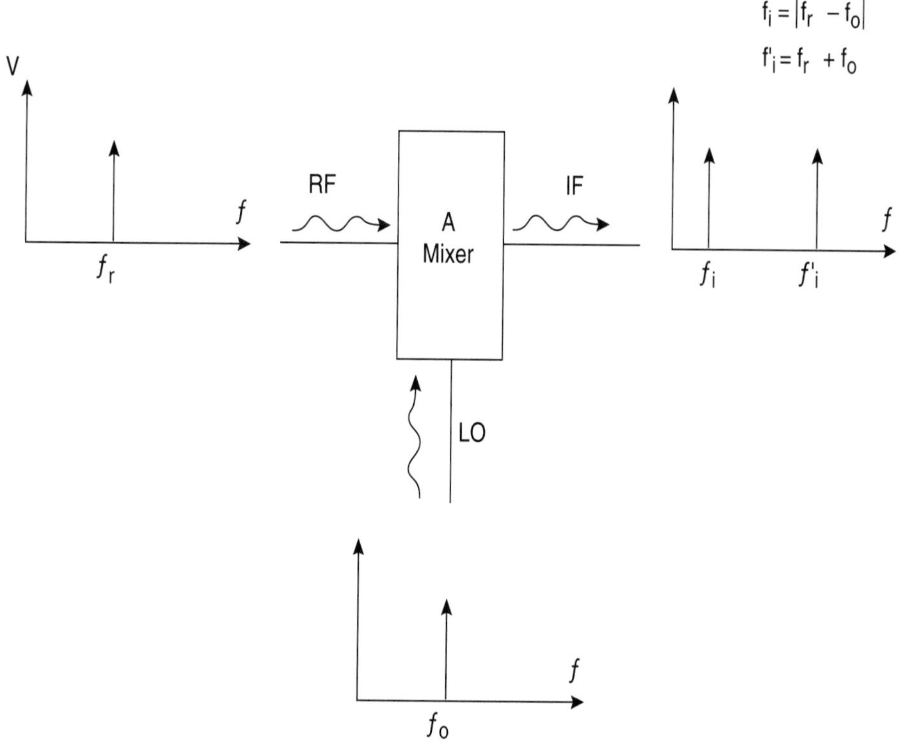

FIGURE 18.3 A mixer response.

Mixers are used, for example, in transmitters to up-convert the signal for transmission purposes or in receivers to down-convert the signal for demodulation purposes. Mixers are further studied in the next chapter.

In this chapter we focus on rectifier and detector circuit design. The general circuit consists of a nonlinear device flanked by two lossless matching networks. A general block diagram for a detector is shown in Figure 18.4. The most common nonlinear devices at RF and microwave frequencies are diodes; however, transistors can also be used.

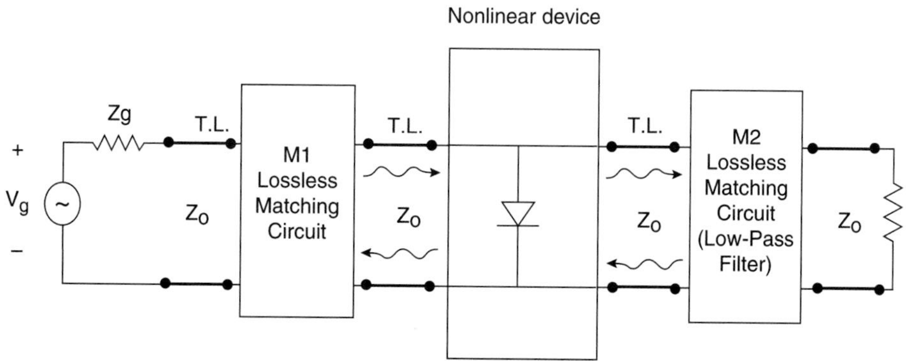

FIGURE 18.4 General diagram of a detector.

Diodes were briefly discussed in Chapter 4, *DC and Low-Frequency Circuits Concepts*, and are further analyzed in the next section. Using a small-signal analysis technique, we will investigate the use of the diode's nonlinear *I-V* characteristic curve in the design of rectifiers or detectors.

18.2 SMALL-SIGNAL ANALYSIS OF A DIODE

In general, a diode can be considered to be a nonlinear resistor with its *I-V* characteristic curve mathematically given by (see Chapter 4):

$$I(V) = I_S(e^{V/nV_T} - 1) \tag{18.1}$$

where:

$V_T = KT/q$ ($V_T = 25$ mV at $T = 293°$ K)
I_S = Diode saturation current
n = Ideality factor ($1 \le n \le 2$), which depends on the material and physical structure of the diode. For example, for a point-contact diode $n = 2$ whereas for a Schottky barrier diode, $n = 1.2$, etc.

Figure 18.5 shows a sketch of a diode *I-V* characteristic curve, as described by Equation 18.1.

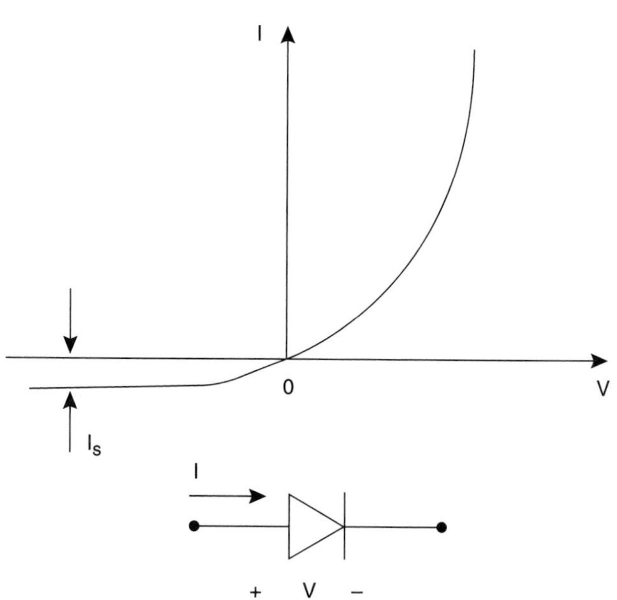

FIGURE 18.5 *I-V* characteristic of a diode.

To perform a small-signal analysis, we assume that the total voltage across the diode (V) is composed of a DC bias voltage (V_o) and a small-signal RF voltage (v), that is:

$$V = V_o + v \tag{18.2}$$

Substituting Equation 18.2 in 18.1 and performing a Taylor series expansion around the Q-point (I_o, V_o), we obtain:

$$I = I(V_o + v) = I(V_o) + v\frac{dI}{dv}\bigg|_{V_o} + \frac{1}{2}v^2\frac{d^2I}{dv^2}\bigg|_{V_o} + \ldots \tag{18.3}$$

where $I(V_o)$ is the DC bias current given by:

$$I_o = I(V_o) = I_S(e^{V_o/nV_T} - 1) \tag{18.4}$$

The first-order derivative corresponds to the dynamic conductance of the diode G_d (the inverse of the junction resistance, R_j) and is given by:

$$G_d = \frac{1}{R_j} = \frac{dI}{dv}\bigg|_{V_o} = \frac{I_S}{nV_T}e^{V_o/nV_T} = \frac{I_o + I_S}{nV_T} \tag{18.5}$$

The second-order derivative is given by:

$$\frac{d^2I}{dv^2}\bigg|_{V_o} = \frac{dG_d}{dv} = G'_d = \frac{I_S}{(nV_T)^2}e^{V_o/nV_T} \tag{18.6}$$

$$G'_d = \frac{(I_S + I_o)}{(nV_T)^2} = \frac{G_d}{nV_T}$$

Equation 18.3 can now be written as a DC current (I_o) and an AC small signal current (i):

$$I(v) = I_o + i \tag{18.7}$$

where

$$i = vG_d + 1/2\,(v^2G'_d) + \ldots. \tag{18.8}$$

The three-term approximation for the diode current, as given by Equation 18.7, is known as the "small-signal approximation." This means that for small signals (i.e., $v/n\,V_T << 1$), higher-order terms (above the second order) for i may be truncated without any loss of accuracy for most of our ensuing analyses and discussions.

NOTE: *As can be observed from Equation 18.8, the AC small signal current (i) is based on derivatives of the I-V curve at the Q-point as given by Equation 18.1. It describes the diode's AC behavior and portrays the ideal nonlinear relationship between i and v as purely resistive, without any considerations for reactive effects. In other words, this analysis is valid for low frequencies.*

In practice, however, the diode's behavior also involves reactive effects caused by junction and parasitic capacitances and lead inductances that are directly related to the packaging as well as the structure of the diode. A typical equivalent circuit for the diode is shown in Figure 18.6. In this figure, (C_p, L_p) are the parasitic elements due to packaging, R_S is the series resistance due to semiconductor neutral regions and the contact areas, and (R_j, C_j) are the junction resistance and capacitance, which are bias-dependent. These parasitic elements cause major deviation from the ideal diode behavior that we have portrayed up to this time in our analysis.

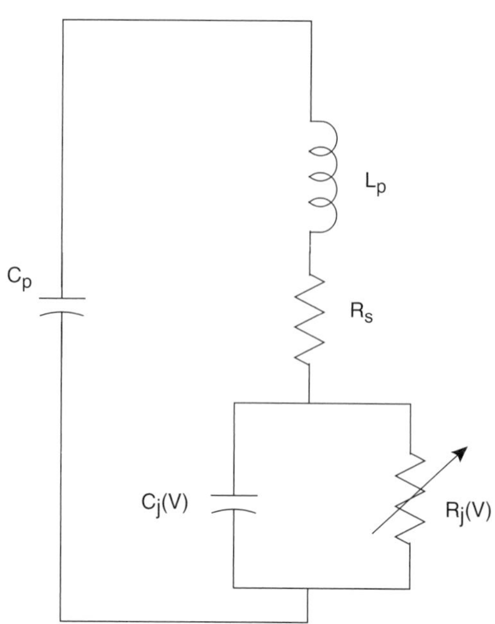

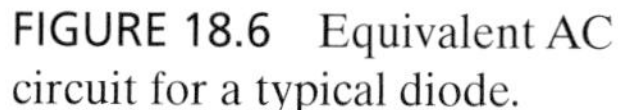
FIGURE 18.6 Equivalent AC circuit for a typical diode.

Even though the deviations from the ideal diode are important in practice and are worth the mention, their consideration and further analysis are relegated to more advanced texts and will not be treated in the rest of this chapter.

18.3 DIODE APPLICATIONS IN DETECTOR CIRCUITS

A typical diode (as analyzed in the previous section) can be used for many different applications, particularly in a rectifier or a detector circuit which is the primary focus of this chapter. The type of circuit used as a detector depends on whether the input RF signal is modulated or not. Thus, we will consider two possible cases.

18.3.1 Unmodulated Signal

As discussed earlier, a diode converts a portion of the input RF energy to a DC current that is proportional to the input RF power. The type of detector circuit that uses an unmodulated RF signal and converts it into a DC output signal may also be referred to as a rectifier. In this section, we will explore the magnitude of the DC current relative to the input RF signal.

Let's consider a diode detector circuit (as shown in Figure 18.7) and assume that the diode is biased at a *Q*-point (I_o, V_o) with an applied input small-signal RF voltage (v) having a frequency (ω_o) and amplitude (v_m) given by:

$$v(t) = v_m \cos(\omega_o t) \tag{18.9}$$

From Equations 18.7 and 18.8, the total current is composed of a DC bias (I_o) and an AC current given by:

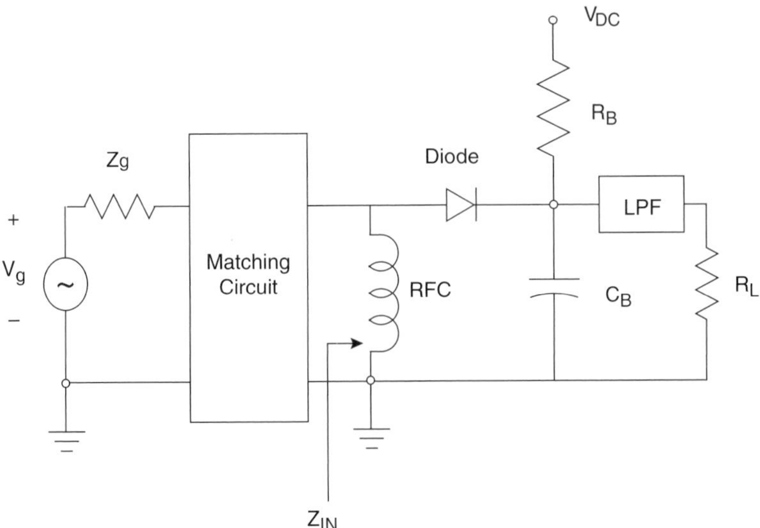

FIGURE 18.7 A typical circuit for a diode rectifier.

$$I = I_o + i \tag{18.10}$$

where

$$i = G_d v_m \cos(\omega_o t) + 1/2[G'_d v_m^2 \cos^2(\omega_o t)] \tag{18.11}$$

Using the identity:

$$\cos^2(\omega_o t) = [1 + \cos(2\omega_o t)]/2,$$

we can write Equation 18.11 as:

$$i = v_m^2 G'_d/4 + v_m G_d \cos(\omega_o t) + v_m^2 G'_d \cos(2\omega_o t)/4 \tag{18.12}$$

Thus, the total DC current is given by:

$$I_{DC} = I_o + v_m^2 G'_d/4 \tag{18.13}$$

If now the output RF signals of frequency (ω_o), $(2\omega_o)$, and other higher-order harmonics, are filtered out using a simple low-pass filter, the remaining output term will be composed of the bias current (I_o) and a term equal to $v_m^2 G'_d/4$. The DC rectified current is proportional to (v_m^2), which is the input RF power, desirable in many applications such as automatic gain control, power monitors, and so on. Under this condition, the detector is said to operate in the square-law region. By proper connection, the resulting current of a detector may be read on a current meter or an oscilloscope.

We need to know a few important detector terms that are related to operation of a rectifier:

- Current sensitivity
- Voltage sensitivity

We will now define each in detail.

a) Current Sensitivity (β_i). An important term describing a rectifier is the current sensitivity (β_i), which is defined to be the ratio of the change in the DC current ΔI_{DC} (due to the RF signal) to the input RF power (P_{IN}):

$$\beta_i = \frac{\Delta I_{DC}}{P_{IN}} \tag{18.14}$$

Considering a purely resistive model for the diode (i.e., neglecting all the parasitics), we can write:

$$\Delta I_{DC} = (v_m^2 G'_d)/4 \tag{18.15}$$

$$P_{IN} = \frac{1}{2}(v_m i_m) = G_d v_m^2/2 \tag{18.16}$$

Thus, we have:

$$\beta_i = \frac{G'_d}{2G_d} = \frac{1}{2nV_T} \text{ (A/W)} \tag{18.17}$$

b) Voltage Sensitivity (β_v). Another useful term is the voltage sensitivity, whose definition varies depending on the value of the load. We will consider two possible situations:

Open-Circuit Voltage Sensitivity The open-circuit voltage sensitivity (β_v) is defined in terms of the voltage drop across the junction resistance (R_j) when the diode is open circuited (i.e., $R_L = \infty$). It can be seen that it is equal to the current sensitivity (β_i) times the junction resistance (R_j):

$$R_j = \left(\frac{dI}{dV}\right)^{-1} \tag{18.18}$$

$$\beta_v = (\beta_i)R_j = nVT(\beta_i)/(I_o + I_S) \tag{18.19}$$

Finite-Load Sensitivity For a finite load (i.e., $R_L \neq \infty$), we need to modify Equation 18.19 by a factor of $R_L/(R_L + R_V)$, where R_V is the video resistance defined by:

$$R_V = R_j + R_S \tag{18.20}$$

Assuming $R_S << R_j$, then:

$$R_V \approx R_j \tag{18.21}$$

Thus, Equation 18.19 for a finite load resistance can be written as:

$$(\beta_v)_{R_L \neq \infty} = \beta_v\left(\frac{1}{1 + R_j/R_L}\right) \tag{18.22}$$

Substituting for β_i in Equation 18.22, we obtain:

$$(\beta_v)_{R_L \neq \infty} = \beta_i\left(\frac{R_j}{1 + R_j/R_L}\right) = \beta_i(1/R_j + 1/R_L)^{-1} \tag{18.23}$$

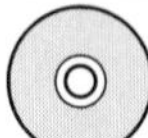

EXAMPLE 18.1

A diode has the following equivalent circuit parameters:

$$C_p = 0.1 \text{ pF}$$

$$L_p = 2.0 \text{ nH}$$

$$C_j = 0.15 \text{ pF}$$
$$R_S = 10\ \Omega$$
$$I_S = 0.1\ \mu\text{A}$$
$$n = 1$$

Assuming the input power (P_{IN}) is 0 dBm, find the junction resistance (R_j), β_i, and β_v for $R_L = \infty$ and $R_L = 10\ k\Omega$ where $I_o = 0$ and $I_o = 60\ \mu A$ at $T = 293^o$ K. For the first order of approximation, ignore all reactive parameters.

Solution:

The junction resistance (R_j) is bias dependent and is given by:

$$R_j = nV_T/(I_o + I_S)$$
$$P_{IN} = 0 \text{ dBm} = 1 \text{ mW}$$
$$V_T = 25 \text{ mV} \quad (T = 293^\text{o} \text{ K})$$

a.

$$I_o = 0$$
$$R_j = 25 \text{ mV}/0.1 \times 10^{-3} \text{ mA} = 250 \text{ k}\Omega$$
$$\beta_i = G'_d/2G_d = 1/2nV_T = 1/2 \times 25 \times 10^{-3} = 20 \text{ A/W}$$
$$\text{For} \quad R_L = \infty \Rightarrow \beta_v = \beta_i R_j = 20 \times 250 \times 10^3 = 5 \times 10^6 \text{ V/W}$$
$$\text{And for} \quad R_L = 10 \text{ k}\Omega\,,\ R_S = 10\ \Omega$$
$$R_V = R_S + R_j \approx R_j = 250 \text{ k}\Omega$$
$$(\beta_v)_{R_L = 10\text{k}\Omega} = \beta_v/(1 + R_V/R_L) = 1.92 \times 10^5 \text{ V/W}$$

b.

$$I_o = 60\ \mu\text{A} >> I_S$$
$$R_j = 25 \text{ mV}/60 \times 10^{-3} \text{ mA} = 417\ \Omega$$

For an ideal diode (i.e., a purely resistive model), β_i remains the same, that is, $\beta_i =$ 20 A/W.

$$\text{For} \quad R_L = \infty \Rightarrow \beta_v = \beta_i R_j = 20 \times 417 = 8340 \text{ V/W}$$
$$\text{And for} \quad R_L = 10 \text{ k}\Omega\,,\ R_S = 10\ \Omega$$
$$R_V = R_S + R_j = 427\ \Omega$$
$$(\beta_v)_{R_L = 10\text{k}\Omega} = \beta_v/(1 + R_V/R_L) = 7998 \text{ V/W}$$

This example shows that there is a tradeoff between the DC bias current and voltage sensitivity. That is, increasing the DC current will reduce the junction resistance and consequently reduce the open-circuit voltage sensitivity.

Furthermore, because R_j is reduced substantially any load value much higher than R_j will have no effect on the (β_v) value and thus the circuit has become load-insensitive, which is a desirable feature for a rectifier.

In the next example, we will analyze a detector using a general model of a diode.

EXAMPLE 18.2

A diode used in a detector has the equivalent circuit model given in Figure 18.8. Assuming the input signal is unmodulated, carry out the following:

1. *Determine β_i and β_v assuming the diode is biased at (I_o, V_o).*
2. *Determine the maximum current when β_v achieves its highest value. Assume $v_j = |v_j|\cos\omega_o t$ is the voltage across the junction, as shown in* Figure 18.8.

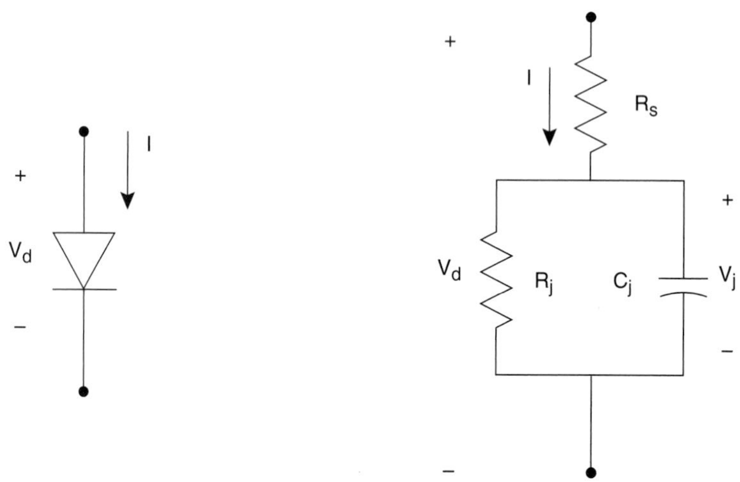

FIGURE 18.8 Circuit for Example 18.2.

Solution:
We know that:

$$\beta_i = \frac{\Delta I_{DC}}{P_{IN}} \tag{18.24a}$$

From Equation 18.24a, we can see that in order to find β_i we need to obtain:

a. ΔI_{DC}, and
b. P_{IN}.

Therefore, we proceed as follows:

1. Assuming i, v_j, and v_d are the small-signal current and voltages in the diode, we can write:

$$v_d = iR_S + v_j \tag{18.24b}$$

$$v_j = i\left(R_j \parallel \frac{1}{j\omega C_j}\right) = \frac{iR_j}{1 + j\omega R_j C_j} \tag{18.25}$$

Eliminating i from Equations 18.24b and 18.25, we obtain:

$$v_j = \frac{v_d}{(1 + R_S/R_j) + j\omega R_S C_j} \tag{18.26}$$

From Equation 18.3, we know that the rectified output DC current is contained in the second order-term and is given by:

$$\Delta i = \frac{v_j^2}{2} \frac{d^2 i}{dv^2}\bigg|_{I_o}$$

where

$$\frac{d^2 i}{dv^2}\bigg|_{I_o} = \frac{1}{nV_T R_j}$$

$$R_j = \frac{nV_T}{I_S + I_o} \tag{18.27}$$

Thus, we can write:

$$\Delta i = \frac{|v_j|^2 (1 + \cos 2\omega_o t)}{4nV_T R_j} \tag{18.28}$$

For an infinite load ($R_L = \infty$) and assuming R_S is negligibly small, ΔI_{DC} can easily be obtained from Equation 18.28 as:

$$\Delta I_{DC} = \frac{|v_j|^2}{4nV_T R_j} \tag{18.29}$$

Adding the effect of R_S, Equation 18.29 can be written as:

$$\Delta I_{DC} = \frac{|v_j|^2}{4nV_T R_j}\left(\frac{R_j}{R_S + R_j}\right) \tag{18.30}$$

2. To find P_{IN}, we note that it is equal to the power absorbed by the diode and is given by:

$$P_{IN} = \frac{1}{2} Re[v_d i_d^*] = \frac{|v_d|^2}{2} Re[Y_d] \tag{18.31}$$

where Y_d is the diode admittance and is given by:

$$Y_d = \frac{1/R_j + j\omega C_j}{(1 + R_S/R_j) + j\omega C_j R_S}$$

$$Re[Y_d] = \frac{(1/R_j)(1 + R_S/R_j) + (\omega C_j)^2 R_S}{(1 + R_S/R_j)^2 + (\omega C_j R_S)^2} \tag{18.32}$$

Using Equations 18.26 and 18.32 in 18.31, we can write:

$$P_{IN} = \frac{|v_j|^2}{2}\left[\frac{(1 + R_S/R_j)}{R_j} + R_S(\omega C_j)^2\right] \tag{18.33}$$

Thus, β_i and β_v are given by:

$$\beta_i = \frac{\Delta I_{DC}}{P_{IN}} = \frac{1}{2nV_T(1 + R_S/R_j)[(1 + R_S/R_j) + (\omega C_j)^2 R_S R_j]} \tag{18.34}$$

$$\beta_v = \beta_i R_j = \frac{nV_T\beta_i}{(I_o + I_S)} \text{ V/W} \qquad (R_L = \infty) \tag{18.35}$$

For non-infinite load values (i.e., $R_L \neq \infty$), β_v must be multiplied by a factor $R_L/(R_L + R_V)$, as discussed earlier.

Effect of Frequency and Bias on βv. Due to the change in voltage across the diode junction, β_v decreases as frequency is increased. The strong frequency dependence of β_v is shown in Figure 18.9. For wideband detector applications, the matching circuit can be designed to compensate for the drop in sensitivity for high-frequency signals in such a way that the overall detector sensitivity remains constant throughout the frequency band.

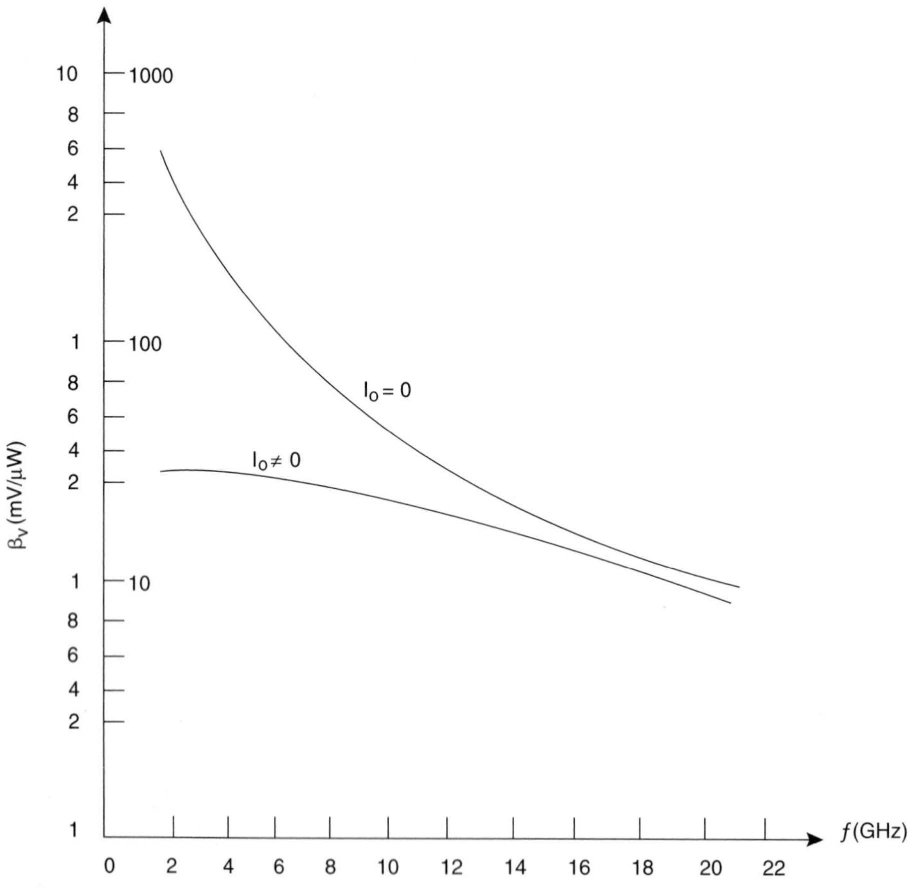

FIGURE 18.9 Typical variation of β_v with frequency and bias current as a parameter.

The diode is usually forward biased to display higher nonlinearity for optimum performance. Typical bias current values (I_o) range from 10 to 100 mA. Typical values for β_v are from 1000 mV/μW to 10 mV/μW.

There is one drawback, though, in biasing the diode; that is a reduction in the variation as well as the value of voltage sensitivity (β_v), particularly at the lower frequencies, as shown in Figure 18.9. This is caused by a lower R_j due to the bias current (I_o), which causes a reduction in voltage across the junction. If we let $I_T = I_o + I_S$, then from Equation 18.35 it can be seen that voltage sensitivity (β_v) is a parabolic-type function of I_T and has a maximum value $(\beta_v)_{max}$ at a given frequency, which can be determined by:

$$I_T = I_o + I_S \tag{18.36}$$

$$\frac{\partial \beta_v}{\partial I_T} = 0 \tag{18.37}$$

Solving Equation 18.37, we obtain $(I_T)_{max}$ given by:

$$(I_T)_{max} = nV_T\omega C_j\sqrt{\frac{R_S}{R_L}} \tag{18.38}$$

$(I_T)_{max}$ as given by Equation 18.38 is a function of frequency and increases monotonically as frequency is increased. This means that in order to operate a detector at $(\beta_v)_{max}$, a higher bias current is needed as frequency is increased.

18.3.2 Amplitude-Modulated (AM) Signal

Nonlinearity of a diode in a detector circuit may be used to demodulate a signal that is amplitude modulated (AM) on a carrier wave. In other words, the detector's function is to separate the envelope (which is the signal) from the carrier wave. In this case, the voltage at the diode's input terminal is given by:

$$\begin{aligned} v_{RF}(t) &= v_m(1 + m\cos\omega_m t)\cos\omega_o t \\ &= v_m\cos\omega_o t + \frac{m}{2}v_m\,[\cos(\omega_o - \omega_m)t + \cos(\omega_o + \omega_m)t\,] \end{aligned} \tag{18.39}$$

where m is called the modulation index ($0 \leq m \leq 1$) and we assume also that $\omega_o >> \omega_m$, which is usually the case for AM. From Equation 18.8, we can write the diode's RF current as:

$$i = vG_d + 1/2\,(v^2G'_d) + \ldots.$$

$$\begin{aligned} i &= (1 + m^2/2)v_m^2\,G'_d/4 + mv_m^2\,G'_d\cos(\omega_m t)/2 + m^2\,v_m^2\,G'_d\cos(2\omega_m t)/8 \\ &+ v_mG_d\cos(\omega_o t) + (1 + m^2/2)v_m^2\,G_d'\cos(2\,\omega_o t)/4 + mv_mG_d\cos[(\omega_o \pm \omega_m)t]/2 \\ &+ mv_m^2\,G'_d\cos[(2\,\omega_o \pm \omega_m)t]/4 + m^2\,v_m^2\,G'_d\cos[2(\omega_o \pm \omega_m)t]/16 \end{aligned} \tag{18.40}$$

The desired output signal (i_{OUT}) is obtained by using a DC block at the output followed by a low-pass filter (with a 3 dB cutoff frequency of ω_m), and is given by:

$$i_{OUT} = mv_m^2\,G'_d\cos(\omega_m t)/2 \tag{18.41}$$

As can be seen, Equation 18.40 provides a plethora of frequencies at 0, ω_m, $2\omega_m$, ω_o, $2\omega_o$, $2\omega_o \pm \omega_m$, and $2(\omega_o \pm \omega_m)$ with relative amplitude of each component as shown in Figure 18.10 and Table 18.1.

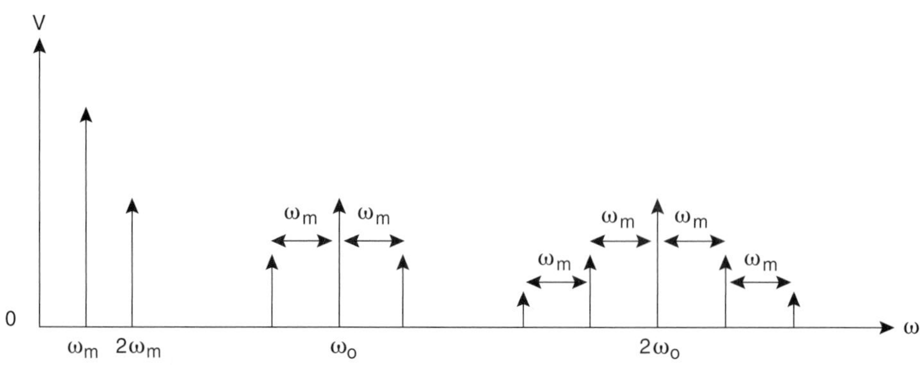

FIGURE 18.10 A diode's output spectrum of a detected AM signal.

TABLE 18.1 Spectrum of Frequencies Generated as a Result of Mixing in Diode

Frequency (rad/sec)	Amplitude $A = v_m G_d$, $B = v_m^2 G'_d/4$
0	$B(1 + m^2/2)$
ω_m	$2mB$
$2\omega_m$	$m^2B/2$
ω_o	A
$2\omega_o$	$(1 + m^2/2)B$
$\omega_o \pm \omega_m$	$mA/2$
$2\omega_o \pm \omega_m$	mB
$2(\omega_o \pm \omega_m)$	$m^2B/4$

The desired demodulated signal at frequency (ω_m) can easily be filtered using a DC block followed by a low-pass filter as discussed earlier. From Table 18.1, we note that the demodulation signal (at $\omega = \omega_m$) is proportional to $v^2{}_m$, which is proportional to the power of the modulating signal (P_{IN}). This is known as operation in the "Square-law" region, used particularly in *VSWR* measurement and signal-level indicator applications as shown in Figure 18.11.

Square-law operation is possible as long as the signal power remains in the small-signal range (typically between –40 dBm and –20 dBm). If the input signal power is gradually increased beyond the small-signal range, then the output will first become linearly proportional to v_m (i.e., $\alpha\sqrt{P_{IN}}$), as shown in Figure 18.11. On further increase of input signal power, the output will saturate and become a constant current regardless of the input power value (typically for $P_{IN} > 10$ to 20 dBm).

18.4 DETECTOR LOSSES

From Chapter 3, *Mathematical Foundation for Understanding Circuits*, we know that in general, the loss of a circuit (e.g., a detector) is defined to be:

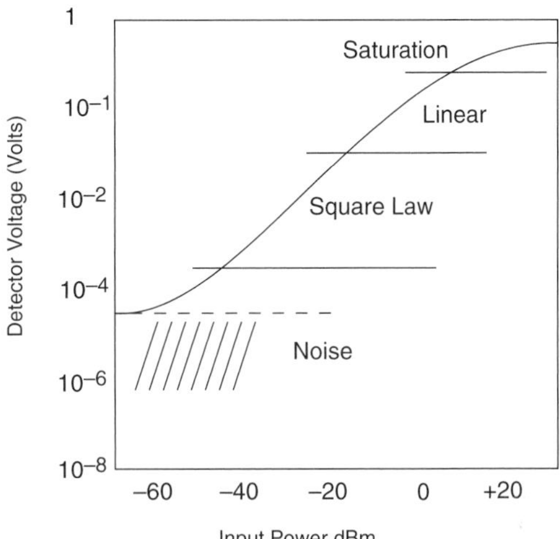

FIGURE 18.11 Typical diode detector output characteristic.

$$L(\text{dB}) = 10\log_{10}\left(\frac{P_{IN}}{P_{OUT}}\right) \tag{18.42}$$

Three types of losses are incurred in a detector while the signal is traveling through the diode for conversion to the proper frequency (i.e., DC for unmodulated or ω_m for AM signals), as discussed in the next section.

18.4.1 Diode Loss

For an unmodulated input signal, we can use the results for P_{IN} of a diode from Equation 18.33, as given in Example 18.2:

$$P_{IN} = \frac{|v_j|^2}{2}\left[\frac{(1 + R_S/R_j)}{R_j} + R_S(\omega C_j)^2\right]$$

To find P_{OUT}, we note that the useful power is obtained at the junction across resistance R_j:

$$P_{OUT} = \frac{1}{2}\frac{|v_j|^2}{R_j}$$

Thus, L_d in dB is found to be:

$$L_d(\text{dB}) = 10\log_{10}\left(1 + \frac{R_S}{R_j} + \omega^2 R_S R_j C_j^2\right) \tag{18.43}$$

18.4.2 Mismatch Loss

From Figure 18.12, we can see that the matching circuit will reflect a portion of the input power (P'_{IN}) equal to:

$$P_{ref} = |\Gamma|^2 P'_{IN}$$

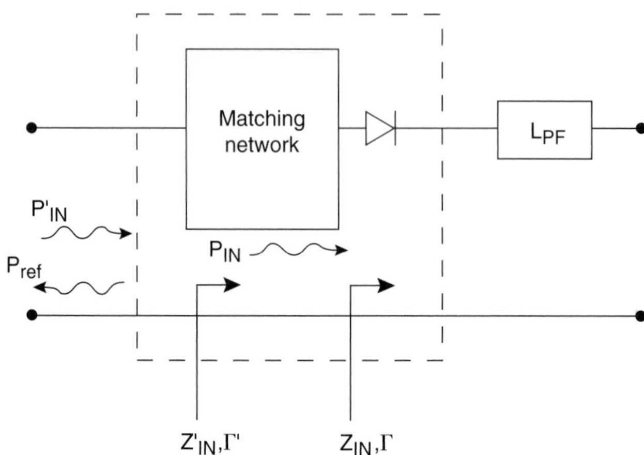

FIGURE 18.12 Power flow in a detector.

Thus, the power reaching the diode (P_{IN}) is the transmitted power, i.e.,

$$P_{IN} = P'_{IN} - P_{ref}$$

or

$$P_{IN} = (1 - |\Gamma'|^2) P'_{IN} \tag{18.44}$$

where

$$\Gamma' = \frac{Z'_{IN} - Z_o}{Z'_{IN} + Z_o}$$

and Z'_{IN} is the diode's impedance transferred to the input of the matching network.

From Equation 18.44, we can see the input power that reaches the diode is reduced by $(1 - |\Gamma'|^2)$. Thus, the mismatch loss is given by:

$$L_m(\text{dB}) = 10 \log_{10} \frac{1}{1 - |\Gamma'|^2}$$

or

$$L_m\,(\text{dB}) = -10 \log_{10} (1 - |\Gamma'|^2) \tag{18.45}$$

18.4.3 Harmonic Loss (Also Called Intrinsic Loss)

Due to the diode's nonlinear *I-V* curve, many signal harmonics are generated that absorb some of the signal power. Therefore, not all of the signal power is converted to the desired signal power, which leads to the concept of harmonic loss (L_h). This loss is normally specified in dB for each detector in the manufacturer's data sheets.

Therefore, the total detector loss (L) in dB, can be summed up as:

$$L(\text{dB}) = L_d + L_m + L_h\ (\text{dB}) \tag{18.46}$$

18.5 EFFECT OF MATCHING NETWORK ON THE VOLTAGE SENSITIVITY

As stated in the previous section, the matching network creates a mismatch loss in the circuit that needs to be taken into account in order to determine the voltage sensitivity (β_v) accurately. From previous discussions, we already know that the detector's voltage sensitivity without the matching network (β_v) is given by:

$$\beta_v = R_j(\Delta I_{DC}/P_{IN}) \tag{18.47}$$

Substituting for P_{IN} from Equation 18.44, we have:

$$\beta_v = R_j \frac{\Delta I_{DC}}{P'_{IN}(1-|\Gamma'|^2)} = \frac{\beta'_v}{(1-|\Gamma'|^2)} \Rightarrow \beta'_v = \beta_v(1-|\Gamma|^2) \tag{18.48}$$

where β'_v is defined by:

$$\beta'_v = R_j \frac{\Delta I_{DC}}{P'_{IN}}$$

which is the detector's voltage sensitivity with the matching section in place.

Equation 18.48 clearly indicates that mismatch loss $(1 - |\Gamma'|^2)$ plays an important role in a detector's overall voltage sensitivity and must be taken into account. Example 18.3 illustrates this point further.

EXAMPLE 18.3

Find the percentage reduction in a detector's sensitivity if the mismatch on the line produces a VSWR = 4:1.

Solution:

$$VSWR = 4 \Rightarrow |\Gamma| = 0.6$$

$$\text{Sensitivity loss factor } = 1 - |\Gamma|^2 = 0.64$$

Thus, $\beta'_v = 0.64\,\beta_v$, and the detector voltage sensitivity is reduced to 64 percent of its original value due to mismatch loss.

18.6 DETECTOR DESIGN

The design of a detector is determined by the trade-offs between many factors such as bandwidth, frequency of operation, and sensitivity. Good guidelines can be established to simplify the design process as described below.

18.6.1 Design Considerations

The following design considerations are of special significance:

1. We know that the diode's impedance is given by:

$$Z_d = R_S + R_j \parallel X_{C_j}$$

To achieve high β_v ($\beta_v = R_j \beta_i$), we require two conditions to be satisfied:

a. The junction shunt resistance (R_j) should be very large, i.e.,

$$R_j >> X_{C_j}$$

$$\text{where} \quad X_{C_j} = \frac{1}{\omega_o C_j}$$

Thus the diode's impedance (Z_d) can be written as:

$$Z_d = R_S + R_j \parallel X_{C_j} \approx R_S + X_{C_j}$$

b. The diode's impedance at the operating frequency (f_o) should be chosen such that it is much higher than R_S, i.e.,

$$|X_{C_j}| > 10 R_S \tag{18.49}$$

2. Furthermore, we need to select a diode with a high 0-V cutoff frequency (f_{co}), which is defined by:

$$f_{co} = \frac{1}{2\pi R_S C_{jo}} \tag{18.50}$$

where C_{jo} is the diode's junction capacitance value at zero bias voltage. Equation 18.50 gives the cutoff frequency (f_{co}) beyond which the diode behaves improperly and is no longer useful.

The diode's junction capacitance (C_j), due to the depletion region for a one-sided junction (e.g., n^+p for microwave frequencies), can be shown to be a function of the applied bias voltage as follows:

$$C_j = \frac{C_{jo}}{\left(1 - \frac{V_a}{V_{bi}}\right)^S} \tag{18.51}$$

where

C_{jo} = The junction capacitance at zero bias voltage
V_a = The applied bias voltage across the junction (for reverse bias $V_a = -V_r$)
V_{bi} = The built-in voltage

$$S = \frac{1}{m+2},$$

and

m is the exponent parameter in the doping distribution function (N) of the lighter side, defined as:

$$N(x) = N_B x^m$$

with N_B being an arbitrary constant established by the doping profile.

3. Combining the preceding two considerations in (1) and (2), we obtain:

$$|X_{C_{jo}}| = \frac{1}{\omega_o C_{jo}} = \frac{1}{2\pi f_o C_{jo}} > 10R_S \Rightarrow \frac{1}{2\pi R_S C_{jo}} > 10f_o \Rightarrow f_{co} > 10f_o \quad (18.52)$$

From Equation 18.52, we can see that f_{co} should be at least 10 times bigger than the highest operating frequency (f_o).

From this discussion, we can conclude that a good diode for a proper detector design should be selected such that:

$$C_{jo} < \frac{1}{20\pi f_o R_S} \quad (18.53)$$

EXAMPLE 18.4

Select a proper diode to be used in a narrowband detector having a high voltage sensitivity and operating at 5 GHz.

Solution:

$$f_o = 5 \text{ GHz}$$

$$f_{co} > 10f_o \Rightarrow f_{co} > 50 \text{ GHz}$$

If we allow maximum R_S to be 15 Ω, then we can write:

$$R_S \le 15\ \Omega$$

$$C_{jo} < \frac{1}{(20\pi)(5\times 10^9)(15)} \Rightarrow C_{jo} < 0.2 \text{ pF}$$

Thus, if we allow the maximum series resistance to be 15 Ω, then the junction capacitance must not exceed 0.2 pF, in order to provide a highly voltage-sensitive detector.

NOTE: *Once the diode's parameters are established, Z_{IN} or Γ_{IN} can subsequently be determined (see Figure 18.7), and the design of the input matching circuit will simply follow the standard procedures set forth in Chapter 10, Applications of the Smith Chart and Chapter 11, Design of Matching Networks.*

LIST OF SYMBOLS/ABBREVIATIONS

A symbol will not be repeated once it has been identified and defined in an earlier chapter, as long as its definition remains unchanged.

C_j	Junction capacitance of a diode
C_{jo}	Junction capacitance of a diode under zero bias
C_P	Parasitic capacitance of a diode
G_d	Dynamic conductance of a diode
G'_d	First-order derivative of the dynamic conductance of a diode
f_{co}	Diode cutoff frequency
I_{DC}	Total diode DC current
I_S	Diode saturation current
L	Detector loss
L_d	Diode loss
L_h	Harmonic loss
L_m	Mismatch loss
L_P	Parasitic inductance of a diode
m	Modulation index
m	The exponent parameter in the doping distribution function of the lighter side of a pn junction, $N(x) = N_B x^m$
n	Ideality factor that varies ($1 \leq n \leq 2$) and is dependent on the material and physical structure of a diode [reference $I(V) = I_S(e^{V/nV_T} - 1)$]
P_{ref}	Reflected power
R_j	Junction resistance of a diode
R_S	Series resistance of a diode
V_a	Applied voltage
V_{bi}	Built-in voltage barrier at the junction
V_j	Junction voltage of a diode
V_r	Reverse bias voltage
β_i	Current sensitivity of a detector (unmodulated signal, m = 0)
β_v	Voltage sensitivity of a detector (unmodulated signal, m = 0)
ω_o	Local oscillator frequency in radians/sec
ω_m	Modulation frequency in radians/sec

PROBLEMS

18.1 Given the following detector diode:

$$I_S = 1 \times 10^{-7}\ \text{A}$$

$$n = 1.05$$

$$C_{jo} = 0.25\ \text{pF}$$

$$R_S = 15\ \Omega$$

$$R_L = 1\ \text{M}\Omega$$

Calculate the voltage sensitivity for each case below:

f (GHz)	T° (K)	I_o (μA)	β_V
2	296	0	?
2	296	40	?
10	296	0	?
10	373	0	?

18.2 Given the following detector diode:

$$I_S = 1 \times 10^{-8}\text{ A}$$
$$n = 1.05$$
$$C_{jo} = 0.20\text{ pF}$$
$$R_S = 10\ \Omega$$
$$R_L = 1\text{ M}\Omega$$
$$L_P = 1\text{ nH}$$
$$C_P = 0.25\text{ pF}$$
$$T = 20°\text{C}$$

Find the diode's impedance in each case below:

a. $I_o = 0, f = 2$ GHz
b. $I_o = 0, f = 6$ GHz
c. $I_o = 50\ \mu\text{A}, f = 2$ GHz

18.3 Find the new voltage sensitivity of a modified detector that has two series transmission lines added at the input terminals, as shown in Figure P18.3. The voltage sensitivity of the unmodified detector is 20 mV/μW with an input impedance $Z_d = 7 - j50\ \Omega$. The input incoming signal is at 5 GHz.

FIGURE P18.3

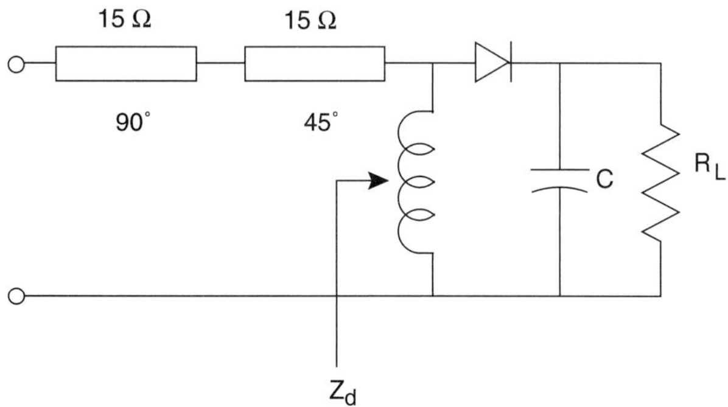

18.4 Given a diode with the following characteristics at 4 GHz:

$$I_S = 1.1 \times 10^{-7} \text{ A}$$
$$n = 1.1$$
$$C_{jo} = 0.20 \text{ pF}$$
$$R_S = 10 \ \Omega$$
$$L_P = 1 \text{ nH}$$
$$C_P = 0.25$$
$$R_L = 10 \text{ K}\Omega$$
$$T = 20°\text{C}$$

Using this diode, design a 4 GHz detector for maximum open-circuit voltage sensitivity.

18.5 Given a diode with the following characteristics at 10 GHz:

$$I_S = 1.0 \times 10^{-7} \text{ A}$$
$$n = 1.1$$
$$C_{jo} = 0.10 \text{ pF}$$
$$R_S = 15 \ \Omega$$
$$L_P = C_P = 0$$

Calculate the open-circuit voltage sensitivity at 10 GHz for $I_o = 0$, 20 and 50 μA. Neglect the effect of bias current on the junction capacitance.

REFERENCES

[18.1] Bahl, I. and P. Bhartia. *Microwave Solid State Circuit Design.* New York: Wiley Interscience, 1988.

[18.2] Bhartia, P. and I. Bahl. *Millimeter Wave Engineering and Application.* New York: John Wiley & Sons, 1984.

[18.3] Chaffin, R. J. *Microwave Semiconductor Devices.* New York: John Wiley & Sons, 1973.

[18.4] Chang, K. *Microwave Solid-State Circuits and Applications.* New York: John Wiley & Sons, 1994.

[18.5] Pozar, D. M. *Microwave Engineering*, 2nd ed. New York: John Wiley & Sons, 1998.

[18.6] Steinbrecher, D. H. "Circuit Aspects of Nonlinear Microwave Networks." *IEEE Proc. Int. Symp. Circuits and Systems*, 1975, pp. 477-9.

[18.7] Watson, H. A. *Microwave Semiconductor Devices and Their Circuit Applications.* New York: McGraw-Hill, 1969.

[18.8] Yngvesson, S. Y. *Microwave Semiconductor Devices.* Norwell, MA: Kluwer Academic Publishers, 1991.

CHAPTER 19

RF/Microwave Frequency Conversion II: Mixer Design

19.1 INTRODUCTION

Having analyzed diodes under small-signal conditions and studied their use in rectifiers and detectors, we are now ready to study another class of frequency conversion devices: mixers. We would like to define the term "mixer" right at the outset of this chapter.

DEFINITION-MIXER: *A nonlinear three-port circuit (two inputs and one output) that generates a spectrum of output frequencies equal to the sum (or difference) of the two input frequencies and their harmonics. The two input ports are referred to as RF and LO whereas the output is called the IF port.*

A mixer uses the nonlinearity of a device to generate a spectrum of frequencies (at the IF port) based on the sum and difference of the harmonics of the RF signal and the local oscillator (also called the pump oscillator) signal frequencies (see Figure 19.1). From this figure, we can see that the nonlinear device is flanked on three sides by three matching circuits, which need to be designed properly for maximum conversion efficiency.

When a mixer is used as an up-converter (e.g., in a transmitter), the "sum frequency" (i.e., $f_i = f_o + f_r$) is utilized and the "difference frequency" is rejected, as shown in Figure 19.2a. In an up-converter, the IF oscillator is modulated with the desired information signal, which when mixed with the LO signal will generate the desired frequency conversion. Use of a mixer as an up-converter particularly in the case of a radar or a transceiver is advantageous because it allows the use of a single local oscillator for both the receiver and the transmitter.

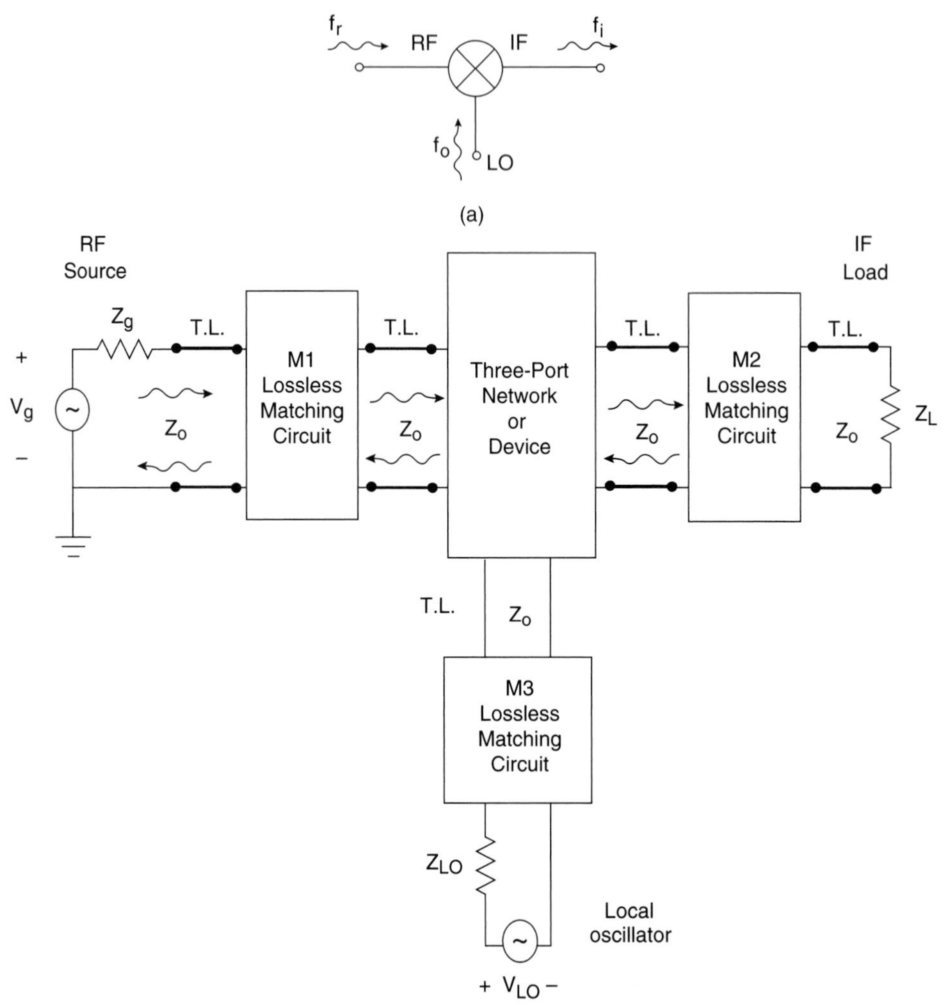

FIGURE 19.1 (a) Mixer symbol, (b) Mixer block diagram.

Vice versa, when the application requires a down-conversion (e.g., in a heterodyne receiver), the difference frequency (i.e., $f_i = |f_o - f_r|$) is used and the sum frequency is filtered out, as shown in Figure 19.2b. Use of a mixer as a down-converter in a heterodyne receiver has several advantages:

- The IF signal, being in the range of 10 MHz $\leq f_{IF} \leq$ 100 MHz, lends itself for low-noise amplification because $1/f$ noise is lower in IF than in RF frequency range.

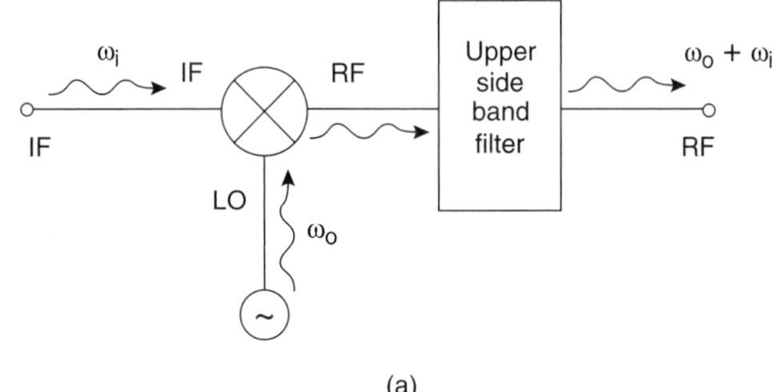

(a)

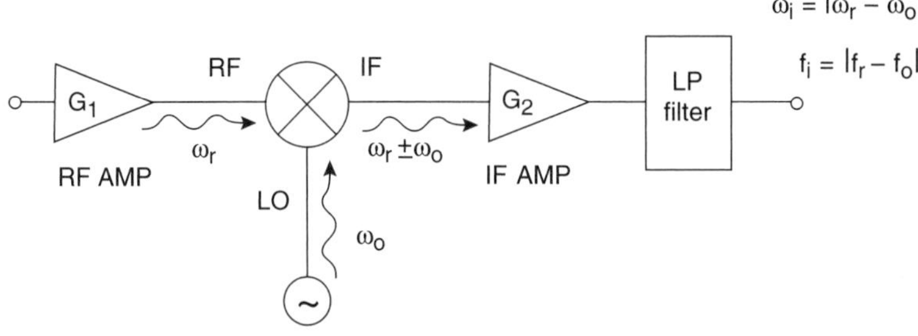

(b)

FIGURE 19.2 Frequency conversion process: (a) Up-conversion in a transmitter, (b) Down-conversion in a heterodyne receiver.

- By changing the LO frequency, a heterodyne receiver can be tuned to receive a wide band of RF frequencies (with the same low-noise quality of IF amplification) without a need for a high-gain wideband RF amplifier, which would have been necessary otherwise.

NOTE: *1/f noise is inversely proportional to the frequency of the signal and is predominant in lower-frequency signals. It is present in a diminishing manner at the rate of -10 dB/decade, from almost DC all the way to a 3-dB cutoff frequency (f_n), of around 300 kHz to 500 kHz. 1/f noise becomes negligibly small and insignificant around 1 MHz. Presence of 1/f noise at the IF frequency band is not desirable because noise is added in a skewed fashion to the signal; that is, lower-frequency signals in the band become noisier than the higher-frequency signals. Such a condition will cause signal distortion and is usually avoided by increasing the IF signal well above the noise cutoff frequency. Beyond the 1/f noise frequency range, white noise becomes predominant. Such a noise is not frequency dependent and is more tolerable for mixer IF needs. Thus, it is a sound design practice to select the IF frequency range (in the final mixer stage) well above 1 MHz, preferably in the 10 MHz to 100 MHz frequency range.*

In general, mixers utilize one or more nonlinear device(s), properly pumped with a relatively large signal (called LO), to mix with an RF signal in order to generate a spectrum of frequencies based on the sum and difference of the harmonics of the RF and LO frequencies. The result can be succinctly be written as:

$$\omega_i = m\,\omega_r \pm n\,\omega_o \tag{19.1}$$

where m and n are positive integers.

The most important terms for mixer operation are those with frequencies at $\omega_r + \omega_o$ and at $\omega_r - \omega_o$ (i.e., $m,n = 1$).

The most common devices used in mixers are diodes, which have already been analyzed in Chapter 18, *RF/Microwave Frequency Conversion I: Rectifier and Detector Design*. The nonlinear behavior of a diode is responsible for the existence of a spectrum of frequencies at the IF port. There are several types of diode mixers, described next.

> **POINT OF CAUTION:** *The pumping signal (or LO) should have a relatively large signal level. The large signal turns the device on and off, thereby causing the device to behave nonlinearly, which is the basis for the mixing action. Therefore, it can be said that the level of the LO signal is responsible for the entire mixing process. In actual practice, care must be taken to see that the LO signal level is high enough so that the mixer operates properly.*

19.2 MIXER TYPES

There are many types of mixers, used for different purposes and needs. The following presents several general categories of these devices along with their possible applications.

19.2.1 Down-Converters

A down-converter is a mixer that, with the help of an LO, shifts the frequency of the RF signal substantially down to an IF signal, ready for further signal processing (see Figure 19.2). The IF signal frequency is given by:

$$\omega_i = |\omega_r - \omega_o| \tag{19.2a}$$

Such a mixer is usually used in a receiver as a demodulator to remove the carrier wave from the transmitted signal in order to obtain the information-carrying signal. For example, a down-converter having an RF signal of 14.1 GHz and an LO signal of 14 GHz will produce an IF frequency of 100 MHz.

> **NOTE:** *Equation 19.2a gives two RF mixing frequencies by which the same IF signal is generated:*

$$\omega_i = +(\omega_r - \omega_o) \Rightarrow \omega_r = \omega_o + \omega_i \tag{19.2b}$$

$$\omega_i = -(\omega'_r - \omega_o) \Rightarrow \omega'_r = \omega_o - \omega_i \tag{19.2c}$$

If we consider one of the two signals (e.g., $\omega_r = \omega_i + \omega_o$) as the "direct frequency" to produce an IF signal then the other ($\omega'_r = \omega_o - \omega_i$), called the image frequency, is also capable of producing the same IF signal (called the image response). The image response is generally undesirable and needs to be eliminated either by proper filtering or sophisticated schemes. A mixer that suppresses the image response is called an image rejection mixer. The generated IF signal, in the absence of image frequency, is usually referred to as a single-sideband (SSB) IF response, whereas a mixer that uses both RF signals (at direct and image frequencies) generates a double-sideband (DSB)

IF response, having twice as much power. Usually, the SSB value for conversion loss or noise figure of a mixer is used and reported in literature, but care must be used to obtain accurate mixer measurements and subsequently to report these quantities properly. In this text, we use SSB values for conversion loss and noise figure as discussed in a later section.

Using the previous example, we can see that if the direct frequency is at 14.1 GHz, then an image frequency (at the RF port) of 13.9 GHz would also produce a 100 MHz IF signal, for an LO signal of 14 GHz.

OBSERVATION: *From the preceding example, we can see that the image frequency at 13.9 GHz is separated from the direct frequency at 14.1 GHz by twice the IF frequency (200 MHz). Thus, in general, we can write:*

$$\omega_r - \omega'_r = 2\,\omega_i$$

CONCLUSION: *Based on this observation, we can define an image signal of a mixer as a spurious signal related to the desired RF signal by twice the IF frequency.*

19.2.2 Up-Converters

An up-converter is a down-converter in reverse order (see Figure 19.2). The desired information signal is modulated on the IF signal, which in turn is mixed with the local oscillator signal (ω_o) producing a signal at $\omega_o + \omega_i$, that is,

$$\omega_r = \omega_o + \omega_i$$

Use of proper filtering or an image rejection mixer is needed to generate the sum frequency ($\omega_r = \omega_o + \omega_i$). An up-converter is used in a transmitter to modulate a carrier wave (LO signal) with an information-bearing signal (IF signal) in order to generate an RF signal for transmission.

For example, an up-converter could have an IF signal of 100 MHz and an LO signal of 10 GHz, with the resulting RF frequency of 10.1 GHz.

NOTE: *In communication engineering jargon, a modulator can be considered to be an up-converter whereas a demodulator is simply a down-converter.*

19.2.3 Harmonic Mixers

A simple method of down-converting a high-frequency RF signal when only a low-frequency LO exists, is the use of "harmonic mixers." Frequency down-conversion is achieved by mixing the high-frequency signal with an appropriate harmonic of the LO frequency, that is,

$$\omega_i = n\omega_o - \omega_r \tag{19.3}$$

where $n = 1, ..., N$ is an integer. As a special case, when $n = 1$, we obtain a fundamental down-converter, described earlier.

It is important to note that because the appropriate higher harmonic of the LO ($n\omega_o$) is generated within the diode mixer itself, the conversion loss is higher for a harmonic mixer than it is for a simple fundamental down-converter. This increase in conversion loss can be approximated by 3 dB for every integral unit increase in n. Possible applications of harmonic mixers are in millimeter-wave instrumentation where the use of a high-frequency LO source, which can generate substantial power to satisfy LO power needs, is impractical or very expensive.

For example, if a fundamental down-converter ($n = 1, f_r = 68$ GHz, $f_o = 66$ GHz, $f_i = 2$ GHz) has a 6 dB conversion loss, then a ×2 harmonic mixer ($n = 2, f_r = 68$ GHz, $f_o = 33$ GHz, $f_i = 2$ GHz) will have about $L_C = 9$ dB and a ×3 harmonic mixer ($n = 3, f_r = 68$ GHz, $f_o = 22$ GHz, $f_i = 2$ GHz) will have approximately $L_C = 12$ dB.

NOTE: *Subharmonic mixers, having an LO frequency of half of the "usual LO frequency," can be considered to be a subset of the general class of harmonic mixers. This is a special type of mixer, which will be discussed in a later section.*

Other Generated Harmonics at the IF Port. In addition to the three primary mixer frequencies (ω_r, ω_o and $\omega_i = \omega_o - \omega_r$) discussed previously, the following frequency components at the IF port must also be taken into consideration for a successful mixer design:

LO harmonics:	$n\omega_o \quad (n = 2,3,...)$
Image response:	$\omega_i = 2\omega_o - \omega_r$ (corresponding to $\omega'_r = \omega_o - \omega_i$)
Harmonic sidebands:	$\omega_{SB} = n\omega_o \pm \omega_r$

These frequencies are the byproducts of the mixing process and exist at the IF port.

19.3 CONVERSION LOSS FOR SSB MIXERS

The conversion loss of an SSB mixer is an important characteristic feature of a mixer that needs to be defined accurately:

DEFINITION-CONVERSION LOSS (L_C): A figure of merit for an SSB mixer defined as the ratio of available RF input power to the IF output power (expressed in dB):

$$L_C(\text{dB}) = 10 \log_{10}\left(\frac{P_{RF}}{P_{IF}}\right)(\text{dB}) \tag{19.4}$$

Typical values for a mixer conversion loss are in the range of 4 to 7 dB.

Similar to a detector (see Chapter 18), the conversion loss of a mixer is made up of three components:

$$L_C = L_d + L_m + L_h \ (\text{dB}) \tag{19.5}$$

where L_d, L_m, and L_h are losses due to diode resistance, port mismatch, and harmonic generation, respectively. These three types of losses are incurred in a mixer while the signal is traveling through the diode for conversion to the desired frequency.

19.3.1 Diode Loss (L_d)

Using the results from Equation 18.43 in Chapter 18, L_d is found to be:

$$L_d(\text{dB}) = 10 \log_{10}\left(1 + \frac{R_S}{R_j} + \omega^2 R_S R_j C_j^2\right) (\text{dB}) \tag{19.6}$$

19.3.2 Mismatch Loss

From Equation 18.44, we can see the input power is reduced by a reduction factor of $(1-|\Gamma'|^2)$. However, because the RF and IF sections are in cascade, the total mismatch loss is given by the multiplication of two reduction factors $(1-|\Gamma_{RF}|^2)(1-|\Gamma_{IF}|^2)$. Thus, the total mismatch loss is given by:

$$L_m(\text{dB}) = -10\log_{10}(1-|\Gamma_{RF}|^2)(1-|\Gamma_{IF}|^2) \quad (\text{dB}) \tag{19.7}$$

19.3.3 Harmonic Loss (Also Called Intrinsic Loss)

Due to the diode's nonlinear *I-V* curve, many signal harmonics are generated but only $\omega_r - \omega_o$ (for down-converters) or $\omega_r + \omega_o$ (for up-converters) is useful. Therefore, not all of the signal power is converted to the desired signal power, which leads to the concept of harmonic loss (L_h) due to mixing action. This loss (also referred to as intrinsic loss) is normally specified in the manufacturer's data sheets for each mixer (e.g., 3 dB). Example 19.1 illustrates the mixer conversion loss further.

> **NOTE:** *It should be noted that one factor that greatly influences the overall conversion loss of a mixer is the power level of the local oscillator signal, as discussed earlier. Usually to obtain a low conversion loss, the LO power level is raised to a large signal level where the small-signal condition is no longer valid and results obtained through small-signal analysis techniques no longer apply. In this case, use of computer aided design (CAD) techniques in solving nonlinear diode equations numerically becomes essential in mixer design.*

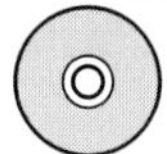

EXAMPLE 19.1

Consider a single-ended mixer having the following port VSWR values at 10 GHz:

$$(VSWR)_{RF} = 2.0$$

$$(VSWR)_{IF} = 3.0$$

$$L_h = 3 \text{ dB}$$

The diode used in the mixer has: $R_j = 100\ \Omega$, $R_S = 2\ \Omega$ and $C_j = 0.2$ pF.

What is the conversion loss of the mixer?

Solution:

From Equation 19.5, we have:

$$L_C = L_d + L_m + L_h \text{ (dB)}$$

$$L_h = 3 \text{ dB}$$

We will calculate each of the remaining two losses as follows:

a.

$$L_d(\text{dB}) = 10\log_{10}\left(1 + \frac{R_S}{R_j} + \omega^2 R_S R_j C_j^2\right) \text{ (dB)}$$

$$L_d(\text{dB}) = 10\log_{10}\left(1 + \frac{2}{100} + (2\pi\times 10^{10})^2 \times 2 \times 100 \times (0.2\times 10^{-12})^2\right) = 0.22 \text{ (dB)}$$

b. $$L_m(\text{dB}) = -10\log_{10}(1-|\Gamma_{RF}|^2)(1-|\Gamma_{IF}|^2)\ \ (\text{dB})$$

We know that:

$$VSWR = \frac{1-|\Gamma|}{1+|\Gamma|} \Rightarrow |\Gamma| = \frac{VSWR-1}{VSWR+1}$$

Thus,

$$(VSWR)_{RF} = 2 \quad \Rightarrow \quad |\Gamma_{RF}| = 1/3 = 0.33$$

$$(VSWR)_{IF} = 3 \quad \Rightarrow \quad |\Gamma_{IF}| = 1/2 = 0.5$$

Now we calculate L_m as:

$$L_m = -10\log_{10}(1-|0.33|^2)(1-|0.5|^2) = 1.76\ (\text{dB})$$

Therefore, the conversion loss is calculated to be:

$$L_C = 0.22 + 1.76 + 3 = 4.98 \approx 5\ \text{dB}$$

A conversion loss of 5 dB means that power at the IF port will be always 5 dB below the power level at the RF port, provided the LO port has sufficient power to drive the diode.

Suppressing harmonic sidebands through the use of resistive terminations results in lost IF signal power that leads to an increase in the mixer's conversion loss (L_C) and consequently its noise figure, which is dependent on the value of L_C. It has been found that reactive terminations with the correct phase at the undesirable frequencies are the key to obtaining low conversion loss. The only drawback will be the resulting mixer performance, which would be sensitive to frequency variations because the added reactive elements are frequency dependent.

19.4 SSB VERSUS DSB MIXERS: CONVERSION LOSS AND NOISE FIGURE

Because a mixer is usually used as the first or second stage in the design of the front end of a receiver system, its noise figure becomes a critical factor. A mixer's noise figure is caused primarily by the conversion loss, which varies depending on whether a DSB or an SSB IF signal is generated.

We need to define these two terms at this point.

DEFINITION-DOUBLE-SIDEBAND (DSB) SIGNAL: *An AM signal that consists of two sidebands—upper and lower sidebands. The upper sideband (USB) is the sum of the LO (or carrier) signal and the RF (or modulating) signal frequencies, and the lower sideband (LSB) is the difference of the LO and RF frequencies.*

DEFINITION-SINGLE-SIDEBAND (SSB) SIGNAL: *An AM signal in which either one of the two sidebands (but not both) of a DSB signal is suppressed.*

19.4.1 SSB versus DSB Conversion Loss

Because a DSB signal has both sidebands available, it has twice as much power available at the IF port compared to an SSB signal. As a result, its conversion loss is 3 dB less than that of an SSB signal, as shown here:

$$(P_{IF})_{DSB} = 2(P_{IF})_{SSB} \tag{19.8}$$

From Equation 19.6, the SSB conversion loss is given by:

$$(L_C)_{SSB}(\text{dB}) = 10\log_{10}\left(\frac{P_{RF}}{P_{IF}}\right) \tag{19.9a}$$

Therefore, based on Equation 19.8 the relation between the conversion loss of a DSB signal versus SSB can be written as:

$$(L_C)_{DSB} = (L_C)_{SSB} - 3 \quad (\text{dB}) \tag{19.9b}$$

or, in terms of ratio, we can write:

$$(L_C)_{DSB} = (L_C)_{SSB}/2 \tag{19.9c}$$

19.4.2 SSB versus DSB Noise Figure

To find the noise figure of a mixer, we recall from Chapter 14, *Noise Considerations in Active Networks*, that the noise figure of a mixer (which is a lossy element), can be written as:

$$F_m = 1 + (L_C - 1)(T/T_o) \tag{19.10a}$$

NOTE: *Equation 19.10a is used for calculation of the mixer's noise figure (F_m), based on its conversion loss and its physical temperature (T), when the IF signal is SSB. Thus, we can rewrite Equation 19.10a as:*

$$(F_m)_{SSB} = 1 + [(L_C)_{SSB} - 1](T/T_o) \tag{19.10b}$$

To find the noise figure $(F_m)_{DSB}$ for a DSB signal, we need to use Equation 19.9c for L_C, which yields:

$$(F_m)_{DSB} = 1 + \left(\frac{T}{T_o}\right)\left[\frac{(L_C)_{SSB}}{2} - 1\right] \tag{19.10c}$$

Special case ($T = T_o$). When the mixer's physical temperature is the same as room temperature, that is,

$$T = T_o$$

then Equations 19.10b and 19.10c simplify and yield the DSB and SSB noise figures as follows:

$$(F_m)_{SSB} = (L_C)_{SSB} \tag{19.11a}$$

$$(F_m)_{DSB} = (L_C)_{SSB}/2 \tag{19.11b}$$

From Equations 19.11a and 19.11b we conclude:

$$(F_m)_{DSB} = (F_m)_{SSB}/2 \tag{19.11c}$$

In terms of dB, from Equation 19.11c we can see that the noise figure for DSB is lower than SSB by 3 dB, i.e.,

$$(F_m)_{DSB} = (F_m)_{SSB} - 3\ (\text{dB}) \tag{19.11d}$$

POINT OF CAUTION: *Equation 19.11d states that the DSB noise figure will be 3 dB lower than the SSB noise figure, which is very important to know when the noise figure of a mixer is being measured. In other words, when the noise figure of a mixer is being measured or reported we need to pay attention to the type of measurement, which is either a DSB or an SSB measurement, and accompany (or annotate) the data with the exact type of measurement to prevent future confusion. Otherwise, we could risk a 3 dB error in all of the related measurement data.*

19.5 ONE-DIODE (OR SINGLE-ENDED) MIXERS

Mixers that use only one diode to generate the IF desired frequency are often referred to as "single-ended mixers." A simple block diagram of a single-ended mixer is shown in Figure 19.3.

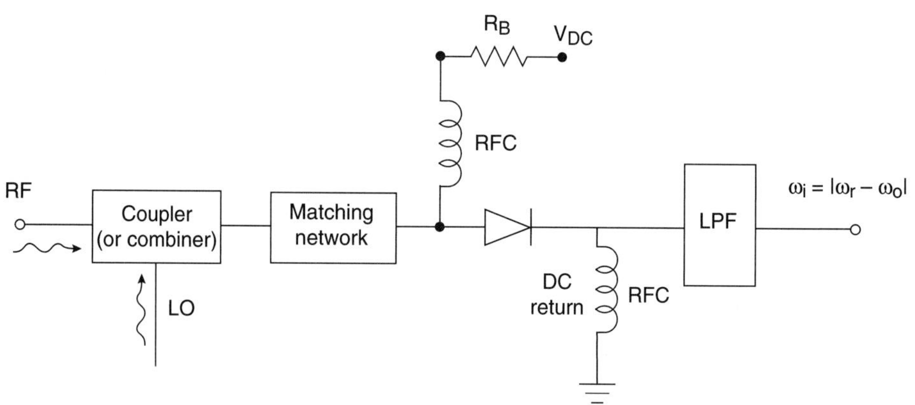

FIGURE 19.3 A typical block diagram of a single-ended mixer.

In this diagram, the coupler (or combiner) combines the two input signals with the help of a matching network (for proper impedance matching) and the combined signal is fed into the diode for mixing. By means of a low-pass filter, the difference signal is obtained. The diode is biased at the desired Q-point by means of RF chokes (RFC) that allow the DC to pass while blocking RF and LO signals.

19.5.1 Single-Ended Mixer Analysis

Let us consider:

$$v_{RF}(t) = V_r \cos(\omega_r t)$$

$$v_{LO}(t) = V_o \cos(\omega_o t)$$

$$v_{tot} = V_r \cos(\omega_r t) + V_o \cos(\omega_o t) \tag{19.12a}$$

The AC current, $i(t)$, in a diode under small-signal conditions, is given from Equation 18.8 as:

$$i(t) = G_d v_{tot} + G'_d v_{tot}^2 / 2 + \ldots \tag{19.12b}$$

By substitution for v_{tot} in Equation 19.12b, we can see that the diode will generate a spectrum of IF frequencies (ω_i) given by:

$$\omega_i = m\,\omega_r \pm n\,\omega_o \tag{19.12c}$$

where m and n are positive integers.

19.5.2 Single-Ended Mixer Design

An ideal mixer consists of a nonlinear element that behaves like an ideal switch with two distinct states: "ON" ($Z_{ON} = 0$, $\Gamma = -1$) and "OFF" ($Z_{OFF} = \infty$, $\Gamma = 1$). From this description, we can see that these two distinct states each have a reflection coefficient magnitude of unity and are 180° apart on the Smith chart, as shown in Figure 19.4a.

Similar to a switch, a diode (when driven by a large LO) also has two impedance states. These two states, unlike an ideal switch, have a reflection coefficient magnitude less than unity and are not separated by 180°, as shown in Figure 19.4b. The conversion loss of a diode is determined by how much these two states are degraded from the ideal case (see Figure 19.4b).

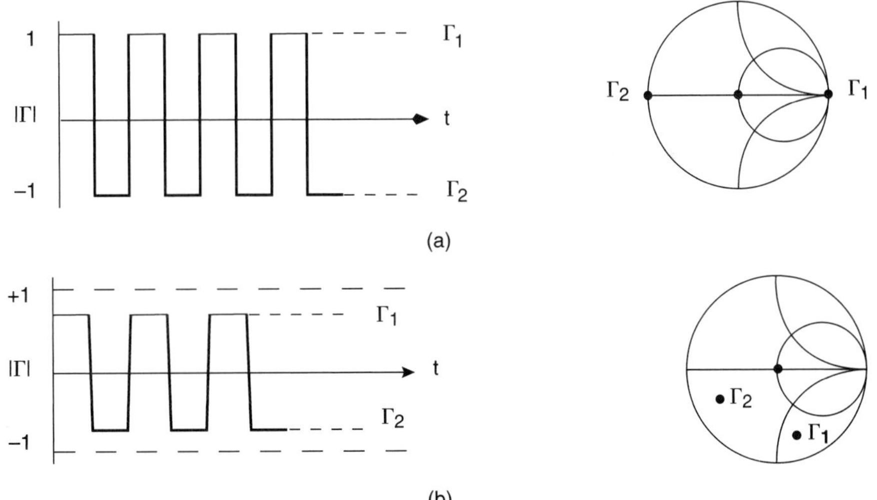

FIGURE 19.4 Comparison of an ideal switch and a diode for mixer application: (a) Reflection coefficient of an ideal switch, (b) Diode reflection coefficient when driven by LO voltage.

In order to obtain a low conversion loss, we need to use a matching circuit to transform the diode's impedance in the two states to two impedances that are not only

180° apart but also have equal and yet the largest possible reflection coefficient magnitude, as shown in Figure 19.5.

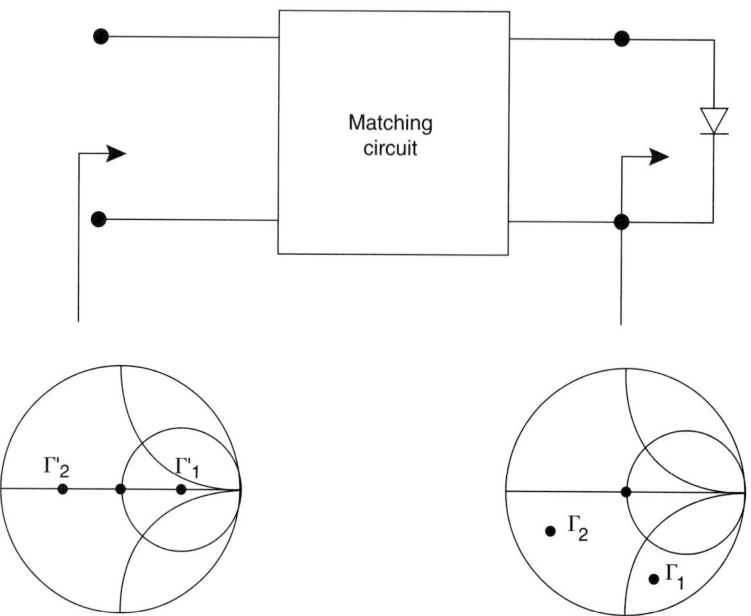

FIGURE 19.5 Mixer matching circuit.

NOTE: *Addition of a matching network will create an additional reflection loss (L_{add}) equal to:*

$$L_{add} = 10\log_{10}(1 - |\Gamma'|^2) \tag{19.13}$$

where $|\Gamma'| = |\Gamma'_1| = |\Gamma'_2|$ *is the magnitude of the transformed reflection coefficient (see Figure 19.5).*

Therefore, the total conversion loss (L'_C)is given by:

$$L'_C = L_C + L_{add} \tag{19.14}$$

To bring forth a procedure for the design of the matching circuit that transforms the two diodes' impedances properly, we use hyperbolic geometry. The steps are delineated below:

STEP 1. Measure the two diodes' impedances at ω_o under large signal conditions.
STEP 2. The two diode states $Z_{ON} = R_1 + jX_1$ and $Z_{OFF} = R_2 + jX_2$ are used to calculate the dynamic Q-factor (Q_d) defined as:

$$Q_d = \sqrt{\frac{(R_1 - R_2)^2 + (X_1 - X_2)^2}{R_1 R_2}} \tag{19.15}$$

STEP 3. Q_d is a measure of the diode's potential conversion loss (where L_C is inversely proportional to Q_d) and is used to compare the mixer performance with itself when different diodes are inserted in the circuit, without actually building the mixer circuit.

STEP 4. Construct the hyperbolic circle C_1 through Z_{ON} and Z_{OFF}. Hyperbolic circles by definition are those circles that are perpendicular to the outside of the Smith chart. For example, the constant reactance circles are hyperbolic circles.

STEP 5. Rotate Z_{ON} and Z_{OFF} until they lie on a hyperbolic circle of constant reactance, C_2 (see Figure 19.6). Q_d is unchanged by this rotation and is given by

$$Q_d = \frac{\sqrt{(R'_1 - R'_2)^2}}{\sqrt{R'_1 R'_2}} = \frac{(R'_1 - R'_2)}{\sqrt{R'_1 R'_2}} \tag{19.16}$$

where

$$Z'_{ON} = R'_1 + jX' \tag{19.17a}$$

$$Z'_{OFF} = R'_2 + jX' \tag{19.17b}$$

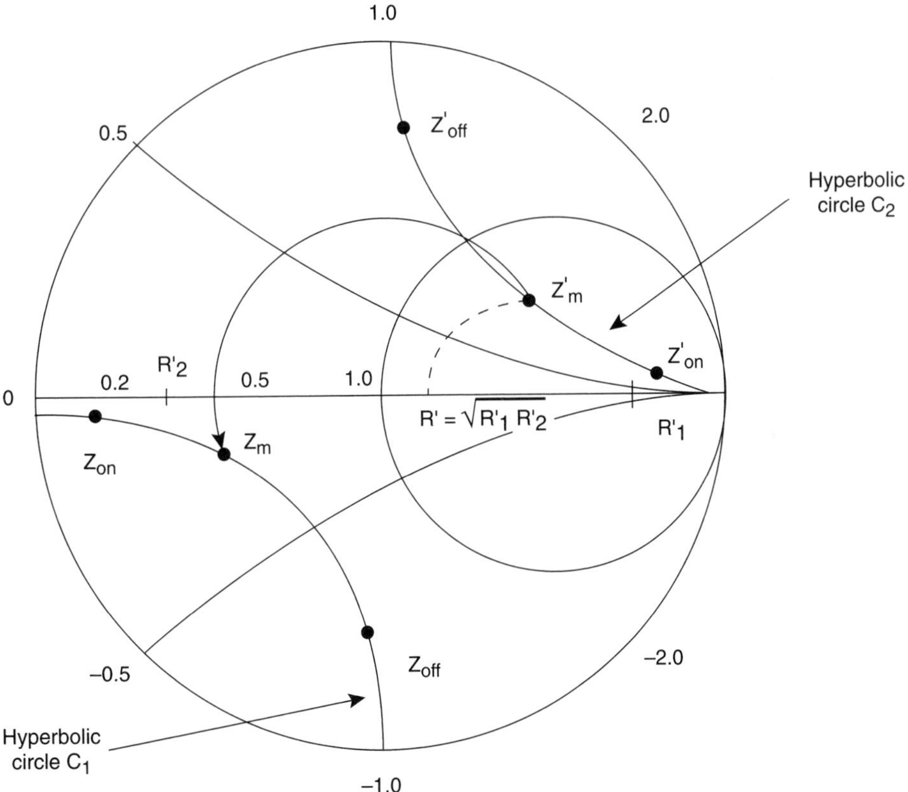

FIGURE 19.6 Diode impedance states.

STEP 6. Calculate the mean impedance (Z'_m) of Z'_{ON} and Z'_{OFF} defined as:

$$Z'_m = R' + jX' \tag{19.18}$$

where $R' = (R'_1 R'_2)^{1/2}$ is the geometric mean of R'_1 and R'_2.

STEP 7. Rotate Z'_m back through the same angle to locate the impedance mean (Z_m) of Z_{ON} and Z_{OFF} on the hyperbolic circle C_1 (see Figure 19.6) given by:

$$Z_m = R_m + jX_m \tag{19.19}$$

where

$$R_m = \sqrt{R_1 R_2 \left(1 + \frac{(X_1 - X_2)^2}{(R_1 + R_2)^2}\right)} \tag{19.20a}$$

$$X_m = X_1 + R_1 \frac{X_2 - X_1}{R_2 - R_1} \tag{19.20b}$$

STEP 8. Having determined Z_m, now we need to design a matching circuit to match Z_m to 50 Ω. This automatically provides maximum $|\Gamma_{ON}|$ and $|\Gamma_{OFF}|$ with 180° phase difference between the two states.

NOTE: *If the LO frequency (ω_o) is varying in a frequency range, a series of Z_m values can be obtained. The locus of all these Z_m values is then matched to 50 Ω to get the lowest conversion loss over the frequency range.*

19.5.3 Design Procedure

Based on the foregoing discussion, the following procedure can be used for a single-ended mixer design:

STEP 1. As discussed in Chapter 18, we need to select a diode with a cutoff frequency (ω_{Co}) at least 10 times the LO frequency (ω_o). This condition yields a zero-Volt junction capacitance (C_{jo}) given by:

$$\omega_{Co} > 10(\omega_o) \Rightarrow C_{jo} < \frac{1}{20\pi f_o R_S} \tag{19.21}$$

STEP 2. To reduce the conversion loss, minimize R_S by choosing a diode with a high cutoff frequency.

STEP 3. Package parasitics should be minimized because they reduce the difference between the two diode states.

STEP 4. Calculate or measure the two impedance states of the diode under large-signal conditions.

STEP 5. Calculate the hyperbolic mean (Z_m) of Z_{ON} and Z_{OFF} and match Z_m to 50 Ω.

STEP 6. Combine the diode and matching circuit along with the short and open stubs and the low-pass filter to create a baseline design as a good starting point. This baseline design may have to be modified to improve impedance match at the RF, LO, and IF frequencies to reduce conversion loss. This design will be the starting point for computer analysis and the optimization process.

Figure 19.7 shows a typical mixer circuit. In this figure, two stubs of 90° electrical length (i.e., θ = 90°) at the LO frequency, straddle the mixer diode. The resonant short stub acts as a bandpass filter allowing only LO and RF signals to enter the diode, whereas the resonant open stub at the output of the diode acts as a notch filter creating a low-impedance path to ground for all LO and RF frequencies at the IF output port. Practical mixer examples are shown in Figure 19.8.

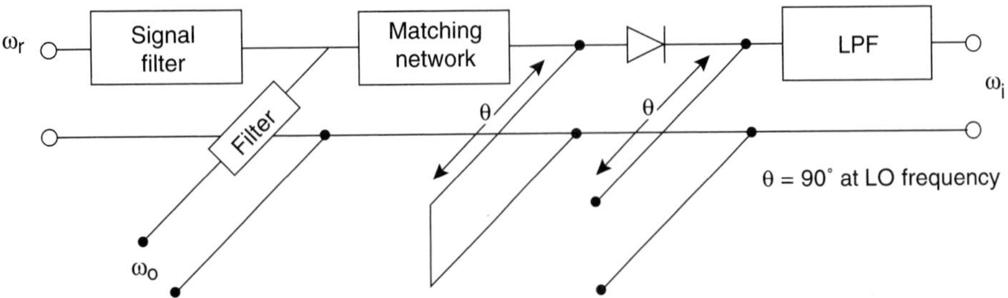

FIGURE 19.7 Mixer design with matching circuit.

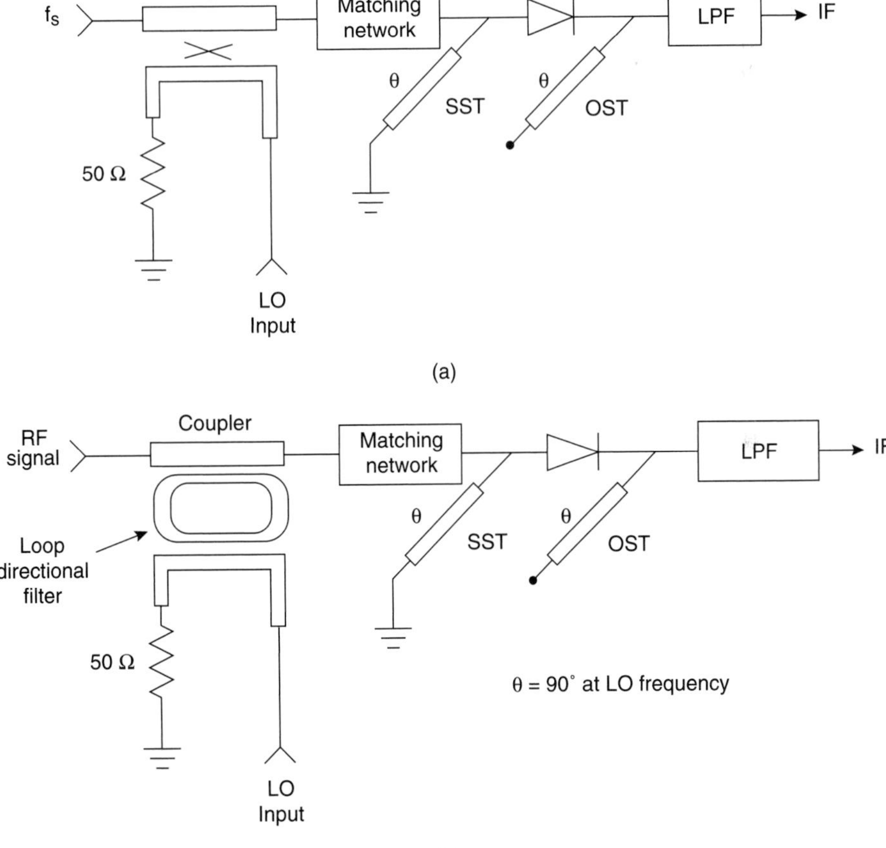

FIGURE 19.8 Single-ended mixers: (a) Broadband-LO mixer configuration, (b) Fixed-LO mixer configuration.

19.5.4 Other Mixer Considerations

Besides conversion loss, other features characterize a mixer, such as:

- Isolation between the RF and the LO ports
- *TOI*: Suppression of higher-order harmonics, expressed in terms of the third-order intercept (*TOI*) point
- *VSWR* at the LO and RF ports
- Noise: cancellation of AM noise at the LO port

The single-ended mixer is a simple mixer with a relatively fair degree of performance across the board in terms of all of these four features. More sophisticated mixers employing more than one diode can be designed that will accentuate one or a few of these features. There are other types of mixers with more complex designs including the balanced mixer (two diodes), the double balanced mixer (four diodes), and double-double balanced mixer with its variation, the image rejection mixer (eight diodes).

Generally speaking, balanced mixers have superior RF and LO port isolation, and they have better rejection of AM noise as well as spurious signals from the LO port. This relatively superior performance is not due to filtering but comes from the proper phase relationship existing between the voltages and currents that leads to cancellation of undesired signals.

NOTE: *Because the input power is divided between two or more devices, a single- or double-balanced mixer has a higher power-handling capability than a single-ended mixer. The only drawback is that the LO power requirements have to increase proportionately as well because the LO power is also divided between the diodes. Whenever, either stronger signals are anticipated at the inputs or better isolation and lower spurious signals are desired, use of balanced mixers is recommended.*

The tradeoff for better mixer performance is sophistication versus price. In the next section, we will study each one of these mixers and identify their main features as well as their strengths and weaknesses as related to their performance.

19.6 TWO-DIODE MIXERS

Types of mixers that use two diodes can be categorized in two classes:

- Single-balanced mixers
- Anti-parallel mixers

These two types are described below.

19.6.1 Single-Balanced Mixers

A single-balanced mixer uses two single-ended mixers in combination with a 3 dB hybrid coupler (also called a hybrid junction or simply a hybrid), as shown in Figure 19.9, with the biasing arrangement for each mixer omitted for simplicity. Due to its importance in this type of mixer, a hybrid coupler needs to be defined now (see also Chapter 16, *RF/Microwave Amplifiers II: Large-Signal Design*, for further details):

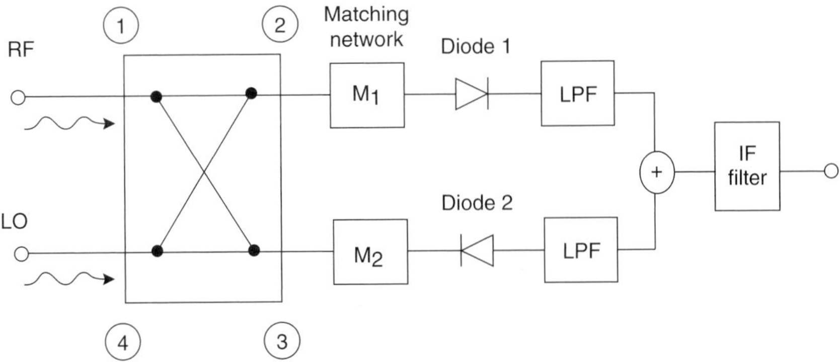

FIGURE 19.9 A single-balanced mixer.

DEFINITION-HYBRID COUPLER (ALSO CALLED HYBRID JUNCTION OR HYBRID): *A component having four ports so arranged that a signal entering at one port divides and emerges from the two adjacent ports but is unable to reach the opposite port.*

The two basic types of hybrids, as shown in a general diagram in Figure 19.10, are these:

- 90° couplers
- 180° hybrid couplers

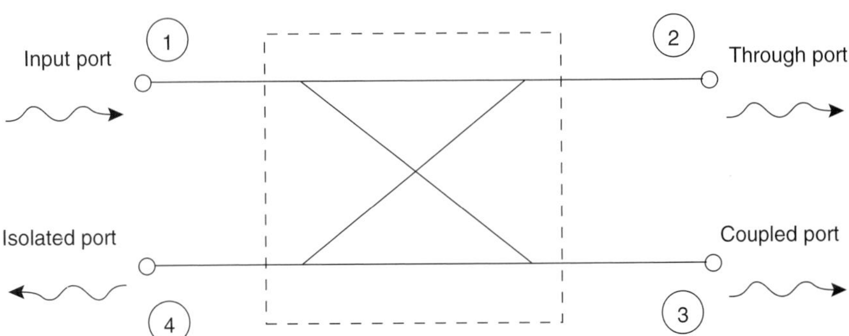

FIGURE 19.10 A hybrid coupler.

The S-parameter matrix for a 90° (quadrature) hybrid coupler is given by:

$$S = \frac{-1}{\sqrt{2}} \begin{bmatrix} 0 & j & 1 & 0 \\ j & 0 & 0 & 1 \\ 1 & 0 & 0 & j \\ 0 & 1 & j & 0 \end{bmatrix} \tag{19.22}$$

A 180° hybrid coupler has the following S-parameters:

$$S = \frac{-j}{\sqrt{2}}\begin{bmatrix} 0 & 1 & 1 & 0 \\ 1 & 0 & 0 & -1 \\ 1 & 0 & 0 & 1 \\ 0 & -1 & 1 & 0 \end{bmatrix} \tag{19.23}$$

Examples of microwave hybrid couplers include 3 dB Lange couplers, 3 dB branch-line couplers and rat-race rings (also called hybrid rings). These are all described in Chapter 16.

Mixer Analysis. We now analyze the effect of AM noise (at the LO port) on the mixer's performance and RF/LO port isolation for two cases of 90° and 180° hybrid couplers for a single-balanced mixer.

a) AM Noise Analysis Assume the RF voltage and LO voltage (plus AM noise) at the inputs of a 90° hybrid coupler are given by:

$$v_{RF}(t) = |V_{RF}|\cos\omega_r t \tag{19.24a}$$

$$v_{LO}(t) = [|V_{LO}| + v_n(t)]\cos\omega_o t \tag{19.24b}$$

where $v_n(t)$ is the AM noise signal at the LO port. In this analysis it is assumed that the RF voltage amplitude and the AM noise amplitude are generally much smaller than the pumping or LO signal amplitude; that is:

$$|V_{RF}| << |V_{LO}|,$$

and

$$|v_n(t)| << |V_{LO}|$$

The desired mixing product is obtained by combining each diode's current to obtain the IF current (i_{IF}) given by:

Diode #1: $$i_1(t) = K_1v_1(t) + K_2v_2(t)^2 \tag{19.25a}$$

Diode #2: $$i_2(t) = -[K_1v_1(t) + K_2v_2(t)^2] \tag{19.25b}$$

Thus,

$$i_{IF}(t) = i_1(t) + i_2(t) = K_2[v_1(t)^2 - v_2(t)^2] \tag{19.26}$$

where

Diode #1: $$v_1(t) = S_{21}V_{RF} + S_{24}V_{LO} \tag{19.27a}$$

Diode #2: $$v_2(t) = S_{31}V_{RF} + S_{34}V_{LO} \tag{19.27b}$$

We now perform noise analysis for 90° and 180° hybrid couplers.

90° Hybrid Coupler For a 90° hybrid coupler, the voltage across each of the two diodes is given by:

$$S_{21} = -j/\sqrt{2},\ S_{24} = -1/\sqrt{2},\ S_{31} = -1/\sqrt{2},\ S_{34} = -j/\sqrt{2}$$

Diode #1: $$v_1(t) = -(jV_{RF} + V_{LO})/\sqrt{2} \tag{19.28a}$$

Diode #2: $$v_2(t) = -(V_{RF} + jV_{LO})/\sqrt{2} \quad (19.28b)$$

Substituting from Equations 19.24 for V_{RF} and V_{LO} in Equations 19.28, we have:

$$v_1(t) = \{|V_{RF}|\cos(\omega_r t - 90°) + [|V_{LO}| + v_n(t)]\cos(\omega_o t - 180°)\}/\sqrt{2} \quad (19.29a)$$

$$v_2(t) = \{|V_{RF}|\cos(\omega_r t - 180°) + [|V_{LO}| + v_n(t)]\cos(\omega_o t - 90°)\}/\sqrt{2} \quad (19.29b)$$

Substituting for v_1 and v_2 from Equations 19.29 in Equation 19.26, and filtering out all the high-frequency terms, we obtain:

$$i_{IF}(t) = -K_2|V_{RF}|[|V_{LO}| + v_n(t)]\sin\omega_i t$$
$$\approx -K_2|v_{RF}||V_{LO}|\sin\omega_i t \quad \text{for} \quad |v_n(t)| << |V_{LO}| \quad (19.30)$$

From Equation 19.30, we can see that using a balanced configuration reduces the effect of noise greatly through the cancellation of the first-order terms (for noise voltage) leading to the reduction of the AM noise up to 30 dB in a practical single-balanced mixer.

180° Hybrid Coupler For a 180° hybrid coupler, the voltages across the two diodes are given by:

$$S_{21} = -j/\sqrt{2}, S_{24} = j/\sqrt{2}, S_{31} = -j/\sqrt{2}, S_{34} = -j/\sqrt{2}$$

Diode #1: $$v_1(t) = j(-V_{RF} + V_{LO})/\sqrt{2} \quad (19.31a)$$

Diode #2: $$v_2(t) = -j(V_{RF} + V_{LO})/\sqrt{2} \quad (19.31b)$$

Substituting from Equation 19.24 for V_{RF} and V_{LO} in Equation 19.31, we have:

$$v_1(t)=\{|V_{RF}|\cos(\omega_r t - 90°)+[|V_{LO}| + v_n(t)]\cos(\omega_o t + 90°)\}/\sqrt{2} \quad (19.32a)$$

$$v_2(t)=\{|V_{RF}|\cos(\omega_r t - 90°)+[|V_{LO}| + v_n(t)]\cos(\omega_o t - 90°)\}/\sqrt{2} \quad (19.32b)$$

Substituting for v_1 and v_2 from Equations 19.32 in 19.28, and filtering out all the high-frequency terms, we obtain:

$$i_{IF}(t) = -K_2|V_{RF}|[|V_{LO}| + v_n(t)]\cos\omega_i t$$
$$\approx -K_2|v_{RF}||V_{LO}|\cos\omega_i t \quad \text{for} \quad |v_n(t)| << |V_{LO}| \quad (19.33)$$

This is very similar to the 90° hybrid coupler results; we can see from Equation 19.33 that using a balanced configuration with a 180° hybrid coupler reduces the effect of noise greatly through the cancellation of the first-order terms (typically 15 to 30 dB of AM noise rejection).

b) VSWR and RF/LO Port Isolation To analyze the isolation between the RF and LO ports, we need to find the reflection coefficients when voltage is applied at these two ports for the case of 90° as well as hybrid couplers as follows:

90° Hybrid Coupler (A Symmetrical Coupler) Using the superposition theorem in the phasor domain,

Let:

$$V_{RF} = |V_{RF}|\angle 0° \quad (19.34)$$

and

$$V_{LO} = 0$$

Using the S-parameters of a 90° hybrid coupler from Equation 19.22, and assuming the reflection coefficient from each diode is Γ, then the reflected voltage phasors from the two diodes are given by:

Diode #1: $$V_{\Gamma 1} = \Gamma V_1 \quad (19.35)$$

Diode #2: $$V_{\Gamma 2} = \Gamma V_2 \quad (19.36)$$

where

$$V_1 = S_{21} V_{RF} = -j|V_{RF}|/\sqrt{2} \quad (19.37)$$

$$V_2 = S_{31} V_{RF} = -|V_{RF}|/\sqrt{2} \quad (19.38)$$

The reflected voltages ($V_{\Gamma 1}$ and $V_{\Gamma 2}$) will then travel back through the coupler and arrive at the RF and LO ports as follows:

$$(V_\Gamma)_{RF} = S_{12} V_{\Gamma 1} + S_{13} V_{\Gamma 2} = (jV_{\Gamma 1} + V_{\Gamma 2})/\sqrt{2} = 0 \text{ (i.e., no reflection)} \quad (19.39\text{a})$$

$$(V_\Gamma)_{LO} = S_{42} V_{\Gamma 1} + S_{43} V_{\Gamma 2} = (V_{\Gamma 1} + jV_{\Gamma 2})/\sqrt{2} = -j\Gamma|V_{RF}|/2 \text{ (i.e., strong coupling)} \quad (19.39\text{b})$$

Now, let:

$$V_{LO} = |V_{LO}|\angle 0° \quad (19.40)$$

and

$$V_{RF} = 0$$

Similarly, we can write the reflected voltage phasors that travel back through the coupler and arrive at the RF and LO ports as follows:

$$(V_\Gamma)'_{RF} = j\Gamma|V_{LO}|/2 \quad \text{(i.e., strong coupling)} \quad (19.41\text{a})$$

$$(V_\Gamma)'_{LO} = 0 \quad \text{(i.e., no reflection)} \quad (19.41\text{b})$$

By combining the results given by Equations 19.39 and 19.41, we can conclude the following for 90° hybrid coupler single-balanced mixers:

1. There are no reflections at the RF and the LO ports; this indicates excellent $VSWR$s, i.e.,

$$(VSWR)_{RF} = 1 \quad (19.42)$$

$$(VSWR)_{LO} = 1 \quad (19.43)$$

2. The isolation between the RF and the LO ports is very poor due to strong cross-coupling that exists between the two ports.

180° Hybrid Coupler (An Anti-Symmetrical Coupler) Employing the superposition theorem in phasor domain as discussed earlier, let:

$$V_{RF} = |V_{RF}|\angle 0° \quad (19.44)$$

and

$$V_{LO} = 0$$

Following the same procedure as in the previous case and using the S-parameters of a 180° hybrid coupler from Equation 19.23, and assuming the reflection coefficient from each diode is Γ, then the reflected voltage phasors back from the two diodes are given by:

Diode #1: $$V_{\Gamma 1} = \Gamma V_1 \quad (19.45\text{a})$$

Diode #2: $$V_{\Gamma 2} = \Gamma V_2 \quad (19.45\text{b})$$

where

$$V_1 = S_{21} V_{RF} = -j\left|V_{RF}\right| / \sqrt{2} \quad (19.46\text{a})$$

$$V_2 = S_{31} V_{RF} = -j\left|V_{RF}\right| / \sqrt{2} \quad (19.46\text{b})$$

The reflected voltages ($V_{\Gamma 1}$ and $V_{\Gamma 2}$) will then travel back through the coupler and arrive at the RF and LO ports as follows:

$$(V_\Gamma)_{RF} = S_{12} V_{\Gamma 1} + S_{13} V_{\Gamma 2} = -jV_{\Gamma 1}/\sqrt{2} - jV_{\Gamma 2}/\sqrt{2} = -\Gamma V_{RF}/2 \quad (19.47\text{a})$$

$$(V_\Gamma)_{LO} = S_{42} V_{\Gamma 1} + S_{43} V_{\Gamma 2} = jV_{\Gamma 1}/\sqrt{2} - jV_{\Gamma 2}/\sqrt{2} = 0 \text{ (i.e., no cross coupling)} \quad (19.47\text{b})$$

Now, let:

$$V_{LO} = |V_{LO}|\angle 0° \quad (19.48)$$

and

$$V_{RF} = 0$$

$$V_1 = S_{24} V_{LO} = j|V_{LO}|/\sqrt{2} \quad (19.49\text{a})$$

$$V_2 = S_{34} V_{LO} = -j|V_{LO}|/\sqrt{2} \quad (19.49\text{b})$$

$$-V_{\Gamma 1} = V_{\Gamma 2} = \Gamma|V_{LO}|/\sqrt{2} \quad (19.50)$$

Similarly, we can write the reflected voltage phasors that travel back through the coupler and arrive at the RF and LO ports as follows:

$$(V_\Gamma)'_{RF} = V_{\Gamma 1}/\sqrt{2} + V_{\Gamma 2}/\sqrt{2} = 0 \text{ (i.e., no cross coupling)} \quad (19.51\text{a})$$

$$(V_\Gamma)'_{LO} = V_{\Gamma 1}/\sqrt{2} - V_{\Gamma 2}/\sqrt{2} = \Gamma|V_{LO}|/2 \quad (19.51\text{b})$$

From the results given by Equation 19.49 and 19.51, we can conclude the following for 180° hybrid coupler single-balanced mixers:

- There are high reflections at the RF and the LO ports that indicate poor *VSWR*s.
- The isolation between the RF and the LO ports is excellent due to no cross-coupling between the two ports.

NOTE: *Due to the manner in which signals combine in a 180° hybrid coupler, the RF and LO ports are also referred to as "the sum, Σ" and "the difference, Δ" ports, respectively.*

19.6.2 Subharmonic (or Anti-Parallel) Diode Mixers

The second type of two-diode mixers is the anti-parallel diode mixer, as shown in Figure 19.11.

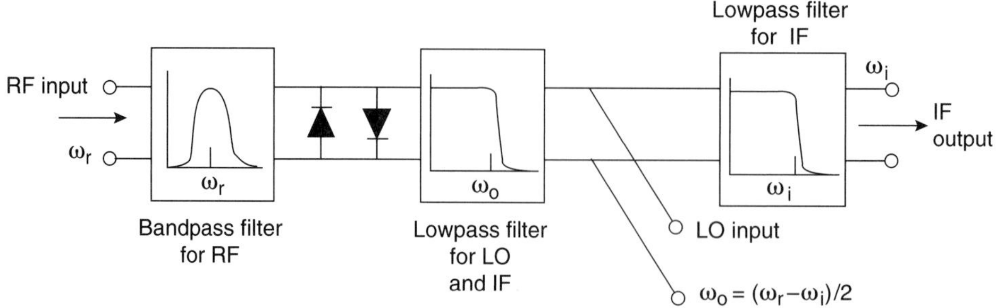

FIGURE 19.11 Subharmonically pumped mixer using an antiparallel diode pair.

This type of configuration is commonly used for subharmonically pumped mixers for millimeter-wave applications. In this type of mixer, the local oscillator signal frequency is half of the "usual LO frequency":

$$\omega_o = \frac{\omega_r - \omega_i}{2} \tag{19.52}$$

or

$$\omega_r = 2\,\omega_o + \omega_i$$

Due to diode nonlinearity, a second harmonic of the LO will be generated that will mix with the RF signal to produce the desired output frequency.

Because an anti-parallel mixer consists of a pair of back-to-back diodes with symmetrical I-V characteristic curves, the following important features become prominent:

Low conversion loss. The fundamental mixing product (of RF and LO) is suppressed, leading to a lower conversion loss. This occurs because only the second harmonic of the LO is required to be mixed with the RF signal and the fundamental mixing product cancels out due to symmetrical *I-V* curves.

Low noise. The AM noise from the LO port cancels out and is thus suppressed, very similar to the single-balanced mixer.

Low LO frequency. This is due to the use of second harmonic of LO.

Large peak inverse-voltage protection. This is a self-protecting mechanism imposed by the diode's anti-parallel configuration.

An example of a typical subharmonic mixer would be:

$$f_r = 64 \text{ GHz},$$

$$f_i = 2 \text{ GHz}$$

$$f_o = (64 - 2)/2 = 31 \text{ GHz},$$

$$L_C = 5 \text{ dB}$$

19.7 FOUR DIODE MIXERS

Increasing the number of diodes from two to four leads to the double-balanced mixer configuration, as shown in Figure 19.12.

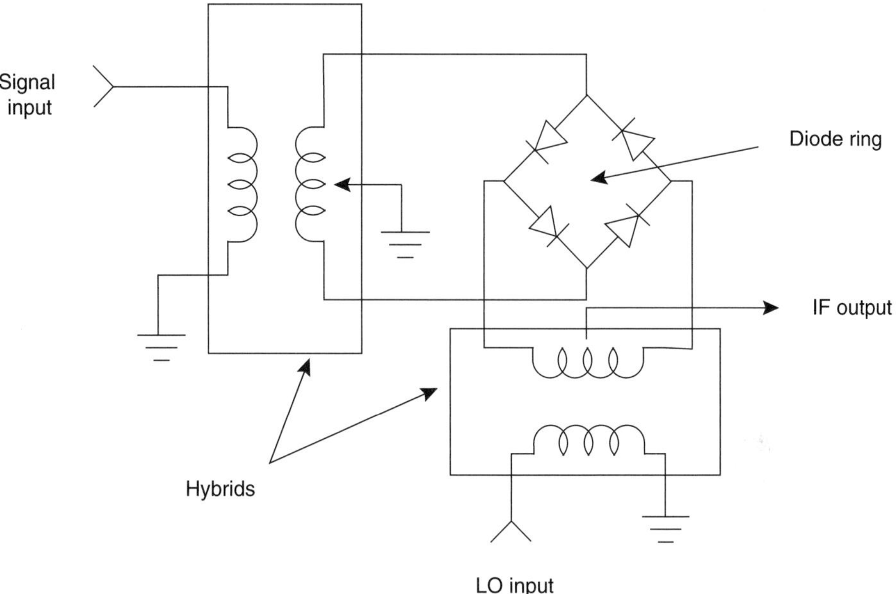

FIGURE 19.12 Double-balanced mixer.

A double-balanced mixer uses a diode quad ring (or sometimes Star) and two 180° hybrid couplers for the RF and the LO ports. From the previous discussion for 180° single-balanced mixers, it can be seen that the double-balanced mixers have good RF/LO isolation but poor RF/LO input *VSWR*.

Although a ring configuration is shown in Figure 19.12, a star arrangement may also be used.

The advantages and disadvantages of this type of mixer can be briefly summarized as follows.

19.7.1 Advantages

Because the double-balanced mixer uses two 180° hybrids for the RF and LO, even harmonics of both the RF and LO signals are suppressed, thus leading to a relatively low conversion loss. This is an improvement compared to a single-balanced mixer, which has only one 180° hybrid and suppresses only the even harmonics of the LO.

19.7.2 Disadvantages

This type of mixer uses two 180° hybrid couplers; therefore it has good RF/LO isolation but poor input *VSWR*. Furthermore, it needs at least four diodes and thus requires a greater LO power in order to drive all of the diodes with sufficient power for proper mixing operation.

19.8 EIGHT-DIODE MIXERS

Use of eight diodes leads to the double-double-balanced mixer configuration as shown in Figure 19.13.

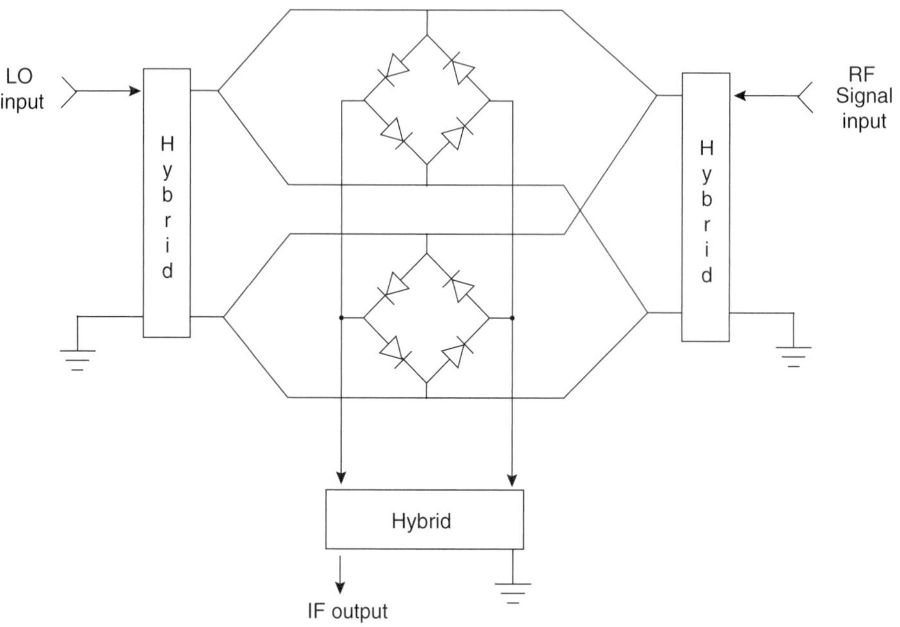

FIGURE 19.13 Double-double-balanced mixer.

In this configuration, two diode quad rings are fed via two RF and LO input hybrid couplers, and the result of mixing is obtained at the output of the IF hybrid. An important application would be configuring a double-double-balanced mixer as an image rejection mixer, as shown in Figure 19.14.

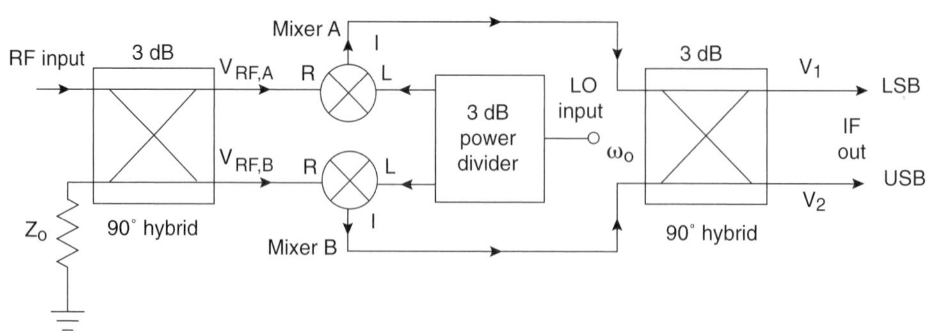

FIGURE 19.14 Image rejection mixer circuit.

This mixer is obtained through intelligent RF and LO signal combination with proper phase such that the image response is isolated and properly suppressed. This circuit consists of two double-balanced mixers, two quadrature (90°) hybrid couplers, and one in-phase hybrid coupler acting as a power divider at the LO input.

19.8.1 Operation

As discussed in an earlier section, because there are two distinct RF signals ($\omega_o \pm \omega_i$) that can produce the same IF frequency when mixed with a local oscillator, one of the RF signals arbitrarily can be considered to be the desired response (real response) while the other is the undesirable response (or image response), as shown in Figure 19.15.

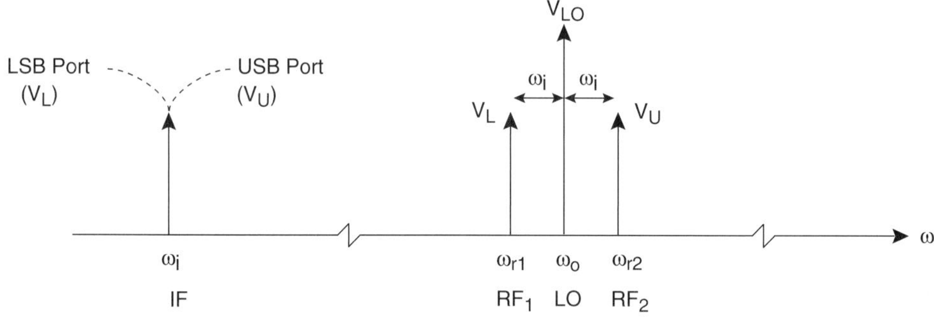

FIGURE 19.15 The two RF signals.

The real response can be selected as either the upper sideband (USB, $\omega_o + \omega_i$) or lower sideband (LSB, $\omega_o - \omega_i$). The image rejection mixer has the capability of separating the IF signal (due to each one of these two RF signals) into USB-IF and LSB-IF.

Analysis. Let the RF signal consist of both the USB signal and LSB signal as follows:

$$V_{RF}(t) = |v_U|\cos(\omega_o + \omega_i)t + |v_L|\cos(\omega_o - \omega_i)t \tag{19.53}$$

Then, the mixers A and B have the following RF inputs:

$$V_{RF,A}(t) = [|v_U|\cos(\omega_o + \omega_i)t + |v_L|\cos(\omega_o - \omega_i)t]/\sqrt{2} \tag{19.54}$$

$$V_{RF,B}(t) = \{|v_U|\cos[(\omega_o + \omega_i)t - 90°] + |v_L|\cos[(\omega_o - \omega_i)t - 90°]\}/\sqrt{2} \tag{19.55}$$

Upon mixing with an LO signal, $V_{LO}(t) = |v_{LO}|\cos(\omega_o)t$, the IF output signal for each mixer can be written as:

$$V_{IF,A}(t) = K[|v_U|\cos(\omega_i t) + |v_L|\cos(\omega_i t)]/2\sqrt{2} \tag{19.56}$$

$$V_{IF,B}(t) = K[|v_U|\cos(\omega_i t - 90°) + |v_L|\cos(\omega_i t + 90°)]/2\sqrt{2} \tag{19.57}$$

Combining the two IF signals at the 90° output hybrid, we obtain the following port signals:

$$\text{LSB port:} \quad V_{LSB} = V_{IF,A}(t - 90°/\omega_i) + V_{IF,B}(t - 180°/\omega_i) = K|v_L|\cos(\omega_i t)/2 \quad (19.58)$$

$$\text{USB port:} \quad V_{LSB} = V_{IF,A}(t - 180°/\omega_i) + V_{IF,B}(t - 90°/\omega_i) = K|v_U|\sin(\omega_i t)/2 \quad (19.59)$$

This analysis clearly shows that the real response can be isolated from the image response, which can be properly terminated and suppressed, if desired.

19.9 MIXER SUMMARY

Table 19.1 summarizes different types of mixers and compares the properties in terms of *VSWR*, RF/LO isolation, conversion loss, and third-order intercept (*TOI*).

TABLE 19.1 Typical Characteristics for Several Classes of Mixers

Mixer Type	# of Diodes	RF VSWR	RF/LO Isolation	Conversion Loss	TOI (dBm)
Single-Ended	1	Poor	Fair	Good	13
Balanced (90°)	2	Good	Poor	Good	13
Balanced (180°)	2	Fair	Excellent	Good	13
Double Balanced	4	Poor	Excellent	Excellent	18
Image Rejection	8	Good	Good	Good	15

From this table, we can see that as the degree of complexity of the mixer is increased, some of the features are more accentuated at the sacrifice of some other features. This is generally valid except for the image rejection mixer, which has good features all across the board with, of course, a higher price tag.

LIST OF SYMBOLS/ABBREVIATIONS

A symbol will not be repeated once it has been identified and defined in an earlier chapter, as long as its definition remains unchanged.

- DSB — Double sideband
- IF — Intermediate frequency, refers to a frequency internal to a receiver or transmitter used in the design to bridge between the RF frequency and the frequency of the signal intelligence
- L_{add} — Reflection loss of an added matching network
- L_C — Conversion loss of a mixer
- LO — Local oscillator
- LSB — Lower sideband
- OST — Open stub
- P_{IF} — Intermediate frequency power
- P_{RF} — RF power
- Q_d — Dynamic Q-factor

RFC	RF Choke
SSB	Single sideband
SST	Shorted stub
USB	Upper sideband
Z_m	Mean impedance of Z_{ON} and Z_{OFF} values
ω_i	Intermediate frequency in radians/sec
ω_o	LO frequency in radians/sec
ω_r	RF frequency in radians/sec

PROBLEMS

19.1 A two-tone input (at ω_1 and ω_2) with a 6 dB difference in the two signal power levels is applied to a nonlinear component. Determine the relative power ratio of the resulting third-order intermodulation products at $2\omega_1 - \omega_2$ and $2\omega_2 - \omega_1$.

19.2 An input level signal composed of two closely spaced frequencies (ω_1, ω_2) is applied to a mixer along with an LO frequency at ω_o as shown in Figure P19.2. Calculate and sketch the resulting output spectrum due to the v^2 term of the mixer response equation.

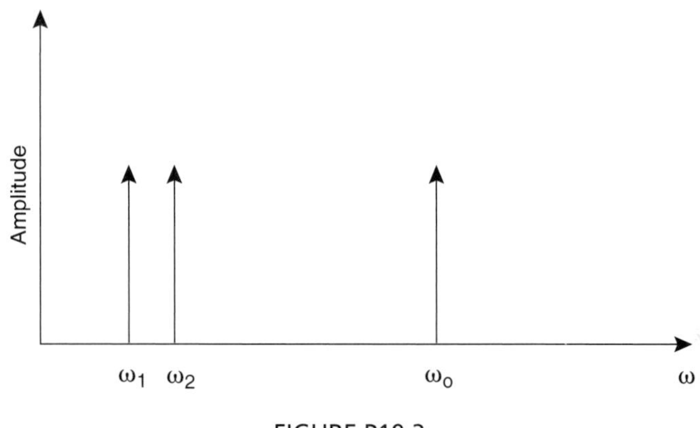

FIGURE P19.2

19.3 Calculate the dynamic Q (Q_d) and hyperbolic mean of the following mixer diode:

$$V_{off} = 0\text{ V}$$
$$V_{on} = 600\text{ mV}$$
$$V_{bi} = 700\text{ mV}$$
$$f_o = 8\text{ GHz}$$
$$C_{jo} = 0.20\text{ PF}$$
$$R_S = 2\ \Omega$$
$$I_S = 1 \times 10^{-9}\text{ A}$$

$$n = 1.1$$
$$L_P = 0.2 \text{ nH}$$
$$C_P = 0.12 \text{ pF}$$

19.4 Design a single-ended mixer using a microwave diode that converts a 4 GHz RF signal into a 500 MHz IF signal. The diode has the following characteristics:

$$f_{co} = 50 \text{ GHz}$$
$$I_S = 1 \times 10^{-9} \text{ A}$$
$$C_{jo} = 0.50 \text{ PF}$$
$$R_S = 10 \ \Omega$$
$$n = 1.05$$
$$L_P = 0.2 \text{ nH}$$
$$C_P = 0.35 \text{ PF}$$

Use a series transmission line to resonate the diode's reactance and a $\lambda/4$ transformer.

NOTE: *The oscillator can be designed using BJT technology (see Chapter 17).*

19.5 A mixer has a parasitic loss of 3 dB and an intrinsic loss of 3 dB. The RF port has *VSWR* of 2 and the IF port has a *VSWR* of 1.5. If an RF signal of 0 dBm is incident into the RF port, what is the signal power at the IF port?

19.6 A mixer has a *VSWR* of 2 at the RF port. Assuming that that the RF and IF and LO ports are perfectly matched, the reflected power detected at the RF port is -10 dBm with the signal at the IF port, -10 dBm. What is the conversion loss (in dB) for the mixer shown in Figure P19.6?

FIGURE P19.6

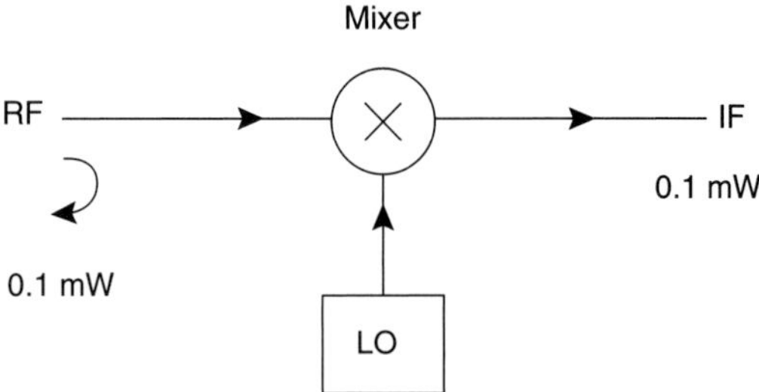

19.7 A mixer when inserted into a certain circuit has a conversion loss of 10 dB and a *VSWR* of 1.5 at both the RF and IF ports. Find the conversion loss of this mixer if the mixer is inserted into a circuit that gives a *VSWR* of 3.0 at both the RF and IF ports.

19.8 A mixer diode having the following *I-V* equation is used as an up-converter:

$$i = a_o + a_1 v + a_2 v^2$$

What are the frequencies of all signals at port A as shown in Figure P19.8?

FIGURE P19.8

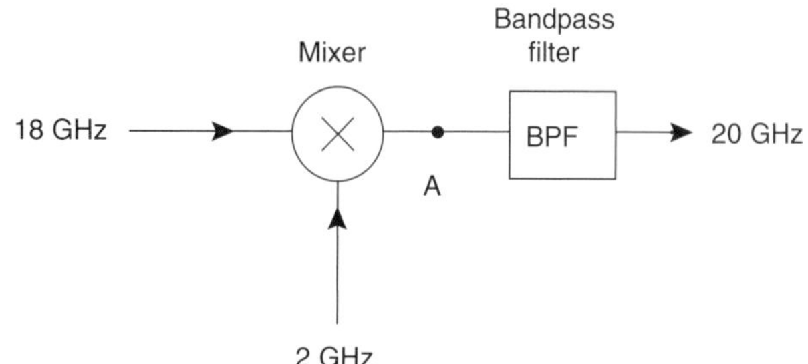

REFERENCES

[19.1] Amaro, L. R. and M. M. Radmanesh. "Re-entrant Filter Analysis Yielding Broadband and Compact Designs." *Microwave Journal*, Vol. 40, No. 7. 1997, pp. 20-34.

[19.2] Arnold, B. W. and M. M. Radmanesh. "Re-entrant Filter Design Using Microstrip-Slotline Transitions." *Microwave Journal*, Vol. 34, No. 4. 1995, pp. 147-51.

[19.3] Bahl, I. and P. Bhartia. *Microwave Solid-State Circuit Design*. New York: Wiley Interscience, 1988.

[19.4] Bhartia, P. and I. Bahl. *Millimeter Wave Engineering and Application*. New York: John Wiley & Sons, 1984.

[19.5] Chang, K. *Microwave Solid-State Circuits and Applications*. New York: John Wiley & Sons, 1994.

[19.6] Held, D. N. and A. R. Kerr. "Conversion Loss and Noise of Microwave and Millimeter-Wave Mixers, Part 1, Theory." *IEEE Transactions on Microwave Theory and Techniques,* Vol. MTT-26., Feb. 1978, pp. 49-55.

[19.7] Kollberg, E. L. "Mixers and Detectors" in *Handbook of Microwave and Optical Components*, Vol. 2. K. Chang, ed. New York: John Wiley & Sons, 1990.

[19.8] Maas, S. A. *Nonlinear Microwave Circuits*. Norwood, MA: Artech House, 1988.

[19.9] Maas, S. A. *Microwave Mixers*, 2nd ed. Norwood, MA: Artech House, 1993.

[19.10] Pozar, D. M. *Microwave Engineering*, 2nd ed. New York: John Wiley & Sons, 1998.

[19.11] Radmanesh, M. M. and N. A. Barakat. "State of the Art S-Band Resistive FET Mixer Design." *IEEE MTT-S International Microwave Symposium Digest*, San Diego, California. 1994, pp. 1435-8.

[19.12] Steinbrecher, D. H. "Circuit Aspects of Nonlinear Microwave Networks." *IEEE Proc. Int. Symp. Circuits and Systems*, 1975, pp. 477-9.

[19.13] Watson, H. A. *Microwave Semiconductor Devices and Their Circuit Applications*. New York: McGraw-Hill, 1969.

[19.14] Yngvesson, S. Y. *Microwave Semiconductor Devices*. Norwell, MA: Kluwer Academic Publishers, 1991.

CHAPTER 20

RF/Microwave Control Circuit Design

20.1 INTRODUCTION

Control circuits are widely used in all aspects of microwave system design and deserve a serious look so that the most important aspects of their design are brought into focus. Before we proceed more deeply into the subject, let's define an important term:

DEFINITION-CONTROL CIRCUIT: *A circuit that directs and regulates a process or sequence of events. In this type of circuit, one signal (or process) is made to control a bigger signal (or process).*

The general concept of "control" in electronics consists of three main actions:

- Starting the flow of a signal
- Changing the signal's property (such as amplitude, phase, or frequency)
- Stopping the flow of a signal

Using the above definition, we can observe that the control of microwave signals can be in terms of a signal's power level, direction of signal flow, phase angle, frequency, and other important characteristics.

Some of the most important microwave control circuits that perform these functions are:

- Microwave switches (start and stop action)
- Microwave phase shifters (change action)
- Microwave attenuators (change action)

These circuits are widely used in all microwave applications such as phased array radar systems, satellite communication systems, and microwave measurement systems. In these applications, many of these circuits are used in combination to handle and process the microwave signal precisely and bring about desired changes in the flow of the signal or any of its properties.

The types of devices that are commonly used in mixer circuit applications are:

- PN junction two-terminal devices such as Varactor diodes, PIN diodes, etc.
- MESFETs

In this chapter, we limit our studies to circuits using PN junction two-terminal devices (e.g., PIN diodes) and relegate MESFET mixer circuit applications to more advanced texts.

20.2 PN JUNCTION DEVICES

Assuming that the reader has a certain familiarity with the physics of PN junction devices, we will focus primarily on circuit applications of these devices. The equivalent circuits of a Varactor diode and of a packaged PIN diode are shown in Figure 20.1 and Figure 20.2, respectively.

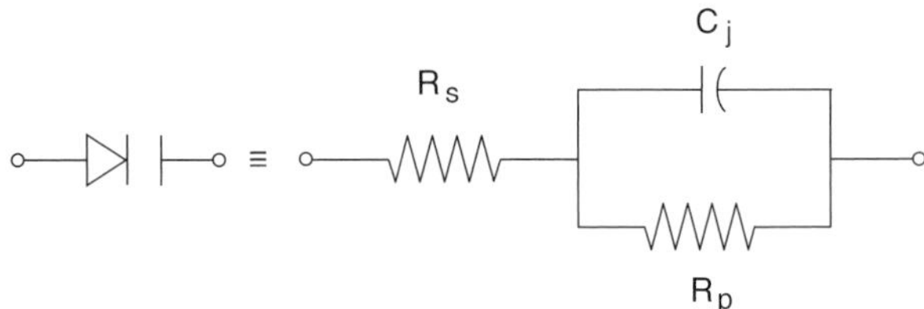

FIGURE 20.1 The equivalent circuit of a Varactor diode.

20.2.1 Varactors

The elements of a Varactor diode as shown in the equivalent circuit of Figure 20.1 are:

- C_j, the junction capacitance
- R_P, parallel equivalent resistance
- R_S, series equivalent resistance

We need to consider the Varactor under two bias conditions:

Reverse Bias. The junction capacitance due to the diode's depletion region, though existing at all voltages, becomes dominant in reverse bias. Assuming a one-sided junction (e.g., n^+p for microwave frequencies), the junction capacitance (C_j) can be shown to be a function of the applied bias voltage as follows:

$$C_j = \frac{C_{jo}}{\left(1 - \frac{V_a}{V_{bi}}\right)^S} \tag{20.1}$$

where

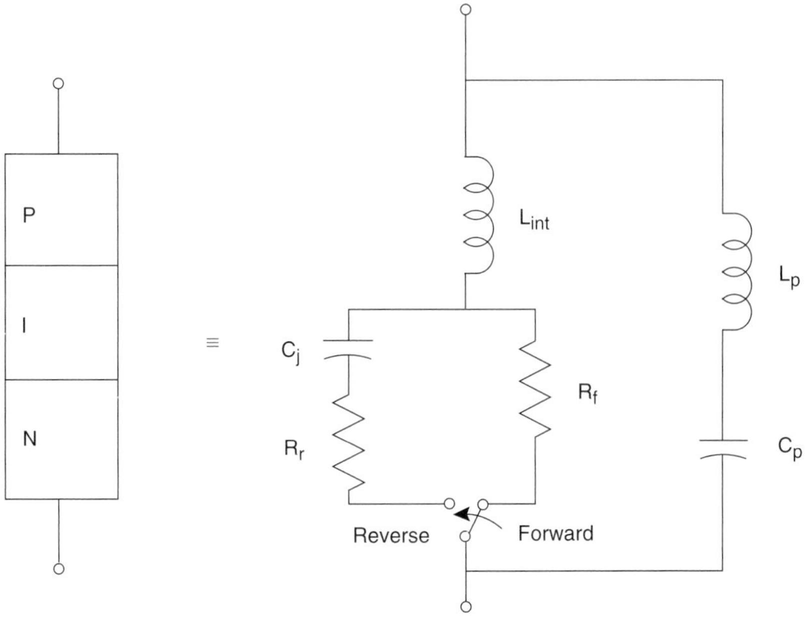

FIGURE 20.2 The equivalent circuit of a packaged PIN diode.

C_{jo} = The junction capacitance at zero bias voltage
V_a= The applied bias voltage across the junction (for reverse bias $V_a = -V_r$)
V_{bi}= The built-in voltage

$$S = \frac{1}{m+2},$$

and
m is the exponent parameter in the doping distribution function (N) of the lighter side, defined as:

$$N(x) = N_B x^m$$

N_B = an arbitrary constant established by the doping profile.

From Equation 20.1, we can see that the larger the S, the larger the variation of the junction capacitance with reverse-bias voltage. For example, for an *abrupt junction* ($m = 0$) under reverse-bias condition ($V_a = -V_r$), we have:

$$m = 0 \Rightarrow S = 1/2$$

$$C_j = \frac{C_{jo}}{\sqrt{1 + \frac{V_r}{V_{bi}}}}$$

However, for an efficient and more sensitive Varactor, we use hyper-abrupt junctions (e.g., $m = -1, -3/2, -5/2$), which gives relatively a high value for S (i.e., $S = 1, 2, 3$).

Forward Bias. As noted earlier, the junction capacitance (due to the diode's transition region) becomes very important under reverse-bias conditions; however, for forward bias the charge storage effects become dominant and the corresponding capacitance (C_S) and conductance (G_S) are given by:

$$G_S = \frac{I_{DC}}{V_T}$$

$$C_S = G_S \tau_n$$

Where I_{DC} is the bias current through the diode, V_T is the thermal voltage and τ_n is the electron's average lifetime in the semiconductor.

The charge storage capacitance (C_S) can be a serious limitation for forward-biased p-n junctions in high frequency applications but in many cases can be overcome by proper device design. For example, the switching performance of an n^+p diode circuit for high-frequency signals can be improved by reducing the electron's carrier lifetime (τ_n). The reduction in carrier lifetime will make the forward-bias storage capacitance (C_S) small enough to be used in many switching applications.

20.2.2 PIN Diodes

The elements of a PIN diode, as shown in the equivalent circuit of Figure 20.2, are:

- C_j, junction capacitance
- R_r, reverse-biased resistance
- R_f, forward-biased resistance
- L_{int}, series inductor
- L_P and C_P, packaging inductor and capacitor, respectively

Under reverse-bias conditions, the junction capacitance of a PIN diode is given by:

$$C_j = \frac{\varepsilon A}{W}$$

where W is the width of the *i*-layer (assuming $W >> W_o$, W_o being the depletion region width), ε is the permittivity of the semiconductor, and A is the cross-sectional area of the diode.

Ignoring the effect of packaging inductor and capacitor, the PIN diode's equivalent circuit under forward and reverse bias is separately shown in Figure 20.3.

Design of Switches. Switching is one of the most important concepts in control of microwave signals and forms one of the basic concepts used in the design of microwave control circuits. We need to define this term at this juncture:

DEFINITION-SWITCHING: *The making, breaking, or changing of connections in an electronic circuit.*

To perform the switching function at microwave frequencies we use a microwave switch, which is a microwave circuit made up of one or more high-speed microwave devices. Microwave switches are used as a basic component in many microwave systems for routing and controlling the direction of flow of the microwave signals among several components. They can be used in the design of other types of control circuits (such as attenuators and phase shifters) as will be discussed shortly.

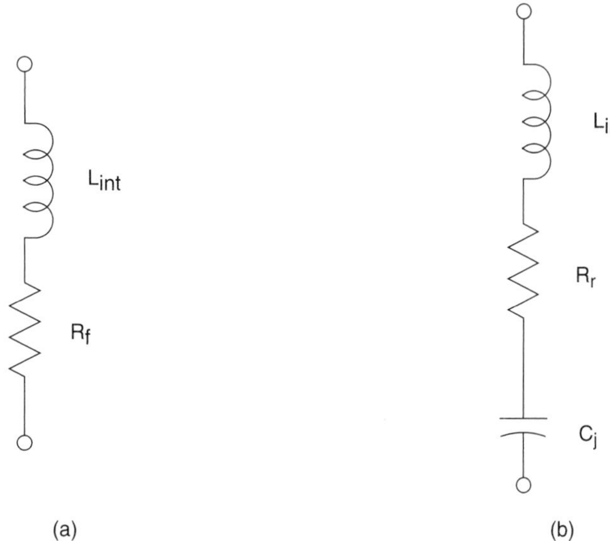

FIGURE 20.3 PIN diode equivalent circuit under a) Forward bias b) Reverse bias.

Use of PIN diodes for high-speed switching applications is very popular because they integrate easily with planar circuits such as microstrip circuits or microwave integrated circuits, and so we will study them in more detail in this section. (**NOTE:** The GaAs MESFET as a switching element may also be employed but is not discussed in this chapter.)

In forward bias, a PIN diode is considered to be on for a forward current of 10 mA to 40 mA typically; whereas in the reverse-bias mode, the PIN diode is assumed to be off for a reverse voltage of 40 V to 60 V.

Before we proceed with the design process of PIN diode switches, let's first define a few important switching terms:

DEFINITION-POLE (OF A SWITCH): *The movable arm of a switch. For example, a single-pole switch has one input terminal with one movable arm (see Figure 20.4)*

FIGURE 20.4 A single-pole single-throw switch (SPST).

DEFINITION-THROW (OF A SWITCH): *The output terminal(s) that each moveable arm of a switch is allowed to make a connection with.*

For example, a single-throw switch has only one output terminal.

20.3 SWITCH CONFIGURATIONS

There are several basic and yet essential switch configurations that need to be treated at this juncture of our study.

20.3.1 Single-Pole Single-Throw (SPST) Switch

To control the flow of microwave signals, a PIN diode can be used in three basic configurations to operate as a single-pole single-throw (SPST) switch as follows:

- A series configuration
- A shunt configuration
- A series-shunt configuration (a combination of the first two)

These three circuit configurations along with the necessary bias networks are shown in Figure 20.5. DC block capacitors in Figure 20.5 should be selected to create very low impedances at the RF operating frequency while the RF choke inductances should have very high RF impedance values. Such RF chokes can be realized by short-circuited quarter-wave transformers at microwave frequencies replacing the ferrite beads, traditional at the lower RF frequencies.

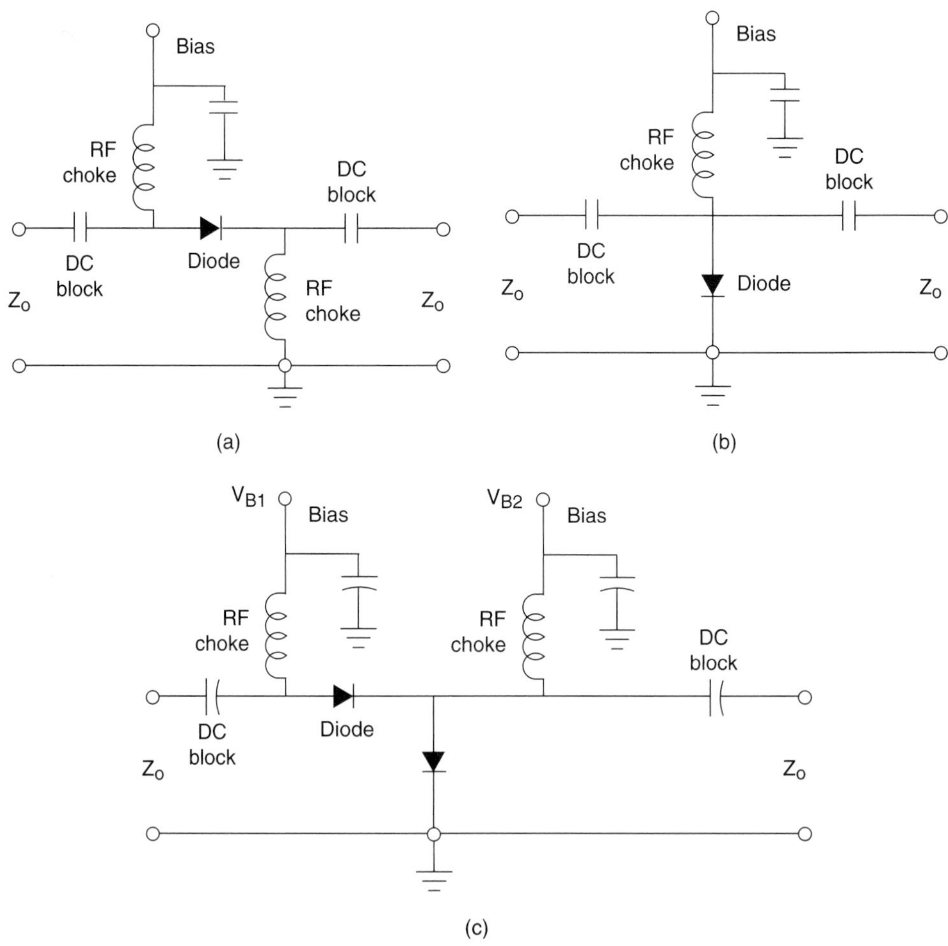

FIGURE 20.5 Single-pole PIN diode switches: a) Series configuration, b) Shunt configuration, c) Series-shunt configuration.

There are several attributes that characterize a switch, such as the *insertion loss* and *isolation*, that should be weighed properly before deciding on a particular configuration. The insertion loss and isolation for each one of these three PIN-diode configurations is presented next.

20.3.2 Switch Insertion Loss (*IL*)

DEFINITION-INSERTION LOSS (IL): *A figure of merit of a switch in the on state defined as the difference (in dB) between the power received at the load before and after the insertion of the switch in the line.*

Based on this definition, we can write insertion loss (IL) for the switch in the on state as:

$$IL = -10\log_{10}\left(\frac{P_L}{P_o}\right) \tag{20.2}$$

where

P_o = Power delivered to the load with an **ideal** switch in the on state.

P_L = Power delivered to the load with the **practical** switch in the on state.

Assuming that the load voltages are V_o (ideal) and V_L (actual) corresponding to P_o and P_L respectively, we can write:

$$P_o = K|V_o|^2 \tag{20.3a}$$

$$P_L = K|V_L|^2 \tag{20.3b}$$

where K is a proportionality constant.

Thus, we can write Equation 20.2 as:

$$IL = 10\log_{10}\left|\frac{V_o}{V_L}\right|^2 = 20\log_{10}\left|\frac{V_o}{V_L}\right| \tag{20.4}$$

Assuming Z_d to be the general impedance of the PIN diode (in the on or off state), the insertion loss (IL) for each case can be derived as shown in the next example.

EXAMPLE 20.1

Derive expressions for the insertion loss of the series, shunt, and series-shunt configuration switching circuits.

Solution:

a. **Series configuration switch**

Consider a series configuration switch as shown in Figure 20.6. The PIN diode should be forward biased to have the switch operate in the on state, i.e.,

$$Z_d = Z_f = R_f + j\omega L_{int} \tag{20.5}$$

where Z_f is the forward-biased impedance of the PIN diode.

Thus, we can write:

$$V_L = \frac{V_S}{2Z_o + Z_f} \tag{20.6}$$

$$V_o = V_S/2 \tag{20.7}$$

FIGURE 20.6 A series switch.

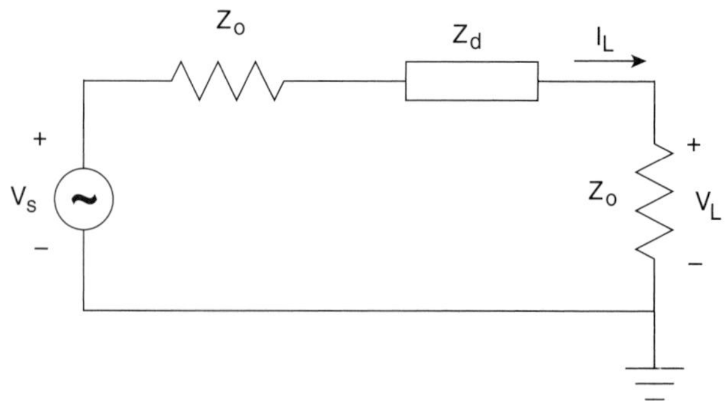

$$IL = 20\log_{10}\left|\frac{2Z_o + Z_f}{2Z_o}\right| \tag{20.8}$$

b. Shunt configuration switch

Consider a shunt configuration switch as shown in Figure 20.7. The PIN diode should be reverse biased to have the switch operate in the on state, i.e.,

$$Z_d = Z_r = R_r + j(\omega L_{int} - 1/\omega C_j) \tag{20.9}$$

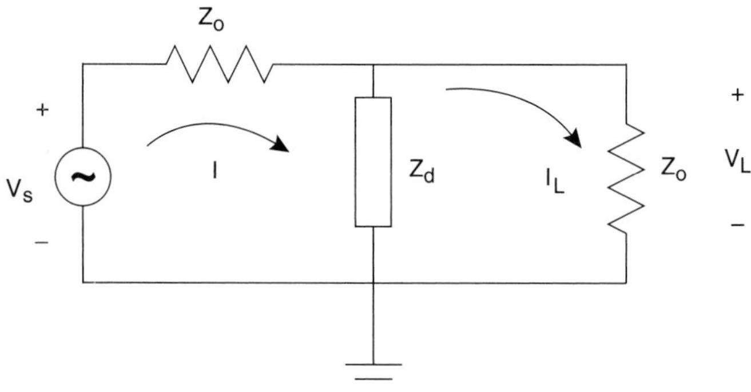

FIGURE 20.7 A shunt switch.

where Z_r is the reverse-biased impedance of the PIN diode. The load voltage can be written in terms of the current in the load (I_L) as:

$$V_L = I_L Z_o \tag{20.10}$$

$$V_o = V_S/2 \tag{20.11}$$

The current in each loop (see Figure 20.7) is given by:

$$I = \frac{V_S}{Z_o + Z_r \parallel Z_o} \tag{20.12}$$

$$I_L = \frac{IZ_r}{Z_o + Z_r} \tag{20.13}$$

Thus, we can write:

$$IL = 20 \log_{10} \left|\frac{V_o}{V_L}\right| = 20 \log_{10} \left|\frac{2Z_r + Z_o}{2Z_r}\right| \tag{20.14}$$

c. Series-shunt configuration switch

Consider the circuit shown in Figure 20.8 where two identical PIN diodes are connected as a switch in series-shunt configuration. To have the switch (as a whole) in the on state, the series diode should be turned on while the shunt diode should be turned off, i.e.,

$$(Z_d)_{se} = Z_f$$
$$(Z_d)_{sh} = Z_r.$$

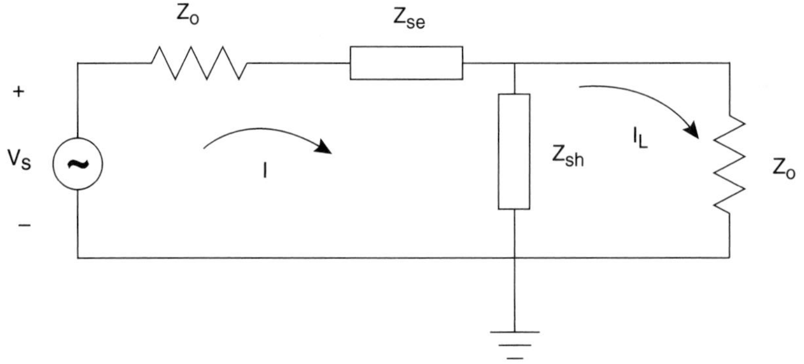

FIGURE 20.8 Series-shunt configuration.

Thus, we have:

$$I = \frac{V_o}{Z_o + Z_f + Z_r \parallel Z_o} \tag{20.15}$$

$$I_L = \frac{IZ_r}{Z_o + Z_r} \tag{20.16}$$

Thus, we can write:

$$V_L = I_L Z_o \tag{20.17}$$

$$V_o = V_S/2 \tag{20.18}$$

$$IL = 20 \log_{10} \left|\frac{V_o}{V_L}\right| = 20 \log_{10} \left|0.5 + \frac{(Z_o + Z_r)(Z_o + Z_f)}{2Z_o Z_r}\right| \tag{20.19}$$

20.3.3 Switch Isolation

DEFINITION-ISOLATION: *A figure of merit of a switch in the off state defined as the difference (in dB) between the power delivered to the load for an ideal switch in the on state and the actual power delivered to the load when the switch is in the off state.*

Based on this definition, we can write isolation as:

$$\text{Isolation} = -10\log_{10}\left(\frac{P'_L}{P_o}\right) \tag{20.20}$$

where

P_o = Power delivered to the load with an **ideal** switch in the on state.
P'_L = Power delivered to the load with the **practical** switch in the off state.

Assuming that the load voltages are V_o (ideal diode on) and V'_L (actual diode off) corresponding to P_o and P'_L respectively, and similar to insertion loss, we can write:

$$\text{Isolation} = 10\log_{10}\left|\frac{V_o}{V'_L}\right|^2 = 20\log_{10}\left|\frac{V_o}{V'_L}\right| \tag{20.21}$$

OBSERVATION: *Comparing Equation 20.4 to 20.21 we can see that isolation is the same as insertion loss except for the switch being in the off state, i.e.,*

$$\text{Insertion Loss} = (IL)_{ON}$$
$$\text{Isolation} = (IL)_{OFF} \tag{20.22}$$

Thus, we expect similar equations for the isolation case, except for changing the diode impedances (in the insertion loss equations) to their opposite values, i.e.,

$$Z_f \rightarrow Z_r$$
$$Z_r \rightarrow Z_f$$

The following example presents the mathematical expressions for the switch isolation.

EXAMPLE 20.2

Using impedance exchange, write the isolation equations for the series, shunt and series-shunt configuration switching circuits.

Solution:

From Equations 20.8, 20.14, and 20.19, we can write:

a. Series configuration

$$\text{Isolation} = (IL)_{OFF} = 20\log_{10}\left|\frac{2Z_o + Z_r}{2Z_o}\right| \tag{20.23}$$

b. Shunt configuration

$$\text{Isolation} = (IL)_{OFF} = 20\log_{10}\left|\frac{Z_o + 2Z_f}{2Z_f}\right| \tag{20.24}$$

c. Series-shunt configuration

$$\text{Isolation} = (IL)_{OFF} = 20\log_{10}\left|0.5 + \frac{(Z_o + Z_f)(Z_o + Z_r)}{2Z_oZ_f}\right| \tag{20.25}$$

SPECIAL CASES: *From Examples 20.1 and 20.2, we can see that the series-shunt configurations is the generalized version of either the series or the shunt configurations and we may obtain the equation for insertion loss $(IL)_{ON}$ or isolation $(IL)_{OFF}$ of these two special cases (series or shunt configuration) by considering them as limiting cases. The next example illustrates this point further.*

EXAMPLE 20.3

Using the results for the series-shunt configuration, derive $(IL)_{ON}$ and $(IL)_{OFF}$ expressions for the series and shunt configurations.

Solution:

a. Series switch

Let the shunt element's impedance (Z_{sh}) approach infinity, i.e.,

$$Z_{sh} = Z_r \rightarrow \infty$$

in both Equations 20.19 and 20.25 to obtain $(IL)_{ON}$ and $(IL)_{OFF}$ exactly as given earlier by Equations 20.8 and 20.23.

b. Shunt switch

Let the series element's impedance (Z_{se}) approach zero, i.e.,

$$Z_{se} = Z_f \rightarrow 0$$

in both Equations 20.19 and 20.25 to obtain $(IL)_{ON}$ and $(IL)_{OFF}$ exactly as given earlier by Equations 20.14 and 20.24.

NOTE: *The on state insertion loss and off state isolation of a switch can be improved by adding external reactances in series or in parallel with the PIN diode. The external reactance will cancel out the diode's reactance and will optimize the switch's performance at the frequency of operation, as desired.*

Combining several SPST switches leads to a variety of multiple-pole multiple-throw switch configurations. For example, a single-pole double-throw (SPDT) switch (as a series or shunt configuration) can be constructed as shown in Figure 20.9.

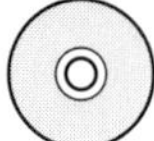

EXAMPLE 20.4

Consider an SPST switch with the following parameters at $f = 3.18$ GHz in a 50 Ω system. The PIN diode has $C_j = 0.1$ pF, $R_f = 1$ Ω, $R_r = 4$ Ω, $L_{int} = 0.3$ nH. Determine the insertion loss and isolation of the switch using a) a series configuration, and b) a shunt configuration.

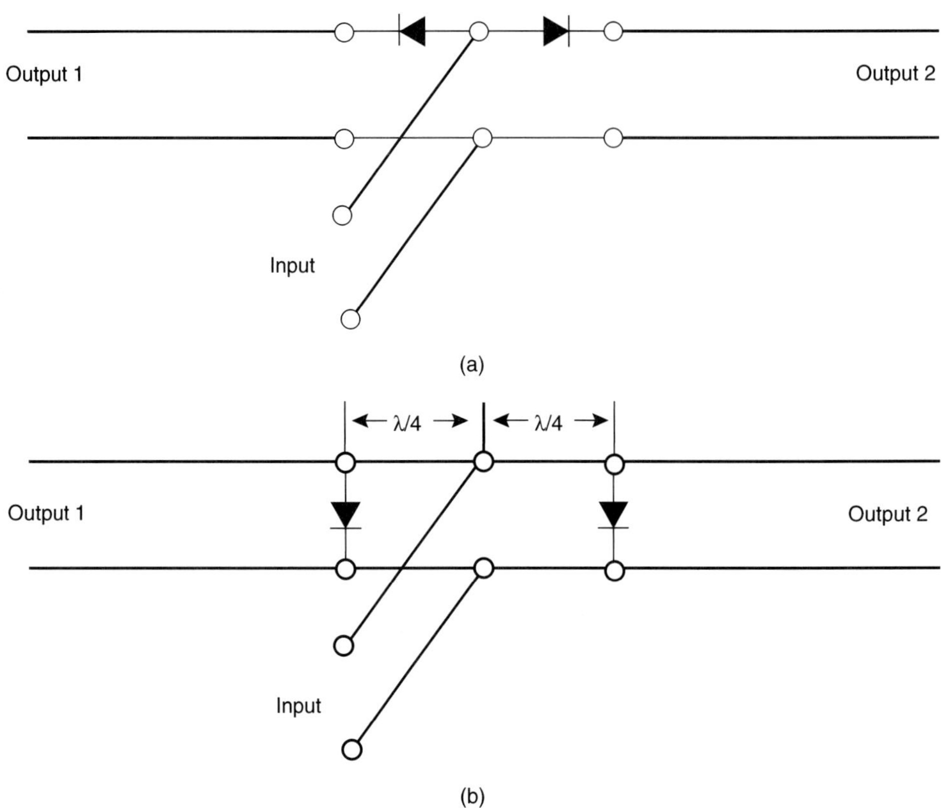

FIGURE 20.9 Circuits for single-pole double-throw (SPDT) PIN diode switch: a) Series b) Shunt.

Solution:

a. Series configuration

$$Z_f = R_f + j\omega L_{int}$$

$$Z_r = R_r + j(\omega L_{int} - 1/\omega C_j)$$

Substituting the PIN diode parameters, we obtain:

$$Z_f = 1 + j6\ \Omega$$

$$Z_r = 4 - j494\ \Omega$$

From Equations 20.8 and 20.23, we have:

$$IL = 20\log_{10}\left|\frac{2Z_o + Z_f}{2Z_o}\right| = 20\log_{10}\left|\frac{101 + j6}{100}\right| = 0.1\text{ dB}$$

$$\text{Isolation} = (IL)_{OFF} = 20\log_{10}\left|\frac{2Z_o + Z_r}{2Z_o}\right| = 20\log_{10}\left|\frac{104 - j494}{100}\right| = 14.1\text{ dB}$$

b. Shunt configuration

Similarly, from Equations 20.14 and 20.24 we have:

$$IL = 20\log_{10}\left|\frac{V_o}{V_L}\right| = 20\log_{10}\left|\frac{2Z_r + Z_o}{2Z_r}\right| = 20\log_{10}\left|\frac{58 - j988}{8 - j988}\right| = 0.015 \text{ dB}$$

$$\text{Isolation} = (IL)_{OFF} = 20\log_{10}\left|\frac{2Z_o + Z_f}{2Z_f}\right| = 20\log_{10}\left|\frac{101 + j6}{2 + j12}\right| = 18.4 \text{ dB}$$

The series configuration has a higher insertion loss (0.1 dB) but a lower isolation value (14.1 dB) compared with the shunt configuration where the insertion loss is 0.015 dB and isolation is 18.4 dB. The attenuation difference between the on and off states is 14 dB for the series configuration, whereas the shunt configuration has higher attenuation difference of 18.385 dB between the two states. For this example, the shunt configuration overall seems to be superior to the series configuration in terms insertion loss, isolation, and attenuation difference between the two states.

EXERCISE 20.1

Using the diode in Example 20.4 in a series-shunt configuration, determine its insertion loss, isolation, and attenuation between the two states.

How does this configuration's performance compare with series or shunt configuration?

20.4 PHASE SHIFTERS

Let's first define what is meant by a phase shifter:

DEFINITION-PHASE SHIFTER: *A two-port network in which the output voltage (or current) may be adjusted to have some desired phase relationship to the input voltage (or current) by using a control signal (e.g., a DC bias, etc.).*

As can be seen from Figure 20.10, phase shifters may be classified into two major categories:

Analog phase shifters are components in which the differential phase shift can be varied in a continuous manner using a control signal.

Digital phase shifters are components in which the differential phase shift can be changed by only a few predetermined phase-shift values such as 45°, 90°, 180°, etc. These networks have extensive applications in phased-array antenna systems where the beam scanning is computer controlled.

In this chapter, we will focus on digital phase shifters primarily.

20.5 DIGITAL PHASE SHIFTERS

There are two methods of designing digital phase shifters at microwave frequencies:

- Ferrite phase shifters
- Semiconductor phase shifters

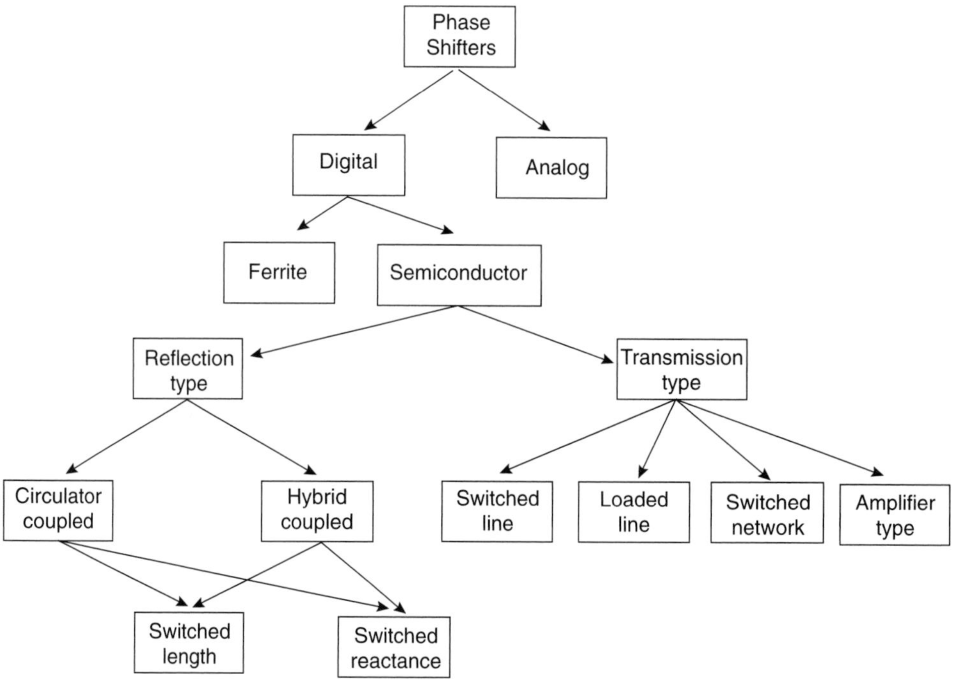

FIGURE 20.10 Various types of microwave phase shifters.

There is a general trend toward semiconductor phase shifters rather than ferrite phase shifters, due to the former's obvious advantages such as smaller size, higher speed, lower drive power, and easier integration with planar circuitry. For this reason, we will only focus on and discuss semiconductor phase shifters in the remainder of this chapter.

The only disadvantage of semiconductor phase shifters is the requirement of a continuous bias current while the ferrite phase shifters only require a pulsed current to change states.

20.6 SEMICONDUCTOR PHASE SHIFTERS

In general, a semiconductor phase shifter can be classified into one of the following two types:

- A transmission type phase shifter
- A reflection type phase shifter

We will now discuss these two types of semiconductor phase shifters separately and will analyze some basic configurations for each as shown in Figure 20.10.

20.6.1 Transmission-Type Phase Shifters

In the first category of phase shifters, the transmission-type phase shifters, we find several design subgroups as follows:

- Switched-line type
- Loaded-line type
- Switching network type
- Amplifier type

In this section, we will discuss the first two subgroups and leave the other two to more advanced books.

I. Switched-Line Phase Shifters. The switched-line phase shifters have a simple and straightforward design. They consist of two SPDT switches and two transmission lines of different lengths, as shown in Figure 20.11. These switched-line phase shifters are normally designed for fixed binary differential phase shifts of $\Delta\phi = 22.5°$, $45°, 90°, 180°$, etc. The differential phase shift between the two paths is given by:

$$\Delta\phi = \beta(\ell_2 - \ell_1) = \beta\Delta\ell \tag{20.26a}$$

where

$$\Delta\ell = \ell_2 - \ell_1 \tag{20.26b}$$

is the physical length difference between the transmission lines and β is the propagation constant of each transmission line. For TEM or quasi-TEM transmission lines, the propagation constant is a linear function of frequency because:

$$\beta = \omega/V_P \tag{20.27a}$$

Thus,

$$\Delta\phi = \omega\Delta\ell/V_P \tag{20.27b}$$

where V_P is the phase velocity on the transmission line.

The linear relationship given by Equation 20.27b reduces the amount of distortion that can occur in a broadband design. The time delay (τ_d) for this type of phase shifters is simply given by:

$$\tau_d = \Delta\phi/\omega = \Delta\ell/V_P \tag{20.28}$$

In actual operation, one line is on (i.e., allowing signal transmission) while the other is off (i.e., no signal transmission); then upon switching of the PIN diodes the on and off lines will reverse and thus a differential phase shift is generated at the output.

POINT OF CAUTION: *One potential problem in switched-line phase shifters, however, is the resonance in the line that is off if its length is near a multiple of $\lambda/2$ at the frequency of operation. Judicious selection of ℓ_1 and ℓ_2 for proper operation is essential.*

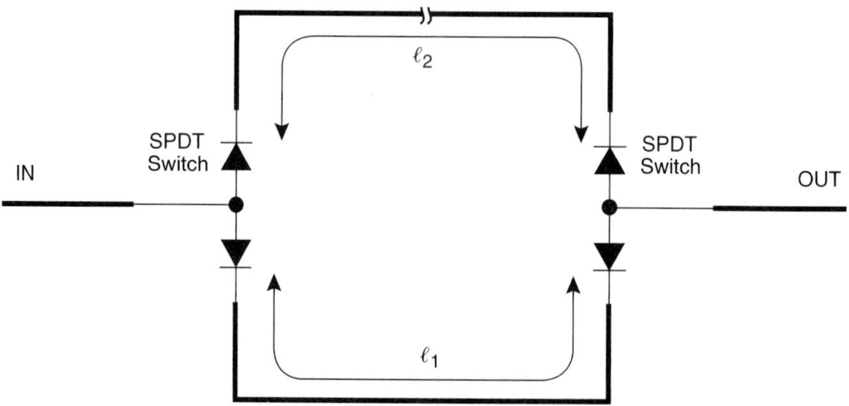

FIGURE 20.11 A switched-line phase shifter.

II. Loaded-Line Phase Shifters. For phase shifts less than or equal to 45°, loaded-line phase shifters are used as shown in Figure 20.12. These circuits are based on the principle of loading a uniform transmission line by a small reactance such that the transmitted wave undergoes a phase shift that depends on the normalized shunt element value. The shunt element can be implemented by means of a lumped lossless element (e.g., an inductor or a capacitor) or a transmission line stub. The following analyzes the loaded-line phase shifters in more exact detail.

FIGURE 20.12 A loaded-line phase shifter.

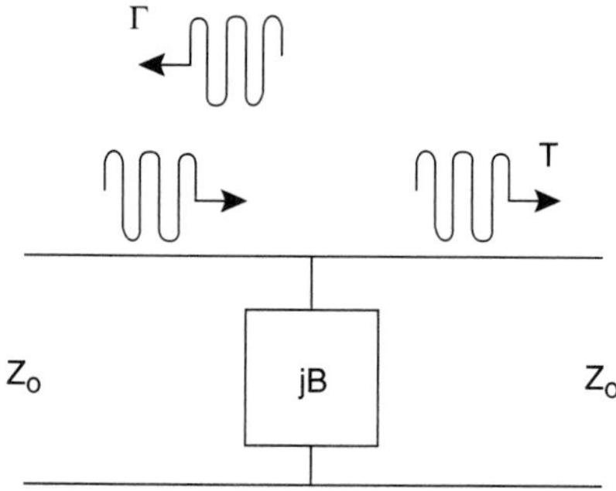

Analysis Consider a transmission line loaded with a normalized shunt susceptance given by:

$$jb = jB/Y_o = jBZ_o \tag{20.29}$$

If a wave of magnitude V_i is incident on this shunt susceptance, then the reflected voltage wave (V_R) and transmitted voltage wave (V_T) are given by:

$$V_R = \Gamma V_i \tag{20.30}$$

$$V_T = \text{T} V_i \tag{20.31}$$

where Γ and T are the reflection and transmission coefficients on the transmission lines, respectively, and are given by:

$$\Gamma = \frac{Z_{IN} - Z_o}{Z_{IN} + Z_o} = \frac{Y_o - Y_{IN}}{Y_o + Y_{IN}} = \frac{1 - (Y_{IN})_N}{1 + (Y_{IN})_N} \tag{20.32}$$

$$T = 1 + \Gamma \tag{20.33}$$

$(Y_{IN})_N$ is the normalized input admittance of the phase shifter as shown in Figure 20.13 and is given by:

$$(Y_{IN})_N = 1 + jb \tag{20.34}$$

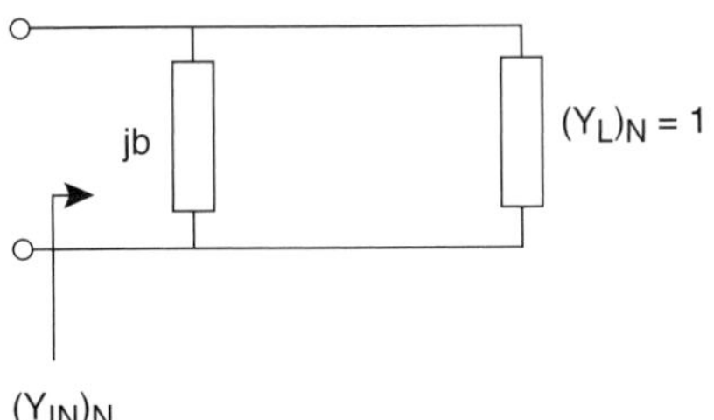

FIGURE 20.13 A shunt element.

Therefore, Γ and T are given by:

$$\Gamma = \frac{-jb}{2 + jb} \tag{20.35}$$

$$T = \frac{2}{2 + jb} = \frac{2}{\sqrt{4 + b^2}} \angle -\tan^{-1}(b/2) \tag{20.36}$$

Thus, the phase shift between the input and the output transmitted signal is given by:

$$\Delta\phi = -\tan^{-1}(b/2) \tag{20.37}$$

The phase shift ($\Delta\phi$) can be positive or negative depending on the sign of b.

Insertion Loss of a Loaded-Line Phase Shifter. The insertion loss (IL) of a phase shifter is defined to be negative of the ratio (in dB) of the output voltage (or in this case the transmitted voltage V_T) and the input voltage (V_i) as follows:

$$V_T/V_i = \frac{1}{\sqrt{1 + b^2/4}},$$

Or, in dB, we have:

$$IL = -20 \log_{10}\left(\frac{V_T}{V_i}\right) = 10 \log_{10}\left(1 + \frac{b^2}{4}\right) \text{ (dB)} \tag{20.38}$$

Disadvantage of Loaded-Line Phase Shifters. We can see from Equation 20.37 that in order to get higher phase shifts, we require a higher value of b. However, we can see from Equation 20.38 that increasing b will increase the insertion loss which means that more power is reflected back to the source because the shunt element is lossless. This is a disadvantage of the loaded-line phase shifter circuits that should be considered when designing these circuits (see Figure 20.12).

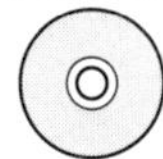

EXAMPLE 20.5

A switched-line phase shifter has a shunt susceptance of jB = j40 Ω in a 50 Ω system. Determine the phase shift and the insertion loss.

If the shunt susceptance value is doubled, determine the new values of the phase shift and the insertion loss.

Solution:

a.

$$b = 40/50 = 0.8$$

$$\Delta\phi = -\tan^{-1}(0.8/2) = -21.8°$$

$$IL = 10\log_{10}\left(1 + \frac{0.8^2}{4}\right) = 0.64 \text{ (dB)}$$

b.

$$b' = 2b = 1.6$$

$$\Delta\phi = -\tan^{-1}(1.6/2) = -38.7°$$

$$IL = 10\log_{10}\left(1 + \frac{1.6^2}{4}\right) = 2.15 \text{ (dB)}$$

From this example, we can see that increasing the phase shift magnitude from (21.8°) to (38.7°) will cause the insertion loss to increase dramatically from 0.64 dB to 2.15 dB (by 1.51 dB!), which indicates that there is an increase in the reflected voltage waves back to the source.

III. Modified Loaded-Line Phase Shifters. To remedy the drawbacks of reflection from a single susceptance, we use two identical susceptances separated by a quarter-wavelength transmission line as shown in Figure 20.14. The partial reflection from the first susceptance will be 180° out of phase with the partial reflection from the second susceptance and the two reflections are almost equal in magnitude, causing cancellation of the signals reflected back to the source. This configuration leads to a very low-input *VSWR*, as shown by the analysis that follows.

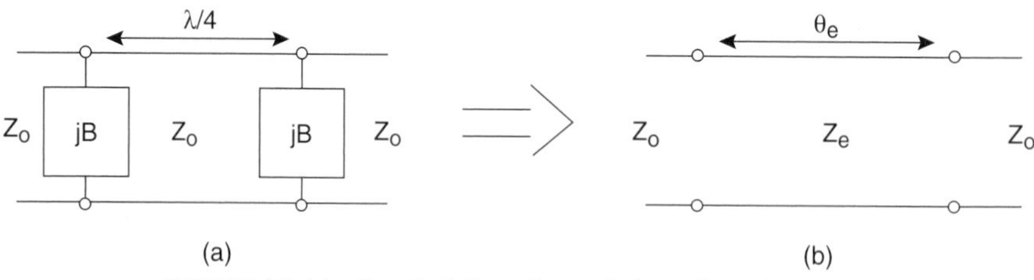

FIGURE 20.14 Loaded-line phase shifter: a) Basic circuit, b) Practical loaded-line phase shifter and its equivalent circuit.

Analysis

1. To analyze this circuit, we first calculate the overall $ABCD$ matrix of the phase shifter as follows:

$$\begin{bmatrix} A & B \\ C & D \end{bmatrix} = \begin{bmatrix} 1 & 0 \\ jB & 1 \end{bmatrix} \begin{bmatrix} 0 & jZ_o \\ j/Z_o & 0 \end{bmatrix} \begin{bmatrix} 1 & 0 \\ jB & 1 \end{bmatrix} = \begin{bmatrix} -BZ_o & jZ_o \\ j(1/Z_o - B^2 Z_o) & -BZ_o \end{bmatrix} \tag{20.39}$$

2. Now we compare the phase shifter's matrix (20.39) with an equivalent transmission line which performs the same function, i.e., giving the same phase shift or time-delay as shown in Figure 20.15. The $ABCD$ matrix of an equivalent and yet fictitious transmission line of electrical length (θ_e) and characteristic impedance (Z_e) performing the same function is given by (see Example 8.5 for details):

$$\begin{bmatrix} A & B \\ C & D \end{bmatrix} = \begin{bmatrix} \cos\theta_e & jZ_e \sin\theta_e \\ j\sin\theta_e / Z_e & \cos\theta_e \end{bmatrix} \tag{20.40}$$

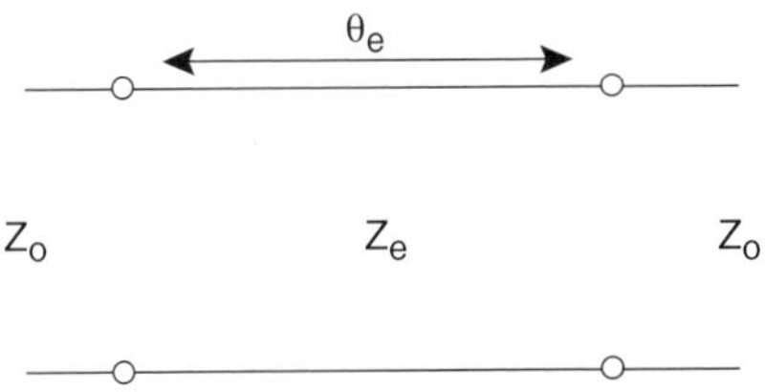

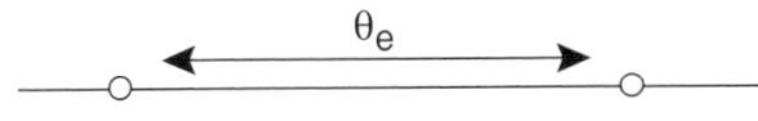

Zo Ze Zo

FIGURE 20.15 Equivalent transmission line.

3. Comparing Equation 20.39 with 20.40, we obtain:

$$\cos\theta_e = -BZ_o = -b \tag{20.41}$$

$$Z_e \sin\theta_e = Z_o \tag{20.42}$$

4. From Equations 20.41 and 20.42, Z_e and θ_e are obtained:

$$Z_e = \frac{Z_o}{\sqrt{1-b^2}} \tag{20.43a}$$

$$\theta_e = \cos^{-1}(-b) \tag{20.43b}$$

From Equation 20.43, we note that to obtain a real and positive number for Z_e, the magnitude of b has to be less than one, i.e.,

$$|b|<1 \Rightarrow Z_e > Z_o$$

Furthermore, the magnitude of Z_e is independent of the sign of b which means that we can keep $|b|$ constant, change its sign, and obtain a differential phase shift for the two states while the circuit remains completely matched because Z_e remains the same. The susceptance b (or B) can be implemented with a lumped lossless element (i.e., an inductor or a capacitor) or a distributed element (such as a short or open stub) and switching between the two states can be done by means of an SPST switch. This concept is illustrated in the next example.

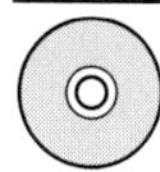

EXAMPLE 20.6

Find the differential phase shift and the VSWR on the line for a modified loaded-line phase shifter which has two shunt inductors as shown in Figure 20.16. Switching the two SPST PIN-diode switches on and off creates two states:

$$B_1 = 20\ \Omega$$

$$B_2 = -20\ \Omega$$

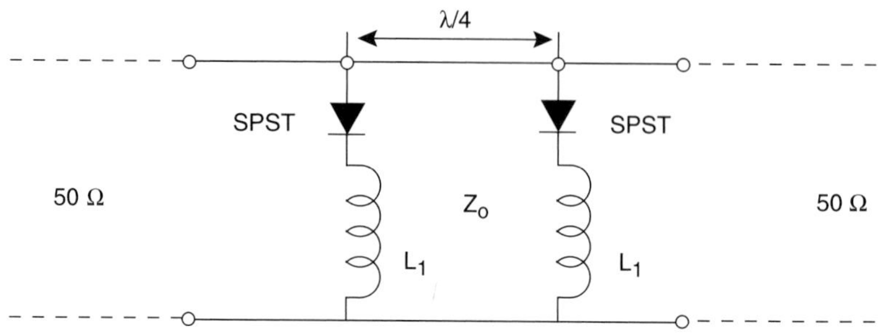

FIGURE 20.16 Circuit for Example 20.6.

Determine the differential phase shift and the equivalent transmission line if the phase shifter operates in a 50 Ω system.

Solution:

$$b_1 = 20/50 = 0.4$$

$$b_2 = -20/50 = -0.4$$

$$\cos\theta_{e1} = -b_1 = -0.4 \Rightarrow \theta_{e1} = 113.6°$$

$$\cos\theta_{e2} = -b_2 = 0.4 \Rightarrow \theta_{e2} = 66.4°$$

The differential phase shift is given by:

$$\Delta\theta_e = \theta_{e2} - \theta_{e1} = -47.2°$$

The equivalent characteristic impedance of a fictitious transmission line having an electrical length of θ_{e1} or θ_{e2} is given by:

$$Z_e = \frac{Z_o}{\sqrt{1-b^2}} = \frac{50}{\sqrt{1-0.4^2}} = 54.6\ \Omega$$

The circuit remains matched to 50 Ω line in both states of the phase shifter.

20.6.2 Reflection-Type Phase Shifters

The second category of phase shifters are reflection-type phase shifters. These phase shifters are based on the concept of switchable terminations, which create switchable reflection coefficients. The reflection coefficient is switched from $\Gamma_1 = |\Gamma_1|\angle\phi_1$ in one state to $\Gamma_2 = |\Gamma_2|\angle\phi_2$ in another state.

Ideally, we would like to have $|\Gamma_1|$ and $|\Gamma_2|$ close to unity (i.e., small or no resistive losses) so that the insertion loss is kept to a minimum. Thus, the reflected signal undergoes a differential phase shift as follows:

$$\Gamma_1 = |\Gamma_1|\angle\phi_1 = \frac{Y_o - Y_{ON}}{Y_o + Y_{ON}}$$

$$\Gamma_2 = |\Gamma_2|\angle\phi_2 = \frac{Y_o - Y_{OFF}}{Y_o + Y_{OFF}}$$

$$\Delta\phi = \phi_2 - \phi_1 \tag{20.44}$$

Implementing such a concept in the design of a practical reflection-type phase shifter leads to two different designs as follows:

1. **Switchable-reactance type phase shifters:** This type of phase shifters can be implemented by using a PIN diode as a switchable termination as shown in Figures 20.17 and 20.18.

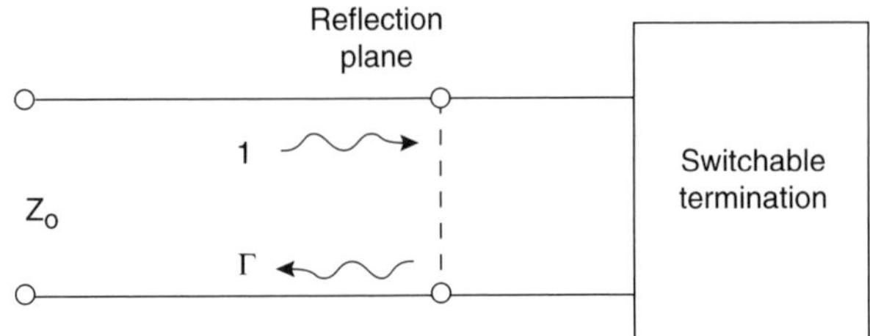

FIGURE 20.17 Basic concept of reflection type phase shifters.

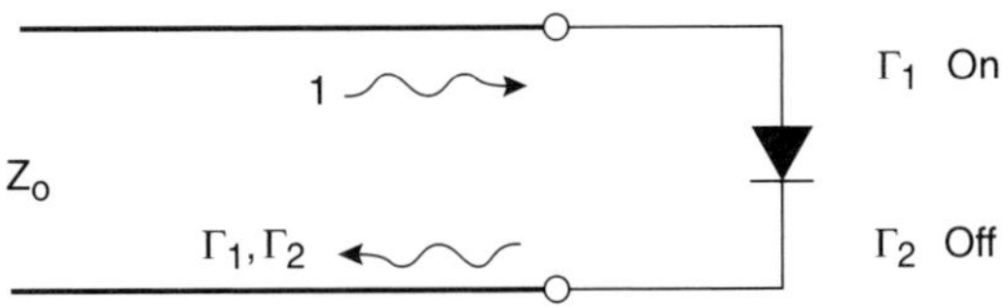

FIGURE 20.18 A switchable-reactance type phase shifter.

2. **Switchable-length type reflection phase shifters:** The second group of reflection type phase shifters is the switchable-length type phase shifters which add an additional line length at the reflection plane by using an SPST PIN-diode switch (see Figure 20.19) very similar to switched line phase shifters discussed earlier.

When the SPST switch is open, the signal travels a longer length of the line than when the SPST switch is closed, creating a total differential phase shift of $\Delta\phi$. The total differential phase shift is twice the electrical length of the additional transmission line (which is being switched in by the SPST switch) because the signal has to travel round-trip to get to the output. Any insertion loss is due to the forward resistance of the PIN diode.

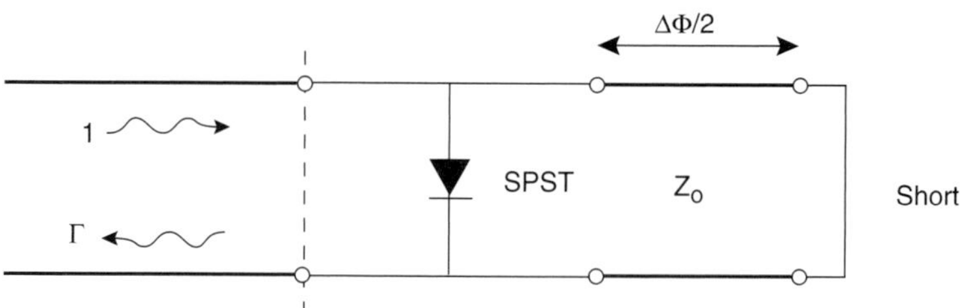

FIGURE 20.19 A switchable-length type phase shifter.

A simple practical design of this case would consist of a 90° hybrid coupler at the input followed by two SPST PIN diode switches to control the path length of the reflected signal as shown in Figure 20.20. The 90° hybrid coupler divides the power equally to the two ports of the phase shifter. Turning the diodes on or off changes the total path length to create a differential phase shift of $\Delta\phi$ at the output. There are infinite number of choices for the line length to obtain the desired $\Delta\phi$.

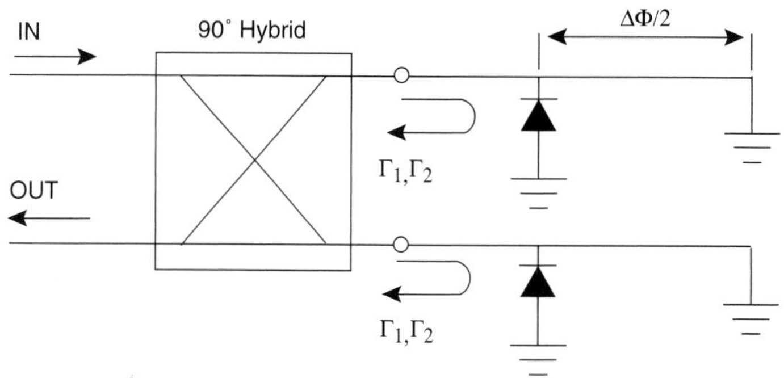

FIGURE 20.20 A switchable-length type phase shifter using a quadrature hybrid.

It can be shown that in order to optimize the bandwidth, the reflection coefficients for the two states (i.e., Γ_1 and Γ_2) must be mutually phase conjugate as follows:

$$\Gamma_1 = e^{j\phi} \qquad \text{diode ON}$$

$$\Gamma_2 = e^{j(\phi+\Delta\phi)} \qquad \text{diode OFF}$$

To provide optimum bandwidth, we require:

$$\Gamma_1 = \Gamma_2^* \Rightarrow e^{j\phi} = e^{-j(\phi+\Delta\phi)} \tag{20.45}$$

This gives:

$$\phi = -(\phi + \Delta\phi) + 2k\pi \Rightarrow 2\phi = 2k\pi - \Delta\phi, \qquad k = 0,1,2,\ldots\ldots$$

$$\Rightarrow \phi = k\pi - \Delta\phi/2 \tag{20.46}$$

This requirement makes the value of ϕ a degree of freedom, which means that any diode will serve the purpose as long as the two PIN diodes are well matched to yield the optimum performance. It should be noted that the 90° hybrid adds an insertion loss additional to the two diodes' resistive losses in the two states.

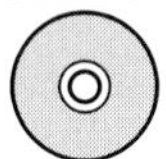

EXAMPLE 20.7

A switchable-length type phase shifter generates a 180° phase shift between the two states. If Γ_1 and Γ_2 for the two states are given by:

$$\Gamma_1 = |\Gamma_o|\angle\phi_1$$

$$\Gamma_2 = |\Gamma_o|\angle\phi_2$$

Find the phase values ϕ_1 and ϕ_2 for best bandwidth.

Solution:

We are given:

$$\Delta\phi = \phi_2 - \phi_1 = 180°$$

For best bandwidth, we require that Γ_1 and Γ_2 be phase conjugates of each other. Thus, from Equation 20.46 we can write:

$$\phi_1 = k\pi - \Delta\phi/2 = \pi(k - 1/2)$$

for $k = 0$

$$\phi_1 = -90°,$$

$$\phi_2 = \phi_1 + \Delta\phi = -90 + 180 = 90°$$

20.7 PIN DIODE ATTENUATORS

By definition, an attenuator is an adjustable or a fixed network of resistors and capacitors (or both) that provide a reduction of the amplitude of an electrical signal without introducing appreciable phase or frequency distortion. A simple example of an attenuator is a potentiometer.

In a broader arena, a voltage-controlled variable attenuator as a control element is widely used for automatic gain control (AGC). PIN diodes (as well as GaAs MESFETs) can be used effectively for the design of variable attenuators. The resistance of a PIN diode is a strong function of the DC bias current and can easily be employed for designing current-controlled variable attenuators at RF and microwave frequencies.

20.7.1 Analysis

An important feature of an attenuator is that it should remain matched to the line as the attenuation varies over its operating range. Let us consider a variable attenuator operating in a circuit as shown in Figure 20.21. One possible design is the use of three PIN diodes in a π-network configuration as shown in Figure 20.22. The equivalent resistive π-network configuration is shown in Figure 20.23.

To have a good match (i.e., $\Gamma = 0$), we require:

$$Z_{IN} = Z_o \tag{20.47}$$

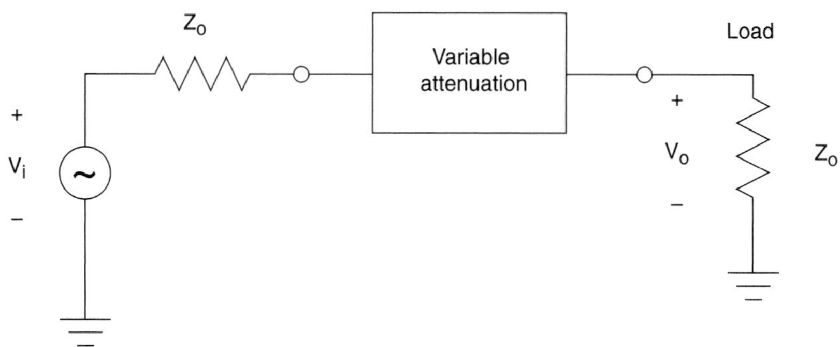

FIGURE 20.21 An attenuator operating in a circuit.

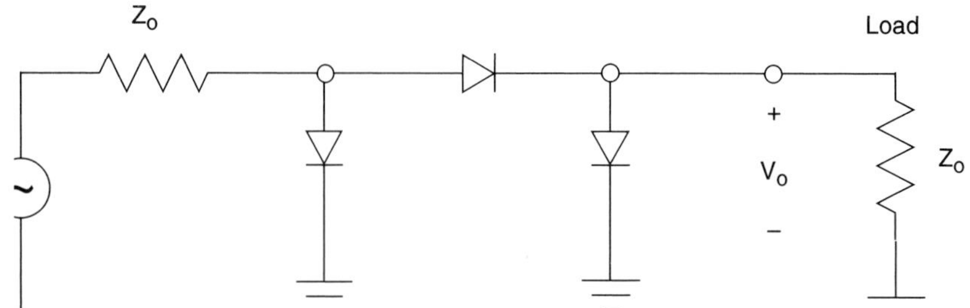

FIGURE 20.22 A PIN diode attenuator in a π-network configuration.

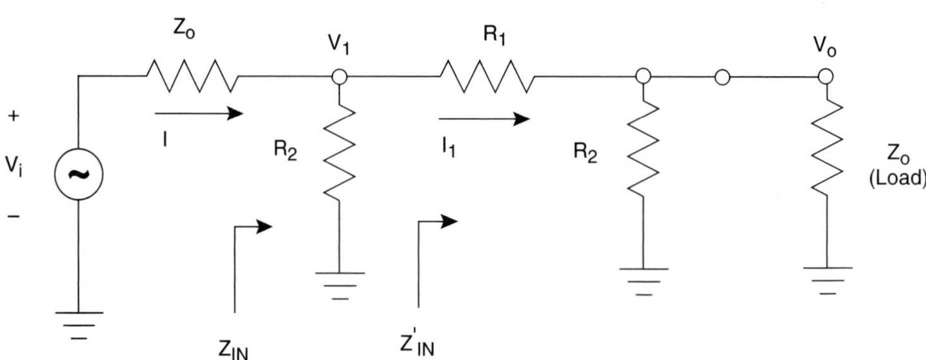

FIGURE 20.23 π-network equivalent circuit.

The input impedance (Z_{IN}) is given by:

$$Z_{IN} = R_2 \parallel Z'_{IN} \tag{20.48}$$

Where Z'_{IN} is given by:

$$Z'_{IN} = R_1 + R_2 \| Z_o = R_1 + \frac{R_2 Z_o}{R_2 + Z_o} \tag{20.49}$$

The values of R_1 and R_2 that satisfy Equation 20.47 can be found in terms of the attenuation ratio (inverse of the voltage gain) of a lossy attenuator. The attenuation ratio (K) is simply given by (see Figure 20.23):

$$K = \left(\frac{V_o}{V_i}\right)^{-1} \quad (K > 1) \tag{20.50}$$

where

$$V_i = I_1 Z'_{IN} = I_1\left(R_1 + \frac{R_2 Z_o}{R_2 + Z_o}\right) \tag{20.51}$$

$$V_o = I_1(R_2 \| Z_o) = I_1\left(\frac{R_2 Z_o}{R_2 + Z_o}\right) \tag{20.52}$$

Using Equations 20.51 and 20.52 in Equation 20.50, we have:

$$K = \frac{V_i}{V_o} = \frac{R_2 + Z_o}{R_2 - Z_o} \tag{20.53}$$

Thus, the values of R_1 and R_2 from Equations 20.47 and 20.53 in terms of K are given as follows:

$$R_1 = \frac{Z_o}{2}\left(K - \frac{1}{K}\right) \tag{20.54a}$$

$$R_2 = Z_o\left(\frac{K+1}{K-1}\right) \tag{20.54b}$$

The PIN diodes have to be biased properly such that the necessary values of resistance as calculated in Equations 20.54 are obtained.

NOTE: *Rather than a π-network of PIN diodes, a T-network of PIN diodes may have equally been employed effectively to obtain a matched attenuator over the operating frequency range (see Figure 20.24). By using a similar analysis, we would be able to obtain very similar results for the resistance values of the PIN diodes.*

EXERCISE 20.2

Using the analysis given above, derive the necessary PIN diode resistor values for a *T*-network PIN diode attenuator having a known attenuation ratio (K'), where K' is the attenuation ratio for the whole attenuator (see Figure 20.24).

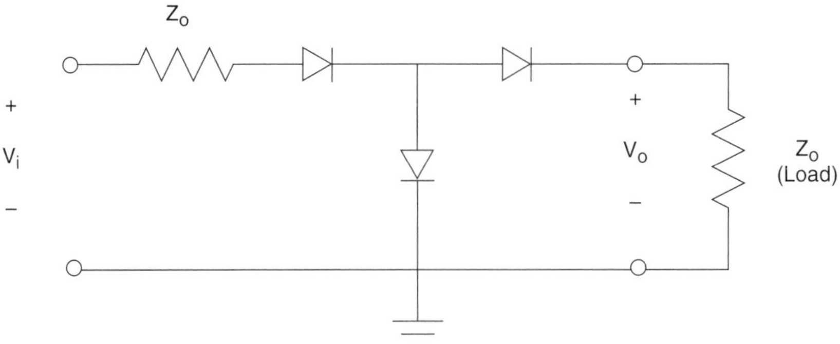

FIGURE 20.24 PIN diode attenuator in a T-network configuration.

LIST OF SYMBOLS/ABBREVIATIONS

A symbol/abbreviation will not be repeated again once it has been identified and defined in an earlier chapter, as long as its definition remains unchanged.

AGC	Automatic gain control
DPDT	Double-pole double-throw switch
IL	Insertion loss
L_{int}	Series inductance of a PIN diode
R_f	Forward-biased resistance
R_r	Reverse-biased resistance
SPDT	Single-pole double-throw switch
SPST	Single-pole single-throw switch
T	Transmission coefficient
V_P	Phase velocity on a transmission line
Z_e	Characteristic impedance for an equivalent transmission line
β	Propagation constant of the transmission line
$\Delta\phi$	Phase shift
θ_e	Transmission line electrical length
τ_d	Time delay

PROBLEMS

20.1 An SPST switch employs two PIN diodes in the series-shunt configuration as shown in Figure P20.1. Calculate the insertion loss and isolation at 6 GHz if each PIN diode has the following characteristics:

$$C_j = 0.1 \text{ pF},$$
$$R_f = 1\ \Omega,$$
$$R_r = 5\ \Omega,$$
$$L_{int} = 0.3 \text{ nH}$$

Ignore the parasitic elements due to the package.

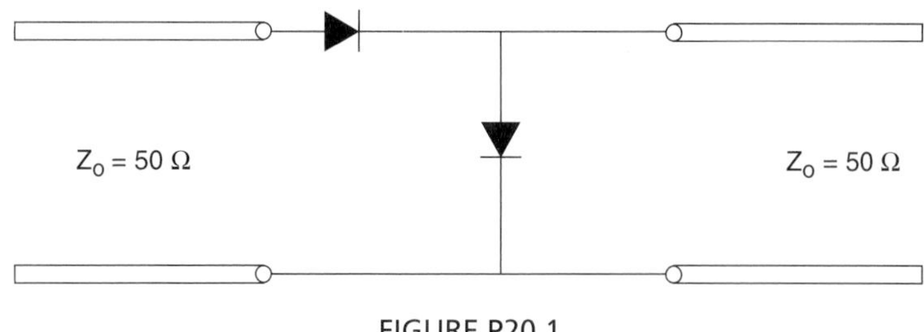

FIGURE P20.1

20.2 An SPST switch is designed using PIN diodes with the following characteristics:

$$C_j = 0.1\ \text{pF},$$
$$R_f = 1\ \Omega,$$
$$R_r = 4\ \Omega,$$
$$L_{int} = 0.3\ \text{nH}$$

The center frequency of the design is 3 GHz. Calculate and plot the switch performance over the frequency range of 2 to 4 GHz for:

a. Series configuration

b. Shunt configuration

20.3 An SPST switch uses PIN diodes in a shunt configuration. The frequency is 4 GHz, $Z_o = 50\ \Omega$ and the diode parameters are:

$$C_j = 0.5\ \text{pF},$$
$$R_f = 0.3\ \Omega,$$
$$R_r = 0.5\ \Omega,$$
$$L_{int} = 0.3\ \text{nH}$$

Find the electrical length of an open-circuited shunt stub, placed across the diode to minimize the insertion loss for the ON state of the switch. Calculate the resulting insertion loss and isolation of the switch.

20.4 Consider the loaded-line phase shifter shown in Figure P20.4. Find the necessary stub lengths for a differential phase shift of 45°, and calculate the resulting insertion loss for both states of the phase shifter. Assume all lines are lossless with $Z_o = 50\ \Omega$ and each diode can be approximated as an ideal short or open.

20.5 A PIN diode is mounted in series with a 50 Ω line. The PIN diode has:

$$C_j = 0.1\ \text{pF},$$
$$R_f = 1.5\ \Omega,$$
$$R_r = 2\ \Omega,$$
$$L_{int} = C_P = 0$$

FIGURE P20.4

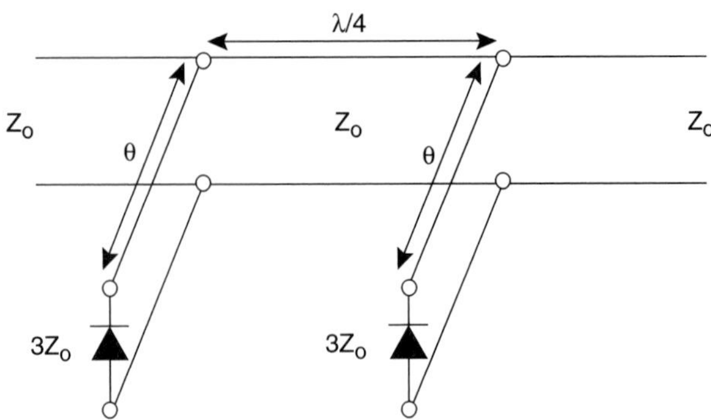

At 1 GHz perform the following calculations:

a. Find the attenuation (in dB) for forward-bias and reverse-bias cases.

b. Assuming there is no mismatch loss and the input power is 10 dBm, find the output power for both the forward- and reverse-bias cases.

20.6 Derive the phase shift and attenuation for the circuits shown in Figure P20.6.

FIGURE P20.6

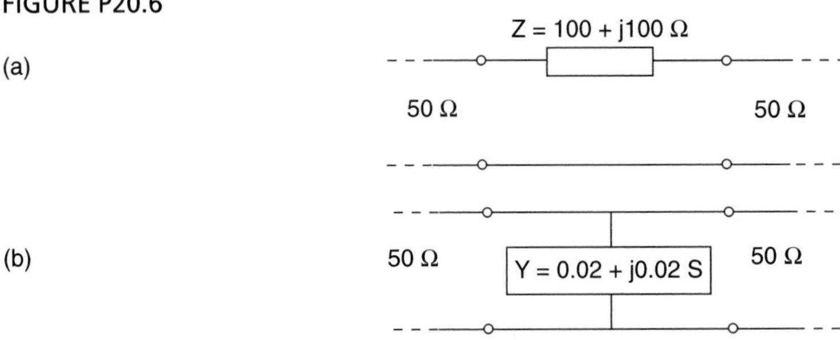

20.7 For the circuit shown in Figure P20.7 (assume $Z_o = 50\ \Omega$):

a. Find the S-parameters.

b. Find the attenuation values (in dB) and the phase shift.

20.8 A PIN diode has:

$$C_j = 0.1\text{ pF},$$
$$R_f = 1.0\ \Omega,$$
$$R_r = 2\ \Omega,$$
$$L_{int} = C_P = 0$$

FIGURE P20.7

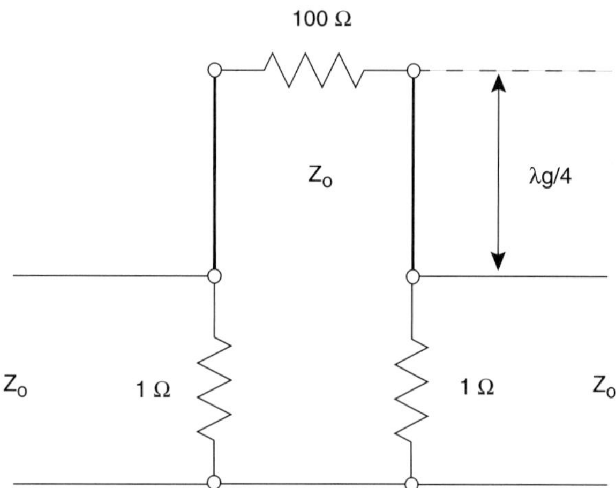

The diode is mounted in series with a 50 Ω line to form a switch as shown in Figure P20.8.

FIGURE P20.8

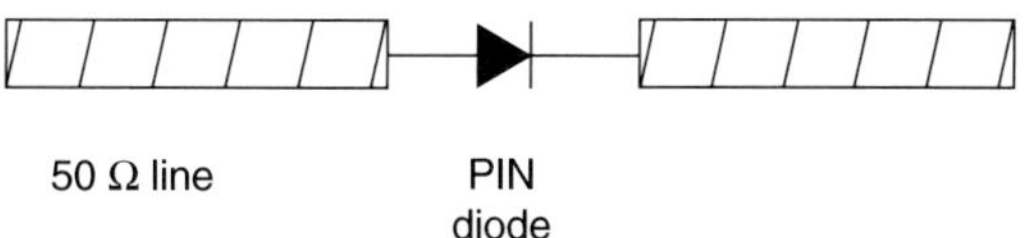

a. Calculate the insertion loss and isolation (in dB) at 5 GHz when the diode is forward and reverse biased.

b. Find the reflection coefficient (Γ) when the diode is reverse biased.

c. What is the insertion loss and isolation if two such identical PIN diodes are connected in series?

20.9 A PIN diode chip is mounted shunt across a 50 Ω microstrip line with two bonding tapes. Each of the bonding tapes can be modeled by an inductance (L) as shown in Figure P20.9. Find the value of the inductance (L') that is needed to optimize the insertion loss at 6 GHz. The PIN diodes have:

$$C_j = 0.1 \text{ pF},$$
$$R_f = 1.0\ \Omega,$$
$$R_r = 2\ \Omega,$$
$$L_{int} = C_P = 0$$

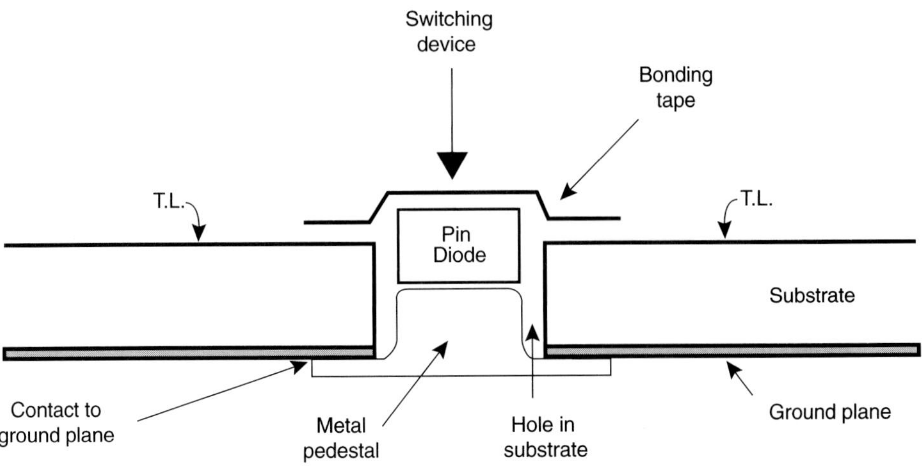

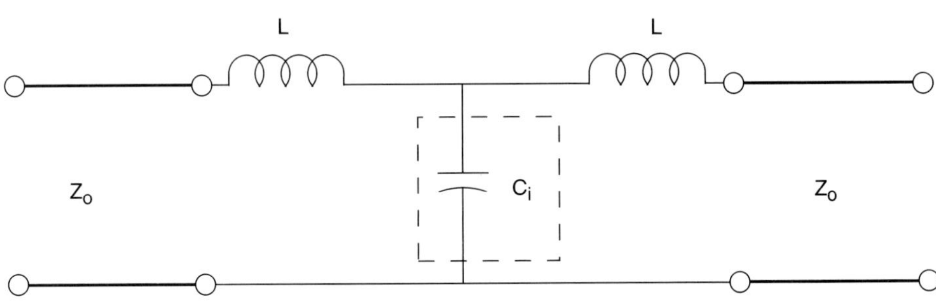

High impedance state

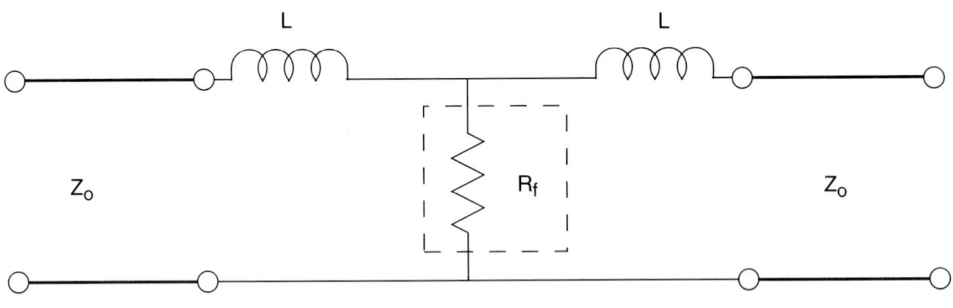

Low impedance state

FIGURE P20.9

REFERENCES

[20.1] Atwater, H. A. "Circuit Designs of the Loaded Line Phase Shifter," *IEEE Transactions on Microwave Theory and Techniques,* Systems, Vol. MTT-33, July 1985, pp. 626-634.

[20.2] Bahl, I. and P. Bhartia. *Microwave Solid State Circuit Design*, New York: John Wiley Interscience, 1988.

[20.3] Bhartia, P. and I. Bahl. *Millimeter Wave Engineering and Application*, New York: John Wiley, 1984.

[20.4] Chang, K. *Microwave Solid-State Circuits and Applications*, New York: John Wiley & Sons, 1994.

[20.5] Chorney, P. "Multi-Octave Multi-Throw, PIN Diode Switches," *Microwave Journal*, Vol. 17, Sept. 1974, pp. 39–42, 48.

[20.6] Hewlett-Packard. "An Attenuator Design Using PIN Diodes," Application Note No. 912.

[20.7] Pozar, D. M. *Microwave Engineering*, 2nd edition, New York: John Wiley & Sons, 1998.

[20.8] Sisson, M. J. et al. "Microstrip Devices for Millimeter-Wave Frequencies," in *IEEE International Microwave Symposium Technical Digest*, 1982, pp. 212–214.

[20.9] Steinbrecher, D. H. "Circuit Aspects of Nonlinear Microwave Networks," *IEEE Proc. Int. Symp. Circuits and Systems*, 1975, pp. 477–479.

[20.10] Watson, H. A. *Microwave Semiconductor Devices and Their Circuit Applications*, New York: McGraw-Hill, 1969.

[20.11] White, J. F. *Microwave Semiconductor Engineering*, Princeton: Van Norstrand, 1990.

[20.12] Yngvesson, S. Y. *Microwave Semiconductor Devices*, Kluwer Academic Publishers, 1991.

CHAPTER 21

RF/Microwave Integrated Circuit Design

21.1 INTRODUCTION

The current trend in circuit design is toward miniaturization and integration. Prime reasons for the trend toward miniaturization include more electronic function in a smaller space and lower weight, especially useful for airborne microwave electronic applications.

RF and microwave integrated circuits (RFICs and MICs) have brought about a revolution in the RF/microwave industry because they have made the concept of a microwave system on a chip a reality. They complement the low-frequency silicon integrated circuit technology, but instead use group III–V semiconductors (e.g., GaAs, InP, etc.).

An RF/microwave integrated circuit consists of an assembly that combines different circuit functions that are connected by microstrip transmission lines. These different circuits all incorporate planar semiconductor devices, passive lumped elements, and distributed elements.

The advantages of RF/microwave integrated circuits compared to traditional circuits using printed circuit technology can be briefly listed as follows:

- Higher reliability
- Reproducibility
- Better performance
- Smaller size
- Lower cost

21.2 MICROWAVE INTEGRATED CIRCUITS

There are two classes of microwave integrated circuits:

- Hybrid Microwave Integrated Circuits (HMICs)
- Monolithic Microwave Integrated Circuits (MMICs)

The first class of MICs can be further subdivided into standard and miniature circuits as shown in Figure 21.1. Each class is described briefly below.

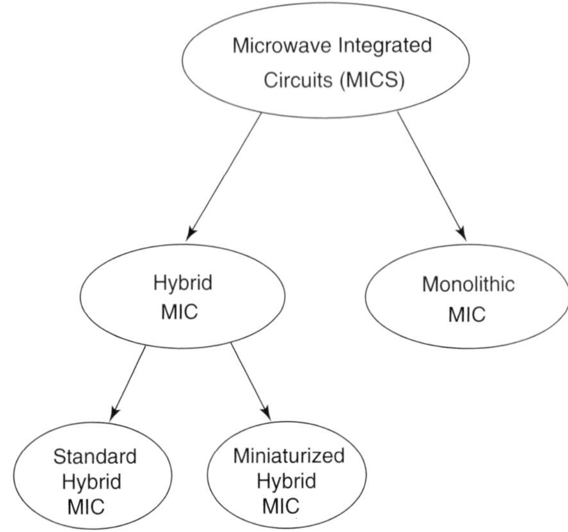

FIGURE 21.1 Types of integrated circuits (MICs).

21.2.1 Hybrid Microwave Integrated Circuits (HMICs)

A hybrid microwave integrated circuit is a type of circuit in which solid state devices and passive elements are bonded to a dielectric substrate. The passive elements (both lumped and distributed) are fabricated by using thick or thin film technology:

- The lumped elements are either bonded in chip form or are fabricated by using multi-level deposition and plating techniques
- The distributed elements are fabricated by using single-level metallization processes

21.2.2 Monolithic Microwave Integrated Circuits (MMICs)

Monolithic microwave integrated circuits are a type of circuit in which all active and passive elements as well as transmission lines are formed into the bulk or onto the surface of a substance by some deposition scheme such as epitaxy, ion implantation, sputtering, evaporation, or diffusion.

We will revisit and study the two types of MICs later in more depth but for now turn to the types of material and basic fabrication steps that are used to create these circuits.

21.3 MIC MATERIALS

There are three general types of circuit elements that either are used in chip form or are fabricated in microwave integrated circuits:

- Distributed transmission lines (e.g., microstrip, strip, coplanar, etc.)
- Lumped elements (R, L, and C)
- Solid state devices (FETs, BJTs, diodes, etc.)

The basic materials for fabrication of RFICs and MICs circuit elements can be categorized in four classes as shown in Figure 21.2:

- Substrate materials
- Conductor materials
- Dielectric materials
- Resistive films

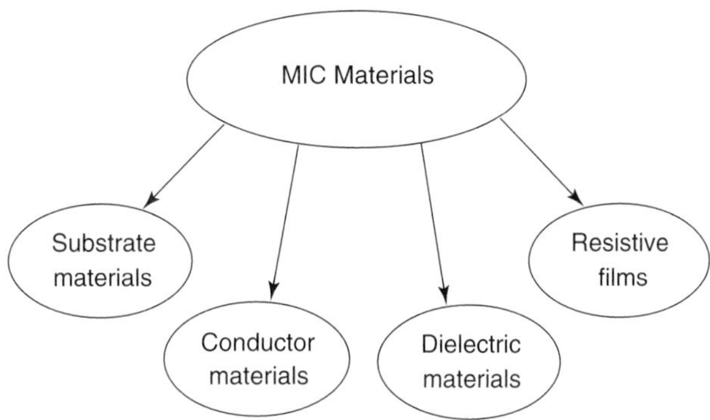

FIGURE 21.2 Types of MIC materials.

Each class will be discussed next.

21.3.1 Substrate materials

This is a class of materials that are used as a substrate or used for active device fabrication. The ideal MIC substrate must have certain features that make it a wise choice for our particular application. These features for an ideal substrate can be summarized as:

- A justifiably low cost
- A suitable dielectric thickness and permittivity to allow useful frequency range and achievable impedance values
- A negligible dielectric loss which corresponds to a low value for loss tangent ($\tan\delta$)
- A good substrate surface finish (0.05–0.1 μm) free of voids to keep conductor loss low with good metal-film adhesion
- A good mechanical strength and thermal conductivity and no deformation occurring during high temperature circuit processing

Table 21.1 summarizes the properties of some of the most commonly used substrate materials:

TABLE 21.1 Types and Properties of MIC Substrate Materials

Types of Material	Surface Roughness (μm)	tanδ @ 10 GHZ ($\times 10^{-4}$)	Relative Dielectric Const. (ε_r)	Thermal Conductivity (KW/cm–°C)	Dielectric Strength (KV/cm) ($\times 10^3$)	MIC Applications
Alumina (99.5%)	2–8	1–2	10	0.3	4	Microstrip lines
Sapphire	1	1	9.3–11.7	0.4	4	Microstrips, Lumped elements
Glass	1	20	5	0.01	_	Lumped elements
Beryllia	2–50	1	6.6	2.5	_	Compound substrate
GaAs	1	6	12.9	0.46	0.35	MMICs, microstrips
Si	1	10–100	12	1.5	0.30	MMICs
Quartz	1	1	3.8	0.01	10	Microstrips

21.3.2 Conductor Materials

The ideal conductor material used for MIC applications should have some desirable features as listed below:

- High conductivity
- High coefficient of thermal expansion
- Low resistance at RF/microwaves
- Good adhesion to the substrate
- Good etchability and solderability
- Easy to deposit or electroplate

Table 21.2 shows the properties of most widely used conductor materials for MIC applications.

TABLE 21.2 Types and Properties of MIC Conductor Materials

Types of Material	Surface Resistivity ($\Omega / sq \times 10^{-7} \sqrt{f}$)	Skin Depth (δ) @ 2 GHZ (μm)	Coefficient of Thermal Expansion ($\alpha_t / °C \times 10^6$)	Adherence to Dielectrics	MIC Deposition Technique
Silver (Ag)	2.5	1.4	21	Poor	(E)*
Copper (Cu)	2.6	1.5	18	Poor	(E), (P)*

TABLE 21.2 Types and Properties of MIC Conductor Materials *(Continued)*

Types of Material	Surface Resistivity $(\Omega/sq \times 10^{-7}\sqrt{f})$	Skin Depth (δ) @ 2 GHZ (μm)	Coefficient of Thermal Expansion $(\alpha_t/°C \times 10^6)$	Adherence to Dielectrics	MIC Deposition Technique
Gold (Au)	3.0	1.7	15	Poor	(E), (P)
Aluminum (Al)	3.3	1.9	26	Poor	(E)
Chromium (Cr)	4.7	2.7	9.0	Good	(E)
Tantalum (Ta)	7.2	4.0	6.6	Good	(EB)*, (S)*
Molybdenum (Mo)	4.7	2.7	6	Fair	(EB), (S), (E)
Tungsten (W)	4.7	2.6	4.6	Fair	(S), (E), (EB), (VPE)*

* Evaporation (E), Electron-beam (EB), Plating (P), Sputtering(S), Vapor phase epitaxy (VPE)

21.3.3 Dielectric Materials

Dielectric materials in MICs are used as insulators for capacitors and as protective or insulating layers for active and passive devices. The following features are desirable in dielectric materials:

- Reproducibility
- High breakdown voltage
- Low loss tangent
- Processability

Table 21.3 shows some of the properties of commonly used dielectric materials for MIC applications.

TABLE 21.3 Types and Properties of MIC Dielectric Materials

Types of Material	Method of Deposition	Relative Dielectric Constant (ε_r)	Dielectric Strength, (V/cm) $(\times 10^5)$	Microwave Q
SiO	(E)*	6–8	4	30
SiO_2	(D)*	4	100	100–1000
Si_3N_4	(S)*	7.6	100	
Al_2O_3	(A)*, (E)	7–10	40	
Ta_2O_5	(A), (E)	22–25	60	100

* Anodization (A), Deposition (D), Evaporation (E), Sputtering (S)

21.3.4 Resistive Films

Resistive films in MICs are needed to fabricate resistors and attenuators. The features desirable in resistive films are:

- Good stability
- Low Temperature Coefficient of Resistance (TCR)
- Sheet resistivity in the range of 10–2000 Ω/square

Temperature coefficient of resistance (TCR) is a figure which states the extent to which the resistance drifts under the influence of temperature, generally expressed in percent per degree (%/°C) or in parts per million per degree (ppm/°C)

Table 21.4 shows some of the properties of commonly used thin-film resistive materials for MIC applications.

TABLE 21.4 Types and Properties of MIC Resistive Materials

Types of Material	Method of Deposition	Resistivity (Ω/square)	TCR (%/°C)	Stability
Cr	(E)*	10–1000	−0.1 to +0.1	Poor
NiCr	(E)	40–400	+0.001 to +0.1	Good
Ta	(S)*	5–100	−0.01 to +0.01	Excellent
Ti	(E)	5–2000	−0.1 to +0.1	Fair

* Evaporation (E), Sputtering (S)

We will now discuss different types of microwave integrated circuits in more detail.

21.4 TYPES OF MICS

As discussed earlier, there are two types of MICs where each one has its own set of applications in the microwave industry. These two types may create some form of competition for each other, but it is a well-known fact that one type will not replace the other despite their obvious advantages or disadvantages.

21.4.1 Hybrid Microwave Integrated Circuits (HMICs)

Hybrid microwave integrated circuits were first developed in the 1960s and ever since have been used extensively in commercial, space, and military applications. Hybrid microwave integrated circuits are subdivided into two categories as shown in Figure 21.1:

Standard Hybrid MICs. Standard hybrid MICs use a single-level metallization for conductors and transmission lines with discrete circuit elements (such as transistors, inductors, capacitors, etc.) bonded to the substrate. Standard hybrid MICs using a single-layer metallization technique to form RF components is a very mature technology. A typical standard hybrid MIC is shown in Figure 21.3.

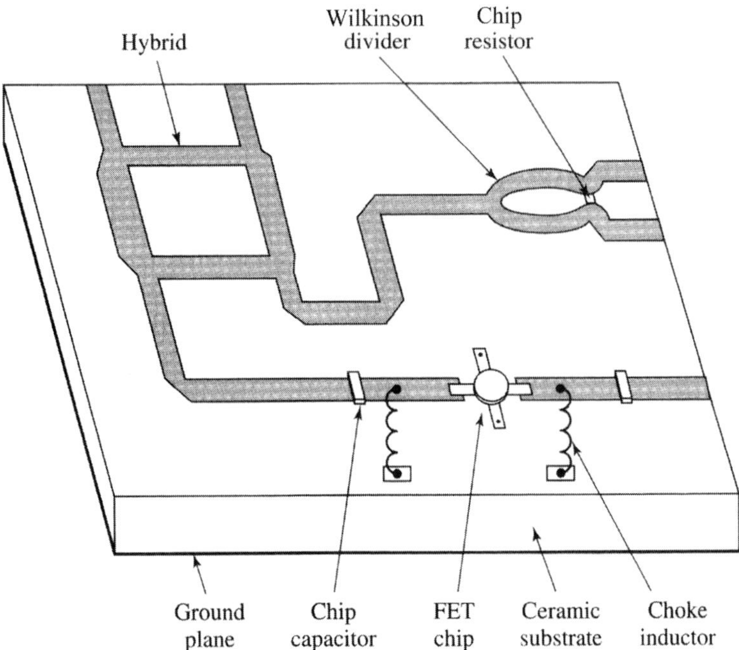

FIGURE 21.3 A typical standard hybrid MIC (HMIC).

Miniature Hybrid MICs. Miniature hybrid MICs use multi-level processes in which passive elements (such as inductors, capacitors, resistors, transmission lines, etc.) are batch deposited on the substrate whereas the semiconductor devices (such as transistors, diodes, etc.) are bonded on the substrate surface. These circuits are smaller than hybrid MICs but are larger than MMICs; therefore miniature hybrid circuit technology can be also called quasi-monolithic.

The advantages of miniature hybrid compared to standard hybrid circuits are:

- Smaller size
- Lighter weight
- Lower loss
- Lower assembly costs (due to batch fabrication process)

The first design consideration is about the substrate material, which depends on frequency. As frequency is increased thinner substrates are required, resulting in smaller sized circuits; for example, in the frequency range of 1–20 GHz, a substrate thickness of 0.635 to 0.254 mm is used whereas for millimeter-wave applications ($f \geq 30$ GHz), a thickness of 0.2 to 0.1 mm is used. Selection of a particular type of substrate depends on specific applications; for example, for high power amplifiers, good heat dissipation is obtained through the use of beryllium oxide (BeO).

Circuit components can be integrated closely together in a small size with minimal coupling (typically around 30 dB). To achieve this small coupling the two adjacent microstrip lines should be separated by at least $3h$, where h is the substrate height or thickness. This is a good guideline to follow when it comes to circuit layout and mask fabrication.

Examples of hybrid MICs widely used in different applications are:

- Electronic systems and instruments
- Satellite communications
- Phased-array radar systems
- Electronic warfare
- Airborne applications

These applications require design and fabrication of components such as amplifiers, mixers, transmit/receive modules, phase shifters, and oscillators, which have been discussed in depth in prior chapters.

Specific and actual examples of hybrid microwave integrated circuits are:

- A 12–Watt GaAs FET amplifier with a gain of 58 dB operating in the 3.7–4.2 GHz range
- A 0–360° analog dual-gate FET phase shifter operating in the 4–8 GHz range
- A 1–Watt Ku-band amplifier with a gain of 4.5 dB operating at 16 GHz with a bandwidth of 400 MHz
- A Ku-band Transmit/Receive (T/R) Module (16.0–16.5 GHz), having a noise figure of 5 dB (as a receiver) and an output power of 3.9–4.4 W (as a transmitter)
- A two-stage LNA with a gain of 20 dB, operating in a 3.7–4.2 GHz range

21.4.2 Microwave Monolithic Integrated Circuits (MMICs)

The concept for MMICs was derived from the low-frequency integrated circuits (ICs). The creation of an integrated circuit is shown in Figure 21.4. Several considerations have emerged as the main incentives for circuit design and fabrication at microwave frequencies:

- The trend in advanced microwave electronic systems is toward increasing integration, reliability, and volume of production with lower costs.
- The new millimeter-wave circuit applications demand the effects of bond-wire parasitics to be minimized and use of discrete elements to be avoided.
- Furthermore, new developments in microwave system designs in the military, commercial, and consumer markets demand a new approach for mass production and for multi-octave bandwidth response in circuits.

The answers to these specific requirements as delineated above lie in the monolithic approach. At this point we need to define an important term:

DEFINITION-MONOLITHIC MICROWAVE INTEGRATED CIRCUITS (MMICs): *Microwave circuits obtained through a multi-level process approach comprising all active and passive circuit elements and interconnections formed into the bulk of or onto the surface of a semi-insulating semiconductor substrate (see Figure 21.5).*

A typical MMIC is shown in Figure 21.5. Even though the first MMIC was reported in 1964, it was not until the late 1970s and early 1980s when MMIC results

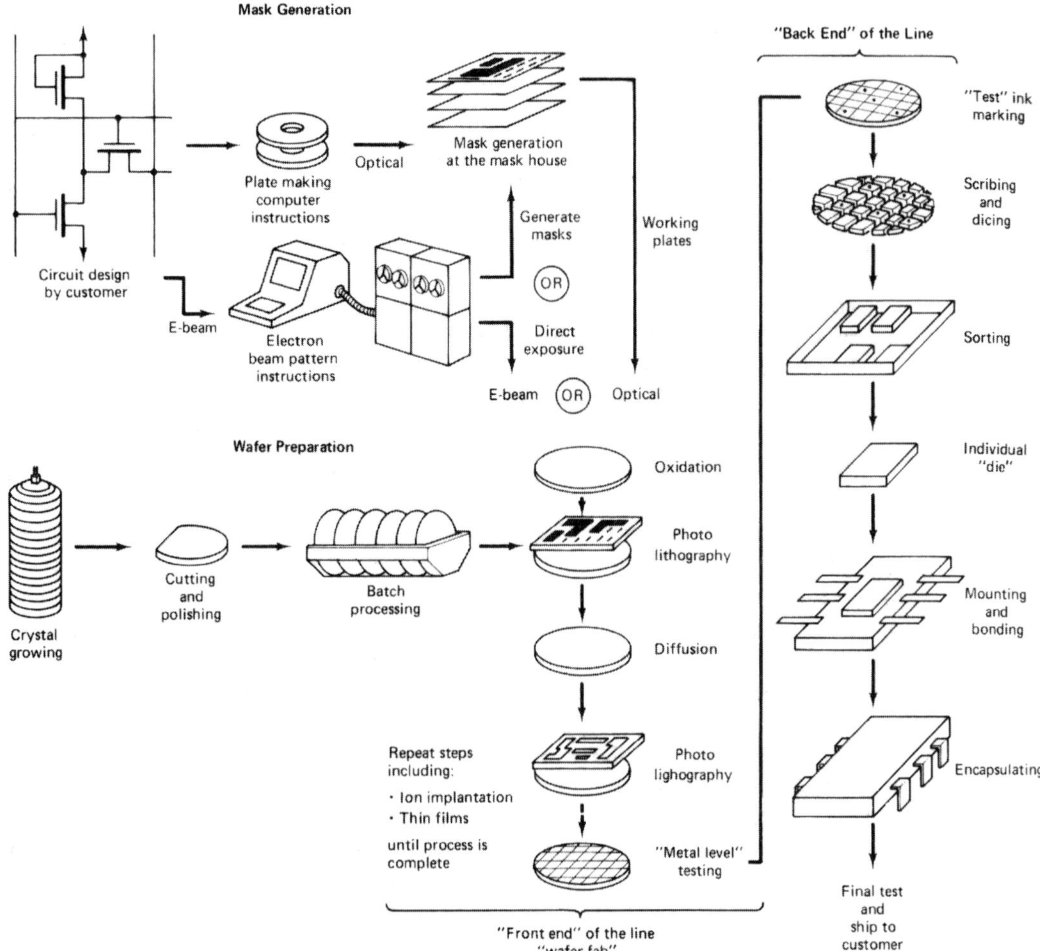

FIGURE 21.4 The creation of an integrated circuit.

using MESFETs became popular. The reason for the increase in MMIC research and development can be summarized briefly as:

- Rapid development of material fabrication technology, such as epitaxy and ion implantation
- Rapid development of low-noise MESFETs up to 60 GHz and power MESFETs up to 30 GHz
- Good ability to fabricate MESFETs, Schottky-Barrier diodes, and switching MESFETS simultaneously using the same process
- Development of semi-insulating GaAs substrate with excellent properties at microwave frequencies (e.g. $\varepsilon_r = 12.9, \tan\delta = 0.0005$)
- Good availability of CAD tools for accurate modeling and optimization of microwave circuits

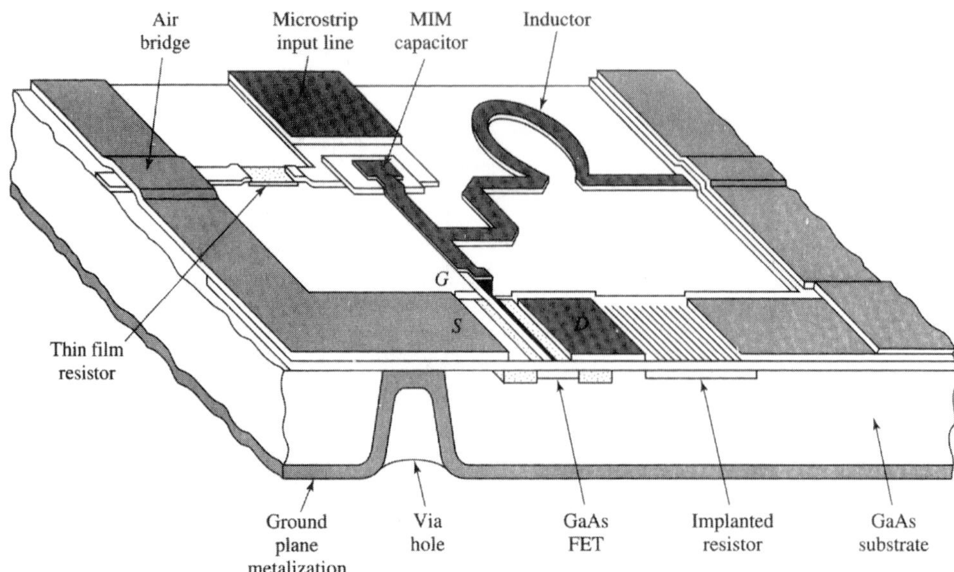

FIGURE 21.5 A typical monolithic MIC (MMIC).

Various substrate materials used for MMICs are bulk silicon, silicon on sapphire (SOS), GaAs, and InP. The comparison of these four types of substrates is summarized in Table 21.5.

TABLE 21.5 Comparison of MMIC Substrate Materials

Property	Si	Si on Sapphire	GaAs	InP
Semi-Insulating	No	Yes	Yes	Yes
Resistivity(Ω–cm)	10^3–10^5	$>10^{14}$	10^7–10^9	10^7
Dielectric constant (ε_r)	11.7	11.6	12.9	14
Mobility (cm^2/V–s)*	700	700	4300	3000
Saturation velocity (cm/s)	9×10^6	9×10^6	1.3×10^7	1.9×10^7
Radiation Hardness	Poor	Poor	Very good	Good
Density (g/cm^3)	2.3	3.9	5.3	4.8
Thermal Conductivity (W/cm–°C)	1.5	0.46	0.46	0.68
Operating Temperature (°C)	250	250	350	300

* Doping level = 10^{17} cm^{-3}

From this table, we can see that GaAs is the substrate of choice for MMIC applications. This is because the electron mobility of GaAs is higher than that of most other substrate materials (for example, it is 6 times higher than silicon's), which leads to higher frequency (or speed) of operation. Another advantage of GaAs is better radiation hardness, important for space-born and military applications. These excellent qualities of GaAs definitely justify the higher cost of the substrate.

Specific examples of MMICs are as follows:

- A 2–40 GHz distributed amplifier with a gain of 4 dB
- A 6–18 GHz two-stage amplifier with a gain of 10 dB and an output power of 20 dBm
- An X-band balanced mixer with a conversion loss of 6.5 dB or better
- A C-band receiver for avionics applications, consisting of three amplifiers, a double-balanced mixer, a limiter, and two LO buffer stages

21.5 HYBRID VERSUS MONOLITHIC MICS

Having gained some familiarity with the two types of MICs, we need to gain a broader perspective on the merits of each by comparing the two MICs in the following areas:

- Cost
- Size and weight
- Design flexibility
- Circuit tweaking
- Broadband performance
- Reproducibility
- Reliability

Each one of these areas is discussed next.

21.5.1 Cost

Hybrid MICs manufacturing is very labor intensive because it involves individual components whereas monolithic MICs are fabricated on wafers in batches. With an acceptable yield of 40–50%, MMICs have a great advantage in cost per unit (for example, currently a 1 mm^2 MMIC chip costs in the range of tens of dollars). However, one of MMIC's drawbacks in costs is the fact that only a very small percentage of the expensive GaAs substrate is occupied by the active circuit elements; for example, in a MMIC balanced amplifier 99% of the chip area is used by the 3 dB hybrid couplers and matching networks and only 1% is used for active circuits.

21.5.2 Size and Weight

In spaceborn system applications, monolithic MICs clearly have a definite edge over bulky hybrid MICs. Therefore, size and weight as a whole is an important consideration in these type of applications. As a point of comparison, it is interesting to note that the weight of a discrete chip resistor or a chip capacitor is more than that of an average monolithic MIC chip.

Table 21.6 summarizes the differences in characteristics of typical high-gain amplifiers fabricated using the hybrid and monolithic MIC technologies.

TABLE 21.6 Numerical Comparison of Hybrid and Monolithic MIC Technologies for Typical Amplifiers

Parameter	Typical Hybrid MIC	Typical Monolithic MIC
Frequency	8–18 GHz	7.5–18.5 GHz
Noise Figure	4.5 dB	5.2 dB
Gain	40±2 dB	57±1.5 dB
No. of wire bonds	400	14
No. of FETs	16	16
No. of substrates	16	1
No. of chip capacitors	40	0
Typical size	64 × 8 × 4 mm (unpackaged)	20 × 10 × 5 mm (packaged)

21.5.3 Design Flexibility

Monolithic MICs use GaAs MESFETs to perform a wide variety of functions such as amplification, oscillation, mixing, multiplication, division, switching, and phase shifting. Monolithic MICs have advantages over hybrid MICs in using FET's geometry in different circuit configurations to provide ample design flexibility. However, there are certain cases (such as in LNAs and narrowband filters) where the superior Q factors of hybrid circuit techniques are advantageous.

Hybrid MICs use several types of transmission lines, such as microstrip, slotline, coplanar lines, and suspended microstrip. They can use high (e.g., Alumina) or low (e.g., Quartz) dielectric constant substrates, or materials chosen for good thermal conductivity (e.g., beryllia). In contrast, the current state-of-the-art monolithic MICs primarily use microstrip lines on GaAs or InP substrates.

21.5.4 Circuit Tweaking

Hybrid MICs have an advantage over monolithic MICs in their amenability to post-manufacturing tuning or tweaking to obtain optimum performance by varying lumped-element values or using microstrip patches which may be bonded in as needed. This is almost impossible to do in monolithic MICs.

Therefore, very accurate modeling of each component is necessary in monolithic MIC circuit design and particularly use of CAD becomes a necessity. Furthermore, careful design is needed in these circuits so that the final performance is not overly sensitive to manufacturing tolerances.

21.5.5 Broadband Performance

The broadband performance is most dependent on the parasitics associated with any of the following factors:

- The lumped elements
- The active devices
- The connecting bond wires or ribbons

In monolithic MICs, active devices and passive elements of circuits are embedded into or onto the substrate, thus the above three parasitics are minimized or accounted for in the design process. Furthermore, with the advent of via-hole technology, only input and output wire connections are required and therefore the number of bond wires is greatly reduced, limited only to those at the periphery of the circuit. Thus, MMICs tend to give superior broadband performance when compared to hybrid MICs.

For example, a 2–40 GHz amplifier is not possible with hybrid MICs, but using monolithic MIC technology, a typical distributed amplifier with a 2–40 GHz bandwidth and 4 dB gain can be built. This is a multi-octave bandwidth design.

It is interesting to note that the performance of single-ended, balanced, low-noise, and multi-stage monolithic amplifiers is somewhat comparable to the performance of their hybrid MIC amplifier counterparts.

21.5.6 Reproducibility

One of the most important advantages of monolithic over hybrid MICs is excellent reproducibility and consistent performance from chip to chip.

Reproducible monolithic MIC chips come from well-controlled manufacturing environments and good circuit designs that can allow small variations in the fabrication process (e.g., doping level, channel depth, substrate thickness, etc.) from batch to batch without any noticeable change in performance.

Because monolithic MICs are smaller in size, the phase angle repeatability and performance (e.g., in phased-array radars) is often superior to hybrid MICs. However, hybrid MICs have improved over the years with the help of semi-automatic equipment (e.g., semi-automatic wire bonders, chip placement, etc.) with a correspondingly higher degree of repeatability.

21.5.7 Reliability

Both monolithic MIC and hybrid MIC technologies are considered to be reliable. However, monolithic MICs are regarded to have the highest reliability, particularly in the areas of lumped-element bonding, thin-film resistors, and wire bonding where hybrid MICs are weak.

On the other hand, circuits using hybrid MIC technology can easily be reworked if failure occurs, thus they avail a degree of recovery in case of nonperformance.

21.5.8 Summary of Monolithic and Hybrid MIC Comparison

We can now recapitulate the foregoing discussions on monolithic versus hybrid technologies in a succinct and easy-to-read table. Table 21.7 summarizes the comparison between monolithic and hybrid MICs:

TABLE 21.7 Summary of Comparison between Monolithic and Hybrid MICs

Features	MMIC	Hybrid MIC
Substrate	Semi-insulator	Insulator
Interconnections	(D)*	(D), wire bonded

TABLE 21.7 Summary of Comparison between Monolithic and Hybrid MICs *(Continued)*

Features	MMIC	Hybrid MIC
Transmission lines	(M)*	(M), Coplanar
Lumped elements	(D)	(D), Discrete
Solid-state devices	(D)	Discrete
Parasitics	Controlled	Uncontrolled
Labor intensive	No	Yes
Repairability	No	Yes
Mass production	Yes	No
Debugging	No	Yes
Cost	Low	High
Size and weight	Smaller	Larger
Design flexibility	Very good	Good
Circuit tweaking	Impossible	Yes
Broadband performance	Good	Limited
Reproducibility	Excellent	Good
Reliability	Excellent	Good

* Deposited (D), Microstrip (M)

21.6 CHIP MATHEMATICS

Radio frequency integrated circuits (RFICs) and microwave monolithic integrated circuits (MMICs) have come in time to have a major impact on our society both economically and socially. High-frequency electronic components containing RFICs and MMICs have been able to provide more functions per dollar, and thus have gone down in price even in periods of economic inflation.

21.6.1 Cost of Processing

The costs of processing an RFIC or MMIC chip may be separated into three areas:

- Initial costs
- Wafer fabrication costs
- Post processing costs

We now describe each of the above areas in the following sections.

21.6.2 Initial Costs

Initial costs are due to research and development (R&D) of a concept, designing a functional circuit, creating the technology, designing and testing the prototype circuits, and other startup costs that need to be spread out over the first batches of chips. Making the first chip is a very expensive process.

At first the yield may be poor and the first few chips very expensive to manufacture. However, as time goes on, the manufacturer learns new ways of improving yield by getting all the outpoints corrected and all of the bugs out of the processing line.

Therefore, yield improvement reduces the cost per chip and, in a competitive market, the manufacturer will reduce the price of the chip at some point in this learning curve and the consumer benefits accordingly.

21.6.3 Wafer Fabrication Costs

In the previous section, we discussed that the yield improvement will result in a decrease in the cost per chip, in contrast with the cost to process a wafer, which remains the same. The cost to process a wafer is a direct manufacturing cost. This is a fixed hard cost, regardless of the yield of the process.

A measure of the capacity of a manufacturing firm is the number of wafers that may start in the processing line each day. To increase the firm's manufacturing capacity, most facilities work up to three shifts per day to maximize equipment usage and tool productivity. For wafer processing steps, see Figure 21.4.

21.6.4 Post Processing Costs

Post processing refers to the steps taken after the chip is manufactured. These steps include scribing, sorting, bonding, and encapsulation. Post processing introduces its own yield detractors because at every step of the process, errors that may cause chip failure can occur.

Comparing the three costs, it turns out that wafer fabrication cost is the least per chip. This is due to automation of the processing lines that allow many wafers to be handled and processed simultaneously. All of the chips undergo processing in a batch (or a run), all being still together on the same wafer. For post-processing steps, see Figure 21.4.

21.6.5 Yield Detractors

There are several factors that detract from the yield of a manufacturing process of a chip among which are:

Dust particles: One of the major elements in reducing yield is dust. Today's devices are many times smaller than a dust particle, thus the entire circuit may become dysfunctional due to one dust particle. This is where the fabrication room (also referred to as the clean room) becomes the focal point of operation in terms of purity and cleanliness of the air. The air filtering system and thus the room is usually classified in terms of number of dust particles per million; for example, a class 100 room indicates a highly sophisticated air filtering system whereby only 100 particles per million are allowed to exist in the air. Thus, in order to have a successful wafer production line and a higher overall yield, very clean rooms and tight controls are needed all along the way.

Physical circuit size: There is a tradeoff between chip size and yield. This relationship is best understood by an example as diagrammed in Figure 21.6. From this figure we can see that for the same defect pattern, the yield can be dramatically improved from 0% (for large chips) to 66% (smaller chips) and finally to 95.6% for even smaller chips.

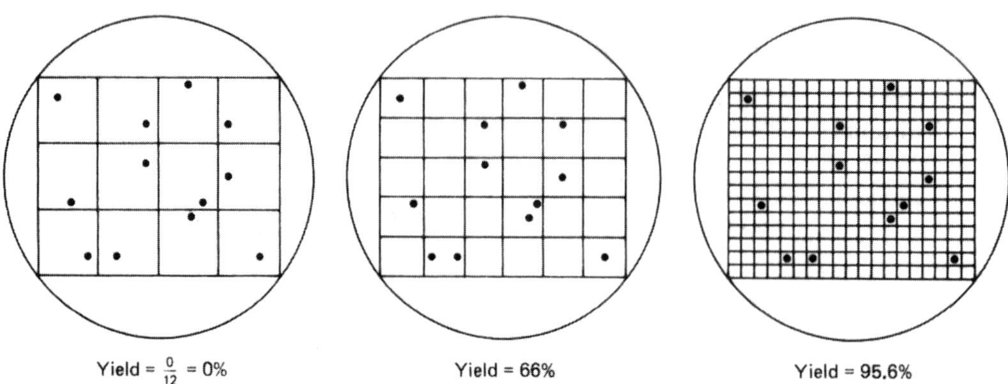

FIGURE 21.6 A dramatic example of how chip size affects yield. For the same defect pattern, the large chips on the left wafer have zero yield, a slightly smaller chip size has a remarkable yield, and an extremely small chip has an extraordinary yield.

21.6.6 Chip Manufacturing Break-Even Analysis

We have used the term yield loosely in the previous sections; we will now define it formally:

DEFINITION-YIELD (Y): *The ratio of the number of usable chips at the end of the manufacturing process shipped to the market (N_{mkt}) to the number of chips initially submitted for processing (N_{IN}), i.e.,*

$$Y_{TOT} = N_{mkt} / N_{IN} \tag{21.1}$$

or

$$N_{mkt} = Y_{TOT} N_{IN} \tag{21.2}$$

The market price of a chip reflects not only its manufacturing costs (hard costs) but also R&D, overhead, and profits (soft costs). Thus to do a break-even analysis, we need to find the number of chips per wafer that will pay for the hard costs. This gives us the largest feasible chip size to manufacture in order to be economically viable.

To produce more chips per wafer, we must make them smaller. The extra chips made beyond the break-even point increase the degree of viability and add to the profits and earnings of the chip manufacturer.

For example, manufacturing a chip of a given complexity that requires a certain surface area (i.e., chip size) may be prevented if the yield is low. Using a certain fabrication technology, the production of the chip may not be economically feasible because the manufacturer is unable to break even.

In other words, cramming in all the necessary functions and still having a break-even chip size may not be physically possible and thus it may be an economically unsound decision for the firm to proceed with the chip's manufacturing. The following example will illustrate this point further.

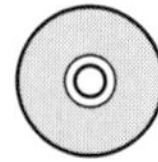

EXAMPLE 21.1

A certain MMIC chip has wafer fabrication costs that add up to \$100/wafer with post processing adding another \$100/wafer to the hard costs. Assuming that the MMIC chip has a minimum selling price of \$10 per chip of which \$5 is attributable to the direct manufacturing costs, find the following:

a. *The optimum number of chips per wafer in order to break even, assuming a yield of 100%.*

b. *Using part (a), what is the optimum number of chips per wafer and the corresponding chip size to break even, assuming a 3-inch (in diameter) wafer?*

c. *Repeat (a) and (b) if the yield drops to 40%.*

Solution:

a. Minimum number of initial chips $(N_{IN})_{min}$ to break even with a yield of 100% is related to the cost per wafer (C_W) and inversely to the manufacturing costs per chip (C_C) and is given by:

$$(N_{IN})_{min} = C_W / C_C$$

$$C_C = \$5$$

$$C_W = 100 + 100 = \$200$$

Thus, $(N_{IN})_{min} = 200/5 = 40$ chips/wafer
This means that 40 chips per wafer are needed to break even at a yield of 100% and to stay economically viable.

b. Using a 3-inch diameter circular wafer ($d = 3''$), we need to have a 7×7 pattern in order to obtain 40 chips as shown in Figure 21.7. Assuming square chips, the size of each side of the square (S) can easily be found by considering the right triangle with the hypotenuse equal to the wafer's diameter (d) as shown in Figure 21.8:

$$d = 7S\sqrt{2} = 3'' = 3 \times 2.54 = 7.62 \text{ cm}$$

Thus, $S = 0.769$ cm
The surface area of each chip (A) is given by:

$$A = S^2 = 0.59 \text{ cm}^2/\text{chip}$$

c. If the yield (Y) drops to 40%, then the number of chips $(N'_{IN})_{min}$ to break even should be increased as follows:

$$(N'_{IN})_{min} = C_W / C_C \times 1/Y$$

$$(N'_{IN})_{min} = 200/5 \times 1/0.4 = 100 \text{ chips/wafer}$$

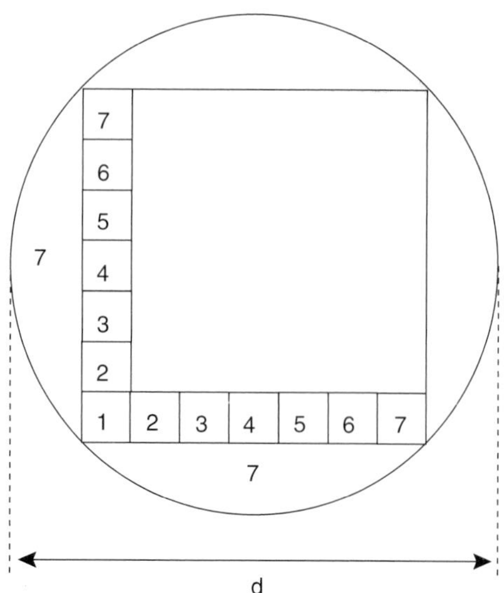

FIGURE 21.7 A circular wafer.

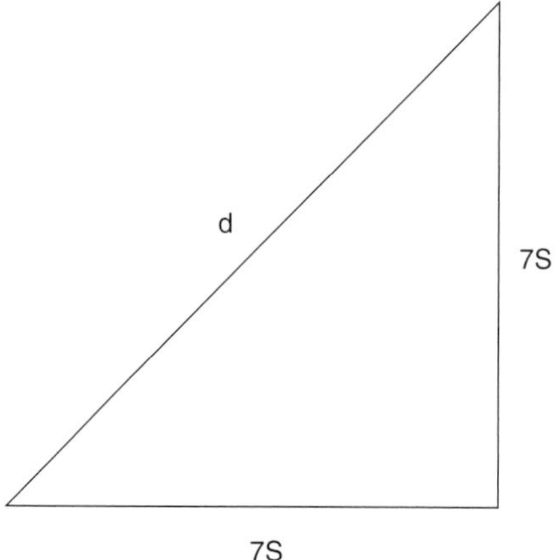

FIGURE 21.8 A right triangle inside a chip with diameter (d).

Thus, a 10×10 pattern is needed to break even with a chip size given by:

$$d = 10S\sqrt{2} \Rightarrow S = 0.53 \text{ cm}$$

and surface area (A) of each chip is given by:

$$A = 0.53 \times 0.53 = 0.29 \text{ cm}^2/\text{chip}$$

OBSERVATION: *From the above example, we can write the following general equations purely by observation:*

$$(N_{IN})_{min} = C_W / C_C \tag{21.3}$$

$$(N'_{IN})_{min} = (N_{IN})_{min} / Y_{TOT} \tag{21.4}$$

or

$$(N'_{IN})_{min} = \frac{C_W}{C_C Y_{TOT}} \tag{21.5}$$

where

$(N_{IN})_{min}$ = Minimum number of initial chips/wafer to break even (where $Y_{TOT} = 100\%$)
$(N'_{IN})_{min}$ = Minimum number of initial chips/wafer to break even (where $Y_{TOT} \neq 100\%$)
Y_{TOT} = Total yield of the fabrication process
C_W = Manufacturing cost per wafer
C_C = Manufacturing cost per chip

Now, considering a general diagram for a circular wafer with $n \times n$ chips (as shown in Figure 21.9), we can write:

$$d = nS\sqrt{2} \tag{21.6}$$

where

d = Wafer diameter
S = Each side of a square chip
n = An integer representing the number of chips per side of wafer

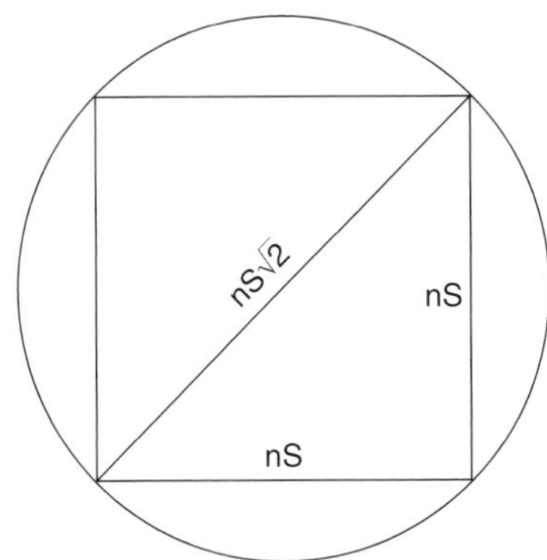

FIGURE 21.9 A circular wafer showing n × n chips.

NOTE: *The total number of chips (N_{IN}) ideally existing per wafer initially submitted for processing (with a yield of 100%) is:*

$$N_{IN} = n^2$$

The number of good chips per wafer shipped to market (N_{mkt}) is always smaller than N_{IN} by a factor of Y_{TOT}:

$$N_{mkt} = N_{IN} Y_{TOT} = n^2 Y_{TOT} \tag{21.7a}$$

For economic viability and solvency, a manufacturer needs to have the number of good chips shipped to the market (N_{mkt}) always greater than (or at worst equal to) the minimum number of initial chips, ($N'_{IN})_{min}$, needed for break-even, i.e.,

$$N_{mkt} \geq (N'_{IN})_{min}$$

The following example clarifies this concept further.

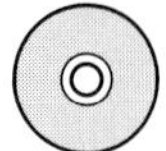

EXAMPLE 21.2

Determine the minimum yield (Y_{min}) necessary to manufacture a 6 mm Square shape MMIC chip on a 3-inch GaAs wafer if the same manufacturing environment of Example 21.1 exists.

Solution:

Given:

$$S = 6 \text{ mm} = 0.6 \text{ cm}$$

$$d = 3'' = 7.62 \text{ cm}$$

Using Equation 21.6, we find:

$$n = d/(S\sqrt{2}) = 7.62/(0.6 \times \sqrt{2}) = 8.98$$

To keep the chip size at S = 6 mm or higher (i.e., S ≥ 6), we have to round down n to the nearest integer, thus,

$$\Rightarrow n = 8$$

The minimum number of good chips per wafer which corresponds to the minimum number of chips to break even is given by:

$$N_{IN} = (N'_{IN})_{min} = n^2 = 64$$

$$(N_{IN})_{min} = C_W/C_C = 200/5 = 40$$

Thus, from (21.4), we find Y_{min} to be:

$$(Y_{TOT})_{min} = (N_{IN})_{min}/(N'_{IN})_{min} = 40/64 = 0.625 = 62.5\%$$

As can be noticed from the previous two examples, the smaller the chips are, the more viable the chip manufacturing operation will be. This line of thinking has pushed for higher wafer diameter size and smaller chips to increase the number of chips per wafer.

From Example 21.1, we can observe that the yield plays an important role in the number of chips produced. This observation makes us look at the total yield of a process more closely.

It turns out that the total yield of a fabrication process can be broken down into three sub-yields where each sub-yield describes the efficiency of a certain stage of the chip manufacturing process. The total yield is a product of these three sub-yields as follows:

$$Y_{TOT} = Y_W Y_A Y_{FT} \tag{21.7b}$$

where

Y_W = Wafer yield, which is the ratio of the number of good wafers at the end of the wafer fabrication process to the initial number of wafers at the start of the process.

Y_A = Assembly yield which is the ratio of the number of good chips after assembly and packaging to the number of chips after the fabrication process.

Y_{FT} = Final test yield which is the ratio of the number of good packaged chips making it to the market to the initial number of packaged chips at the start of the testing process.

Based on Equations 21.2 and 21.7b, the number of good chips per wafer (N_{mkt}) making it to the end is given by:

$$N_{mkt} = Y_W Y_A Y_{FT} N_{IN} \tag{21.8}$$

Because there is usually more than one wafer at the start of the process, N_{mkt} must be scaled up by the number of wafers (N_W) giving the total number of packaged chips shipped to the market (N_{TOT}) as:

$$N_{TOT} = N_W N_{mkt} \tag{21.9}$$

The following example will illustrate this concept further.

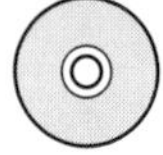

EXAMPLE 21.3

Assume that the wafer yield is 40%, the assembly yield is 80%, and the final test yield is 85%. If the minimum chip surface area required for an MMIC chip of certain complexity is 4mm × 4mm:

a. *Determine the number of good chips that will be shipped to the market if initially we start with 10,000 wafers of 3" in diameter?*

b. *Through debugging of the assembly line, the yield for each stage is increased to 50%, 90%, and 90%, respectively. Determine the number of good chips that will be shipped to the market if a 4" wafer is used?*

c. *Using the yields for the step (b), find the total number of chips to the market if the minimum chip size is increased to 6.3mm × 6.3mm (use a 4" wafer)?*

Solution:

a.

$$Y_{TOT} = 0.4 \times 0.8 \times 0.85 = 0.272$$

$$n = d/(S\sqrt{2}) = 3 \times 2.54/(0.4 \times \sqrt{2}) = 13.6$$

To keep the chip size at S = 4 mm or higher (i.e., S ≥ 4), we have to round down n to the nearest integer, thus:

$$n = 13.$$

Having found n, we find N_{TOT} as follows:

$$N_{IN} = n^2 = 169 \text{ chips/wafer}$$

$$N'_{IN} = n^2\, Y_{TOT} = 169 \times 0.272 = 45 \text{ chips/wafer}$$

Thus, the total number of good chips to market from Equation 21.9 is given by:

$$N_W = 10{,}000$$

$$N_{TOT} = 10{,}000 \times 45 = 450{,}000 \text{ good chips}$$

b. The new total yield is now given by;

$$Y_{TOT} = 0.5 \times 0.9 \times 0.9 = 0.405$$

Using a 4" wafer, we obtain:

$$n = d/(S\sqrt{2}) = 4 \times 2.54/(0.4 \times \sqrt{2}) = 17.96$$

To keep the chip size at S = 4 mm or higher (i.e. S ≥ 4), we have to round down n to the nearest integer, thus:

$$n = 17.$$

Thus, there are more chips per wafer proportionately, i.e.,

$$N_{IN} = n^2 = 289 \text{ chips/wafer}$$

$$N_{mkt} = 289 \times 0.405 = 117 \text{ chips/wafer}$$

Thus, the total number of good chips to market from Equation 21.9 is given by:

$$N_{TOT} = 10{,}000 \times 117 = 1{,}170{,}000 \text{ good chips}$$

c. Going to a bigger chip size of 6.3mm × 6.3mm, we have:

$$n = d/(S\sqrt{2}) = 4 \times 2.54/(0.63 \times \sqrt{2}) = 11.4$$

To keep the chip size at S = 6.3 mm or higher (i.e. S ≥ 6.3), we have to round down n to the nearest integer, thus:

$$n = 11$$

$$N_{IN} = n^2 = 121 \text{ chips/wafer}$$

Using the total yield value from step (b), we have:

$$N_{mkt} = 121 \times 0.405 = 49 \text{ good chips/wafer}$$

Thus, the total number of good chips to market from Equation 21.9 is given by:

$$N_{TOT} = 10{,}000 \times 49 = 490{,}000 \text{ good chips}$$

Thus, going to the bigger chip size reduces the number of good chips by a ratio of (49/117 = 41.9%) which is a substantial decrease. To compensate for this decrease, the sales price per chip should be increased by approximately a factor of 2.4 times (i.e., 1/41.9%) to keep the operations at break-even point and viable.

It is interesting to note that as time goes on, innovation, learning, and increase in wafer size are among many factors that contribute to bring about lower chip prices, all to the benefit of consumers.

EXERCISE 21.1

Using the results from Example 21.3, predict the reduction of the price of a chip for the 5" wafer, provided all manufacturing costs are kept the same as given in the Example 21.3.

LIST OF SYMBOLS/ABBREVIATIONS

A symbol/abbreviation will not be repeated once it has been identified and defined in an earlier chapter, as long as its definition remains unchanged.

C_C	Cost per wafer
C_W	Cost per chip
d	Wafer diameter
MIC	Microwave Integrated Circuit
MMIC	Monolithic Microwave Integrated Circuit
n	Integer representing the number of chips per side of wafer such that $n^2 \geq N_{min}$
N_{IN}	Number of chips submitted for processing
N_{min}	Number of chips needed to break even assuming 100% yield
N_{mkt}	Number of chips available for the market
N_{TOT}	Total number of packaged chips shipped to market
N_W	Number of wafers
S	Side dimension for a square chip
TCR	Thermal coefficient of resistance
Y	Yield
Y_A	Assembly yield
Y_{FT}	Final test yield
Y_{min}	Minimum yield required to manufacture chips in order to be viable
Y_{TOT}	Total yield
Y_W	Wafer yield

PROBLEMS

21.1 Describe the advantages of microwave integrated circuits over conventional circuits.

21.2 Describe the advantages of monolithic millimeter wave integrated circuits over hybrid integrated circuits.

21.3 List the basic properties required for an ideal a) substrate material, b) conductor material, c) resistive film, and d) capacitor dielectric film used in an MIC.

21.4 What is the relationship between yield and chip size?

21.5 Consider that it costs \$250/wafer to process a certain MMIC chip with \$10/chip direct manufacturing cost allowance. How many good chips per wafer must be produced to break even if the overall yield is 30%?

21.6 Assume a certain MMIC processing line has 33% wafer fabrication yield on 4" wafers. The cost of processing is $500 and a $5 allowance per chip has been made.

a. Can a 5 × 5 mm chip be economically produced?

b. What is the size of the largest square chip that can be made?

21.7 For a certain MMIC manufacturing process, the cost per wafer is $300 with a $10 direct allowance per chip for the manufacturing.

a. Calculate the minimum number of chips per wafer to break even for a yield of 50%.

b. What must the yield be if 4 × 4 mm chips are to be manufactured on 3" wafers?

c. If the yield is improved to 90%, repeat the above two steps.

REFERENCES

[21.1] Bahl, I. and P. Bhartia. *Microwave Solid State Circuit Design*, New York: Wiley Interscience, 1988.

[21.2] Bhartia, P. and I. Bahl. *Millimeter Wave Engineering and Application*, New York: John Wiley & Sons, 1984.

[21.3] Caulton, M. and H. Sobol. "Microwave Integrated Circuit Technology—A Survey," *IEEE J. On Solid State Circuits*, Vol. SC-5, Dec. 1970, pp. 292–303.

[21.4] Decker, D. R. "Are MMICs a Fad or a Fact?," *Microwave System News*, Vol. 13, July 1983, pp. 84–92.

[21.5] Fortino, Andres G. *Fundamentals of Integrated Circuit Technology*, Reston Publishing Company (Prentice-Hall), 1984.

[21.6] Howes, M. J. and P. Shepard. "Passive Components Essential to Microwave Integrated Circuit Design," *Microwave System News*, Vol. 14, 1984, pp. 113–122.

[21.7] McQuiddy, Jr., D. N., J. W. Wassel, J. B. Lagrange and W. R. Wisseman. "Monolithic Microwave Integrated Circuits: An Historical Perspective," *IEEE Transactions on Microwave Theory and Techniques*, Vol. MTT-32, Sept. 1984, pp. 997–1008.

[21.8] Mehal, E. and R. W. Wacker. "GaAs Integrated Microwave Circuits," *IEEE Transactions on Microwave Theory and Techniques*, Vol. MTT-16, July 1968, pp. 451–454.

[21.9] Pengelly, R. S. "Hybrid vs. Monolithic Microwave Circuits—A Matter of Cost," *Microwave System News*, Vol. 13, Jan. 1983, pp. 77–114.

[21.10] Pozar, D. M. "Microwave Engineering," 2nd edition, New York: John Wiley & Sons, 1998.

[21.11] Pucel, R. A. (Ed.) *Monolithic Microwave Integrated Circuits*, Piscataway, NJ: IEEE Press, 1985.

[21.12] Pucel, R. A. "Design Considerations for Monolithic Microwave Circuits," *IEEE Transactions on Microwave Theory and Techniques*, Vol. MTT-29, June 1981, pp. 513–534.

[21.13] Shillady R. W. "Microwave Integrated Circuits Offer Alternatives to the Radar System Designer," *Microwave System News*, Vol. 13, Aug. 1983, pp. 100–119.

[21.14] Sobol, H. "Technology and design of Hybrid Integrated Circuits," Solid State Technol., Vol. 13, Feb. 1970, pp. 49–57.

[21.15] Sobol, H. "Applications of Integrated Circuit Technology to Microwave Frequencies," Proc. *IEEE*, Vol. 59, Aug. 1971, pp. 1200–1211.

[21.16] Yamaskai, H. and D. Maki. "Hybrid vs. Monolithic—Is More Monolithic Better," *Microwave Journal*, Vol. 25, Nov. 1982, pp. 95–100.

PART VI

APPENDICES

APPENDIX A

List of Symbols & Abbreviations

NOTE: *Our symbol convention for DC analysis uses capital letters for both the quantity (voltage or current) and its subscript. Small-signal AC analysis uses similar designations but with small letters for both the quantity and its subscript. Total signal variables use small letters for the quantity with capital letter subscripts.*

$\angle$	Angle symbol
α	Common-base current gain
β	Common-emitter current gain
β	Propagation constant of a lossless transmission line (also called Phase constant)
β_i	Current sensitivity of a detector
β_v	Voltage sensitivity of a detector
δ	Skin depth or depth of penetration
Δ	Calculated value used to evaluate stability (K-Δ test)
$\Delta\phi$	Phase shift
γ	Propagation constant
γ	Sensitivity of a varactor diode
Γ	Reflection coefficient
$\Gamma(x)$	Reflection coefficient at point "x" on a transmission line
Γ_D	Reflection coefficient of the device
Γ_g	Reflection coefficient at the generator/source
$\Gamma_{IN,M1}$	Reflection coefficient for input of stage 1 of a multistage amplifier
Γ_L	Reflection coefficient at the load
Γ_{LP}	Large-signal reflection coefficient at the load
Γ_{ML}	Reflection coefficient for conjugate match value of Γ_L
Γ_{MS}	Reflection coefficient for conjugate match value of Γ_S
Γ_{opt} or Γ_o	Reflection coefficient for minimum noise figure of a device
$\Gamma_{OUT,M2}$	Output reflection coefficient of stage 2 of a multistage amplifier
Γ_{SP}	Large signal reflection coefficient at the source
Γ_T	Reflection coefficient at the terminating network
ε	Dielectric permittivity
ε_o	Permittivity of vacuum (8.854×10^{-12} F/m)
ε_r	Relative dielectric constant of a material
θ	Phase angle of a complex number
θ_e	Transmission line electrical length

λ	Wavelength
λ_o	Wavelength in free space
μ	Permeability
μ	Calculated value used to evaluate the stability of a network (the μ-parameter test)
ρ	Resistivity
σ	Conductivity
τ_d	Time delay
φ	Phase of a function
ϕ	Phase angle
Φ	Magnetic flux
ω	Angular frequency ($\omega = \beta v$) in radians/s
ω_i	Intermediate frequency (IF) in radians/s
ω_m	Modulating signal frequency in radians/s
ω_o	Local oscillator frequency in radians/s
ω_r	RF frequency in radians/s
A	Amplitude
A	A complex number representing a phasor
A	Surface area
AC	Alternating current
AGC	Automatic gain control
AWG	American Wire Gauge
B	Used to represent Susceptance
B	Frequency bandwidth
B	Base of a BJT
b	Normalized susceptance, $b = B/Y_o$
b	Base of a transistor (small-signal equivalent circuit model)
Bel	An international unit for measuring attenuation
B_g	Susceptance of the generator-tuning network (shunt configuration)
BJT	Bipolar Junction Transistor
b_L	Normalized susceptance at the load
b_P	Normalized susceptance of a parallel element
BW	Frequency bandwidth
c	Speed of light
C	Capacitance
C	Coulomb, SI unit for measuring charge
C	Capacitance per unit length
C_μ	Capacitance between collector and base in hybrid-π model
C_π	Capacitance between emitter and base terminals in hybrid-π model
CBJ	Collector Base junction of a transistor

C_C	Cost per wafer
C_d	Distributed capacitance
C_{eq}	Equivalent lumped capacitance
cf.,	Compare
C_j	Junction capacitance of a diode
C_P	Parasitic capacitance of a diode
C_T	Total capacitance of a center tapped capacitor
C_W	Cost per chip
d	Wafer diameter
dB	Decibel—a tenth of Bel
dBm	Decibels referenced to one milliWatt
dBW	Decibels referenced to one Watt
dBmW	Decibels referenced to one microWatt
DC	Direct Current
DR	Dynamic Range (expressed in dB)
DSB	Double-Side Band
e.g.	For example
E	Emitter
E	Energy
EBJ	Emitter Base junction of a transistor
E-Field	Electric Field
EM	Electromagnetic
F	Force
F	Noise figure
f_o	Frequency where Q is maximum (Center of pass band)
f_1, f_2	Frequency of a signal
F_{cas}	Noise figure for cascaded components
f_{co}	Diode cutoff frequency
FET	Field Effect Transistor
F_{min}	Minimum noise figure
f_r	Resonance frequency
G	Conductance
g	Normalized conductance, $g = G/Y_o$
G_o	Small-signal transistor power gain
G_o(dB)	Small-signal power gain in dB
G_{1dB}	1 dB gain compression point
G_A	Available power gain
GaAs	Gallium Arsenide
G_D	Dynamic conductance of a diode
G'_D	First order derivative of the dynamic conductance of a diode
G_g	Conductance of the generator-tuning network (shunt configuration)
g_L	Normalized conductance of the load

G_L	Output matching network power gain
$G_{L,MAX}$	Maximum output matching network power gain
G_{LU}	Output matching network power gain for unilateral transistors
g_m	Transconductance
G_{MSG}	Maximum stable gain
g_P	Normalized conductance of a parallel element
G_P	Operating power gain
G_S	Power gain of the input matching network
$G_{S,MAX}$	Maximum power gain of the input matching network
G_{SU}	Input matching network power gain for unilateral transistors
G_T	Transducer power gain
$G_{T,max}$	Maximum transducer power gain
G_{TU}	Transducer power gain for unilateral transistors
HAWK	Honesty, Actions, Work, Knowledge
$H(\omega)$	Voltage gain (also called transfer function)
$h_{11}, h_{12}, h_{21}, h_{22}$	Hybrid or *h*-parameters
I	Current
i.e.	That is, namely
$i(t)$	Instantaneous current
i_1	Current into port 1 of a network
i_2	Current into port 2 of a network
i_b, I_B	Base current of a transistor
i_c, I_C	Collector current of a transistor
i_e, I_E	Emitter current for a transistor
I_{CQ}	Collector current at the quiescent (Q) point on the transistor *I-V* curve characteristic graph
I_D	Drain current, FET
I_{DC}	Total DC current
I_{DSQ}	Drain to source quiescent (Q) point current, FET
I_{DSS}	Drain current at the onset of saturation when $V_{GS} = 0$, FET
IF	Intermediate frequency signal
I_G	Gate current, FET
i_k	Current at branch *k*
I_L	Current through the load
IL	Insertion loss
Im	The imaginary portion of a complex number
I_{max}	Maximum current on a transmission line
I_{min}	Minimum current on a transmission line
I_o	Oscillator current amplitude
$I_{o,max}$	Oscillator current amplitude under maximum power condition
I_S	Diode saturation current
I_S	Source current, FET

I_{SC}	Short-circuit current
j	An imaginary number where $j = \sqrt{-1}$
JFET	Junction Field Effect Transistor
k	Arbitrary integer constant
k	Boltzmann's constant ($= 1.38 \times 10^{-23}$ J/°K)
K	A parameter used to evaluate stability ($K - \Delta$ Test)
K	Overall voltage gain (used in Miller theorem)
K	Attenuation ratio (used in PIN diode attenuators)
KCL	Kirchoff's Current Law
KE	Kinetic Energy
KVL	Kirchoff's Voltage Law
ℓ	Length of a circuit
ℓ_{max}	Location of Z_{max} on a transmission line
ℓ_{min}	Location of Z_{min} on a transmission line
L	Inductance
L_{add}	Reflection loss of an added matching network
L_C	Conversion loss of a mixer
L_{eq}	Equivalent lumped inductance
L_{int}	Series inductance of a PIN diode
LNA	Low Noise Amplifier
LO	Local Oscillator
L_P	Parasitic inductance of a diode
LSB	Lower Side Band
m	Mass
m	Modulation index
m	Exponent parameter in the doping distribution function: $N(x) = N_B x^m$
M	Noise measure
MESFET	Metal Semiconductor Field Effect Transistor
MIC	Microwave Integrated Circuit
MKSA	A system of measurement having four basic units: meter (m), kilo gram (Kg), second (s), and ampere (A)
M_L	Load mismatch factor
MMIC	Monolithic Microwave Integrated Circuit
M_{min}	Minimum Noise measure
MNA	Minimum Noise Amplifier
MOSFET	Metal-Oxide Semiconductor Field Effect Transistor
M_S	Source mismatch factor
n	Diode ideality factor, which varies $1 < n < 2$ dependent on the material and physical structure of a diode [reference $I(V) = I_S(e^{V/nV_T} - 1)$]
n	Integer representing the number of chips per side of wafer
N	Noise figure parameter

N	Used to represent the number of turns of wire in an inductor
Np	Used to represent the unit Neper, which measures attenuation
N_{IN}	Number of chips submitted for processing
$(N_{IN})_{min}$	Number of chips needed to break even assuming 100% yield
$(N'_{IN})_{min}$	Number of chips needed to break even assuming the yield is not 100%
N_{mkt}	Number of chips shipped to the market
NPN	In a BJT: n-type collector, p-type base, and n-type emitter
NR	Negative Resistance
N_{TOT}	Total number of packaged chips shipped to market
N_W	Number of wafers
OST	Open Stub
P	Power
$P_{1\text{dB}}$	Output power at the 1 dB gain compression point
P_{2f1-f2}	Output power at the third-order intermodulation product frequency ($2f_1 - f_2$)
P_{av}	Average power
$(P_{av})_{max}$	Maximum average power
P_{AVN}	Power available from the transistor under matched condition
P_{AVS}	Power available from the source under matched condition
PE	Potential Energy
PF	Power Factor
P_{f1}	Output power at frequency (f_1)
P_i	Power incident
$P_{i,mds}$	Minimum detectable signal input power
P_{IF}	Intermediate frequency power
P_{IN}	Input power for network under consideration
$P_{IN,1\text{dB}}$	Input power at the 1 dB gain compression point
P_{inc}	Incident power
P_{IP}	Power at the third-order intercept point
P_L	Power delivered to the load
P_N	Noise power
$P_{NO,tot}$	Total output noise power
P_{NI}	Input thermal noise power
P_{NO}	Output thermal noise power
PNP	In a BJT: p-type collector, n-type base, and p-type emitter
$P_{o,mds}$	Output power level for minimum detectable signal
P_{OUT}	Output power from network under consideration
P_r, P_{ref}	Reflected power
P_{RF}	Power at the RF port of a mixer
P_{TOI}	Output power at the third-order intercept point (same as P_{IP})
PWR	Power reflection coefficient

P^+	Incident power
P^-	Reflected power
q	Magnitude of electric charge
q	Electric charge of an electron
Q	Quality Factor
Q	Quiescent point subscript
Q	Total Charge
Q_o	Maximum Quality Factor at $f = f_o$
Q_d	Dynamic Q-factor
r	Normalized resistance ($r = R/Z_o$)
R	Resistance
r_π	Resistance between emitter and base in hybrid-π model
r_μ	Resistance between collector and base in hybrid-π model
R_B	Base resistor
R_C	Collector resistor
$R_{D,max}$	Device negative resistance that provides maximum power in an oscillator design
Re	The real portion of a complex number
REFL	Reflection loss or mismatch loss
REFL COEF	Reflection coefficient
R_{eq}	Equivalent lumped resistance
RETN	Return loss
R_f	Forward-biased resistance of a diode
RF	Radio Frequency
RFC	RF Choke
R_g	Resistance of the generator-tuning network (series configuration)
R_i	Used to represent one resistor out of several (i=1,...,N)
R_j	Junction resistance of a diode
R_L	Load resistance
R_{loss}	Loss resistance
R_N	A noisy resistor
R_n	Noise resistance of a transistor
R_{No}	Noiseless resistor
r_o	Shunt resistance between the collector and emitter in a hybrid-π model
R_r	Reverse-biased resistance of a diode
R_S	Series resistance of a diode
R_S	Source resistance in an FET circuit
R_S	Input matching network resistance
S	Sensitivity parameter of a varactor
S	Number of stages in a power combiner or divider
S	Side dimension of a square chip

$S_{11}, S_{12}, S_{21}, S_{22}$	Scattering or *S*-parameters
SF	Shape Factor
SI	International system of units
SPST	Single-Pole Single-Throw
SSB	Single-Side Band
SST	Shorted stub
t	Time
t_d	Time delay
T	Temperature (use degrees Kelvin for computation)
T	Transmission coefficient
T_o	Standard room temperature (290° K)
$T_{11}, T_{12}, T_{21}, T_{22}$	Transmission or *T*-parameters
TCR	Thermal coefficient of resistance
T_e	Equivalent noise temperature
$T_{e,cas}$	Equivalent noise temperature for cascaded components
TEM	Transverse Electromagnetic wave
TL	Transmission Line
TOI	Third-order intercept point
T_S	Source temperature
TV	Television
USB	Upper sideband
n	Speed of propagation
v_1	Voltage at port 1 of a network
$v_{12}(t)$	Voltage difference between two designated points, 1 and 2, in a circuit
v_2	Voltage at port 2 of a network
V	Voltage
$\bar{V}$	Velocity
$V(t)$	Voltage with respect to time
V^+	Incident voltage
V^-	Reflected voltage
V_{BE}	Voltage from the base to emitter of a transistor
V_{CC}	Collector power supply voltage
V_{CE}	Collector to emitter voltage in a BJT
V_{CEQ}	Collector to emitter voltage at the quiescent point
V_D	Drain voltage, FET
V_{DD}	Drain power supply voltage, FET
V_{DS}	Drain to source voltage, FET
V_{DSQ}	Drain to source quiescent voltage, FET
V_G	Gate voltage, FET
V_{GS}	Gate to source voltage, FET
V_{GSQ}	Gate to source quiescent voltage, FET
V_j	Junction voltage of a diode

v_k	Voltage at branch k
V_{max}	Maximum voltage, on a transmission line
V_{min}	Minimum voltage, on a transmission line
V_{oc}	Open-circuit voltage
VOL	Voltage reflection coefficient magnitude
$V_{n,rms}$	Root-mean-square (rms) of noise signal
$V_n(t)$	Instantaneous noise voltage
V_P	Phase velocity
V_S	Source AC voltage
V_S	Voltage at the source of an FET
$VSWR$	Voltage Standing Wave Ratio
V_T	Thermal voltage defined as $V_T = KT/q$, normally 25 mV ($T = 293°$ K)
V_{zk}	Voltage at the Zener knee of the I-V characteristic of a diode
W	Work
X	Reactance
x	Normalized reactance ($x = X/Z_o$)
X_g	Reactance of the generator-tuning network (series configuration)
X_N	Normalized reactance
Y	Admittance
Y	Yield
Y_o	Characteristic admittance
$Y_{11}, Y_{12}, Y_{21}, Y_{22}$	Admittance or Y-parameters
Y_A	Assembly yield
Y_{FT}	Final test yield
Y_g	Admittance of the generator-tuning network (shunt configuration)
Y_{min}	Minimum yield required to manufacture chips
Y_T	Thevenin input admittance (used in Norton's theorem)
Y_T	Total admittance
Y_{TOT}	Total yield
Y_W	Wafer yield
Z	Impedance
$\|Z\|$	Absolute value (or magnitude) of the variable inside
$Z_{\lambda/4}$	Characteristic impedance of a $\lambda/4$ transmission line
Z_o	Characteristic impedance
$Z_{11}, Z_{12}, Z_{21}, Z_{22}$	Impedance or Z-parameters
Z_D	Device impedance
Z_g	Impedance of the generator-tuning network (series configuration)
Z_{IN}	Input impedance
$(Z_{IN})_N$	Normalized input impedance
Z_L	Load impedance
Z_m	Mean impedance of two states of a diode (Z_{ON}, and Z_{OFF})
Z_{max}	Maximum impedance, on a transmission line

Z_{min}	Minimum impedance, on a transmission line
Z_N	Normalized impedance
Z_{OC}	Open-circuit impedance
$Z_{opt}(Y_{opt})$	Normalized source impedance (or admittance) for minimum noise figure
Z_{SC}	Short-circuit impedance
Z_T	Total impedance
Z_T	Thevenin input impedance
Z_T	Characteristic impedance of a quarter-wave transformer

A P P E N D I X B

Physical Constants

Quantity	Symbol	Value
Angstrom unit	$A°$	$1\ A° = 10^{-4}\ \mu m = 10^{-10}$ m
Avogadro constant	N_{AVO}	$6.02204 \times 10^{23}\ mol^{-1}$
Boltzmann constant	k	1.38066×10^{-23} J/K
Charge of electron	q_e	1.60218×10^{-19} C
Electron charge/mass ratio	q_e/m_e	1.75880×10^{11} C/kg
Electron rest mass	m_e	9.1095×10^{-31} kg
Electron volt	eV	$1\ eV = 1.60218 \times 10^{-19}$ J
Intrinsic impedance (vacuum)	η_o	$120\pi\ (=377)\ \Omega$
Neutron rest mass	m_n	1.67495×10^{-27} kg
Permeability (vacuum)	$\mu_o\ (=1/\varepsilon_o c^2)$	$4\pi \times 10^{-7} = 1.25663 \times 10^{-6}$ H/m
Permittivity (vacuum)	$\varepsilon_o\ (=1/\mu_o c^2)$	8.85418×10^{-12} F/m
Planck constant	h	6.62617×10^{-34} J-s
Proton rest mass	M_p	1.67264×10^{-27} kg
Speed of light (vacuum)	c	2.99792×10^{8} m/s
Thermal voltage (at 293° K)	$V_T = kT/q$	$0.0252\ V \approx 25$ mV

APPENDIX C

International System Of Units (SI)

Quantity	Unit	Symbol	Dimension	Type
Capacitance	Farad	F	C/V	Derived unit
Charge	Coulomb	C	A-s	Derived unit
Conductance	Siemens	S	A/V	Derived unit
Current	Ampere	A	Basic dimension	Base unit
Energy	Joule	J	N-m	Derived unit
Force	Newton	N	$Kg\text{-}m/s^2$	Derived unit
Frequency	Hertz	Hz	1/s	Derived unit
Inductance	Henry	H	Wb/A	Derived unit
Length	Meter	m	Basic dimension	Base unit
Magnetic flux	Weber	Wb	V-s	Derived unit
Magnetic induction	Tesla	T	Wb/m^2	Derived unit
Mass	Kilogram	kg	Basic dimension	Base unit
Potential	Volt	V	J/C	Derived unit
Power	Watt	W	J/s	Derived unit
Pressure	Pascal	Pa	N/m^2	Derived unit
Resistance	Ohm	Ω	V/A	Derived unit
Temperature	Kelvin	K	Basic dimension	Base unit
Time	Second	s	Basic dimension	Base unit

Conversion Factors between Units

1 angstrom	10^{-4} micron $=10^{-8}$ cm
1 cm	393.7 mils
1 dB	0.115 nepers
1 foot	0.305 m
1 gauss	10^{-4} tesla
1 inch	2.54 cm = 25.4 mm
1 kg	2.2 lb = 1000 g
1 lb	453.6 g
1 micron	10^{-6} m $= 10^{-8}$ cm
1 mil	10^{-3} inch $= 2.54 \times 10^{-3}$ cm
1 mile	1.61 km
1 neper	8.686 dB

APPENDIX D

Unit Prefixes

Multiple	Prefix	Symbol
10^{18}	exa-	E
10^{15}	peta-	P
10^{12}	tera-	T
10^{9}	giga-	G
10^{6}	mega-	M
10^{3}	kilo-	k
10^{2}	hecta-	h
10	deka-	da
10^{-1}	deci-	d
10^{-2}	centi-	c
10^{-3}	milli-	m
10^{-6}	micro-	μ
10^{-9}	nano-	n
10^{-12}	pico-	p
10^{-15}	femto-	f
10^{-18}	atto-	a

APPENDIX E

Greek Alphabet

Letter	Lowercase	Uppercase
Alpha	α	Α
Beta	β	Β
Gamma	γ	Γ
Delta	δ	Δ
Epsilon	ε	Ε
Zeta	ζ	Ζ
Eta	η	Η
Theta	θ	Θ
Iota	ι	Ι
Kappa	κ	Κ
Lambda	λ	Λ
Mu	μ	Μ
Nu	ν	Ν
Xi	ξ	Ξ
Omicron	ο	Ο
Pi	π	Π
Rho	ρ	Ρ
Sigma	σ	Σ
Tau	τ	Τ
Upsilon	υ	Υ
Phi	φ	Φ
Chi	χ	Χ
Psi	ψ	Ψ
Omega	ω	Ω

APPENDIX F

Classical Laws of Electricity, Magnetism and Electromagnetics

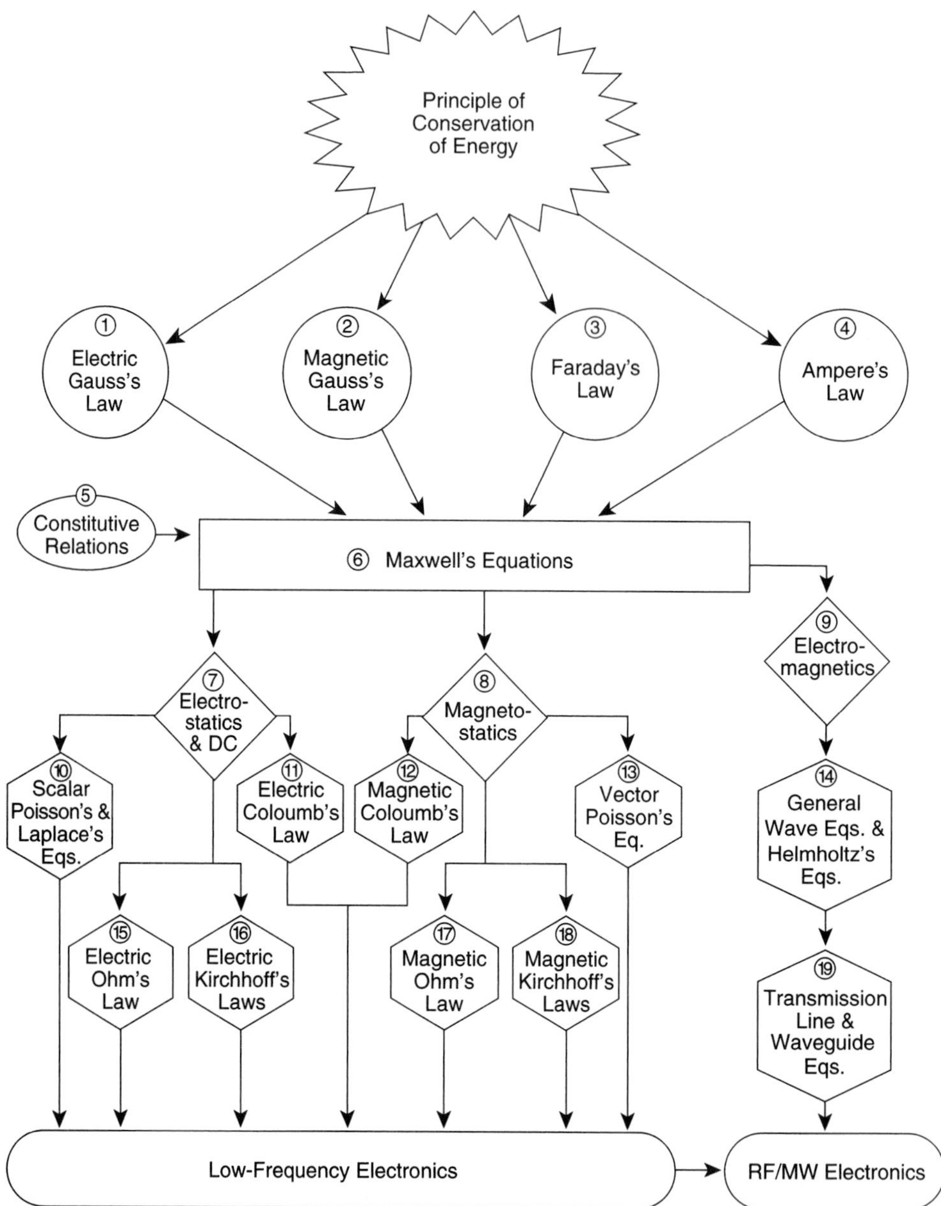

FIGURE F.1 All fundamental classical laws of electricity and magnetism as they apply to the field of electronics.

SYMBOL LIST

$\hat{a}_{R_{12}}$	A unit vector in the direction of R_{12}
$\overline{A}$	Magnet potential vector
$\overline{B}$	Magnetic flux density vector
C	Capacitance per unit length
$d\ell$	An infinitesimal length
$\overline{D}$	Electric displacement vector
emf	Electromotive force
$\overline{E}$	Electric field intensity vector
$\overline{F}$	Lorentz force vector
$\overline{H}$	Magnetic field intensity vector
G	Conductance per unit length
j	Unity imaginary number
$\overline{J}$	Current density vector
k	Wave propagation constant in unbounded space ($k = \omega/c$)
k_C	Eigenvalue (or characteristic root) which determines β ($k_C{}^2 = k^2 - \beta^2$)
ℓ	Path length
L	Inductance per unit length
m	Number of branches
mmf	Magnetomotive force
M	Magnetic pole
Q	Electric charge
R	Resistance per unit length
R_{12}	Distance between point 1 and point 2
t	Time
$\overline{V}$	Velocity
V_P	Phase velocity
V_{12}	Voltage difference between point 1 and point 2
V	Voltage (or potential difference) function
x	A length variable
∇	Gradient operator
$\nabla\cdot$	Divergence operator
$\nabla\times$	Curl operator
∇^2	Laplacian operator
$\equiv$	Defined as
$\parallel$	Parallel to

*	Complex conjugate
$\approx$	Approximately equal to
Re[ ]	Real part of [ ]
Im[ ]	Imaginary part of [ ]

GREEKS

α	Attenuation constant
β	Phase constant of a propagating wave ($e^{\pm j\beta x}$)
ε	Permittivity
γ	Propagation constant
ΔX	Difference in value of a quantity (X) at two points (i.e., $\Delta X = X_2 - X_1$)
ε_o	Free space permittivity
λ	Wavelength
μ	Permeability
μ_o	Free space permeability
ϕ	Potential function
Φ	Magnetic flux
ρ	Volume charge density
$\Re$	Magnetic reluctance
σ	Electrical conductivity
ω	Frequency in radians

GENERAL LAWS AND EQUATIONS

1. Electric Gauss's Law:

$$\nabla \cdot \overline{D} = \rho$$

2. Magnetic Gauss's Law:

$$\nabla \cdot \overline{B} = 0$$

3. Faraday's Law:

$$\nabla \times \overline{E} = -\frac{\partial \overline{B}}{\partial t}$$

4. Ampere's Law:

$$\nabla \times \overline{H} = \frac{\partial \overline{D}}{\partial t} + \bar{J}$$

5. Constitutive Relations:

$$\bar{D} = \varepsilon\bar{E}$$

$$\bar{B} = \mu\bar{H}$$

$$\bar{J} = \sigma\bar{E}$$

6. Maxwell's Equations:

$$\nabla \times \bar{E} = -\frac{\partial \bar{B}}{\partial t}$$

$$\nabla \times \bar{H} = \frac{\partial \bar{D}}{\partial t} + \bar{J}$$

$$\nabla \cdot \bar{D} = \rho$$

$$\nabla \cdot \bar{B} = 0$$

Under steady-state sinusoidal time dependence, the Maxwell's equations become:

$$\nabla \times \bar{E} = -j\omega\mu\bar{H}$$

$$\nabla \times \bar{H} = j\omega\varepsilon\bar{E} + \bar{J}$$

$$\nabla \cdot \bar{D} = \rho$$

$$\nabla \cdot \bar{B} = 0$$

where

$$\bar{J} = \sigma\bar{E}$$

$$\bar{D} = \varepsilon\bar{E}$$

$$\bar{B} = \mu\bar{H}$$

7. Electrostatics and DC:

$$\nabla \times \bar{E} = 0$$

$$\nabla \cdot \bar{D} = \rho$$

$$\bar{E} = -\nabla V$$

$$\bar{F} = q\bar{E}$$

$$\bar{D} = \varepsilon\bar{E}$$

$$\bar{J} = \sigma\bar{E}$$

8. Magnetostatics:

$$\nabla \times \bar{H} = \bar{J}$$

$$\nabla \cdot \bar{B} = 0$$

$$\bar{B} = \nabla \times \bar{A}$$

$$\bar{F} = Q(\bar{V} \times \bar{B})$$

$$\bar{B} = \mu \bar{H}$$

($\bar{F}$ is the Lorentz Force, which can also be written as: $\bar{F} = I\bar{dl} \times \bar{B}$)

9. Electromagnetics:

$$\nabla \times \bar{E} = -\frac{\partial \bar{B}}{\partial t}$$

$$\nabla \times \bar{H} = \frac{\partial \bar{D}}{\partial t} + \bar{J}$$

$$\nabla \cdot \bar{D} = \rho$$

$$\nabla \cdot \bar{B} = 0$$

Under steady-state sinusoidal time dependence, the Maxwell's equations become:

$$\nabla \times \bar{E} = -j\omega\mu \bar{H}$$

$$\nabla \times \bar{H} = j\omega\varepsilon \bar{E} + \bar{J}$$

$$\nabla \cdot \bar{D} = \rho$$

$$\nabla \cdot \bar{B} = 0$$

where

$$\bar{J} = \sigma \bar{E}$$

$$\bar{D} = \varepsilon \bar{E}$$

$$\bar{B} = \mu \bar{H}$$

10. Scalar Poisson's & Laplace's Equations:

a. Poisson's Equation:

$$\nabla^2 \phi = -\frac{\rho}{\varepsilon}$$

b. Laplace's Equation:

$$\nabla^2 \phi = 0$$

11. Electric Coulomb's Law:

$$\overline{F_2} = \left(\frac{Q_1 Q_2}{4\pi\varepsilon R_{12}^2}\right)\hat{a}_{R_{12}}$$

Or, in terms of field quantity (E), we can write Electric Coulomb's Law as:

$$\overline{dE_2} = \frac{dQ_1 \hat{a}_{R_{12}}}{4\pi\varepsilon R_{12}^2}$$

12. Magnetic Coulomb's Law:

$$\overline{F_2} = \left(\frac{M_1 M_2}{4\pi\mu R_{12}^2}\right)\hat{a}_{R_{12}}$$

Also, the Biot-Savart Law, considered by some to be the Coulomb's Law of Magnetostatics, is expressed in terms of field quantitiy ($\overline{H}$) as:

$$\overline{dH_2} = \frac{I_1 \overline{d\ell_1} \times \hat{a}_{R_{12}}}{4\pi R_{12}^2}$$

13. Vector Poisson's Equation:

$$\nabla^2 \overline{A} = -\mu \overline{J}$$

14. General and Helmholtz wave equations

a. General Wave Equations in a nonconducting medium ($\sigma = 0$):

$$\nabla^2 \overline{E} - \mu\varepsilon \frac{\partial^2 \overline{E}}{\partial t^2} = 0$$

$$\nabla^2 \overline{H} - \mu\varepsilon \frac{\partial^2 \overline{H}}{\partial t^2} = 0$$

General Wave Equations in a lossy medium ($\sigma \neq 0$):

$$\nabla^2 \overline{E} - \mu\varepsilon \frac{\partial^2 \overline{E}}{\partial t^2} - \mu\sigma \frac{\partial \overline{E}}{\partial t} = 0$$

$$\nabla^2 \overline{H} - \mu\varepsilon \frac{\partial^2 \overline{H}}{\partial t^2} = 0$$

b. Helmholtz Equations (Sinusoidal Waves):
Homogeneous Vector Equations:

$$\nabla^2 \overline{E} + k^2 \overline{E} = 0,$$

$$\nabla^2 \overline{H} + k^2 \overline{H} = 0$$

Inhomogeneous Scalar Equation:

$$\nabla^2 \phi + k^2 \phi = -\frac{\rho}{\varepsilon}$$

Inhomogeneous Vector Equation:

$$\nabla^2\bar{A} + k^2\bar{A} = -\mu\bar{J}$$

Lorentz Condition:

$$\nabla \cdot \bar{A} = -j\omega\mu\varepsilon\phi$$

Wave Number:

a. Non-Conducting Media:

$$k = \omega\sqrt{\mu\varepsilon} = \omega/V_P = 2\pi/\lambda$$

b. Conducting Media:

$$k = \omega\sqrt{\mu_o\varepsilon}, \qquad \varepsilon = \varepsilon_o\left(1 - j\frac{\sigma}{\omega\varepsilon_o}\right)$$

15. Electric Ohm's Law:

$$V_{12} = RI$$

16. Electric Kirchhoff's Laws (EKVL & EKCL):
EKVL:

$$\sum_{i=1}^{m} \Delta V_i = 0$$

or

$$\sum_{i=1}^{m_1} emf_i = \sum_{j=1}^{m_2} \Delta V_j$$

where $m = m_1 + m_2$.
EKCL:

$$\sum_{j=1}^{n} I_j = 0$$

17. Magnetic Ohm's Law:

$$mmf_{12} = \Re\Phi$$

18. Magnetic Kirchhoff's Laws (MKVL & MKCL):
MKVL:

$$\sum_{i=1}^{m} \Delta(H_i\ell_i) = 0$$

or

$$\sum_{k=1}^{m_1} mmf_k = \sum_{i=1}^{m_2} \Delta(H_i\ell_i)$$

where $m = m_1 + m_2$.
MKCL:

$$\sum_{j=1}^{n} \Phi_j = 0$$

19. Waves under Steady-State Sinusoidal conditions:

A1. TEM Waves: Field Equations; $E_z = 0,\ H_z = 0$

$$\nabla_t^2 \overline{E}_t + (k^2 - \beta^2)\overline{E}_t = 0$$

$$\nabla_t^2 \phi = 0$$

$$\overline{E} = -\nabla_t \phi e^{\mp jkz}$$

$$\overline{H} = \pm\frac{1}{\eta}\hat{z} \times \overline{E}$$

where

$$\eta = \sqrt{\frac{\mu}{\varepsilon}}$$

$$\beta = k = \omega\sqrt{\mu\varepsilon} = \omega/V_P = 2\pi/\lambda$$

A2. Transmission Line Waves: Current and Voltage Equations (Sinusoidal Waves)—TEM Waves

$$I(x) = \left(\frac{-1}{R + j\omega L}\right)\frac{dV(x)}{dx},$$

$$\gamma = \sqrt{(R + j\omega L)(G + j\omega C)} = \alpha + j\beta$$

Lossless Equations ($R, G = 0$):

$$\frac{d^2V(x)}{dx^2} + \beta^2 V(x) = 0,$$

$$\frac{d^2I(x)}{dx^2} + \beta^2 I(x) = 0,$$

$$I(x) = \left(\frac{-1}{j\omega L}\right)\frac{dV(x)}{dx}$$

$$\beta = \omega\sqrt{LC}$$

B. TE Waves: $E_z = 0,\ H_z \neq 0$

$$(\nabla_t^2 - \beta^2)(\overline{H}_t + \overline{H}_z) = 0$$

or,

$$\nabla_t^2 \overline{H}_z + K_C^2 \overline{H}_z = 0$$

$$\nabla_t^2 \overline{H}_t + K_C^2 \overline{H}_t = 0$$

C. TM Waves: $E_z \neq 0,\ H_z = 0$

$$(\nabla_t^2 - \beta^2)(\overline{E}_t + \overline{E}_z) = 0$$

or,

$$\nabla_t^2 \overline{E}_z + K_C^2 \overline{E}_z = 0$$
$$\nabla_t^2 \overline{E}_t + K_C^2 \overline{E}_t = 0$$

where

$$\nabla = \nabla_t - j\beta\hat{z}$$
$$\nabla^2 = \nabla_t^2 - \beta^2$$
$$\overline{E} = \overline{E}_t + \overline{E}_z$$
$$\overline{H} = \overline{H}_t + \overline{H}_z$$
$$k_C^2 = k^2 - \beta^2$$

APPENDIX G

Materials Constants & Frequency Bands

TABLE G.1 Conductivity σ (S/m)

Material	σ
Aluminum	3.82×10^7
Bakelite	10^{-9}
Brass	1.50×10^7
Bronze	1.00×10^7
Clay	10^{-4}
Copper	5.80×10^7
Diamond	10^{-13}
Ferrite	10^{-2}
GaAs	2.5×10^{-7}
Germanium (Ge)	2.3
Glass	10^{-12}
Gold	4.10×10^7
Ground (wet)	$10^{-2} - 10^{-3}$
Iron	1.03×10^7
Marble	10^{-8}
Mica	10^{-14}
Nichrome	0.10×10^7
Nickel	1.45×10^7
Polystrene	10^{-16}
Porcelain	10^{-13}
Quartz	10^{-17}
Rubber (hard)	10^{-15}
Silicon	4.0×10^{-4}
Silver	6.17×10^7
Soil (sandy)	10^{-5}
Solder	0.70×10^7
Steel (stainless)	0.11×10^7
Tungsten	1.82×10^7

TABLE G.1 Conductivity σ (S/m) *(Continued)*

Material	σ
Water (distilled)	2×10^{-4}
Water (fresh)	10^{-3}
Water (sea)	3–5
Zinc	1.67×10^{7}

TABLE G.2 Relative Permittivity (ε_r) (Also Called Dielectric Constant)

Material	ε_r
Air	1
Alcohol	25
Bakelite	4.8
Gallium Arsenide(GaAs)	12.9
Germanium (Ge)	15.8
Glass	4–7
Ground (dry)	2–5
Ice	4.2
Indium Phosphide (InP)	14
Mica (ruby)	5.4
Nylon	4
Paper	2–4
Plexiglass	2.6–3.5
Polyethylene	2.25
Polystrene	2.55–6
Porcelain	6
Quartz (fused)	3.8
Rubber	2.5–4
Salt (NaCl)	5.9
Sand (dry)	4
Silica (fused)	3.8
Silicon	11.7
Snow	3.3
Soil (dry)	2.8
Styrofoam	1.03
Teflon	2.1
Water (Distilled)	80
Water (fresh)	80
Water (Sea)	20
Wood (dry)	1.5–4

TABLE G.3 Relative Permeability (μ_r)

Dielectric	μ_r
Aluminum	1.00000065
Beryllium	1.00000079
Bismuth	0.99999860
Cast Iron	60
Cobalt	60
Ferrite	1,000
Iron (pure)	4,000
Iron (transformer)	3,000
Machine Steel	300
Mumetal	20,000
Nickel	50
Parafin	0.99999942
Silicon Iron	4,000
Silver	0.99999981
Supermalloy	100,000
Wood	0.99999950

TABLE G.4 Semiconductor Substrate Material Constants

Property	Si	Si on Sapphire	GaAs	InP
Density (g/cm^3)	2.3	3.9	5.3	4.8
Dielectric constant (ε_r)	11.7	11.6	12.9	14
Mobility (cm^2/V–s)*	700	700	4300	3000
Operating Temperature (°C)	250	250	350	300
Radiation Hardness	Poor	Poor	Very good	Good
Resistivity (intrinsic) (Ω–cm)	10^3–10^5	$>10^{14}$	10^7–10^9	10^7
Saturation velocity (cm/s)	9×10^6	9×10^6	1.3×10^7	1.9×10^7
Semi-Insulating	No	Yes	Yes	Yes
Thermal Conductivity (W/cm–°C)	1.5	0.46	0.46	0.68

* Doping level = 10^{17} cm^{-3}

TABLE G.5 Commercial Radio Frequency/Microwave Band Subdivisions

Name of Band	Abbreviation	Frequency Range
Very low freq.	VLF	3–30 kHz
Low freq.	LF	30–300 kHz
Medium freq.	MF	300 kHz–3 MHz
High freq.	HF	3–30 MHz
Very high freq.	VHF	30–300 MHz
Ultra-high freq.	UHF	0.3–3 GHz
Super-high freq.	SHF	3–30 GHz
Extra-high freq.	EHF	30–300 GHz

TABLE G.6 IEEE and Commercial Microwave Band Symbol Designations

Band Designation (by Symbol)	Frequency Range (GHz)
L band	1.0–2.0
S band	2.0–4.0
C band	4.0–8.0
X band	8.0–12.0
Ku band	12.0–18.0
K band	18.0–26.5
Ka band (mmw)	26.5–40.0
Q band (mmw)	33.0–50.0
U band (mmw)	40.0–60.0
V band (mmw)	50.0–75.0
E band (mmw)	60.0–90.0
W band (mmw)	75.0–110.0
F band (mmw)	90.0–140.0
D band (mmw)	110.0–170.0
G band (mmw)	140.0–220.0

APPENDIX H

Conversion Among Two-Port Network Parameters

Source: Microwave Transistor Amplifiers: Analysis and Design. By Guillermo Gonzalez. Prentice Hall, Upper Saddle River, 1997. Used with permission.

	S	z	y	h	$ABCD$
S	$\begin{matrix} S_{11} & S_{12} \\ S_{21} & S_{22} \end{matrix}$	$S_{11} = \frac{(z'_{11} - 1)(z'_{22} + 1) - z'_{12} z'_{21}}{\Delta_1}$ $S_{12} = \frac{2z'_{12}}{\Delta_1}$ $S_{21} = \frac{2z'_{21}}{\Delta_1}$ $S_{22} = \frac{(z'_{11} + 1)(z'_{22} - 1) - z'_{12} z'_{21}}{\Delta_1}$	$S_{11} = \frac{(1 - y'_{11})(1 + y'_{22}) + y'_{12} y'_{21}}{\Delta_2}$ $S_{12} = \frac{-2y'_{12}}{\Delta_2}$ $S_{21} = \frac{-2y'_{21}}{\Delta_2}$ $S_{22} = \frac{(1 + y'_{11})(1 - y'_{22}) + y'_{12} y'_{21}}{\Delta_2}$	$S_{11} = \frac{(h'_{11} - 1)(h'_{22} + 1) - h'_{12} h'_{21}}{\Delta_3}$ $S_{12} = \frac{2h'_{12}}{\Delta_3}$ $S_{21} = \frac{-2h'_{21}}{\Delta_3}$ $S_{22} = \frac{(1 + h'_{11})(1 - h'_{22}) + h'_{12} h'_{21}}{\Delta_3}$	$\begin{matrix} \frac{A' + B' - C' - D'}{\Delta_4} & \frac{2(A'D' - B'C')}{\Delta_4} \\ \frac{2}{\Delta_4} & \frac{-A' + B' - C' + D'}{\Delta_4} \end{matrix}$
z	$z'_{11} = \frac{(1 + S_{11})(1 - S_{22}) + S_{12} S_{21}}{\Delta_5}$ $z'_{12} = \frac{2S_{12}}{\Delta_5}$ $z'_{21} = \frac{2S_{21}}{\Delta_5}$ $z'_{22} = \frac{(1 - S_{11})(1 + S_{22}) + S_{12} S_{21}}{\Delta_5}$	$\begin{matrix} z_{11} & z_{12} \\ z_{21} & z_{22} \end{matrix}$	$\begin{matrix} \frac{y_{22}}{\lvert y \rvert} & \frac{-y_{12}}{\lvert y \rvert} \\ \frac{-y_{21}}{\lvert y \rvert} & \frac{y_{11}}{\lvert y \rvert} \end{matrix}$	$\begin{matrix} \frac{\lvert h \rvert}{h_{22}} & \frac{h_{12}}{h_{22}} \\ \frac{-h_{21}}{h_{22}} & \frac{1}{h_{22}} \end{matrix}$	$\begin{matrix} \frac{A}{C} & \frac{\Delta_8}{C} \\ \frac{1}{C} & \frac{D}{C} \end{matrix}$
y	$y'_{11} = \frac{(1 - S_{11})(1 + S_{22}) + S_{12} S_{21}}{\Delta_6}$ $y'_{12} = \frac{-2S_{12}}{\Delta_6}$ $y'_{21} = \frac{-2S_{21}}{\Delta_6}$ $y'_{22} = \frac{(1 + S_{11})(1 - S_{22}) + S_{12} S_{21}}{\Delta_6}$	$\begin{matrix} \frac{z_{22}}{\lvert z \rvert} & \frac{-z_{12}}{\lvert z \rvert} \\ \frac{-z_{21}}{\lvert z \rvert} & \frac{z_{11}}{\lvert z \rvert} \end{matrix}$	$\begin{matrix} y_{11} & y_{12} \\ y_{21} & y_{22} \end{matrix}$	$\begin{matrix} \frac{1}{h_{11}} & \frac{-h_{12}}{h_{11}} \\ \frac{h_{21}}{h_{11}} & \frac{\lvert h \rvert}{h_{11}} \end{matrix}$	$\begin{matrix} \frac{D}{B} & \frac{-\Delta_8}{B} \\ \frac{-1}{B} & \frac{A}{B} \end{matrix}$
h	$h'_{11} = \frac{(1 + S_{11})(1 + S_{22}) - S_{12} S_{21}}{\Delta_7}$ $h'_{12} = \frac{2S_{12}}{\Delta_7}$ $h'_{21} = \frac{-2S_{21}}{\Delta_7}$ $h'_{22} = \frac{(1 - S_{11})(1 - S_{22}) - S_{12} S_{21}}{\Delta_7}$	$\begin{matrix} \frac{\lvert z \rvert}{z_{22}} & \frac{z_{12}}{z_{22}} \\ \frac{-z_{21}}{z_{22}} & \frac{1}{z_{22}} \end{matrix}$	$\begin{matrix} \frac{1}{y_{11}} & \frac{-y_{12}}{y_{11}} \\ \frac{y_{21}}{y_{11}} & \frac{\lvert y \rvert}{y_{11}} \end{matrix}$	$\begin{matrix} h_{11} & h_{12} \\ h_{21} & h_{22} \end{matrix}$	$\begin{matrix} \frac{B}{D} & \frac{-\Delta_8}{D} \\ \frac{-1}{D} & \frac{C}{D} \end{matrix}$
$ABCD$	$A' = \frac{(1 + S_{11})(1 - S_{22}) + S_{12} S_{21}}{2S_{21}}$ $B' = \frac{(1 + S_{11})(1 + S_{22}) - S_{12} S_{21}}{2S_{21}}$ $C' = \frac{(1 - S_{11})(1 - S_{22}) - S_{12} S_{21}}{2S_{21}}$ $D' = \frac{(1 - S_{11})(1 + S_{22}) + S_{12} S_{21}}{2S_{21}}$	$\begin{matrix} \frac{z_{11}}{z_{21}} & \frac{\lvert z \rvert}{z_{21}} \\ \frac{1}{z_{21}} & \frac{z_{22}}{z_{21}} \end{matrix}$	$\begin{matrix} \frac{-y_{22}}{y_{21}} & \frac{-1}{y_{21}} \\ \frac{-\lvert y \rvert}{y_{21}} & \frac{-y_{11}}{y_{21}} \end{matrix}$	$\begin{matrix} \frac{-\lvert h \rvert}{h_{21}} & \frac{-h_{11}}{h_{21}} \\ \frac{-h_{22}}{h_{21}} & \frac{-1}{h_{21}} \end{matrix}$	$\begin{matrix} A & B \\ C & D \end{matrix}$

$\Delta_1 = (z'_{11} + 1)(z'_{22} + 1) - z'_{12} z'_{21}$
$\Delta_2 = (1 + y'_{11})(1 + y'_{22}) - y'_{12} y'_{21}$
$\Delta_3 = (h'_{11} + 1)(h'_{22} + 1) - h'_{12} h'_{21}$
$\Delta_4 = A' + B' + C' + D'$
$\Delta_5 = (1 - S_{11})(1 - S_{22}) - S_{12} S_{21}$
$\Delta_6 = (1 + S_{11})(1 + S_{22}) - S_{12} S_{21}$
$\Delta_7 = (1 - S_{11})(1 + S_{22}) + S_{12} S_{21}$
$\Delta_8 = AD - BC$

$z'_{11} = z_{11}/Z_o,\ z'_{12} = z_{12}/Z_o,\ z'_{21} = z_{21}/Z_o,\ z'_{22} = z_{22}/Z_o$
$y'_{11} = y_{11}Z_o,\ y'_{12} = y_{12}Z_o,\ y'_{21} = y_{21}Z_o,\ y'_{22} = y_{22}Z_o$
$h'_{11} = h_{11}/Z_o,\ h'_{12} = h_{12},\ h'_{21} = h_{21},\ h'_{22} = h_{22}Z_o$
$A' = A,\ B' = B/Z_o,\ C' = CZ_o,\ D' = D$
$\lvert z \rvert = z_{11}z_{22} - z_{12}z_{21}$
$\lvert y \rvert = y_{11}y_{22} - y_{12}y_{21}$
$\lvert h \rvert = h_{11}h_{22} - h_{12}h_{21}$

APPENDIX I

Conversion among the Y-Parameters of a Transistor (Three Configurations: CE, CB, and CC)

Let

e ≡ Common emitter (CE) configuration
b ≡ Common base (CB) configuration
c ≡ Common collector (CC) configuration

The conversions among the three configurations are given as follows:

1. CB & CC to CE

$$Y_{11,e} = Y_{11,b} + Y_{12,b} + Y_{21,b} + Y_{22,b} = Y_{11,c}$$
$$Y_{12,e} = -(Y_{12,b} + Y_{22,b}) = -(Y_{11,c} + Y_{12,c})$$
$$Y_{21,e} = -(Y_{21,b} + Y_{22,b}) = -(Y_{11,c} + Y_{21,c})$$
$$Y_{22,e} = Y_{22,b} = Y_{11,c} + Y_{12,c} + Y_{21,c} + Y_{22,c}$$

2. CE & CC to CB

$$Y_{11,b} = Y_{11,e} + Y_{12,e} + Y_{21,e} + Y_{22,e} = Y_{22,c}$$
$$Y_{12,b} = -(Y_{12,e} + Y_{22,e}) = -(Y_{21,c} + Y_{22,c})$$
$$Y_{21,b} = -(Y_{21,e} + Y_{22,e}) = -(Y_{12,c} + Y_{22,c})$$
$$Y_{22,b} = Y_{22,e} = Y_{11,c} + Y_{12,c} + Y_{21,c} + Y_{22,c}$$

3. CE & CB to CC

$$Y_{11,c} = Y_{11,e} = Y_{11,b} + Y_{12,b} + Y_{21,b} + Y_{22,b}$$
$$Y_{12,c} = -(Y_{11,e} + Y_{12,e}) = -(Y_{11,b} + Y_{21,b})$$
$$Y_{21,c} = -(Y_{11,e} + Y_{21,e}) = -(Y_{11,b} + Y_{12,b})$$
$$Y_{22,c} = Y_{11,e} + Y_{12,e} + Y_{21,e} + Y_{22,e} = Y_{11,b}$$

NOTE: *If other parameters other than Y-parameters (e.g., S-parameters, etc.) are desired, Appendix H can be used effectively to accomplish the conversion (e.g., from Y- to S-parameters, etc.).*

APPENDIX J

Useful Mathematical Formulas

1. Binomial Formulas

$$(x \pm y)^2 = x^2 \pm 2xy + y^2$$

$$(x \pm y)^3 = x^3 \pm 3x^2y + 3xy^2 \pm y^3$$

$$(x \pm y)^4 = x^4 \pm 4x^3y + 6x^2y^2 \pm 4xy^3 + y^4$$

Or, in general (for + sign):

$$(x+y)^n = x^n + nx^{n-1} + \frac{n(n-1)}{2!}x^{n-2}y^2 + \frac{n(n-1)(n-2)}{3!}x^{n-3}y^3 + \ldots + y^n$$

Where factorial n (or n!) is defined by:

n! = 1·2·3·...·n

NOTE: *Zero factorial is defined by:* $0! = 1$.

2. Special Formulas

$$x^2 - y^2 = (x-y)(x+y)$$

$$x^3 - y^3 = (x-y)(x^2 + xy + y^2)$$

$$x^3 + y^3 = (x+y)(x^2 - xy + y^2)$$

$$x^4 - y^4 = (x^2 - y^2)(x^2 + y^2) = (x-y)(x+y)(x^2+y^2)$$

$$\frac{1-x^n}{1-x} = 1 + x + x^2 + \ldots + x^{n-1}, \ |x| < 1$$

$$\sum_{n=0}^{\infty} x^n = 1 + x + x^2 + \ldots + x^n + \ldots = \frac{1}{1-x}, \ |x| < 1$$

3. Trigonometric Functions

$$\sin(-x) = -\sin x$$

$$\cos(-x) = \cos x$$

$$\sin\left(\frac{\pi}{2} \pm x\right) = \cos x$$

$$\cos\left(\frac{\pi}{2} \pm x\right) = \mp \sin x$$

$$\sin(\pi \pm x) = \mp \sin x$$

$$\cos(\pi \pm x) = -\cos x$$

$$\tan x = \frac{\sin x}{\cos x}$$

$$\cot x = \frac{\cos x}{\sin x}$$

$$\sec x = \frac{1}{\cos x}$$

$$\csc x = \frac{1}{\sin x}$$

$$\sin^2 x + \cos^2 x = 1$$

$$\tan 2x = \frac{2\tan x}{1 - \tan^2 x}$$

$$\sin 3x = 3\sin x - 4\sin^3 x$$

$$\cos 3x = -3\cos x + 4\cos^3 x$$

$$\sin^2 x = \frac{1 - \cos 2x}{2}$$

$$\cos^2 x = \frac{1 + \cos 2x}{2}$$

$$\sin^3 x = \frac{3\sin x - \sin 3x}{4}$$

$$\cos^3 x = \frac{3\cos x + \cos 3x}{4}$$

$$\sin x \pm \sin y = 2\sin\left(\frac{x \pm y}{2}\right)\cos\left(\frac{x \mp y}{2}\right)$$

$$\cos x + \cos y = 2\cos\left(\frac{x + y}{2}\right)\cos\left(\frac{x - y}{2}\right)$$

$$\cos x - \cos y = -2\sin\left(\frac{x + y}{2}\right)\sin\left(\frac{x - y}{2}\right)$$

$$\sin(x \pm y) = \sin x \cos y \pm \cos x \sin y$$

$$\cos(x \pm y) = \cos x \cos y \mp \sin x \sin y$$

$$\tan(x \pm y) = \frac{\tan x \pm \tan y}{1 \mp \tan x \tan y}$$

$$\cot(x \pm y) = \frac{\cot x \cot y \mp 1}{\cot x \pm \cot y}$$

$$\sin x \sin y = \frac{1}{2}[\cos(x-y) - \cos(x+y)]$$

$$\cos x \cos y = \frac{1}{2}[\cos(x-y) + \cos(x+y)]$$

$$\sin x \cos y = \frac{1}{2}[\sin(x-y) + \sin(x+y)]$$

4. Hyperbolic Functions

$$\sinh x = \frac{e^x - e^{-x}}{2}$$

$$\cosh x = \frac{e^x + e^{-x}}{2}$$

$$\tanh x = \frac{\sinh x}{\cosh x} = \frac{e^x - e^{-x}}{e^x + e^{-x}}$$

$$\coth x = \frac{1}{\tanh x}$$

$$\operatorname{sech} x = \frac{1}{\cosh x}$$

$$\operatorname{csch} x = \frac{1}{\sinh x}$$

$$\cosh^2 x - \sinh^2 x = 1$$

$$\sinh(-x) = -\sinh x$$

$$\cosh(-x) = \cosh x$$

$$\tanh(-x) = -\tanh x$$

$$\sinh(x \pm y) = \sinh x \cosh y \pm \cosh x \sinh y$$

$$\cosh(x \pm y) = \cosh x \cosh y \pm \sinh x \sinh y$$

$$\tanh(x \pm y) = \frac{\tanh x \pm \tanh y}{1 \pm \tanh x \tanh y}$$

$$\sinh 2x = 2 \sinh x \cosh x$$

$$\cosh 2x = \cosh^2 x + \sinh^2 x = 2\cosh^2 x - 1 = 1 + 2\sinh^2 x$$

$$\tanh 2x = \frac{2\tanh x}{1 + \tanh^2 x}$$

$$\sinh^2 x = \frac{1}{2}[\cosh 2x - 1]$$

$$\cosh^2 x = \frac{1}{2}[\cosh 2x + 1]$$

$$\sinh x \pm \sinh y = 2\sinh\frac{(x \pm y)}{2}\cosh\frac{(x \mp y)}{2}$$

$$\cosh x + \cosh y = 2\cosh\frac{(x+y)}{2}\cosh\frac{(x-y)}{2}$$

$$\cosh x - \cosh y = 2\sinh\frac{(x+y)}{2}\sinh\frac{(x-y)}{2}$$

$$\sinh x \sinh y = \frac{1}{2}[\cosh(x+y) - \cosh(x-y)]$$

$$\cosh x \cosh y = \frac{1}{2}[\cosh(x+y) + \cosh(x-y)]$$

$$\sinh x \cosh y = \frac{1}{2}[\sinh(x+y) + \sinh(x-y)]$$

5. Logarithmic Functions

$$\log_a xy = \log_a x + \log_a y$$

$$\log_a \frac{x}{y} = \log_a x - \log_a y$$

$$\log_a x^y = y\log_a x$$

$$\log_a x = \frac{\log_b x}{\log_b a}$$

$$\log_a a = 1$$

6. Complex Numbers
Rectangular to polar representation:

$$x + jy = re^{j\theta}$$

where:

$$j = \sqrt{-1},$$

$$r = \sqrt{x^2 + y^2},$$

$$\theta = \tan^{-1}(y/x)$$

$$(re^{j\theta})^n = r^n e^{jn\theta}$$

$$(r_1 e^{j\theta_1})(r_2 e^{j\theta_2}) = r_1 r_2 e^{j(\theta_1 + \theta_2)}$$

7. Relationship between exponential, trigonometric and hyperbolic functions

$$e^{\pm j\pi} = -1$$

$$e^{\pm j\pi/2} = \pm j$$

$$e^{\pm j(x + 2k\pi)} = e^{\pm jx}$$

$$e^{\pm j[x + (2k+1)\pi]} = -e^{\pm jx}$$

$$e^{\pm jx} = \cos x \pm j\sin x$$

$$\sin x = \frac{e^{jx} - e^{-jx}}{2j}$$

$$\cos x = \frac{e^{jx} + e^{-jx}}{2}$$

$$\tan x = -j\left(\frac{e^{jx} - e^{-jx}}{e^{jx} + e^{-jx}}\right)$$

$$\sin(jx) = j\sinh x$$

$$\cos(jx) = \cosh x$$

$$\tan(jx) = j\tanh x$$

$$\sinh(jx) = j\sin x$$

$$\cosh(jx) = \cos x$$

$$\tanh(jx) = j\tan x$$

where $j = \sqrt{-1}$ and k is a positive integer.

8. Derivatives

$$\frac{d}{dx}(u^n) = nu^{n-1}\frac{du}{dx}$$

$$\frac{d}{dx}(uv) = u\frac{dv}{dx} + v\frac{du}{dx} \quad \text{(chain rule)}$$

$$\frac{d}{dx}(u/v) = \frac{v(du/dx) - u(dv/dx)}{v^2}$$

$$\frac{d}{dx}\sin u = \cos u\frac{du}{dx}$$

$$\frac{d}{dx}\cos u = -\sin u\frac{du}{dx}$$

$$\frac{d}{dx}\tan u = \sec^2 u\frac{du}{dx}$$

$$\frac{d}{dx}\cot u = -\csc^2 u\frac{du}{dx}$$

$$\frac{d}{dx}\log_a u = \frac{\log_a e}{u}\frac{du}{dx}$$

$$\frac{d}{dx}\log_e u = \frac{1}{u}\frac{du}{dx}$$

$$\frac{d}{dx}a^u = a^u\log_e a\frac{du}{dx}$$

$$\frac{d}{dx}e^u = e^u\frac{du}{dx}$$

$$\frac{d}{dx}\sinh u = \cosh u\frac{du}{dx}$$

$$\frac{d}{dx}\cosh u = \sinh u\frac{du}{dx}$$

$$\frac{d}{dx}\tanh u = \operatorname{sech}^2 u\frac{du}{dx}$$

$$\frac{d}{dx}\coth u = -\operatorname{csch}^2 u\frac{du}{dx}$$

9. Indefinite Integrals

$$\int u^n du = \frac{u^{n+1}}{n+1} + C$$

$$\int u dv = uv - \int v du \text{ (integration by parts)}$$

$$\int \frac{du}{u} = \log_e|u| + C$$

$$\int a^u du = \frac{a^u}{\log_e a} + C \qquad \mathrm{a} > 0, \mathrm{a} \neq 1$$

$$\int e^u du = e^u + C$$

$$\int \log_e u du = u\log_e u - u + C$$

$$\int \sin u du = -\cos u + C$$

$$\int \cos u du = \sin u + C$$

$$\int \tan u du = -\log_e \cos u + C$$

$$\int \cot u du = \log_e \sin u + C$$

$$\int \sinh u du = \cosh u + C$$

$$\int \cosh u du = \sinh u + C$$

$$\int \tanh u du = \log_e \cosh u + C$$

$$\int \coth u du = \log_e \sinh u + C$$

$$\int \sin^2 u du = \frac{u - \sin u \cos u}{2} + C$$

$$\int \cos^2 u du = \frac{u + \sin u \cos u}{2} + C$$

10. Taylor Series Expansion

$$f(x)|_{x=a} = f(a) + f'(a)(x-a) + \frac{f''(a)(x-a)^2}{2!} + \dots + \frac{f^{(n-1)}(a)(x-a)^{(n-1)}}{(n-1)!} + \dots$$

$$e^x\Big|_{x=0} = 1 + x + \frac{x^2}{2!} + \frac{x^3}{3!} + \dots$$

11. Vector Operations and Identities (in Rectangular Coordinate System)

The dot product between vectors U and V is defined by:

$$U \cdot V = |U||V|\cos\theta = U_xV_x + U_yV_y + U_zV_z$$

The cross product between vectors U and V is defined by:

$$V = \hat{i}(U_yV_z - U_zV_y) + \hat{j}(U_zV_x - U_xV_z) + \hat{k}(U_xV_y - U_yV_x)$$

$$|U \times V| = |U||V|\sin\theta$$

$$U \times V = -V \times U$$

The Del operator is defined by:

$$\nabla = \hat{i}\frac{\partial}{\partial x} + \hat{j}\frac{\partial}{\partial y} + \hat{k}\frac{\partial}{\partial z}$$

The gradient of scalar Φ (x,y,z) is defined by:

$$\nabla\Phi = \hat{i}\frac{\partial\Phi}{\partial x} + \hat{j}\frac{\partial\Phi}{\partial y} + \hat{k}\frac{\partial\Phi}{\partial z}$$

The divergence of vector U (x,y,z) is defined by:

$$\nabla \cdot U = \frac{\partial U_x}{\partial x} + \frac{\partial U_y}{\partial y} + \frac{\partial U_z}{\partial z}$$

The curl of vector U(x,y,z) is defined by:

$$\nabla \times U = \hat{i}\left(\frac{\partial U_z}{\partial y} - \frac{\partial U_y}{\partial z}\right) + \hat{j}\left(\frac{\partial U_x}{\partial z} - \frac{\partial U_z}{\partial x}\right) + \hat{k}\left(\frac{\partial U_y}{\partial x} - \frac{\partial U_x}{\partial y}\right)$$

The laplacian of a scalar Φ is defined by:

$$\nabla \cdot \nabla\Phi = \nabla^2\Phi = \frac{\partial^2\Phi}{\partial x^2} + \frac{\partial^2\Phi}{\partial y^2} + \frac{\partial^2\Phi}{\partial z^2}$$

The laplacian of a vector U is defined by:

$$\nabla^2 U = \hat{i}\,\nabla^2 U_x + \hat{j}\,\nabla^2 U_y + \hat{k}\,\nabla^2 U_z$$

Vector identities:

$$\nabla \cdot (\Phi U) = (\nabla\Phi) \cdot U + \Phi(\nabla \cdot U)$$

$$\nabla(\Phi\Psi) = \Psi(\nabla\Phi) + \Phi(\nabla\Psi)$$

$$\nabla \times (\Phi U) = (\nabla\Phi) \times U + \Phi(\nabla \times U)$$

$$\nabla \times \nabla\Phi = 0$$

$$\nabla \cdot (\nabla \times U) = 0$$

$$\nabla \times (\nabla \times U) = \nabla(\nabla \cdot U) - \nabla^2 U$$

where Φ and Ψ are scalar functions.

12. Equation of a Circle

In complex number plane, the equation of a circle is given by:

$$|z - z_o| = R$$

where ($z = x + jy$) is any point on the circle, with ($z_o = x_o + jy_o$) as the center and R as the radius.

In a Cartesian plane, the equivalent equation of a circle centered at (x_o, y_o) having a radius (R) is given by:

$$(x - x_o)^2 + (y - y_o)^2 = R^2$$

APPENDIX K

DC Bias Networks for an FET

There are five basic DC bias networks for an FET that uses one or two power supplies.[1]

[1] Vendelin, G. D. *Five Basic Bias Design for GaAs FET Amplifiers, Microwaves & RF*, February 1978. (Reproduced with permission of Microwaves & RF.)

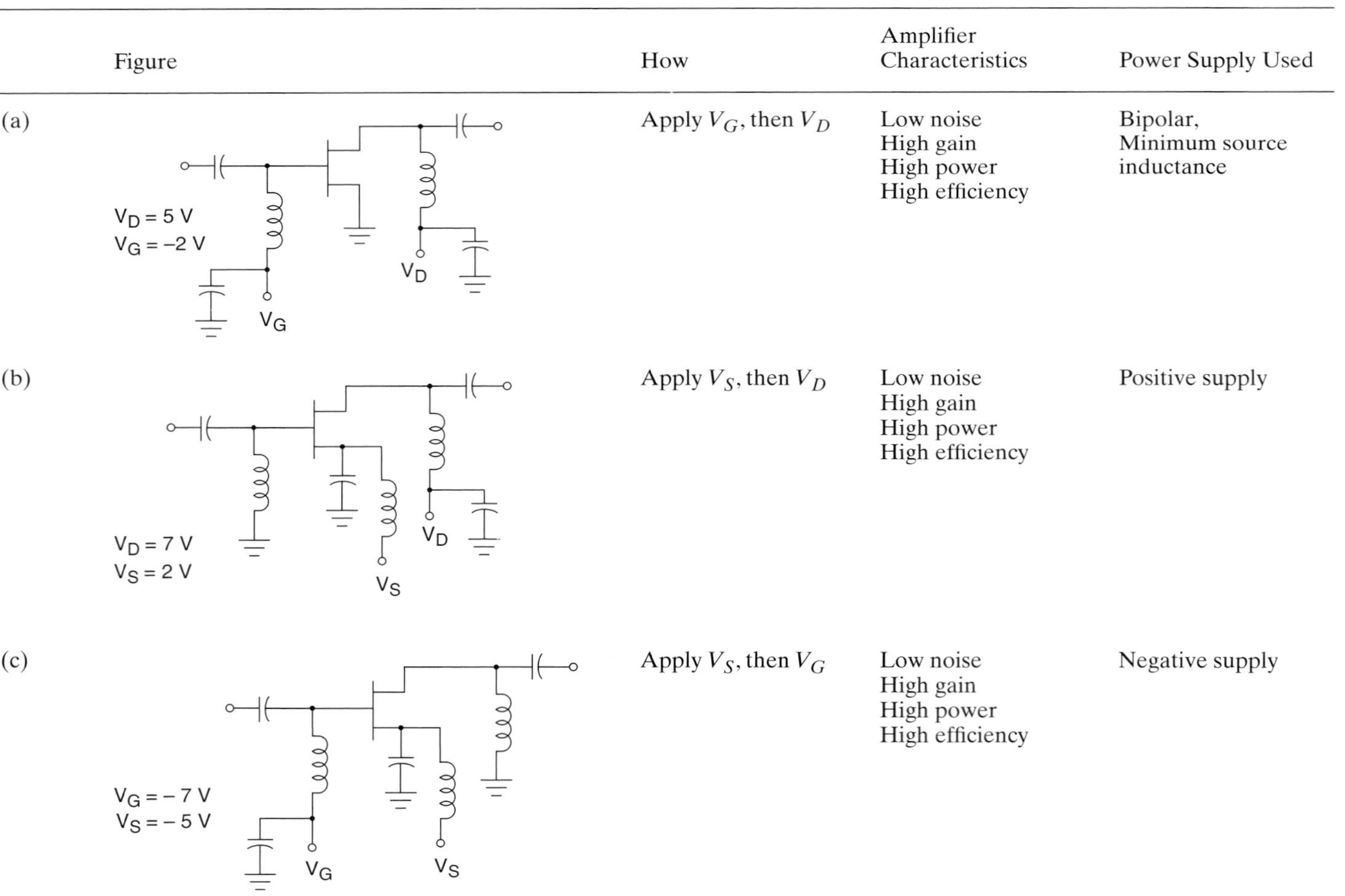

	Figure	How	Amplifier Characteristics	Power Supply Used
(a)		Apply V_G, then V_D	Low noise High gain High power High efficiency	Bipolar, Minimum source inductance
(b)		Apply V_S, then V_D	Low noise High gain High power High efficiency	Positive supply
(c)		Apply V_S, then V_G	Low noise High gain High power High efficiency	Negative supply

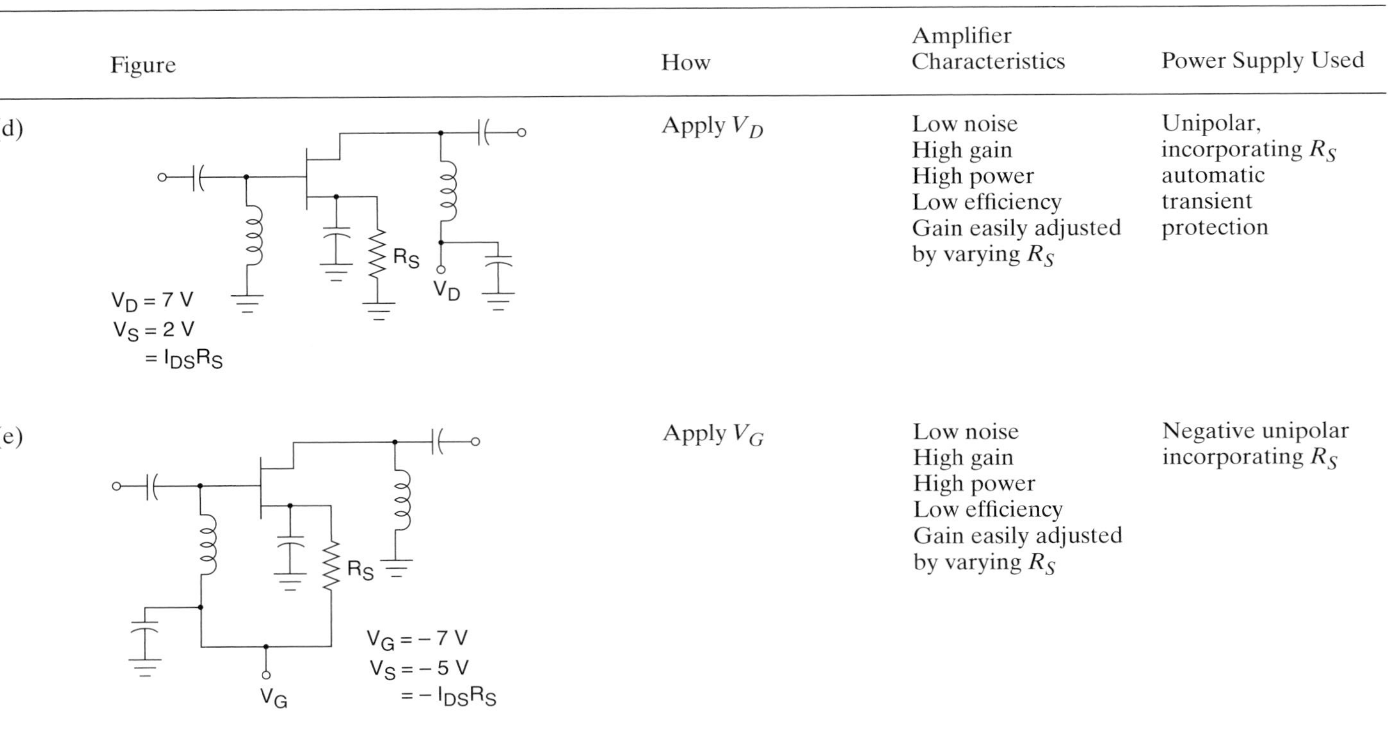

	Figure	How	Amplifier Characteristics	Power Supply Used
(d)		Apply V_D	Low noise High gain High power Low efficiency Gain easily adjusted by varying R_S	Unipolar, incorporating R_S automatic transient protection
(e)		Apply V_G	Low noise High gain High power Low efficiency Gain easily adjusted by varying R_S	Negative unipolar incorporating R_S

APPENDIX L

Computer Aided Design (CAD) Examples

Computer aided design (CAD) programs are currently available for use with microwave design. The examples presented in this appendix were completed using HP-EEsof Libra, Series IV, Version 6.1. EEsof Libra operates in the UNIX environment and is accessed using HP3410/3420 workstations. Use of the equipment was available through workstations available at the RF/Microwave/lightwave laboratory at California State University, Northridge.

The three examples in this Appendix are: (a) a BJT microwave amplifier design using an *S*-parameter file generated from the 2SC3603 BJT data sheet and also using the EEsof library file for the packaged device, (b) a FET microwave amplifier design, and (c) a microwave mixer design using a silicon diode.

EXAMPLE 1—BJT MICROWAVE AMPLIFIER DESIGN: HGA DESIGN

Design Problem

Design a microwave amplifier at 1 GHz for maximum possible gain utilizing an NEC NE85635 bipolar junction transistor (BJT). The transistor device is the gain stage of the amplifier. The NE85635 device (NEC old part No.) is the 2SC3603 silicon transistor (NEC new part No.). The *S*-parameters for the microwave transistor at $V_{CE} = 10$ V, $I_C = 20$ mA, and $Z_o = 50\ \Omega$ are shown in Table L.1.

TABLE L.1 *S*-Parameters for the Microwave Transistor: NEC NE85635

f (GHz)	$\lvert S_{11} \rvert$	∠ S11°	$\lvert S_{21} \rvert$	∠ S21°	$\lvert S_{12} \rvert$	∠ S12°	$\lvert S_{22} \rvert$	∠ S22°
0.5	0.629	–160.8	10.10	92.60	0.040	41.50	0.256	–49.00
1.0	0.631	175.8	5.411	75.10	0.048	51.40	0.244	–57.20
1.5	0.628	162.5	3.565	60.60	0.070	59.20	0.232	–66.80
2.0	0.646	152.2	2.720	48.40	0.086	56.00	0.220	–77.40
2.5	0.659	142.1	2.161	38.80	0.105	52.20	0.213	–89.10
3.0	0.677	132.0	1.916	25.7	0.127	45.10	0.217	–103.10
3.5	0.695	123.8	1.585	14.30	0.151	39.70	0.232	–119.50
4.0	0.713	116.5	1.392	05.30	0.168	34.80	0.254	–134.0

Design Procedure

The requirement for this project is to design a maximum gain amplifier using an NEC NE85635 BJT at 1 GHz. The design procedure used for this activity began with defining the biasing network for the transistor. Next the stability conditions (K and Δ) were calculated for the active device from the manufacturer's S-parameter information. From the stability calculation, the approach for choosing the matching network design was determined.

Matching networks are not limited to a single configuration of elements. Designs can be developed and simulated for lumped element implementation and transmission line implementations. Both lumped and distributed element configurations can be developed and simulated using EEsof Libra software.

(A) DC Bias Design. Two resistors R_1 and R_2 compose the voltage divider for the base voltage. R_C is the current limit resistor. R_E is the emitter resistor. The manufacturer's performance information was given for bias conditions of $I_C = 7.0$ mA.

A voltage of $V_{CC} = 30$ V was chosen. $V_{CE} = 10\text{ V} = 1/3\, V_{CC}$ was given as a rule-of-thumb in Chapter 4, Example 4.3. The following is a list of assumptions made to calculate the bias circuitry:

$$V_B = 1/3\, V_{CC} = 10\text{ V}$$

$\beta = 120$, from manufacturer's data sheets. From Figure 4.25, we have:

$$I_1 = I_2 \text{ if } I_B << I_1\text{, choose } I_1 = I_C / 10 = 0.7\text{ mA}$$

$$V_{BE} = 0.7\text{ V}$$

First, the values of R_C and R_E were calculated as:

$$V_E = V_B - V_{BE} = 10.0 - 0.7\text{ V} = 9.3\text{ V}$$

$$R_C = (V_{CC} - V_{CE} - V_E) / I_{CQ} = (30 - 10 - 9.3)\text{ V}/ 0.7\text{ mA} = 1{,}428\ \Omega \approx 1.5\text{K}\Omega$$

$$R_E = V_{CE} / I_E = V_{CE} / (I_{CQ} / \alpha) = 9.3 / (0.007/0.99) = 1{,}315\ \Omega$$

Next, the R_1 and R_2 resistors were determined:

$$I_1 \approx V_{CC} / (R_1 + R_2) \Rightarrow (R_1 + R_2) = 30\text{ V} / 0.7\text{ mA} = 42.85\text{ K}\Omega$$

$$V_B = V_{CC}\, R_2 / (R_1 + R_2) \Rightarrow R_2 = 10\text{ V}\ (42.85\text{ K}\Omega) / 30\text{ V} = 14.28\text{ K}\Omega$$

$$R_1 = (R_1 + R_2) - R_2 = 28.5\text{ K}\Omega$$

Finally, we need to confirm the transistor is operating in the active region:

$$V_{CE} = V_{CB} + V_{BE} \Rightarrow V_{CB} = 10 - 0.7 = 9.3\text{ V}$$

The CB junction is reverse biased. The BE junction is forward biased. The transistor is operating in the active mode.

A bypass capacitor was used with an emitter resistor to improve bias stability at the lower microwave frequencies.

(B) RF Design. The first step after selecting the device is to assess the stability conditions at the desired frequency.

Following the stability determination, the unilateral figure of merit, U, is checked. This design will require a bilateral analysis because $S_{12} \neq 0$. However, the book recommends a look at U to assess the error.

Finally, the input and output filter networks will be determined by using a Smith chart.

(1) STABILITY CONDITIONS The design procedure used to calculate the matching networks is chosen by the stability conditions at the desired frequency. This is done by calculating values of K and Δ. The values were calculated as follows:

$$K = (1 - |S_{11}|^2 - |S_{22}|^2 - |\Delta|^2) = 1.067$$

$$\Delta = |S_{11}S_{22} - S_{12}S_{21}| = 0.109$$

Because $K > 1$ and $\Delta < 1$ the device is unconditionally stable. This agrees with the graphic method. Because $S_{12} \neq 0$, the unilateral figure of merit needs to be calculated.

(2) ERROR RANGE The unilateral figure of merit, U, varies with frequency due to S-Parameter dependence and determines if the unilateral design procedure is applicable. U must be small to allow the simplification. U for the amplifier in the design is given by:

$$U = |S_{12}|\,|S_{21}|\,|S_{11}|\,|S_{22}| / |(1 - |S_{11}|^2)(1 - |S_{22}|^2) = 0.706$$

The maximum error using the unilateral assumption is bounded by:

$$1/(1+U)^2 < G_T / G_{TU,MAX} < 1/(1-U)^2$$

$$-0.59 \text{ dB} < G_T / G_{TU,MAX} < 0.63 \text{ dB}$$

$$G_{S,MAX} = 2.2 \text{ dB}$$

$$G_o = 14.7 \text{ dB}$$

$$G_{L,MAX} = 0.2 \text{ dB}$$

$$G_{TU,MAX} = 17.1 \text{ dB}$$

In this case, the unilateral assumption is valid because an error of –0.59 dB to 0.63 dB is acceptable. Thus, the unilateral design method can be used.

(3) REFLECTION COEFFICIENT DETERMINATION The unilateral design uses Γ_L and Γ_S, which are chosen equal to S^*_{11} and S^*_{22}. These are:

$$\Gamma_S = S^*_{11} = 0.631\angle{-175.8^\circ}$$

$$\Gamma_L = S^*_{22} = 0.244\angle 57.2^\circ$$

(4) MISMATCH FACTOR The mismatch factor for the source and the load should be near unity since the source and load are matched to the device. They are given by:

$$M_S = (1 - |\Gamma_S|^2)(1 - |\Gamma_{IN}|^2)/|1 - \Gamma_S \Gamma_{IN}|^2$$

$$M_L = (1 - |\Gamma_L|^2)(1 - |\Gamma_{OUT}|^2)/|1 - \Gamma_L \Gamma_{OUT}|^2$$

(5) DESIGN #1: LUMPED ELEMENTS First a design using lumped elements was developed. For the input matching network $\Gamma_S = 0.631 \angle{-175.8^\circ}$ was used. Starting from the source at O on the Smith chart, first a shunt inductor, L_1, $Y_{L1} = -j1.8$, was added to arrive at $Y_N = 1 - j1.82$. Then, a series capacitor, C_1, $Z_{C1} = -j0.642$, was added to arrive at $Z_N = 0.231 - j0.03$. All values are given as normalized.

For the output matching network $\Gamma_L = 0.244 \angle 57.2^\circ$ was used. Starting from the source at O on the Smith chart, first a series inductor, L_2, $Y_{L2} = j0.65$, was added to get to $Z_N = 1{+}j0.65$. Then, a shunt C_2, $Y_{C2} = j0.136$ was added (from $0.703 - j0.456$ to $Y_N = 0.703 - j0.320$). All values are normalized.

Next, the component values were calculated for the impedance and admittance values obtained from the Smith chart:

$$J\omega L_1 = (50)/(-j1.8) \Rightarrow L_1 = (50)/[2\pi(1\times 10^9)(1.8)] = 4.42 \text{ nH}$$

$$J\omega C_1 = -j0.642(50) \Rightarrow C_1 = (0.425)(50)/[2\pi(1\times 10^9)] = 7.49 \text{ pF}$$

$$J\omega L_2 = j0.65(50) \Rightarrow L_2 = 0.65(50)/[2\pi(1\times 10^9)] = 5.17 \text{ nH}$$

$$J\omega C_2 = (50)/(j0.136) \Rightarrow C_2 = (50)/[(0.65)2\pi(1\times 10^9)(50)] = 0.43 \text{ pF}$$

(6) DESIGN #2: DISTRIBUTED ELEMENTS (SHUNT/SERIES STUBS) As an alternate possibility to lumped elements, a design using distributed elements was developed. The intent was to use open-circuit stubs with minimum lengths and transmission lines for ease of construction. A number of designs can be developed and modeled using distributed elements.

First, design an input open shunt stub and an output series open stub.

For the input matching network, $\Gamma_S = 0.631 \angle{-175.8^\circ}$ was used. Starting from the source at O on the Smith chart, an open circuit shunt stub, $\ell_1 = 0.159\lambda$, $Y_{\ell 1} = j1.55$, was added to arrive at $Y_N = 1 + j1.55$. Then, a transmission line, ℓ_2, was added to move at a constant *VSWR* for 0.066λ to arrive at $\Gamma_S = 0.631 \angle{-175.8^\circ}$. All values are normalized using $Z_o = 50\ \Omega$.

For the output matching network $\Gamma_L = 0.244 \angle 57.2^\circ$ was used. Starting from the source at O on the Smith chart, a series open-circuit shunt stub, $\ell_3 = 0.325\lambda$, $Z_{l3} = j0.52$, was added to arrive at $Z_N = 1{+}j0.52$. Then, a transmission line, ℓ_4, was added to move at a constant *VSWR* for 0.171λ to arrive at $\Gamma_L = 0.244 \angle 57.2^\circ$.

(7) DESIGN #3: DISTRIBUTED ELEMENTS (SHUNT/SHUNT STUBS)—SECOND DESIGN The configuration of the microstrip open series stub in design #2 was not available in the EEsof program. As a result, the output matching network was designed using an open shunt stub. No changes were necessary for the input matching network, where a value of $\Gamma_S = 0.631 \angle{-175.8^\circ}$ was used.

However, for the output matching network ($\Gamma_L = 0.244 \angle 57.2°$) a shunt stub (instead of a series stub) was used. Starting from the source at O on the Smith chart an open-circuit shunt stub, $\ell_3 = 0.100\lambda$, $Y_{\ell 3} = j0.52$, was added to arrive at $Y_N = 1 + j0.52$. Then, a transmission line, ℓ_4, was added to move on a constant *VSWR* for a distance of 0.275λ to arrive at $\Gamma_S = 0.631 \angle -175.8°$.

(8) Microstrip Dimensions & Calculations The microstrip elements used in the design have to be defined by length and width when placed in the schematic. The dimensions for the elements were calculated for FR4 material (FR4 is the epoxy glass laminated dielectric which is manufactured by Isola/ADI) with a dielectric thickness of 0.062 inches (62 mils) and $\varepsilon_r = 4.81$, clad in one-ounce copper. The one-ounce copper clad has a thickness of 1.4 mil. Specific data were entered in the EEsof program for the microstrip elements so that the transmission lines could be correctly modeled by the program. The microstrip information is entered through the MSUB datum in the element definition, and this in turn is taken from the master default window into the MSUB Data Default. The MSUB Data Default information block is:

$$ER = 4.81$$

$$H = 62$$

$$T = 1.40$$

$$RHO = 1$$

$$RGH = 0$$

$$COND1 = cond$$

$$COND2 = cond2$$

$$DIEL1 = diel1$$

$$DIEL2 = diel2$$

$$RES = resi$$

Next, the dimensions of the individual microstrip elements were calculated.

A value of $Z_o = 50\ \Omega$ was selected for the characteristic impedance of the transmission lines that were to be used in the design. The ratio of W/h (W is the width of the transmission line and h is the dielectric thickness) was determined by using two equations.

For $W/h \le 2$:

$$W/h = 8e^{A}/(e^{2A} - 2)$$

$$A = (Z_o/60)\{\sqrt{(\varepsilon_r + 1)/2} + [(\varepsilon_r - 1)/(\varepsilon_r + 1)](0.23 + 0.11/\varepsilon_r)\}$$

For $W/h \ge 2$:

$$W/h = (2/\pi)\{B - 1 - \ln(2B - 1) + [(\varepsilon_r - 1)/2\varepsilon_r][\ln(B - 1) + 0.39 - 0.61/\varepsilon_r]\}$$

$$B = 377\pi / 2 Z_o \sqrt{\varepsilon_r}$$

The calculations resulted in a value of $W/h = 1.787$ using the $W/h \leq 2$ method and a value of 1.788 using the $W/h \geq 2$ method.

Wave propagation on the microstrip line varies from wave propagation in free space. The wavelength, λ, on the microstrip equals the wavelength in free space divided by the square root of the effective relative dielectric constant. The following calculations specify the lengths of the microstrip sections for the input and output networks.

$$\lambda = \lambda_o / \sqrt{\varepsilon_{ff}}$$

$$\varepsilon_{ff} = [(\varepsilon_r + 1)/2] + [(\varepsilon_r - 1)/2]\,[(1 + 12\,h/W)^{-1/2}] = 3.638$$

$$\lambda / \lambda_o = 1/\sqrt{\varepsilon_{ff}} = 0.525$$

Once the wavelength for the microstrip transmission lines was established, the values for the lengths of the microstrip sections used for the matching networks were calculated.

$$\ell_1 = 0.159\,\lambda$$

$$L_1 = 0.159\,\lambda = 0.159\,(c/\mathrm{f})\,(\lambda/\lambda_o)$$

$$L_1 = 0.159\,(3 \times 10^{10}\ \mathrm{cm/sec})\,(393.7\ \mathrm{mil/cm})\,(0.525)/10^9 = 985.9\ \mathrm{mil}$$

$$\ell_2 = 0.066\,\lambda$$

$$L_2 = 409.3\ \mathrm{mil}$$

$$\ell_3 = 0.325\,\lambda$$

$$L_3 = 2015.3\ \mathrm{mil}$$

$$\ell_4 = 0.024\,\lambda$$

$$L_4 = 148.8\ \mathrm{mil}$$

The EEsof Libra program is restricted to only two connections per node when the automated routing feature is used for layout of the board design. A layout can be developed manually if this design rule is violated. Design synchronization should be done before the simulation of the final layout is run.

(9) RF/Biasing Circuitry Isolation The final consideration for the amplifier design was the isolation of the bias circuitry from the RF signal. RF chokes are included to isolate the bias power supply from the high frequency RF signals. These components are designated RFC in the schematic. The inductance was chosen to be much greater than $|Z_C| = 367\Omega$, which was the impedance of the capacitor filter in the lumped element design.

$$X_{RFC} = 10 \times |Z_C|$$

$$X_{RFC} = \omega L_{RFC} \Rightarrow L_{RFC} = X_{RFC}/\omega$$

$$L_{RFC} = 3670 / 2\pi\,(10^9)\ \mathrm{H}$$

$$L_{RFC} = 584 \text{ nH}$$

DC blocking capacitors were added to prevent the bias current from flowing into the source and the load.

$$-jX_{DC\text{-}BLK} = 1/(j\omega C_{DC\text{-}BLK}) \Rightarrow C_{DC\text{-}BLK} = 1/\omega X_{DC\text{-}BLK}$$

$$C_{DC\text{-}BLK} = 1/2\pi\,(10^9)\,(0.3)\text{ F}$$

$$C_{DC\text{-}BLK} = 497\text{ pF}$$

A bypass capacitor, $X_{BP} << 32\ \Omega$, was added across R_E to improve the voltage gain:

$$-jX_{BP} = 1/(j\omega C_{BP}) \Rightarrow C_{BP} = 1/\omega X_{BP}$$

$$C_{BP} = 1/2\pi\,(10^9)\,(2.12)\ \text{ pF}$$

$$C_{BP} = 75\text{ pF}$$

(10) Final Circuit Configuration A stylized schematic for the Board 1 configuration is shown in Figure L.1.

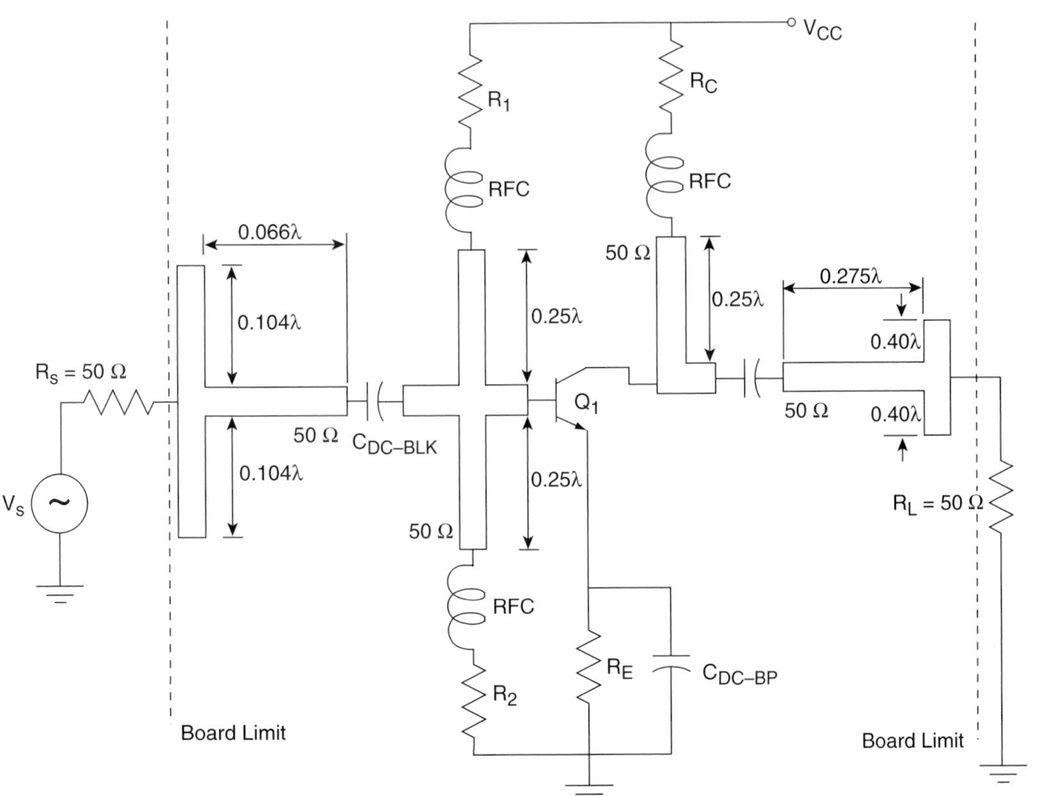

FIGURE L.1 Example 1—Circuit configuration.

We will now simulate this circuit using HP-EEsof Libra. The following figures show examples of the EEsof documentation. The first figure shown is the EEsof Default Values window (see Figure L.2). The data elements in this window define the units used in the program. Once the user moves to the schematic window or test bench window the units for circuit elements are not displayed, so this window is important to ensure that the results of the analysis-run for the circuit are correctly interpreted.

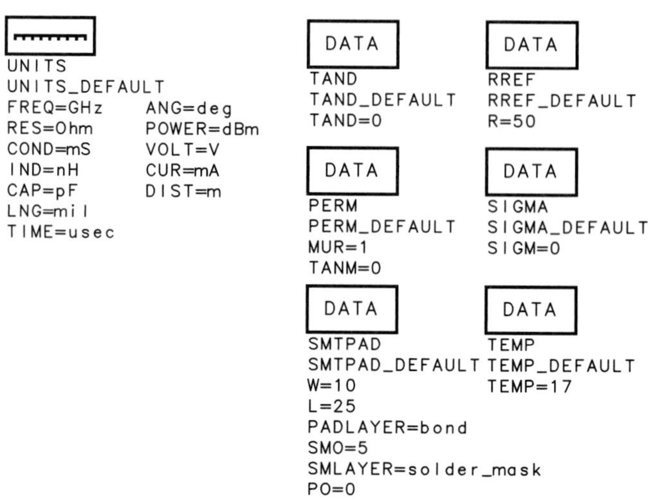

FIGURE L.2 Example 1—Default Window.

The EEsof Schematic window is used to capture the circuit schematic (see Figure L.3). The circuit elements are individually defined in the schematic window. After the element is defined, it is connected to capture the circuit configuration.

The EEsof Test Bench Parameters are shown in the Test Bench window (see Figure L.4). To evaluate the circuit, performance-specific measurements are identified in the Test Bench window prior to running an analysis. Each test measurement is placed in the window along with the circuit that is being tested.

A graph window displays performance information as specified by the user. For each selected measurement that is displayed, the program automatically calculates the proper scaling factors. A readable format is obtained by grouping measurements according to the output range. A linear scale was chosen for the graph. Figure L.5 shows a plot of S_{11} and S_{21}. Figure L.6 shows a plot of S_{12} and S_{22}.

A tabular form for display of performance data is also available. This format is available in the table window, and units are selected for a clear display of performance values.

The user selects the measurements for display in the Table L.2.

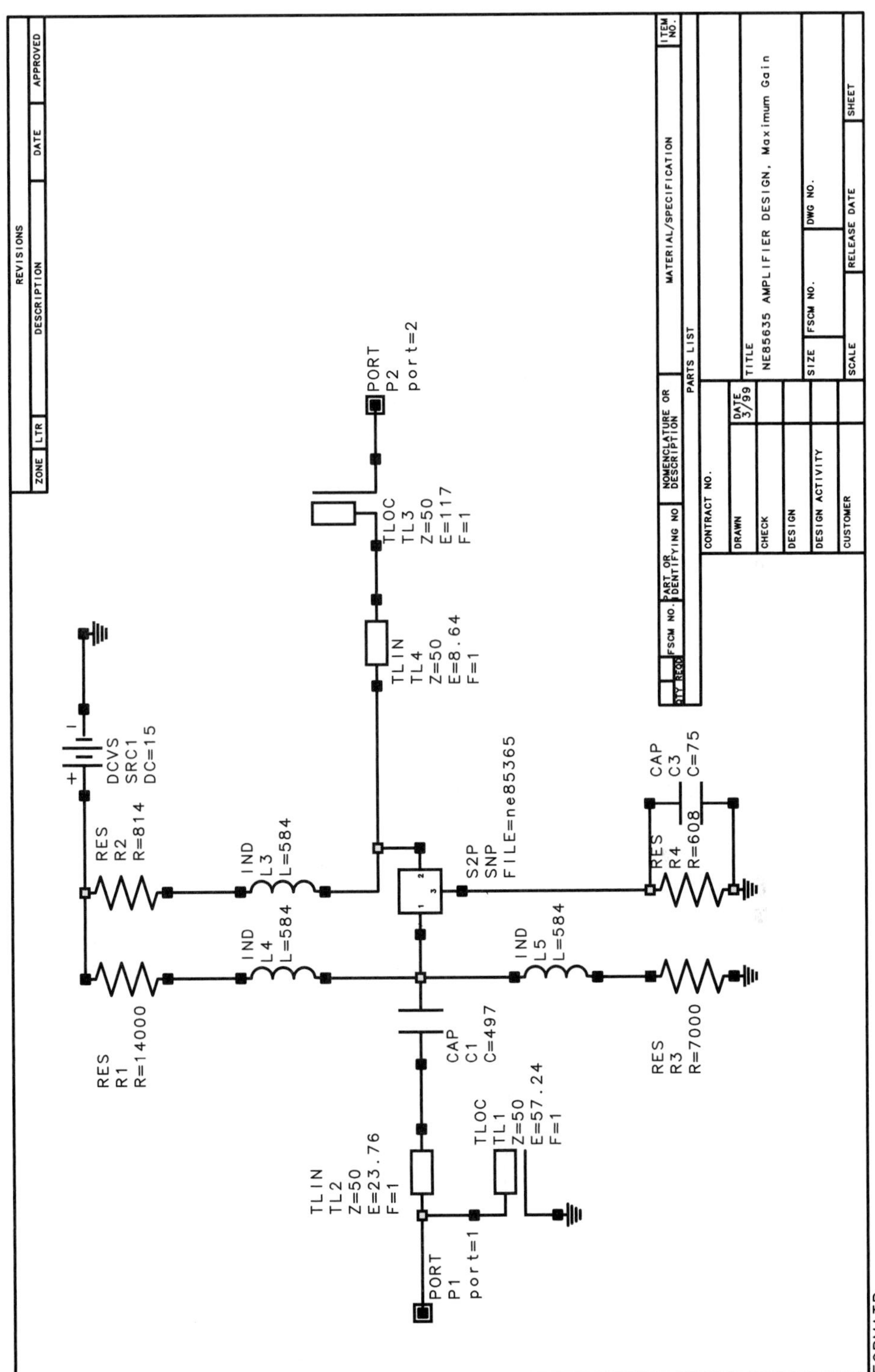

FIGURE L.3 Example 1—Schematic Capture.

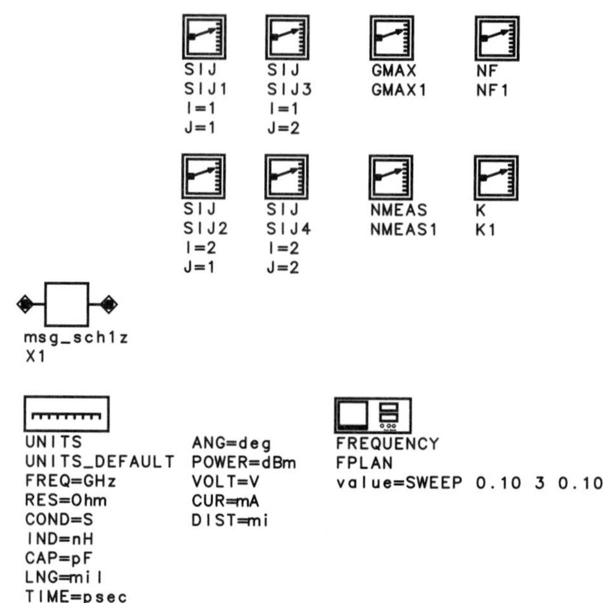

FIGURE L.4 Example 1—Test Bench Window.

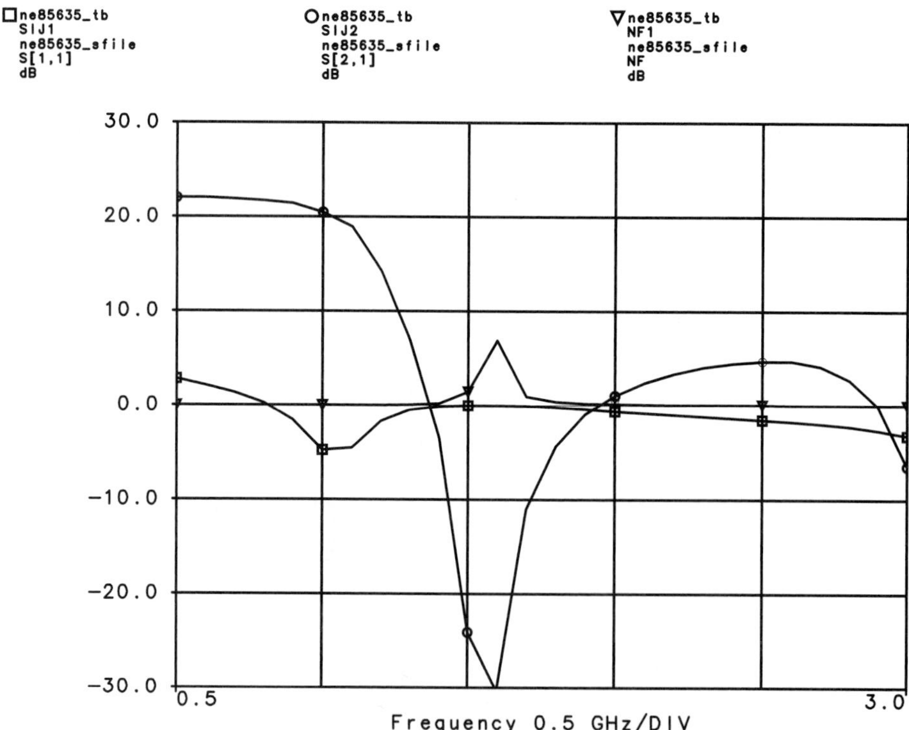

FIGURE L.5 Example 1—S_{11} and S_{21} Performance.

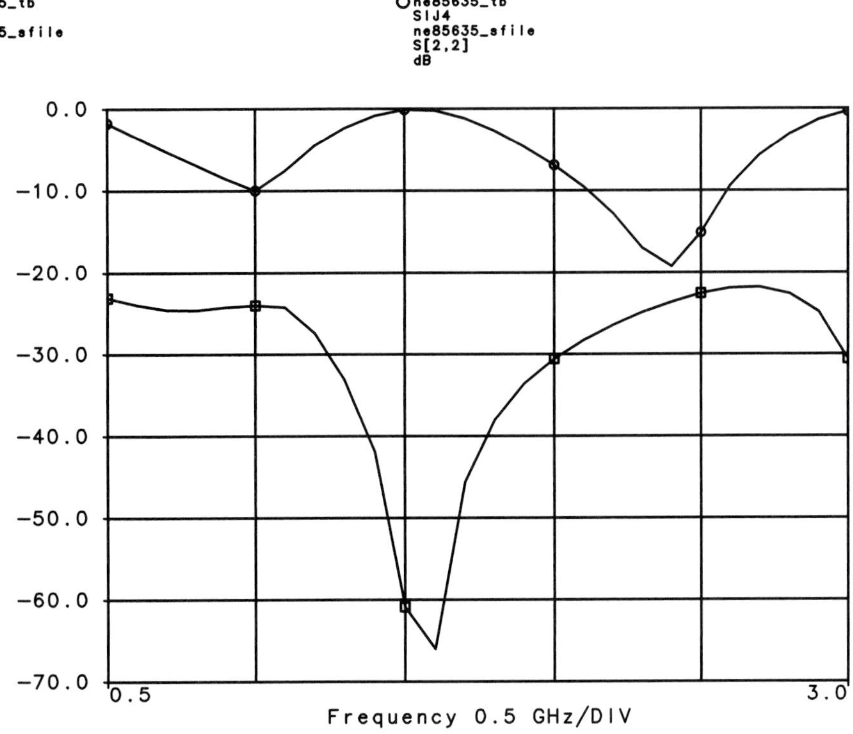

FIGURE L.6 Example 1—S_{12} and S_{22} Performance.

TABLE L.2 S-Parameter File Distributed Element Design Performance Measurements

ne85635_transistor

Freq GHz	SIJ1 S[1,1] dB	SIJ2 S[2,1] dB	GMAX1 GMAX dB	NF1 NF dB	SIJ3 S[1,2] dB	SIJ4 S[2,2] dB
0.50	2.8060	22.0201	22.5731	0.0398	–23.1260	–1.7493
0.60	2.1510	22.0314	23.0211	0.0343	–24.0109	–3.4854
0.70	1.3324	21.8883	23.2219	0.0286	–24.5554	–5.2293
0.80	0.2464	21.7088	23.1647	0.0228	–24.6206	–6.9095
0.90	–1.5229	21.4192	22.8377	0.0172	–24.2562	–8.5495
1.00	–4.7340	20.4516	22.2384	0.0123	–24.0252	–9.9786
1.10	–4.5383	18.9592	21.6152	0.0101	–24.2713	–7.5911
1.20	–1.6540	14.3337	20.8723	0.0155	–27.4109	–4.4877
1.30	–0.4758	7.0245	20.0713	0.0463	–33.1182	–2.3411
1.40	–0.1358	–3.4326	17.5267	0.1932	–41.9171	–0.8559
1.50	–0.0234	–24.0825	15.9258	1.4655	–60.8711	–0.0695
1.60	–0.0037	–30.3030	15.3503	6.8858	–66.0058	–0.1749

TABLE L.2 S-Parameter File Distributed Element Design Performance Measurements *(Continued)*

ne85635_transistor

Freq GHz	SIJ1 S[1,1] dB	SIJ2 S[2,1] dB	GMAX1 GMAX dB	NF1 NF dB	SIJ3 S[1,2] dB	SIJ4 S[2,2] dB
1.70	−0.0781	−10.9960	14.7661	0.9346	−45.6467	−1.1306
1.80	−0.2279	−4.3545	14.1665	0.3901	−37.9765	−2.6965
1.90	−0.4127	−0.9532	13.5463	0.2285	−33.5605	−4.6521
2.00	−0.6051	1.0171	12.9012	0.1554	−30.5809	−6.9121
2.10	−0.7978	2.4287	12.5051	0.1144	−28.2844	−9.5696
2.20	−0.9859	3.3977	12.0885	0.0882	−26.4478	−12.8585
2.30	−1.1694	4.0676	11.6493	0.0701	−24.9222	−17.0153
2.40	−1.3502	4.5000	11.1855	0.0568	−23.6408	−19.2733
2.50	−1.5316	4.6984	10.6955	0.0467	−22.5954	−15.1225
2.60	−1.7126	4.6981	10.4136	0.0388	−21.9472	−9.3880
2.70	−1.9338	4.1436	10.0814	0.0325	−21.8677	−5.6539
2.80	−2.2301	2.7558	9.7092	0.0276	−22.6299	−3.0752
2.90	−2.6324	−0.0574	9.3104	0.0237	−24.8211	−1.3207
3.00	−3.1795	−6.4398	8.8984	0.0208	−30.5817	−0.2755

EXAMPLE 1A—LIBRARY FILE AMPLIFIER DESIGN

The NEC NE85635 bipolar junction transistor (BJT) is available in the parts library for the EEsof. Use of the library simplifies the design tasks required for placing elements in the schematic. All of the device information is defined in the library entry for the device, so all the designer has to do is select the component from the appropriate library file.

The EEsof Default Values window for the library design is shown in Figure L.7. The data elements in this window are used to define the units used in the program. EEsof Schematic window displays the circuit schematic (see Figure L.8). The EEsof Test Bench Parameters are shown in the Test Bench window (see Figure L.9).

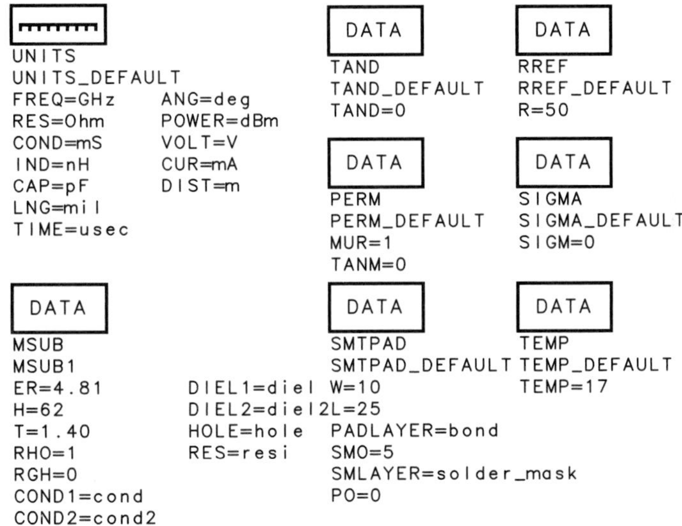

FIGURE L.7 Example 1A-Default Window.

Figure L.10 shows the amplifier circuit layout. A graph window displays performance information plotted on a linear scaled graph. Figure L.11 shows a plot of S_{11} and S_{21}. Figure L.12 shows a plot of S_{12} and S_{22}. A tabular display of performance data clearly shows performance values (see Table L.3).

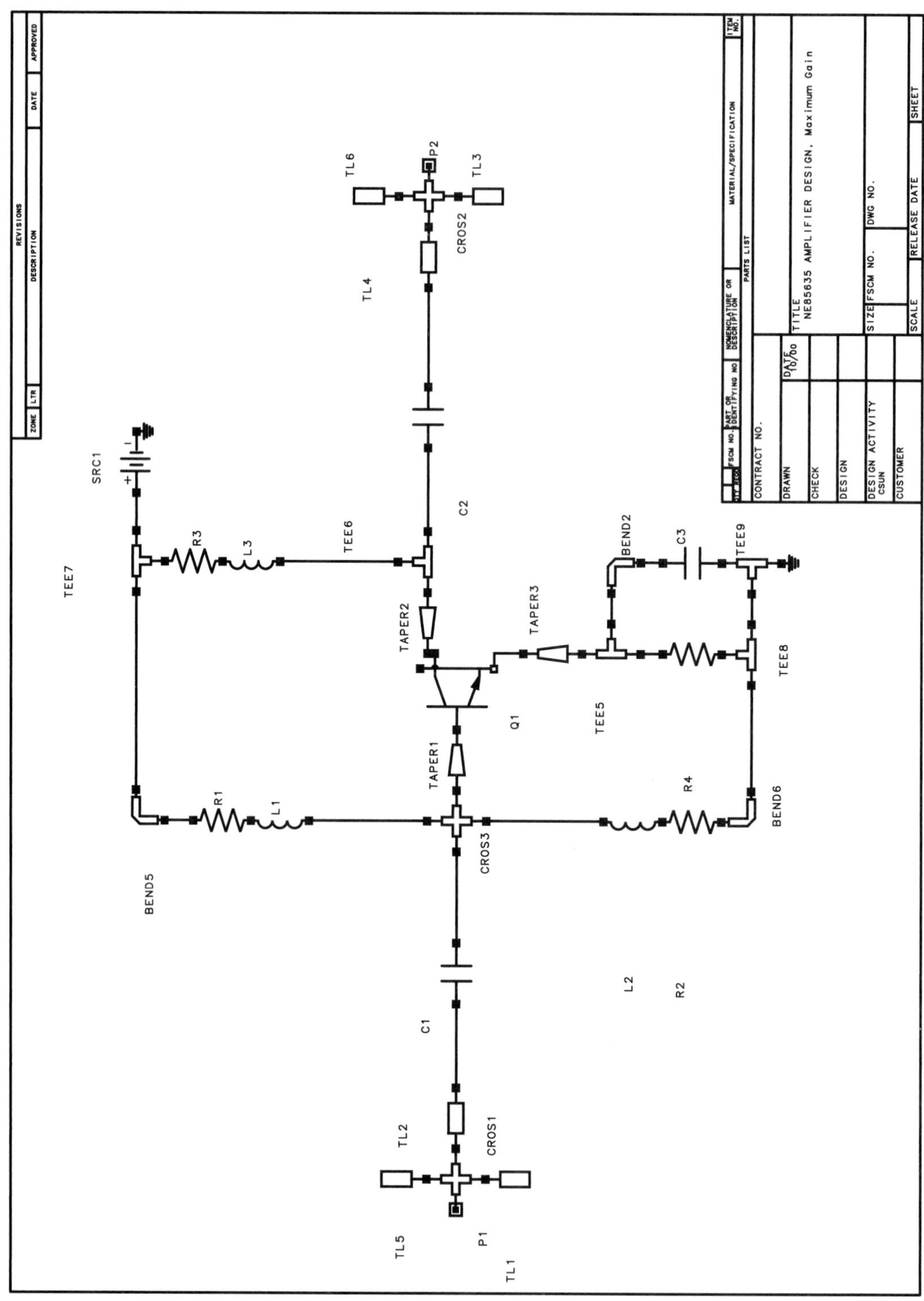

FIGURE L.8 Example 1A—Schematic Capture.

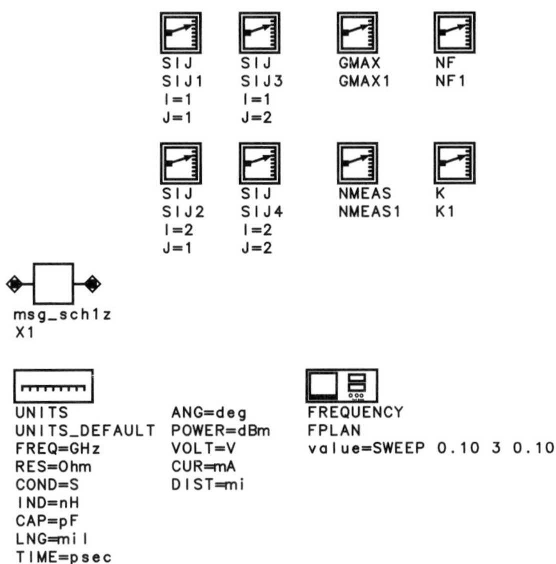

FIGURE L.9 Example 1A—Test Bench Window.

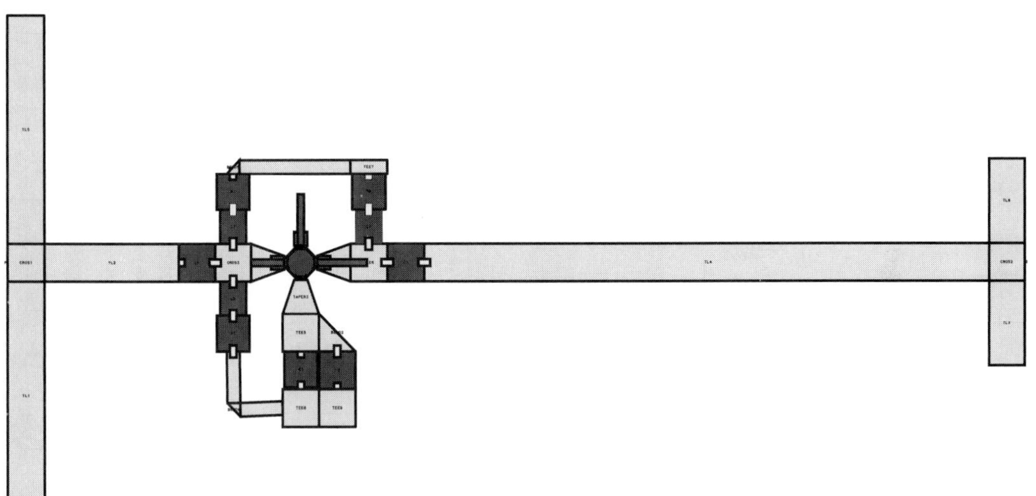

FIGURE L.10 Example 1A—Layout.

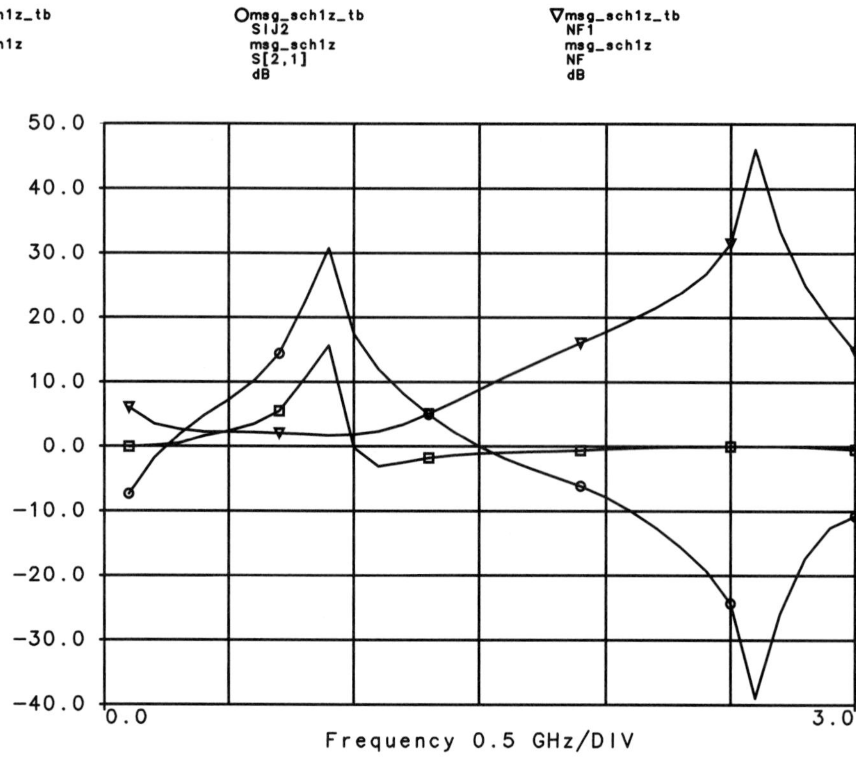

FIGURE L.11 Example 1A—S_{11}, S_{21} and NF Performance.

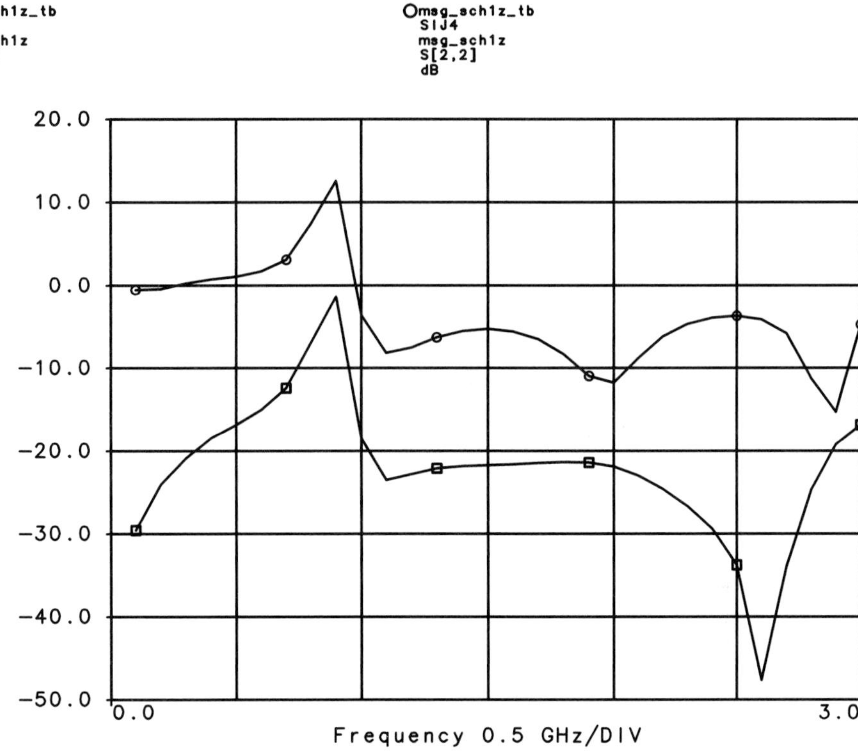

FIGURE L.12 Example 1A—S_{12} and S_{22} Performance.

TABLE L.3 Package Library Distributed Element Design Performance Measurements

Freq GHz	SIJ1 S[1,1] dB	SIJ2 S[2,1] dB	GMAX1 GMAX dB	NF1 NF dB	SIJ3 S[1,2] dB	SIJ4 S[2,2] dB
0.10	−0.0915	−7.3949	11.0816	6.0897	−29.5581	−0.5634
0.20	0.1409	−1.7544	11.1462	3.5094	−24.0468	−0.4637
0.30	0.5075	1.8307	11.3408	2.6621	−20.8508	0.2550
0.40	1.6891	4.8282	11.6401	2.2867	−18.4520	0.7385
0.50	2.4280	7.2709	12.0598	2.2783	−16.8487	1.0781
0.60	3.5213	10.2322	12.6317	2.2222	−15.0313	1.6831
0.70	5.4981	14.4281	13.4107	2.0878	−12.3932	3.0884
0.80	10.4393	22.1237	14.4920	1.9016	−6.8602	7.5062
0.90	15.6580	30.7260	16.0299	1.7560	−1.3337	12.5545
1.00	−0.2285	17.4274	17.9428	1.8333	−18.4582	−3.6191
1.10	−3.1121	11.9943	16.8090	2.3655	−23.5111	−8.1587
1.20	−2.4559	8.1246	12.5333	3.4783	−22.7805	−7.5089
1.30	−1.7792	4.9454	10.6434	5.0713	−22.1010	−6.2716
1.40	−1.3598	2.2617	9.3056	6.9201	−21.8299	−5.5213
1.50	−1.1100	0.0176	8.2475	8.8387	−21.7102	−5.2975
1.60	−0.9528	−1.8345	7.3622	10.7282	−21.5940	−5.5934
1.70	−0.8370	−3.3737	6.5944	12.5540	−21.4449	−6.5145
1.80	−0.7239	−4.7361	5.9102	14.3178	−21.3273	−8.3045
1.90	−0.5847	−6.1338	5.2857	16.0421	−21.4062	−10.9652
2.00	−0.4175	−7.8186	4.7006	17.7660	−21.9013	−11.7337
2.10	−0.2569	−9.9582	4.1334	19.5493	−22.9566	−8.8180
2.20	−0.1399	−12.5537	3.5522	21.4879	−24.5564	−6.2069
2.30	−0.0714	−15.5626	2.8923	23.7551	−26.6448	−4.6744
2.40	−0.0361	−19.1580	1.9743	26.7240	−29.3849	−3.9145
2.50	−0.0186	−24.2966	0.0527	31.4864	−33.7251	−3.7314
2.60	−0.0115	−38.9522	−11.2205	46.0948	−47.6330	−4.1727
2.70	−0.0206	−25.9907	−2.0667	33.4115	−33.9690	−5.7822
2.80	−0.0878	−17.3025	0.5056	25.0341	−24.6191	−11.2769
2.90	−0.2884	−12.5319	0.9582	19.4994	−19.2244	−15.2994
3.00	−0.5056	−10.7558	0.9725	14.7406	−16.8582	−4.7290

EXAMPLE 2—A FET MICROWAVE AMPLIFIER DESIGN: MNA DESIGN

Design Problem

Design a field effect transistor (FET) amplifier for minimum noise at 3 GHz. The FET *S*-parameters are shown in Table L.4. Open stub and quarter-wavelength transformers are to be used for the matching networks. The *S*-parameters for the device are at 3.0 GHz over a 100 MHz bandwidth. The low noise bias point is at $V_{DS} = 5$ V and $I_D = 10$ mA; with $Z_o = 50\ \Omega$, $F_{min} = 3$ dB, $\Gamma_{opt} = 0.5\angle{-135^\circ}$, and $R_n = 4$. Assume FET has a threshold voltage of $V_t = 2$ V.

TABLE L.4 *S*-Parameters for the Microwave Transistor

F (GHz)	$\lvert S_{11}\rvert$	∠ S11°	$\lvert S_{21}\rvert$	∠ S21°	$\lvert S_{12}\rvert$	∠ S12°	$\lvert S_{22}\rvert$	∠ S22°
3.0	0.900	−60.0	2.000	90.00	0.000	00.00	0.500	−45.00

Design Procedure

The requirements for this project were to design a minimum noise amplifier (MNA) using the FET characterized in Table L.4 at 3 GHz. From this starting point, the design procedure used for this activity began with defining the biasing network for the transistor. Next, the stability conditions (K and Δ) for the active device were calculated from the manufacturer's *S*-parameter information. From the stability calculations, the approach for choosing the matching network design was determined.

Matching networks are not limited to a single configuration of elements. Designs can be developed and simulated for lumped element implementation as well as distributed element implementation using transmission lines. Both types of design can be developed and simulated using EEsof.

(A) DC Biasing. We use a common-source configuration with $R_S = 0$ (see Chapter 4, Example 4.7). Two resistors R_{G1} and R_{G2} compose the voltage divider for the gate voltage. R_D is the current limiting resistor between the power supply and the drain terminal. The minimum noise performance was given for bias conditions of $V_{DS} = 5$ V and $I_D = 10.0$ mA.

$V_{DD} = 15$ V was chosen for the supply voltage. The FET is operated in the saturation region of the I_D - V_{DS} characteristics. The following is a list of assumptions made to calculate the bias circuitry:

$$V_{GS} \geq V_t$$

$$V_{GD} < V_t$$

$$V_{GS} = V_{DD}/3 = 5.0\text{ V} = V_{DS}$$

$$V_{GS} + V_{SD} + V_{DG} = 0 \Rightarrow V_{GS} - V_{DS} - V_{GD} = 0 \Rightarrow V_{GD} = 0 < V_t$$

$$R_D = (V_{DD} - V_D)/I_D = (15 - 5)/10 \text{ mA} = 1 \text{ K}\Omega$$

Biasing the device such that $V_G = V_D$ means that $V_{GD} \approx 0$ and $V_{GD} < V_t$ so the device will always be in saturation. From this biasing arrangement, the values of R_{G1} and R_{G2} were calculated:

$$V_G = 5.0 \text{ V}$$

$$V_G = R_{G2} / (R_{G1} + R_{G2})\, V_{DD} \Rightarrow R_{G2} / (R_{G1} + R_{G2}) = 1/3$$

$$\text{Use } R_{G1} + R_{G2} = 1.5 \text{ M}\Omega$$

$$R_{G1} = 1.0 \text{ M}\Omega \text{ and } R_{G2} = 0.5 \text{ M}\Omega$$

(B) RF Design. The first step after selecting the device is to assess the stability conditions at the desired frequency.

Following the stability determination, we will proceed with the RF circuit design procedure to determine the input and output matching networks. This design will not require a bilateral analysis because $S_{12} = 0$.

(1) STABILITY CONDITIONS The design procedure used to calculate the matching networks is determined by the stability conditions at the desired frequency. This is done by calculating values of K and Δ. The values are calculated from the following equations:

$$K = (1 - |S_{11}|^2 - |S_{22}|^2 - |\Delta|^2 / |S_{12}S_{21}| = \infty$$

$$\Delta = |S_{11}S_{22} - S_{12}S_{21}| = 0.45$$

Because $K >> 1$ and $\Delta < 1$, the device is unconditionally stable. This agrees with the graphical method. Furthermore, because $S_{12} = 0$, this is a unilateral case and thus the unilateral figure of merit will not be calculated.

(2) REFLECTION COEFFICIENT DETERMINATION The unilateral maximum gain design uses Γ_L and Γ_S chosen equal to S^*_{11} and S^*_{22}, respectively. However, for a minimum noise design $\Gamma_S = \Gamma_{OPT}$ is used to ensure a minimum noise where Γ_{OPT} is given by the device manufacturer, or in this case is stipulated in the problem. Thus,

$$\Gamma_S = \Gamma_{OPT} = 0.5\angle{-135^\circ}$$

$$\Gamma_L = \Gamma^*_{OUT} = S^*_{22} = (0.5\angle{-45^\circ})^* = 0.5\angle 45^\circ$$

(3) MISMATCH FACTORS The mismatch factor for the source will be high because Γ_{OPT} is not close to a conjugate match of S_{11}. The load will be a close match because S^*_{22} was used for the output matching network design. M_S *and* M_L are calculated as:

$$M_S = (1 - |\Gamma_S|^2)(1 - |\Gamma_{IN}|^2) / |1 - \Gamma_S \Gamma_{IN}|^2 = 0.0828 = -10.8 \text{ dB}$$

$$M_L = (1 - |\Gamma_L|^2)(1 - |\Gamma_{OUT}|^2) / |1 - \Gamma_L \Gamma_{OUT}|^2 = 0.976 = -0.10 \text{ dB}$$

(4) INPUT AND OUTPUT MATCHING NETWORKS USING DISTRIBUTED ELEMENTS A design using distributed elements was developed. The intent was to use open-circuit stubs and transmission lines with minimum lengths for ease of construction.

(5) λ/4 TRANSFORMER AND OPEN SHUNT STUB For the input matching network, $\Gamma_S = 0.5\ \angle{-135°}$ was used. From the origin "O" on the Smith chart, a series λ/4 transformer was employed effectively to arrive at $r = 0.33$ or $R = 0.33(50) = 16.5\Omega$. So $(Z_o)_{\lambda/4} = \sqrt{50(16.5)} = 28.7\,\Omega$ for a 50 Ω system. Then, we use an open shunt stub, $\ell_1 = 0.438 - 0.25 = 0.188\ \lambda$, to move to Γ_S.

For the output matching network, $\Gamma_L = 0.5\angle 45°$ was used. Starting from the source at O on the Smith chart, a λ/4 transformer was used to arrive at $r = 2.85$ or $R = 2.85(50) = 142.3\Omega$. So $Z_{OC1} = \sqrt{50(142.3)} = 80.62\,\Omega$ for a 50 Ω load. Then we use an open λ/8 shunt stub, $\ell_4 = 0.125$ (where $Z_{\lambda/8} = 1/B = 1/.0072 = 138.8\ \Omega$) to move to Γ_L.

(6) MICROSTRIP DIMENSIONS The microstrip elements were designed using FR4 material (FR4 is the epoxy glass laminated dielectric which is manufactured by Isola/ADI) with a dielectric thickness of 0.062 inches (62 mils) and $\varepsilon_r = 4.81$, clad in one ounce copper. Specific data were entered in the EEsof program for the microstrip elements through the MSUB Data Default information block.

Next, the dimensions of the individual microstrip elements were calculated. A value of $Z_o = 50\ \Omega$ was selected for the characteristic impedance of the transmission lines that will be used in the design. The ratio of W/h is calculated as per Example 1.

$$\varepsilon_{ff} = [(\varepsilon_r + 1)/2] + [(\varepsilon_r - 1)/2]\,[(1 + 12\,h/W)^{-1/2}] = 3.638$$

$$\lambda/\lambda_o = 1/\sqrt{\varepsilon_{ff}} = 0.525$$

Once the wavelengths for the microstrip transmission lines are established, the values for the lengths of the microstrip sections used for the matching networks are calculated.

$$\ell_1 = 0.159\ \lambda$$

$$L_1 = 0.159\ (c/\mathrm{f})\ (\lambda/\lambda_0)$$

$$L_1 = 0.159\ (3 \times 10^{10}\ \mathrm{cm/sec})\ (393.7\ \mathrm{mil/cm})\ (0.525)\ /\ (10^9\ \mathrm{cycles/sec}) = 985.9\ \mathrm{mil}$$

$$\ell_2 = 0.066\ \lambda$$

$$L_2 = 409.3\ \mathrm{mil}$$

$$\ell_3 = 0.325\ \lambda$$

$$L_3 = 2015.3\ \mathrm{mil}$$

$$\ell_4 = 0.024\ \lambda$$

$$L_4 = 148.8\ \text{mil}$$

(7) RF/Biasing Circuitry Isolation The final consideration for the amplifier design was isolation of the bias circuitry from the RF signal. In the complete schematic, RF chokes are included to isolate the bias power supply from the high frequency RF. These components are designated RFC in the schematic. The inductance was chosen for X_{RFC} much greater than 50 Ω, which is the characteristic impedance of the transmission lines in the final design.

$$jX_{RFC} = j\omega L_{RFC} \Rightarrow L_{RFC} = X_{RFC}/\omega,$$

$$\text{For } X_{RFC} = 5\text{K}\Omega \Rightarrow L_{RFC} = 265\ \text{nH}$$

DC blocking capacitors were added to isolate the bias current from the source and the load. The capacitance was chosen for X_{RFC} much less than 50Ω, which is the characteristic impedance of the transmission lines in the final design.

$$-jX_{DC\text{-}BLK} = 1/(j\omega C_{DC\text{-}BLK}) \Rightarrow C_{DC\text{-}BLK} = 1/\omega X_{DC\text{-}BLK},$$

$$\text{For } X_{DC\text{-}BLK} = 1\Omega \Rightarrow C_{DC\text{-}BLK} = 53\ \text{pF}$$

TABLE L.5 Example 2—Distributed Element Design Performance Measurements

Freq GHz	SIJ1 S[1,1] dB	NF1 NF dB	SIJ2 S[2,1] dB	SIJ3 S[1,2] dB	SIJ4 S[2,2] dB
2.950000000	−0.3479	3.1501	1.4610	−235.6574	−5.3309
2.960000000	−0.3454	3.1053	1.3936	−235.7372	−5.2450
2.970000000	−0.3430	3.0710	1.3269	−235.7725	−5.1608
2.980000000	−0.3407	3.0471	1.2611	−235.8706	−5.0782
2.990000000	−0.3385	3.0336	1.1962	−235.9281	−4.9973
3.000000000	−0.3365	3.0301	1.1321	−235.9737	−4.9181
3.010000000	−0.3345	3.0363	1.0688	−236.0707	−4.8404
3.020000000	−0.3326	3.0519	1.0064	−236.0877	−4.7643
3.030000000	−0.3308	3.0763	0.9448	−236.1601	−4.6898
3.040000000	−0.3292	3.1091	0.8840	−236.2233	−4.6168
3.050000000	−0.3276	3.1497	0.8241	−236.2775	−4.5454

Similar to the Examples 1 and 1A, we obtained the results as shown in Table L.2 and Figures L.13 through L.17.

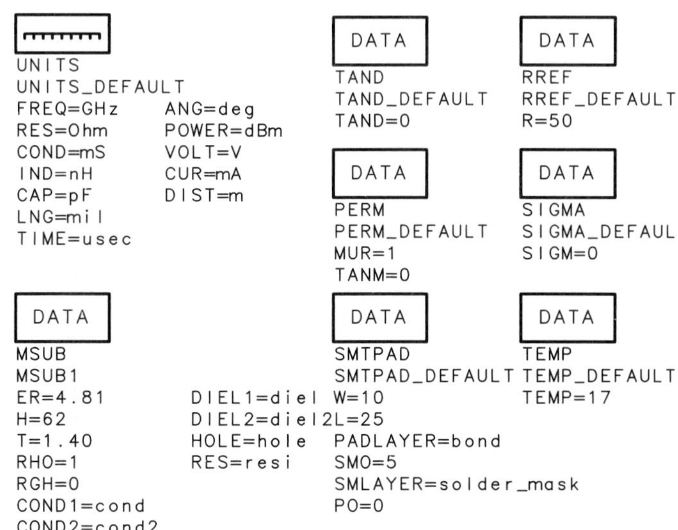

FIGURE L.13 Example 2—Default Window.

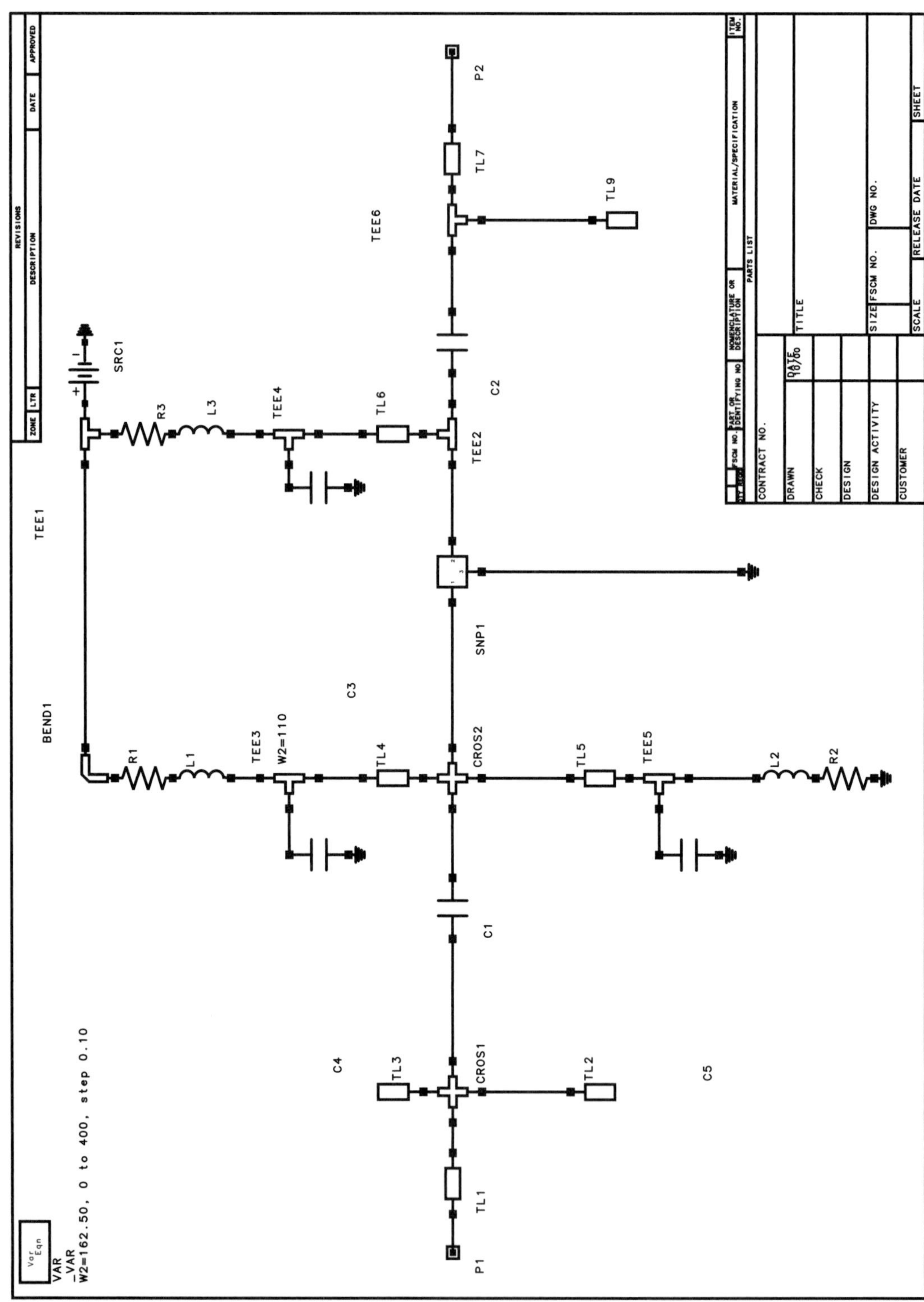

FIGURE L.14 Example 2—Schematic Capture.

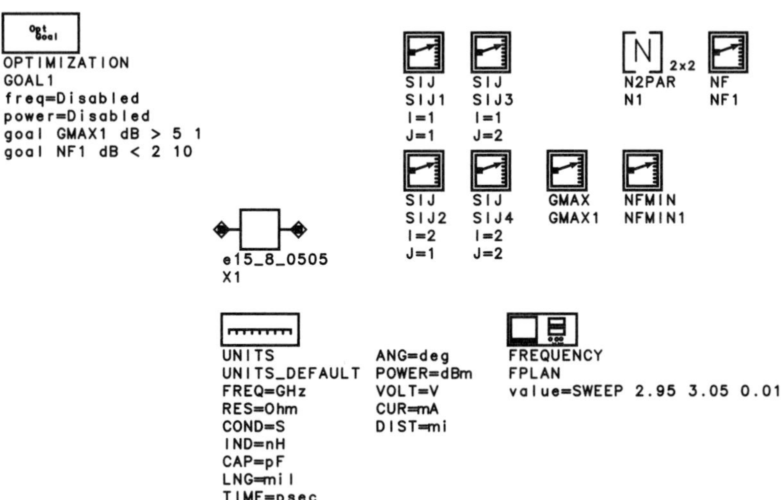

FIGURE L.15 Example 2—Test Bench Window.

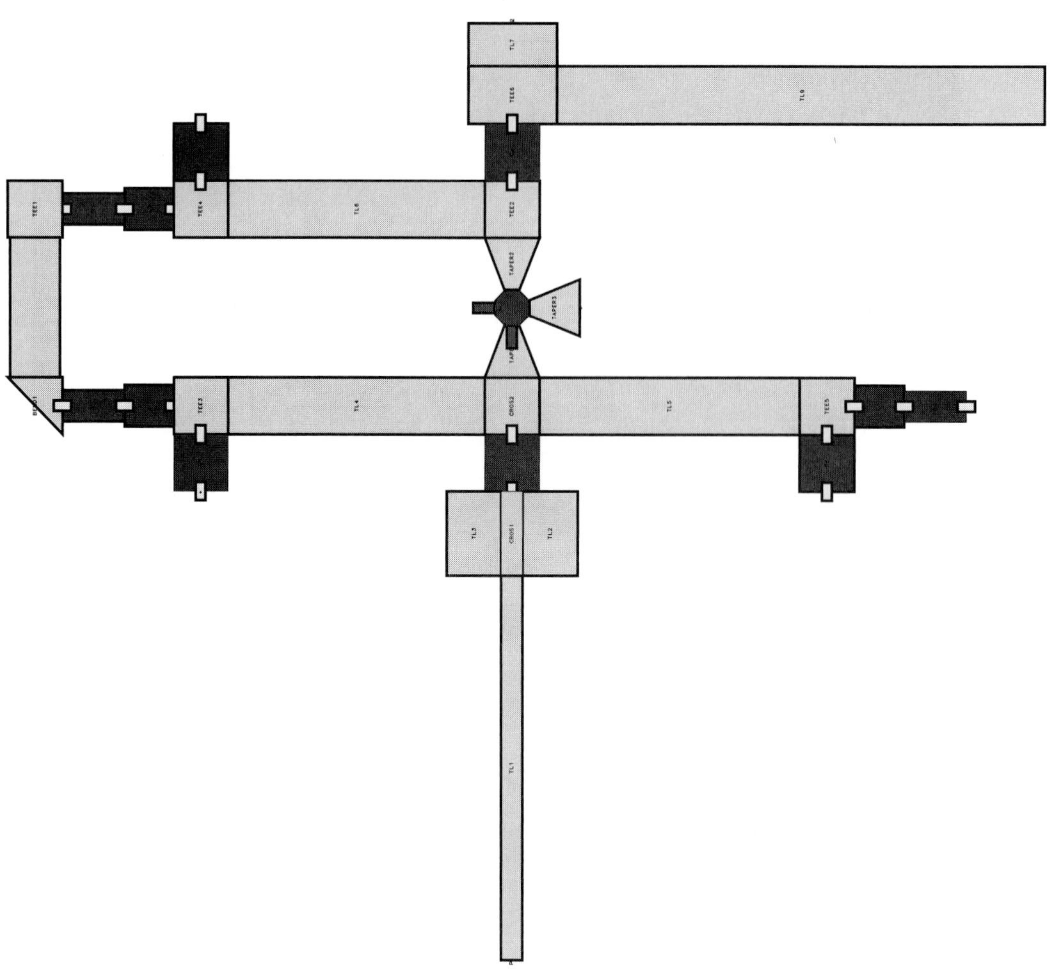

FIGURE L.16 Example 2—Layout.

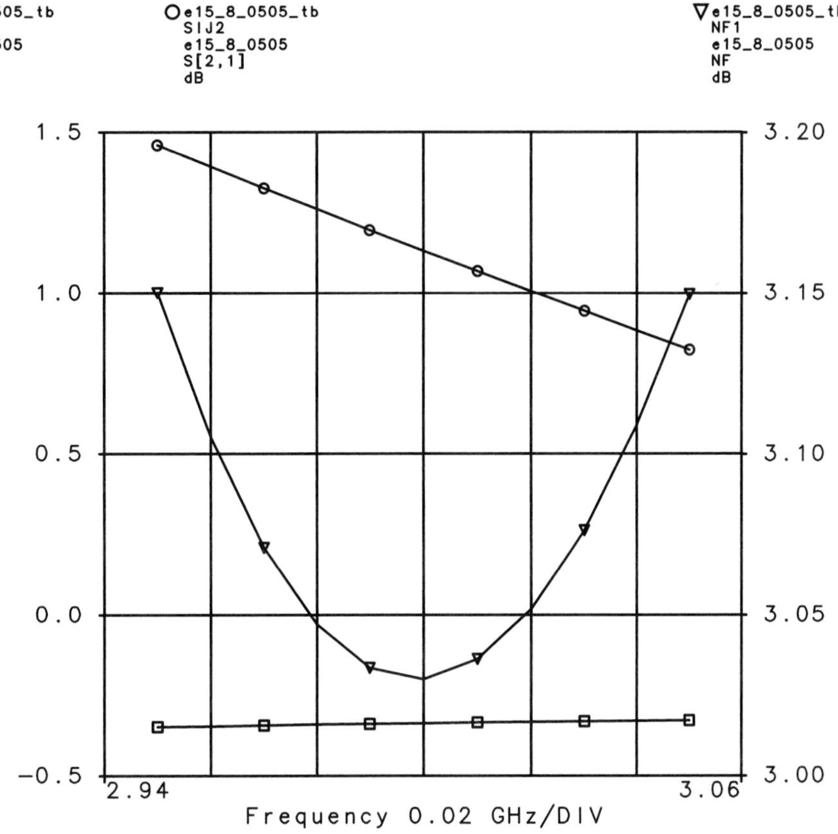

FIGURE L.17 Example 2—S_{11}, S_{21} and NF Performance.

EXAMPLE 3—MICROWAVE MIXER DESIGN

Design Problem

Design a microwave diode mixer circuit to convert a 4 GHz RF signal into a 500 MHz IF signal. The diode characteristics are given by:

$$f_{co} = 50 \text{ GHz}$$

$$I_s = 1 \times 10^{-9} \text{ A}$$

$$R_s = 15 \ \Omega \text{ Max}$$

$$n = 1.05$$

$$C_j = 0.5 \text{ pF}$$

$$L_p = 0.2 \text{ nH}$$

$$C_p = 0.35 \text{ pF}$$

Use series transmission line to resonate the diode's reactance and a $\lambda/4$ transformer.

Design Procedure

The design procedure used for this activity begins with defining the diode ON and OFF impedances.

A) DC Biasing. Resistor R_B provides current limiting for the diode bias circuitry (see Chapter 19, Figure 19.3). $V_{DC} = 15$ V was chosen for the supply voltage. The main concern is to ensure that the diode voltage is high enough so that the diode is turned ON. The following is a list of assumptions made to calculate the bias circuitry:

$$I_o = 60\ \mu\text{A}$$

$$V_{CC} = 15\ \text{V} \quad R_B = V_{DC}/I_o = 15 / 60\ \mu\text{A} = 250\ \text{K}\Omega$$

B) Mixer RF Design. The RF circuit is designed by using distributed elements. The first step after selecting the device is to determine the impedance for the OFF and ON states. This requires defining the values of the elements in the model.

Assuming an abrupt PN junction diode, we can write:

$$\text{Diode OFF, } I_o = 0\ \text{A and } V_o = 0\ \text{V} \Rightarrow C_j = C_{jo}$$

$$\text{Diode ON, } I_o \neq 0\ \text{A and } V_o \approx 0.7\ \text{V} \Rightarrow C_j = \infty$$

$$R_j = nV_T / (I_o + I_S)$$

$$\text{Diode OFF, } I_o = 0\ \text{A and } R_j = 1.05(25 \times 10^{-3})/10^{-9} = 26.25\ \text{M}\Omega$$

Diode ON, $I_o \neq 0$ A and $C_j = \infty$ creates a short circuit across the resistance (R_j)

$$f_{co} \geq 10\ \text{f}_o \Rightarrow \text{f}_{co} \geq 40\ \text{GHz}$$

$$\Rightarrow f_{co} = 1/2\pi R_S C_{jo} = 40\ \text{GHz} \Rightarrow C_{jo} = 1/2\pi(15)(40 \times 10^9) = 0.265\ \text{pF}$$

$$jX_{Cjo} = 1/j2\pi\omega C_{jo} = -j/2\pi(4 \times 10^9)(0.265 \times 10^{-12}) = -j150\,\Omega$$

$$jX_{LP} = j2\pi\omega L_P = j2\pi(4 \times 10^9)(0.2 \times 10^{-9}) = j5\,\Omega$$

$$jX_{CP} = 1/j2\pi\omega C_P = j/2\pi(4 \times 10^9)(0.35 \times 10^{-12}) = -j113.7\Omega$$

$$(Z_d)_{OFF} = -j133.7 \,||\, (5 - j145) = 2.87 - j63.7\,\Omega$$

$$(Z_d)_{ON} = -j113.7 \,||\, (15 + j5) = 16.1 - j3\,\Omega$$

$$Q_d = \{[(R_1 - R_2)^2 + (X_1 - X_2)^2]/(R_1 R_2)\}^{1/2}$$

$$= \{[(16.1 - 2.87)^2 + (3 - (-63.7))^2]/16.1 \times 2.87\}^{1/2} = 15.6$$

$$R_m = [(R_1 R_2)[1 + (X_1 - X_2)^2]/(R_1 + R_2)^2]^{1/2}$$

$$\Rightarrow R_m = \{ (16.1)(2.87) [1 + (3 - (-63.7)]^2/(16.1 + 2.87)^2\}^{1/2} = 105.7\ \Omega$$

$$X_m = X_1 + R_1 [(X_2 - X_1)/(R_2 + R_1)] = 3 + 16.1 (-63.7 - 3)/(2.87+ 16.1) = -53.52\Omega$$

Thus, we have:

$$Z_m = 105.7 - j53.52\ \Omega \Rightarrow (Z_m)_N = 2.1 - j1.07$$

INPUT MATCHING NETWORK USING DISTRIBUTED ELEMENTS The matching network design was defined using distributed elements. The problem calls for the use of a series transmission line to resonate the diode's reactance and a λ/4 transformer.

For the input matching network, the diode impedance of $Z_m = 105.7 - j53.52\ \Omega$ was matched with a 50 Ω source. Starting from the source at O on the Smith chart and moving to the normalized impedance of $(Z_m)_N = 2.1 - j1.07\ \Omega$, we use a λ/4 transformer then a series 50 Ω TL. For the λ/4 transformer, we move from the origin "O" to $r = 4.8$. or $R = 4.8(50) = 240\Omega$. Therefore, for the λ/4 transformer we have: $(Z_o)_{\lambda/4} = \sqrt{50(240)} = 109.5\ \Omega$. Now we move to $(Z_m)_N$ on a constant *VSWR* circle, to obtain a series TL having a length of: $\ell_1 = 0.36 - 0.25 = 0.11\lambda$.

Figure L.18 shows the library design in the EEsof Default Values window. The data elements in this window are used to define the units used in the program. EEsof schematic window displays the circuit schematic (see Figure L.19). The EEsof Test Bench Parameters are shown in the Test Bench window (see Figure L.20).

A graph window displays the power spectrum performance plotted on a log scale. Figure L.21 shows a plot of the model diode. Figure L.22 is a plot of the power spectrum for a 1N5456 diode installed in the circuit in place of the model diode.

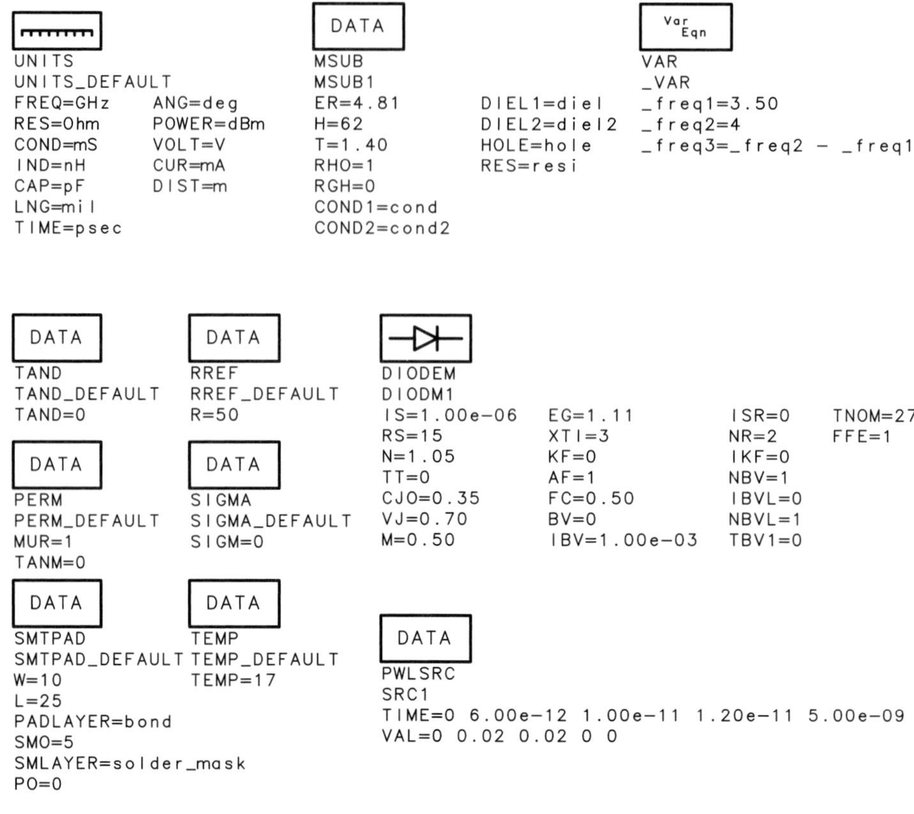

FIGURE L.18 Example 3—Default Window.

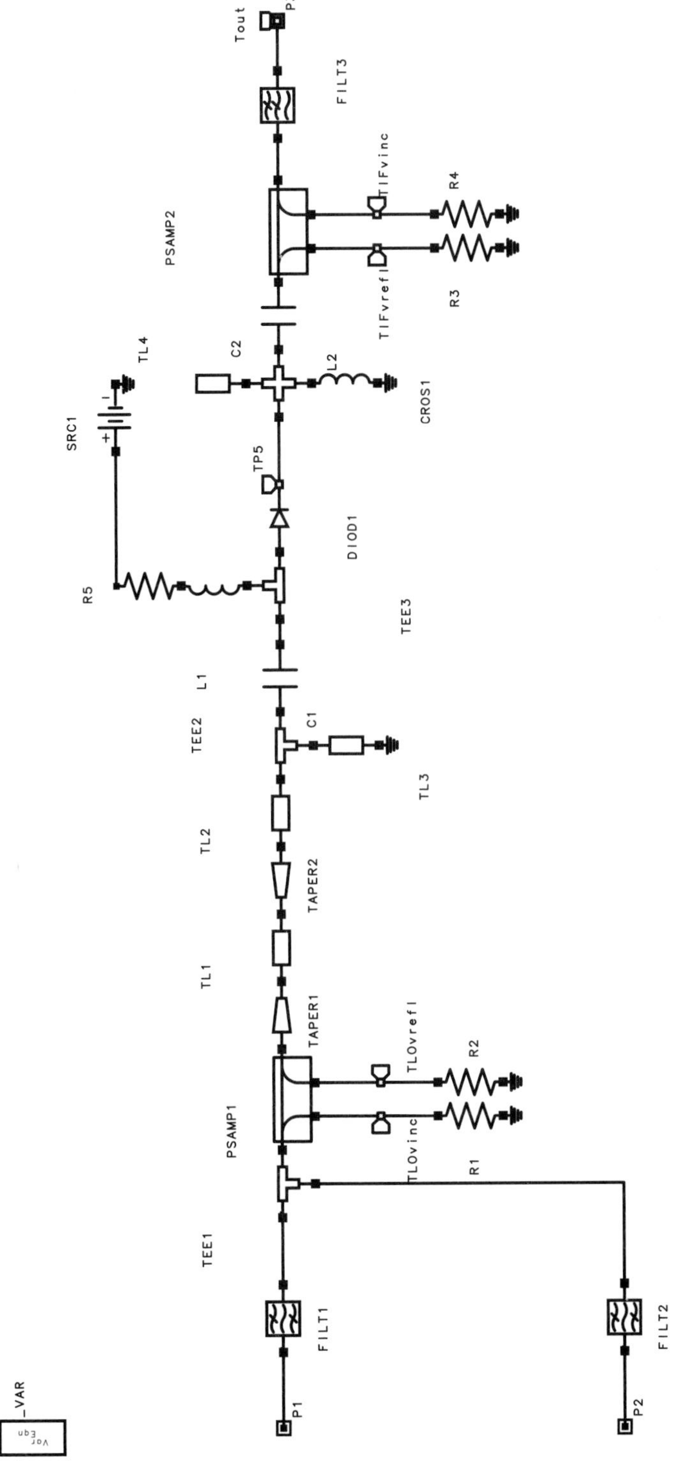

FIGURE L.19 Example 3—Schematic Capture.

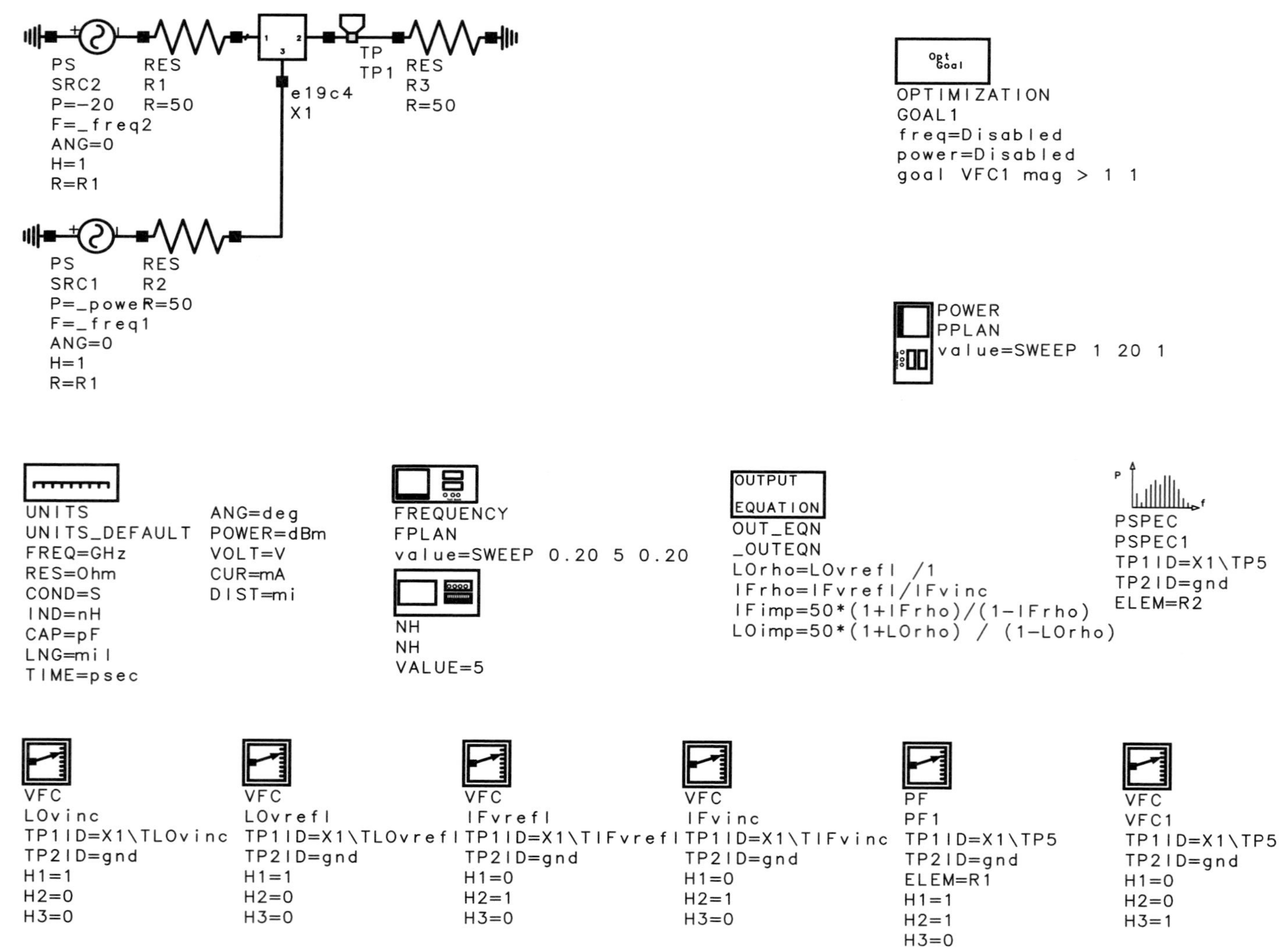

FIGURE L.20 Example 3—Test Bench Window.

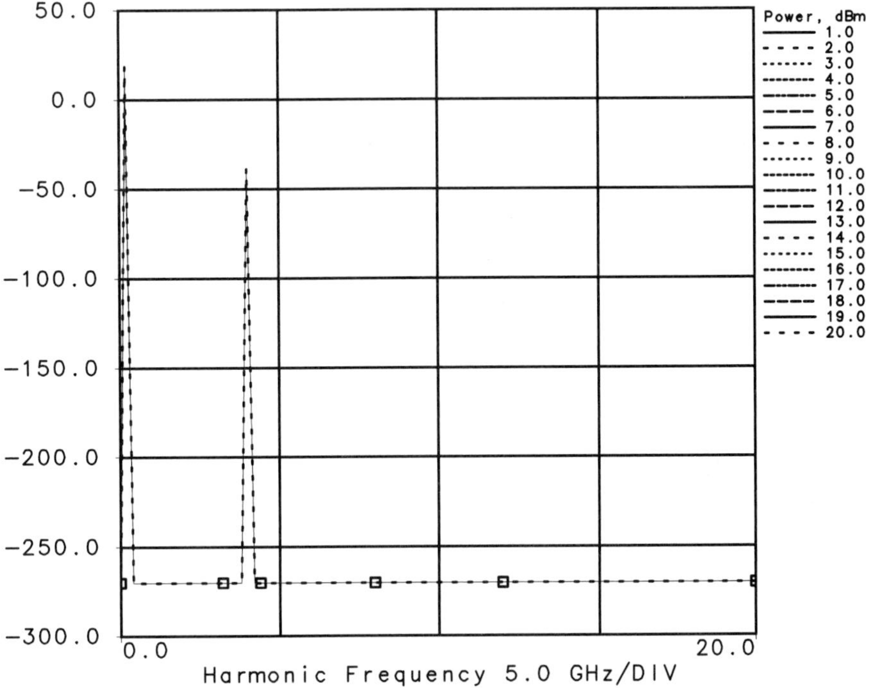

FIGURE L.21 Example 3—Frequency Spectrum, Diode Model.

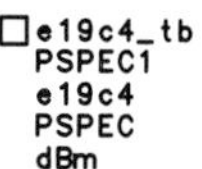

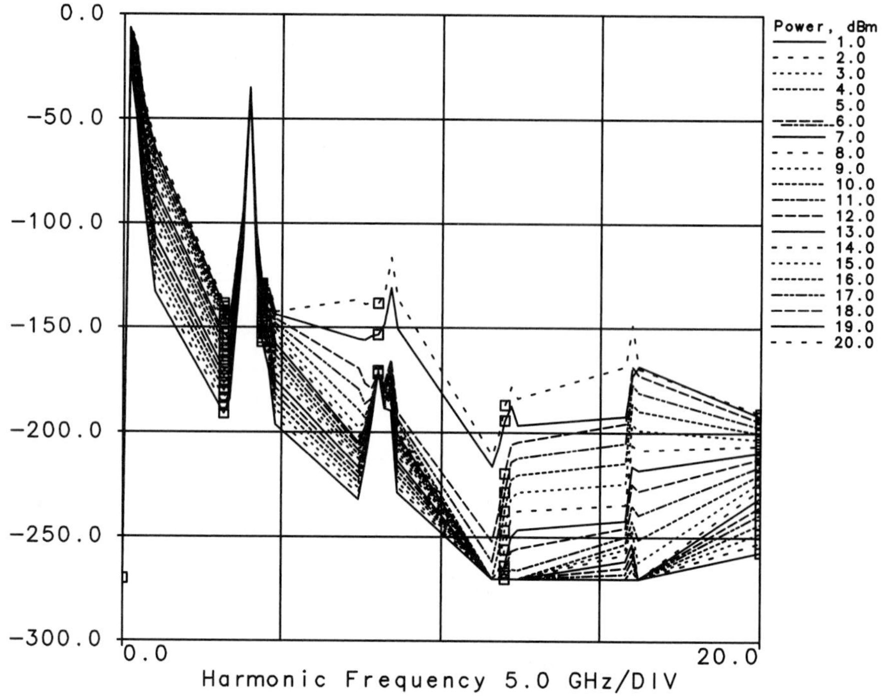

FIGURE L.22 Example 3—Frequency Spectrum, 1N5654 Diode.

APPENDIX M

Derivation of the Constant Gain and Noise Figure Circles

From Chapter 13, the unilateral gain equations are given by:

$$g_S = \frac{G_S}{G_{S,max}} = \frac{1-|\Gamma_S|^2}{|1-S_{11}\Gamma_S|^2}(1-|S_{11}|^2) \quad \text{(M.1)}$$

$$g_L = \frac{G_L}{G_{L,max}} = \frac{1-|\Gamma_L|^2}{|1-S_{22}\Gamma_L|^2}(1-|S_{22}|^2) \quad \text{(M.2)}$$

We will derive the constant gain circle equation in the Γ_S plane using Equation M.1. A similar procedure could be utilized for Equation M.2 in the Γ_L plane.

First, we will present the general equation of a circle in part (1) and then proceed to obtain the desired derivation from Equation M.1 in part (2). In the last segment (3), we show that the constant noise figure circles can also be derived using the same procedure as outlined in parts (1) and (2).

1. **General equation of a circle**

   The general equation for a circle in the Γ_S plane, with radius (R_{gS}) and a center at (C_{gS}), is given by:

$$|\Gamma_S - C_{gS}| = R_{gS} \quad \text{(M.3)}$$

   Squaring both sides of Equation M.3, we have:

$$|\Gamma_S - C_{gS}|^2 = R_{gS}^2 \quad \text{(M.4)}$$

   Equation M.4 can be rewritten as:

$$\Rightarrow (\Gamma_S - C_{gS})(\Gamma_S - C_{gS})^* = R_{gS}^2 \quad \text{(M.5)}$$

   Applying the conjugate operation and multiplying the terms in Equation M.5 yields the desired equation form for a circle as:

$$|\Gamma_S|^2 - C_{gS}^*\Gamma_S - C_{gS}\Gamma_S^* = R_{gS}^2 - |C_{gS}|^2 \quad \text{(M.6)}$$

2. **Constant gain circles**

   Multiplying the terms in Equation M.1, we have:

$$g_S 1 + |S_{11}\Gamma_S|^2 - S_{11}\Gamma_S - S_{11}^*\Gamma_S^*) = 1 - |\Gamma_S|^2 - |S_{11}|^2 + |\Gamma_S|^2|S_{11}|^2 \quad \text{(M.7)}$$

Separating the various terms, we can write Equation M.7 as:

$$(1-|S_{11}|^2+g_S|S_{11}|^2)|\Gamma_S|^2-g_S S_{11}\Gamma_S-g_S S_{11}^*\Gamma_S^*=1-g_S-|S_{11}|^2 \qquad \text{(M.8)}$$

Dividing both sides by the coefficient of $|\Gamma_S|^2$, Equation M.8 can be recast as:

$$|\Gamma_S|^2-\left(\frac{g_S S_{11}}{1-|S_{11}|^2(1-g_S)}\right)\Gamma_S-\left(\frac{g_S S_{11}^*}{1-|S_{11}|^2(1-g_S)}\right)\Gamma_S^*=\frac{1-g_S-|S_{11}|^2}{1-|S_{11}|^2(1-g_S)} \qquad \text{(M.9)}$$

Comparing Equations M.9 to M.6, we obtain:

$$C_{gS}=\frac{g_S S_{11}^*}{1-|S_{11}|^2(1-g_S)} \qquad \text{(M.10)}$$

$$R_{gS}^2-|C_{gS}|^2=\frac{1-g_S-|S_{11}|^2}{1-|S_{11}|^2(1-g_S)} \qquad \text{(M.11)}$$

Substituting for C_{gS} from Equation M.10 in M.11, we obtain R_{gS} as:

$$R_{gS}=\frac{\sqrt{1-g_S}\,(1-|S_{11}|^2)}{1-|S_{11}|^2(1-g_S)} \qquad \text{(M.12)}$$

Equations M.10 and M.12 provide the equations for C_{gS} and R_{gS} exactly as given in Chapter 13.

3. **Constant noise figure circles**

The preceding procedure can also be used successfully to derive the constant noise figure family of circles' center and radius (C_F, R_F). From Chapter 14 we have:

$$N=\frac{|\Gamma_S-\Gamma_{opt}|^2}{1-|\Gamma_S|^2} \qquad \text{(M.13)}$$

Upon multiplication and rearrangement of terms, we can rewrite Equation M.13 in terms of the variable Γ_S as:

$$|\Gamma_S|^2-\frac{(\Gamma_S\Gamma_{opt}^*+\Gamma_S^*\Gamma_{opt})}{1+N}=\frac{N}{1+N}-\frac{|\Gamma_{opt}|^2}{1+N} \qquad \text{(M.14)}$$

Comparing Equation M.14 to M.6, C_F and R_F are found to be:

$$C_F=\frac{\Gamma_{opt}}{1+N} \qquad \text{(M.15)}$$

$$R_F=\frac{\sqrt{N^2+N(1-|\Gamma_{opt}|^2)}}{1+N} \qquad \text{(M.16)}$$

APPENDIX N

About the Software...

A CD containing software in the form of an electronic book (E-book), which contains all numerical examples from the text, is bound into the back of each textbook. The solutions are programmed using Visual Basic software, which is built into the "Microsoft Excel®" application software.

I MAIN FEATURES

The main features of this CD are as follows:

1. It is a powerful interactive tool for learning the textbook content and also for solving the numerical problems.
2. The software includes 90 solved problems based on the numerical examples in the book.
3. The big advantage of the interactive software tool is its use of live math. Every number and formula is interactive. The reader can change the starting parameters of a problem and watch as the final results change before his or her eyes. This feature allows the reader to experiment with every number, formula, etc.
4. Each solved problem becomes a worksheet that the reader can modify to solve dozens of related problems.
5. The electronic book takes advantage of the powerful Microsoft Excel® environment to perform many tedious and complicated RF and microwave design calculations (usually using complex numbers), allowing the student to focus on the essential concepts.
6. This is an excellent tool for students, engineers, and educators, to:
   a. Understand the fundamentals and practical concepts of RF and Microwaves and,
   b. Encourage applications and new RF/Microwave circuit designs using the concepts presented in the book.

II HOW TO START THE PROGRAM

The following steps need to be carried out before the software is ready to use:

1. Either one of the following two methods may be used to utilize the contents of the CD-ROM:
   a. Read all the files directly from the CD-ROM or,
   b. Create a folder called "E-book" in the C: Drive. Copy the entire content of the CD-ROM into the folder entitled "E-book."
2. Open Microsoft Excel 2000 software (or Excel 97 with SR-1 or SR-2 revision) and open the "E-book" folder.
3. You may begin the program by double clicking on the "Table of Contents" file.

   NOTE: *You may double click on the "Table of Contents" file directly as a shortcut without opening the Excel software.*

4. Once the "table of contents" file opens up, you need to open up the "about the CD" file by clicking on it.
5. Carefully read all the information in the "about the CD" file and close it by either
   a. holding the ALT + TAB keys and choosing EXCEL or,
   b. clicking "return to table of contents" ARROW to return to the E-book table of contents.
6. Turn on both *Analysis Toolpack* and *Analysis Toolpack—VBA* as discussed below.
7. Proceed to the desired example by clicking on it and selecting "Enable Macros."

III HOW TO USE THE E-BOOK SOFTWARE

Before proceeding to the worked-out examples, we need to select from the toolbar menu "Tools," "Add-ins," and, from the dialog box, select both "Analysis Toolpack" and "Analysis Toolpack—VBA" in order to set up the software properly.

When a particular example is selected and clicked for interactive use, the user will encounter a dialogue box where "Enable Macros" must be selected. When the example is opened, the user will observe that each numerical example consists of several sections, which can be briefly summarized as:

1. **Problem statement:** Word for word text taken from the book that describes the nature of the problem.
2. **Input data:** This section provides all the manipulatable data, which the reader may have at his or her own disposal to vary interactively and experiment with, in order to examine different scenarios and obtain answers to "what if" questions. *Inside the input data box, the user may change the values of the parameters only, and not any of the units. The user should type the new value in the appropriate box and press enter/return to observe the desired change. This is the only place where the user is allowed to make any changes to the software.*

3. **Output data:** This section contains a step-by-step solution of the problem as well as easy explanations provided for quick assimilation of the results. Most of the complicated calculations are done in complex numbers using the Visual Basic programming technique, which is part of the Microsoft Excel® software.
4. **Problem format and color codes:** All problems are formatted and color-coded in the same manner throughout the software. This is done for the user's easy recognition and reference, and is delineated as follows:

Example xx.xx	Cyan
Problem text	Tan
Solution:	Navy Blue
Input data (interactive part)	
Heading	Red
Content	Turquoise/Brown
Output data	
Heading	Sky Blue
Content	Yellow
Interactive Answers	Green
Caution	Red
Note/Conclusion	Pink
Reference	Violet

IV SOFTWARE KNOWLEDGE REQUIRED

A rudimentary knowledge of Microsoft Excel® is required to operate the software successfully. The user does *not* need to know Visual Basic programming techniques to work with the examples' solutions interactively.

V MINIMUM SOFTWARE/HARDWARE REQUIREMENTS

The user needs to have the following:

1. **Hardware requirements:** A personal computer (PC) with a Pentium chip, preferably.
2. **Software requirements:** Windows 95/98/NT operating system and Microsoft Excel 2000 (or Excel 97 with: SR-1 or SR-2 revisions).

   NOTE 1: *To obtain the Service Release 1 or 2 (SR-1 or SR-2), the user should download the required software from the following Website:*

   http://www.microsoft.com/

   Go to the search link and find the SR-1 or SR-2 revision, which is suitable to the version of Microsoft Office that you own. Download and install the SR-1 or SR-2 upgrade to repair all known bugs in Microsoft Excel 97. Without this correction, Excel 97 gives incomplete values for the worked-out examples in the textbook CD.

NOTE 2: *If you own Microsoft Excel 2000, please ignore "Note 1." Install the textbook CD directly without any changes to the Microsoft Excel software using the procedure outlined in the previous section. If you experience any problems, you need to download and install the SR-1A revision for Microsoft Excel 2000 from the site mentioned previously.*

VI TROUBLESHOOTING PROBLEMS

If the following problems occur, you may correct them as follows:

1. If "######" appears in place of a numerical answer, it means that the cell is too small and you have to resize that cell in order to display the final numerical result correctly. To resize the cell, go to the Excel toolbar menu and select **Format**, **Column**, and **Autofit Selection**.
2. If "#VALUE!" appears, it means that any of the following conditions may have occurred:
   a. Divide by zero.
   b. Negative number under a square root.
   c. The number is out of range.
   d. Excel 97 software is not used with SR-1 or SR-2 revision.
   e. The *Analysis Toolpack* and *Analysis Toolpack—VBA* are not turned on.*

* You need to select from the toolbar menu "Tools," "Add-ins," and from the dialog box select *Analysis Toolpack* and *Analysis Toolpack—VBA* in order to set up the software properly.

Glossary of Technical Terms

The following glossary of terms is provided for easy reference and supplements the presented materials in the text. This brief summary of technical terms does not supersede or replace the use of a technical dictionary, which is highly recommended and heavily encouraged for deeper understanding of electronics.

Absolute temperature The temperature scale measured relative to absolute zero which, in Celsius degrees, is at –273.15 °C (giving rise to the Kelvin scale), or in Fahrenheit degrees, is at –459.67 °F (giving rise to the Rankine scale). *See also* **temperature**.

AC load line A straight line drawn over a series of current-voltage characteristic curves on a graph to show current changes with voltage for a specified load resistance under AC conditions. Slope of the line depends on the AC load resistance. *See also* **load line**.

Active device An electronic component such as a transistor, that can be used to produce amplification (or gain) in a circuit.

Active mode (in a BJT) In this mode the emitter-base junction (EBJ) is forward biased and the collector-base junction (CBJ) is reverse biased.

Admittance The measure of ease of AC current flow in a circuit, the reciprocal of impedance expressed in Siemens (symbol Y or y).

Ampere (A) The unit of electric current defined as the flow of one coulomb of charge per second. Alternately, it can also be defined as the constant current that would produce a force of 2×10^{-7} Newton per meter of length in two straight parallel conductors of infinite length, of negligible cross section, placed one meter apart in a vacuum.

Ampere's Law Current (either conduction or displacement) flowing in a wire generates a magnetic flux that encircles the wire in a clockwise direction when the current is moving away from the observer. The direction of the magnetic field follows the right hand rule. (This law may also be referred to as the law of magnetic field generation).

Differential form: $\nabla \times \overline{H} = \bar{J} + \dfrac{\partial \overline{D}}{\partial t}$,

Integral form: $\oint_C \overline{H} \cdot \overline{d\ell} = I + \int_S \dfrac{\partial \overline{D}}{\partial t} \cdot \overline{dS}$

Amplifier A linear device that draws power from a DC source and produces as an output an amplified or magnified reproduction of the essential features of the input signal (such as amplitude or power), without distorting the wave shape of the signal.

Amplitude The extent to which an alternating current or pulsating current or voltage swings from zero or a mean value.

Analog Pertaining to the general class of devices or circuits in which the output varies as a continuous function of the input.

Anode The positive electrode of a device (such as a diode, etc.) toward which the electrons move during current flow.

Anodization An electrolytic process in which a protective oxide film is deposited on the surface of a metallic body, which is acting temporarily as the anode of an electrolytic cell.

Anti-parallel mixer (also called *subharmonically pumped mixer*) A configuration used for sub-harmonically pumped mixers for millimeter-wave applications. The LO signal frequency is half of the usual LO frequency, and due to non-linearity, a second harmonic of the LO will be generated that will mix with the RF signal to produce the desired output frequency.

Attenuation The decrease in amplitude of a signal during its transmission from one point to another.

Attenuation constant The real component of the propagation constant.

Attenuator A resistive network that provides reduction of the amplitude of an electrical signal without introducing appreciable phase or frequency distortion.

Automatic Gain Control (AGC) A type of circuit that holds the gain and accordingly the output of a receiver or amplifier substantially constant in spite of input-signal amplitude fluctuations. A rectifier samples the AC signal output and delivers a DC voltage proportional to that output. That DC voltage is then applied in correct polarity as bias to the early stage(s) of the receiver or amplifier to reduce the gain when the output swings beyond a predetermined level, and vice versa.

Available power gain (G_A) The ratio of the output power available from the network under matched condition to the power available from the driving source (also under matched condition).

Average power The power delivered averaged over one cycle.

Axiom A self-evident truth, accepted without proof.

Balanced amplifier An amplifier circuit with two identical signal branches, connected so as to operate with the inputs in phase opposition and with the output connections in phase. Each branch is balanced to ground (i.e., each has the same input impedance with respect to ground).

Balanced mixer The input power is divided between two or more devices providing more power handling capability, and better rejection of AM noise and spurious outputs form the LO port.

Bandwidth (B) The difference between upper (f_2) and lower (f_1) frequencies at which the amplitude response is 3 dB below the passband response value (also called the *half-power bandwidth*).

Base The region between the emitter and collector of a transistor that receives minority carriers injected from the emitter.

Bel A dimensionless unit expressing the ratio of two powers, such that the number of bels is the common logarithm of this ratio (named after Alexander Graham Bell). One bel is equivalent to a power ratio of +10.

Beta (β) The current gain of a transistor connected as a common emitter (grounded emitter) amplifier. It is the ratio of the change in collector current to a change in base current: $\beta = dI_C/dI_B$.

Bias The steady and constant current or voltage applied to an electrical device to establish an operating point for proper operation of the device.

Bias current A steady and constant current supplied to the electrode of a semiconductor device to preset the operating point for optimal performance.

Bias voltage A steady voltage, usually DC, that presets the operating point of a circuit or device such as a transistor.

Bidirectional Responsive in both directions.

Bilateral Having a voltage-current characteristic curve that is symmetrical with respect to the origin. If a positive voltage produces a positive current magnitude, then an equal negative voltage produces a negative current of the same magnitude.

Bipolar Junction Transistor (BJT) An active semiconductor device (made of Silicon, Germanium, gallium arsenide, etc.) having three terminals. Conduction is by electrons and holes. Current through the collector and emitter can be controlled by small changes in current at the base terminal.

Bonding A method used to produce good electrical contact between parts of a device.

Breakdown A large usually abrupt rise in electrical current in the presence of a small increase in voltage, for example between two electrodes in a confined gas. In a reverse biased semiconductor diode, this is the region of transition from a high dynamic resistance to a lower dynamic resistance that results in a significant increase in current through the device.

Breakdown voltage The voltage at which the reverse current of a reverse biased semiconductor junction suddenly rises to a high value (non-destructive if current is limited).

Break-even analysis A cost analysis done to determine the minimum number of components that must be produced and sold to offset the cost of development and production setup.

Broad-Band Amplifier (BBA) An amplifier having essentially a flat response (i.e., no resonant element) over a wide range of frequencies—one to several octaves of bandwidth.

Broadband Means that the bandwidth, in dB ($20\log_{10}B$), spans approximately 6 dB per octave or more.

Capacitance The property that permits the storage of electrically separated charges when a potential difference exists between two conductors. The capacitance of a capacitor is defined as the ratio between the electric charge that has been transferred from one electrode to the other, and the difference in potential between the electrodes.

Capacitor A device consisting essentially of two conducting surfaces separated by an insulating material (or a dielectric) such as air, paper, mica, etc., that can store electric charge.

Cathode The portion or element of a two-terminal device that is the primary source of electrons during operation.

Cavity (also called a *cavity resonator*) A metallic enclosure inside which resonant fields at microwave frequencies are excited in such a way that it becomes a source of electromagnetic oscillations.

Cell A single and basic unit for producing electricity by electrochemical or biochemical action. For example, a battery consists of a series of connected cells.

Celsius (°C) $1/100^{th}$ of the temperature difference between the freezing point of water (0°C) and the boiling point of water (100°C) on the Celsius temperature scale.

$$T(°C) = T(K) - 273.15, \quad T(°C) = \frac{5}{9}\{T(°F) - 32\}$$

Characteristic curve The family of I-V curves that define the operating characteristics of a device plotted for several parameter values. Example: the collector voltage-collector current ($V_{CE} - I_C$) characteristic curve of a transistor.

Characteristic impedance The driving-point impedance of a transmission line if it were of infinite length. This can also be defined as the ratio of the voltage to current at every point along a transmission line on which there are no standing waves. It is given in general by: $Z_o = \sqrt{(R + j\omega C)/(G + j\omega C)}$.

Charge A basic property of elementary particles of matter (electrons, protons, etc.) that is capable of creating a force field in its vicinity. The built-in force field is a result of stored electric energy. *See also* **electric charge**.

Chip A single substrate upon which all the active and passive circuit elements are fabricated using one or all of the semiconductor techniques of diffusion, passivation, masking, photoresist, epitaxial growth, etc.

Circuit The interconnection of a number of devices in one or more closed paths to perform a desired electrical or electronic function.

Circulation (of a vector field) The scalar line integral of a vector over a closed path.

Clapp oscillator A Colpitts type oscillator that uses a series-resonant tank circuit for improved stability.

Class A amplifier A mode of operation in which each transistor in the amplifier operates in its active region for the entire cycle.

Class AB amplifier A mode of operation in which an amplifier operates in class A for small signals, and in class B for large signals.

Class B amplifier In this mode of operation, each transistor is in its active region for approximately half of the signal cycle.

Class C amplifier In this mode of operation, each transistor is in its active region for significantly less than half of the signal cycle.

Closed loop circuit The output of the circuit is continuously fed back to the input for constant comparison and for control.

Coaxial transmission line (also called *coaxial cable*) A concentric transmission line in which one conductor completely surrounds the other, the two being separated by a continuous solid dielectric or by dielectric spacers. Such a line has no external field and is not susceptible to external fields.

Collector The region into which majority carriers in a transistor flow from the base under the influence of a reverse bias across the two regions.

Colpitt's oscillator A sinusoidal oscillator using a three-terminal active element (transistor) and a feedback loop containing a parallel LC circuit.

Coulomb (C) The unit of electric charge defined as the charge transported across a surface in one second by an electric current of one ampere. An electron has a charge of 1.602×10^{-19} coulomb.

Coulomb's Laws The laws that state that the force (F) of attraction or repulsion between two electric charges (or magnetic poles) is directly proportional to the product of the magnitude of charges, Q (or magnetic pole strengths, M), and inversely proportional to the square of distance (d) between them; i.e.,

$$\text{Electric: } F = \frac{Q_1 Q_2}{4\pi\varepsilon d^2},$$

$$\text{Magnetic: } F = \frac{M_1 M_2}{4\pi\mu d^2}$$

The force between unlike charges, Q_1 and Q_2 (or poles, M_1 and M_2) is an attraction, and between like charges (or poles) is a *repulsion*.

Common-base configuration A circuit configuration in which the base terminal is common to the input circuit and the output circuit, and in which the input terminal is the emitter terminal and the output terminal is the collector terminal.

Common-collector configuration A circuit configuration in which the collector terminal is common to the input circuit and the output circuit, and in which the input terminal is the base terminal and the output terminal is the emitter terminal.

Common-drain configuration A circuit configuration in which the drain terminal of a FET is common to the input circuit and the output circuit, and in which the input terminal is the gate terminal and the output terminal is the source terminal.

Common-emitter configuration A circuit configuration in which the emitter terminal is common to the input circuit and the output circuit, and in which the input terminal is the base terminal and the output terminal is the collector terminal.

Common-gate configuration A circuit configuration in which the gate terminal of a FET is common to the input circuit and the output circuit, and in which the input terminal is the source terminal and the output terminal is the drain terminal.

Common-source configuration A circuit configuration in which the source terminal of a FET is common to the input circuit and the output circuit, and in which the input terminal is the gate terminal and the output terminal is the drain terminal.

Communication principle A fundamental concept in existence (that is intertwined throughout the entire field of RF/microwaves) that states for communication to take place between two or more entities, three elements must be present: a source point, a receipt point, and a spacing or distance between the two.

Complex power Power calculated based on the impedance of a component.

Component A packaged functional unit consisting of one or more circuits made up of devices, which in turn may be part of an operating system or subsystem.

Conductivity The ratio of the current density (J) to the electric field (E) in a material. It represents the ability to conduct or transmit electricity.

Conductor a) A material that conducts electricity with ease, such as metals, electrolytes, and ionized gases. b) An individual metal wire in a cable, insulated or uninsulated.

Control circuit A circuit that directs and regulates a process or sequence of events. In this type of circuit, one signal (or process) is made to control a bigger signal (or process).

Coupler A passive device that divides a signal to feed two or more circuits, or that combines two or more signals to feed a single circuit.

Curl operation ($\nabla \times \bar{A}$) A vector whose magnitude measures the maximum net circulation per unit area and has a direction perpendicular to the area, as the area tends to zero. The cause of the curl of a vector field is a vortex source.

Current Net transfer of electrical charges across a surface per unit time, usually represented by (I). Current density (J) is current per unit area.

Current gain The ratio of the magnitude of the current in the input circuit of a device to the magnitude of the current in a specified load impedance connected to the output.

Current source An electronic circuit that generates a constant direct current in to or out of a high-impedance output node.

Cut-off The condition where the emitter base junction (EBJ) of a transistor (BJT) has zero bias or is reverse biased and the collector-base junction (CBJ) is reverse biased. Under these voltage conditions, there is no collector, emitter or base current (i.e., $I_C = I_E = I_B = 0$).

Cutoff frequency The frequency at which transmission or rejection of signals begins in a filter, amplifier, a transmission line and so on.

DC (also called *direct current*) A current which always flows in one direction (e.g., a current delivered by a battery).

DC biasing The setting of the DC voltages at each of the two transistor junctions (EBJ or CBJ), such that the transistor will be stable in the intended mode (e.g., active mode for amplifiers, etc.).

DC load line A straight line drawn over a series of current-voltage characteristic curves of a device on which the operating point moves as the voltage is varied. The slope of the line depends on the load resistance under DC conditions.

DC analysis In this kind of analysis, the goal is to find the Q-point of a circuit containing a nonlinear device. Since the equations to be solved are nonlinear, exact solutions are very difficult and normally quick approximate answers are found using iterative numerical analysis techniques.

Decibel The ratio of two powers or intensities, or the ratio of a power to a reference power. It is one-tenth of an international unit known as *bel.*

Depletion layer The region near the pn junction in a semiconductor in which there are no current carriers (either free electrons or holes), unless biased by a voltage.

Depletion mode An FET (normally on) that passes maximum current at zero gate potential and has decreasing current with applied gate potential, is operating in the depletion mode.

Depletion-type FET An FET in which a conducting channel exists at zero gate voltage (i.e. device is normally "on") and a voltage with proper polarity must be applied to the gate to eliminate the channel (i.e., to turn the device off).

Deposition The application of a layer of one substance (usually a metal) to the surface of a substrate, as in evaporation, sputtering, etc.

Detector (also called *demodulator*) A device that recovers the modulating signal from an RF carrier.

Device A single discrete conventional electronic part such as a resistor, a transistor, etc.

Device bandwidth The range of frequencies in a device, within which it functions optimally. It is usually taken between the 3 dB points of the plot of the transfer function $H(\omega) = V_o(\omega)/V_i(\omega)$ versus angular frequency (ω).

Die (also called *chip*) A single substrate on which all the active and passive elements of an electronic circuit have been fabricated. This is one portion taken from a wafer bearing many chips, but it is not ready for use until it is packaged and provided with terminals for connection to the outside world.

Dielectric A material that is a non-conductor of electricity. It is characterized by a parameter called *dielectric constant* or *relative permittivity* (ε_r).

Dielectric constant The property of a dielectric defined as the ratio of the capacitance of a capacitor (with the given dielectric) to the capacitance with air as the dielectric, but otherwise identical.

Differential amplifier An amplifier with two similar input circuits so connected as to respond to the difference between the voltages (or currents), and suppress voltages or currents that are alike in the two input circuits.

Diffusion Spatial variation in carrier concentration resulting in a net motion of carriers from regions of high concentration to regions of low concentration.

Digital Circuitry in which data-carrying signals are restricted to either of two voltage levels.

Diode, ideal A nonlinear two terminal device which conducts electricity completely one way acting as a short circuit (forward biased), while no current passes through the opposite way behaving like an open circuit (reverse biased).

Diode, real (also called a *crystal diode*) A two-electrode device consisting of n-type and p-type semiconductor material that makes use of the rectifying properties of a p-n junction; namely, it passes current (i) in the forward direction (from anode to cathode) for a positive voltage (v) and suppresses current in the reverse direction, in the following manner:

$$i = I_s(e^{v/nV_T} - 1)$$

I_S is the saturation current, V_T is the thermal voltage and n is a constant between 1 and 2 depending on the material and physical structure.

Discrete device An individual electrical component such as a resistor, capacitor, or transistor as opposed to an integrated circuit that consists of several discrete components.

Dispersive medium A medium in which the phase velocity (V_P) of a wave is a function of its frequency.

Distributed element An element whose property is spread out over an electrically significant length or area of a circuit instead of being concentrated at one location or within a specific component.

Divergence ($\nabla \cdot \bar{A}$) a) The emanation of many flows from a single point, or reversely, the convergence of many flows to one point. b) (of a vector quantity) The flux per unit volume leaving an infinitesimal element of volume at a point in a vector field. The cause or source of the divergence of a vector field is called a flow source.

Doping The process of adding controlled amounts of impurities or dopants to a semiconductor crystal in order to control its resistivity.

Double balanced mixer The mixer uses four diodes and two 180° hybrid couplers for the RF and LO ports, providing more power handling capability, and better rejection of AM noise and spurious outputs from the LO port.

Double-double balanced mixer The mixer uses eight diodes and three 180° hybrid couplers for the RF and LO ports, providing more power handling capability, and better rejection of AM noise and spurious outputs from the LO port.

Duality theorem States that when a theorem is true, it will remain true if each quantity and operation is replaced by its dual quantity and operation. In circuit theory, the dual quantities are "voltage and current" and "impedance and admittance." The dual operations are "series and parallel" and "meshes and nodes."

Dynamic range The difference between the maximum acceptable signal level and the minimum acceptable signal level. Both signals are measured at the output.

Electric charge (or *charge*) A basic property of elementary particles of matter (e.g., electron, protons, etc.) that is capable of creating a force field in its vicinity. This built-in force field is a result of stored electric energy. The charge of an object is the algebraic sum of the charges of its constituents (such as electrons, protons, etc.), and may be zero, or a positive or a negative number.

Electric current (or *current*) The net transfer of electric charges (Q) across a surface per unit time.

Electric displacement vector ($\bar{D}$) The electric field intensity vector multiplied by the permittivity of the medium.

Electric field ($\bar{E}$) The region about a charged body capable of exerting force. The intensity of the electric field at any point is defined to be the force that would be exerted on a unit positive charge at that point.

Electric field intensity The electric force on a stationary positive unit charge at a point in an electric field (also called *electric field strength*, *electric field vector*, and *electric vector*).

Electrical noise (or *noise*) Any unwanted electrical disturbance or spurious signal. These unwanted signals are random in nature, and are generated either internally in the electronic components or externally through impinging electromagnetic radiation.

Electrolysis The action whereby a current passing through a conductive solution (called an *electrolyte*) produces a chemical change in the solution and the electrodes.

Electrolyte A substance that ionizes when dissolved in a solution. Electrolytes conduct electricity, and in batteries they are instrumental in producing electricity by chemical action.

Electrolytic cell In general, a cell containing an electrolyte and at least two electrodes. Examples include voltaic cells, electrolytic capacitors, and electrolytic resistors.

Electromagnetic (EM) wave A radiant energy flow produced by oscillation of an electric charge as the source of radiation. In free space and away from the source, EM rays of waves consist of vibrating electric and magnetic fields that move at the speed of light (in vacuum), and are at right angles to each other and to the direction of motion. EM waves propagate with no actual transport of matter, and grow weaker in amplitude as they travel farther in space. EM waves include radio, microwaves, infrared, visible/ultraviolet light waves, X, gamma, and cosmic rays.

Electron One of the natural elementary constituents of matter. It carries a negative electric charge of one electronic unit.

Electronics The study, control, and application of the conduction of electricity through different media (e.g., semiconductors, conductors, gases, vacuum, etc.).

Electroplating Depositing one metal on the surface of another by electrolytic action.

Electrostatics The branch of physics concerned with electricity at rest.

Emitter A region in a transistor from which charge carriers (that are minority carriers in the base) are injected into the base.

Energy The capacity or ability of a body to perform work. Energy of a body is either potential motion (called *potential energy*) or due to its actual motion (called *kinetic energy*).

Enhancement mode A FET device that is normally off with zero gate voltage. A threshold voltage is then required to turn the device on.

Enhancement-type FET An FET which requires a gate voltage equal to or larger than the threshold voltage (V_t) in order to induce a conducting channel between source and drain (i.e., to turn the transistor "on"). The device is normally "off" when the gate voltage is zero.

Epitaxy The controlled growth of a layer of semiconductor material on the surface of a substance in such a way that the orientation of the atoms in the original substrate is preserved; i.e., the atoms in the grown layer are aligned to the orientation of the substrate atoms.

Equivalent circuit A simplified circuit that has the same response to changing voltage and frequency as the original more complex circuit. It is based upon specific models employed for each of the elements and devices in the circuit.

Equivalent noise temperature The absolute temperature at which a perfect resistor with the same resistance (as the component) would generate the same amount of noise as the component at room temperature.

Etch a) The selective removal of unwanted material on a printed circuit board to form the circuit connections for components. Usually done by chemical means. b) Refers to the material left on the board serving as transmission lines for the circuit.

Evaporation A technique for electrically depositing a film of selected metal on a metallic or non-metallic surface. In this technique, a filament of the desired metal is heated by an electric current in a vacuum chamber, which makes filament particles evaporate and condense as a film on the desired surface.

Excess Noise Ratio (ENR) A common measure of effective noise power for a noise source, defined as:

$$\text{ENR} = 10\log_{10}\left(\frac{P_N - P_o}{P_o}\right) = 10\log_{10}\left(\frac{T_N - T_o}{T_o}\right)$$

Where P_N and T_N are the noise power and the effective noise temperature of the active noise source, and P_o and T_o are the noise power and temperature of the matched load, respectively.

Fall time The time during which a pulse decreases from 90% (high level) to 10% of its maximum value (low level).

Fahrenheit (°F) 1/180th of the temperature difference between the freezing point of water (32°F) and the boiling point of water (212°F) on the Fahrenheit temperature scale.

$$T(°F) = T(°R) - 459.67, \quad T(°F) = \frac{9}{5}T(°C) + 32$$

°R and °C are symbols for degrees Rankin and Celsius, respectively.

Farad (F) The unit of capacitance in the MKS system of units equal to the capacitance of a capacitor that has a charge of one coulomb when a potential difference of one volt is applied.

Faraday's Law (also called the law of electromagnetic induction) When a magnetic field cuts a conductor, or when a conductor cuts a magnetic field, an electrical current will flow through the conductor if a closed path is provided over which the current can circulate; i.e.,

$$\text{Differential form: } \nabla \times \bar{E} = -\frac{\partial \bar{B}}{\partial t},$$

$$\text{Integral form: } \oint_C \bar{E} \cdot \overline{d\ell} = -\int_S \frac{\partial \bar{B}}{\partial t} \cdot \overline{dS} = -\frac{d\Phi}{dt}$$

Feedback The process of coupling some of the output of an amplifier back to its input. Negative feedback reduces the gain of an amplifier, and positive gain can be used to boost the gain.

Field a) A volume of space in which force is operative. b) An entity that acts as an intermediary agent in interactions between particles, is distributed over a region of space, and whose properties are a function of space and time in general.

Field-Effect Transistor (FET) A transistor in which current carriers are injected at one terminal (the source) and pass to another (the drain) through a channel of semiconductor material whose resistivity depends mainly on the applied voltage to the gate.

Flicker noise (or 1/*f* noise) Small vibrations of a current due to the following factors: a) random injection or recombination of charge carriers at an interface, such as at a metal semiconductor interface (in semiconductor devices), and b) random changes in cathode emissions of electric charges such as at cathode-air interface (in a thermionic tube).

Flow The passage of particles (e.g., electrons, etc.) between two points. Example: electrons moving from one terminal of a battery to the other terminal through a conductor.

Force The agency that accomplishes work.

Forward bias An external voltage applied in the conducting direction of a pn junction (i.e., positive terminal to p and negative terminal to n).

Frequency The number of complete cycles in one second of a time harmonic signal (e.g., an AC current or voltage, electromagnetic waves, etc).

Frequency response A measure of how effectively a circuit or device transmits the different frequencies applied to it. It is a phasor with magnitude and phase equal to:

$$H(\omega) = |V_o / V_i| \angle \theta_o - \theta_i$$

Gain The ratio of output to input that identifies the increase in signal or amplification when the input signal passes through a circuit.

Gallium Arsenide (GaAs) A binary III-V compound semiconductor material used to produce high-frequency devices.

Gate One of the electrodes of a field effect transistor used to control the resistance of the source to drain current path by the application of a voltage that causes the depletion layer under the gate to vary, thus reducing or increasing the conductance of the path. It is analogous to the base of a transistor.

Gauss The unit of magnetic induction (also called *magnetic flux density*) in the CGS system of units equal to one line per square centimeter, which is the magnetic flux density of one maxwell per square centimeter, or 10^{-4} tesla.

Gauss's Law (electric) The summation of the normal component of the electrical displacement vector over any closed surface is equal to the electric charges within the surface, which means that the source of the electric flux lines is the electric charge; i.e.,

Differential form: $\nabla \cdot \overline{D} = \rho$,

Integral form: $\oint_C \overline{D} \cdot \overline{dS} = \int_V \rho dv = Q$

Gauss's Law (magnetic) The summation of the normal component of the magnetic flux density vector over any closed surface is equal to zero, which in essence means that the magnetic flux lines have no source (or magnetic charge); i.e.,

Differential form: $\nabla \cdot \overline{B} = 0$,

Integral form: $\oint_S \overline{B} \cdot \overline{dS} = 0$

Generalized Ohm's Law When dealing with linear circuits under the influence of time harmonic signals, Ohm's law can be restated under the steady-state condition in the phasor domain as $V = ZI$.

Gilbert (Gi) The unit of magnetomotive force in the CGS system of units, equal to the magnetomotive force of a closed loop of one turn in which there is a current of $10/4\pi$ amperes. One Gilbert equals $10/4\pi$ Ampere-turn.

Ground (a) A metallic connection with the earth to establish zero potential (used for protection against short circuit). (b) The voltage reference point in a circuit. There may or may not be an actual connection to earth but it is understood that a point in the circuit said to be at ground potential could be connected to earth without disturbing the operation of the circuit in any way.

Hartley oscillator An oscillator in which a parallel-tuned tank circuit is connected between the collector and base of a transistor, the inductive element of the tank having an intermediate tap at the emitter potential.

Henry (H) The unit of self and mutual inductance in the MKS system of units equal to the inductance of a closed loop that gives rise to a magnetic flux of one weber for each ampere of current that flows through.

Hertz (Hz) The unit of frequency equal to the number of cycles of a periodic function that occur in one second.

High-Gain Amplifier (HGA) A general class of amplifiers with a high gain value that is not necessarily equal to the maximum possible gain available from the amplifier.

High-Power Amplifier (HPA) A large signal amplifier designed to maximize power output while generally ignoring the small-signal amplifier considerations.

Hole A vacant electron energy state near the top of the valence band in a semiconductor. It behaves as a positively charged particle having mass and mobility.

Hybrid circuit An integrated microelectronic circuit in which each component is fabricated on a separate chip or substrate, interconnected by means of lead wires so that each component can be individually optimized for performance.

Hybrid-π model (of a BJT) The BJT is modeled as a resistor between the base and emitter along with a current-controlled current source for small AC signal analysis at the collector branch.

Impedance The total opposition that a circuit presents to an AC signal, and a complex number equal to the ratio of the voltage phasor (V) to the current phasor (I).

Incident waves In a medium of certain propagation characteristics, a wave that encounters a discontinuity in the medium, or encounters a medium having different propagation characteristics, is said to be incident upon the discontinuity or medium.

Inductance (L) The inertial property of an element (caused by an induced reverse voltage), which opposes the flow of current when a voltage is applied; it opposes a change in current that has been established.

Inductor A conductor used to introduce inductance into an electric circuit, normally configured as a coil to maximize the inductance value.

Input The current, voltage, power, or other driving force applied to a circuit or device.

Insertion loss The attenuation resulting from inserting a circuit between source and load.

Insulator A material in which the outer electrons are tightly bound to the atom and are not free to move. Thus, there is negligible current through the material when a voltage is applied.

Integrated Circuit (IC) An electrical network composed of two or more circuit elements on a single semiconductor substrate.

Intermediate Frequency (IF) A frequency to which a signal wave is shifted locally as an intermediate step in transmission or reception.

Intermodulation products The additional frequencies at the output of a non-linear network (e.g., a non-linear amplifier, etc.) when two or more sinusoidal signals are applied at the input.

Isolation Electrical separation between two points.

Joule(J) The unit of energy or work in the MKS system of units, which is equal to the work performed as the point of application of a force of one Newton moves the object through a distance of one meter in the direction of the force.

Junction A joining of two different semiconductors or of semiconductor and metal.

Junction capacitance The capacitance associated with a region of transition between p- and n-type semiconductor materials.

Junction Field-Effect Transistor (JFET) A transistor in which current carriers are injected at one terminal (the source) and pass to another (the drain) though a channel of semiconductor material whose resistivity depends mainly on the applied voltage to the gate (which is a Pn junction).

Kelvin (K) The unit of measurement of temperature in the absolute scale (based on Celsius temperature scale), in which the absolute zero is at –273.15 °C. It is precisely equal to a value of 1/273.15 of the absolute temperature of the triple point of water, which is a particular pressure and temperature (273.15 K) point at which three different phases of water (i.e., vapor, liquid, and ice) can coexist in equilibrium. *See also* **temperature**.

Kinetic Energy *(K.E.)* The energy of a particle due to its motion. This motion is caused by a force on the particle.

Kirchhoff's Current Law (KVL) The law of conservation of charge that states that the total current flowing to a given point in a circuit is equal to the total current flowing away from that point.

Kirchhoff's Voltage Law (KCL) An electrical version of the law of conservation of energy that states that the algebraic sum of the voltage drops in any closed path in a circuit is equal to the algebraic sum of the electromotive forces in that path.

Large signal A signal that is large enough to transverse a large portion of the operating characteristic curves of the device, such that the device operates in a non-linear fashion.

Large signal amplifier (or power amplifier) A large signal amplifier designed to maximize power output while generally ignoring the small-signal amplifier considerations.

Large signal analysis A method of analysis of an active circuit under high amplitude signals that traverse such a large part of the operating characteristics of a device that non-linear portions of the characteristic are usually encountered, causing non-linear operation of the device.

Law of conservation of energy (excluding all metaphysical sources of energy) This fundamental law simply states that any form of energy in the physical universe can neither be created nor destroyed, but only converted into another form of energy (also called principle of conservation of energy).

Life time (also called mean life) The average time during which a hole or electron exists before recombination.

Light-Emitting Diode (LED) A pn junction that emits light when biased in the forward direction.

Linear network A network in which the parameters of resistance, inductance, and capacitance of the lumped elements are constant with respect to current or voltage, and in which the voltage or current sources are independent of or directly proportional to other voltages and currents or their derivatives, in the network.

Load A device that absorbs power and converts it into the desired form. This is the impedance to which energy is being supplied.

Load line A straight line drawn over a series of current-voltage characteristic curves on a graph to show current changes with voltage for a specified load resistance. The slope of the line depends on the load resistance under DC conditions (for DC load line) or under AC conditions (for AC load line).

Local Oscillator (LO) An oscillator used to generate a signal that, when combined with an RF signal, reproduces a sum or difference frequency equal to the intermediate frequency (or IF signal).

Lossless A theoretically perfect component that has no loss and hence, transmits all of the energy fed to it.

Low-Noise Amplifier (LNA) An amplifier that is designed for a given low noise figure value and a certain gain requirement. This is a tradeoff targeted toward a specific application, since maximum power gain and minimum noise figure cannot be obtained simultaneously. A Minimum-Noise Amplifier (MNA) is a special case of this category of amplifiers.

Lumped element A self-contained and localized element that offers one particular electrical property throughout the frequency range of interest.

Magnetic field The space surrounding a magnetic pole, a current-carrying conductor, or a magnetized body that is permeated by magnetic energy and is capable of exerting a magnetic force. This space can be characterized by magnetic lines of force.

Magnetic field intensity (H) The force that a magnetic field would exert on a unit magnetic pole placed at a point of interest, which expresses the free space strength of the magnetic field at that point (also called *magnetic field strength*, *magnetic intensity*, *magnetic field*, *magnetic force*, and *magnetizing force*).

Magnetic flux density vector (B) The flux per unit area. It is equal to that field which causes a force ($\bar{F}$) on a charged particle (Q) travelling through with a velocity ($\bar{V}$) such that $\bar{F} = Q\bar{V} \times \bar{B}$ (also called magnetic induction vector).

Magnetostatics The study of magnetic fields that are neither moving nor changing direction.

Matching circuit A passive circuit designed to connect two networks such that the maximum transfer of energy occurs.

Matching The concept of connecting two circuits (source and load) or two networks together via a coupling device or network in such a way that the maximum transfer of energy (or power) occurs between the two circuits or networks.

Mathematics Mathematics are short-hand methods of stating, analyzing, or resolving real or abstract problems and expressing their solutions by symbolizing data, decisions, conclusions, and assumptions.

Matter Matter particles are a condensation of energy particles into a very small volume.

Maximum-Gain Amplifier (MGA) An amplifier that is designed to provide maximize possible gain. The conditions $\Gamma_S = \Gamma^*_N$ and $\Gamma_L = \Gamma^*_{OUT}$ are the conditions that are to be met for this design.

Maximum Power Transfer Theorem The Maximum Power Transfer Theorem states that the maximum power that can be delivered to a load is only feasible when the load has an optimum impedance value $(Z_L)_{opt}$ equal to the complex conjugate of the source impedance value (Z_S); i.e., $(Z_L)_{opt}=Z^*_S$.

Maxwell (Mx) The unit for magnetic flux in the CGS system of units, equal to 10^{-8} weber.

Maxwell's Equations A series of four advanced classical equations developed by James Clerk Maxwell between 1864 and 1873, which describe the behavior of electromagnetic fields and waves in all practical situations. They relate the vector quantities for electric and magnetic fields as well as electric charges existing (at any point or in a volume), and set forth stringent requirements that the fields must satisfy. These celebrated equations are given as follows:

	Differential form	Integral form
a. Ampere's Law	$\nabla \times \bar{H} = \bar{J} + \frac{\partial \bar{D}}{\partial t}$,	$\oint_C \bar{H} \cdot \overline{d\ell} = I + \int_S \frac{\partial \bar{D}}{\partial t} \cdot \overline{dS}$
b. Faraday's Law	$\nabla \times \bar{E} = -\frac{\partial \bar{B}}{\partial t}$,	$\oint_C \bar{E} \cdot \overline{d\ell} = -\int_S \frac{\partial \bar{B}}{\partial t} \cdot \overline{dS} = -\frac{d\Phi}{dt}$
c. Gauss's Law (electric)	$\nabla \cdot \bar{D} = \rho$,	$\oint_S \bar{D} \cdot \overline{dS} = \int_V \rho dv = Q$
d. Gauss's Law (magnetic)	$\nabla \cdot \bar{B} = 0$,	$\oint_S \bar{B} \cdot \overline{dS} = 0$

From these equations, Maxwell predicted the existence of electromagnetic waves whose later discovery made radio possible. He showed that where a varying electric field exists, it is accompanied by a varying magnetic field induced at right angles, and vice versa, and the two form an electromagnetic field that could propagate as a transverse wave. He calculated that in a vacuum, the speed of the wave was given by $1/\sqrt{\varepsilon_o \mu_o}$, where ε_o and μ_o are the permittivity and permeability of vacuum. The calculated value for this speed was in remarkable agreement with the measured speed of light, and Maxwell concluded that light is propagated as electromagnetic waves.

Metal-Oxide-Semiconductor Field Effect Transistor (MOSFET) A device consisting of diffused source and drain regions on either side of a p or n type channel region and a gate electrode insulated from the channel by silicon oxide. When a control voltage is applied to the gate, the channel is converted to the same type of semiconductor as the source and the drain. This permits current to flow between the source and drain.

Metal-Semiconductor Field Effect Transistor (MESFET) A field-effect transistor that uses a thin film of III-V semiconductor (such as gallium arsenide, etc.) with a Schottky barrier gate formed by depositing a layer of metal directly onto the surface of the film.

Microelectronics The technology of electronics that is connected with or applied to the realization of electronic systems from extremely small electronic parts.

Microstrip line A microwave transmission line that is a single conductor supported above a ground plane.

Microwave Integrated Circuit (MIC) Consists of an assembly that combines different circuit functions that are connected by Microstrip transmission lines. These different circuits all incorporate planar semiconductor devices, passive lumped elements, and distributed elements.

Microwaves Waves in the frequency range of 1 GHz to 300 GHz.

Miller's Theorem Consider a linear two-port network where the feedback (or the bridging) element with admittance (Y_f) is connected from the output to the input. The overall voltage gain ($v_o/v_i = K$) is expected to be known already through independent means. With this assumption in mind, Miller's Theorem states that, the feedback element can be equivalently replaced by two admittances, Y_i and Y_o, as follows a) Y_i between input node (1) and ground having a value of $Y_i = Y_f(1 - K)$ b) Y_o between output node (2) and ground with a value of $Y_o = Y_f(1 - 1/K)$.

Millimeter wave Electromagnetic radiation in the frequency range of 30 to 300 GHz. (Corresponds to 10 mm to 1 mm in wavelength.)

Minimum Noise Amplifier (MNA) A special case of low-noise amplifiers (LNAs) that are designed to provide the minimum possible noise. This is obtained by designing the input matching network using the optimum reflection coefficient, $\Gamma_S = \Gamma_{opt}$. The output matching network is designed for the best VSWR such that

$$\Gamma_L = \Gamma_{OUT}^* = \left(S_{22} + \frac{S_{12}S_{21}\Gamma_o}{1 - S_{11}\Gamma_o}\right)^*$$

Mismatch factor (also called ***mismatch loss***) The ratio P_2/P_1 (always less than one) for a load that is mismatched to a source, where P_1 is the power a matched load would absorb from the source, and P_2 is the power actually absorbed by the mismatched load.

Mixer A non-linear three-port circuit (two inputs and one output) that generates a spectrum of output frequencies equal to the sum (or difference) of the two input frequencies and their harmonics. The two input ports are referred to as RF and LO, whereas the output is called the IF port.

Model A simplified mathematical representation of a process, device, circuit, or system employed to facilitate their analysis. It is based on a series of assumptions, which limits its usefulness to only a specific range of operation. Therefore, the assumptions (or limitations) are part of the description of the model and should always be kept in mind when using the model.

Monolithic circuit Is an integrated circuit built entirely on a single chip of semiconductor.

Monolithic integrated circuit An integrated circuit that is formed in a single block or wafer of semiconductor materials. The name is derived from the Greek monolithos, which means "one stone."

Monolithic Microwave Integrated Circuit (MMIC) A microwave circuit obtained through a multilevel process approach comprising of all active and passive circuit elements as well as interconnecting transmission lines formed into the bulk or onto the surface of a semi-insulating semiconductor substrate by some deposition scheme such as epitaxy, ion implantation, sputtering, evaporation, diffusion, etc.

Multistage amplifier (MSA) A number of amplifiers connected in cascade to form a single amplification unit. Proper matching between stages is crucial for optimum performance.

Narrow-Band Amplifier (NBA) An amplifier designed to operate over a narrow bandwidth that is greater than 1 percent of the center frequency and less than one-third octave.

N-channel device A device constructed on p-type semiconductor substrate whose drain and source components are n-type silicon.

Negative feedback Part of the output signal of an amplifying circuit feeds back into the input of the circuit to control the overall circuit response.

Negative resistance The resistance of a device where current decreases with an increase in the voltage across the device, instead of the increase in current as it normally occurs.

Negative Resistance (NR) device A device in which the voltage and current are 180° out of phase with respect to each other, which means that an increase in the applied voltage causes a decrease in the current.

Neper A unit of attenuation used for expressing the ratio of two currents , voltages, or fields by taking the natural logarithm (logarithm to base e) of this ratio. If voltage V_1 is attenuated to V_2 so that $V_2/V_1 = e^{-N}$, then N is attenuation in nepers (always a positive number) and is defined by: $N = \log_e(V_1/V_2) = \ln(V_1/V_2)$.

Network A collection of electric devices and elements (such as resistors, capacitors, etc.) connected together to form several interrelated circuits.

Network parameters A set of parameters used to characterize a circuit in terms of voltage and current conditions at the network input and output ports. For Z-, Y-, H-, and $ABCD$- parameters, the net values of current and voltage are used. However, for S- and T-parameters, the incident and reflected values of current and voltage are used.

Neutron One of uncharged elementary particles of an atom having the same mass as a proton. A free neutron decomposes into a proton, an electron, and a neutrino. A neutrino is a neutral uncharged particle having a mass that approaches zero very rapidly (a half-life of about 13 minutes).

Newton (N) The unit of force in MKS system of units equal to the force that imparts an acceleration of one meter per second^2 to a mass of one kilogram.

NMOS (or *N-channel MOS*) Metal Oxide Silicon (MOS) devices made on p-type substrates in which the active carriers are electrons that flow between n-type source and drain contacts.

Noise Random unwanted electrical signals that cause unwanted and false output signals in a circuit.

Noise figure The ratio of the total available noise power at the output, $(P_o)_{tot}$, to the output available noise power $(P_o)_i$ due to thermal noise coming only from the input resistor at the standard room temperature ($T_o = 290$ °K).

Nonlinear Having an output that does not rise and fall in direct proportion to the input.

Norton's Theorem (is the dual of Thevenin's theorem) It states that the voltage across a load element (Z_L) that is connected to the two terminals of a linear network is equal to the short-circuit current (I_{sc}) between these terminals in the absence of the load element divided by the sum of the load admittance(Y_L) and the admittance (Y_T) of the network when looking back into these terminals while setting all independent sources to zero.

NPN transistor A transistor with a p-type base and an n-type collector and emitter.

N-type (in semiconductors) Is a semiconductor material that has been doped with an excess of donor impurity atoms, so that the free charge carriers are electrons.

Nucleus The core of an atom containing most of the mass composed of protons and neutrons, having a positive charge equal to the charge of the number of protons that it contains.

Octave The interval between any two frequencies having a ratio of 2 to 1.

Oersted (Oe) The unit of magnetic field in the CGS system of units equal to the field strength at the center of a plane circular coil of one turn and 1-cm radius when there is a current of $10/2\pi$ ampere in the coil.

Ohm (Ω) The unit of resistance in the MKS system of units equal to the resistance between two points on a conductor through which a current of one ampere flows as a result of a potential difference of one volt applied between the two points.

Ohm's Law The potential difference V across the resistor terminals is directly proportional to the electrical current flowing through the resistor. The proportionality constant is called resistance (R); i.e., $V = RI$. Ohm's Law can also be interpreted as the conversion of potential energy (V) into kinetic energy (I), which is a simple statement expressing the principle of conservation of energy.

One-dB gain compression point The point (on the P_{OUT} vs. P_{IN} plot) at which the power gain of the transistor due to non-linearities is reduced by 1-dB from its small signal linear power gain value.

Open loop gain The ratio of the output signal of an amplifier (without any feedback) to its input signal at any given frequency.

Operating point (or *quiescent point*) On a family of characteristic curves, the point indicating the quiescent level of operation determined by applying a fixed bias voltage or current. This point represents the condition existing when only DC values are present. An AC signal applied at this point oscillates above and below the point as a mean.

Operating power gain (G_p) The ratio of the input power to the output power actually reaching the load.

Oscillator An electronic device that generates alternating-current power at a frequency determined by constants in its circuits.

Output The current, voltage, power, or driving force delivered by a circuit or device.

Packaging The physical process of connecting, protecting, and enclosing electronic devices, components, etc.

Passive A component that may control but does not create or amplify electrical energy.

P-channel A p-type material conduction channel in a semiconductor device.

Penetration depth See skin depth.

Phase The angular relationship of a wave to some time reference or other wave.

Phase constant The imaginary component of the propagation constant for a traveling wave at a given frequency.

Phase shifter A two-port network in which the output voltage (or current) may be adjusted to have some desired phase relationship with the input voltage (or current) by using a control signal (e.g., a DC bias, etc.).

Phase velocity The velocity at which a point of constant phase is propagated in a propagating sinusoidal wave.

Phasor A result of a mathematical transformation of a sinusoidal waveform (voltage, current, or EM wave) from the time domain into the complex number domain (or frequency domain) whereby the magnitude and phase angle information of the sinusoid is retained.

Pin diode A diode made by diffusing the semiconductor with p dopant from one side and n dopant from the opposite side, with the process so controlled that a thin intrinsic region separates the *n* and *p* regions.

Pinch-off condition (in FETs) The condition in which the current flow between the source and the drain (I_D) no longer increases with further increase of V_{DS}. See also saturation region (FET).

Pinch-off region In a FET, the region of the V-I characteristic where the gate voltage causes the depletion region to extend completely across the channel, resulting in an almost constant current flow at the drain. See also saturation region (FET).

Plane wave A wave where the wavefronts are normal to the direction of propagation.

Plating *See* **electroplating**.

PMOS Metal Oxide Silicon (MOS) devices made on n-type substrates in which the active carriers are holes that flow between p-type source and drain contacts.

Pn junction A transition between *p*-type and *n*-type semiconductor regions within a crystal lattice. Such a junction possesses specific electrical properties such as the ability to conduct in only one direction, and is used as the basis for semiconductor devices.

Pnp transistor A transistor consisting of two p-type regions separated by an n-type region. When a forward bias is applied to the first junction and a reverse bias to the second junction, the device operates in the active mode and functions as an amplifier.

Pole (of a switch) The movable arm of a switch. For example, a single-pole switch has one input terminal with one movable arm.

Port Access point to a system or circuit, consisting of two terminals.

Positive feedback The process by which amplification is increased by having part of the power or signal in the output circuit returned to the input circuit in order to increase the output.

Potential difference (or *voltage*) The electrical pressure or force between any two points caused by accumulation of charges at one point relative to another, which has the capability of creating a current between the two points.

Potential Energy (P.E.) Any form of stored energy that has the capability of performing work when released. This energy is due to the position of particles relative to each other.

Potential unstability The characteristics of an amplifier evaluated at the desired frequency of operation and at gain levels that can cause the device to oscillate uncontrollably. This condition can be influenced by the operating environment; temperature, frequency, source, or load conditions. Proper design of the amplifier will establish circuit conditions for the device that prevent this condition.

Power The rate at which work is performed; i.e., the rate at which energy is being either generated or absorbed.

Power combiner A device used to combine the signal from two or more sources such that it can be used to feed a single input.

Power divider A device that provides a desired distribution of power at a branch point in a microwave system. For example, for a half-power coupler, the resulting signal level is reduced by 3dB at each branch.

Power factor The ratio of resistance to impedance that indicates a measure of loss in an inductor, capacitor, or insulator. This can be found to be the cosine of phase angle between the voltage applied to an element and the current passing through it.

Power gain The ratio of the signal power developed at the output of a device to the signal power applied at the input.

Principle of conservation of energy See law of conservation of energy.

Processing The act of converting material from one form into another more desired form, such as in integrated circuit fabrication where one starts with a wafer and through many steps ends up with a functional circuit on a chip.

Propagation The travel of electromagnetic waves through a medium.

Propagation constant A number showing the effect (such as losses, wave velocity, etc.) a transmission line has on a wave as it propagates along the line. It is a complex term having a real term called the *attenuation factor* and an imaginary term called the *phase constant*.

Proton One of the three basic subatomic particles, with a positive charge magnitude equivalent to the negative charge of an electron.

P-type (in semiconductors) Is a semiconductor material that has been doped with an excess of acceptor impurity atoms, so that the free charge carriers are holes.

Pulse A variation of a quantity whose value is normally constant. The variation is characterized by a rise to a certain level (amplitude), a finite duration, and a decay back to the normal level.

Quality factor (Q) A measure of the ability of an element (or circuit) with periodic behavior to store energy equal to 2π times the average energy stored divided by the energy dissipated per cycle.

Quiescent point (also called *Q-point* or *bias point*) See operating point.

Radio Frequency (RF) Any frequency at which coherent electromagnetic radiation of energy is possible.

Rankine (°*R*) The unit of measurement of temperature in the absolute scale (based on Fahrenheit temperature scale), in which the absolute zero is at –459.67 °F. *See also* **temperature.**

Reactance Symbolized by X. This parameter is the measure of the opposition to the flow of alternating current.

Reactive element An inductor or a capacitor are reactive elements.

Reciprocal network A network that satisfies the reciprocity theorem. If the excitation to the network is interchanged with the response, the ratio of the excitation to the response will remain the same.

Reciprocity theorem A theorem stating that the interchange of electromotive force at one point (e.g., in branch k, v_k) in a *passive linear network*, with the current produced at any other point (e.g., branch m, i_m) results in the same current (in branch k, i_k) when the same electromotive force is applied in the new location (branch m, v_m), i.e. $v_k/i_m = v_m/i_k$.

Recombination The combination and resultant neutralization of particles having unlike charges (e.g., an electron and a hole).

Rectifier A device having an asymmetrical conduction characteristic such that current can flow in only one direction through the device.

Reflected waves The waves reflected from a discontinuity in the medium they are traveling in.

Reflection coefficient The ratio of the reflected wave phasor to the incident wave phasor.

Resistance A property of a resistor dependent upon dimensions, material, and temperature that determines the amount of current flowing through when a voltage is applied.

Resistor A lumped bilateral and linear element that impedes the flow of current, $i(t)$, through it when a potential difference, $V_{12}(\mathrm{t})$, is imposed between its two terminals. The resistor's value is found by: $R = V_{12}(t)/i(t)$.

Resonant frequency The frequency at which a given system or circuit will respond with maximum amplitude when driven by an external sinusoidal force.

Reverse bias An external voltage applied to a semiconductor pn junction to reduce the current across the junction and thereby widen the depletion region. This voltage reinforces the internal potential barrier set up in equilibrium.

Right-hand rule (also called *Fleming's rule*) For a current-carrying wire, the rule that if the fingers of the right hand are placed around the wire so that the thumb points in the direction of the current flow, the curling fingers will be pointing in the direction of the magnetic field produced by the wire.

Ripple A measure of the flatness of the frequency response of the resonance circuit and defined to be the attenuation difference (in dB) of the maximum value from the minimum in the passband; i.e., $\varepsilon = |\text{max. attenuation} - \text{min. attenuation}|$ (in dB).

Root mean square value The square root of the average of the square of the values of a period quantity taken throughout one complete period.

Saturation In general, is a circuit condition whereby an increase in the driving or input signal (e.g., voltage or current) no longer produces a change in the output.

Saturation region (in a BJT) The operating condition of a transistor when an increase in base current produces no further increase in collector current (cf., triode region in FET).

Saturation region or pinch-off region (in a FET) Is a region in the FET characteristic curves plot (drawn in $I_D - V_{DS}$ plane), where the current flow between the source and drain (I_D) no longer increases with further increase of V_{DS} (cf., active region in BJT).

Scattering parameters (or *S-parameters*) The network representation of a multiport network at high RF/microwave frequencies. These parameters are defined in terms of reflected and incident voltages and currents at the ports under specified load conditions.

Semiconductor A material having a resistance between that of conductors and insulators, and usually having a negative temperature coefficient of resistance.

Shape Factor (SF) of a resonant circuit The ratio of the 60-dB bandwidth to the 3-dB bandwidth.

Shot noise (or *Schottky noise*) Caused by random passage of discrete charge carriers in a solid state device while crossing a junction or other discontinuities, causing a variable current I due to random motion of electrons or holes. It is commonly found in a semiconductor device (e.g., in a pn junction of a diode or a transistor) and is proportional to $(I)^{1/2}$.

Siemens (S) (also called *mho*) The unit for conductance, susceptance, and admittance in the MKS system of units, and is equal to the reciprocal of the resistance of an element that has a resistance of one ohm.

Signal An electrical quantity (such as a current or voltage) that can be used to convey information for communication, control, etc.

Signal Flow Graph (SFG) An abbreviated block diagram consisting of small circles (called *nodes*) representing the variables that are connected by several directed lines (called *branches*) representing one-way signal multipliers. An arrow on the line indicates direction of signal flow, and a letter near the arrow indicates the multiplication factor.

Silicon (Si) A semiconductor material element in column IV of the periodic table used in device fabrication.

Single-ended circuit A circuit with one terminal grounded so that operation is asymmetric with respect to ground.

Sinusoidal Varying in proportion to the sine or cosine of an angle (or time function). For example, the ordinary AC signal is a sinusoidal.

Skin depth (also called *penetration depth*) The depth beneath the surface of a current-carrying conductor, where the current density amplitude is $1/e$ or 36.8 percent of its surface amplitude (i.e. depth at 1 neper down). The skin depth is given by:

$$\delta = \sqrt{\frac{1}{\pi f \mu \sigma}}$$

where μ is the permeability (H/m) and σ is the conductivity of the conductor

Skin effect The tendency of RF currents to flow closer to the surface of a conductor as the frequency is increased. Since the current is then restricted to a smaller cross sectional area, this effectively increases the resistance.

Small signal A low-amplitude signal that covers such a small part of the operating characteristic curve of a device that operation is nearly always linear.

Small-signal amplifier An amplifier that operates in the linear region of the V-I Characteristic curves.

Small-signal analysis Consideration of only small excursions from the no-signal bias, so that a device can be represented by a linear equivalent circuit.

Smith Chart A graphical display of impedance to reflection coefficient transformation in which all passive values of impedance are mapped one-to-one into a circle of radius one in the reflection coefficient plane.

Solid-state device Any element that can control current without moving parts, heated filaments, or vacuum gaps. All semiconductors are solid-state devices, although not all solid-state devices (such as transformers, etc.) are semiconductors.

Space (also called *created space*) The continuous three-dimensional expanse extending in all directions, within which all things exist.

Sputtering A technique used to deposit a thin layer of metal on a glass, plastic, metal, or other surfaces in vacuum. In a vacuum chamber, a piece of desired metal (to be deposited) is made the cathode of a high-voltage circuit with respect to a nearby anode plate. The high voltage causes metal atoms to be ejected from the surface of the cathode and strike the surface of the desired object placed in their path, thus becoming deposited on it as a film.

Stability In general, the ability of an amplifier to maintain effectiveness in its nominal operating characteristics in spite of large changes in the environment such as physical temperature, signal frequency, source or load conditions, etc.

Standing wave A standing, apparent motionless-ness, of particles causing an apparent no out-flow, no in-flow. A standing wave is caused by two energy flows, impinging against one another, with the same frequency and comparable magnitudes to cause a suspension of energy particles in space, enduring with a duration longer than the duration of the flows themselves.

Standing Wave Ratio (SWR) The ratio of current or voltage on a transmission line that results from two waves having the same frequency and traveling in opposite directions meeting and creating a standing wave.

Steady state response Is the response of a system that either does not change with time or maintains a state of relative equilibrium after all initial transients have disappeared.

Stub A short section of a transmission line, open or shorted at the far end, connected in series or in parallel with a transmission line to match the impedance of the primary line to that of a source or load.

Subharmonically pumped mixer *See* **anti-parallel mixer**.

Substrate A single body of material on or in which one or more electronic circuit elements or integrated circuits are fabricated.

Superposition Theorem This theorem states that in a linear network, the voltage or current in any element resulting from several sources acting together is the sum of the voltages or currents resulting from each source acting alone, while all other independent sources are set to zero; i.e.,

$$f(v_1 + v_2 + \ldots\ldots\ldots\ldots + v_n) = f(v_1) + f(v_2) + \ldots\ldots\ldots\ldots + f(v_n)$$

Switch A mechanical or electrical device that completes or breaks the path of the current or sends it over a different path.

Switching Making, breaking, or changing the connections in an electric circuit.

Switch Insertion Loss (IL) A figure of merit of a switch in the "on" state and defined to be the difference (in dB) between the power received at the load before and after the insertion of the switch in the line.

Switch isolation A figure of merit of a switch in the "off" state and defined to be the difference (in dB) between the power delivered to the load for an ideal switch in the "on" state and the actual power delivered to the load when the switch is in the "off" state.

Temperature The degree of hotness or coldness measured with respect to an arbitrary zero or an absolute zero, and expressed on a degree scale. Examples of arbitrary-zero degree scales are Celsius scale (°C) and Fahrenheit scale (°F); and examples of absolute-zero degree scales are Kelvin degree scale (based on Celsius degree scale) and Rankine degree scale (based on Fahrenheit degree scale).

Terminal One of the electric input or output points of a circuit or component.

Tesla (T) The unit of magnetic field in the MKS system of units equal to one weber per square meter.

Thermal noise (or *Johnson noise* or *Nyquist noise*) The most basic type of noise that is caused by thermal vibration of bound charges and thermal agitation of electrons in a conductive material. This is common to all passive or active devices.

Thevenin's Theorem (also known as *Helmholt's Theorem*) At any given frequency, the current (I_L) that will flow through a load impedance (Z_L) when connected to any two terminals of a linear network is equal to the open-circuit voltage (V_{oc} or

V_T), with the load (Z_L) removed, and divided by the sum of the load impedance and the impedance (Z_T) obtained by looking back from the open terminals into the network with all independent sources reduced to zero (i.e., replacing each independent source by its internal impedance).

Third-Order Intercept (TOI) point The intersection of the P_{f1} plot with the P_{2f1-f2} on a P_{IN} versus P_{OUT} plot for an amplifier. The third-order intercept output power is used to evaluate the spurious-free dynamic range of an amplifier.

Threshold voltage (also referred to as *pinch -off voltage, Vp,* for JFETs) The minimum gate voltage in an FET required to induce a channel between source and drain in an enhancement type FET (i.e., to turn the transistor "on") when $V_{DS} = 0$.

Throw (of a switch) The output terminal(s) that each moveable arm of a switch is allowed to make a connection with. For example, a single throw switch has only one output terminal.

Time That characteristic of the physical universe at a given location (on a macroscopic and/or microscopic level) that orders the sequence of events. It proceeds from the interaction of matter and energy and is merely an "index of change," used to keep track of change of particle location. The fundamental unit of time measurement is supplied by the earth's rotation on its axis while orbiting around the sun. It can also alternately be defined as the co-motion and co-action of moving particles relative to one another in space.

Time harmonic signal A sinusoidal signal shown as: $A\cos(\omega t + \theta)$ or $A\sin(\omega t + \theta)$.

T model (of a BJT) A model used to evaluate the AC characteristics of a BJT device. This model is obtained from: $v_{be} = r_e i_e$, $r_e = r_\pi/(1 + \beta)$, and re $= V_T/I_{EQ}$.

Transducer power gain (G_T) The ratio of power that the transducer delivers to the load under specified operating conditions, to the power available from the source.

Transfer function (in linear circuits) Is a complex function expressing the mathematical relationship between the output (of a circuit or system) to its input, in phasor domain (i.e., under steady state condition).

Transformer An electrical device that, by electromagnetic induction, transforms electric energy from one (or more) circuit(s) to one (or more) other circuit(s) at the same frequency, but usually at a different voltage and current value.

Transistor In general, a non-linear, three-terminal, active semiconductor device where the flow of the electrical current between two of the terminals is controlled by the third terminal. The name is an acronym for transfer resistor.

Transistor circuit configurations There are three major circuit configurations for each type of transistor (BJT or FET). These are common emitter (CE), common base (CB) and common collector or emitter follower (CC) for BJTs, and common source (CS), common gate (CG) and common drain or source follower (CD) for FETs. In each of these circuit configurations the stated "common terminal" is common to both the input and the output circuit-terminals. The "common terminal" need not be directly connected to ground, even though AC-wise it usually is. For transistor amplifier design, the CE and CS configurations are usually preferred and are of primary focus in the text.

Transmission Line (T.L.) Any system of conductors suitable for conducting TEM-mode electromagnetic energy efficiently between two or more terminals.

Transmission parameters (or *T-parameters*) The network representation of a multi-port network at high RF/microwave frequencies. These parameters are defined in terms of the input reflected and incident voltage and current as independent variables, and the output incident and reflected waves as the dependent variables. This matrix allows cascaded networks to be easily combined and represented with T parameters using a matrix multiplication.

Transmitted wave That portion of an incident wave that is not reflected at the interface, but actually travels from one medium to another.

Transverse Electro-Magnetic (TEM) wave Waves having the electric and magnetic fields perpendicular to each other and to the direction of propagation. These waves have no field components in the direction of propagation.

Triode region (of an FET) The linear region of operation below pinch-off. Also called Voltage Controlled Resistance (VCR) region.

Tuned circuit A circuit consisting of inductance and capacitance that can be adjusted for resonance at the desired frequency.

Tuning The adjustment relating to frequency of a circuit or system to secure optimum performance.

Tunnel diode (or *esaki diode*) A heavily doped pn junction diode. The tunnel diode uses the tunneling effect for its operation and has a negative resistance region above a minimum level of applied forward bias. With the addition of suitable external circuits, it can be used as an oscillator or amplifier.

Tunneling The observed effect of the ability of certain atomic particles to pass through a barrier that they cannot cross over because of the required energy level. The tunneling effect uses a low-energy electron to pierce the potential barrier.

Two-port network A network that has only two access ports, one for input or excitation, and one for output or response.

Ultimate attenuation The final minimum attenuation that the resonance circuit presents outside of the specified pass-band.

Unconditional stability The ability of an amplifier to maintain effectiveness regardless of changes to the operating environment; temperature, frequency, source, or load conditions. For example, requirements for unconditional stability under various source or load conditions are:

$$|\Gamma_{IN}| < 1 \text{ and } |\Gamma_{OUT}| < 1 \text{ for all } |\Gamma_S| < 1 \text{ and } |\Gamma_L| < 1.$$

Unidirectional Flowing in only one direction (e.g., direct current).

Unilateral A non-reciprocal characteristic.

Unilateral figure of merit (U) A calculated parameter used to evaluate the error involved in using the transistor amplifier approximate gain ($G_{TU,max}$, obtained by assuming $|S_{12}|$ as zero) rather than its actual gain (G_T):

$$\frac{1}{(1+U)^2} < \frac{G_T}{G_{TU,max}} < \frac{1}{(1-U)^2}$$

Unipolar transistor (also called a *field effect transistor*) A transistor in which charge carriers are of only one polarity.

Vapor-Phase Epitaxy (VPE) The use of chemical vapor deposition to grow epitaxial layers on a substrate. In this technique, by introduction of specific substances into a high temperature reactor, a chemical reaction is brought about that causes the desired substance to be produced and deposited epitaxially onto the substrate.

Varactor (also called *voltage-variable capacitor*) A two-terminal solid-state device that utilizes the voltage-variable capacitance of a pn junction. In the normal semiconductor diode, efforts are made to minimize inherent capacitance, while in the varactor, this capacitance is emphasized. The capacitance varies with the applied voltage and allows the device to be used for various applications such as tuning, switching, etc.

Volt (V) The unit of potential difference (or electromotive force) in the MKS system of units equal to the potential difference between two points for which one coulomb of charge will do one joule of work in going from one point to the other.

Voltage Voltage or potential difference between two points is defined to be the amount of work done against an electric field in order to move a unit charge from one point to the other.

Voltage Controlled Oscillator (VCO) Any oscillator for which a change in the tuning voltage (usually DC) results in a change in the oscillation frequency.

Voltage divider A resistor or reactor (a capacitor or inductor) connected across a voltage and tapped to make a fixed or variable fraction of the applied voltage available.

Voltage follower amplifier An amplifier with a high input impedance having an output voltage characteristic that is the same as the input. It is normally used as a buffer stage to prevent the loading effect, especially when the load value is very small compared to the output impedance of the amplifier.

Voltage gain The ratio of the voltage across the specified load impedance of a circuit to the voltage across the input of the circuit.

Voltage source The device or generator connected to the input of a network or circuit.

Voltage Standing Wave Ratio (VSWR) The ratio of maximum voltage to the minimum voltage on a transmission. The standing wave on a line results from two voltage (or current) waves having the same frequency, and traveling in opposite directions.

Wafer A thin semiconductor slice of silicon or germanium on which matrices of microcircuits can be formed using manufacturing processes. After processing, the wafer is separated into chips (or *die*) containing individual circuits.

Watt (W) The unit of power in MKS system of units defined as the work of one joule done in one second.

Wave A disturbance that propagates from one point in a medium to other points without giving the medium as a whole any permanent displacement.

Wavefront A surface of constant phase.

Waveguide A transmission line comprised of a hollow conducting tube within which electromagnetic waves are propagated.

Wavelength The physical distance between two points having the same phase in two consecutive cycles of a periodic wave along a line in the direction of propagation.

Wave propagation The travel of electromagnetic waves through a medium.

Weber (Wb) The unit of magnetic flux in the MKS system of units equal to the magnetic flux which, linking a circuit of one turn, produces an electromotive force of one volt when the flux is reduced to zero at a uniform rate in one second.

White noise Random noise that has a constant energy per unit bandwidth at every frequency in the range of interest.

Work The advancement of the point of application of a force on a particle.

Y factor The ratio of two output noise powers of an amplifier under two physical temperatures, normally chosen far apart for practical reasons.

Yield (Y) The ratio of the number of usable chips at the end of the manufacturing process shipped to the market (N_{mkt}) to the number of chips initially submitted for processing (N_{IN}); i.e., $Y = N_{mkt}/N_{IN}$.

Zener diode A two-layer device that, if forward biased, operates as a normal diode, but when reverse biased, the diode exhibits a knee or sharp break in its current-voltage characteristic curve. The voltage remains essentially constant for any further increase of reverse current, up to the allowable dissipation rating.

ZY Smith Chart A graphical display of impedance-to-reflection coefficient transformation in which all passive values of impedance and admittance are mapped one to one into a circle of radius one in the reflection coefficient plane. The grids for both impedance and admittance are coexistent, allowing either value to be read from the chart.

Index

C

D

M

N

O

P

Q

R

S

T

About the Author

MATTHEW M. RADMANESH received his MSEE and Ph.D. degrees from the University of Michigan at Ann Arbor. He has taught at GMI Engineering & Management (presently Kettering University) and has worked in the RF and Microwave industry for Hughes Aircraft Co., Maury Microwave Corp., and Boeing Aircraft Co. He is currently a faculty member in the Electrical and Computer Engineering department at California State University, Northridge, CA. He is a senior member of IEEE, Eta Kappa Nu Honor Society, and a Past President (for three years) of the SFV Chapter of the IEEE Microwave Theory and Technique (MTT) Society. His many years of experience in both microwave industry and academia has led to several dozen technical papers in national and international journals. His current research interests include design of RF and microwave devices and circuits, millimeter-wave circuit applications, and integrated optics. Dr. Radmanesh won the MPD divisional award while at Hughes Aircraft Co. and a similar award from Boeing Aircraft Co. He holds two patents for his pioneering work and novel designs of two millimeter-wave noise sources.

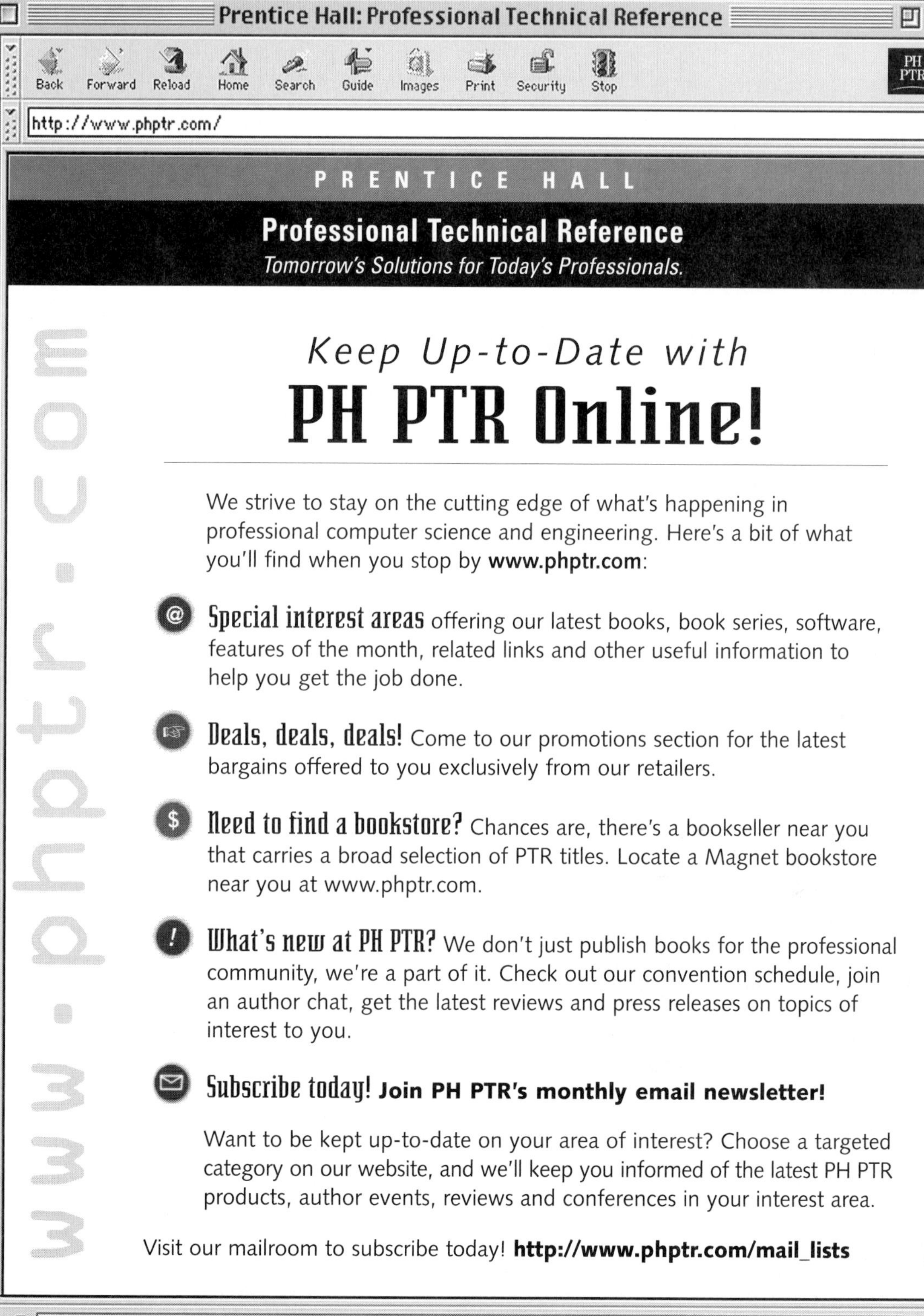

Prentice Hall: Professional Technical Reference
Back Forward Reload Home Search Guide Images Print Security Stop
PH PTR
http://www.phptr.com/
PRENTICE HALL
Professional Technical Reference
Tomorrow's Solutions for Today's Professionals.
www.phptr.com
Keep Up-to-Date with
PH PTR Online!
We strive to stay on the cutting edge of what's happening in professional computer science and engineering. Here's a bit of what you'll find when you stop by www.phptr.com:
Special interest areas offering our latest books, book series, software, features of the month, related links and other useful information to help you get the job done.
Deals, deals, deals! Come to our promotions section for the latest bargains offered to you exclusively from our retailers.
Need to find a bookstore? Chances are, there's a bookseller near you that carries a broad selection of PTR titles. Locate a Magnet bookstore near you at www.phptr.com.
What's new at PH PTR? We don't just publish books for the professional community, we're a part of it. Check out our convention schedule, join an author chat, get the latest reviews and press releases on topics of interest to you.
Subscribe today! Join PH PTR's monthly email newsletter!
Want to be kept up-to-date on your area of interest? Choose a targeted category on our website, and we'll keep you informed of the latest PH PTR products, author events, reviews and conferences in your interest area.
Visit our mailroom to subscribe today! http://www.phptr.com/mail_lists

LICENSE AGREEMENT AND LIMITED WARRANTY

READ THE FOLLOWING TERMS AND CONDITIONS CAREFULLY BEFORE OPENING THIS SOFTWARE PACKAGE. THIS LEGAL DOCUMENT IS AN AGREEMENT BETWEEN YOU AND PRENTICE-HALL, INC. (THE "COMPANY"). BY OPENING THIS SEALED SOFTWARE PACKAGE, YOU ARE AGREEING TO BE BOUND BY THESE TERMS AND CONDITIONS. IF YOU DO NOT AGREE WITH THESE TERMS AND CONDITIONS, DO NOT OPEN THE SOFTWARE PACKAGE. PROMPTLY RETURN THE UNOPENED SOFTWARE PACKAGE AND ALL ACCOMPANYING ITEMS TO THE PLACE YOU OBTAINED THEM FOR A FULL REFUND OF ANY SUMS YOU HAVE PAID.

1. **GRANT OF LICENSE:** In consideration of your payment of the license fee, which is part of the price you paid for this product, and your agreement to abide by the terms and conditions of this Agreement, the Company grants to you a nonexclusive right to use and display the copy of the enclosed software program (hereinafter the "software") on a single computer (i.e., with a single CPU) at a single location so long as you comply with the terms of this Agreement. The Company reserves all rights not expressly granted to you under this Agreement.

2. **OWNERSHIP OF SOFTWARE:** You own only the magnetic or physical media (the enclosed software) on which the software is recorded or fixed, but the Company retains all the rights, title, and ownership to the software recorded on the original software copy(ies) and all subsequent copies of the software, regardless of the form or media on which the original or other copies may exist. This license is not a sale of the original software or any copy to you.

3. **COPY RESTRICTIONS:** This software and the accompanying printed materials and user manual (the "Documentation") are the subject of copyright. You may not copy the Documentation or the software, except that you may make a single copy of the software for backup or archival purposes only. You may be held legally responsible for any copying or copyright infringement which is caused or encouraged by your failure to abide by the terms of this restriction.

4. **USE RESTRICTIONS:** You may not network the software or otherwise use it on more than one computer or computer terminal at the same time. You may physically transfer the software from one computer to another provided that the software is used on only one computer at a time. You may not distribute copies of the software or Documentation to others. You may not reverse engineer, disassemble, decompile, modify, adapt, translate, or create derivative works based on the software or the Documentation without the prior written consent of the Company.

5. **TRANSFER RESTRICTIONS:** The enclosed software is licensed only to you and may not be transferred to any one else without the prior written consent of the Company. Any unauthorized transfer of the software shall result in the immediate termination of this Agreement.

6. **TERMINATION:** This license is effective until terminated. This license will terminate automatically without notice from the Company and become null and void if you fail to comply with any provisions or limitations of this license. Upon termination, you shall destroy the Documentation and all copies of the software. All provisions of this Agreement as to warranties, limitation of liability, remedies or damages, and our ownership rights shall survive termination.

7. **MISCELLANEOUS:** This Agreement shall be construed in accordance with the laws of the United States of America and the State of New York and shall benefit the Company, its affiliates, and assignees.

8. **LIMITED WARRANTY AND DISCLAIMER OF WARRANTY:** The Company warrants that the software, when properly used in accordance with the Documentation, will operate in substantial conformity with the description of the software set forth in the Documentation. The Company does not warrant that the software will meet your requirements or that the operation of the software will be uninterrupted or error-free. The Company warrants that the media on which the software is delivered shall be free from defects in materials and workmanship under normal use

for a period of thirty (30) days from the date of your purchase. Your only remedy and the Company's only obligation under these limited warranties is, at the Company's option, return of the warranted item for a refund of any amounts paid by you or replacement of the item. Any replacement of software or media under the warranties shall not extend the original warranty period. The limited warranty set forth above shall not apply to any software which the Company determines in good faith has been subject to misuse, neglect, improper installation, repair, alteration, or damage by you. EXCEPT FOR THE EXPRESSED WARRANTIES SET FORTH ABOVE, THE COMPANY DISCLAIMS ALL WARRANTIES, EXPRESS OR IMPLIED, INCLUDING WITHOUT LIMITATION, THE IMPLIED WARRANTIES OF MERCHANTABILITY AND FITNESS FOR A PARTICULAR PURPOSE. EXCEPT FOR THE EXPRESS WARRANTY SET FORTH ABOVE, THE COMPANY DOES NOT WARRANT, GUARANTEE, OR MAKE ANY REPRESENTATION REGARDING THE USE OR THE RESULTS OF THE USE OF THE SOFTWARE IN TERMS OF ITS CORRECTNESS, ACCURACY, RELIABILITY, CURRENTNESS, OR OTHERWISE.

IN NO EVENT, SHALL THE COMPANY OR ITS EMPLOYEES, AGENTS, SUPPLIERS, OR CONTRACTORS BE LIABLE FOR ANY INCIDENTAL, INDIRECT, SPECIAL, OR CONSEQUENTIAL DAMAGES ARISING OUT OF OR IN CONNECTION WITH THE LICENSE GRANTED UNDER THIS AGREEMENT, OR FOR LOSS OF USE, LOSS OF DATA, LOSS OF INCOME OR PROFIT, OR OTHER LOSSES, SUSTAINED AS A RESULT OF INJURY TO ANY PERSON, OR LOSS OF OR DAMAGE TO PROPERTY, OR CLAIMS OF THIRD PARTIES, EVEN IF THE COMPANY OR AN AUTHORIZED REPRESENTATIVE OF THE COMPANY HAS BEEN ADVISED OF THE POSSIBILITY OF SUCH DAMAGES. IN NO EVENT SHALL LIABILITY OF THE COMPANY FOR DAMAGES WITH RESPECT TO THE SOFTWARE EXCEED THE AMOUNTS ACTUALLY PAID BY YOU, IF ANY, FOR THE SOFTWARE.

SOME JURISDICTIONS DO NOT ALLOW THE LIMITATION OF IMPLIED WARRANTIES OR LIABILITY FOR INCIDENTAL, INDIRECT, SPECIAL, OR CONSEQUENTIAL DAMAGES, SO THE ABOVE LIMITATIONS MAY NOT ALWAYS APPLY. THE WARRANTIES IN THIS AGREEMENT GIVE YOU SPECIFIC LEGAL RIGHTS AND YOU MAY ALSO HAVE OTHER RIGHTS WHICH VARY IN ACCORDANCE WITH LOCAL LAW.

ACKNOWLEDGMENT

YOU ACKNOWLEDGE THAT YOU HAVE READ THIS AGREEMENT, UNDERSTAND IT, AND AGREE TO BE BOUND BY ITS TERMS AND CONDITIONS. YOU ALSO AGREE THAT THIS AGREEMENT IS THE COMPLETE AND EXCLUSIVE STATEMENT OF THE AGREEMENT BETWEEN YOU AND THE COMPANY AND SUPERSEDES ALL PROPOSALS OR PRIOR AGREEMENTS, ORAL, OR WRITTEN, AND ANY OTHER COMMUNICATIONS BETWEEN YOU AND THE COMPANY OR ANY REPRESENTATIVE OF THE COMPANY RELATING TO THE SUBJECT MATTER OF THIS AGREEMENT.

Should you have any questions concerning this Agreement or if you wish to contact the Company for any reason, please contact in writing at the address below.

Robin Short
Prentice Hall PTR
One Lake Street
Upper Saddle River, New Jersey 07458

About the CD-ROM

Bound into the back of each textbook, is a CD containing software in the form of an electronic book (E-book), which contains all of the numerical examples from the text. The solutions are programmed using Visual Basic software, a built-in feature of the Microsoft Excel® application software.

FEATURES

The CD-ROM included with *Radio Frequency and Microwave Electronics Illustrated* contains a powerful interactive tool for learning the textbook content, as well as for solving numerical problems. The software includes 90 solved problems based on numerical examples in the book. They enable the user to analyze or design RF and microwave circuits with ease, accuracy and great speed. The big advantage of the interactive software tool is its use of live math.

HOW TO START THE PROGRAM

The following steps need to be carried out before we can utilize the contents of the CD-ROM:

1. Either
   a. Read all the files directly from the CD-ROM or,
   b. Create a folder called "E-book" in the C: drive. Copy the entire contents of the CD-ROM into a folder entitled "E-book."
2. Open the Microsoft Excel 2000 software (or Excel 97 with SR-1 or SR-2 revision). Then, open the "E-book" folder from the CD-ROM or C: drive.
3. The user may begin the program by double clicking on the "Table of Contents" file. As a shortcut, double click on the "Table of Contents" file directly without opening the Microsoft Excel software.
4. Once the "Table of Contents" file opens, the user needs to click on the "About the CD" file to open it.
5. Carefully read all of the detailed information in the "About the CD" file, and close it by either
   a. holding ALT + TAB keys and choosing EXCEL, or
   b. clicking the "Return to Table of Contents" ARROW to return to the E-book table of contents.
6. Before proceeding to the completed examples, the user needs to select from the toolbar menu "Tools," "Add-ins." From the dialog box, select both *Analysis Toolpack* and *Analysis Toolpack—VBA* to set up the software properly.
7. Proceed to the desired example by clicking it and selecting the "Enable Macros" from the dialogue box.
8. For in-depth information on how to use the CD-ROM, please refer to Appendix N.

LICENSE AGREEMENT

Use of the software accompanying *Radio Frequency and Microwave Electronics Illustrated* is subject to the terms of the License Agreement and Limited Warranty, found on the previous two pages.

TECHNICAL SUPPORT

Prentice Hall does not offer technical support for any of the programs on the CD-ROM. However, if the CD-ROM is damaged, you may obtain a replacement copy by sending an email that describes the problem to: disc_exchange@prenhall.com.